FUNDAMENTALS OF THERMODYNAMICS

FIFTH EDITION

RICHARD E. SONNTAG
CLAUS BORGNAKKE
University of Michigan

GORDON J. VAN WYLEN
Hope College

JOHN WILEY & SONS, INC.
New York • Chichester • Weinheim
Brisbane • Singapore • Toronto

Cover Photo: Courtesy Norman Chigier, Department of Mechanical Engineering, Carnegie Mellon University.

The cover photograph shows a flame emerging from a nozzle into a turbulent air flow field. The methane gas flow is initially laminar as it emerges from the nozzle. The turbulent eddies in the air flow are entrained and penetrate into the gas flow. The brightest colors represent the highest temperature regions. Carbon, in the form of soot, is formed in the fuel-rich regions, resulting in radiation from the flame (the yellow regions).

ACQUISITIONS EDITOR   Joseph Hayton

MARKETING MANAGER   Karen Allman

SENIOR PRODUCTION MANAGER   Lucille Buonocore

SENIOR PRODUCTION EDITOR   Monique Calello

COVER & TEXT DESIGNER   Dawn L. Stanley

ILLUSTRATION COORDINATOR   Jaime Perea

ILLUSTRATION STUDIO   Precision Graphics

SENIOR PHOTO EDITOR   Hilary Newman

MANUFACTURING MANAGER   Monique Calello

This book was set in Times Roman by GGS Information Services and printed and bound by Von Hoffman Press. The cover was printed by Phoenix Color Corporation.

This book is printed on acid-free paper.

The paper in this book was manufactured by a mill whose forest management programs include sustained yield harvesting of its timberlands. Sustained yield harvesting principles ensure that the numbers of trees cut each year does not exceed the amount of new growth.

Library of Congress Cataloging-in-Publication Data:
Sonntag, Richard Edwin.
    Fundamentals of thermodynamics / Richard E. Sonntag, Claus
Borgnakke, Gordon J. Van Wylen. — 5th ed.
        p.   cm.
    Prev. eds. under title: Fundamentals of classical thermodynamics /
Gordon J. Van Wylen
    Includes index.
    ISBN 0-471-18361-X (cloth, alk. paper)
    1. Thermodynamics.   I. Borgnakke, C. (Claus)   II. Van Wylen
Gordon John.   III. Van Wylen, Gordon John   Fundamentals of
classical thermodynamics.   IV. Title
TJ265.S66   1998
621.402′-dc21   97-45721
                                                                                              CIP

Printed in the United States of America

10 9 8 7 6 5 4 3 2 1

PREFACE

In this fifth edition we have retained the basic objective of the first four editions: to present a comprehensive and rigorous treatment of classical thermodynamics while retaining an engineering perspective and, in doing so, to lay the groundwork for subsequent studies in such fields as fluid mechanics, heat transfer, and statistical thermodynamics, and also to prepare the student to effectively use thermodynamics in the practice of engineering. However, in this edition, we have departed from our earlier practice of including both undergraduate- and graduate-level material and problems. This edition is intended to provide strong emphasis as an undergraduate textbook.

We have deliberately directed our presentation to students. New concepts and definitions are presented in the context where they are first relevant. The first thermodynamic properties to be defined (Chapter 2) are those that can be readily measured: pressure, specific volume, and temperature. In Chapter 3, tables of thermodynamic properties are introduced, but only in regard to these measurable properties. Internal energy and enthalpy are introduced in connection with the first law, entropy with the second law, and the Helmholtz and Gibbs functions in the chapter on thermodynamic relations. Many examples have been included in the book to assist the student in gaining an understanding of thermodynamics, and the problems at the end of each chapter have been carefully sequenced to correlate with the subject matter and provide some progression in difficulty.

We have attempted to cover fairly comprehensively the basic subject matter of classical thermodynamics, and believe that the book provides adequate preparation for study of the application of thermodynamics to the various professional fields as well as for study of more advanced topics in thermodynamics, such as those related to materials, surface phenomena, plasmas, and cryogenics. We also recognize that a number of colleges offer a single introductory course in thermodynamics for all departments, and we have tried to cover those topics that the various departments might wish to have included in such a course. However, since specific courses vary considerably in prerequisites, specific objectives, duration, and background of the students, we have arranged the material, particularly in the later chapters, so that there is considerable flexibility in the amount of material that may be covered.

The principal change of our philosophy is in presenting this edition as an undergraduate text. In that regard, we have deleted variables, material, and problems that would be used only in an advanced course. The heart of the book concerns the introduction and application of the first and second laws. The first law presentation is now divided into two separate chapters, and the two second law chapters are now presented as three, with control volume analysis in each case rating its own chapter. Development of the control volume is now much simplified and less mathematical than in earlier editions. Description and explanation of example processes for steady flow devices and machines is greatly expanded, to assist the undergraduate students in coming to understand the nature of these processes. There are 25 percent more problems overall in these five (first and second law) chapters, in spite of the deletion of most of the more difficult problems from the previous edition. A few advanced problems have been retained, but are now located together at the end of the SI problem set and are clearly marked as such.

Chapter 10 is a completely new presentation of the material on irreversibility and availability (exergy), and is at a much more basic, and less mathematical, level. Chapter 11 is essentially the same presentation of power and refrigeration systems that was extensively reorganized and expanded in the Fourth Edition. This chapter includes a number of modern and relevant applications: heat pumps, cogeneration, topping and bottoming cycles, two-stage systems, combined cycles, and many others. We have also included comments and problems that relate thermodynamics to environmental concerns.

Advanced material has been deleted from the final five chapters of this edition. Chapter 12 now contains material only on mixtures of ideal gases, including air-water vapor mixtures. The presentation of thermodynamic relations in Chapter 13 has been reorganized, with determination of change-of-phase properties now at the beginning, followed by homogeneous phase property changes. This approach is smoother and more natural than the previous organization, and the introduction of Helmholtz and Gibbs functions follows as a logical necessity for the use of measurable properties to calculate entropy changes. The variables fugacity, activity, and acentric factor have been dropped as being unnecessary at the undergraduate level. Finally, a basic model for dealing with real gas mixtures in addition to pure substances is introduced at the end of this chapter. Advanced material has also been dropped from Chapters 14 and 15, including Gibbs function of formation, which is not needed for the calculation of chemical equilibrium constants, and also phase equilibrium in real multicomponent systems. The sections on flow through blade passages has been deleted from Chapter 16, which now contains only material on compressible flow. We recognize that in many cases this chapter is not included in the thermodynamics courses, but may instead by covered elsewhere in the curriculum.

A new topic in this edition is a brief introduction to heat transfer modes, following the definition of heat in Chapter 4. We have included this material for those who wish to give a slightly expanded discussion of the nature of heat transfer without going into much detail at this level. This section is independent and need not be covered by those who do not choose to discuss the topic in an introductory thermodynamics course.

The Appendix Tables have been revised and updated, and are compatible with the more extensive set of tables in our 1997 book *Thermodynamic and Transport Properties*. The disk included with this book contains the same information as that in the printed Appendix. Two new advantages are that the programs can now be run under Windows, and calculated properties can also be exported to spreadsheets.

Throughout the book we have attempted to maintain an engineering perspective, particularly through the choice of examples and problems. There are many new problems throughout the book, in addition to the large number in the chapters on the first and second laws. In each chapter, the presentation of problems begins with a large number of undergraduate problems in SI units, coordinated to the chapter material. This is followed by a small set of advanced problems for those who wish to include some of the more difficult applications, and then a set of English unit problems for those who wish to include this material. The problem set concludes with a section on computer, design, and open-ended problems, which has largely been carried over from the Fourth Edition. We find these problems to be particularly useful as projects in the second, or application, course on thermodynamics.

In regard to symbols used in this text, we have tried to maintain consistency throughout the book. In a limited number of cases, we have used a given symbol for more than one purpose. We believe, however, that the context will clarify the meaning of the sumbol in these cases.

Our philosophy regarding units in this edition has been to organize the book so that the course or sequence can be taught entirely in SI units (Le Système International d'Unités). Thus, all the text examples are in SI units, as are the complete problem sets and the thermodynamic tables. In recognition, however, of the continuing need for engineering graduates to be familiar with English Engineering units, we have included an introduction to this system in Chapter 2. We have also repeated a sufficient number of examples, problems, and tables in these units, which should allow for suitable practice for those who wish to use these units. For dealing with English units, there is a problem of distinguishing between force and mass units. The symbols lbf and lbm have been used for pound force and pound mass to emphasize this distinction in the English system. Such symbols as lbf/in.^{2} have been used for pressure (rather than psi) and ft^3/lbm for specific volume (rather than cu ft/lb) in order to stress the fundamental units involved in the various parameters. The force–mass conversion constant g_c is treated as being implicit, to enable equations including kinetic or potential energy terms to be handled in a consistent manner with regard to units. In dealing with SI units, we have commonly used the basic units for pressure (pascal) and volume (cubic meter), althouh we have used the liter extensively as a convenient volume unit. Others may wish to use the bar more extensively as the pressure unit, and we trust that such flexibility in these units will present no particular difficulties to the student. Concerning the extensive properties, a lowercase letter (v, u, h, s) designates the property per unit mass (either pound mass or kilogram); an uppercase letter (V, U, H, S) the property for the entire system; a lowercase letter with a bar ($\bar{v}$, $\bar{u}$, $\bar{h}$, $\bar{s}$) the property per unit mole (either the pound mole or lb mol, or the kilogram mole or kmol, in this text); and an uppercase letter with a bar ($\bar{V}$, $\bar{U}$, $\bar{H}$, $\bar{S}$) the partial molal property in a mixture. Following this pattern, we have found it convenient to designate the total heat transfer as Q, the heat transfer per unit mass of the system as q, the total work as W, and the work per unit mass of the system as w.

Furthermore, we represent the rate of flow across a system boundary or control surface by a dot over the given quantity. Thus, $\dot{Q}$ represents a rate of heat transfer across the system boundary; $\dot{W}$ the rate at which work crosses the system boundary (that is, the power); and $\dot{m}$ the mass rate of flow across a control surface ($\dot{n}$ is used when the mass rate of flow is expressed in moles per unit time). We realize that we have departed from the usual mathematical use of a dotted symbol, in which the dotted symbol typically refers to a derivative with respect to time. However, we have used the dotted symbol to indicate only a flow of heat and work across a system boundary and a flow of heat, work, and mass across a control surface, and not in any other context. We believe that these designations have contributed to a simple and consistent use of symbols for this book.

We acknowledge with appreciation the suggestions, counsel, and encouragement of many colleagues, both at the University of Michigan and elsewhere. This assistance has been very helpful to us during the writing of this edition, as it was with the earlier editions of the book. Both undergraduate and graduate students have been of particular assistance, for their perceptive questions have often caused us to rewrite or rethink a given portion of the text, or to try to develop a better way of presenting the material in order to anticipate such questions or difficulties. We must single out three former doctoral students at the University of Michigan who have been of particular help over the years with this ongoing project, especially in the development of the computer programs for the last two editions: Dr. Young Moo Park, Dr. Kyoung Kuhn Park, and Dr. Youngil Kim. We appreciate their many valuable contributions to this material. Finally, for each of us, the encouragement and patience of our wives and families have been indispensable, and have made this time of writing pleasant and enjoyable, in spite of the pressures of the project.

We would also like to thank those many survey respondents who took the time to fill out a market survey for the fifth edition of the text. Valuable information has been gleaned from these responses and was used to help modify the fifth edition of the text. The time and effort spent on these surveys is much appreciated.

Our hope is that this book will contribute to the effective teaching of thermodynamics to students who face very significant challenges and opportunities during their professional careers. Your comments, criticism, and suggestions will also be appreciated.

RICHARD E. SONNTAG
CLAUS BORGNAKKE
GORDON J. VAN WYLEN
Ann Arbor, Michigan
September 1997

CONTENTS

SYMBOLS

a	acceleration
a, A	specific Helmholtz function and total Helmholtz function
AF	air-fuel ratio
c	velocity of sound
c	mass fraction
C_D	coefficient of discharge
C_p	constant-pressure specific heat
C_v	constant-volume specific heat
C_{po}	zero-pressure constant-pressure specific heat
C_{vo}	zero-pressure constant-volume specific heat
e, E	specific energy and total energy
F	force
FA	fuel-air ratio
g	acceleration due to gravity
g, G	specific Gibbs function and total Gibbs function
g_c	a constant that relates force, mass, length, and time
h, H	specific enthalpy and total enthalpy
i	electrical current
I	irreversibility
J	proportionality factor to relate units of work to units of heat
k	specific heat ratio: C_p/C_v
K	equilibrium constant
KE	kinetic energy
L	length
m	mass
$\dot{m}$	mass rate of flow
M	molecular weight
M	Mach number
n	number of moles
n	polytropic exponent
P	pressure
P_i	partial pressure of component i in a mixture
PE	potential energy
P_r	relative pressure as used in gas tables
q, Q	heat transfer per unit mass and total heat transfer
$\dot{Q}$	rate of heat transfer
Q_H, Q_L	heat transfer with high-temperature body and heat transfer with low-temperature body; sign determined from context

R	gas constant
$\bar{R}$	universal gas constant
s, S	specific entropy and total energy
S_{gen}	entropy generation
$\dot{S}_{\text{gen}}$	rate of entropy generation
t	time
T	temperature
u, U	specific internal energy and total internal energy
v, V	specific volume and total volume
v_r	relative specific volume as used in gas tables
$\mathbf{V}$	velocity
w, W	work per unit mass and total work
$\dot{W}$	rate of work, or power
w^{rev}	reversible work between two states
x	quality
y	gas-phase mole fraction
Z	elevation
Z	compressibility factor
Z	electrical charge

Script Letters

$\mathscr{A}$	area
$\mathscr{E}$	electrical potential
$\mathscr{S}$	surface tension
$\mathscr{T}$	tension

Greek Letters

α	residual volume
α_P	volume expansivity
β	coefficient of performance for a refrigerator
β'	coefficient of performance for a heat pump
β_S	adiabatic compressibility
β_T	isothermal compressibility
η	efficiency
μ	chemical potential
μ_J	Joule-Thomson coefficient
ν	stoichiometric coefficient
ρ	density
Φ	equivalence ratio
ϕ	relative humidity
ϕ	availability for a control mass
ψ	availability associated with a steady-state, steady-flow process
ω	humidity ratio or specific humidity

Subscripts

c	property at the critical point
c.v.	control volume
e	state of a substance leaving a control volume
f	formation

f	property of saturated liquid
fg	difference in property for saturated vapor and saturated liquid
g	property of saturated vapor
i	state of a substance entering a control volume
i	property of saturated solid
if	difference in property for saturated liquid and saturated solid
ig	difference in property for saturated vapor and saturated solid
r	reduced property
s	isentropic process
0	property of the surroundings
0	stagnation property

SUPERSCRIPTS

$-$	bar over symbol denotes property on a molal basis (over V, H, S, U, A, G, the bar denotes partial molal property)
$\circ$	property at standard-state condition
$*$	ideal gas
$*$	property at the throat of a nozzle
rev	reversible

SOME INTRODUCTORY COMMENTS 1

In the course of our study of thermodynamics, a number of the examples and problems presented refer to processes that occur in equipment such as a steam power plant, a fuel cell, a vapor-compression refrigerator, a thermoelectric cooler, a rocket engine, and an air separation plant. In this introductory chapter, a brief description of this equipment is given. There are at least two reasons for including such a chapter. First, many students have had limited contact with such equipment, and the solution of problems will be more meaningful when they have some familiarity with the actual processes and the equipment. Second, this chapter will provide an introduction to thermodynamics, including the use of certain terms (which will be more formally defined in later chapters), some of the problems to which thermodynamics can be applied, and some of the things that have been accomplished, at least in part, from the application of thermodynamics.

Thermodynamics is relevant to many other processes than those cited in this chapter. It is basic to the study of materials, chemical reactions, and plasmas. The student should bear in mind that this chapter is only a brief and necessarily very incomplete introduction to the subject of thermodynamics.

1.1 THE SIMPLE STEAM POWER PLANT

A schematic diagram of a simple steam power plant is shown in Fig. 1.1. High-pressure superheated steam leaves the boiler, which is also referred to as a steam generator, and enters the turbine. The steam expands in the turbine and, in doing so, does work, which enables the turbine to drive the electric generator. The low-pressure steam leaves the turbine and enters the condenser, where heat is transferred from the steam (causing it to condense) to the cooling water. Because large quantities of cooling water are required, power plants are frequently located near rivers or lakes. This transfer of heat to the water in lakes and rivers leads to the thermal pollution problem, which has been studied extensively in recent years. During our study of thermodynamics we will gain an understanding of why this heat transfer is necessary and how it can be minimized. When the supply of cooling water is limited, a cooling tower may be used. In the cooling tower some of the cooling water evaporates in such a way that it lowers the temperature of the water that remains as a liquid.

The pressure of the condensate leaving the condenser is increased in the pump and thus enables the condensate to flow into the steam generator. In many steam generators an economizer is used. An economizer is simply a heat exchanger in which heat is trans-

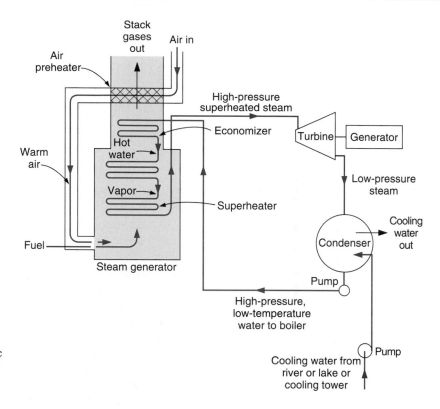

FIGURE 1.1 Schematic diagram of a steam power plant.

ferred from the products of combustion (just before they leave the steam generator) to the condensate. The temperature of the condensate is increased, but no evaporation takes place. In other sections of the steam generator, heat is transferred from the products of combustion to the water, causing it to evaporate. The temperature at which evaporation occurs is called the saturation temperature. The steam then flows through another heat exchanger, known as a superheater, where the temperature of the steam is increased well above the saturation temperature.

In many power plants the air that is used for combustion is preheated in the air preheater by transferring heat from the stack gases as they are leaving the furnace. This air is then mixed with fuel—which might be coal, fuel oil, natural gas, or other combustible material—and combustion takes place in the furnace. As the products of combustion pass through the furnace, heat is transferred to the water in the superheater, the boiler, the economizer, and to the air in the air preheater. The products of combustion from power plants are discharged to the atmosphere, which is one of the facets of the air pollution problem we now face.

A large power plant has many other pieces of equipment, some of which will be considered in later chapters.

Figure 1.2 shows a steam turbine and the generator that it drives. Steam turbines vary in capacity from less than 10 kilowatts to more than 1 000 000 kilowatts.

Figure 1.3 shows a large steam generator. The flow of air and products of combustion are indicated. The condensate, also called the boiler feedwater, enters at the economizer inlet, and the superheated steam leaves at the superheater outlet.

The number of operating nuclear power plants has increased substantially in recent

FIGURE 1.2 A large steam turbine. (Courtesy General Electric Co.)

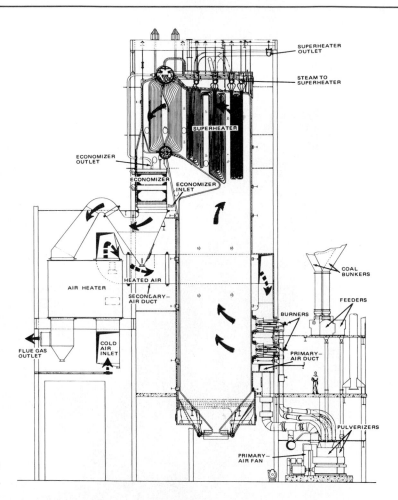

FIGURE 1.3 A large steam generator (Courtesy Babcock and Wilcox Co.).

years. In these power plants the reactor replaces the steam generator of the conventional power plant, and the radioactive fuel elements replace the coal, oil, or natural gas.

There are several different reactor designs in current use. One of these is the boiling-water reactor, such as the system shown in Fig. 1.4. In other nuclear power plants a secondary fluid circulates from the reactor to the steam generator, where heat is transferred from the secondary fluid to the water, which in turn goes through a conventional steam cycle. Safety considerations and the necessity to keep the turbine, condenser, and related equipment from becoming radioactive are always major considerations in the design and operation of a nuclear power plant.

1.2 FUEL CELLS

When a conventional power plant is viewed as a whole, as shown in Fig. 1.5, fuel and air enter the power plant and products of combustion leave the unit. There is also a transfer of heat to the cooling water, and work is done in the form of the electrical energy leaving the power plant. The overall objective of a power plant is to convert the availability (to do work) of the fuel into work (in the form of electrical energy) in the most efficient manner, taking into consideration cost, space, safety, and environmental concerns.

We might well ask whether all the equipment in the power plant, such as the steam generator, the turbine, the condenser, and the pump, is necessary. Is it possible to produce electrical energy from the fuel in a more direct manner?

The fuel cell accomplishes this objective. Figure 1.6 shows a schematic arrangement of a fuel cell of the ion-exchange membrane type. In this fuel cell, hydrogen and oxygen react to form water. Let us consider the general features of the operation of this type of fuel cell.

The flow of electrons in the external circuit is from anode to cathode. Hydrogen enters at the anode side, and oxygen enters at the cathode side. At the surface of the ion-exchange membrane the hydrogen is ionized according to the reaction

$$2H_2 \rightarrow 4H^+ + 4e^-$$

The electrons flow through the external circuit and the hydrogen ions flow through the membrane to the cathode, where the following reaction takes place.

$$4H^+ + 4e^- + O_2 \rightarrow 2H_2O$$

There is a potential difference between the anode and cathode, and thus there is a flow of electricity through a potential difference; this, in thermodynamic terms, is called work. There may also be a transfer of heat between the fuel cell and the surroundings.

At the present time the fuel used in fuel cells is usually either hydrogen or a mixture of gaseous hydrocarbons and hydrogen. The oxidizer is usually oxygen. However, current development is directed toward the production of fuel cells that use hydrocarbon fuels and air. Although the conventional (or nuclear) steam power plant is still used in large-scale power-generating systems, and conventional piston engines and gas turbines are still used in most transportation power systems, the fuel cell may eventually become a serious competitor. The fuel cell is already being used to produce power for space and other special applications.

Thermodynamics plays a vital role in the analysis, development, and design of all power-producing systems, including reciprocating internal-combustion engines and gas turbines. Considerations such as the increase of efficiency, improved design, optimum operating conditions, environmental pollution, and alternate methods of power generation involve, among other factors, the careful application of the fundamentals of thermodynamics.

1.3 THE VAPOR-COMPRESSION REFRIGERATION CYCLE

A simple vapor-compression refrigeration cycle is shown schematically in Fig. 1.7. The refrigerant enters the compressor as a slightly superheated vapor at a low pressure. It then leaves the compressor and enters the condenser as a vapor at some elevated pressure, where the refrigerant is condensed as heat is transferred to cooling water or to the surroundings. The refrigerant then leaves the condenser as a high-pressure liquid. The pressure of the liquid is decreased as it flows through the expansion valve, and as a result, some of the liquid flashes into cold vapor. The remaining liquid, now at a low pressure and temperature, is vaporized in the evaporator as heat is transferred from the refrigerated space. This vapor then reenters the compressor.

In a typical home refrigerator the compressor is located in the rear near the bottom of the unit. The compressors are usually hermetically sealed; that is, the motor and compressor

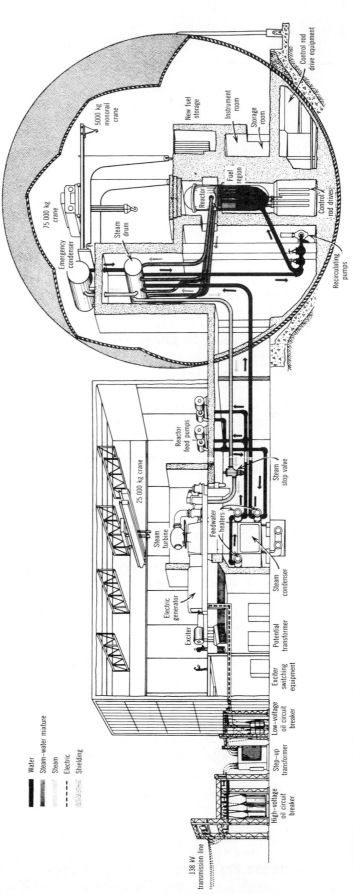

FIGURE 1.4 Schematic diagram of the Big Rock Point nuclear plant of Consumers Power Company at Charlevoix, Michigan (Courtesy Consumers Power Company).

Water

Steam–water mixture

Steam

Electric

Shielding

138 kV transmission line

High-voltage oil circuit breaker

Step-up transformer

Low-voltage oil circuit breaker

Exciter switching equipment

Potential transformer

Exciter

Electric generator

Steam turbine

25 000 kg crane

Steam condenser

Feedwater heaters

Reactor feed pumps

Steam stop valve

Emergency condenser

75 000 kg crane

Steam drum

5000 kg monorail crane

New fuel storage

Instrument room

Storage room

Reactor

Fuel region

Control rod drives

Recirculating pumps

Control rod drive equipment

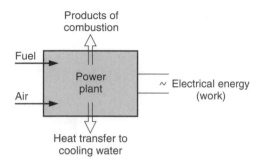

FIGURE 1.5 Schematic diagram of a power plant.

are mounted in a sealed housing, and the electric leads for the motor pass through this housing. This seal prevents leakage of the refrigerant. The condenser is also located at the back of the refrigerator and is arranged so that the air in the room flows past the condenser by natural convection. The expansion valve takes the form of a long capillary tube, and the evaporator is located around the outside of the freezing compartment inside the refrigerator.

Figure 1.8 shows a large centrifugal unit that is used to provide refrigeration for an air-conditioning unit. In this unit, water is cooled and then circulated to provide cooling where needed.

1.4 THE THERMOELECTRIC REFRIGERATOR

We may well ask the same question about the vapor-compression refrigerator that we asked about the steam power plant—is it possible to accomplish our objective in a more direct manner? Is it possible, in the case of a refrigerator, to use the electrical energy

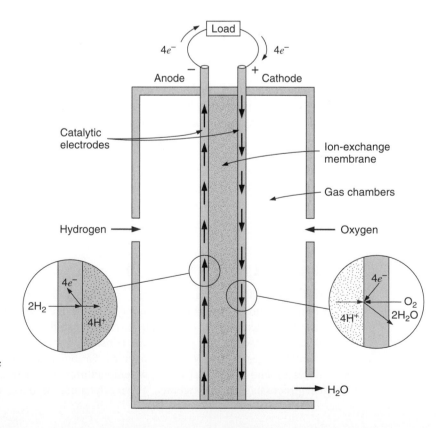

FIGURE 1.6 Schematic arrangement of an ion-exchange membrane type of fuel cell.

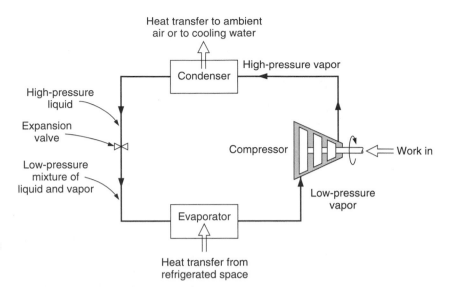

FIGURE 1.7 Schematic diagram of a simple refrigeration cycle.

(which goes to the electric motor that drives the compressor) to produce cooling in a more direct manner, and to avoid the cost of the compressor, condenser, evaporator, and all the related piping?

The thermoelectric refrigerator is such a device. This is shown schematically in Fig. 1.9. The thermoelectric device, like the conventional thermocouple, uses two dissimilar materials. There are two junctions between these two materials in a thermoelectric refrigerator. One is located in the refrigerated space and the other in ambient surroundings. When a potential difference is applied, as indicated, the temperature of the junction located in the refrigerated space will decrease and the temperature of the other junction will increase. Under steady-state operating conditions, heat will be transferred from the refrigerated space to the cold junction. The other junction will be at a temperature above the ambient, and heat will be transferred from the junction to the surroundings.

A thermoelectric device can also be used to generate power by replacing the refrigerated space with a body that is at a temperature above the ambient. Such a system is shown in Fig. 1.10.

The thermoelectric refrigerator cannot yet compete economically with the conventional vapor-compression units. However, in certain special applications, the thermoelectric refrigerator is already in use and, in view of research and development efforts under way in this field, it is quite possible that thermoelectric refrigerators will be much more extensively used in the future.

1.5 THE AIR SEPARATION PLANT

One process of great industrial significance is the air separation plant, in which air is separated into its various components. The oxygen, nitrogen, argon, and rare gases so produced are used extensively in various industrial, research, space, and consumer-goods applications. The air separation plant can be considered an example from two major fields, the chemical process industry and cryogenics. Cryogenics is a term applied to technology, processes, and research at very low temperatures (in general, below 150 K). In both chemical processing and cryogenics, thermodynamics is basic to an understanding of

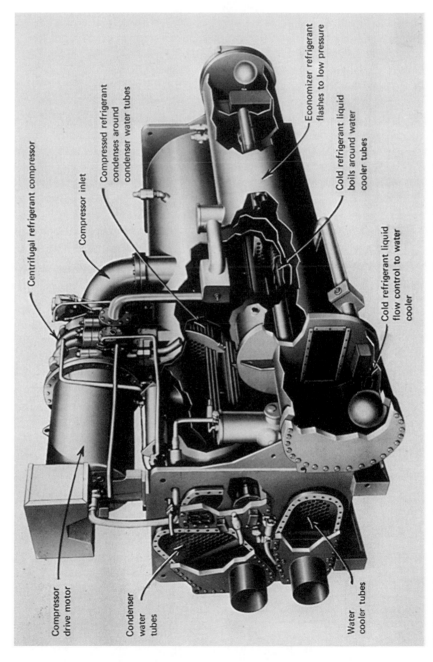

Centrifugal refrigerant compressor

Compressor inlet

Compressed refrigerant condenses around condenser water tubes

Economizer refrigerant flashes to low pressure

Cold refrigerant liquid boils around water cooler tubes

Cold refrigerant liquid flow control to water cooler

Compressor drive motor

Condenser water tubes

Water cooler tubes

FIGURE 1.8   A refrigeration unit for an air-conditioning system (Courtesy Carrier Air Conditioning Co.).

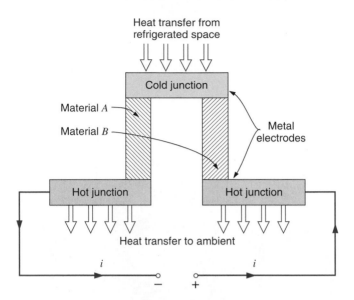

FIGURE 1.9 A themo-electric refrigerator.

many phenomena that occur and to the design and development of processes and equipment.

A number of different designs of air separation plants have been developed. Consider Fig. 1.11, which shows a somewhat simplified sketch of a type of plant that is frequently used. Air from the atmosphere is compressed to a pressure of 2 to 3 MPa. It is then purified, particularly to remove carbon dioxide (which would plug the flow passages as it solidifies when the air is cooled to its liquefaction temperature). The air is then compressed to a pressure of 15 to 20 MPa, cooled to the ambient temperature in the aftercooler, and dried to remove the water vapor (which would also plug the flow passages as it freezes).

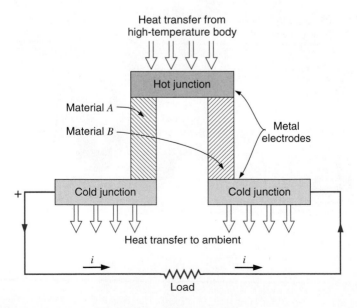

FIGURE 1.10 A thermoelectric power generation device.

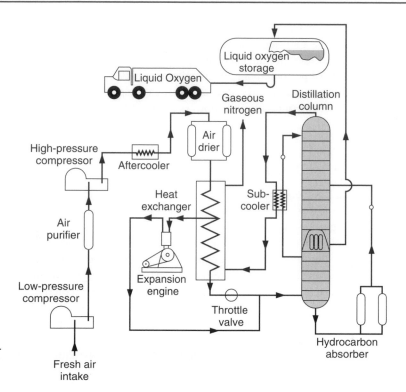

FIGURE 1.11 A simplified diagram of a liquid oxygen plant.

The basic refrigeration in the liquefaction process is provided by two different processes. In one process the air in the expansion engine expands. During this process the air does work and as a result the temperature of the air is reduced. In the other refrigeration process air passes through a throttle valve that is so designed and so located that there is a substantial drop in the pressure of the air and, associated with this, a substantial drop in the temperature of the air.

As shown in Fig. 1.11, the dry, high-pressure air enters a heat exchanger. The air temperature drops as it flows through the heat exchanger. At some intermediate point in the heat exchanger, part of the air is bled off and flows through the expansion engine. The remaining air flows through the rest of the heat exchanger and through the throttle valve. The two streams join (both are at the pressure of 0.5 to 1 MPa) and enter the bottom of the distillation column, which is referred to as the high-pressure column. The function of the distillation column is to separate the air into its various components, principally oxygen and nitrogen. Two streams of different composition flow from the high-pressure column through throttle valves to the upper column (also called the low-pressure column). One of these streams is an oxygen-rich liquid that flows from the bottom of the lower column, and the other is a nitrogen-rich stream that flows through the subcooler. The separation is completed in the upper column. Liquid oxygen leaves from the bottom of the upper column, and gaseous nitrogen leaves from the top of the column. The nitrogen gas flows through the subcooler and the main heat exchanger. It is the transfer of heat to this cold nitrogen gas that causes the high-pressure air entering the heat exchanger to become cooler.

Not only is a thermodynamic analysis essential to the design of the system as a whole, but essentially every component of such a system, including the compressors, the expansion engine, the purifiers and driers, and the distillation column, operates according to the principles of thermodynamics. In this separation process we are also concerned with the thermodynamic properties of mixtures and the principles and procedures by which these mixtures can be separated. This is the type of problem encountered in the refining of petroleum and many other chemical processes. It should also be noted that cryogenics is particularly relevant to many aspects of the space program, and a thorough knowledge of thermodynamics is essential for creative and effective work in cryogenics.

1.6 THE GAS TURBINE

The basic operation of a gas turbine is similar to the steam power plant, except that air is used instead of water. Fresh atmospheric air flows through a compressor that brings it to a high pressure. Energy is then added by spraying fuel into the air and igniting it so the combustion generates a high temperature flow. This high-temperature, high-pressure gas enters a turbine, where it expands down to the exhaust pressure, producing a shaft work output in the process. The turbine shaft work is used to drive the compressor and other devices, such as an electric generator that may be coupled to the shaft. The energy that is not used for shaft work comes out in the exhaust gases, so these have either a high temperature or a high velocity. The purpose of the gas turbine determines the design so that the most desirable energy form is maximized. An example of a large gas turbine for stationary power generation is shown in Fig. 1.12; the unit has 16 stages of compression, 4 stages in the turbine and is rated at 150 MW. Notice that since the combustion of fuel uses the oxygen in the air, the exhaust gases cannot be recirculated as the water is in the steam power plant.

A gas turbine is often the preferred power generating device where a large amount of power is needed, but only a small physical size is possible. Examples are jet engines, turbofan jet engines, offshore oilrig power plants, ship engines, helicopter engines, smaller local power plants, or peak load power generators in larger power plants. Since the gas turbine has relatively high exhaust temperatures, it can also be arranged so the exhaust gases are used to heat water that runs in a steam power plant before it exhausts to the atmosphere.

In the examples mentioned previously, the jet engine and turboprop application utilize part of the power to discharge the gases at high velocity. This is what generates the thrust of the engine that moves the airplane forward. The gas turbines in these applica-

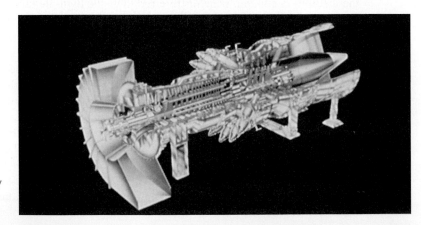

FIGURE 1.12 A 150-MW gas turbine (Courtesy Westinghouse Electric Corporation).

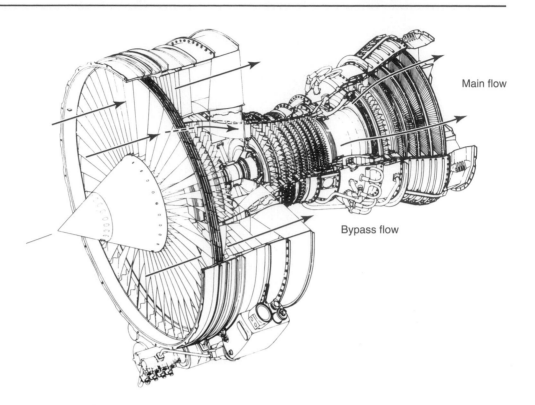

Main flow

Bypass flow

FIGURE 1.13 A turbofan jet engine (Courtesy General Electric Aircraft Engines).

tions are therefore designed differently than for the stationary power plant, where the energy is taken out as shaft work to an electric generator. An example of a turbofan jet engine used in a commercial airplane is shown in Fig. 1.13. The large front end fan also blows air past the engine providing cooling and gives additional thrust.

1.7 THE CHEMICAL ROCKET ENGINE

The advent of missiles and satellites brought to prominence the use of the rocket engine as a propulsion power plant. Chemical rocket engines may be classified as either liquid propellant or solid propellant, according to the fuel used.

Figure 1.14 shows a simplified schematic diagram of a liquid propellant rocket. The oxidizer and fuel are pumped through the injector plate into the combustion chamber where combustion takes place at high pressure. The high-pressure, high-temperature products of combustion expand as they flow through the nozzle, and as a result they leave the nozzle with a high velocity. The momentum change associated with this increase in velocity gives rise to the forward thrust on the vehicle.

The oxidizer and fuel must be pumped into the combustion chamber, and some auxiliary power plant is necessary to drive the pumps. In a large rocket this auxiliary power plant must be very reliable and have a relatively high power output, yet it must be light in weight. The oxidizer and fuel tanks occupy the largest part of the volume of an actual rocket, and the range and payload of a rocket are determined largely by the amount of oxidizer and fuel that can be carried. Many different fuels and oxidizers have been considered and tested, and much effort has gone into the development of fuels and oxidizers that will give a higher thrust per unit mass rate of flow of reactants. Liquid oxygen

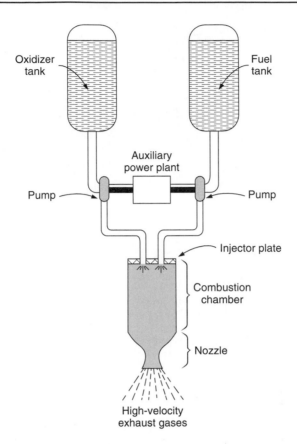

FIGURE 1.14
Simplified schematic
diagram of a liquid-
propellant rocket engine.

is frequently used as the oxidizer in liquid-propellant rockets, and liquid hydrogen is frequently used as the fuel.

Much work has also been done on solid-propellant rockets. They have been very successfully used for jet-assisted takeoffs of airplanes, military missiles, and space vehicles. They are much simpler in both the basic equipment required for operation and the logistic problems involved in their use, but they are more difficult to control.

1.8 ENVIRONMENTAL ISSUES

In the first seven sections of this chapter, we have introduced and discussed a number of devices, each of which produces certain effects for the use and convenience of humankind. For example, the construction and operation of the steam power plant creates electricity, which is so deeply entrenched in our society that we take its ready availability for granted. In recent years, however, it has become increasingly apparent that we need to consider seriously the effects of such an operation on our environment. Combustion of hydrocarbon fuels releases carbon dioxide into the atmosphere, where its concentration is increasing. Carbon dioxide, as well as other gases, absorbs infrared radiation from the surface of the earth, holding it close to the planet and creating the "greenhouse effect," which in turn is believed to cause global warming and critical climatic changes around the earth. Power plant combustion, particularly of coal, releases sulfur dioxide, which is absorbed in clouds and later falls as acid rain in many areas. Combustion processes in power plants and gasoline and diesel engines generate pollutants other than the carbon dioxide and sulfur oxides mentioned. Species such as carbon monoxide, nitric oxides, and partly burned fuels together with particulates all contribute to atmospheric pollution and are regulated by law for many applications. Refrigeration and air-conditioning systems, as well as other industrial processes, use certain chlorofluorocarbons that eventually find their way to the upper atmosphere and destroy the protective ozone layer.

These are only some of the many environmental problems caused by our efforts to produce goods and effects intended to improve our way of life. During our study of thermodynamics, which is the science of the conversion of energy from one form to another, we must continue to reflect on these issues. We must consider how we can eliminate or at least minimize damaging effects, as well as use our natural resources, efficiently and responsibly.

2 Some Concepts and Definitions

One excellent definition of thermodynamics is that it is the science of energy and entropy. Since we have not yet defined these terms, an alternate definition in already familiar terms is: thermodynamics is the science that deals with heat and work and these properties of substances that bear a relation to heat and work. Like all sciences, the basis of thermodynamics is experimental observation. In thermodynamics these findings have been formalized into certain basic laws, which are known as the first, second, and third laws of thermodynamics. In addition to these laws, the zeroth law of thermodynamics, which in the logical development of thermodynamics precedes the first law, has been set forth.

In the chapters that follow, we will present these laws and the thermodynamic properties related to these laws, and apply them to a number of representative examples. The objective of the student should be to gain both a thorough understanding of the fundamentals and an ability to apply these fundamentals to thermodynamic problems. The examples and problems further this twofold objective. It is not necessary for the student to memorize numerous equations, for problems are best solved by the application of the definitions and laws of thermodynamics. In this chapter some concepts and definitions basic to thermodynamics are presented.

2.1 A THERMODYNAMIC SYSTEM AND THE CONTROL VOLUME

A thermodynamic system is a device or combination of devices containing a quantity of matter that is being studied. To define this more precisely a control volume is chosen, so that it contains the matter and devices inside a control surface. Everything external to the control volume is the surroundings, with the separation given by the control surface. The surface may be open or closed to mass flows and it may have flows of energy in terms of heat transfer and work across it. The boundaries may be movable or stationary. In the case of a control surface that is closed to mass flow, so that no mass can escape or enter the control volume, it is called a **control mass** containing the same amount of matter at all times.

Selecting the gas in the cylinder of Fig. 2.1 as a control volume by placing a control surface around it, we recognize this as a control mass. If a Bunsen burner is placed under the cylinder, the temperature of the gas will increase and the piston will rise. As the piston rises, the boundary of the control mass moves. As we will see later, heat and work

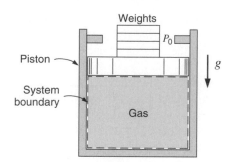

FIGURE 2.1 Example of a control mass.

cross the boundary of the control mass during this process, but the matter that comprises the control mass can always be identified and remains the same.

An isolated system is one that is not influenced in any way by the surroundings. This means that no mass, heat, or work cross the boundary of the system. In many cases a thermodynamic analysis must be made of a device, such as an air compressor, which has a flow of mass into it, out of it, or both, as shown schematically in Fig. 2.2. The procedure followed in such an analysis is to specify a control volume that surrounds the device under consideration. The surface of this control volume is the control surface, which may have mass momentum, and also heat and work, cross it.

Thus the more general control surface defines a control volume, where mass may flow in or out, with a control mass as the special case of no mass flow in or out, so the control mass contains a fixed mass at all times, hence its name. The difference in the formulation of the analysis is considered in detail in Chapter 6. The terms *closed system* (fixed mass) and *open system* (involving a flow of mass) are sometimes used to make this distinction. Here, we use the term *system* as a more general and loose description for a mass, device, or combination of devices that then is more precisely defined, when a control volume is selected. The procedure that will be followed in the presentation of the first and the second laws of thermodynamics is first to present these laws for a control mass and then to extend the analysis to the more general control volume.

2.2 MACROSCOPIC VERSUS MICROSCOPIC POINT OF VIEW

An investigation into the behavior of a system may be undertaken from either a microscopic or a macroscopic point of view. Let us briefly describe a system from a microscopic point of view. Consider a system consisting of a cube 25 mm on a side and con-

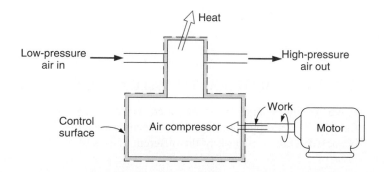

FIGURE 2.2 Example of a control volume.

taining a monatomic gas at atmospheric pressure and temperature. This volume contains approximately 10^{20} atoms. To describe the position of each atom, we need to specify three coordinates; to describe the velocity of each atom, we specify three velocity components.

Thus, to describe completely the behavior of this system from a microscopic point of view we must deal with at least 6×10^{20} equations. Even with a large digital computer, this is a quite hopeless computational task. However, there are two approaches to this problem that reduce the number of equations and variables to a few that can be computed relatively easily. One approach is the statistical approach, in which, on the basis of statistical considerations and probability theory, we deal with "average" values for all particles under consideration. This is usually done in connection with a model of the atom under consideration. This is the approach used in the disciplines known as kinetic theory and statistical mechanics.

The other approach that reduces the number of variables to a few that can be handled is the macroscopic point of view of classical thermodynamics. As the word macroscopic implies, we are concerned with the gross or average effects of many molecules. These effects can be perceived by our senses and measured by instruments. However, what we really perceive and measure is the time-averaged influence of many molecules. For example, consider the pressure a gas exerts on the walls of its container. This pressure results from the change in momentum of the molecules as they collide with the wall. From a macroscopic point of view, however, we are not concerned with the action of the individual molecules but with the time-averaged force on a given area, which can be measured by a pressure gage. In fact, these macroscopic observations are completely independent of our assumptions regarding the nature of matter.

Although the theory and development in this book is presented from a macroscopic point of view, a few supplementary remarks regarding the significance of the microscopic perspective are included as an aid to the understanding of the physical processes involved. Another book in this series, *Introduction to Thermodynamics: Classical and Statistical,* by R. E. Sonntag and G. J. Van Wylen, includes thermodynamics from the microscopic and statistical point of view.

A few remarks should be made regarding the continuum. From the macroscopic view, we are always concerned with volumes that are very large compared to molecular dimensions and, therefore, with systems that contain many molecules. Because we are not concerned with the behavior of individual molecules, we can treat the substance as being continuous, disregarding the action of individual molecules; this is called a continuum. The concept of a continuum, of course, is only a convenient assumption that loses validity when the mean free path of the molecules approaches the order of magnitude of the dimensions of the vessel, as, for example, in high-vacuum technology. In much engineering work the assumption of a continuum is valid and convenient, and goes hand in hand with the macroscopic view.

2.3 Properties and State of a Substance

If we consider a given mass of water, we recognize that this water can exist in various forms. If it is a liquid initially, it may become a vapor when it is heated or a solid when it is cooled. Thus, we speak of the different phases of a substance. A phase is defined as a quantity of matter that is homogeneous throughout. When more than one phase is present, the phases are separated from each other by the phase boundaries. In each phase the substance may exist at various pressures and temperatures or, to use the thermodynamic term, in various states. The state may be identified or described by certain observable, macroscopic properties; some familiar ones are temperature, pressure, and density. In later chapters other properties will be introduced. Each of the properties of a substance in a given state has only one definite value, and these properties always have the same value for a given state, regardless of how the substance arrived at the state. In fact, a property can be defined as any quantity that depends on the state of the system and is independent of the path (that is, the prior history) by which the system arrived at the given state. Conversely, the state is specified or described by the properties, and later we will consider the number of independent properties a substance can have, that is, the minimum number of properties that must be specified to fix the state of the substance.

Thermodynamic properties can be divided into two general classes, intensive and extensive properties. An intensive property is independent of the mass; the value of an extensive property varies directly with the mass. Thus, if a quantity of matter in a given state is divided into two equal parts, each part will have the same value of intensive properties as the original, and half the value of the extensive properties. Pressure, temperature, and density are examples of intensive properties. Mass and total volume are examples of extensive properties. Extensive properties per unit mass, such as specific volume, are intensive properties.

Frequently we will refer not only to the properties of a substance but to the properties of a system. When we do so we necessarily imply that the value of the property has significance for the entire system, and this implies equilibrium. For example, if the gas that comprises the system (control mass) in Fig. 2.1 is in thermal equilibrium, the temperature will be the same throughout the entire system, and we may speak of the temperature as a property of the system. We may also consider mechanical equilibrium, which is related to pressure. If a system is in mechanical equilibrium, there is no tendency for the pressure at any point to change with time as long as the system is isolated from the surroundings. There will be a variation in pressure with elevation because of the influence of gravitational forces, although under equilibrium conditions there will be no tendency for the pressure at any location to change. However, in many thermodynamic problems, this variation in pressure with elevation is so small that it can be neglected. Chemical equilibrium is also important and will be considered in Chapter 15. When a system is in equilibrium regarding all possible changes of state, we say that the system is in thermodynamic equilibrium.

2.4 PROCESSES AND CYCLES

Whenever one or more of the properties of a system change, we say that a change in state has occurred. For example, when one of the weights on the piston in Fig. 2.3 is removed, the piston rises and a change in state occurs, for the pressure decreases and the specific volume increases. The path of the succession of states through which the system passes is called the process.

Let us consider the equilibrium of a system as it undergoes a change in state. The moment the weight is removed from the piston in Fig. 2.3, mechanical equilibrium does

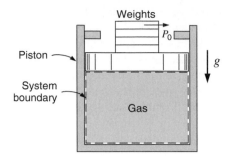

FIGURE 2.3 Example of a system that may undergo a quasi-equilibrium process.

not exist, and as a result the piston is moved upward until mechanical equilibrium is again restored. The question is this: Since the properties describe the state of a system only when it is in equilibrium, how can we describe the states of a system during a process if the actual process occurs only when equilibrium does not exist? One step in the answer to this question concerns the definition of an ideal process, which we call a quasi-equilibrium process. A quasi-equilibrium process is one in which the deviation from thermodynamic equilibrium is infinitesimal, and all the states the system passes through during a quasi-equilibrium process may be considered equilibrium states. Many actual processes closely approach a quasi-equilibrium process, and may be so treated with essentially no error. If the weights on the piston in Fig. 2.3 are small and are taken off one by one, the process could be considered quasi-equilibrium. On the other hand, if all the weights were removed at once, the piston would rise rapidly until it hit the stops. This would be a nonequilibrium process, and the system would not be in equilibrium at any time during this change of state.

For nonequilibrium processes, we are limited to a description of the system before the process occurs and after the process is completed and equilibrium is restored. We are unable to specify each state through which the system passes or the rate at which the process occurs. However, as we will see later, we are able to describe certain overall effects that occur during the process.

Several processes are described by the fact that one property remains constant. The prefix iso- is used to describe this process. An isothermal process is a constant-temperature process, an isobaric (sometimes called isopiestic) process is a constant-pressure process, and an isochoric process is a constant-volume process.

When a system in a given initial state goes through a number of different changes of state or processes and finally returns to its initial state, the system has undergone a cycle. Therefore, at the conclusion of a cycle, all the properties have the same value they had at the beginning. Steam (water) that circulates through a steam power plant undergoes a cycle.

A distinction should be made between a thermodynamic cycle, which has just been described, and a mechanical cycle. A four-stroke cycle internal-combustion engine goes through a mechanical cycle once every two revolutions. However, the working fluid does not go through a thermodynamic cycle in the engine, since air and fuel are burned and changed to products of combustion that are exhausted to the atmosphere. In this text the term cycle will refer to a thermodynamic "cycle" unless otherwise designated.

2.5 UNITS FOR MASS, LENGTH, TIME, AND FORCE

Since we are considering thermodynamic properties from a macroscopic perspective, we are dealing with quantities that can, either directly or indirectly, be measured and counted. Therefore, the matter of units becomes an important consideration. In the remaining sections of this chapter we will define certain thermodynamic properties and the basic units. The relation between force and mass is often a difficult matter for students, and is considered in this section in some detail.

Force, mass, length, and time are related by Newton's second law of motion, which states that the force acting on a body is proportional to the product of the mass and the acceleration in the direction of the force.

$$F \propto ma$$

The concept of time is well established. The basic unit of time is the second (s), which in the past was defined in terms of the solar day, the time interval for one complete revolution of the earth relative to the sun. Since this period varies with the season of the year, an average value over a one-year period is called the mean solar day, and the mean solar second is 1/86 400 of the mean solar day. (The measurement of the earth's rotation is sometimes made relative to a fixed star, in which case the period is called a sidereal day.) In 1967, the General Conference of Weights and Measures (CGPM) adopted a definition of the second as the time required for a beam of cesium-133 atoms to resonate 9 192 631 770 cycles in a cesium resonator.

For periods of time less than a second, the prefixes milli, micro, nano, or pico, as listed in Table 2.1, are commonly used. For longer periods of time, the units minute (min), hour (h), or day (day) are frequently used. It should be pointed out that the prefixes in Table 2.1 are used with many other units as well.

The concept of length is also well established. The basic unit of length is the meter (m). For many years the accepted standard was the International Prototype Meter, the distance between two marks on a platinum–iridium bar under certain prescribed conditions. This bar is maintained at the International Bureau of Weights and Measures, Sevres, France. In 1960, the CGPM adopted a definition of the meter as a length equal to 1 650 763.73 wavelengths in a vacuum of the orange-red line of krypton-86. Then in 1983, the CGPM adopted a more precise definition of the meter in terms of the speed of light (which is now a fixed constant): the meter is the length of the path traveled by light in a vacuum during a time interval of 1/299 792 458 of a second.

The fundamental unit of mass is the kilogram (kg). As adopted by the first CGPM in 1889 and restated in 1901, it is the mass of a certain platinum–iridium cylinder maintained under prescribed conditions at the International Bureau of Weights and Measures.

TABLE 2.1
Unit Prefixes

Factor	Prefix	Symbol	Factor	Prefix	Symbol
10^{12}	tera	T	10^{-3}	milli	m
10^{9}	giga	G	10^{-6}	micro	μ
10^{6}	mega	M	10^{-9}	nano	n
10^{3}	kilo	k	10^{-12}	pico	p

A related unit that is used frequently in thermodynamics is the mole (mol), defined as an amount of substance containing as many elementary entities as there are atoms in 0.012 kg of carbon-12. These elementary entities must be specified, and may be atoms, molecules, electrons, ions, or other particles or specific groups. For example, one mole of diatomic oxygen, having a molecular weight of 32 (compared to 12 for carbon), is a mass of 0.032 kg. The mole is often termed a gram mole, since it is an amount of substance in grams numerically equal to the molecular weight. In this text, when using the metric SI system we will find it preferable to use the kilomole (kmol), the amount of substance in kilograms numerically equal to the molecular weight, rather than the mole.

The system of units in use presently throughout most of the world is the metric International System, commonly referred to as SI units (from Le Système International d'Unités). In this system, the second, meter, and kilogram are the basic units for time, length, and mass, respectively, as just defined, and the unit of force is defined directly from Newton's second law.

Therefore, a proportionality constant is unnecessary, and we may write that law as an equality:

$$F = ma$$

The unit of force is the newton (N), which by definition is the force required to accelerate a mass of one kilogram at the rate of one meter per second per second.

$$1 \, \mathrm{N} = 1 \, \mathrm{kg \, m / s^2}$$

It is worth noting that SI units derived from proper nouns use capital letters for symbols; others use the lowercase letters. The liter, with the symbol L, is an exception.

The traditional system of units used in the United States is the English Engineering System. In this system the unit of time is the second, which has been discussed earlier. The basic unit of length is the foot (ft), which at present is defined in terms of the meter as

$$1 \, \mathrm{ft} = 0.3048 \, \mathrm{m}$$

The inch (in.) is defined in terms of the foot

$$12 \, \mathrm{in.} = 1 \, \mathrm{ft}$$

The unit of mass in this system is the pound mass (lbm). It was originally the mass of a certain platinum cylinder kept in the Tower of London, but now it is defined in terms of the kilogram as

$$1 \, \mathrm{lbm} = 0.453 \, 592 \, 37 \, \mathrm{kg}$$

A related unit is the pound mole (lb mol), which is an amount of substance in pounds mass numerically equal to the molecular weight of that substance. It is important to distinguish between a pound mole and a mole (gram mole).

In the English Engineering System of Units, the concept of force is not defined from Newton's second law, but is instead established as an independent quantity. Thus, for this system, unlike other systems of units, a conversion constant g_c is necessary in Newton's second law. The unit for force is defined in terms of an experimental procedure as follows. Let the standard pound mass be suspended in the earth's gravitational field at a location where the acceleration due to gravity is 32.1740 ft/s². The force with which the standard pound mass is attracted to the earth (the buoyant effects of the atmosphere on the standard pound mass must also be standardized) is defined as the unit for force and is termed a pound force (lbf). We must be careful to distinguish between a lbm and a lbf, and we do not use the term pound alone. Since we have independently defined the units for force, mass, length, and time, we have the relation, from Newton's second law,

$$F = \frac{ma}{g_c}$$

or

$$1 \text{ lbf} = \frac{1 \text{ lbm} \times 32.174 \text{ ft} / \text{s}^2}{g_c}$$

or

$$g_c = 32.174 \frac{\text{lbm ft}}{\text{lbf s}^2}$$

Note that the conversion constant g_c has both a numerical value and dimensions. We also note that for gravitational accelerations that are not too different from the standard value (32.174 ft/s^2 or 9.806 65 m/s^2), the lbf is approximately numerically equal to the lbm. This is the reason that people often drop the designation, which can lead to confusion and inconsistencies. To illustrate, let us calculate the force due to gravity on a one-pound mass at a location where the acceleration due to gravity is 32.14 ft/s^2 (about 10 000 ft above sea level).

$$F = \frac{ma}{g_c} = \frac{1 \text{ lbm} \times 32.14 \text{ ft} / \text{s}^2}{32.174 \text{ lbm ft} / \text{lbf s}^2}$$
$$= 0.999 \text{ lbf}$$

We further note that the pound force could very well have been defined as

$$1 \text{ lbf} = 32.174 \text{ lbm ft} / \text{s}^2$$

in a manner analogous to that for the newton in the SI system. It follows that

$$1 = 32.174 \frac{\text{lbm ft}}{\text{lbf s}^2}$$

Comparing this with the conversion constant g_c, we have

$$g_c = 32.174 \frac{\text{lbm ft}}{\text{lbf s}^2} = 1$$

Since a pure number can be substituted into an equation at any point, we can therefore substitute this unit conversion constant g_c as necessary. Throughout the remainder of this text, we will follow this philosophy, and will not include the constant g_c explicitly in our equations, even when our attention is focused on the English Engineering system of units.

The term *weight* is often used with respect to a body, and is sometimes confused with mass. Weight is really correctly used only as a force. When we say a body weighs so much, we mean that this is the force with which it is attracted to the earth (or some other body), that is, the product of its mass and the local gravitational acceleration. The mass of a substance remains constant with elevation, but its weight varies with elevation.

2.6 Energy

One of the very important concepts in a study of thermodynamics is the concept of energy. Energy is a fundamental concept, such as mass or force and, as is often the case with such concepts, is very difficult to define. Energy has been defined as the capability to produce an effect. Fortunately the word *energy* and the basic concept that this word represents are familiar to us in everyday usage, and a precise definition is not essential at this point.

It is important to note that energy can be stored within a system and can be transferred (as heat, for example) from one system to another. In a study of statistical thermodynamics we would examine, from a molecular view, the ways in which energy can be stored. Because it is helpful in a study of classical thermodynamics to have some notion of how this energy is stored, a brief introduction is presented here.

Consider as a system a certain gas at a given pressure and temperature contained within a tank or pressure vessel. When considered from the molecular view, we identify three general forms of energy.

1. Intermolecular potential energy, which is associated with the forces between molecules.
2. Molecular kinetic energy, which is associated with the translational velocity of individual molecules.
3. Intramolecular energy (that within the individual molecules), which is associated with the molecular and atomic structure and related forces.

The first of these forms of energy, the intermolecular potential energy, depends on the magnitude of the intermolecular forces and the position the molecules have relative to each other at any instant of time. It is impossible to determine accurately the magnitude of this energy because we do not know either the exact configuration and orientation of the molecules at any time, or the exact intermolecular potential function. However, there are two situations for which we can make good approximations. The first situation is at low or moderate densities. In this case the molecules are relatively widely spaced, so that only two-molecule or two- and three-molecule interactions contribute to the potential energy. At these low and moderate densities, techniques are available for determining, with reasonable accuracy, the potential energy of a system composed of reasonably simple molecules. The second situation is at very low densities; under these conditions the average intermolecular distance between molecules is so large that the potential energy may be assumed to be zero. Consequently, we have in this case a system of independent particles (an ideal gas) and, therefore, from a statistical point of view, we are able to concentrate our efforts on evaluating the molecular translational and internal energies.

The translational energy, which depends only on the mass and velocities of the molecules, is determined by using the equations of mechanics—either quantum or classical.

The intramolecular internal energy is more difficult to evaluate because, in general, it may result from a number of contributions. Consider a simple monatomic gas such as helium. Each molecule consists of a helium atom. Such an atom possesses electronic energy as a result of both orbital angular momentum of the electrons about the nucleus and angular momentum of the electrons spinning on their axes. The electronic energy is commonly very small compared with the translational energies. (Atoms also possess nuclear

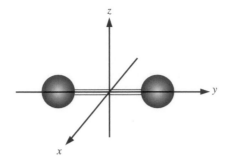

FIGURE 2.4 The coordinate system for a diatomic molecule.

energy, which, except in the case of nuclear reactions, is constant. We are not concerned with nuclear energy at this time.) When we consider more complex molecules, such as those comprised of two or three atoms, additional factors must be considered. In addition to having electronic energy, a molecule can rotate about its center of gravity and thus have rotational energy. Furthermore, the atoms may vibrate with respect to each other and have vibrational energy. In some situations there may be an interaction between the rotational and vibrational modes of energy.

In the evaluation of the energy of a molecule, reference is often made to the degree of freedom, f, for these energy modes. For a monatomic molecule, such as helium, $f = 3$, and this represents the three directions, x, y, and z in which the molecule can move. For a diatomic molecule, such as oxygen, $f = 6$. Three of these are the translation of the molecule as a whole in the x, y, and z directions, and two are for rotation. The reason why there are only two modes of rotational energy is evident from Fig. 2.4, where we take the origin of the coordinate system at the center of gravity of the molecule, and the y-axis along the molecule's internuclear axis. The molecule will then have an appreciable moment of inertia about the x-axis and the z-axis, but not about the y-axis. The sixth degree of freedom of the molecule is vibration, which relates to stretching of the bond joining the atoms.

For a more complex molecule such as H_2O, there are additional vibrational degrees of freedom. Figure 2.5 shows a model of the H_2O molecule. From this diagram it is evident that there are three vibrational degrees of freedom. It is also possible to have rotational energy about all three axes. Thus, for the H_2O molecule, there are nine degrees of freedom ($f = 9$), three translational, three rotational, and three vibrational.

This general discussion can be summarized by referring to Fig. 2.6. Let heat be transferred to the water. During this process the temperature of the liquid and vapor (steam) will increase, and eventually all the liquid will become vapor. From the macroscopic view we are concerned only with the energy that is transferred as heat, the change in properties, such as temperature and pressure, and the total amount of energy (relative to some base) that the H_2O contains at any instant. Thus, questions about how energy is

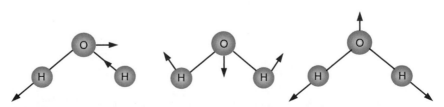

FIGURE 2.5 The three principal vibrational modes for the water molecule.

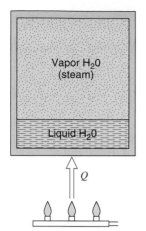

FIGURE 2.6 Heat transfer to water.

stored in the H_2O do not concern us. From a microscopic viewpoint we are concerned about the way in which energy is stored in the molecules. We might be interested in developing a model of the molecule so that we could predict the amount of energy required to change the temperature a given amount. Although the focus in this book is on the macroscopic or classical viewpoint, it is helpful to keep in mind the microscopic or statistical perspective as well, as the relationship between the two helps us in understanding basic concepts such as energy.

2.7 SPECIFIC VOLUME

The specific volume of a substance is defined as the volume per unit mass, and is given the symbol v. The density of a substance is defined as the mass per unit volume, and is therefore the reciprocal of the specific volume. Density is designated by the symbol ρ. Specific volume and density are intensive properties.

The specific volume of a system in a gravitational field may vary from point to point. For example, if the atmosphere is considered a system, the specific volume increases as the elevation increases. Therefore, the definition of specific volume involves the specific volume of a substance at a point in a system.

Consider a small volume δV of a system, and let the mass be designated δm. The specific volume is defined by the relation

$$v = \lim_{\delta V \to \delta V'} \frac{\delta V}{\delta m}$$

where $\delta V'$ is the smallest volume for which the mass can be considered a continuum. Volumes smaller than this will lead to the recognition that mass is not evenly distributed in space, but concentrated in particles as molecules, atoms, electrons, etc. This is tentatively indicated in Fig. 2.7, where in the limit of a zero volume the specific volume may be infinite (the volume does not contain any mass) or very small (the volume is part of a nucleus).

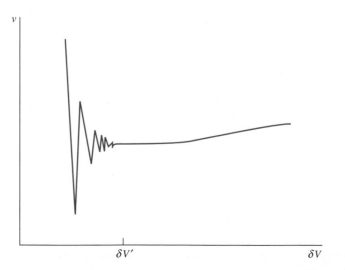

FIGURE 2.7 The continuum limit for the specific volume.

Thus, in a given system, we should speak of the specific volume or density at a point in the system, and recognize that this may vary with elevation. However, most of the systems that we consider are relatively small, and the change in specific volume with elevation is not significant. Therefore, we can speak of one value of specific volume or density for the entire system.

In this text, the specific volume and density will be given either on a mass or on a mole basis. A bar over the symbol (lowercase) will be used to designate the property on a mole basis. Thus, $\bar{v}$ will designate molal specific volume and $\bar{\rho}$ will designate the molal density. In SI units, those for specific volume are m^3/kg and m^3/mol (or $m^3/kmol$); for density the corresponding units are kg/m^3 and mol/m^3 (or $kmol/m^3$). In English units, those for specific volume are ft^3/lbm and $ft^3/lb\ mol$; the corresponding units for density are lbm/ft^3 and $lb\ mol/ft^3$.

Although the SI unit for volume is the cubic meter, a commonly used volume unit is the liter (L), which is a special name given to a volume of 0.001 cubic meters, that is, 1 L = 10^{-3} m^3.

2.8 PRESSURE

When dealing with liquids and gases, we ordinarily speak of pressure; for solids we speak of stresses. The pressure in a fluid at rest at a given point is the same in all directions, and we define pressure as the normal component of force per unit area. More specifically, if $\delta\mathcal{A}$ is a small area, $\delta\mathcal{A}'$ the smallest area over which we can consider the fluid a continuum, and δF_n the component of force normal to $\delta\mathcal{A}$, we define pressure, P, as

$$P = \lim_{\delta\mathcal{A}\to\delta\mathcal{A}'} \frac{\delta F_n}{\delta\mathcal{A}}$$

where the lower limit corresponds to sizes as mentioned for the specific volume, shown in Fig. 2.7. The pressure P at a point in a fluid in equilibrium is the same in all directions. In a viscous fluid in motion, the variation in the state of stress with orientation becomes an important consideration. These considerations are beyond the scope of this book, and we will consider pressure only in terms of a fluid in equilibrium.

The unit for pressure in the International System is the force of one newton acting on a square meter area, which is called the pascal (Pa). That is,

$$1\ Pa = 1\ N/m^2$$

In general, the unit for pressure that is consistent with the English units is pounds force per square foot (lbf/ft^2). On the other hand, in common parlance and general experimental work, pressures are often measured in pounds force per square inch ($lbf/in.^2$). Therefore, the student should be careful in numerical calculations to introduce the conversion 144 $in.^2$ = 1 ft^2 as necessary.

Two other units, not part of the International System, continue to be widely used. These are the bar, where

$$1\ bar = 10^5\ Pa = 0.1\ MPa$$

and the standard atmosphere, where

$$1\ atm = 101\,325\ Pa = 14.696\ lbf/in.^2$$

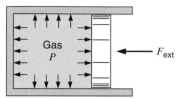

FIGURE 2.8 The balance of forces on a movable boundary relates to inside gas pressure.

which is slightly larger than the bar. In this text, we will normally use the SI unit, the pascal, and especially the multiples of kilopascal and megapascal. The bar will be utilized often in the examples and problems, but the atmosphere will not be used, except in specifying certain reference points.

Consider a gas contained in a cylinder fitted with a movable piston, as shown in Fig. 2.8. The pressure exerted by the gas on all its boundaries is the same, assuming that the gas is in an equilibrium state. This pressure is fixed by the external force acting on the piston, since there must be a balance of forces in order that the piston remain stationary. Thus, the product of the pressure and the movable piston area must be equal to the external force. If the external force is now changed, in either direction, the gas pressure inside must accordingly adjust, with appropriate movement of the piston, to establish a force balance at a new equilibrium state. As another example, if the gas in the cylinder is heated by an outside body, which tends to increase the gas pressure, the piston will move instead, such that the pressure remains equal to whatever value is required by the external force.

In most thermodynamic investigations we are concerned with absolute pressure. Most pressure and vacuum gauges, however, read the difference between the absolute pressure and the atmospheric pressure existing at the gauge. This referred to as gauge pressure. This is shown graphically in Fig. 2.9, and the following examples illustrate the

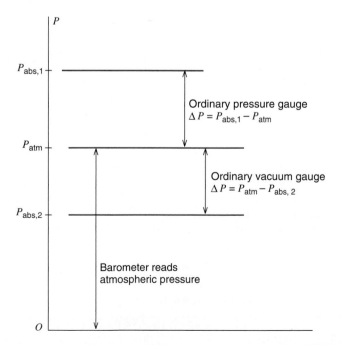

FIGURE 2.9 Illustration of terms used in pressure measurement.

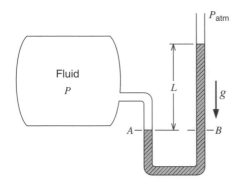

FIGURE 2.10 Example of pressure measurement using a column of fluid.

principles. Pressures below atmospheric and slightly above atmospheric, and pressure differences (for example, across an orifice in a pipe) are frequently measured with a manometer, which contains water, mercury, alcohol, oil, or other fluids. From the principles of hydrostatics one concludes that for a difference in level of L meters, the pressure difference in pascals is calculated by the relation

$$\Delta P = \rho L g$$

where ρ is the density of the fluid, and g is the local acceleration due to gravity. The accepted standard value for gravitational acceleration is

$$g = 9.806\ 65\ \mathrm{m/s}^2 = 32.174\ \mathrm{ft/s}^2$$

but the value varies with location and elevation. The use of a manometer is illustrated in Fig. 2.10.

For distinguishing between absolute and gauge pressure in this text, the term pascal or lbf/in.^{2} will always refer to absolute pressure. Any gauge pressure will be indicated as such.

EXAMPLE 2.1 A mercury (Hg) manometer is used to measure the pressure in a vessel as shown in Fig. 2.10. The mercury has a density of 13590 kg/m^3 and the height difference between the two columns is measured to be 24 cm. We want to determine the pressure inside the vessel.

Solution:
The manometer measures the gauge pressure as a pressure difference

$$\Delta P = P_{gauge} = \rho L g = 13590 \times 0.24 \times 9.80665$$

$$= 31985 \frac{kg}{m^3} m \frac{m}{s^2} = 31985\ Pa = 31.985\ kPa$$

$$= 0.316\ atm$$

In order to get the absolute pressure inside the vessel we have

$$P_A = P_{vessel} = P_B = \Delta P + P_{atm}$$

so we need to know the atmospheric pressure measured by a barometer (absolute pressure). Assume this pressure is known as 750 mm Hg, measured similar to the above setup with one side open to the atmosphere and the other side closed so there is mercury vapor with a very small pressure on top of the liquid column. The absolute pressure in the vessel becomes

$$P_{vessel} = \Delta P + P_{atm} = 13590 \times 0.750 \times 9.80665 + 31985$$
$$= 99954 + 31985 = 131940 \ Pa = 1.302 \ atm$$

2.9 EQUALITY OF TEMPERATURE

Although temperature is a familiar property, an exact definition of it is difficult. We are aware of "temperature" first of all as a sense of hotness or coldness when we touch an object. We also learn early that when a hot body and a cold body are brought into contact, the hot body becomes cooler and the cold body becomes warmer. If these bodies remain in contact for some time, they usually appear to have the same hotness or coldness. However, we also realize that our sense of hotness or coldness is very unreliable. Sometimes very cold bodies may seem hot, and bodies of different materials that are at the same temperature appear to be at different temperatures.

Because of these difficulties in defining temperature, we define equality of temperature. Consider two blocks of copper, one hot and the other cold, each of which is in contact with a mercury-in-glass thermometer. If these two blocks of copper are brought into thermal communication, we observe that the electrical resistance of the hot block decreases with time and that of the cold block increases with time. After a period of time has elapsed, however, no further changes in resistance are observed. Similarly, when the blocks are first brought in thermal communication, the length of a side of the hot block decreases with time, but the length of a side of the cold block increases with time. After a period of time, no further change in length of either of the blocks is perceived. In addition, the mercury column of the thermometer in the hot block drops at first and that in the cold block rises, but after a period of time no further changes in height are observed. We may say, therefore, that two bodies have equality of temperature if, when they are in thermal communication, no change in any observable property occurs.

2.10 THE ZEROTH LAW OF THERMODYNAMICS

Now consider the same two blocks of copper and another thermometer. Let one block of copper be brought into contact with the thermometer until equality of temperature is established, and then remove it. Then let the second block of copper be brought into contact with the thermometer. Suppose that no change in the mercury level of the thermometer occurs during this operation with the second block. We then can say that both blocks are in thermal equilibrium with the given thermometer.

The zeroth law of thermodynamics states that when two bodies have equality of temperature with a third body, they in turn have equality of temperature with each other. This seems obvious to us because we are so familiar with this experiment. Because the principle is not derivable from other laws, and because it precedes the first and second laws of thermodynamics in the logical presentation of thermodynamics, it is called the zeroth law of thermodynamics. This law is really the basis of temperature measurement. Every time a body has equality of temperature with the thermometer, we can say that the body has the temperature we read on the thermometer. The problem remains how to relate temperatures that we might read on different mercury thermometers or obtain from different temperature-measuring devices, such as thermocouples and resistance thermometers. This observation suggests the need for a standard scale for temperature measurements.

2.11 TEMPERATURE SCALES

Two scales are commonly used for measuring temperature, namely the Fahrenheit (after Gabriel Fahrenheit, 1686–1736) and the Celsius. The Celsius scale was formerly called the centigrade scale but is now designated the Celsius scale after Anders Celsius (1701–1744), the Swedish astronomer who devised this scale.

The Fahrenheit temperature scale is used with the English Engineering system of units, and the Celsius scale with the SI unit system. Until 1954 both of these scales were based on two fixed, easily duplicated points—the ice point and the steam point. The temperature of the ice point is defined as the temperature of a mixture of ice and water that is in equilibrium with saturated air at a pressure of 1 atm. The temperature of the steam point is the temperature of water and steam, which are in equilibrium at a pressure of 1 atm. On the Fahrenheit scale these two points are assigned the numbers 32 and 212, respectively, and on the Celsius scale the points are 0 and 100, respectively. Why Fahrenheit chose these numbers is an interesting story. In searching for an easily reproducible point, Fahrenheit selected the temperature of the human body and assigned it the number 96. He assigned the number 0 to the temperature of a certain mixture of salt, ice, and salt solution. On this scale the ice point was approximately 32. When this scale was slightly revised and fixed in terms of the ice point and steam point, the normal temperature of the human body was found to be 98.6 F.

In this text the symbols F and °C will denote the Fahrenheit and Celsius scales, respectively. The symbol T will refer to temperature on all temperature scales.

At the tenth CGPM in 1954, the Celsius scale was redefined in terms of a single fixed point and the ideal-gas temperature scale. The single fixed point is the triple point of water (the state in which the solid, liquid, and vapor phases of water exist together in equilibrium). The magnitude of the degree is defined in terms of the ideal-gas temperature scale, which is discussed in Chapter 7. The essential features of this new scale are a single fixed point and a definition of the magnitude of the degree. The triple point of water is assigned the value of 0.01°C. On this scale the steam point is experimentally found to be 100.00°C. Thus, there is essential agreement between the old and new temperature scales.

We have not yet considered an absolute scale of temperature. The possibility of such a scale comes from the second law of thermodynamics and is discussed in Chapter 7. On the basis of the second law of thermodynamics, a temperature scale that is independent of any thermometric substance can be defined. This absolute scale is usually referred to as the thermodynamic scale of temperature. However, it is very complicated to use this scale directly, and therefore, a more practical scale, the International Temperature Scale, which closely represents the thermodynamic scale, has been adopted.

The absolute scale related to the Celsius scale is the Kelvin scale (after William Thomson, 1824–1907, who is also known as Lord Kelvin), and is designated K (without the degree symbol). The relation between these scales is

$$K = {}^{\circ}C + 273.15$$

In 1967, the CGPM defined the kelvin as 1/273.16 of the temperature at the triple point of water. The Celsius scale is now defined by this equation instead of by its earlier definition.

The absolute scale related to the Fahrenheit scale is the Rankine scale and is designated R. The relation between these scales is

$$R = F + 459.67$$

A number of empirically based temperature scales, to standardize temperature measurement and calibration, have been in use over the last 70 years. The most recent of these is the International Temperature Scale of 1990, or ITS-90. It is based on a number of fixed and easily reproducible points that are assigned definite numerical values of temperature, and on specified formulas relating temperature to the readings on certain temperature-measuring instruments for the purpose of interpolation between the defining fixed points. Details of the ITS-90 are not considered further in this text. It is noted that this scale is a practical means for establishing measurements that conform closely to the absolute thermodynamic temperature scale.

PROBLEMS

2.1 The "standard" acceleration (at sea level and 45° latitude) due to gravity is 9.80665 m/s². What is the force needed to hold a mass of 2 kg at rest in this gravitational field? How much mass can a force of 1 N support?

2.2 A model car rolls down an incline with a slope so the gravitational "pull" in the direction of motion is one-third of the standard gravitational force (see Problem 2.1). If the car has a mass of 0.45 kg, find the acceleration.

2.3 A car drives at 60 km/h and is brought to a full stop with constant deceleration in 5 seconds. If the total car and driver mass is 1075 kg, find the necessary force.

2.4 A washing machine has 2 kg of clothes spinning at a rate that generates an acceleration of 24 m/s². What is the force needed to hold the clothes?

2.5 A 1200-kg car moving at 20 km/h is accelerated at a constant rate of 4 m/s² up to a speed of 75 km/h. What are the force and total time required?

2.6 A steel plate of 950 kg accelerates from rest with 3 m/s² for a period of 10s. What force is needed and what is the final velocity?

2.7 A 15-kg steel container has 1.75 kilomoles of liquid propane inside. A force of 2 kN now accelerates this system. What is the acceleration?

2.8 A rope hangs over a pulley with the two equally long ends down. On one end you attach a mass of 5 kg and on the other end you attach 10 kg. Assuming standard gravitation and no friction in the pulley, what is the acceleration of the 10 kg mass when released?

2.9 A bucket of concrete of total mass 200 kg is raised by a crane with an acceleration of 2 m/s² relative to the ground at a location where the local gravitational acceleration is 9.5 m/s². Find the required force.

2.10 On the moon the gravitational acceleration is approximately one-sixth that on the surface of the earth. A 5-kg mass is "weighed" with a beam balance on the surface on the moon. What is the expected reading? If this mass is weighed with a spring scale that reads correctly for standard gravity on earth (see Problem 2.1), what is the reading?

2.11 One kilogram of diatomic oxygen (O_2 molecular weight 32) is contained in a 500-L tank. Find the specific volume on both a mass and mole basis (v and $\bar{v}$).

2.12 A 5 m³ container is filled with 900 kg of granite (density 2400 kg/m³) and the rest of the volume is air with density 1.15 kg/m³. Find the mass of air and the overall (average) specific volume.

2.13 A 15-kg steel gas tank holds 300 L of liquid gasoline, having a density of 800 kg/m³. If the system is decelerated with 6 m/s² what is the needed force?

2.14 A vertical hydraulic cylinder has a 125-mm-diameter piston with hydraulic fluid inside the cylinder and an ambient pressure of 1 bar. Assuming standard gravity, find the piston mass that will create a pressure inside of 1500 kPa.

2.15 A barometer to measure absolute pressure shows a mercury column height of 725 mm. The temperature is such that the density of the mercury is 13550 kg/m³. Find the ambient pressure.

2.16 A cannon-ball of 5 kg acts as a piston in a cylinder of 0.15-m diameter. As the gunpowder is burned a pressure of 7 MPa is created in the gas behind the ball. What is the acceleration of the ball if the cylinder (cannon) is pointing horizontally?

2.17 Repeat the previous problem for a cylinder (cannon) pointing 40 degrees up relative to the horizontal direction.

2.18 A piston/cylinder with a cross-sectional area of 0.01 m² has a piston mass of 100 kg resting on the stops, as shown in Fig. P2.18. With an outside atmospheric pressure of 100 kPa, what should the water pressure be to lift the piston?

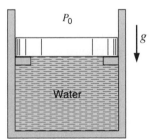

P_0

g

Water

FIGURE P2.18

2.19 The hydraulic lift in an auto-repair shop has a cylinder diameter of 0.2 m. To what pressure should the hydraulic fluid be pumped to lift 40 kg of piston/arms and 700 kg of a car?

2.20 A differential pressure gauge mounted on a vessel shows 1.25 MPa and a local barometer gives atmospheric pressure as 0.96 bar. Find the absolute pressure inside the vessel.

2.21 The absolute pressure in a tank is 85 kPa and the local ambient absolute pressure is 97 kPa. If a U-tube with mercury, density 13550 kg/m³, is attached to the tank to measure the vacuum, what column height difference would it show?

2.22 A 5-kg piston in a cylinder with diameter of 100 mm is loaded with a linear spring and the outside atmospheric pressure of 100 kPa. The spring exerts no force on the piston when it is at the bottom of the cylinder and for the state shown, the pressure is 400 kPa with volume 0.4 L. The valve is opened to let some air in, causing the piston to rise 2 cm. Find the new pressure.

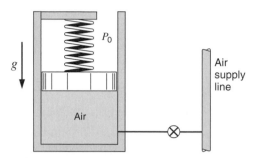

FIGURE P2.22

2.23 A U-tube manometer filled with water, density 1000 kg/m^3, shows a height difference of 25 cm. What is the gauge pressure? If the right branch is tilted to make an angle of 30° with the horizontal, as shown in Fig. P2.23, what should the length of the column in the tilted tube be relative to the U-tube?

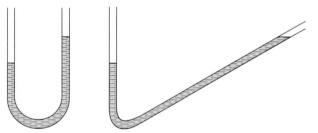

FIGURE P2.23

2.24 The difference in height between the columns of a manometer is 200 mm with a fluid of density 900 kg/m^3. What is the pressure difference? What is the height difference if the same pressure difference is measured using mercury, density 13600 kg/m^3, as manometer fluid?

2.25 Two reservoirs, A and B, open to the atmosphere, are connected with a mercury manometer. Reservoir A is moved up/down so the two top surfaces are level at h_3 as shown in Fig. P2.25. Assuming that you know ρ_A, ρ_{Hg} and measure the heights h_1, h_2, and h_3, find the density ρ_B.

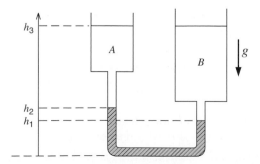

FIGURE P2.25

2.26 Two vertical cylindrical storage tanks are full of liquid water, density 1000 kg/m³, the top open to the atmoshere. One is 10 m tall, 2 m diameter, the other is 2.5 m tall with diameter 4 m. What is the total force from the bottom of each tank to the water and what is the pressure at the bottom of each tank?

2.27 The density of mercury changes approximately linearly with temperature as

$$\rho_{Hg} = 13595 - 2.5\,T \text{ kg/m}^3. \qquad T \text{ in Celsius}$$

so the same pressure difference will result in a manometer reading that is influenced by temperature. If a pressure difference of 100 kPa is measured in the summer at 35°C and in the winter at −15°C, what is the difference in column height between the two measurements?

2.28 Liquid water with density ρ is filled on top of a thin piston in a cylinder with cross-sectional area A and total height H, as shown in Fig. P2.28. Air is let in under the piston so it pushes up, spilling the water over the edge. Deduce the formula for the air pressure as a function of the piston elevation from the bottom, h.

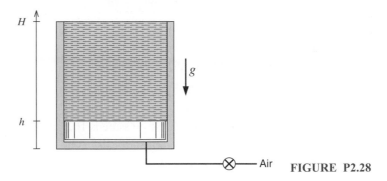

FIGURE P2.28

2.29 A piston, $m_p = 5$ kg, is fitted in a cylinder, $A = 15$ cm², that contains a gas. The setup is in a centrifuge that creates an acceleration of 25 m/s² in the direction of piston motion towards the gas. Assuming standard atmospheric pressure outside the cylinder, find the gas pressure.

2.30 A piece of experimental apparatus is located where $g = 9.5$ m/s² and the temperature is 5°C. An air flow inside the apparatus is determined by measuring the pressure drop across an orifice with a mercury manometer (see Problem 2.27 for density) showing a height difference of 200 mm. What is the pressure drop in kPa?

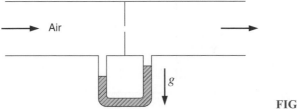

FIGURE P2.30

2.31 Repeat the previous problem if the flow inside the apparatus is liquid water, $\rho \simeq$ 1000 kg/m³, instead of air. Find the pressure difference between the two holes flush with the bottom of the channel. You cannot neglect the two unequal water columns.

2.32 Two piston/cylinder arrangements, A and B, have their gas chambers connected by a pipe, as shown in Fig. P2.32. Cross-sectional areas are $A_A = 75$ cm^2 and $A_B = 25$ cm^2 with the piston mass in A being $m_A = 25$ kg. Outside pressure is 100 kPa and standard gravitation. Find the mass m_B so that none of the pistons have to rest on the bottom.

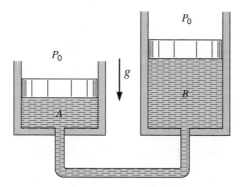

FIGURE P2.32

2.33 Two hydraulic piston/cylinders are of same size and setup as in Problem 2.32, but with negligible piston masses. A single point force of 250 N presses down on piston A. Find the needed extra force on piston B so that none of the pistons have to move.

2.34 At the beach, atmospheric pressure is 1025 mbar. You dive 15 m down in the ocean and you later climb a hill up to 250 m elevation. Assume the density of water is about 1000 kg/m^3 and the density of air is 1.18 kg/m^3. What pressure do you feel at each place?

2.35 In the city water tower, water is pumped up to a level 25 m above ground in a pressurized tank with air at 125 kPa over the water surface. This is illustrated in Fig. P2.35. Assuming the water density is 1000 kg/m^3 and standard gravity, find the pressure required to pump more water in at ground level.

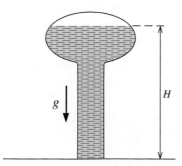

FIGURE P2.35

2.36 Two cylinders are connected by a piston, as shown in Fig. P2.36. Cylinder A is used as a hydraulic lift and pumped up to 500 kPa. The piston mass is 25 kg and there is standard gravity. What is the gas pressure in cylinder B?

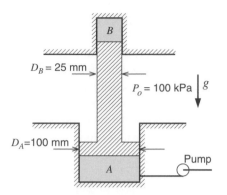

FIGURE P2.36

2.37 Two cylinders are filled with liquid water, $\rho \approx 1000$ kg/m^3, and connected by a line with a closed valve, as shown in Fig. P2.37. A has 100 kg and B has 500 kg of water, their cross-sectional areas are $A_A = 0.1$ m^2 and $A_B = 0.25$ m^2 and the height h is 1 m. Find the pressure on each side of the valve. The valve is opened and water flows to an equilibrium. Find the final pressure at the valve location.

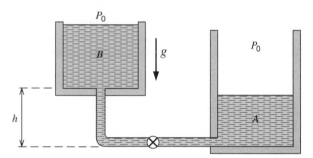

FIGURE P2.37

2.38 Using the freezing and boiling point temperatures for water in both Celsius and Fahrenheit scales, develop a conversion formula between the scales. Find the conversion formula between Kelvin and Rankine temperature scales.

ENGLISH UNIT PROBLEMS

2.39E A 2500-lbm car moving at 15 mi/h is accelerated at a constant rate of 15 ft/s^2 up to a speed of 50 mi/h. What are the force and total time required?

2.40E Two pound moles of diatomic oxygen gas are enclosed in a 20-lbm steel container. A force of 2000 lbf now accelerates this system. What is the acceleration?

2.41E A bucket of concrete of total mass 400 lbm is raised by a crane with an acceleration of 6 ft/s^2 relative to the ground at a location where the local gravitational acceleration is 31 ft/s^2. Find the required force.

2.42E One pound-mass of diatomic oxygen (O_2 molecular weight 32) is contained in a 100-gal tank. Find the specific volume on both a mass and mole basis (v and $\overline{v}$).

2.43E A 30-lbm steel gas tank holds 10 ft^3 of liquid gasoline, having a density of 50 lbm/ft^3. What force is needed to accelerate this combined system at a rate of 15 ft/s^2?

2.44E A differential pressure gauge mounted on a vessel shows 185 lbf/in.^{2} and a local barometer gives atmospheric pressure as 0.96 atm. Find the absolute pressure inside the vessel.

2.45E A U-tube manometer filled with water, density 62.3 lbm/ft^3, shows a height difference of 10 in. What is the gauge pressure? If the right branch is tilted to make an angle of 30° with the horizontal, as shown in Fig. P2.23, what should the length of the column in the tilted tube be relative to the U-tube?

2.46E A piston/cylinder with cross-sectional area of 0.1 ft^2 has a piston mass of 200 lbm resting on the stops, as shown in Fig. P2.18. With an outside atmospheric pressure of 1 atm, what should the water pressure be to lift the piston?

2.47E The density of mercury changes approximately linearly with temperature as

$$\rho_{Hg} = 851.5 - 0.086\ T \quad lbm/ft^3 \qquad T \text{ in degrees Fahrenheit}$$

so the same pressure difference will result in a manometer reading that is influenced by temperature. If a pressure difference of 14.7 lbf/in.^{2} is measured in the summer at 95 F and in the winter at 5 F, what is the difference in column height between the two measurements?

2.48E A piston, $m_p = 10$ lbm, is fitted in a cylinder, $A = 2.5$ in.^{2}, that contains a gas. The setup is in a centrifuge that creates an acceleration of 75 ft/s^2. Assuming standard atmospheric pressure outside the cylinder, find the gas pressure.

2.49E At the beach, atmospheric pressure is 1025 mbar. You dive 30 ft down in the ocean and you later climb a hill up to 300-ft elevation. Assume the density of water is about 62.3 lbm/ft^3 and the density of air is 0.0735 lbm/ft^3. What pressure do you feel at each place?

COMPUTER, DESIGN, AND OPEN-ENDED PROBLEMS

2.50 Write a program to list corresponding temperatures in °C, K, F, and R from −50°C to 100°C in increments of 10 degrees.

2.51 Write a program that will input pressure in kPa or atm or lbf/in.^{2} and write the pressure out in: kPa, atm, bar, and lbf/in.^{2}.

2.52 Write a program to do the temperature correction on a mercury barometer reading (see Problem 2.27). Input reading and temperature and output corrected reading at 20°C and pressure in kPa.

2.53 Make a list of different weights and scales that are used to measure mass directly or indirectly. Investigate the ranges of mass and the accuracy that can be obtained.

2.54 Thermometers are based on several principles. Expansion of a liquid with a rise in temperature is used in many applications. Electrical resistance, thermistors, and thermocouples are common in instrumentation and remote probes. Investigate a variety of thermometers and make a list of their range, accuracy, advantages, and disadvantages.

2.55 Collect information for a resistance, thermistor, and thermocouple based thermometer suitable for the range of temperatures from 0°C to 200°C. For each of the three types list the accuracy and response of the transducer (output per degree change). Is any calibration or corrections necessary when it is used in an instrument?

2.56 A thermistor is used as a temperature transducer. Its resistance changes with temperature approximately as

$$R = R_o \exp[\alpha(1/T - 1/T_o)]$$

where it has the resistance, R_o, at the temperature, T_o. Select the constants as $R_o = 3000\Omega$, $T_o = 298$ K and compute α so it has the resistance of 200 Ω at 100°C.

Write a program to convert a measured resistance, R, into information about the temperature. Find information for actual thermistors and plot the calibration curves with the above formula and the recommended correction given by the manufacturer.

2.57 Investigate possible transducers for the measurement of temperature in a flame with temperatures near 1000 K. Are any available for a temperature of 2000 K?

2.58 Devices to measure pressure are available as differential or absolute pressure transducers. Make a list of 5 different differential pressure transducers to measure pressure differences in the order of 100 kPa. Note their accuracy, response (linear or ?), and price.

2.59 A micromanometer uses a fluid with density 1000 kg/m^3 and it is able to measure the height difference with an accuracy of ±0.5 mm. Its range is a maximum height difference of 0.5 m. Investigate if any transducers are available to replace the micromanometer.

2.60 An experiment involves the measurements of temperature and pressure of a gas flowing in a pipe at 300°C and 250 kPa. Write a report with a suggested set of transducers (at least two alternatives for each) and give the expected accuracy and cost.

3 PROPERTIES OF A PURE SUBSTANCE

In the previous chapter we considered three familiar properties of a substance—specific volume, pressure, and temperature. We now turn our attention to pure substances and consider some of the phases in which a pure substance may exist, the number of independent properties a pure substance may have, and methods of presenting thermodynamic properties.

3.1 THE PURE SUBSTANCE

A pure substance is one that has a homogeneous and invariable chemical composition. It may exist in more than one phase, but the chemical composition is the same in all phases. Thus, liquid water, a mixture of liquid water and water vapor (steam), and a mixture of ice and liquid water are all pure substances; every phase has the same chemical composition. On the other hand, a mixture of liquid air and gaseous air is not a pure substance because the composition of the liquid phase is different from that of the vapor phase.

Sometimes a mixture of gases, such as air, is considered a pure substance as long as there is no change of phase. Strictly speaking, this is not true. As we will see later, we should say that a mixture of gases such as air exhibits some of the characteristics of a pure substance as long as there is no change of phase.

In this text the emphasis will be on simple compressible substances. This term designates substances whose surface effects, magnetic effects, and electrical effects are insignificant when dealing with the substances. On the other hand, changes in volume, such as those associated with the expansion of a gas in a cylinder, are very important. Reference will be made, however, to other substances for which surface, magnetic, and electrical effects are important. We will refer to a system consisting of a simple compressible substance as a simple compressible system.

3.2 VAPOR–LIQUID–SOLID-PHASE EQUILIBRIUM IN A PURE SUBSTANCE

Consider as a system 1 kg of water contained in the piston-cylinder arrangement shown in Fig. 3.1a. Suppose that the piston and weight maintain a pressure of 0.1 MPa in the cylinder, and that the initial temperature is 20°C. As heat is transferred to the water, the temperature increases appreciably, the specific volume increases slightly, and the pres-

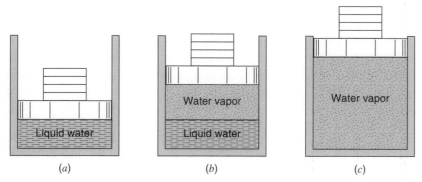

FIGURE 3.1 Constant-pressure change from liquid to vapor phase for a pure substance.

sure remains constant. When the temperature reaches 99.6°C, additional heat transfer results in a change of phase, as indicated in Fig. 3.1*b*. That is, some of the liquid becomes vapor, and during this process both the temperature and pressure remain constant, but the specific volume increases considerably. When the last drop of liquid has vaporized, further transfer of heat results in an increase in both temperature and specific volume of the vapor, as shown in Fig. 3.1*c*.

The term *saturation temperature* designates the temperature at which vaporization takes place at a given pressure. This pressure is called the saturation pressure for the given temperature. Thus, for water at 99.6°C the saturation pressure is 0.1 MPa, and for water at 0.1 MPa the saturation temperature is 99.6°C. For a pure substance there is a definite relation between saturation pressure and saturation temperature. A typical curve, called the vapor-pressure curve, is shown in Fig. 3.2.

If a substance exists as liquid at the saturation temperature and pressure, it is called saturated liquid. If the temperature of the liquid is lower than the saturation temperature for the existing pressure, it is called either a subcooled liquid (implying that the temperature is lower than the saturation temperature for the given pressure) or a compressed liquid (implying that the pressure is greater than the saturation pressure for the given temperature). Either term may be used, but the latter term will be used in this text.

When a substance exists as part liquid and part vapor at the saturation temperature, its quality is defined as the ratio of the mass of vapor to the total mass. Thus, in Fig. 3.1*b*, if the mass of the vapor is 0.2 kg and the mass of the liquid is 0.8 kg, the quality is 0.2 or 20%. The quality may be considered an intensive property and has the symbol x. Quality has meaning only when the substance is in a saturated state, that is, at saturation pressure and temperature.

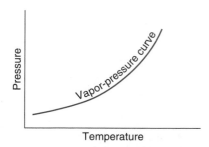

FIGURE 3.2 Vapor-pressure curve of a pure substance.

If a substance exists as vapor at the saturation temperature, it is called saturated vapor. (Sometimes the term *dry saturated vapor* is used to emphasize that the quality is 100%.) When the vapor is at a temperature greater than the saturation temperature, it is said to exist as superheated vapor. The pressure and temperature of superheated vapor are independent properties, since the temperature may increase while the pressure remains constant. Actually, the substances we call gases are highly superheated vapors.

Consider Fig. 3.1 again. Let us plot on the temperature–volume diagram of Fig. 3.3 the constant-pressure line that represents the states through which the water passes as it is heated from the initial state of 0.1 MPa and 20°C. Let state *A* represent the initial state, *B* the saturated-liquid state (99.6°C), and line *AB* the process in which the liquid is heated from the initial temperature to the saturation temperature. Point *C* is the saturated-vapor state, and line *BC* is the constant-temperature process in which the change of phase from liquid to vapor occurs. Line *CD* represents the process in which the steam is superheated at constant pressure. Temperature and volume both increase during this process.

Now let the process take place at a constant pressure of 1 MPa, starting from an initial temperature of 20°C. Point *E* represents the initial state, in which the specific volume is slightly less than that at 0.1 MPa and 20°C. Vaporization begins at point *F*, where the temperature is 179.9°C. Point *G* is the saturated-vapor state, and line *GH* the constant-pressure process in which the steam is superheated.

In a similar manner, a constant pressure of 10 MPa is represented by line *IJKL,* for which the saturation temperature is 311.1°C.

At a pressure of 22.09 MPa, represented by line *MNO*, we find, however, that there is no constant-temperature vaporization process. Instead, point *N* is a point of inflection with a zero slope. This point is called the critical point. At the critical point the saturated-liquid and saturated-vapor states are identical. The temperature, pressure, and specific volume at the critical point are called the critical temperature, critical pressure, and critical volume. The critical-point data for some substances are given in Table 3.1. More extensive data are given in Table A.2 in the Appendix.

A constant-pressure process at a pressure greater than the critical pressure is represented by line *PQ*. If water at 40 MPa, 20°C, is heated in a constant-pressure process in a cylinder as shown in Fig. 3.1, there will never be two phases present and the state shown

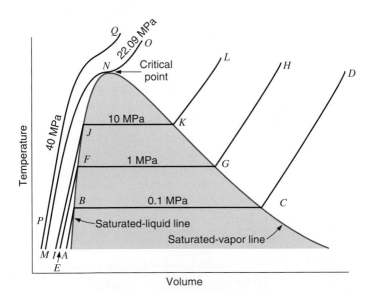

FIGURE 3.3
Temperature–volume diagram for water showing liquid and vapor phases (not to scale).

TABLE 3.1

Some Critical-Point Data

	Critical Temperature, °C	Critical Pressure, MPa	Critical Volume, m^3/kg
Water	374.14	22.09	0.003 155
Carbon dioxide	31.05	7.39	0.002 143
Oxygen	−118.35	5.08	0.002 438
Hydrogen	−239.85	1.30	0.032 192

in Fig. 3.1*b* will never exist. Instead, there will be a continuous change in density and at all times there will be only one phase present. The question then is when do we have a liquid and when do we have a vapor? The answer is that this is not a valid question at supercritical pressures. We simply term the substance a fluid. However, rather arbitrarily, at temperatures below the critical temperature we usually refer to it as a compressed liquid and at temperatures above the critical temperature as a superheated vapor. It should be emphasized, however, that at pressures above the critical pressure we never have a liquid and vapor phase of a pure substance existing in equilibrium.

In Fig. 3.3, line *NJFB* represents the saturated-liquid line and line *NKGC* represents the saturated-vapor line.

Let us consider another experiment with the piston–cylinder arrangement. Suppose that the cylinder contains 1 kg of ice at −20°C, 100 kPa. When heat is transferred to the ice, the pressure remains constant, the specific volume increases slightly, and the temperature increases until it reaches 0°C, at which point the ice melts and the temperature remains constant. In this state the ice is called a saturated solid. For most substances the specific volume increases during this melting process, but for water the specific volume of the liquid is less than the specific volume of the solid. When all the ice has melted, a further heat transfer causes an increase in temperature of the liquid.

If the initial pressure of the ice at −20°C is 0.260 kPa, heat transfer to the ice results in an increase in temperature to −10°C. At this point, however, the ice passes directly from the solid phase to the vapor phase in the process known as sublimation. Further heat transfer results in superheating of the vapor.

Finally, consider an initial pressure of the ice of 0.6113 kPa and a temperature of −20°C. Through heat transfer let the temperature increase until it reaches 0.01°C. At this point, however, further heat transfer may cause some of the ice to become vapor and some to become liquid, for at this point it is possible to have the three phases in equilibrium. This point is called the triple point, which is defined as the state in which all three phases may be present in equilibrium. The pressure and temperature at the triple point for a number of substances are given in Table 3.2.

This whole matter is best summarized by the diagram of Fig. 3.4, which shows how the solid, liquid, and vapor phases may exist together in equilibrium. Along the sublimation line the solid and vapor phases are in equilibrium, along the fusion line the solid and liquid phases are in equilibrium, and along the vaporization line the liquid and vapor phases are in equilibrium. The only point at which all three phases may exist in equilibrium is the triple point. The vaporization line ends at the critical point because there is no distinct change from the liquid phase to the vapor phase above the critical point.

TABLE 3.2
Some Solid-Liquid-Vapor Triple-Point Data

	Temperature, °C	Pressure, kPa
Hydrogen (normal)	−259	7.194
Oxygen	−219	0.15
Nitrogen	−210	12.53
Carbon dioxide	−56.4	520.8
Mercury	−39	0.000 000 13
Water	0.01	0.6113
Zinc	419	5.066
Silver	961	0.01
Copper	1083	0.000 079

Consider a solid in state *A*, as shown in Fig. 3.4. When the temperature increases but the pressure (which is less than the triple-point pressure) is constant, the substance passes directly from the solid to the vapor phase. Along the constant-pressure line *EF*, the substance passes from the solid to the liquid phase at one temperature, and then from the liquid to the vapor phase at a higher temperature. Constant-pressure line *CD* passes through the triple point, and it is only at the triple point that the three phases may exist together in equilibrium. At a pressure above the critical pressure, such as *GH*, there is no sharp distinction between the liquid and vapor phases.

Although we have made these comments with rather specific reference to water (only because of our familiarity with water), all pure substances exhibit the same general behavior. However, the triple-point temperature and critical temperature vary greatly from one substance to another. For example, the critical temperature of helium, as given

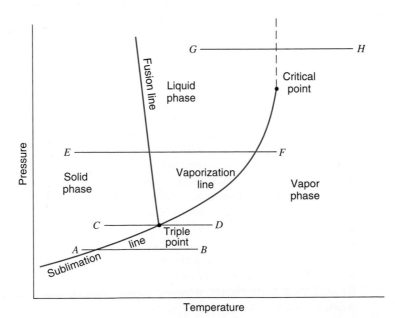

FIGURE 3.4
Pressure–temperature diagram for a substance such as water.

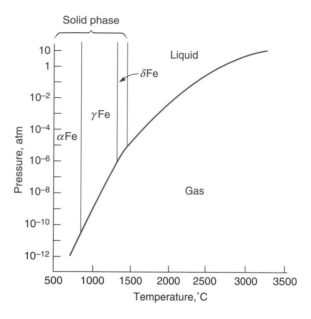

FIGURE 3.5 Estimated pressure–temperature diagram for iron (From *Phase Diagrams in Metallurgy* by F. N. Rhines, copyright 1956, McGraw-Hill Book Company; used by permission).

in Table A.2, is 5.3 K. Therefore, the absolute temperature of helium at ambient conditions is over 50 times greater than the critical temperature. On the other hand, water has a critical temperature of 374.14°C (647.29 K), and at ambient conditions the temperature of water is less than one-half the critical temperature. Most metals have a much higher critical temperature than water. When we consider the behavior of a substance in a given state, it is often helpful to think of this state in relation to the critical state or triple point. For example, if the pressure is greater than the critical pressure, it is impossible to have a liquid phase and a vapor phase in equilibrium. Or, to consider another example, the states at which vacuum-melting a given metal is possible can be ascertained by a consideration of the properties at the triple point. Iron at a pressure just above 5 Pa (the triple-point pressure) would melt at a temperature of about 1535°C (the triple-point temperature).

It should also be pointed out that a pure substance can exist in a number of different solid phases. A transition from one solid phase to another is called an allotropic transformation. Figure 3.5 is a pressure–temperature diagram for iron that shows three solid phases, the liquid phase, and the vapor phase. Figure 3.6 shows a number of solid phases for water. A pure substance can have a number of triple points, but only one triple point has a solid, liquid, and vapor equilibrium. Other triple points for a pure substance can have two solid phases and a liquid phase, two solid phases and a vapor phase, or three solid phases.

3.3 INDEPENDENT PROPERTIES OF A PURE SUBSTANCE

One important reason for introducing the concept of a pure substance is that the state of a simple compressible pure substance (that is, a pure substance in the absence of motion, gravity, and surface, magnetic, or electrical effects) is defined by two independent properties. For example, if the specific volume and temperature of superheated steam are specified, the state of the steam is determined.

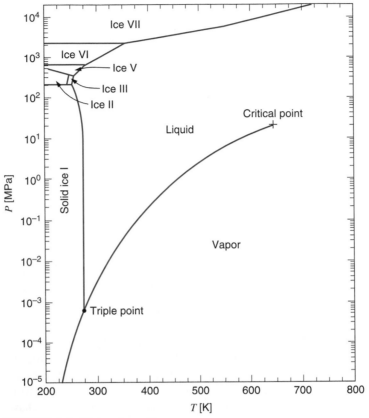

FIGURE 3.6 Water phase diagram.

To understand the significance of the term independent property, consider the saturated-liquid and saturated-vapor states of a pure substance. These two states have the same pressure and the same temperature, but they are definitely not the same state. In a saturation state, therefore, pressure and temperature are not independent properties. Two independent properties such as pressure and specific volume or pressure and quality are required to specify a saturation state of a pure substance.

The reason for mentioning previously that a mixture of gases, such as air, has the same characteristics as a pure substance as long as only one phase is present, concerns precisely this point. The state of air, which is a mixture of gases of definite composition, is determined by specifying two properties as long as it remains in the gaseous phase. Air then can be treated as a pure substance.

3.4 Equations of State for the Vapor Phase of a Simple Compressible Substance

From experimental observations it has been established that the P–v–T behavior of gases at low density is closely given by the following equation of state,

$$P\overline{v} = \overline{R}T \tag{3.1}$$

where $\bar{R}$ is the universal gas constant. The value of $\bar{R}$ is

$$\bar{R} = 8.3145 \frac{kN\ m}{kmol\ K} = 8.3145 \frac{kJ}{kmol\ K}$$

In the English Engineering system,

$$\bar{R} = 1545 \frac{ft\ lbf}{lb\ mol\ R}$$

Dividing Eq. 3.1 by M, the molecular weight, we have the equation of state on a unit mass basis,

$$\frac{P\bar{v}}{M} = \frac{\bar{R}T}{M}$$

or

$$Pv = RT \tag{3.2}$$

where

$$R = \frac{\bar{R}}{M} \tag{3.3}$$

Here R is a constant for a particular gas. The value of R for a number of substances is given in Table A.5 of the Appendix. It follows from Eqs. 3.1 and 3.2 that this equation of state can be written in terms of the total volume:

$$PV = n\bar{R}T$$
$$PV = mRT \tag{3.4}$$

It should also be noted that Eq. 3.4 can be written alternately in the form

$$\frac{P_1 V_1}{T_1} = \frac{P_2 V_2}{T_2} \tag{3.5}$$

That is, gases at low density closely follow the well-known Boyle's and Charles' laws. Boyle and Charles, of course, based their statements on experimental observations. (Strictly speaking, neither of these statements should be called a law, since they are only approximately true and even then only under conditions of low density.)

EXAMPLE 3.1   What is the mass of air contained in a room 6 m × 10 m × 4 m if the pressure is 100 kPa and the temperature is 25°C? Assume air to be an ideal gas. By using Eq. 3.4 and the value of R from Table A.5, we have

$$m = \frac{PV}{RT} = \frac{100\ kN/m^2 \times 240\ m^3}{0.287\ kN\ m/kg\ K \times 298.2\ K} = 280.5\ kg$$

EXAMPLE 3.2 A tank has a volume of 0.5 m³ and contains 10 kg of an ideal gas having a molecular weight of 24. The temperature is 25°C. What is the pressure?

The gas constant is determined first:

$$R = \frac{\overline{R}}{M} = \frac{8.3145 \text{ kN m} / \text{kmol K}}{24 \text{ kg} / \text{kmol}}$$

$$= 0.346\ 44 \text{ kN m} / \text{kg K}$$

We now solve for P:

$$P = \frac{mRT}{V} = \frac{10 \text{ kg} \times 0.346\ 44 \text{ kN m} / \text{kg K} \times 298.2 \text{ K}}{0.5 \text{ m}^3}$$

$$= 2066 \text{ kPa}$$

EXAMPLE 3.2E A tank has a volume of 15 ft³ and contains 20 lbm of an ideal gas having a molecular weight of 24. The temperature is 80 F. What is the pressure?

The gas constant is determined first:

$$R = \frac{\overline{R}}{M} = \frac{1545 \text{ ft lbf} / \text{lb mole R}}{24 \text{ lbm} / \text{lb mole}} = 64.4 \text{ ft lbf} / \text{lbm R}$$

We now solve for P.

$$P = \frac{mRT}{V} = \frac{20 \text{ lbm} \times 64.4 \text{ ft lbf} / \text{lbm R} \times 540 \text{R}}{144 \text{ in.}^2 / \text{ft}^2 \times 15 \text{ ft}^3} = 321 \text{ lbf} / \text{in.}^2$$

The equation of state given by Eq. 3.1 (or Eq. 3.2) is referred to as the ideal-gas equation of state. At very low density, all gases and vapors approach ideal-gas behavior, with the P–v–T relationship being given by the ideal-gas equation of state. At higher densities the behavior may deviate substantially from the ideal-gas equation of state.

Because of its simplicity, the ideal-gas equation of state is very convenient to use in thermodynamic calculations. However, two questions are now appropriate. The ideal-gas equation of state is a good approximation at low density. But what constitutes low density? Or, expressed in other words, over what range of density will the ideal-gas equation of state hold with accuracy? The second question is, how much does an actual gas at a given pressure and temperature deviate from ideal-gas behavior?

To answer both questions, we introduce the concept of the compressibility factor, Z, which is defined by the relation

$$Z = \frac{P\overline{v}}{\overline{R}T}$$

or

$$P\overline{v} = Z\overline{R}T \tag{3.6}$$

Note that for an ideal gas $Z = 1$, and the deviation of Z from unity is a measure of the deviation of the actual relation from the ideal-gas equation of state.

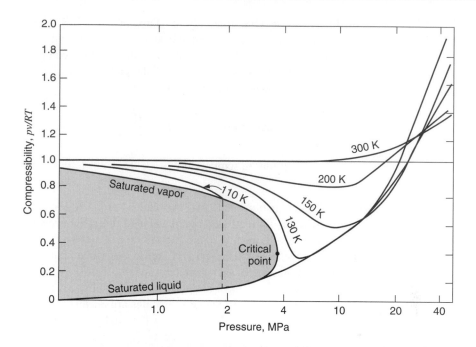

FIGURE 3.7
Compressibility of nitrogen.

Figure 3.7 shows a skeleton compressibility chart for nitrogen. From this chart we make three observations. The first is that at all temperatures $Z \to 1$ as $P \to 0$. That is, as the pressure approaches zero, the P–v–T behavior closely approaches that predicted by the ideal-gas equation of state. Note also that at temperatures of 300 K and above (that is, room temperature and above) the compressibility factor is near unity up to pressure of about 10 MPa. This means that the ideal-gas equation of state can be used for nitrogen (and, as it happens, air) over this range with considerable accuracy.

Now suppose we reduce the temperature from 300 K but keep the pressure constant at 4 MPa. The density will increase, and we note a sharp decrease below unity in the value of the compressibility factor. Values of $Z < 1$ mean that the actual density is greater than would be predicted by ideal-gas behavior. The physical explanation of this is as follows. As the temperature is reduced from 300 K and the pressure remains constant at 4 MPa, the molecules are brought closer together. In this range of intermolecular distances, and at this pressure and temperature, there is an attractive force between the molecules. The lower the temperature the greater is this intermolecular attractive force. This attractive force between the molecules means that the density is greater than would be predicted by the ideal-gas behavior, which assumes no intermolecular forces. Note also from the compressibility chart that at very high densities, for pressures above 30 MPa, the compressibility factor is always greater than unity. In this range the intermolecular distances are very small, and there is a repulsive force between the molecules. This factor tends to make the density less than would otherwise be expected.

The precise nature of intermolecular forces is a rather complex matter. These forces are a function of the temperature as well as the density. The preceding discussion should be considered a qualitative analysis to assist in gaining some understanding of the ideal-gas equation of state and how the P–v–T behavior of actual gases deviates from this equation.

If we examine compressibility diagrams for other pure substances, we find that the diagrams are all similar in the characteristics described above for nitrogen, at least in a qualitative sense. Quantitatively the diagrams are all different, since the critical temperatures and pressures of different substances vary over wide ranges, as evidenced from the values listed in Table A.2. Is there a way in which we can put all of these substances on a common basis? To do so, we "reduce" the properties with respect to the values at the critical point. The reduced properties are defined as

$$\text{Reduced pressure} \ = P_r = \frac{P}{P_c} \qquad P_c = \text{Critical pressure}$$

$$\text{Reduced temperature} \ = T_r = \frac{T}{T_c} \qquad T_c = \text{Critical temperature}$$

These equations state that the reduced property for a given state is the value of this property in this state divided by the value of this same property at the critical point.

If lines of constant T_r are plotted on a Z versus P_r diagram, a plot such as that in Fig. D.1 is obtained. The striking fact is that when such Z versus P_r diagrams are prepared for a number of different substances, all of them very nearly coincide, especially when the substances have simple, essentially spherical molecules. Correlations for substances with more complicated molecules are reasonably close, except near or at saturation or at high density. Thus, Fig. D.1 is actually a generalized diagram for simple molecules, which means that it represents the average behavior for a number of different simple substances. When such a diagram is used for a particular substance, the results will generally be somewhat in error. On the other hand, if P–v–T information is required for a substance in a region where no experimental measurements have been made, this generalized compressibility diagram will give reasonably accurate results. We need know only the critical pressure and critical temperature to use this basic generalized chart.

Extensive use of Fig. D.1 will be made in Chapter 13. For the present, this chart will be used primarily to help us decide whether, in a given circumstance, it is reasonable to assume ideal-gas behavior as a model. For example, we note from the chart that if the pressure is very low (that is, $\ll P_c$), the ideal-gas model can be assumed with good accuracy, regardless of the temperature. Furthermore, at high temperatures (that is, greater than about twice T_c), the ideal-gas model can be assumed with good accuracy to pressures as high as four or five times P_c. When the temperature is less than about twice the critical temperature and the pressure is not extremely low, we are in a region, commonly termed superheated vapor, in which the deviation from ideal-gas behavior may be considerable. In this region it is preferable to use tables of thermodynamic properties or charts for a particular substance. These tables are considered in the following section.

The value of compressibility factor from the generalized chart, Fig. D.1, also gives us a good estimate of the amount of error resulting from the use of the ideal gas model. The following example illustrates this point.

EXAMPLE 3.3   Calculate the specific volume of propane at a pressure of 7 MPa and a temperature of 150°C, and compare this with the specific volume given by the ideal-gas equation of state.

For propane

$$T_c = 369.8 \text{ K} \qquad P_c = 4.25 \text{ MPa}$$
$$R = 0.188\ 55 \text{ kJ / kg K}$$

$$T_r = \frac{423.2}{369.8} = 1.144 \qquad P_r = \frac{7}{4.25} = 1.647$$

From the compressibility chart

$$Z = 0.523$$

(This is the value from the software at the given reduced temperature and pressure. A value read from Fig. D.1 will be slightly different.)

$$v = \frac{ZRT}{P} = \frac{0.523 \times 0.188\ 55 \times 423.2}{7000} = 0.005\ 96 \text{ m}^3 / \text{kg}$$

The ideal-gas equation would give the value

$$v = \frac{0.188\ 55 \times 423.2}{7000} = 0.0114 \text{ m}^3 / \text{kg}$$

Instead of the ideal-gas model to represent gas behavior, or even the generalized compressibility chart, which is approximate, it is desirable to have an equation of state that accurately represents the P–v–T behavior for a particular gas over the entire superheated vapor region. Such an equation is necessarily more complicated and consequently more difficult to use. Many such equations have been proposed and used to correlate the observed behavior of gases. To illustrate the nature and complexity of these equations, we present one of the best known, the Benedict–Webb–Rubin equation of state:

$$P = \frac{RT}{v} + \frac{RTB_0 - A_0 - C_0 / T^2}{v^2} + \frac{RTb - a}{v^3} + \frac{a\alpha}{v^6} + \frac{c}{v^3 T^2}\left(1 + \frac{\gamma}{v^2}\right)e^{-\gamma / v^2} \qquad (3.7)$$

This equation contains eight empirical constants and is accurate to densities of about twice the critical density. The empirical constants for the Benedict–Webb–Rubin equation for a number of substances are given in Appendix D.

The matter of equations of state will be discussed further in Chapter 13. Note that an equation of state that accurately describes the relation between pressure, temperature, and specific volume is rather cumbersome and that the solution requires considerable time. When we use a digital computer, it is often most convenient to determine the thermodynamic properties in a given state from such equations. However, in hand calculations, it is much more convenient to tabulate values of pressure, temperature, specific volume, and other thermodynamic properties for various substances. The Appendix includes summary tables and graphs of the thermodynamic properties of water, ammonia, chlorofluorocarbon refrigerants R-12 and R-22, the new refrigerant R-134a, nitrogen, and methane. The tables of the properties of water are usually referred to as the "steam tables," where the first set was presented by Keenan, Keyes, Hill, and Moore. The method for compiling the P–v–T data for such a table is to find an equation of state that accurately fits the experimental data and then to solve the equation of state for the values listed in the table.

3.5 TABLES OF THERMODYNAMIC PROPERTIES

Tables of thermodynamic properties of many substances are available, and in general, all these tables have the same form. In this section we will refer to the steam tables. The steam tables are selected both because they are a vehicle for presenting thermodynamic tables and because steam is used extensively in power plants and industrial processes. Once the steam tables are understood, other thermodynamic tables can be readily used.

Before we introduce and discuss the actual steam tables, it is worthwhile to examine the ideas behind thermodynamic tables and to explore some of the initial difficulties that students frequently encounter. We therefore introduce these concepts on an artificial, oversimplified, and somewhat abstract basis, after which we will proceed to the actual tables. These "steam tables" actually consist of four separate tables, say A, B, C, and D, each of which is associated with a different region (phase) of the possible combinations of the two variables, T and P, as shown in Fig. 3.8. Each table is listed according to values of T and P. For each set of T and P (a state), the table contains values for four other variables, v, u, h, and s.

If the values of T and P are given, then by comparing them with the boundary values of T' and P', we are able to choose which of the four tables (A, B, C, and D) has the desired values of v, u, h, and s. Table A is the correct choice, for example, only if $T < T'$ and $P > P'$ are both true.

The principal difficulty in first learning how to use the tables is that the state may actually be specified by any two of the variables (assuming that T, P, v, u, h, and s are all independent). We then wish to find the other four variables from the appropriate table. If the two given values are not T and P, it may not be obvious which table to choose to find the other four variables. Even when we have had experience comparing the appropriate values at the table boundaries and understand through practice in which directions the changes go, we must go through the process of finding the appropriate table each and every time.

In addition to finding the correct table, the other nuisance of everyday use of the tables is interpolation, that is, when one (or both) of the stated values is not exactly equal to a value listed in the table.

Computerized tables eliminate both of these problems, but the student nevertheless must learn to understand the significance, construction, and limitations of the tables because there will be occasions when it will be necessary to use printed tables.

Let us use this background, to introduce the actual steam tables, as presented in the Appendix, Table B.1. Table B.1 is a summary table based on a complicated curve fit to the behavior of water. It is very similar to the *Steam Tables,* by Keenan, Keyes, Hill, and Moore, published in 1969 and 1978, revisions of the original tables published by Keenan and Keyes in 1936. We will concentrate on the three properties already discussed in Chapter 2, namely T, P, and v, and note that the other properties listed in Table B.1, u, h, and s, will be discussed later. We further note that the separation of phases in terms of val-

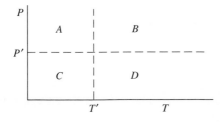

FIGURE 3.8 Schematic diagram of "steam tables" principle.

ues of T and P is actually described by the relations illustrated in Fig. 3.4, not Fig. 3.8. The region of superheated vapor in Fig. 3.4 is given in Table B.1.3, and that of compressed liquid is given in Table B.1.4. The compressed-solid region shown in Fig. 3.4 is not listed in the Appendix. The saturated-liquid and saturated-vapor region, as seen in the T and v diagram of Fig. 3.3 (and as the vaporization line in Fig. 3.4), is listed according to the values of T in Table B.1.1 and according to the values of P (T and P are not independent in the two-phase regions) in Table B.1.2. Similarly, the saturated-solid and saturated-vapor region is listed according to T in Table B.1.5, but the saturated-solid and saturated-liquid region, the third phase boundary line shown in Fig. 3.4, is not listed in the Appendix.

In Table B.1.1, the first column after the temperature gives the corresponding saturation pressure in kilopascals or megapascals. The next two columns give specific volume in cubic meters per kilogram. The first of these columns gives the specific volume of the saturated liquid, v_f; the second column gives the specific volume of saturated vapor, v_g. The difference between these two, $v_g - v_f$, represents the increase in specific volume when the state changes from saturated liquid to saturated vapor and is designated v_{fg}.

The specific volume of a substance having a given quality can be found by using the definition of quality. Quality has already been defined as the ratio of the mass of vapor to the total mass of liquid plus vapor when a substance is in a saturation state. Let us consider a mass m having a quality x. The volume is the sum of the volume of the liquid and the volume of the vapor:

$$V = V_{\text{liq}} + V_{\text{vap}} \qquad (3.8)$$

In terms of the masses, Eq. 3.8 can be written in the form

$$mv = m_{\text{liq}} v_f + m_{\text{vap}} v_g$$

Dividing by the total mass and introducing the quality x, we have

$$v = (1-x)v_f + xv_g \qquad (3.9)$$

Using the definition

$$v_{fg} = v_g - v_f$$

Equation 3.9 can then be written in the form

$$v = v_f + xv_{fg} \qquad (3.10)$$

As an example, let us calculate the specific volume of saturated steam at 200°C having a quality of 70%. Using Eq. 3.9 gives

$$v = 0.3(0.001\ 157) + 0.7(0.127\ 36)$$

$$= 0.0895\ \text{m}^3 / \text{kg}$$

In Table B.1.2, the first column after the pressure lists the saturation temperature for each pressure. The next column lists specific volume in a manner similar to Table B.1.1. When necessary, v_{fg} can readily be found by subtracting v_f from v_g.

Table 3 of the steam tables, which is summarized in Table B.1.3, gives the properties of superheated vapor. In the superheated region, pressure and temperature are independent properties and, therefore, for each pressure a large number of temperatures is given, and for each temperature four thermodynamic properties are listed, the first one being specific volume. Thus, the specific volume of steam at a pressure of 0.5 MPa and 200°C is 0.4249 m³/kg.

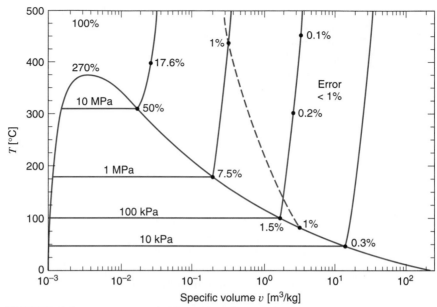

FIGURE 3.9 Temperature-specific volume diagram for water.

Figure 3.9 shows a $T - v$ plot for water, with an indication of percent error in assuming ideal gas behavior along the saturated vapor curve and also in several areas of the superheated vapor region.

Table 4 of the steam tables, summarized in Table B.1.4, gives the properties of the compressed liquid. To demonstrate the use of this table, consider a piston and a cylinder (as shown in Fig. 3.10) that contains 1 kg of saturated-liquid water at 100°C. Its properties are given in Table B.1.1, and we note that the pressure is 0.1013 MPa and the specific volume is 0.001 044 m³/kg. Suppose the pressure is increased to 10 MPa while the temperature is held constant at 100°C by the necessary transfer of heat, Q. Since water is slightly compressible, we would expect a slight decrease in specific volume during this process. Table B.1.4 gives this specific volume as 0.001 039 m³/kg. This is only a slight decrease, and only a small error would be made if one assumed that the volume of a compressed liquid was equal to the specific volume of the saturated liquid at the same temperature. In many situations this is the most convenient procedure, particularly when compressed-liquid data are not available.

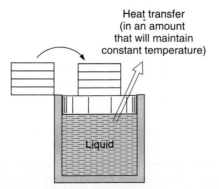

FIGURE 3.10 Illustration of compressed-liquid state.

Table B.1.5 of the steam tables gives the properties of saturated solid and saturated vapor that are in equilibrium. The first column gives the temperature and the second column gives the corresponding saturation pressure. As would be expected, all these pressures are less than the triple-point pressure. The next two columns give the specific volume of the saturated solid and saturated vapor (note that the tabulated value is $v_i \times 10^3$).

EXAMPLE 3.4

Determine the specific volume for R-134a at 100°C, 3.0 MPa, for the following models:

1. The R-134a tables, B.5.

2. Ideal gas.

3. The generalized chart, Fig. D.1.

From Table B.5.2 at 100°C, 3 MPa,

$$v = 0.006\ 65\ \text{m}^3/\text{kg} \qquad \text{(most accurate value)}$$

Assuming ideal gas, $\quad R = \dfrac{\overline{R}}{M} = \dfrac{8.3145}{102.03} = 0.081\ 49 \dfrac{\text{kJ}}{\text{kg K}}$

$$v = \frac{RT}{P} = \frac{0.08149 \times 373.2}{3000} = 0.010\ 14\ \text{m}^3/\text{kg}$$

which is more than 50% too large.
Use the generalized chart, Fig. D.1.

$$T_r = \frac{373.2}{374.2} = 1.0, \quad P_r = \frac{3}{4.06} = 0.74, \quad Z = 0.67$$

$$v = Z \times \frac{RT}{P} = 0.67 \times 0.010\ 14 = 0.006\ 79\ \text{m}^3/\text{kg}$$

which is only 2% too large.

3.6 COMPUTERIZED TABLES

Most of the tables in the Appendix are supplied in a computer program on the disk accompanying this book. The main program operates with a visual interface in the Windows environment on a PC-type computer and is generally self-explanatory. A secondary program operates in the DOS environment with a simple-text menu-driven interface.

The main program covers the full set of tables for water, refrigerants, and cryogenic fluids, as in Tables B.1 to B.7 including the compressed liquid region, which is only printed for water. For these substances a small graph with the *P-v* diagram shows the region around the critical point down toward the triple line covering the compressed liquid, two-phase liquid-vapor, dense fluid, and superheated vapor regions. As a state is selected and the properties computed, a thin crosshair set of lines indicates the state in the diagram so this can be seen with a visual impression of the state's location.

Ideal gases are covered corresponding to the Tables A.7 for air and A.8 for other ideal gases. You are able to select the substance and the units to work in for all the various table sections giving a wider choice than the printed tables. Metric units (SI) or standard English units for the properties can be used as well as a mass basis (kg or lbm) or a mole basis, satisfying the need for the most common applications.

The generalized chart, Fig. D.1, with the compressibility factor, is included so it is possible to get the value of Z a little more accurately than reading the graph. This is particularly useful for the case of a two-phase mixture where the saturated liquid and saturated vapor values are needed. Besides the compressibility factor, this part of the program includes correction terms beyond ideal gas approximations for changes in the other thermodynamic properties, which will be explained in Chapter 13.

The remaining tables are available with a text-based, menu-driven interface in the DOS environment. These tables cover the single-state properties, as in Tables A.2 to A.6, the combustion-related tables A.9 and A.10 and the compressible flow tables A.11 and A.12.

3.7 THERMODYNAMIC SURFACES

The matter discussed in this chapter can be well summarized by a consideration of a pressure-specific volume–temperature surface. Two such surfaces are shown in Figs. 3.11 and 3.12. Figure 3.11 shows a substance such as water in which the specific volume

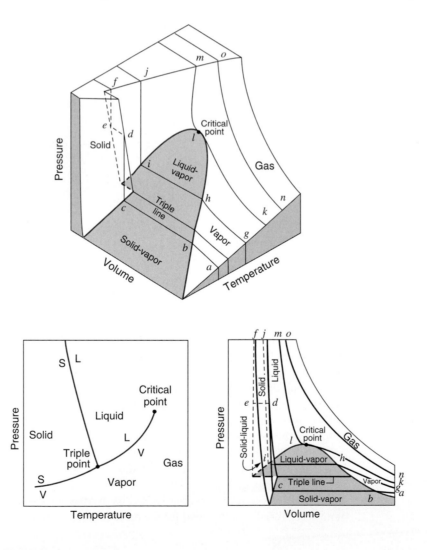

FIGURE 3.11
Pressure–volume–temperature surface for a substance that expands on freezing.

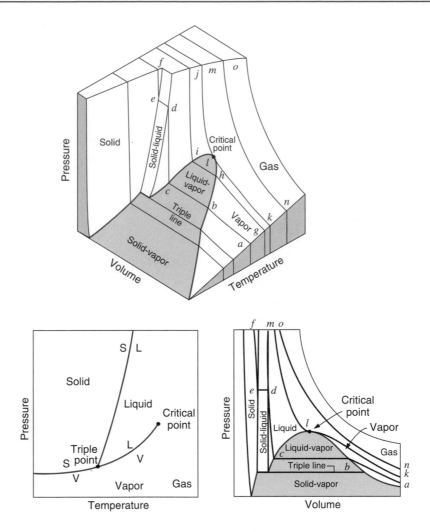

FIGURE 3.12
Pressure–volume–temperature surface for a substance that contracts on freezing.

increases during freezing. Figure 3.12 shows a substance in which the specific volume decreases during freezing.

In these diagrams the pressure, specific volume, and temperature are plotted on mutually perpendicular coordinates, and each possible equilibrium state is thus represented by a point on the surface. This follows directly from the fact that a pure substance has only two independent intensive properties. All points along a quasi-equilibrium process lie on the P–v–T surface, since such a process always passes through equilibrium states.

The regions of the surface that represent a single phase—the solid, liquid, and vapor phases—are indicated. These surfaces are curved. The two-phase regions—the solid–liquid, solid–vapor, and liquid–vapor regions—are ruled surfaces. By this we understand that they are made up of straight lines parallel to the specific-volume axis. This, of course, follows from the fact that in the two-phase region, lines of constant pressure are also lines of constant temperature, although the specific volume may change. The triple point actually appears as the triple line on the P–v–T surface, since the pressure and temperature of the triple point are fixed, but the specific volume may vary, depending on the proportion of each phase.

It is also of interest to note the pressure–temperature and pressure–volume projections of these surfaces. We have already considered the pressure–temperature diagram for a substance such as water. It is on this diagram that we observe the triple point. Various lines of constant temperature are shown on the pressure–volume diagram, and the corresponding constant-temperature sections are lettered identically on the P–v–T surface. The critical isotherm has a point of inflection at the critical point.

One notices that for a substance such as water, which expands on freezing, the freezing temperature decreases with an increase in pressure. For a substance that contracts on freezing, the freezing temperature increases as the pressure increases. Thus, as the pressure of vapor is increased along the constant-temperature line *abcdef* in Fig. 3.11, a substance that expands on freezing first becomes solid and then liquid. For the substance that contracts on freezing, the corresponding constant-temperature line, Fig. 3.12 indicates that as the pressure on the vapor is increased, it first becomes liquid and then solid.

EXAMPLE 3.5 A vessel having a volume of 0.4 m³ contains 2.0 kg of a liquid water and water vapor mixture in equilibrium at a pressure of 600 kPa. Calculate

1. The volume and mass of liquid.
2. The volume and mass of vapor.

The specific volume is calculated first:

$$v = \frac{0.4}{2.0} = 0.20 \text{ m}^3 / \text{kg}$$

From the steam tables (Table B.1.2),

$$v_{fg} = 0.3157 - 0.001\,101 = 0.3146$$

The quality can now be calculated using Eq. 3.10:

$$0.20 = 0.001\,101 + x0.3146$$

$$x = 0.6322$$

Therefore, the mass of liquid is

$$2.0(0.3678) = 0.7356 \text{ kg}$$

The mass of vapor is

$$2.0(0.6322) = 1.2644 \text{ kg}$$

The volume of liquid is

$$V_{\text{liq}} = m_{\text{liq}} v_f = 0.7356(0.001\,101) = 0.0008 \text{ m}^3$$

The volume of vapor is

$$V_{\text{vap}} = m_{\text{vap}} v_g = 1.2644(0.3157) = 0.3992 \text{ m}^3$$

EXAMPLE 3.6   A rigid vessel contains saturated ammonia vapor at 20°C. Heat is transferred to the system until the temperature reaches 40°C. What is the final pressure?

Since the volume does not change during this process, the specific volume also remains constant. From the ammonia tables, Table B.2,

$$v_1 = v_2 = 0.149 \ 28 \ \text{m}^3 / \text{kg}$$

Since v_g at 40°C is less than 0.149 28 m³/kg, it is evident that in the final state the ammonia is superheated vapor. By interpolating between the 900- and 1000-kPa columns of Table B.2.2, we find that

$$P_2 = 936 \ \text{kPa}$$

EXAMPLE 3.6E   A rigid vessel contains saturated ammonia vapor at 70F. Heat is transferred to the system until the temperature reaches 120F. What is the final pressure?

Since the volume does not change during this process, the specific volume also remains constant. From the ammonia tables, Table C.9,

$$v_1 = v_2 = 2.311 \ \text{ft}^3 / \text{lbm}$$

Since v_g at 120F is less than 2.311 ft³/lbm, it is evident that in the final state the ammonia is superheated vapor. By interpolating between the 140- and 180- lbf/in.² columns of Table C.9.2, we find that

$$P_2 = 145 \ \text{lbf} / \text{in.}^2$$

PROBLEMS

3.1 Water at 27°C can exist in different phases dependent on the pressure. Give the approximate pressure range in kPa for water being in each one of the three phases, vapor, liquid, or solid.

3.2 Find the lowest temperature at which it is possible to have water in the liquid phase. At what pressure must the liquid exist?

3.3 If density of ice is 920 kg/m³, find the pressure at the bottom of a 1000-m-thick ice cap on the North Pole. What is the melting temperature at that pressure?

3.4 A substance is at 2 MPa, 17°C in a rigid tank. Using only the critical properties, can the phase of the mass be determined if the substance is nitrogen, water or propane?

3.5 A cylinder fitted with a frictionless piston contains butane at 25°C, 500 kPa. Can the butane reasonably be assumed to behave as an ideal gas at this state?

3.6 A 1-m³ tank is filled with a gas at room temperature 20°C and pressure 100 kPa. How much mass is there if the gas is a) air, b) neon, or c) propane?

3.7 A cylinder has a thick piston initially held by a pin as shown in Fig. P3.7. The cylinder contains carbon dioxide at 200 kPa and ambient temperature of 290 K. The metal piston has a density of 8000 kg/m³ and the atmospheric pressure is 101 kPa. The pin is now removed, allowing the piston to move and after a while the gas returns to ambient temperature. Is the piston against the stops?

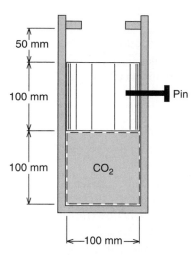

50 mm

100 mm — Pin

100 mm CO_2

←100 mm→

FIGURE P3.7

3.8 A cylindrical gas tank 1 m long, inside diameter of 20 cm, is evacuated and then filled with carbon dioxide gas at 25°C. To what pressure should it be charged if there should be 1.2 kg of carbon dioxide?

3.9 A 1-m^3 rigid tank with air at 1 MPa, 400 K is connected to an air line as shown in Fig. P3.9. The valve is opened and air flows into the tank until the pressure reaches 5 MPa, at which point the valve is closed and the temperature inside is 450K.

  a. What is the mass of air in the tank before and after the process?

  b. The tank eventually cools to room temperature, 300 K. What is the pressure inside the tank then?

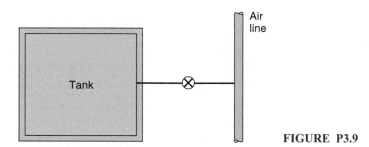

Air line

Tank

FIGURE P3.9

3.10 A hollow metal sphere of 150-mm inside diameter is weighed on a precision beam balance when evacuated and again after being filled to 875 kPa with an unknown gas. The difference in mass is 0.0025 kg, and the temperature is 25°C. What is the gas, assuming it is a pure substance listed in Table A.5?

3.11 A piston/cylinder arrangement, shown in Fig. P3.11, contains air at 250 kPa, 300°C. The 50-kg piston has a diameter of 0.1 m and initially pushes against the stops. The atmosphere is at 100 kPa and 20°C. The cylinder now cools as heat is transferred to the ambient.

  a. At what temperature does the piston begin to move down?

  b. How far has the piston dropped when the temperature reaches ambient?

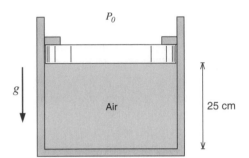

P_0

g

Air

25 cm

FIGURE P3.11

3.12 Air in a tank is at 1 MPa and room temperature of 20°C. It is used to fill an initially empty balloon to a pressure of 200 kPa, at which point the diameter is 2 m and the temperature is 20°C. Assume the pressure in the balloon is linearly proportional to its diameter and that the air in the tank also remains at 20°C throughout the process. Find the mass of air in the balloon and the minimum required volume of the tank.

3.13 A vacuum pump is used to evacuate a chamber where some specimens are dried at 50°C. The pump rate of volume displacement is 0.5 m³/s with an inlet pressure of 0.1 kPa and temperature 50°C. How much water vapor has been removed over a 30-min period?

3.14 An initially deflated and flat balloon is connected by a valve to a 12 m³ storage tank containing helium gas at 2 MPa and ambient temperature, 20°C. The valve is opened and the balloon is inflated at constant pressure, $P_o = 100$ kPa, equal to ambient pressure, until it becomes spherical at $D_1 = 1$ m. If the balloon is larger than this, the balloon material is stretched giving a pressure inside as

$$P = P_o + C\left(1 - \frac{D_1}{D}\right)\frac{D_1}{D}$$

The balloon is inflated to a final diameter of 4 m, at which point the pressure inside is 400 kPa. The temperature remains constant at 20°C. What is the maximum pressure inside the balloon at any time during this inflation process? What is the pressure inside the helium storage tank at this time?

3.15 The helium balloon described in Problem 3.14 is released into the atmosphere and rises to an elevation of 5000 m, with a local ambient pressure of $P_o = 50$ kPa and temperature of −20°C. What is then the diameter of the balloon?

3.16 A cylinder is fitted with a 10-cm-diameter piston that is restrained by a linear spring (force proportional to distance), as shown in Fig. P3.16. The spring force

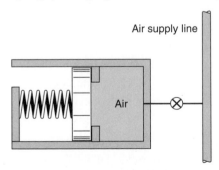

Air supply line

Air

FIGURE P3.16

is 80 kN/m and the piston initially rests on the stops, with a cylinder volume of 1 L. The valve to the air line is opened and the piston begins to rise when the cylinder pressure is 150 kPa. When the valve is closed, the cylinder volume is 1.5 L and the temperature is 80°C. What mass of air is inside the cylinder?

3.17 Air in a tire is initially at $-10°C$, 190 kPa. After driving awhile, the temperature goes up to 10°C. Find the new pressure. You must make one assumption on your own.

3.18 A substance is at 2 MPa, 17°C in a 0.25-m^3 rigid tank. Estimate the mass from the compressibility factor if the substance is a) air, b) butane, or c) propane.

3.19 Argon is kept in a rigid 5 m^3 tank at $-30°C$, 3 MPa. Determine the mass using the compressibility factor. What is the error (%) if the ideal gas model is used?

3.20 A bottle with a volume of 0.1 m^3 contains butane with a quality of 75% and a temperature of 300 K. Estimate the total butane mass in the bottle using the generalized compressibility chart.

3.21 A mass of 2 kg of acetylene is in a 0.045 m^3 rigid container at a pressure of 4.3 MPa. Use the generalized charts to estimate the temperature. (This becomes trial and error.)

3.22 Is it reasonable to assume that at the given states the substance behaves as an ideal gas?

    a. Oxygen at 30°C, 3 MPa   d. R-134a at 30°C, 3 MPa

    b. Methane at 30°C, 3 MPa   e. R-134a at 30°C, 100 kPa

    c. Water at 30°C, 3 MPa

3.23 Determine whether water at each of the following states is a compressed liquid, a superheated vapor, or a mixture of saturated liquid and vapor.

    a. 10 MPa, 0.003 m^3/kg   d. 10 kPa, 10°C

    b. 1 MPa, 190°C   e. 130°C, 200 kPa

    c. 200°C, 0.1 m^3/kg   f. 70°C, 1 m^3/kg

3.24 Determine whether refrigerant R-22 in each of the following states is a compressed liquid, a superheated vapor, or a mixture of saturated liquid and vapor.

    a. 50°C, 0.05 m^3/kg   d. 50°C, 0.3 m^3/kg

    b. 1.0 MPa, 20°C   e. $-20°C$, 200 kPa

    c. 0.1 MPa, 0.1 m^3/kg   f. 2 MPa, 0.012 m^3/kg

3.25 Verify the accuracy of the ideal gas model when it is used to calculate specific volume for saturated water vapor as shown in Fig. 3.9. Do the calculation for 10 kPa and 1 MPa.

3.26 Determine the quality (if saturated) or temperature (if superheated) of the following substances at the given two states:

    a. Water at   1: 120°C, 1 m^3/kg;   2: 10 MPa, 0.01 m^3/kg

    b. Nitrogen at   1: 1 MPa, 0.03 m^3/kg;   2: 100 K, 0.03 m^3/kg

    c. Ammonia at   1: 400 kPa, 0.327 m^3/kg;   2: 1000 kPa, 0.1 m^3/kg

    d. R-22 at   1: 130 kPa, 0.1 m^3/kg;   2: 150 kPa, 0.17 m^3/kg

3.27 Calculate the following specific volumes:

    a. R-134a   50°C, 80% quality

    b. Water   4 MPa, 90% quality

    c. Methane   140 K, 60% quality

    d. Ammonia   60°C, 25% quality

3.28 Give the phase and the specific volume.

a. H_2O	$T = 275°C$	$P = 5$ MPa
b. H_2O	$T = -2°C$	$P = 100$ kPa
c. CO_2	$T = 267°C$	$P = 0.5$ MPa
d. Air	$T = 20°C$	$P = 200$ kPa
e. NH_3	$T = 170°C$	$P = 600$ kPa

3.29 Give the phase and the specific volume.

a. R-22	$T = -25°C$	$P = 100$ kPa
b. R-22	$T = -25°C$	$P = 300$ kPa
c. R-12	$T = 5°C$	$P = 300$ kPa
d. Ar	$T = 200°C$	$P = 200$ kPa
e. NH_3	$T = 20°C$	$P = 100$ kPa

3.30 Find the phase, quality x if applicable, and the missing property P or T.

a. H_2O	$T = 120°C$	$v = 0.5$ m³/kg
b. H_2O	$P = 100$ kPa	$v = 1.8$ m³/kg
c. H_2O	$T = 263$ K	$v = 200$ m³/kg
d. Ne	$P = 750$ kPa	$v = 0.2$ m³/kg
e. NH_3	$T = 20°C$	$v = 0.1$ m³/kg

3.31 Give the phase and the missing properties of P, T, v, and x.

a. R-22	$T = 10°C$	$v = 0.01$ m³/kg
b. H_2O	$T = 350°C$	$v = 0.2$ m³/kg
c. CO_2	$T = 800$ K	$P = 200$ kPa
d. N_2	$T = 200$ K	$P = 100$ kPa
e. CH_4	$T = 190$ K	$x = 0.75$

3.32 Give the phase and the missing properties of P, T, v, and x. These may be a little more difficult if the appendix tables are used instead of the software.

a. R-22	$T = 10°C$	$v = 0.036$ m³/kg
b. H_2O	$v = 0.2$ m³/kg	$x = 0.5$
c. H_2O	$T = 60°C$	$v = 0.001016$ m³/kg
d. NH_3	$T = 30°C$	$P = 60$ kPa
e. R-134a	$v = 0.005$ m³/kg	$x = 0.5$

3.33 What is the percent error in specific volume if the ideal gas model is used to represent the behavior of superheated ammonia at 40°C, 500 kPa? What if the generalized compressibility chart, Fig. D.1, is used instead?

3.34 What is the percent error in pressure if the ideal gas model is used to represent the behavior of superheated vapor R-22 at 50°C, 0.03082 m³/kg? What if the generalized compressibility chart, Fig. D.1, is used instead (iterations needed)?

3.35 Determine the mass of methane gas stored in a 2 m³ tank at −30°C, 3 MPa. Estimate the percent error in the mass determination if the ideal gas model is used.

3.36 A water storage tank contains liquid and vapor in equilibrium at 110°C. The distance from the bottom of the tank to the liquid level is 8 m. What is the absolute pressure at the bottom of the tank?

3.37 A sealed rigid vessel has volume of 1 m³ and contains 2 kg of water at 100°C. The vessel is now heated. If a safety pressure valve is installed, at what pressure should the valve be set to have a maximum temperature of 200°C?

3.38 A 500-L tank stores 100 kg of nitrogen gas at 150 K. To design the tank the pressure must be estimated and three different methods are suggested. Which is the most accurate, and how different in percent are the other two?

   a. Nitrogen tables, Table B.6

   b. Ideal gas

   c. Generalized compressibility chart, Fig. D.1

3.39 A 400-m³ storage tank is being constructed to hold LNG, liquified natural gas, which may be assumed to be essentially pure methane. If the tank is to contain 90% liquid and 10% vapor, by volume, at 100 kPa, what mass of LNG (kg) will the tank hold? What is the quality in the tank?

3.40 A storage tank holds methane at 120 K, with a quality of 25%, and it warms up by 5°C per hour due to a failure in the refrigeration system. How much time will it take before the methane becomes single phase and what is the pressure then?

3.41 Saturated liquid water at 60°C is put under pressure to decrease the volume by 1% keeping the temperature constant. To what pressure should it be compressed?

3.42 Saturated water vapor at 60°C has its pressure decreased to increase the volume by 10% keeping the temperature constant. To what pressure should it be expanded?

3.43 A boiler feed pump delivers 0.05 m³/s of water at 240°C, 20 MPa. What is the mass flowrate (kg/s)? What would be the percent error if the properties of saturated liquid at 240°C were used in the calculation? What if the properties of saturated liquid at 20 MPa were used?

3.44 A glass jar is filled with saturated water at 500 kPa, quality 25%, and a tight lid is put on. Now it is cooled to −10°C. What is the mass fraction of solid at this temperature?

3.45 A cylinder/piston arrangement contains water at 105°C, 85% quality with a volume of 1 L. The system is heated, causing the piston to rise and encounter a linear spring, as shown in Fig. P3.45. At this point the volume is 1.5 L, the piston diameter is 150 mm, and the spring constant is 100 N/mm. The heating continues, so the piston compresses the spring. What is the cylinder temperature when the pressure reaches 200 kPa?

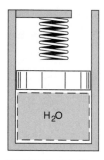

FIGURE P3.45

3.46 Saturated (liquid + vapor) ammonia at 60°C is contained in a rigid steel tank. It is used in an experiment, where it should pass through the critical point when the system is heated. What should the initial mass fraction of liquid be?

3.47 For a certain experiment, R-22 vapor is contained in a sealed glass tube at 20°C. It is desired to know the pressure at this condition, but there is no means of measuring it, since the tube is sealed. However, if the tube is cooled to −20°C small droplets of liquid are observed on the glass walls. What is the initial pressure?

3.48 A steel tank contains 6 kg of propane (liquid + vapor) at 20°C with a volume of 0.015 m³. The tank is now slowly heated. Will the liquid level inside eventually rise to the top or drop to the bottom of the tank? What if the initial mass is 1 kg instead of 6 kg?

3.49 A cylinder containing ammonia is fitted with a piston restrained by an external force that is proportional to cylinder volume squared. Initial conditions are 10°C,

90% quality and a volume of 5 L. A valve on the cylinder is opened and additional ammonia flows into the cylinder until the mass inside has doubled. If at this point the pressure is 1.2 MPa, what is the final temperature?

3.50 A container with liquid nitrogen at 100 K has a cross-sectional area of 0.5 m² as shown in Fig. P3.50. Due to heat transfer, some of the liquid evaporates and in one hour the liquid level drops 30 mm. The vapor leaving the container passes through a valve and a heater and exits at 500 kPa, 260 K. Calculate the volume rate of flow of nitrogen gas exiting the heater.

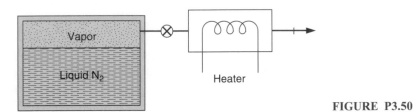

Heater

FIGURE P3.50

3.51 A pressure cooker (closed tank) contains water at 100°C with the liquid volume being 1/10 of the vapor volume. It is heated until the pressure reaches 2.0 MPa. Find the final temperature. Has the final state more or less vapor than the initial state?

3.52 Ammonia in a piston/cylinder arrangement is at 700 kPa, 80°C. It is now cooled at constant pressure to saturated vapor (state 2) at which point the piston is locked with a pin. The cooling continues to −10°C (state 3). Show the processes 1 to 2 and 2 to 3 on both a P-v and T-v diagram.

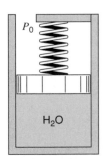

FIGURE P3.53

3.53 A piston/cylinder arrangement is loaded with a linear spring and the outside atmosphere. It contains water at 5 MPa, 400°C with the volume being 0.1 m³, as shown in Fig. P3.53. If the piston is at the bottom, the spring exerts a force such that $P_{\text{lift}} = 200$ kPa. The system now cools until the pressure reaches 1200 kPa. Find the mass of water, the final state (T_2, v_2) and plot the P-v diagram for the process.

3.54 Water in a piston/cylinder is at 90°C, 100 kPa, and the piston loading is such that pressure is proportional to volume, $P = CV$. Heat is now added until the temperature reaches 200° C. Find the final pressure and also the quality if in the two-phase region.

3.55 A spring-loaded piston/cylinder contains water at 500°C, 3 MPa. The setup is such that pressure is proportional to volume, $P = CV$. It is now cooled until the water becomes saturated vapor. Sketch the P-v diagram and find the final pressure.

3.56 Refrigerant-12 in a piston/cylinder arrangement is initially at 50° C, $x = 1$. It is then expanded in a process so that $P = Cv^{-1}$ to a pressure of 100 kPa. Find the final temperature and specific volume.

3.57 A sealed rigid vessel of 2 m³ contains a saturated mixture of liquid and vapor R-134a at 10°C. If it is heated to 50°C, the liquid phase disappears. Find the pressure at 50°C and the initial mass of the liquid.

3.58 Two tanks are connected as shown in Fig. P3.58, both containing water. Tank A is at 200 kPa, $v = 0.5$ m³/kg, $V_A = 1$ m³ and tank B contains 3.5 kg at 0.5 MPa, 400°C. The valve is now opened and the two come to a uniform state. Find the final specific volume.

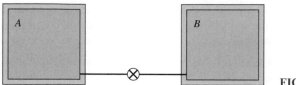

FIGURE P3.58

3.59 A tank contains 2 kg of nitrogen at 100 K with a quality of 50%. Through a volume flowmeter and valve, 0.5 kg is now removed while the temperature remains constant. Find the final state inside the tank and the volume of nitrogen removed if the valve/meter is located at

a. The top of the tank b. The bottom of the tank

3.60 Consider two tanks, A and B, connected by a valve, as shown in Fig. P3.60. Each has a volume of 200 L and tank A has R-12 at 25°C, 10% liquid and 90% vapor by volume, while tank B is evacuated. The valve is now opened and saturated vapor flows from A to B until the pressure in B has reached that in A, at which point the valve is closed. This process occurs slowly such that all temperatures stay at 25°C throughout the process. How much has the quality changed in tank A during the process?

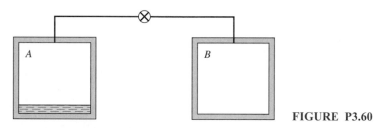

FIGURE P3.60

3.61E A substance is at 300 lbf/in.^{2}, 65 F in a rigid tank. Using only the critical properties, can the phase of the mass be determined if the substance is nitrogen, water, or propane?

3.62E A cylindrical gas tank 3 ft long, inside diameter of 8 in., is evacuated and then filled with carbon dioxide gas at 77 F. To what pressure should it be charged if there should be 2.6 lbm of carbon dioxide?

3.63E A vacuum pump is used to evacuate a chamber where some specimens are dried at 120 F. The pump rate of volume displacement is 900 ft^3/min with an inlet pressure of 1 mm Hg and temperature 120 F. How much water vapor has been removed over a 30-min period?

3.64E A cylinder is fitted with a 4-in.-diameter piston that is restrained by a linear spring (force proportional to distance) as shown in Fig. P3.16. The spring force constant is 400 lbf/in. and the piston initially rests on the stops, with a cylinder volume of 60 in.^{3}. The valve to the air line is opened and the piston begins to rise when the cylinder pressure is 22 lbf/in.^{2}. When the valve is closed, the cylinder volume is 90 in.^{3} and the temperature is 180 F. What mass of air is inside the cylinder?

3.65E A substance is at 70 F, 300 lbf/in.^{2} in a 10 ft^3 tank. Estimate the mass from the compressibility chart if the substance is a) air, b) butane, or c) propane.

3.66E Determine the mass of an ethane gas is stored in a 25 ft^3 tank at 250 F, 440 lbf/in.^{2} using the compressibility chart. Estimate the error (%) if the ideal gas model is used.

3.67E Argon is kept in a rigid 100 ft^3 tank at -30 F, 450 lbf/in.^{2}. Determine the mass using the compressibility factor. What is the error (%) if the ideal gas model is used?

3.68E Determine whether water at each of the following states is a compressed liquid, a superheated vapor, or a mixture of saturated liquid and vapor.

a. 1800 lbf/in.^{2}, 0.03 ft^3/lbm   d. 2 lbf/in.^{2}, 50 F

b. 150 lbf/in.^{2}, 320 F   e. 270 F, 30 lbf/in.^{2}

c. 380 F, 3 ft^3/lbm   f. 160 F, 10 ft^3/lbm

3.69E Give the phase and the specific volume.

a. H$_2$O $T = 520$F $P = 700$ lbf/in.^{2}

b. H$_2$O $T = 30$ F $P = 15$ lbf/in.^{2}

c. CO$_2$ $T = 510$ F $P = 75$ lbf/in.^{2}

d. Air $T = 68$ F $P = 2$ atm

e. NH$_3$ $T = 290$ F $P = 90$ lbf/in.^{2}

3.70E Give the phase and the specific volume.

a. R-22 $T = -10$ F $P = 30$ lbf/in.^{2}

b. R-22 $T = -10$ F $P = 40$ lbf/in.^{2}

c. H$_2$O $T = 280$ F $P = 35$ lbf/in.^{2}

d. Ar $T = 300$ F $P = 30$ lbf/in.^{2}

e. NH$_3$ $T = 60$ F $P = 15$ lbf/in.^{2}

3.71E Give the phase and the missing properties of P, T, v, and x. These may be a little more difficult if the appendix tables are used instead of the software.

a. R-22 $T = 50$ F $v = 0.6$ ft^3/lbm

b. H$_2$O $v = 2$ ft^3/lbm $x = 0.5$

c. H$_2$O $T = 150$ F $v = 0.01632$ ft^3/lbm

d. NH$_3$ $T = 80$ F $P = 13$ lbf/in.^{2}

e. R-134a $v = 0.08$ ft^3/lbm $x = 0.5$

3.72E What is the percent error in specific volume if the ideal gas model is used to represent the behavior of superheated ammonia at 100 F, 80 lbf/in.^{2}? What if the generalized compressibility chart, Fig. D.1, is used instead?

3.73E A water storage tank contains liquid and vapor in equilibrium at 220 F. The distance from the bottom of the tank to the liquid level is 25 ft. What is the absolute pressure at the bottom of the tank?

3.74E A sealed rigid vessel has volume of 35 ft^3 and contains 2 lbm of water at 200 F. The vessel is now heated. If a safety pressure valve is installed, at what pressure should the valve be set to have a maximum temperature of 400 F?

3.75E Saturated liquid water at 200 F is put under pressure to decrease the volume by 1%, keeping the temperature constant. To what pressure should it be compressed?

3.76E Saturated water vapor at 200 F has its pressure decreased to increase the volume by 10%, keeping the temperature constant. To what pressure should it be expanded?

3.77E A boiler feed pump delivers 100 ft^3/min of water at 400 F, 3000 lbf/in.^{2}. What is the mass flowrate (lbm/s)? What would be the percent error if the properties of saturated liquid at 400 F were used in the calculation? What if the properties of saturated liquid at 3000 lbf/in.^{2} were used?

3.78E Saturated (liquid + vapor) ammonia at 140 F is contained in a rigid steel tank. It is used in an experiment, where it should pass through the critical point when the system is heated. What should the initial mass fraction of liquid be?

3.79E A steel tank contains 14 lbm of propane (liquid + vapor) at 70 F with a volume of 0.25 ft^3. The tank is now slowly heated. Will the liquid level inside eventually rise to the top or drop to the bottom of the tank? What if the initial mass is 2 lbm instead of 14 lbm?

3.80E A pressure cooker (closed tank) contains water at 200 F with the liquid volume being 1/10 of the vapor volume. It is heated until the pressure reaches 300 lbf/in.^{2}. Find the final temperature. Has the final state more or less vapor than the initial state?

3.81E Two tanks are connected together as shown in Fig. P3.58, both containing water. Tank A is at 30 lbf/in.^{2}, $v = 8$ ft^3/lbm, $V = 40$ ft^3, and tank B contains 8 lbm at 80 lbf/in.^{2}, 750 F. The valve is now opened and the two come to a uniform state. Find the final specific volume.

3.82E A spring-loaded piston/cylinder contains water at 900 F, 450 lbf/in.^{2}. The setup is such that pressure is proportional to volume, $P = CV$. It is now cooled until the water becomes saturated vapor. Find the final pressure.

3.83E Refrigerant-22 in a piston/cylinder arrangement is initially at 120 F, $x = 1$. It is then expanded in a process so that $P = Cv^{-1}$ to a pressure of 30 lbf/in.^{2}. Find the final temperature and specific volume.

Computer, Design, and Open-Ended Problems

3.84 Write a program that lists the saturated P, T for ammonia with pressure as an entry, beginning with $P = 100$ kPa and ending with $P = 500$ kPa, in steps of 25 kPa.

3.85 Write a program that tabulates values of P and T along a constant specific volume line for water. The starting state is 100 kPa, quality of $x = 0.5$, and the ending state is 1 MPa.

3.86 Write a computer program that lists the states P, T, and v along the process curves in Problem 3.11.

3.87 Use the computer software to sketch the variation of pressure with temperature in Problem 3.40. Extend the curve a little into the single phase region.

3.88 By the use of the computer software find a few of the states between the beginning and end states and show the variation of pressure and temperature as a function of volume for Problem 3.54.

3.89 In Problem 3.56, we wish to follow the path of the process for the R-12 for any state between the initial and final states inside the cylinder.

3.90 For any specified substance in Tables B.1–B.7, fit a polynomial equation of degree n to tabular data for pressure as a function of density along any given isotherm in the superheated vapor region.

3.91 The refrigerant fluid in a household refrigerator changes phase from liquid to vapor at the low temperature in the refrigerator. It changes phase from vapor to liquid at the higher temperature in the heat exchanger that gives the energy to the room air.

Measure or otherwise estimate these temperatures. Based on these temperatures make a table with the refrigerant pressures for the refrigerants for which tables are available in Appendix B. Discuss the results and the requirements for a substance to be a potential refrigerant.

3.92 Repeat the previous problem for refrigerants that are listed in Table A.2 and use the compressibility chart Fig. D.1 to estimate the pressures.

3.93 The saturated pressure as a function of temperature follows the correlation developed by Wagner as

$$\ln P_r = [w_1 \tau + w_2 \tau^{1.5} + w_3 \tau^3 + w_4 \tau^6]/T_r$$

where the reduced pressure and temperature are $P_r = P/P_c$ and $T_r = T/T_c$. The temperature variable is $\tau = 1 - T_r$. The parameters are found for R-12 and R-134a as

	w_1	w_2	w_3	w_4
R-12	-6.91826	1.49560	-2.65015	-0.63170
R-134a	-7.59884	1.48886	-3.79873	1.81379

Compare these correlations to the tables in Appendix B.

3.94 Find the constants in the curve fit for the saturation pressure using Wagner's correlation as shown in the previous problem for water and methane. Find other correlations in the literature and compare them to the tables and give the maximum deviation.

3.95 The specific volume of saturated liquid can be approximated by the Rackett equation as

$$v_f = \frac{\overline{R}\,T_c}{M P_c} Z_c^n \; ; \; n = 1 + (1 - T_r)^{2/7}$$

with the reduced temperature, $T_r = T/T_c$, and the compressibility factor, $Z_c = P_c v_c/RT_c$. Using values from Table A.2 with the critical constants, compare the formula to the tables for substances where the saturated specific volume is available.

4 WORK AND HEAT

In this chapter we consider work and heat. It is essential for the student of thermodynamics to understand clearly the definitions of both work and heat, because the correct analysis of many thermodynamic problems depends on distinguishing between them.

4.1 DEFINITION OF WORK

Work is usually defined as a force F acting through a displacement x, the displacement being in the direction of the force. That is,

$$W = \int_1^2 F\,dx \tag{4.1}$$

This is a very useful relationship because it enables us to find the work required to raise a weight, to stretch a wire, or to move a charged particle through a magnetic field.

However, when treating thermodynamics from a macroscopic point of view, it is advantageous to tie in the definition of work with the concepts of systems, properties, and processes. We therefore define work as follows: work is done by a system if the sole effect on the surroundings (everything external to the system) could be the raising of a weight. Notice that the raising of a weight is in effect a force acting through a distance. Notice also that our definition does not state that a weight was actually raised or that a force actually acted through a given distance, but that the sole effect external to the system could be the raising of a weight. Work done *by* a system is considered positive and work done *on* a system is considered negative. The symbol W designates the work done by a system.

In general, we will speak of work as a form of energy. No attempt will be made to give a rigorous definition of energy. Since the concept is familiar, the term *energy* will be used as appropriate, and various forms of energy will be identified. Work is the form of energy that fulfills the definition.

Let us illustrate this definition of work with a few examples. Consider as a system the battery and motor of Fig. 4.1a and let the motor drive a fan. Does work cross the boundary of the system? To answer this question using the definition of work given earlier, replace the fan with the pulley and weight arrangement shown in Fig. 4.1b. As the motor turns, the weight is raised, and the sole effect external to the system is the raising of a weight. Thus, for our original system of Fig. 4.1a, we conclude that work is crossing the boundary of the system since the sole effect external to the system could be the raising of a weight.

Let the boundaries of the system be changed now to include only the battery shown in Fig. 4.2. Again we ask the question, does work cross the boundary of the system? To

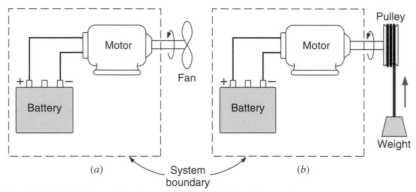

FIGURE 4.1 Example of work crossing the boundary of a system.

answer this question, we need to ask a more general question. Does the flow of electrical energy across the boundary of a system constitute work?

The only limiting factor in having the sole external effect be the raising of a weight is the inefficiency of the motor. However, as we design a more efficient motor, with lower bearing and electrical losses, we recognize that we can approach a certain limit that meets the requirement of having the only external effect be the raising of a weight. Therefore, we can conclude that when there is a flow of electricity across the boundary of a system, as in Fig. 4.2, it is work.

4.2 UNITS FOR WORK

As already noted, work done *by* a system, such as that done by a gas expanding against a piston, is positive, and work done *on* a system, such as that done by a piston compressing a gas, is negative. Thus, positive work means that energy leaves the system, and negative work means that energy is added to the system.

Our definition of work involves the raising of a weight, that is, the product of a unit force (one newton) acting through a unit distance (one meter). This unit for work in SI units is called the joule (J).

$$1\,J = 1\,N\,m$$

Power is the time rate of doing work and is designated by the symbol $\dot{W}$:

$$\dot{W} \equiv \frac{\delta W}{dt}$$

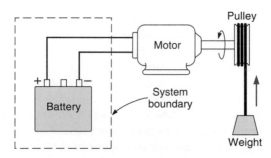

FIGURE 4.2 Example of work crossing the boundary of a system because of a flow of an electric current across the system boundary.

The unit for power is a rate of work of one joule per second, which is a watt (W):

$$1 \text{ W} = 1 \text{ J} / \text{s}$$

A familiar unit for power in English units is the horsepower (hp), where

$$1 \text{ hp} = 550 \text{ ft lbf} / \text{s}$$

It is often convenient to speak of the work per unit mass of the system, often termed "specific work." This quantity is designated w and is defined

$$w \equiv \frac{W}{m}$$

4.3 Work Done at the Moving Boundary of a Simple Compressible System

We have already noted that there are a variety of ways in which work can be done on or by a system. These include work done by a rotating shaft, electrical work, and the work done by the movement of the system boundary, such as the work done in moving the piston in a cylinder. In this section we will consider in some detail the work done at the moving boundary of a simple compressible system during a quasi-equilibrium process.

Consider as a system the gas contained in a cylinder and piston, as in Fig. 4.3. Let one of the small weights be removed from the piston, which will cause the piston to move upward a distance dL. We can consider this quasi-equilibrium process and calculate the amount of work W done by the system during this process. The total force on the piston is $P\mathscr{A}$, where P is the pressure of the gas and $\mathscr{A}$ is the area of the piston. Therefore, the work δW is

$$\delta W = P\mathscr{A} \, dL$$

But $\mathscr{A} \, dL = dV$, the change in volume of the gas. Therefore,

$$\delta W = P \, dV \tag{4.2}$$

The work done at the moving boundary during a given quasi-equilibrium process can be found by integrating Eq. 4.2. However, this integration can be performed only if we know the relationship between P and V during this process. This relationship may be expressed in the form of an equation, or it may be shown in the form of a graph.

Let us consider a graphical solution first. We use as an example a compression process such as occurs during the compression of air in a cylinder, Fig. 4.4. At the beginning of the process the piston is at position 1, and the pressure is relatively low. This state is represented on a pressure–volume diagram (usually referred to as a P–V diagram). At the conclusion of the process the piston is in position 2, and the corresponding state of the gas is shown at point 2 on the P–V diagram. Let us assume that this compression was a quasi-equilibrium process, and that during the process the system passed through the states shown by the line connecting states 1 and 2 on the P–V diagram. The assumption of a quasi-equilibrium process is essential here because each point on line 1–2 represents a definite state, and these states will correspond to the actual state of the system only if

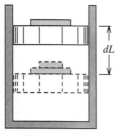

FIGURE 4.3 Example of work done at the moving boundary of a system in a quasi-equilibrium process.

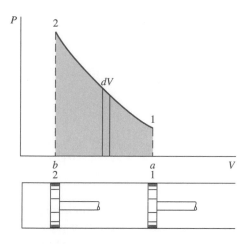

FIGURE 4.4 Use of pressure–volume diagram to show work done at the moving boundary of a system in a quasi-equilibrium process.

the deviation from equilibrium is infinitesimal. The work done on the air during this compression process can be found by integrating Eq. 4.2:

$$_1W_2 = \int_1^2 \delta W = \int_1^2 P\, dV \tag{4.3}$$

The symbol $_1W_2$ is to be interpreted as the work done during the process from state 1 to state 2. It is clear from examining the P–V diagram that the work done during this process,

$$\int_1^2 P\, dV$$

is represented by the area under the curve 1–2, area a–1–2–b–a. In this example the volume decreased, and the area a–1–2–b–a represents work done on the system. If the process had proceeded from state 2 to state 1 along the same path, the same area would represent work done by the system.

Further consideration of a P–V diagram, such as Fig. 4.5, leads to another important conclusion. It is possible to go from state 1 to state 2 along many different quasi-equilibrium paths, such as A, B, or C. Since the area underneath each curve represents the work for each process, it is evident that the amount of work done during each process not only is a function of the end states of the process, but depends on the path that is followed in going from one state to another. For this reason work is called a path function or, in mathematical parlance, δW is an inexact differential.

This concept leads to a brief consideration of point and path functions or, to use another term, exact and inexact differentials. Thermodynamic properties are point functions, a name that comes from the fact that for a given point on a diagram (such as Fig. 4.5) or surface (such as Fig. 3.11), the state is fixed, and thus there is a definite value of each property corresponding to this point. The differentials of point functions are exact differentials, and the integration is simply

$$\int_1^2 dV = V_2 - V_1$$

Thus, we can speak of the volume in state 2 and the volume in state 1, and the change in volume depends only on the initial and final states.

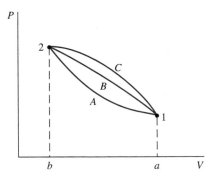

FIGURE 4.5 Various quasi-equilibrium processes between two given states, indicating that work is a path function.

Work, on the other hand, is a path function, for, as has been indicated, the work done in a quasi-equilibrium process between two given states depends on the path followed. The differentials of path functions are inexact differentials, and the symbol δ will be used in this text to designate inexact differentials (in contrast to d for exact differentials). Thus, for work, we write

$$\int_1^2 \delta W = {}_1W_2$$

It would be more precise to use the notation, ${}_1W_{2A}$, which would indicate the work done during the change from state 1 to state 2 along path A. However, it is implied in the notation ${}_1W_2$ that the process between states 1 and 2 has been specified. It should be noted that we never speak about the work in the system in state 1 or state 2, and thus we would never write $W_2 - W_1$.

In evaluating the integral of Eq. 4.3, we should always keep in mind that we wish to determine the area under the curve on Fig. 4.5. In connection with this point, we identify the following two classes of problems:

1. The relationship between P and V is given in terms of experimental data or in graphical form (as, for example, the trace on an oscilloscope). Therefore, we may evaluate the integral, Eq. 4.3, by graphical or numerical integration.

2. The relationship between P and V makes it possible to fit an analytical relationship between them. We may then integrate directly.

One common example of this second type of functional relationship is a process called a polytropic process, one in which

$$PV^n = \text{constant}$$

throughout the process. The exponent n may possibly be any value from $-\infty$ to $+\infty$, depending on the particular process. For this type of process, we can integrate Eq. 4.3 as follows:

$$PV^n = \text{constant} = P_1V_1^n = P_2V_2^n$$

$$P = \frac{\text{constant}}{V^n} = \frac{P_1V_1^n}{V^n} = \frac{P_2V_2^n}{V^n}$$

$$\int_1^2 P\,dV = \text{constant} \int_1^2 \frac{dV}{V^n} = \text{constant} \left(\frac{V^{-n+1}}{-n+1} \right) \Bigg|_1^2$$

$$\int_1^2 P\,dV = \frac{\text{constant}}{1-n}(V_2^{1-n} - V_1^{1-n}) = \frac{P_2 V_2^n V_2^{1-n} - P_1 V_1^n V_1^{1-n}}{1-n}$$

$$= \frac{P_2 V_2 - P_1 V_1}{1-n} \tag{4.4}$$

Note that the resulting Eq. 4.4 is valid for any exponent n, except $n = 1$. Where $n = 1$,

$$PV = \text{constant} = P_1 V_1 = P_2 V_2$$

and

$$\int_1^2 P\,dV = P_1 V_1 \int_1^2 \frac{dV}{V} = P_1 V_1 \ln \frac{V_2}{V_1} \tag{4.5}$$

Note that in Eqs. 4.4 and 4.5 we did not say that the work is equal to the expressions given in these equations. These expressions give us the value of a certain integral, that is, a mathematical result. Whether or not that integral equals the work in a particular process depends on the result of a thermodynamic analysis of that process. It is important to keep the mathematical result separate from the thermodynamic analysis, for there are many situations in which work is not given by Eq. 4.3.

The polytropic process as described demonstrates one special functional relationship between P and V during a process. There are many other possible relations, some of which will be examined in the problems at the end of this chapter.

EXAMPLE 2.1   Consider as a system the gas in the cylinder shown in Fig. 4.6; the cylinder is fitted with a piston on which a number of small weights are placed. The initial pressure is 200 kPa, and the initial volume of the gas is 0.04 m³.

1. Let a Bunsen burner be placed under the cylinder, and let the volume of the gas increase to 0.1 m³ while the pressure remains constant. Calculate the work done by the system during this process.

$$_1W_2 = \int_1^2 P\,dV$$

Since the pressure is constant, we conclude from Eq. 4.3 that

$$_1W_2 = P\int_1^2 dV = P(V_2 - V_1)$$

$$_1W_2 = 200 \text{ kPa} \times (0.1 - 0.04)\text{m}^3 = 12.0 \text{ kJ}$$

2. Consider the same system and initial conditions, but at the same time the Bunsen burner is under the cylinder and the piston is rising, let weights be removed from the piston at such a rate that, during the process, the temperature of the gas remains constant.

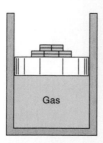

Gas

FIGURE 4.6
Sketch for Example 4.1.

If we assume that the ideal-gas model is valid, then, from Eq. 3.4,

$$PV = mRT$$

We note that this is a polytropic process with exponent $n = 1$. From our analysis, we conclude that the work is given by Eq. 4.3 and that the integral in this equation is given by Eq. 4.5. Therefore,

$$_1W_2 = \int_1^2 P\,dV = P_1V_1 \ln \frac{V_2}{V_1}$$

$$= 200 \text{ kPa} \times 0.04 \text{ m}^3 \times \ln \frac{0.10}{0.04} = 7.33 \text{ kJ}$$

3. Consider the same system, but during the heat transfer let the weights be removed at such a rate that the expression $PV^{1.3} = $ constant describes the relation between pressure and volume during the process. Again the final volume is 0.1 m³. Calculate the work.

   This is a polytropic process in which $n = 1.3$. Analyzing the process, we conclude again that the work is given by Eq. 4.3, and that the integral is given by Eq. 4.4. Therefore,

$$P_2 = 200 \left(\frac{0.04}{0.10} \right)^{1.3} = 60.77 \text{ kPa}$$

$$_1W_2 = \int_1^2 P\,dV = \frac{P_2V_2 - P_1V_1}{1 - 1.3} = \frac{60.77 \times 0.1 - 200 \times 0.04}{1 - 1.3}$$

$$= 6.41 \text{ kJ}$$

4. Consider the system and initial state given in the first three examples, but let the piston be held by a pin so that the volume remains constant. In addition, let heat be transferred from the system until the pressure drops to 100 kPa. Calculate the work.

Since $\delta W = P\,dV$ for a quasi-equilibrium process, the work is zero, because there is no change in volume.

The process for each of the four examples is shown on the P–V diagram of Fig. 4.7. Process 1–2a is a constant-pressure process, and area 1–2a–f–e–1 represents the work. Similarly, line 1–2b represents the process in which $PV = $ constant, line 1–2c the process in which $PV^{1.3} = $ constant, and line 1–2d the constant-volume process. The student should compare the relative areas under each curve with the numerical results obtained for the amounts of work done.

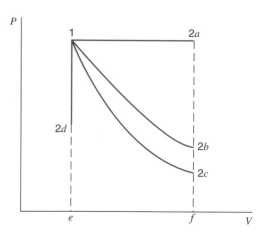

FIGURE 4.7 Pressure–volume diagram showing work done in the various processes of Example 4.1.

EXAMPLE 4.2 Consider a slightly different piston cylinder arrangement as shown in Fig. 4.8. In this example the piston is loaded with a mass, m_p, the outside atmosphere P_0, a linear spring and a single point force F_1. The piston is restricted in its motion by lower and upper stops trapping the gas with a pressure P. A force balance on the piston in the direction of motion yields

$$m_p a > 0 = \sum F_\uparrow - \sum F_\downarrow$$

with a zero acceleration in a quasi-equilibrium process. The forces, when the piston is between the stops, are

$$\sum F_\uparrow = P\mathcal{A} \qquad \sum F_\downarrow = m_p g + P_0 \mathcal{A} + k_s (x - x_0) + F_1$$

with the linear spring constant, k_s. The piston position for a relaxed spring is x_0, which depends on how the spring is installed. The force balance then gives the gas pressure by division with the area, $\mathcal{A}$, as

$$P = P_0 + \left[m_p g + F_1 + k_s (x - x_0) \right] / \mathcal{A}$$

To illustrate the process in a P–V diagram, the distance x is converted to volume by division and multiplication with $\mathcal{A}$:

$$P = P_0 + \frac{m_p g}{\mathcal{A}} + \frac{F_1}{\mathcal{A}} + \frac{k_s}{\mathcal{A}^2}(V - V_0) = C_1 + C_2 V$$

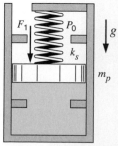

FIGURE 4.8 Sketch of physical system for Example 4.2.

This relation gives the pressure as a linear function of the volume, with the line having a slope of $C_2 = k_s / \mathcal{A}^2$. With the stops installed, the minimum and maximum volumes limit the possible states of the system to the combination of P and V as shown in Fig. 4.9. Regardless of what substance is inside, any process must proceed along the lines in the P–V diagram. The work term in a quasi-equilibrium process then follows as

$$_1W_2 = \int_1^2 P \, dV = \text{AREA under the process curve}$$

$$_1W_2 = \frac{1}{2}\left(P_1' + P_2'\right)\left(V_2 - V_1\right)$$

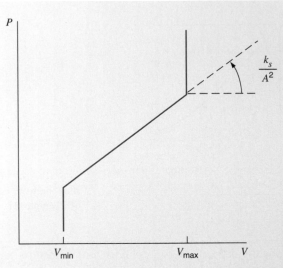

FIGURE 4.9 The process curve showing possible P–V combinations for Example 4.2.

with $P_1' = P_1$ and $P_2' = P_2$, subject to the constraint that

$$P_{\min} \le P_1', \ P_2' \le P_{\max}$$

These limits show that only the part of the process that follows the sloped line, when the piston moves, contributes to the work. Any part of the process with a pressure smaller than $P_{\min}$ or larger than $P_{\max}$ does not involve work as the piston is held in a fixed position by one of the stops. The maximum work then arises, when the piston travels the total distance between the two stops.

In this section we have discussed boundary movement work in a quasi-equilibrium process. We should also realize that there may very well be boundary movement work in a nonequilibrium process. Then the total force exerted on the piston by the gas inside the cylinder, $P\mathcal{A}$, does not equal the external force, F_{ext}, and the work is not given by Eq. 4.2. The work can, however, be evaluated in terms of F_{ext} or, dividing by area, an equivalent external pressure, P_{ext}. The work done at the moving boundary in this case is

$$\delta W = F_{\text{ext}} \, dL = P_{\text{ext}} \, dV \qquad (4.6)$$

Evaluation of Eq. 4.6 in any particular instance requires a knowledge of how the external force or pressure changes during the process.

4.4 OTHER SYSTEMS THAT INVOLVE WORK

In the preceding section we considered the work done at the moving boundary of a simple compressible system during a quasi-equilibrium process and also during a nonequilibrium process. There are other types of systems in which work is done at a moving boundary. In this section we briefly consider three such systems, a stretched wire, a surface film, and electrical work.

Consider as a system a stretched wire that is under a given tension $\mathcal{T}$. When the length of the wire changes by the amount dL, the work done by the system is

$$\delta W = -\mathcal{T}\, dL \tag{4.7}$$

The minus sign is necessary because work is done by the system when dL is negative. This equation can be integrated to give

$$_1W_2 = -\int_1^2 \mathcal{T}\, dL \tag{4.8}$$

The integration can be performed either graphically or analytically if the relation between $\mathcal{T}$ and L is known. The stretched wire is a simple example of the type of problem in solid-body mechanics that involves the calculation of work.

EXAMPLE 4.3 A metallic wire of initial length L_0 is stretched. Assuming elastic behavior, determine the work done in terms of the modulus of elasticity and the strain.

Let σ = stress, e = strain, and E = the modulus of elasticity.

$$\sigma = \frac{\mathcal{T}}{\mathcal{A}} = Ee$$

Therefore,

$$\mathcal{T} = \mathcal{A}Ee$$

From the definition of strain,

$$de = \frac{dL}{L_0}$$

Therefore,

$$\delta W = -\mathcal{T}\, dL = -\mathcal{A}EeL_0\, de$$

$$W = -\mathcal{A}EL_0 \int_{e=0}^e e\, de = -\frac{\mathcal{A}EL_0}{2}(e)^2$$

Now consider a system that consists of a liquid film having a surface tension $\mathcal{S}$. A schematic arrangement of such a film, maintained on a wire frame, one side of which can be moved, is shown in Fig. 4.10. When the area of the film is changed, for example, by sliding the movable wire along the frame, work is done on or by the film. When the area changes by an amount $d\mathcal{A}$, the work done by the system is

$$\delta W = -\mathcal{S}\, d\mathcal{A} \tag{4.9}$$

For finite changes,

$$_1W_2 = -\int_1^2 \mathcal{S}\, d\mathcal{A} \tag{4.10}$$

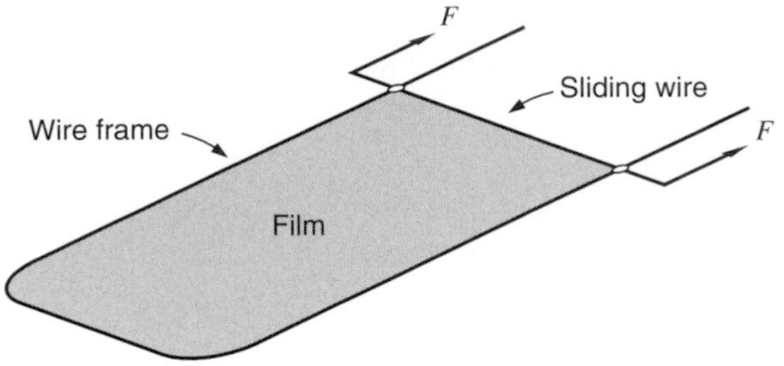

FIGURE 4.10 Schematic arrangement showing work done on a surface film.

We have already noted that electrical energy flowing across the boundary of a system is work. We can gain further insight into such a process by considering a system in which the only work mode is electrical. As examples of such a system, we can think of a charged condenser, an electrolytic cell, and the type of fuel cell described in Chapter 1. Consider a quasi-equilibrium process for such a system, and during this process let the potential difference be $\mathscr{E}$ and the amount of electrical charge that flows into the system be dZ. For this quasi-equilibrium process the work is given by the relation

$$\delta W = -\mathscr{E}\,dZ \tag{4.11}$$

Since the current, i, equals dZ/dt (where t = time), we can also write

$$\delta W = -\mathscr{E}i\,dt$$

$$_1W_2 = -\int_1^2 \mathscr{E}i\,dt \tag{4.12}$$

Equation 4.12 may also be written as a rate equation for work (the power).

$$\frac{\delta W}{dt} = -\mathscr{E}i \tag{4.13}$$

Since the ampere (electric current) is one of the fundamental units in the International System, and the watt has been defined previously, this relation serves as the definition of the unit for electric potential, the volt (V), which is one watt divided by one ampere.

4.5 CONCLUDING REMARKS REGARDING WORK

The similarity of the expressions for work in the three processes discussed in Section 4.4 and in the processes in which work is done at a moving boundary should be noted. In each of these quasi-equilibrium processes, the work is given by the integral of the product of an intensive property and the change of an extensive property. The following is a summary list of these processes and their work expressions

Simple compressible system $\qquad _1W_2 = \int_1^2 P\,dV$

Stretched wire $\qquad _1W_2 = -\int_1^2 \mathscr{T}\,dL$

Surface film $$_1W_2 = -\int_1^2 \mathcal{S}\, d\mathcal{A} \qquad (4.14)$$

System in which the work is
completely electrical $$_1W_2 = -\int_1^2 \mathcal{E}\, dZ$$

Although we will deal primarily with systems in which there is only one mode of work, it is quite possible to have more than one work mode in a given process. Thus, we could write

$$\delta W = P\, dV - \mathcal{T}\, dL - \mathcal{S}\, d\mathcal{A} - \mathcal{E}\, dZ + \cdots \qquad (4.15)$$

where the dots represent other products of an intensive property and the derivative of a related extensive property. In each term the intensive property can be viewed as the driving force that causes a change to occur in the related extensive property, which is often termed the displacement.

It should also be noted that many other forms of work can be identified in processes that are not quasi-equilibrium processes. For example, there is the work done by shearing forces in the friction in a viscous fluid or the work done by a rotating shaft that crosses the system boundary.

The identification of work is an important aspect of many thermodynamic problems. We have already noted that work can be identified only at the boundaries of the system. For example, consider Fig. 4.11, which shows a gas separated from the vacuum by a membrane. Let the membrane rupture and the gas fill the entire volume. Neglecting any work associated with the rupturing of the membrane, we can ask whether work is done in the process. If we take as our system the gas and the vacuum space, we readily conclude that no work is done because no work can be identified at the system boundary. If we take the gas as a system, we do have a change of volume, and we might be tempted to calculate the work from the integral

$$\int_1^2 P\, dV$$

However, this is not a quasi-equilibrium process, and therefore the work cannot be calculated from this relation. Because there is no resistance at the system boundary as the volume increases, we conclude that for this system no work is done in this process of filling the vacuum.

Another example can be cited with the aid of Fig. 4.12. In Fig. 4.12a the system consists of the container plus the gas. Work crosses the boundary of the system at the

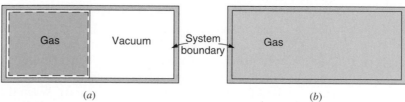

(a) (b)

FIGURE 4.11 Example of process involving a change of volume for which the work is zero.

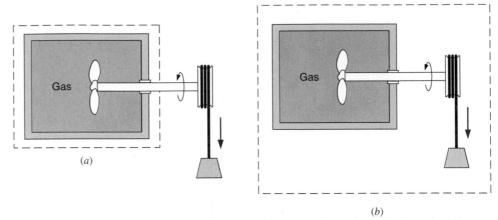

FIGURE 4.12 Example showing how selection of the system determines whether work is involved in a process.

point where the system boundary intersects the shaft, and can be associated with the shearing forces in the rotating shaft. In Fig. 4.12b the system includes shaft and weight as well as the gas and the container. Therefore, no work crosses the system boundary as the weight moves downward. As we will see in the next chapter, we can identify a change of potential energy within the system, but this should not be confused with work crossing the system boundary.

4.6 DEFINITION OF HEAT

The thermodynamic definition of heat is somewhat different from the everyday understanding of the word. It is essential to understand clearly the definition of heat given here, because it plays a part in so many thermodynamic problems.

If a block of hot copper is placed in a beaker of cold water, we know from experience that the block of copper cools down and the water warms up until the copper and water reach the same temperature. What causes this decrease in the temperature of the copper and the increase in the temperature of the water? We say that it is the result of the transfer of energy from the copper block to the water. It is out of such a transfer of energy that we arrive at a definition of heat.

Heat is defined as the form of energy that is transferred across the boundary of a system at a given temperature to another system (or the surroundings) at a lower temperature by virtue of the temperature difference between the two systems. That is, heat is transferred from the system at the higher temperature to the system at the lower temperature, and the heat transfer occurs solely because of the temperature difference between the two systems. Another aspect of this definition of heat is that a body never contains heat. Rather, heat can be identified only as it crosses the boundary. Thus, heat is a transient phenomenon. If we consider the hot block of copper as one system and the cold water in the beaker as another system, we recognize that originally neither system contains any heat (they do contain energy, of course). When the copper block is placed in the water and the two are in thermal communication, heat is transferred from the copper to the water until equilibrium of temperature is established. At this point we no longer have heat transfer, because there is no temperature difference. Neither of the systems contains heat at the conclusion of the process. It also follows that heat is identified at the boundary of the system, for heat is defined as energy being transferred across the system boundary.

4.7 UNITS OF HEAT

Heat, like work, is a form of energy transfer to or from a system. Therefore, the units for heat, and to be more general, for any other form of energy as well, are the same as the units for work, or are at least directly proportional to them. In the International System the unit for heat (energy) is the joule. Similarly, in the English System, the foot pound force is an appropriate unit for heat. However, another unit came to be used naturally over the years, the result of an association with the process of heating water, such as that used in connection with defining heat in the previous section. Consider as a system 1 lbm of water at 59.5F. Let a block of hot copper of appropriate mass and temperature be placed in the water, so that when thermal equilibrium is established the temperature of the water is 60.5F. This unit amount of heat transferred from the copper to the water in this process is called the British thermal unit (Btu). More specifically, it is called the 60-degree Btu, defined as the amount of heat required to raise 1 lbm of water from 59.5F to 60.5F. (The Btu as used today is actually defined in terms of the standard SI units.) It is worth noting here that a unit of heat in metric units, the calorie, originated naturally in a manner similar to the origin of the Btu in the English system. The calorie is defined as the amount of heat required to raise 1 gram of water from 14.5°C to 15.5°C.

Heat transferred *to* a system is considered positive, and heat transferred *from* a system is negative. Thus, positive heat represents energy transferred to a system, and negative heat represents energy transferred from a system. The symbol Q represents heat. A process in which there is no heat transfer ($Q = 0$) is called an adiabatic process.

From a mathematical perspective, heat, like work, is a path function and is recognized as an inexact differential. That is, the amount of heat transferred when a system undergoes a change from state 1 to state 2 depends on the path that the system follows during the change of state. Since heat is an inexact differential, the differential is written δQ. On integrating, we write

$$\int_1^2 \delta Q = {_1}Q_2$$

In words, ${_1}Q_2$ is the heat transferred during the given process between states 1 and 2.

The rate at which heat is transferred to a system is designated by symbol $\dot{Q}$.

$$\dot{Q} \equiv \frac{\delta Q}{dt}$$

It is also convenient to speak of the heat transfer per unit mass of the system, q, often termed "specific heat transfer," which is defined as

$$q \equiv \frac{Q}{m}$$

4.8 HEAT TRANSFER MODES

Heat transfer is the transport of energy due to a temperature difference between different amounts of matter. We know that an ice cube taken out of the freezer will melt as it is placed in a warmer environment such as a glass of liquid water or on a plate with room air around it. From the discussion about energy in Section 2.6 we realize that molecules of matter have translational (kinetic), rotational, and vibrational energy. Energy in these modes can be transmitted to the nearby molecules by interactions (collisions) or by exchange of molecules such that energy is given out by molecules that have more in the average (higher temperature) to those that have less in the average (lower temperature). This energy exchange between molecules is heat transfer by **conduction** and it increases with the temperature difference and the ability of the substance to make the transfer. This is expressed in Fourier's law of conduction

$$\dot{Q} = -kA\frac{dT}{dx} \quad [\text{W}]$$

giving the rate of heat transfer as proportional to the conductivity, k, the total area, A, and the temperature gradient. The minus sign gives a direction of the heat transfer from a higher temperature to a lower temperature region. Often the gradient is evaluated as a temperature difference divided by a distance when an estimate has to be done if a mathematical or numerical solution is not available.

Values for the conductivity, k, ranges from the order of 100 W/mK for metals, 1 to 10 for nonmetallic solids as glass, ice and rock, from 0.1 to 10 for liquids, around 0.1 for insulation materials, and from 0.1 down to less than 0.01 for gases.

A different mode of heat transfer takes place when a media is flowing, called **convective** heat transfer. In this mode the bulk motion of a substance moves matter with a certain energy level over or near a surface with a different temperature. Now the heat transfer by conduction is dominated by the manner in which the bulk motion brings the two substances in contact or close proximity. Examples of this are the wind blowing over a building, flow through heat exchangers, which can be air flowing over/through a radiator with water flowing inside the radiator piping. The overall heat transfer is typically correlated with Newton's law of cooling as

$$\dot{Q} = Ah\Delta T$$

where the transfer properties are lumped into the heat transfer coefficient, h, which then becomes a function of the media properties, the flow and geometry. A more detailed study of fluid mechanics and heat transfer aspects of the overall process is necessary to evaluate the heat transfer coefficient for a given situation.

Typical values for the convection coefficient (all in W/m^2) are

Natural convection	$h = 5$–25	gas	$h = 50$–1000	liquid
Forced convection	$h = 25$–250	gas	$h = 50$–20000	liquid
Boiling phase change	$h = 2500$–$100\,000$			

The final mode of heat transfer is **radiation,** which transmits energy as electromagnetic waves in space. The transfer can happen in empty space and does not require any matter, but the emission (generation) of the radiation and the absorption does require a substance to be present. Surface emission is usually written as a fraction, emissivity ε, of a perfect black body emission as

$$\dot{Q} = \varepsilon\sigma A T_s^4 \quad [\text{W/m}^2]$$

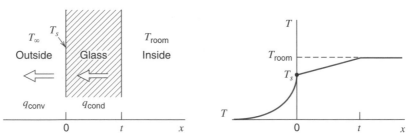

FIGURE 4.13 Conduction and convection heat transfer through a window pane.

with the surface temperature, T_s, and Stefan-Boltzmann constant, σ. Typical values of the emissivity range from 0.92 for nonmetalic surfaces, to 0.6 to 0.9 for nonpolished metalic surfaces, to less than 0.1 for highly polished metal surfaces. Radiation is distributed over a range of wavelengths and it is emitted and absorbed differently for different surfaces, but such a description is beyond the scope of the present text.

EXAMPLE 4.4 Consider the constant transfer of energy from a warm room at 20°C inside a house to the colder ambient at −10°C through a single pane window as shown in Fig. 4.13. The temperature variation with distance from the outside glass surface is shown with an outside convection heat transfer layer, but no such layer inside the room as a simplification. The glass pane has a thickness of 5 mm (0.005 m) with a conductivity of 1.4 W/mK and a total surface area of 0.5 m². The outside wind is blowing so the convective heat transfer coefficient is 100 W/m²K. With an outer glass surface temperature of 12.1°C we would like to know the rate of heat transfer in the glass and the convective layer.

For the conduction through the glass we have

$$\dot{Q} = -kA\frac{dT}{dx} = -kA\frac{\Delta T}{\Delta x} = -1.4 \times 0.5\frac{20 - 12.1}{0.005} = -1106 \text{ W}$$

and the negative sign shows that energy is leaving the room. For the outside convection layer we have

$$\dot{Q} = hA\Delta T = 100 \times 0.5\left[12.1 - (-10)\right] = 1105 \text{ W}$$

with a direction from the higher to the lower temperature, i.e., toward the outside.

4.9 COMPARISON OF HEAT AND WORK

At this point it is evident that there are many similarities between heat and work.

1. Heat and work are both transient phenomena. Systems never possess heat or work, but either or both cross the system boundary when a system undergoes a change of state.

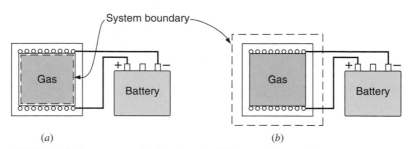

(a)   (b)

FIGURE 4.14 An example showing the difference between heat and work.

2. Both heat and work are boundary phenomena. Both are observed only at the boundaries of the system, and both represent energy crossing the boundary of the system.

3. Both heat and work are path functions and inexact differentials.

It should also be noted that in our sign convention, $+Q$ represents heat transferred *to* the system and thus is energy added to the system, and $+W$ represents work done *by* the system and thus represents energy leaving the system.

A final illustration may help explain the difference between heat and work. Figure 4.14 shows a gas contained in a rigid vessel. Resistance coils are wound around the outside of the vessel. When current flows through the resistance coils, the temperature of the gas increases. Which crosses the boundary of the system, heat or work?

In Fig. 4.14*a* we consider only the gas as the system. The energy crosses the boundary of the system because the temperature of the walls is higher than the temperature of the gas. Therefore, we recognize that heat crosses the boundary of the system.

In Fig. 4.14*b* the system includes the vessel and the resistance heater. Electricity crosses the boundary of the system and, as indicated earlier, this is work.

Consider a gas in a cylinder fitted with a movable piston, as shown in Fig. 4.15. There is a positive heat transfer to the gas, which tends to make the temperature increase. It also tends to increase the gas pressure. However, the pressure is dictated by the external force acting on its movable boundary, as discussed in Section 2.8. If this remains constant, then the volume increases instead. There are also the opposite tendencies for a negative heat transfer, that is, one out of the gas. Consider again the positive heat transfer, except that in this case the external force simultaneously decreases. This causes the gas pressure to decrease, such that the temperature tends to go down. In this case, there are simultaneous tendencies for temperature change in the opposite direction, which effectively decouples directions of heat transfer and temperature change.

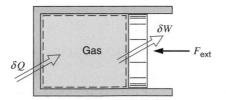

FIGURE 4.15 The effects of heat addition to a control volume that also can give out work.

PROBLEMS

4.1 A piston of mass 2 kg is lowered 0.5 m in the standard gravitational field. Find the required force and work involved in the process.

4.2 An escalator raises a 100-kg bucket of sand 10 m in 1 minute. Determine the total amount of work done and the instantaneous rate of work during the process.

4.3 A linear spring, $F = k_s(x - x_o)$, with spring constant $k_s = 500$ N/m, is stretched until it is 100 mm longer. Find the required force and work input.

4.4 A nonlinear spring has the force versus displacement relation of $F = k_{ns}(x - x_o)^n$. If the spring end is moved to x_1 from the relaxed state, determine the formula for the required work.

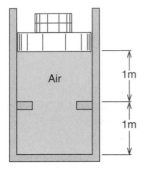

FIGURE P4.6

4.5 A cylinder fitted with a frictionless piston contains 5 kg of superheated refrigerant R-134a vapor at 1000 kPa, 140°C. The setup is cooled at constant pressure until the R-134a reaches a quality of 25%. Calculate the work done in the process.

4.6 A piston/cylinder arrangement shown in Fig. P4.6 initially contains air at 150 kPa, 400°C. The setup is allowed to cool to the ambient temperature of 20°C.

a. Is the piston resting on the stops in the final state? What is the final pressure in the cylinder?

b. What is the specific work done by the air during this process?

4.7 The refrigerant R-22 is contained in a piston/cylinder as shown in Fig. P4.7, where the volume is 11 L when the piston hits the stops. The initial state is −30°C, 150 kPa with a volume of 10 L. This system is brought indoors and warms up to 15°C.

a. Is the piston at the stops in the final state?

b. Find the work done by the R-22 during this process.

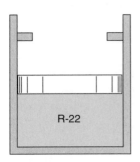

FIGURE P4.7

4.8 Consider a mass going through a polytropic process where pressure is directly proportional to volume ($n = -1$). The process starts with $P = 0$, $V = 0$ and ends with $P = 600$ kPa, $V = 0.01$ m³. The physical setup could be as in Problem 2.22. Find the boundary work done by the mass.

4.9 A piston/cylinder contains 50 kg of water at 200 kPa with a volume of 0.1 m³. Stops in the cylinder restrict the enclosed volume to 0.5 m³, similar to the setup in Problem 4.7. The water is now heated to 200°C. Find the final pressure, volume, and work done by the water.

4.10 A piston/cylinder contains 1 kg of liquid water at 20°C and 300 kPa. Initially the piston floats, similar to the setup in Problem 4.7, with a maximum enclosed volume of 0.002 m³ if the piston touches the stops. Now heat is added so a final pressure of 600 kPa is reached. Find the final volume and the work in the process.

4.11 A piston/cylinder contains butane, C_4H_{10}, at 300°C, 100 kPa with a volume of 0.02 m³. The gas is now compressed slowly in an isothermal process to 300 kPa.

a. Show that it is reasonable to assume that butane behaves as an ideal gas during this process.

b. Determine the work done by the butane during the process.

4.12 The piston/cylinder shown in Fig. P4.12 contains carbon dioxide at 300 kPa, 100°C with a volume of 0.2 m³. Weights are added to the piston such that the gas compresses according to the relation $PV^{1.2} = $ constant to a final temperature of 200°C. Determine the work done during the process.

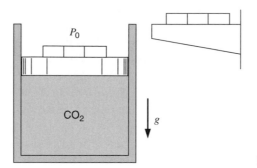

FIGURE P4.12

4.13 Air in a spring loaded piston/cylinder has a pressure that is linear with volume, $P = A + BV$. With an initial state of $P = 150$ kPa, $V = 1$ L and a final state of 800 kPa and volume 1.5 L it is similar to the setup in Problem 3.16. Find the work done by the air.

4.14 A gas initially at 1 MPa, 500°C is contained in a piston and cylinder arrangement with an initial volume of 0.1 m³. The gas is then slowly expanded according to the relation $PV =$ constant until a final pressure of 100 kPa is reached. Determine the work for this process.

4.15 Consider a two-part process with an expansion from 0.1 to 0.2 m³ at a constant pressure of 150 kPa followed by an expansion from 0.2 to 0.4 m³ with a linearly rising pressure from 150 kPa ending at 300 kPa. Show the process in a P-V diagram and find the boundary work.

4.16 A cylinder fitted with a piston contains propane gas at 100 kPa, 300 K with a volume of 0.2 m³. The gas is now slowly compressed according to the relation $PV^{1.1}$ = constant to a final temperature of 340 K. Justify the use of the ideal gas model. Find the final pressure and the work done during the process.

4.17 The gas space above the water in a closed storage tank contains nitrogen at 25°C, 100 kPa. Total tank volume is 4 m³, and there is 500 kg of water at 25°C. An additional 500 kg of water is now forced into the tank. Assuming constant temperature throughout, find the final pressure of the nitrogen and the work done on the nitrogen in this process.

4.18 A steam radiator in a room at 25°C has saturated water vapor at 110 kPa flowing through it, when the inlet and exit valves are closed. What is the pressure and the quality of the water, when it has cooled to 25°C? How much work is done?

4.19 A balloon behaves such that the pressure inside is proportional to the diameter squared. It contains 2 kg of ammonia at 0°C, 60% quality. The balloon and ammonia are now heated so that a final pressure of 600 kPa is reached. Considering the ammonia as a control mass, find the amount of work done in the process.

4.20 Consider a piston/cylinder with 0.5 kg of R-134a as saturated vapor at −10°C. It is now compressed to a pressure of 500 kPa in a polytropic process with $n = 1.5$. Find the final volume and temperature, and determine the work done during the process.

4.21 A cylinder having an initial volume of 3 m³ contains 0.1 kg of water at 40°C. The water is then compressed in an isothermal quasi-equilibrium process until it has a quality of 50%. Calculate the work done in the process. Assume the water vapor is an ideal gas.

4.22 Consider the nonequilibrium process described in Problem 3.7. Determine the work done by the carbon dioxide in the cylinder during the process.

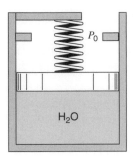

FIGURE P4.23

4.23 Two kilograms of water are contained in a piston/cylinder (Fig. P4.23) with a massless piston loaded with a linear spring and the outside atmosphere. Initially the spring force is zero and $P_1 = P_0 = 100$ kPa with a volume of 0.2 m³. If the piston just hits the upper stops the volume is 0.8 m³ and $T = 600°C$. Heat is now added until the pressure reaches 1.2 MPa. Find the final temperature, show the P-V diagram and find the work done during the process.

4.24 A piston/cylinder (Fig. P4.24) contains 1 kg of water at 20°C with a volume of 0.1 m³. Initially the piston rests on some stops with the top surface open to the atmosphere, P_0, and a mass so a water pressure of 400 kPa will lift it. To what temperature should the water be heated to lift the piston? If it is heated to saturated vapor, find the final temperature, volume, and work, $_1W_2$.

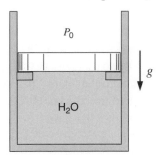

FIGURE P4.24

4.25 Assume the same system as in the previous problem, but let the piston be locked with a pin. If the water is heated to saturated vapor, find the final temperature, volume, and work, $_1W_2$.

4.26 A piston/cylinder setup similar to Problem 4.24 contains 0.1 kg saturated liquid and vapor water at 100 kPa with quality 25%. The mass of the piston is such that a pressure of 500 kPa will float it. The water is heated to 300°C. Find the final pressure, volume, and work, $_1W_2$.

4.27 A 400-L tank, A (see Fig. P4.27) contains argon gas at 250 kPa, 30°C. Cylinder B, having a frictionless piston of such mass that a pressure of 150 kPa will float it, is initially empty. The valve is opened and argon flows into B and eventually reaches a uniform state of 150 kPa, 30°C throughout. What is the work done by the argon?

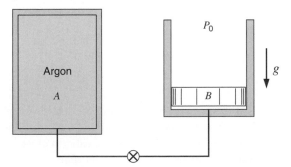

FIGURE P4.27

4.28 Air at 200 kPa, 30°C is contained in a cylinder/piston arrangement with initial volume 0.1 m³. The inside pressure balances ambient pressure of 100 kPa plus an externally imposed force that is proportional to $V^{0.5}$. Now heat is transferred to the system to a final pressure of 225 kPa. Find the final temperature and the work done in the process.

4.29 A spring-loaded piston/cylinder arrangement contains R-134a at 20°C, 24% quality with a volume 50 L. The setup is heated and thus expands, moving the piston. It is noted that when the last drop of liquid disappears the temperature is 40°C. The heating is stopped when $T = 130$°C. Verify that the final pressure is about 1200 kPa by iteration and find the work done in the process.

4.30 A cylinder containing 1 kg of ammonia has an externally loaded piston. Initially the ammonia is at 2 MPa, 180°C and is now cooled to saturated vapor at 40°C, and then further cooled to 20°C, at which point the quality is 50%. Find the total work for the process, assuming a piecewise linear variation of P versus V.

4.31 A vertical cylinder (Fig. P4.31) has a 90-kg piston locked with a pin trapping 10 L of R-22 at 10°C, 90% quality inside. Atmospheric pressure is 100 kPa, and the cylinder cross-sectional area is 0.006 m². The pin is removed, allowing the piston to move and come to rest with a final temperature of 10°C for the R-22. Find the final pressure, final volume, and work done by the R-22.

4.32 A piston/cylinder has (Fig. P4.32) 1 kg of R-134a at state 1 with 110°C, 600 kPa, and is then brought to saturated vapor, state 2, by cooling while the piston is locked with a pin. Now the piston is balanced with an additional constant force and the pin is removed. The cooling continues to a state 3 where the R-134a is saturated liquid. Show the processes in a P-V diagram and find the work in each of the two steps, 1 to 2 and 2 to 3.

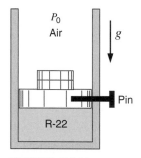

FIGURE P4.31

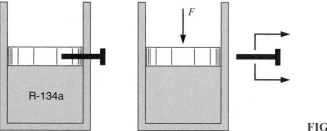

FIGURE P4.32

4.33 Consider the process described in Problem 3.49. With the ammonia as a control mass, determine the boundary work during the process.

4.34 Find the specific work for Problem 3.52.

4.35 Find the work for Problem 3.53.

4.36 Find the specific work for Problem 3.55.

4.37 Find the work for Problem 3.56.

4.38 A spherical elastic balloon initially containing 5 kg of ammonia as saturated vapor at 20°C is connected by a valve to a 3-m³ evacuated tank. The balloon is made such that the pressure inside is proportional to the diameter. The valve is now opened, allowing ammonia to flow into the tank until the pressure in the balloon has dropped to 600 kPa, at which point the valve is closed. The final temperature in both the balloon and the tank is 20°C. Determine

a. The final pressure in the tank

b. The work done by the ammonia in the process

4.39 A 0.5-m-long steel rod with a 1-cm diameter is stretched in a tensile test. What is the work required to obtain a relative strain of 0.1%? The modulus of elasticity of steel is 2×10^8 kPa.

4.40 A film of ethanol at 20°C has a surface tension of 22.3 mN/m and is maintained on a wire frame as shown in Fig. P4.40. Consider the film with two surfaces as a control mass and find the work done when the wire is moved 10 mm to make the film 20 × 40 mm.

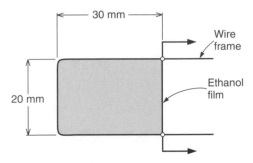

FIGURE P4.40

4.41 A simple magnetic substance is one involving only magnetic work, that is, a change in magnetization of a substance in the presence of a magnetic field. For such a substance undergoing a quasiequilibrium process at constant volume, the work is

$$\delta W = -\, C_o \mathcal{H} d\mathfrak{M}$$

where $\mathcal{H}$ = magnetic field intensity, $\mathfrak{M}$ = magnetization, and C_o = a proportionality constant. For a first approximation, assume that magnetization is proportional to the magnetic field intensity divided by the temperature of the magnetic substance. Determine the work done in an isothermal process during a change of magnetization from $\mathfrak{M}_1$ to $\mathfrak{M}_2$.

4.42 For the magnetic substance described in Problem 4.41, determine the work done in a process at constant magnetic field intensity (temperature varies), instead of one at constant temperature.

4.43 A battery is well insulated while being charged by 12.3 V at a current of 6 A. Take the battery as a control mass and find the instantaneous rate of work and the total work done over 4 hours.

4.44 Two springs with the same spring constant are installed in a massless piston/cylinder with the outside air at 100 kPa. If the piston is at the bottom, both springs are relaxed, and the second spring comes in contact with the piston at $V = 2$ m³. The cylinder (Fig. P4.44) contains ammonia initially at $-2°C$, $x = 0.13$, $V = 1$ m³, which is then heated until the pressure finally reaches 1200 kPa. At what pressure will the piston touch the second spring? Find the final temperature and the total work done by the ammonia.

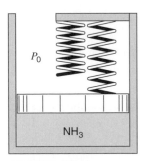

FIGURE P4.44

4.45 Consider the process of inflating a helium balloon, as described in Problem 3.14. For a control volume that consists of the space inside the balloon, determine the work done during the overall process.

4.46 A cylinder (Fig. P4.46), $A_{cyl} = 7.012$ cm², has two pistons mounted, the upper one, $m_{p1} = 100$ kg, initially resting on the stops. The lower piston, $m_{p2} = 0$ kg, has 2 kg water below it, with a spring in vacuum connecting the two pistons. The spring force is zero when the lower piston stands at the bottom, and when the lower piston hits the stops the volume is 0.3 m³. The water, initially at 50 kPa, $V = 0.00206$ m³, is then heated to saturated vapor.

a. Find the initial temperature and the pressure that will lift the upper piston.

b. Find the final T, P, v and the work done by the water.

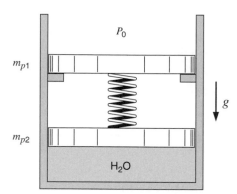

FIGURE P4.46

4.47 The sun shines on a 150 m^2 road surface so it is at 45°C. Below the 5-cm-thick asphalt, average conductivity of 0.06 W/m K, is a layer of compacted rubble at a temperature of 15°C. Find the rate of heat transfer to the rubble.

4.48 A steel pot, conductivity 50 W/m K, with a 5-mm-thick bottom is filled with 15°C liquid water. The pot has a diameter of 20 cm and is now placed on an electric stove that delivers 250 W as heat transfer. Find the temperature on the outer pot bottom surface assuming the inner surface is at 15°C.

4.49 A water-heater is covered up with insulation boards over a total surface area of 3 m^2. The inside board surface is at 75°C, the outside surface is at 20°C, and the board material has a conductivity of 0.08 W/m K. How thick should the board be to limit the heat transfer loss to 200 W ?

4.50 You drive a car on a winter day with the atmospheric air at −15°C, and you keep the outside front windshield surface temperature at +2°C by blowing hot air on the inside surface. If the windshield is 0.5 m^2 and the outside convection coefficient is 250 W/m^{2}K, find the rate of energy loss through the front windshield.

4.51 A large condenser (heat exchanger) in a power plant must transfer a total of 100 MW from steam running in a pipe to sea water being pumped through the heat exchanger. Assume the wall separating the steam and seawater is 4 mm of steel, conductivity 50 W/m K, and that a maximum of 5°C difference between the two fluids is allowed in the design. Find the required minimum area for the heat transfer neglecting any convective heat transfer in the flows.

4.52 The black grille on the back of a refrigerator has a surface temperature of 35°C with a total surface area of 1 m^2. Heat transfer to the room air at 20°C takes place with an average convective heat transfer coefficient of 15 W/m^2 K. How much energy can be removed during 15 minutes of operation?

4.53 Due to a faulty door contact the small light bulb (25 W) inside a refrigerator is kept on and limited insulation lets 50 W of energy from the outside seep into the refrigerated space. How much of a temperature difference to the ambient at 20°C must the refrigerator have in its heat exchanger with an area of 1 m^2 and an average heat transfer coefficient of 15 W/m^2 K to reject the leaks of energy?

4.54 The brake shoe and steel drum on a car continuously absorb 25 W as the car slows down. Assume a total outside surface area of 0.1 m^2 with a convective heat transfer coefficient of 10 W/m^2 K to the air at 20°C. How hot does the outside brake and drum surface become when steady conditions are reached?

4.55 A wall surface on a house is at 30°C with an emissivity of $\varepsilon = 0.7$. The surrounding ambient to the house is at 15°C, average emissivity of 0.9. Find the rate of radiation energy from each of those surfaces per unit area.

4.56 A log of burning wood in the fireplace has a surface temperature of 450°C. Assume the emissivity is 1 (perfect black body) and find the radiant emission of energy per unit surface area.

4.57 A radiant heat lamp is a rod, 0.5 m long and 0.5 cm in diameter, through which 400 W of electric energy is deposited. Assume the surface has an emissivity of 0.9 and neglect incoming radiation. What will the rod surface temperature be?

4.58 Consider a window-mounted air conditioning unit used in the summer to cool incoming air. Examine the system boundaries for rates of work and heat transfer, including signs.

4.59 Consider a hot-air heating system for a home. Examine the following systems for heat transfer.

a. The combustion chamber and combustion gas side of the heat transfer area

b. The furnace as a whole, including the hot- and cold-air ducts and chimney

4.60 Consider a household refrigerator that has just been filled up with room-temperature food. Define a control volume (mass) and examine its boundaries for rates of work and heat transfer, including sign.

a. Immediately after the food is placed in the refrigerator

b. After a long period of time has elapsed and the food is cold

4.61 A room is heated with an electric space heater on a winter day. Examine the following control volumes, regarding heat transfer and work, including sign.

a. The space heater   c. The space heater and the room together

b. Room

ENGLISH UNIT PROBLEMS

4.62E A cylinder fitted with a frictionless piston contains 10 lbm of superheated refrigerant R-134a vapor at 100 lbf/in.^{2}, 300 F. The setup is cooled at constant pressure until the water reaches a quality of 25%. Calculate the work done in the process.

4.63E An escalator raises a 200 lbm bucket of sand 30 ft in 1 minute. Determine the total amount of work done and the instantaneous rate of work during the process.

4.64E A linear spring, $F = k_s(x - x_o)$, with spring constant $k_s = 35$ lbf/ft, is stretched until it is 2.5 in. longer. Find the required force and work input.

4.65E The piston/cylinder shown in Fig. P4.12 contains carbon dioxide at 50 lbf/in.^{2}, 200 F with a volume of 5 ft^3. Mass is added at such a rate that the gas compresses according to the relation $PV^{1.2} = $ constant to a final temperature of 350 F. Determine the work done during the process.

4.66E Consider a mass going through a polytropic process where pressure is directly proportional to volume ($n = -1$). The process starts with $P = 0$, $V = 0$ and ends with $P = 90$ lbf/in.^{2}, $V = 0.4$ ft^3. The physical setup could be as in Problem 2.22. Find the boundary work done by the mass.

4.67E The gas space above the water in a closed storage tank contains nitrogen at 80 F, 15 lbf/in.^{2}. Total tank volume is 150 ft^3 and there is 1000 lbm of water at 80 F.

An additional 1000 lbm water is now forced into the tank. Assuming constant temperature throughout, find the final pressure of the nitrogen and the work done on the nitrogen in this process.

4.68E A steam radiator in a room at 75 F has saturated water vapor at 16 lbf/in.^{2} flowing through it, when the inlet and exit valves are closed. What is the pressure and the quality of the water when it has cooled to 75 F? How much work is done?

4.69E Consider a two-part process with an expansion from 3 to 6 ft^3 at a constant pressure of 20 lbf/in.^{2} followed by an expansion from 6 to 12 ft^3 with a linearly rising pressure from 20 lbf/in.^{2} ending at 40 lbf/in.^{2}. Show the process in a P-V diagram and find the boundary work.

4.70E A cylinder having an initial volume of 100 ft^3 contains 0.2 lbm of water at 100 F. The water is then compressed in an isothermal quasi-equilibrium process until it has a quality of 50%. Calculate the work done in the process assuming water vapor is an ideal gas.

4.71E Air at 30 lbf/in.^{2}, 85 F is contained in a cylinder/piston arrangement with initial volume 3.5 ft^3. The inside pressure balances ambient pressure of 14.7 lbf/in.^{2} plus an externally imposed force that is proportional to $V^{0.5}$. Now heat is transferred to the system to a final pressure of 40 lbf/in.^{2}. Find the final temperature and the work done in the process.

4.72E A cylinder containing 2 lbm of ammonia has an externally loaded piston. Initially the ammonia is at 280 lbf/in.^{2}, 360 F and is now cooled to saturated vapor at 105 F, and then further cooled to 65 F, at which point the quality is 50%. Find the total work for the process, assuming a piecewise linear variation of P versus V.

4.73E A piston/cylinder has 2 lbm of R-134a at state 1 with 200 F, 90 lbf/in.^{2}, and is then brought to saturated vapor, state 2, by cooling while the piston is locked with a pin. Now the piston is balanced with an additional constant force and the pin is removed. The cooling continues to a state 3 where the R-134a is saturated liquid. Show the processes in a P-V diagram and find the work in each of the two steps, 1 to 2 and 2 to 3.

4.74E Find the work for Problem 3.82.

4.75E Find the specific work for Problem 3.83.

4.76E A 1-ft-long steel rod with a 0.5-in. diameter is stretched in a tensile test. What is the required work to obtain a relative strain of 0.1%? The modulus of elasticity of steel is 30 × 10^6 lbf/in.^{2}.

4.77E The sun shines on a 1500 ft^2 road surface so it is at 115 F. Below the 2-in.-thick asphalt, average conductivity of 0.035 Btu/h ft F, is a layer of compacted rubble at a temperature of 60 F. Find the rate of heat transfer to the rubble.

4.78E A water-heater is covered up with insulation boards over a total surface area of 30 ft^2. The inside board surface is at 175 F, the outside surface is at 70 F and the board material has a conductivity of 0.05 Btu/h ft F. How thick should the board be to limit the heat transfer loss to 720 Btu/h?

4.79E The black grille on the back of a refrigerator has a surface temperature of 95 F with a total surface area of 10 ft^2. Heat transfer to the room air at 70 F takes place with an average convective heat transfer coefficient of 3 Btu/h ft^2 R. How much energy can be removed during 15 minutes of operation?

COMPUTER, DESIGN, AND OPEN-ENDED PROBLEMS

4.80 In Problem 4.12, determine the work done by the carbon dioxide at any point during the process.

4.81 In Problem 4.28, determine the work done by the air at any point during the process.

4.82 A piston/cylinder arrangement of initial volume 0.025 m³ contains saturated water vapor at 200°C. The steam now expands in a quasi-equilibrium isothermal process to a final pressure of 200 kPa, while it does work against the piston. Determine the work done in this process by a numerical integration (summation) of the area below the *P-V* process curve. Compute about 10 points along the curve by using the computerized software to get the volume at 200°C and the various pressures. How different is the work calculated if ideal gas is assumed?

4.83 Consider the process described in Problem 4.38, in which ammonia is transferred from a balloon to a rigid tank. Follow the process from the initial state in 10 steps until the pressure in the balloon and tank eventually equalize. Get the properties from the computerized tables.

4.84 Reconsider the process in Problem 4.30 in which three states were specified. Solve the problem by fitting a single smooth curve (*P* versus *v*) through the three points. Map out the path followed (including temperature and quality) during the process.

4.85 Write a computer program to determine the boundary movement work for a specified substance undergoing a process for a given set of data (values of pressure and corresponding volume during the process).

4.86 Ammonia vapor is compressed inside a cylinder by an external force acting on the piston. The ammonia is initially at 30°C, 500 kPa, and the final pressure is 1400 kPa. The following data have been measured for the process:

Pressure, kPa	500	653	802	945	1100	1248	1400
Volume, L	1.25	1.08	0.96	0.84	0.72	0.60	0.50

Determine the work done by the ammonia by summing the area below the *P-V* process curve. As you plot it *P* is the height and the change in volume is the base of a number of rectangles.

4.87 A substance is brought from a state of P_1, v_1 to a state P_2, v_2 in a piston/cylinder arrangement. Assume that the process can be approximated as a polytropic process. Write a program that will find the polytropic exponent, *n,* and the boundary work per unit mass. The four state properties are input variables. Check the program with cases that you can easily hand calculate.

4.88 Assume that you have a plate of $A = 1$ m² with thickness $L = 0.02$ m over which there is a temperature difference of 20°C. Find the conductivity, κ, from the literature and compare the heat transfer rates if the plate substance is a metal like aluminum or steel, or wood, foam insulation, air, argon, or liquid water. Assume the average substance temperature is 25°C.

4.89 Make a list of household appliances such as refrigerators, electric heaters, vacuum cleaners, hair dryers, TVs, stereo sets, and any others you may think of. For each, list its energy consumption and explain where you have energy transfer as work and where there is heat transfer.

5 THE FIRST LAW OF THERMODYNAMICS

Having completed our consideration of basic definitions and concepts, we are ready to proceed to a discussion of the first law of thermodynamics. This law is often called the law of the conservation of energy and, as we will see later, this is essentially true. Our procedure will be to state this law for a system (control mass) undergoing a cycle, and then for a change of state of a system. The law of the conservation of matter will also be considered in this chapter.

5.1 THE FIRST LAW OF THERMODYNAMICS FOR A CONTROL MASS UNDERGOING A CYCLE

The first law of thermodynamics states that during any cycle a system (control mass) undergoes, the cyclic integral of the heat is proportional to the cyclic integral of the work.

To illustrate this law, consider as a control mass the gas in the container shown in Fig. 5.1. Let this system go through a cycle that is made up of two processes. In the first process work is done on the system by the paddle that turns as the weight is lowered. Let the system then return to its initial state by transferring heat from the system until the cycle has been completed.

Historically, work was measured in mechanical units of force times distance, such as foot pounds force or joules, and heat was measured in thermal units, such as the British thermal unit or the calorie. Measurements of work and heat were made during a cycle for a wide variety of systems and for various amounts of work and heat. When the amounts of work and heat were compared, it was found that they were always proportional. Such observations led to the formulation of the first law of thermodynamics, which in equation form is written

$$ J \oint \delta Q = \oint \delta W \tag{5.1} $$

The symbol $\oint \delta Q$, which is called the cyclic integral of the heat transfer, represents the net heat transfer during the cycle, and $\oint \delta W$, the cyclic integral of the work, represents the net work during the cycle. Here, J is a proportionality factor that depends on the units used for work and heat.

The basis of every law of nature is experimental evidence, and this is true also of the first law of thermodynamics. Many different experiments have been conducted on the first law, and every one thus far has verified it either directly or indirectly. The first law has never been disproved.

96

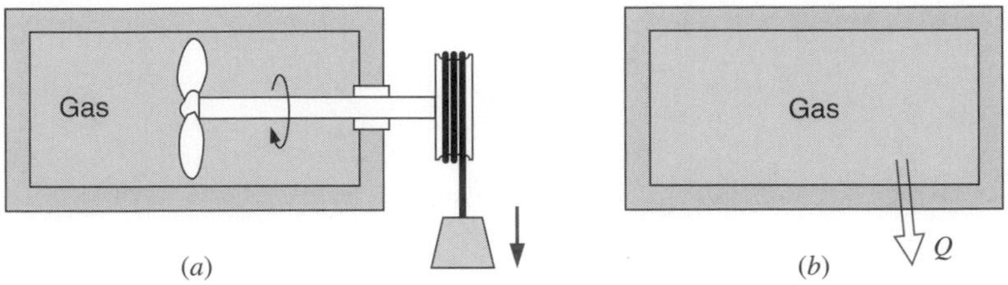

FIGURE 5.1 Example of a control mass undergoing a cycle.

(a)

(b) Q

As was discussed in Chapter 4, the units for work and heat or for any other form of energy either are the same or are directly proportional. In SI units, the joule is used as the unit for both work and heat and for any other energy unit. In English units, the basic unit for work is the foot pound force, and the basic unit for heat is the British thermal unit (Btu). James P. Joule (1818–1889) did the first accurate work in the 1840s on measurement of the proportionality factor J, which relates these units. Today, the Btu is defined in terms of the basic SI metric units,

$$1 \text{ Btu} = 778.17 \text{ ft lbf}$$

This unit is termed the International British thermal unit. For much engineering work, the accuracy of other data does not warrant more accuracy than the relation 1 Btu = 778 ft lbf, which is the value used with English units in the problems in this text. Because these units are equivalent, it is not necessary to include the factor J explicitly in Eq. 5.1, but simply to recognize that for any system of units, each equation must have consistent units throughout. Therefore, we may write Eq. 5.1 as

$$\oint \delta Q = \oint \delta W \tag{5.2}$$

which can be considered the basic statement of the first law of thermodynamics.

5.2 THE FIRST LAW OF THERMODYNAMICS FOR A CHANGE IN STATE OF A CONTROL MASS

Equation 5.2 states the first law of thermodynamics for a control mass during a cycle. Many times, however, we are concerned with a process rather than a cycle. We now consider the first law of thermodynamics for a control mass that undergoes a change of state. We begin by introducing a new property, the energy, which is given the symbol E. Consider a system that undergoes a cycle, in which it changes from state 1 to state 2 by process A, and returns from state 2 to state 1 by process B. This cycle is shown in Fig. 5.2 on a pressure (or other intensive property)–volume (or other extensive property) diagram. From the first law of thermodynamics, Eq. 5.2,

$$\oint \delta Q = \oint \delta W$$

Considering the two separate processes, we have

$$\int_1^2 \delta Q_A + \int_2^1 \delta Q_B = \int_1^2 \delta W_A + \int_2^1 \delta W_B$$

Now consider another cycle in which the control mass changes from state 1 to state 2 by

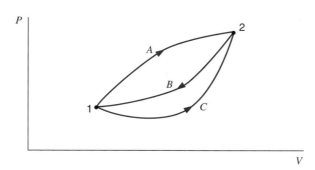

FIGURE 5.2 Demonstration of the existence of thermodynamic property E.

process C and returns to state 1 by process B, as before. For this cycle we can write

$$\int_1^2 \delta Q_C + \int_2^1 \delta Q_B = \int_1^2 \delta W_C + \int_2^1 \delta W_B$$

Subtracting the second of these equations from the first, we have

$$\int_1^2 \delta Q_A - \int_1^2 \delta Q_C = \int_1^2 \delta W_A - \int_1^2 \delta W_C$$

or, by rearranging,

$$\int_1^2 \left(\delta Q - \delta W \right)_A = \int_1^2 \left(\delta Q - \delta W \right)_C \qquad (5.3)$$

Since A and C represents arbitrary processes between states 1 and 2, the quantity $\delta Q - \delta W$ is the same for all processes between states 1 and 2. Therefore, $\delta Q - \delta W$ depends only on the initial and final states and not on the path followed between the two states. We conclude that this is a point function, and therefore it is the differential of a property of the mass. This property is the energy of the mass and is given the symbol E. Thus we can write

$$dE = \delta Q - \delta W$$

$$\delta Q = dE + \delta W \qquad (5.4)$$

Because E is a property, its derivative is written dE. When Eq. 5.4 is integrated from an initial state 1 to a final state 2, we have

$$_1Q_2 = E_2 - E_1 + {}_1W_2 \qquad (5.5)$$

where E_1 and E_2 are the initial and final values of the energy E of the control mass, $_1Q_2$ is the heat transferred to the control mass during the process from state 1 to state 2 and $_1W_2$ is the work done by the control mass during the process.

Note that a control mass may be made up of several different subsystems, such as shown in Fig. 5.3. In this case, each part must be analyzed and included separately in applying the first law, Eq. 5.5

The physical significance of the property E is that it represents all the energy of the system in the given state. This energy might be present in a variety of forms, such as the kinetic or potential energy of the system as a whole with respect to the chosen coordinate frame, energy associated with the motion and position of the molecules, energy associated with the structure of the atom, chemical energy present in a storage battery, energy present in a charged condenser, or in any of a number of other forms.

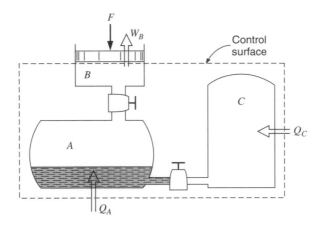

FIGURE 5.3 A control mass with several different subsystems.

In the study of thermodynamics, it is convenient to consider the bulk kinetic and potential energy separately and then to consider all the other energy of the control mass in a single property that we call the internal energy, and to which we give the symbol U. Thus, we would write

$$E = \text{Internal energy} + \text{Kinetic energy} + \text{Potential energy}$$

or

$$E = U + \text{KE} + \text{PE}$$

The kinetic and potential energy of the control mass are associated with the coordinate frame that we select and can be specified by the macroscopic parameters of mass, velocity, and elevation. The internal energy U includes all other forms of energy of the control mass and is associated with the thermodynamic state of the system.

Since the terms comprising E are point functions, we can write

$$dE = dU + d(\text{KE}) + d(\text{PE}) \tag{5.6}$$

The first law of thermodynamics for a change of state may therefore be written

$$\delta Q = dU + d(\text{KE}) + d(\text{PE}) + \delta W \tag{5.7}$$

In words this equation states that as a control mass undergoes a change of state, energy may cross the boundary as either heat or work, and each may be positive or negative. The net change in the energy of the system will be exactly equal to the net energy that crosses the boundary of the system. The energy of the system may change in any of three ways—by a change in internal energy, in kinetic energy, or in potential energy.

This section concludes by deriving an expression for the kinetic and potential energy of a control mass. Consider a mass that is initially at rest relative to the earth, which is taken as the coordinate frame. Let this system be acted on by an external horizontal force F that moves the mass a distance dx in the direction of the force. Thus, there is no change in potential energy. Let there be no heat transfer and no change in internal energy. Then from the first law, Eq. 5.7,

$$\delta W = -F\, dx = -d\text{KE}$$

But

$$F = ma = m\frac{d\mathbf{V}}{dt} = m\frac{dx}{dt}\frac{d\mathbf{V}}{dx} = m\mathbf{V}\frac{d\mathbf{V}}{dx}$$

Then

$$d\text{KE} = F\,dx = m\mathbf{V}\,d\mathbf{V}$$

Integrating, we obtain

$$\int_{\text{KE}=0}^{\text{KE}} d\text{KE} = \int_{\mathbf{V}=0}^{\mathbf{V}} m\mathbf{V}\,d\mathbf{V}$$

$$\text{KE} = \frac{1}{2}m\mathbf{V}^2 \tag{5.8}$$

A similar expression for potential energy can be found. Consider a control mass that is initially at rest and at the elevation of some reference level. Let this mass be acted on by a vertical force F of such magnitude that it raises (in elevation) the mass with constant velocity an amount dZ. Let the acceleration due to gravity at this point be g. From the first law, Eq. 5.7,

$$\delta W = -F\,dZ = -d\text{PE}$$

$$F = ma = mg$$

Then

$$d\text{PE} = F\,dZ = mg\,dZ$$

Integrating gives

$$\int_{\text{PE}_1}^{\text{PE}_2} d\,\text{PE} = m\int_{Z_1}^{Z_2} g\,dZ$$

Assuming that g does not vary with Z (which is a very reasonable assumption for moderate changes in elevation),

$$\text{PE}_2 - \text{PE}_1 = mg(Z_2 - Z_1) \tag{5.9}$$

Substituting the expressions for kinetic and potential energy into Eq. 5.6, we have

$$dE = dU + m\mathbf{V}\,d\mathbf{V} + mg\,dZ$$

Integrating for a change of state from state 1 to state 2 with constant g, we have

$$E_2 - E_1 = U_2 - U_1 + \frac{m\mathbf{V}_2^2}{2} - \frac{m\mathbf{V}_1^2}{2} + mgZ_2 - mgZ$$

Similarly, substituting these expressions for kinetic and potential energy into Eq. 5.7, we have

$$\delta Q = dU + \frac{d(m\mathbf{V}^2)}{2} + d(mgZ) + \delta W \tag{5.10}$$

Assuming g is a constant, in the integrated form of this equation,

$$_1Q_2 = U_2 - U_1 + \frac{m\left(\mathbf{V}_2^2 - \mathbf{V}_1^2\right)}{2} + mg\left(Z_2 - Z_1\right) + {}_1W_2$$

(5.11)

Three observations should be made regarding this equation. The first observation is that the property E, the energy of the control mass, was found to exist, and we were able to write the first law for a change of state using Eq. 5.5. However, rather than deal with this property E, we find it more convenient to consider the internal energy and the kinetic and potential energies of the mass. In general, this procedure will be followed in the rest of this book.

The second observation is that Eqs. 5.10 and 5.11 are in effect a statement of the conservation of energy. The net change of the energy of the control mass is always equal to the net transfer of energy across the boundary as heat and work. This is somewhat analogous to a joint checking account shared by a husband and wife. There are two ways in which deposits and withdrawals can be made—either by the husband or by the wife—and the balance will always reflect the net amount of the transaction. Similarly, there are two ways in which energy can cross the boundary of a control mass—either as heat or as work—and the energy of the mass will change by the exact amount of the net energy crossing the boundary. The concept of energy and the law of the conservation of energy are basic to thermodynamics.

The third observation is that Eqs. 5.10 and 5.11 can give only changes in internal energy, kinetic energy, and potential energy. We can learn nothing about absolute values of these quantities from these equations. If we wish to assign values to internal energy, kinetic energy, and potential energy, we must assume reference states and assign a value to the quantity in this reference state. The kinetic energy of a body with zero velocity relative to the earth is assumed to be zero. Similarly, the value of the potential energy is assumed to be zero when the body is at some reference elevation. With internal energy, therefore, we must also have a reference state if we wish to assign values of this property. This matter is considered in the following section.

5.3 INTERNAL ENERGY—A THERMODYNAMIC PROPERTY

Internal energy is an extensive property, because it depends on the mass of the system. Similarly, the kinetic and potential energies are extensive properties.

The symbol U designates the internal energy of a given mass of a substance. Following the convention used with other extensive properties, the symbol u designates the internal energy per unit mass. We could speak of u as the specific internal energy, as we do with specific volume. However, because the context will usually make it clear whether u or U is referred to, we will simply use the term internal energy to refer to both internal energy per unit mass and the total internal energy.

In Chapter 3 it was noted that in the absence of the effects of motion, gravity, surface effects, electricity, or other effects, the state of a pure substance is specified by two independent properties. It is very significant that with these restrictions, the internal energy may be one of the independent properties of a pure substance. This means, for example, that if we specify the pressure and internal energy (with reference to an arbitrary base) of superheated steam, the temperature is also specified.

Thus, in a table of thermodynamic properties such as the steam tables, the value of internal energy can be tabulated along with other thermodynamic properties. Tables 1 and 2 of the steam tables by Keenan and his colleagues (Tables B.1.1 and B.1.2) list the internal energy for saturated states. Included are the internal energy of saturated liquid u_f, the internal energy of saturated vapor u_g, and the difference between the internal energy of saturated liquid and saturated vapor u_{fg}. The values are given in relation to an arbitrarily assumed reference state, which will be discussed later. The internal energy of saturated steam of a given quality is calculated in the same way as specific volume. The relations are

$$U = U_{liq} + U_{vap}$$

or

$$mu = m_{liq}u_f + m_{vap}u_g$$

Dividing by m and introducing the quality x, gives

$$u = (1-x)u_f + xu_g$$

$$u = u_f + xu_{fg}$$

As an example, the specific internal energy of saturated steam having a pressure of 0.6 MPa and a quality of 95% can be calculated as

$$u = u_f + xu_{fg} = 669.9 + 0.95(1897.5) = 2472.5 \text{ kJ / kg}$$

EXAMPLE 5.1

A tank containing a fluid is stirred by a paddle wheel. The work input to the paddle wheel is 5090 kJ. The heat transfer from the tank is 1500 kJ. Consider the tank and the fluid inside a control surface and determine the change in internal energy of this control mass.

The first law of thermodynamics is (Eq. 5.11)

$$_1Q_2 = U_2 - U_1 + \frac{1}{2}m(\mathbf{V}_2^2 - \mathbf{V}_1^2) + mg(Z_2 - Z_1) + {}_1W_2$$

Since there is no change in kinetic and potential energy, this reduces to

$$_1Q_2 = U_2 - U_1 + {}_1W_2$$

$$U_2 - U_1 = -1500 - (-5090) = 3590 \text{ kJ}$$

EXAMPLE 5.2

Consider a stone having a mass of 10 kg and a bucket containing 100 kg of liquid water. Initially the stone is 10.2 m above the water, and the stone and the water are at the same temperature, state 1. The stone then falls into the water.

Determine $\Delta U, \Delta KE, \Delta PE, Q$, and W for the following changes of state, assuming standard gravitational acceleration of 9.806 65 m/s².

a. The stone is about to enter the water, state 2.

b. The stone has just come to rest in the bucket, state 3.

c. Heat has been transferred to the surroundings in such an amount that the stone and water are at the temperature, T_1, state 4.

Analysis and Solution

The first law for any of the steps is

$$Q = \Delta U + \Delta KE + \Delta PE + W$$

and each term can be identified for each of the changes of state.

1. The stone has fallen from Z_1 to Z_2, and we assume no heat transfer as it falls. The water has not changed state, thus

$$\Delta U = 0; \quad {}_1Q_2 = 0; \quad {}_1W_2 = 0$$

and the first law reduces to

$$\Delta KE + \Delta PE = 0$$

$$\Delta KE = -\Delta PE = -mg\left(Z_2 - Z_1\right)$$

$$= -10 \text{ kg} \times 9.806\ 65 \text{ m/s}^2 \times \left(-10.2 \text{ m}\right)$$

$$= 1000 \text{ J} = 1 \text{ kJ}$$

That is, for the process from state 1 to state 2,

$$\Delta KE = 1 \text{ kJ} \quad \text{and} \quad \Delta PE = -1 \text{ kJ}$$

2. For the process from state 2 to state 3 with zero kinetic energy, we have

$$\Delta PE = 0 \qquad {}_2Q_3 = 0 \qquad {}_2W_3 = 0$$

Then

$$\Delta U + \Delta KE = 0$$

$$\Delta U = -\Delta KE = 1 \text{ kJ}$$

3. In the final state, there is no kinetic, nor potential energy and the internal energy is the same as in state 1.

$$\Delta U = -1 \text{ kJ} \qquad \Delta KE = 0 \qquad \Delta PE = 0 \qquad {}_3W_4 = 0$$

$$_3Q_4 = \Delta U = -1 \text{ kJ}$$

5.4 PROBLEM ANALYSIS AND SOLUTION TECHNIQUE

At this point in our study of thermodynamics, we have progressed sufficiently far (that is, we have accumulated sufficient tools to work with) that it is worthwhile to develop a somewhat formal technique or procedure for analyzing and solving thermodynamic problems. For the time being it may not seem entirely necessary to use such a rigorous procedure for many of our problems, but we should keep in mind that as we acquire more analytical tools the problems that we are capable of dealing with will become much more complicated. Thus, it is appropriate that we begin to practice this technique now in anticipation of these future problems.

Our problem analysis and solution technique is contained within the framework of the following set of questions that must be answered in the process of an orderly solution of a thermodynamic problem.

1. What is the control mass or control volume? Is it useful, or necessary, to choose more than one? It may be helpful to draw a sketch of the system at this point, illustrating all heat and work flows, and indicating forces such as external pressures and gravitation.

2. What do we know about the initial state (which properties)?

3. What do we know about the final state?

4. What do we know about the process that takes place? Is anything constant or zero? Is there some known functional relation between two properties?

5. Is it helpful to draw a diagram of the information in steps 2 to 4 (for example, a T–v or P–v diagram)?

6. What is our thermodynamic model for the behavior of the substance (for example, steam tables, ideal gas, and so on)?

7. What is our analysis of the problem (examine control surfaces for various work modes, first law, conservation of mass)?

8. What is our solution technique (in other words, from what we have done so far in steps 1–7, how do we proceed to find whatever it is that is desired)? Is a trial-and-error solution necessary?

It is not always necessary to write out all these steps, and in the majority of the examples throughout this text we will not do so. However, when faced with a new and unfamiliar problem, the student should always at least think through this set of questions to develop the ability to solve increasingly more challenging problems. In solving the following example, we will use this technique in detail.

EXAMPLE 5.3 A vessel having a volume of 5 m³ contains 0.05 m³ of saturated liquid water and 4.95 m³ of saturated water vapor at 0.1 MPa. Heat is transferred until the vessel is filled with saturated vapor. Determine the heat transfer for this process.

Control mass: All the water inside the vessel.

Sketch: Fig. 5.4.

Initial state: Pressure, volume of liquid, volume of vapor; therefore, state 1 is fixed.

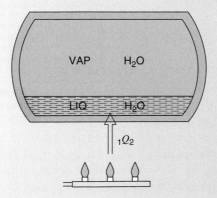

FIGURE 5.4 Sketch for Example 5.3.

Final state: Somewhere along the saturated-vapor curve; the water was heated, so $P_2 > P_1$.

Process: Constant volume and mass; therefore, constant specific volume.

Diagram: Fig. 5.5.

Model: Steam tables.

Analysis:

First law: $$_1Q_2 = U_2 - U_1 + m\frac{\mathbf{V}_2^2 - \mathbf{V}_1^2}{2} + mg(Z_2 - Z_1) + {}_1W_2$$

From examining the control surface for various work modes, we conclude that the work for this process is zero. Furthermore, the system is not moving, so there is no change in kinetic energy. There is a small change in the center of mass of the system but we will assume that the corresponding change in potential energy is negligible (in kilojoules). Therefore,

$$_1Q_2 = U_2 - U_1$$

Solution

The heat transfer will be found from the first law. State 1 is known, so U_1 can be calculated. The specific volume at state 2 is also known (from state 1 and the process). Since state 2 is saturated vapor, state 2 is fixed, as is seen from Fig. 5.5. Therefore, U_2 can also be found.

The solution proceeds as follows:

$$m_{1\,\text{liq}} = \frac{V_{\text{liq}}}{v_f} = \frac{0.05}{0.001\,043} = 47.94\ \text{kg}$$

$$m_{1\,\text{vap}} = \frac{V_{\text{vap}}}{v_g} = \frac{4.95}{1.6940} = 2.92\ \text{kg}$$

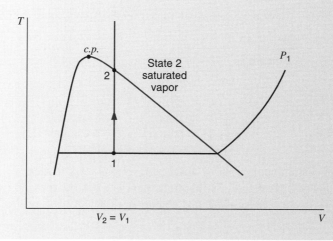

FIGURE 5.5 Diagram for Example 5.3.

Then

$$U_1 = m_{1\,liq}u_{1\,liq} + m_{1\,vap}u_{1\,vap}$$

$$= 47.94(417.36) + 2.92(2506.1) = 27\,326 \text{ kJ}$$

To determine u_2 we need to know two thermodynamic properties, since this determines the final state. The properties we know are the quality, $x = 100$ percent, and v_2, the final specific volume, which can readily be determined.

$$m = m_{1\,liq} + m_{1\,vap} = 47.94 + 2.92 = 50.86 \text{ kg}$$

$$v_2 = \frac{V}{m} = \frac{5.0}{50.86} = 0.098\,31 \text{ m}^3 / \text{kg}$$

In Table 2 of the steam tables we find, by interpolation, that at a pressure of 2.03 MPa, $v_g = 0.098\,31$ m³/kg. The final pressure of the steam is therefore 2.03 MPa. Then

$$u_2 = 2600.5 \text{ kJ} / \text{kg}$$

$$U_2 = mu_2 = 50.86(2600.5) = 132\,261 \text{ kJ}$$

$${}_1Q_2 = U_2 - U_1 = 132\,261 - 27\,326 = 104\,935 \text{ kJ}$$

EXAMPLE 5.3E   A vessel having a volume of 100 ft³ contains 1 ft³ of saturated liquid water and 99 ft³ of saturated water vapor at 14.7 lbf/in². Heat is transferred until the vessel is filled with saturated vapor. Determine the heat transfer for this process.

Control mass: All the water inside the vessel.

Sketch: Fig. 5.4.

Initial state: Pressure, volume of liquid, volume of vapor; therefore, state 1 is fixed.

Final state: Somewhere along the saturated-vapor curve; the water was heated, so $P_2 > P_1$.

Process: Constant volume and mass; therefore, constant specific volume.

Diagram: Fig. 5.5.

Model: Steam tables.

Analysis:

First law: $${}_1Q_2 = U_2 - U_1 + m\frac{(\mathbf{V}_2^2 - \mathbf{V}_1^2)}{2} + mg(Z_2 - Z_1) + {}_1W_2$$

From examining the control surface for various work modes, we conclude that the work for this process is zero. Furthermore, the system is not moving, so there is no change in kinetic energy. There is a small change in center of mass of the system but we will assume that the corresponding change in potential energy is negligible (compared to other terms). Therefore,

$$_1Q_2 = U_2 - U_1$$

Solution

The heat transfer will be found from the first law above. State 1 is known, so U_1 can be calculated. Also, the specific volume at state 2 is known (from state 1 and the process). Since state 2 is saturated vapor, state 2 is fixed, as is seen from Fig. 5.5. Therefore, U_2 can also be found.

The solution proceeds as follows:

$$m_{1\,\text{liq}} = \frac{V_{\text{liq}}}{v_f} = \frac{1}{0.01672} = 59.81 \text{ lbm}$$

$$m_{1\,\text{vap}} = \frac{V_{\text{vap}}}{v_g} = \frac{99}{26.80} = 3.69 \text{ lbm}$$

Then,

$$U_1 = m_{1\,\text{liq}} u_{1\,\text{liq}} + m_{1\,\text{vap}} u_{1\,\text{vap}}$$

$$= 59.81(180.1) + 3.69(1077.6) = 14\,748 \text{ Btu}$$

To determine u_2 we need to know two thermodynamic properties, since this determines the final state. The properties we know are the quality, $x = 100\%$, and v_2, the final specific volume, which can readily be determined.

$$m = m_{1\,\text{liq}} + m_{1\,\text{vap}} = 59.81 + 3.69 = 63.50 \text{ lbm}$$

$$v_2 = \frac{V}{m} = \frac{100}{63.50} = 1.575 \text{ ft}^3 / \text{lbm}$$

In Table 2 of the steam tables we find, by interpolation, that at a pressure of 294 lbf/in.^{2}, $v_g = 1.575$ ft^3/lbm. The final pressure of the steam is therefore 294 lbf/in.^{2}. Then,

$$u_2 = 1117.0 \text{ Btu} / \text{lbm}$$

$$U_2 = mu_2 = 63.50(1117.0) = 70\,930 \text{ Btu}$$

$$_1Q_2 = U_2 - U_1 = 70\,930 - 14\,748 = 56\,182 \text{ Btu}$$

Up to this point, we have discussed how internal energies are listed in the saturation tables of the steam tables. The internal energy of superheated vapor steam is listed as a function of temperature and pressure in Table B.1.3. Similarly, Tables B.1.4 and B.1.5 list values in the compressed-liquid region and in the saturated-solid–saturated-vapor region, respectively. In summary, all the values of internal energy are listed in the tables in the same manner as are values of specific volume.

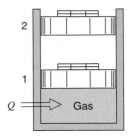

FIGURE 5.6 The constant-pressure quasi-equilibrium process.

5.5 THE THERMODYNAMIC PROPERTY ENTHALPY

In analyzing specific types of processes, we frequently encounter certain combinations of thermodynamic properties, which are therefore also properties of the substance undergoing the change of state. To demonstrate one such situation, let us consider a control mass undergoing a quasi-equilibrium constant-pressure process, as shown in Fig. 5.6. Assume that there are no changes in kinetic or potential energy and that the only work done during the process is that associated with the boundary movement. Taking the gas as our control mass and applying the first law, Eq. 5.11, we have

$$_1Q_2 = U_2 - U_1 + {}_1W_2$$

The work can be calculated from the relation

$$_1W_2 = \int_1^2 P\, dV$$

Since the pressure is constant,

$$_1W_2 = P\int_1^2 dV = P(V_2 - V_1)$$

Therefore,

$$_1Q_2 = U_2 - U_1 + P_2V_2 - P_1V_1$$
$$= (U_2 + P_2V_2) - (U_1 + P_1V_1)$$

We find that in this very restricted case, the heat transfer during the process is given in terms of the change in the quantity $U + PV$ between the initial and final states. Because all these quantities are thermodynamic properties, that is, functions only of the state of the system, their combination must also have these same characteristics. Therefore, we find it convenient to define a new extensive property, the enthalpy,

$$H \equiv U + PV \tag{5.12}$$

or, per unit mass,

$$h \equiv u + Pv \tag{5.13}$$

As for internal energy, we could speak of specific enthalpy, h, and total enthalpy, H. However, we will refer to both as enthalpy, since the context will make it clear which is being discussed.

The heat transfer in a constant-pressure quasi-equilibrium process is equal to the change in enthalpy, which includes both the change in internal energy and the work for this particular process. This is by no means a general result. It is valid for this special case only because the work done during the process is equal to the difference in the PV product for the final and initial states. This would not be true if the pressure had not remained constant during the process.

The significance and use of enthalpy is not restricted to the special process just described. Other cases in which this same combination of properties $u + Pv$ appear will be developed later, notably in Chapter 6 in which we discuss control volume analyses. Our reason for introducing enthalpy at this time is that although the steam tables list values for internal energy, many other tables and charts of thermodynamic properties give values for enthalpy but not for the internal energy. Therefore, it is necessary to calculate the internal energy at a state using the tabulated values and Eq. 5.13:

$$u = h - Pv$$

Students often become confused about the validity of this calculation when analyzing system processes that do not occur at constant pressure, for which enthalpy has no physical significance. We must keep in mind that enthalpy, being a property, is a state or point function, and its use in calculating internal energy at the same state is not related to, or dependent on, any process that may be taking place.

Tabular values of enthalpy, such as those included in Tables B.1 through B.7, are all relative to some arbitrarily selected base. In the steam tables, the internal energy of saturated liquid at 0.01°C is the reference state and is given a value of zero. For refrigerants, such as ammonia and chlorofluorocarbons R-12 and R-22, the reference state is arbitrarily taken as saturated liquid at −40°C. The enthalpy in this reference state is assigned the value of zero. Cryogenic fluids, such as nitrogen or methane, have other arbitrary reference states chosen for enthalpy values listed in their tables. Because each of these reference states is arbitrarily selected, it is always possible to have negative values for enthalpy, as for saturated-solid water in Table B.1.5. When enthalpy and internal energy are given values relative to the same reference state, as they are in essentially all thermodynamic tables, the difference between internal energy and enthalpy at the reference state is equal to Pv. Since the specific volume of the liquid is very small, this product is negligible as far as the significant figures of the tables are concerned, but the principle should be kept in mind, for in certain cases it is significant.

In the superheat region of many thermodynamic tables, values of the specific internal energy u are not given. As mentioned earlier, these values can be readily calculated from the relation $u = h - Pv$, though it is important to keep the units in mind. As an example, let us calculate the internal energy u of superheated R-134a at 0.4 MPa, 70°C.

$$u = h - Pv$$
$$= 460.545 - 400 \times 0.066\,484$$
$$= 433.951 \text{ kJ / kg}$$

The enthalpy of a substance in a saturation state and with a given quality is found in the same way as the specific volume and internal energy. The enthalpy of saturated liquid has the symbol h_f, saturated vapor h_g, and the increase in enthalpy during vaporization h_{fg}. For a saturation state, the enthalpy can be calculated by one of the following relations:

$$h = (1-x)h_f + xh_g$$
$$h = h_f + xh_{fg}$$

The enthalpy of compressed liquid water may be found from Table B.1.4. For substances for which compressed-liquid tables are not available, the enthalpy is taken as that of saturated liquid at the same temperature.

EXAMPLE 5.4 A cylinder fitted with a piston has a volume of 0.1 m³ and contains 0.5 kg of steam at 0.4 MPa. Heat is transferred to the steam until the temperature is 300°C, while the pressure remains constant.

Determine the heat transfer and the work for this process.

Control mass: Water inside cylinder.

Initial state: P_1, V_1, m; therefore v_1 is known, state 1 is fixed (at P_1, v_1, check steam tables—two-phase region).

Final state: P_2, T_2; therefore state 2 is fixed (superheated).

Process: Constant pressure.

Diagram: Fig. 5.7.

Model: Steam tables.

Analysis:

No change in kinetic energy, no change in potential energy. Work is done by movement at the boundary. Assume the process to be quasi-equilibrium. Since the pressure is constant,

$$_1W_2 = \int_1^2 P\,dV = P\int_1^2 dV = P(V_2 - V_1) = m(P_2 v_2 - P_1 v_1)$$

Therefore, the first law is

$$_1Q_2 = m(u_2 - u_1) + {}_1W_2$$
$$= m(u_2 - u_1) + m(P_2 v_2 - P_1 v_1) = m(h_2 - h_1)$$

Solution

There is a choice of procedures to follow. State 1 is known, so v_1 and h_1 (or u_1) can be found. State 2 is also known, so v_2 and h_2 (or u_2) can be found. Using the first law and the work equation, we can calculate the heat transfer and work. Using the enthalpies, we have

$$v_1 = \frac{V_1}{m} = \frac{0.1}{0.5} = 0.2 = 0.001\,084 + x_1 0.4614$$

$$x_1 = \frac{0.1989}{0.4614} = 0.4311$$

$$h_1 = h_f + x_1 h_{fg}$$

$$= 604.74 + 0.4311 \times 2133.8 = 1524.7$$

$$h_2 = 3066.8$$

$$_1Q_2 = 0.5(3066.8 - 1524.7) = 771.1 \text{ kJ}$$

$$_1W_2 = mP(v_2 - v_1) = 0.5 \times 400(0.6548 - 0.2) = 91.0 \text{ kJ}$$

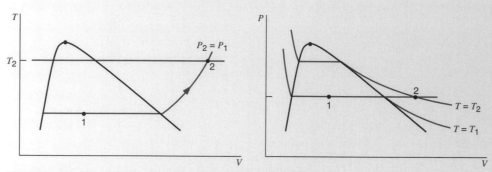

FIGURE 5.7 The constant-pressure quasi-equilibrium process.

Therefore,

$$U_2 - U_1 = {_1}Q_2 - {_1}W_2 = 771.1 - 91.0 = 680.1 \text{ kJ}$$

The heat transfer could also have been found from u_1 and u_2,

$$u_1 = u_f + x_1 u_{fg}$$

$$= 604.31 + 0.4311 \times 1949.3 = 1444.7$$

$$u_2 = 2804.8$$

and

$${_1}Q_2 = U_2 - U_1 + {_1}W_2$$

$$= 0.5(2804.8 - 1444.7) + 91.0 = 771.1 \text{ kJ}$$

5.6 THE CONSTANT-VOLUME AND CONSTANT-PRESSURE SPECIFIC HEATS

In this section we will consider a homogeneous phase of a substance of constant composition. This phase may be a solid, a liquid, or a gas, but no change of phase will occur. We will then define a variable termed the specific heat, the amount of heat required per unit mass to raise the temperature by one degree. Since it would be of interest to examine the relation between the specific heat and other thermodynamic variables, we note first that the heat transfer is given by Eq. 5.10. Neglecting changes in kinetic and potential energies, and assuming a simple compressible substance and quasi-equilibrium process, for which the work in Eq. 5.10 is given by Eq. 4.2, we have

$$\delta Q = dU + \delta W = dU + P\, dV$$

We find that this expression can be evaluated for two separate special cases.

1. Constant volume, for which the work term ($P\, dV$) is zero, so that the specific heat (at constant volume) is

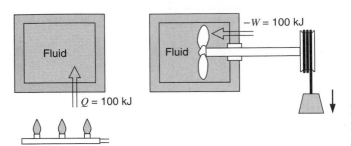

FIGURE 5.8 Sketch showing two ways in which a given ΔU may be achieved.

$$C_v = \frac{1}{m}\left(\frac{\delta Q}{\delta T}\right)_v = \frac{1}{m}\left(\frac{\partial U}{\partial T}\right)_v = \left(\frac{\partial u}{\partial T}\right)_v \qquad (5.14)$$

2. Constant pressure, for which the work term can be integrated and the resulting PV terms at the initial and final states be associated with the internal energy terms, as in Section 5.5, thereby leading to the conclusion that the heat transfer can be expressed in terms of the enthalpy change. The corresponding specific heat (at constant pressure) is

$$C_p = \frac{1}{m}\left(\frac{\delta Q}{\delta T}\right)_p = \frac{1}{m}\left(\frac{\partial H}{\partial T}\right)_p = \left(\frac{\partial h}{\partial T}\right)_p \qquad (5.15)$$

Note that in each of these special cases, the resulting expression, Eq. 5.14 or 5.15, contains only thermodynamic properties, from which we conclude that the constant-volume and constant-pressure specific heats must themselves be thermodynamic properties. This means that, although we began this discussion by considering the amount of heat transfer required to cause a unit temperature change and then proceeded through a very specific development leading to Eq. 5.14 (or 5.15), the result ultimately expresses a relation among a set of thermodynamic properties and therefore constitutes a definition that is independent of the particular process leading to it (in the same sense that the definition of enthalpy in the previous section is independent of the process used to illustrate one situation in which the property is useful in a thermodynamic analysis). As an example, consider the two identical fluid masses shown in Fig. 5.8. In the first system 100 kJ of heat is transferred to it, and in the second system 100 kJ of work is done on it. Thus, the change of internal energy is the same for each, and therefore the final state and the final temperature are the same in each. In accordance with Eq. 5.14, therefore, exactly the same value for the average constant-volume specific heat would be found for this substance for the two processes, even though the two processes are very different as far as heat transfer is concerned.

It should also be noted that this development was carried out for a simple compressible substance, for which the work was given by Eq. 4.2. Where there are different work modes, such as those described in Section 4.4, different specific heats become appropriate. For any particular quasi-equilibrium work mode, there is one specific heat at constant displacement and one at constant driving force, as these terms were discussed in connection with Eq. 4.16. For example, in a system involving surface effects, it is convenient to utilize a specific heat at constant surface tension and one at constant area.

EXAMPLE 5.5 Estimate the constant-pressure specific heat of steam at 0.5 MPa, 375°C.

Solution

If we consider a change of state at constant pressure, Eq. 5.15 may be written

$$C_p \approx \left(\frac{\Delta h}{\Delta T}\right)_p$$

From the steam tables,

$$\text{at } 0.5 \text{ MPa}, \ 350°C, \quad h = 3167.7$$

$$\text{at } 0.5 \text{ MPa}, \ 400°C, \quad h = 3271.8$$

Since we are interested in C_p at 0.5 MPa, 375°C,

$$C_p \simeq \frac{104.1}{50} = 2.082 \text{ kJ / kg K}$$

As a special case, consider either a solid or a liquid. Since both of these phases are nearly incompressible,

$$dh = du + d(Pv) \approx du + vdP \tag{5.16}$$

Also, for both of these phases, the specific volume is very small, such that in many cases

$$dh \approx du \approx C \ dT \tag{5.17}$$

where C is either the constant-volume or the constant-pressure specific heat, as the two would be nearly the same. In many processes involving a solid or a liquid, we might further assume that the specific heat in Eq. 5.17 is constant (unless the process is at low temperature or over a wide range of temperatures). Equation 5.17 can then be integrated to

$$h_2 - h_1 \simeq u_2 - u_1 \simeq C(T_2 - T_1) \tag{5.18}$$

Specific heats for various solids and liquids are listed in Tables A.3 and A.4.

In other processes for which it is not possible to assume constant specific heat, there may be a known relation for C as a function of temperature. Equation 5.17 could then also be integrated.

5.7 THE INTERNAL ENERGY, ENTHALPY, AND SPECIFIC HEAT OF IDEAL GASES

At this point certain comments about the internal energy, enthalpy, and the constant-pressure and constant-volume specific heats of an ideal gas should be made. An ideal gas has been defined in Chapter 3 as a gas at sufficiently low density so that intermolecular forces and the associated energy are negligibly small. Therefore, an ideal gas has the equation of state

$$Pv = RT$$

It can be shown that for an ideal gas, the internal energy is a function of the temperature only. That is, for an ideal gas,

$$u = f(T) \tag{5.19}$$

This means that an ideal gas at a given temperature has a certain definite specific internal energy u, regardless of the pressure. This will be demonstrated mathematically using the methods of classical thermodynamics in Chapter 13.

In 1843, Joule demonstrated this fact when he conducted the following classical experiment in thermodynamics. Two pressure vessels (Fig. 5.9), connected by a pipe and valve, were immersed in a bath of water. Initially vessel A contained air at 22-atm pressure and vessel B was highly evacuated. When thermal equilibrium was attained, the valve was opened, allowing the pressures in A and B to equalize. No change in the temperature of the bath was detected during or after this process. Because there was no change in the temperature of the bath, Joule concluded that no heat had been transferred to the air. Because the work was also zero, he concluded from the first law of thermodynamics that there was no change in the internal energy of the gas. Because the pressure and volume changed during this process, one concludes that internal energy is not a function of pressure and volume. Because air does not conform exactly to the definition of an ideal gas, a small change in temperature will be detected when very accurate measurements are made in Joule's experiment.

The relation between the internal energy u and the temperature can be established by using the definition of constant-volume specific heat given by Eq. 5.14.

$$C_v = \left(\frac{\partial u}{\partial T} \right)_v$$

Since the internal energy of an ideal gas is not a function of volume, for an ideal gas we can write

$$C_{v0} = \frac{du}{dT}$$

$$du = C_{v0} \, dT \tag{5.20}$$

where the subscript 0 denotes the specific heat of an ideal gas. For a given mass m,

$$dU = mC_{v0} \, dT \tag{5.21}$$

From the definition of enthalpy and the equation of state of an ideal gas, it follows that

$$h = u + Pv = u + RT \tag{5.22}$$

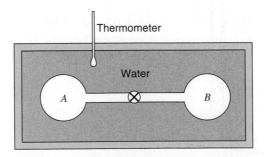

FIGURE 5.9 Apparatus for conducting Joule's experiment.

Since R is a constant and u is a function of temperature only, it follows that the enthalpy, h, of an ideal gas is also a function of temperature only. That is,

$$h = f(T) \tag{5.23}$$

The relation between enthalpy and temperature is found from the constant-pressure specific heat as defined by Eq. 5.15.

$$C_p = \left(\frac{\partial h}{\partial T} \right)_p$$

Since the enthalpy of an ideal gas is a function of the temperature only and is independent of the pressure, it follows that

$$C_{p0} = \frac{dh}{dT}$$

$$dh = C_{p0}\, dT \tag{5.24}$$

For a given mass m,

$$dH = mC_{p0}\, dT \tag{5.25}$$

The consequences of Eqs. 5.20 and 5.24 are demonstrated in Fig. 5.10, which shows two lines of constant temperature. Since internal energy and enthalpy are functions of temperature only, these lines of constant temperature are also lines of constant internal energy and constant enthalpy. From state 1 the high temperature can be reached by a variety of paths, and in each case the final state is different. However, regardless of the path, the change in internal energy is the same, as is the change in enthalpy, for lines of constant temperature are also lines of constant u and constant h.

Because the internal energy and enthalpy of an ideal gas are functions of temperature only, it also follows that the constant-volume and constant-pressure specific heats are also functions of temperature only. That is,

$$C_{v0} = f(T) \qquad C_{p0} = f(T) \tag{5.26}$$

Because all gases approach ideal-gas behavior as the pressure approaches zero, the ideal-gas specific heat for a given substance is often called the zero-pressure specific heat, and the zero-pressure, constant-pressure specific heat is given the symbol C_{p0}. The zero-pressure, constant-volume specific heat is given the symbol C_{v0}. Figure 5.11 shows $\overline{C}_{p0}$ as a function of temperature for a number of different substances. These values are determined by the techniques of statistical thermodynamics and will not be discussed in this

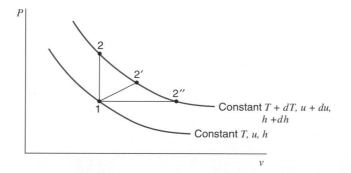

FIGURE 5.10 Pressure-volume diagram for an ideal gas.

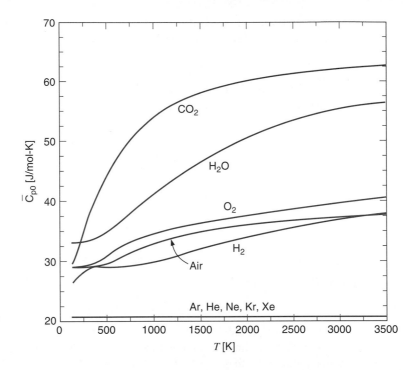

FIGURE 5.11 Heat capacity for some gases as function of temperature.

text. However, a brief qualitative discussion at this point should provide some insight into this behavior and will be beneficial in determining under what conditions an assumption of constant specific heat is justified.

As was discussed in Section 2.6, the energy possessed by molecules may be stored in several forms. The translational and rotational energies increase linearly with temperature, which means that these contributions to the specific heat are not temperature-dependent. Contributions from vibrational and electronic modes, on the other hand, are temperature-dependent (the electronic usually being very small). From Fig. 5.11, it is evident that the specific heat of a diatomic gas (such as hydrogen or oxygen) increases with an increase in temperature, primarily because of the vibration. A polyatomic gas (such as carbon dioxide or water) shows a much greater increase in specific heat as the temperature increases, and this is due to the additional vibrational modes of a polyatomic molecule. A monatomic gas (such as helium, argon, or neon), possessing only translational and electronic energies, shows little or no variation of specific heat over a wide range of temperatures.

A very important relation between the constant-pressure and constant-volume specific heats of an ideal gas may be developed from the definition of enthalpy.

$$h = u + Pv = u + RT$$

Differentiating and substituting Eqs. 5.20 and 5.24, we have

$$dh = du + R\ dT$$

$$C_{p0}\ dT = C_{v0}\ dT + R\ dT$$

Therefore,

$$C_{p0} - C_{v0} = R \tag{5.27}$$

On a mole basis this equation is written

$$\overline{C}_{p0} - \overline{C}_{v0} = \overline{R} \tag{5.28}$$

This tells us that the difference between the constant-pressure and constant-volume specific heats of an ideal gas is always constant, though both are functions of temperature. Thus, we need examine only the temperature dependency of one, and the other is given by Eq. 5.27.

Let us consider the specific heat C_{p0}. There are three possibilities to examine. The situation is simplest if we assume constant specific heat, that is, no temperature dependence. Then it is possible to integrate Eq. 5.24 directly to

$$h_2 - h_1 = C_{p0}(T_2 - T_1) \tag{5.29}$$

We note from Fig. 5.11 the circumstances under which this will be an accurate model. It should be added, however, that it may be a reasonable approximation under other conditions, especially if an average specific heat in the particular temperature range is used in Eq. 5.29. Values of specific heat at room temperature and gas constants for various gases are given in Table A.5.

The second possibility for the specific heat is to use an analytical equation for C_{p0} as a function of temperature. Because the results of specific-heat calculations from statistical thermodynamics do not lend themselves to convenient mathematical forms, these results have been approximated empirically. The equations for C_{p0} as a function of temperature are listed in Table A.6 for a number of gases.

The third possibility is to integrate the results of the calculations of statistical thermodynamics from an arbitrary reference temperature to any other temperature T, and to define a function

$$h_T = \int_{T_0}^{T} C_{p0} \, dT$$

This function can then be tabulated in a single-entry (temperature) table. Then, between any two states 1 and 2,

$$h_2 - h_1 = \int_{T_0}^{T_2} C_{p0} \, dT - \int_{T_0}^{T_1} C_{p0} \, dT = h_{T_2} - h_{T_1} \tag{5.30}$$

and it is seen that the reference temperature cancels out. This function h_T (and a similar function $u_T = h_T - RT$) is listed for air in Table A.7. The enthalpy function (with respect to a reference temperature of 25°C) is listed for other gases in Table A.8.

To summarize the three possibilities, we note that using the ideal-gas tables, Tables A.7 and A.8, gives us the most accurate answer, but that the equations in Table A.6 would give a close empirical approximation. Constant specific heat would be less accurate, except for monatomic gases and gases below room temperature. It should be remembered that all these results are a part of the ideal-gas model, which in many of our problems is not a valid assumption for the behavior of the substance.

EXAMPLE 5.6   Calculate the change of enthalpy as 1 kg of oxygen is heated from 300 to 1500 K. Assume ideal-gas behavior.

Solution

For an ideal gas, the enthalpy change is given by Eq. 5.24. However, we also need to make an assumption about the dependence of specific heat on temperature. Let us solve this problem in several ways and compare the answers.

Our most accurate answer for the ideal-gas enthalpy change for oxygen between 300 and 1500 K would be from the ideal-gas tables, Table A.8. This result is, using Eq. 5.30,

$$h_2 - h_1 = \frac{\overline{h}_{1500} - \overline{h}_{300}}{M} = \frac{40\,600 - 54}{32} = 1267.0 \text{ kJ / kg}$$

The empirical equation from Table A.6 should give a good approximation to this result. Integrating Eq. 5.24, we have

$$\overline{h}_2 - \overline{h}_1 = \int_{T_1}^{T_2} \overline{C}_{p0} \, dT = \int_{\theta_1}^{\theta_2} \overline{C}_{p0}(\theta) \times 100 \, d\theta$$

$$\overline{h}_{1500} - \overline{h}_{300} = 100 \left(37.432\theta + \frac{0.020\,102}{2.5} \theta^{2.5} + \frac{178.57}{0.5} \theta^{-0.5} - 236.88\theta^{-1} \right)\Bigg|_{\theta_1=3}^{\theta_2=15}$$

$$= 40\,525 \text{ kJ / kmol}$$

$$h_2 - h_1 = \frac{\overline{h}_{1500} - \overline{h}_{300}}{M} = \frac{40\,525}{32}$$

$$= 1266.4 \text{ kJ / kg}$$

which is different than the first result by less than 0.1%.

If we assume constant specific heat, we must be concerned about what value we are going to use. If we use the value at 300 K from Table A.5, we find, from Eq. 5.29, that

$$h_2 - h_1 = C_{p0}(T_2 - T_1) = 0.9216 \times 1200 = 1105.9 \text{ kJ / kg}$$

which is low by 12.7%. On the other hand, suppose we assume that the specific heat is constant at its value at 900 K, the average temperature. Substituting 900 K into the equation for specific heat from Table A.6, we have

$$\overline{C}_{p0} = 37.432 + 0.020\,102(9)^{1.5} - 178.57(9)^{-1.5} + 236.88(9)^{-2}$$

$$= 34.2855 \text{ kJ / kmol K}$$

or

$$C_{p0} = \frac{34.2855}{32} = 1.0714 \text{ kJ / kg K}$$

Substituting this value into Eq. 5.29 gives the result

$$h_2 - h_1 = 1.0714 \times 1200 = 1285.7 \text{ kJ / kg}$$

which is high by about 1.5%, a much closer result than the one using the room temperature specific heat. It should be kept in mind that part of the model involving ideal gas with constant specific heat is the choice of what value is to be used.

EXAMPLE 5.7

A cylinder fitted with a piston has an initial volume of 0.1 m³ and contains nitrogen at 150 kPa, 25°C. The piston is moved, compressing the nitrogen until the pressure is 1 MPa and the temperature is 150°C. During this compression process heat is transferred from the nitrogen, and the work done on the nitrogen is 20 kJ. Determine the amount of this heat transfer.

Control mass: Nitrogen.

Initial state: P_1, T_1, V_1; state 1 fixed.

Final state: P_2, T_2; state 2 fixed.

Process: Work input known.

Model: Ideal gas, constant specific heat with value at 300 K, Table A.5.

Analysis:

First law: $_1Q_2 = m(u_2 - u_1) + {_1}W_2$

Solution

The mass of nitrogen is found from the equation of state with the value of R from Table A.5.

$$m = \frac{PV}{RT} = \frac{150 \times 0.1}{0.2968 \times 298.15} = 0.1695 \text{ kg}$$

Assuming constant specific heat as given in Table A.5, we have

$$_1Q_2 = mC_{v0}(T_2 - T_1) + {_1}W_2$$
$$= 0.1695(0.7448)(150 - 25) - 20.0$$
$$= 15.8 - 20.0 = -4.2 \text{ kJ}$$

It would, of course, be somewhat more accurate to use Table A.8 than to assume constant specific heat (room temperature value), but often the slight increase in accuracy does not warrant the added difficulties of manually interpolating the tables.

EXAMPLE 5.7E

A cylinder fitted with a piston has an initial volume of 2 ft³ and contains nitrogen at 20 lbf/in.², 80 F. The piston is moved, compressing the nitrogen until the pressure is 160 lbf/in.² and the temperature is 300F. During this compression process heat is transferred from the nitrogen, and the work done on the nitrogen is 9.15 Btu. Determine the amount of this heat transfer.

Control mass: Nitrogen.

Initial state: P_1, T_1, V_1; state 1 fixed.

Final state: P_2, T_2; state 2 fixed.

Process: Work input known.

Model: Ideal gas, constant specific heat with value at 540 R, Table C.4.

Analysis:

First law:
$$_1Q_2 = m(u_2 - u_1) + {}_1W_2$$

Solution

The mass of nitrogen is found from the equation of state with the value of R from Table C.4.

$$m = \frac{PV}{RT} = \frac{20 \times 144 \times 2}{55.15 \times 540} = 0.1934 \text{ lbm}$$

Assuming constant specific heat as given in Table C.4,

$$_1Q_2 = mC_{v0}(T_2 - T_1) + {}_1W_2$$

$$= 0.1934(0.177)(300 - 80) - 9.15$$

$$= 7.53 - 9.15 = -1.62 \text{ Btu}$$

It would, of course, be somewhat more accurate to use Table C.7 than to assume constant specific heat (room temperature value), but often the slight increase in accuracy does not warrant the added difficulties of manually interpolating the tables.

5.8 THE FIRST LAW AS A RATE EQUATION

We frequently find it desirable to use the first law as a rate equation that expresses either the instantaneous or average rate at which energy crosses the control surface as heat and work and the rate at which the energy of the control mass changes. In so doing we are departing from a strictly classical point of view, because basically classical thermodynamics deals with systems that are in equilibrium, and time is not a relevant parameter for systems that are in equilibrium. However, since these rate equations are developed from the concepts of classical thermodynamics and are used in many applications of thermodynamics, they are included in this book. This rate form of the first law will be used in the development of the first law for the control volume in Section 6.2, and in this form the first law finds extensive applications in thermodynamics, fluid mechanics, and heat transfer.

Consider a time internal δt during which an amount of heat δQ crosses the control surface, an amount of work δW is done by the control mass, the internal energy change is ΔU, the kinetic energy change is ΔKE, and the potential energy change is ΔPE. From the first law we can write

$$\delta Q = \Delta U + \Delta KE + \Delta PE + \delta W$$

Dividing by δt we have the average rate of energy transfer as heat and work and increase of the energy of the control mass.

$$\frac{\delta Q}{\delta t} = \frac{\Delta U}{\delta t} + \frac{\Delta KE}{\delta t} + \frac{\Delta PE}{\delta t} + \frac{\delta W}{\delta t}$$

Taking the limit for each of these quantities as δt approaches zero, we have

$$\lim_{\delta t \to 0} \frac{\delta Q}{\delta t} = \dot{Q}, \quad \text{the heat transfer rate}$$

$$\lim_{\delta t \to 0} \frac{\delta W}{\delta t} = \dot{W}, \quad \text{the power}$$

$$\lim_{\delta t \to 0} \frac{\Delta U}{\delta t} = \frac{dU}{dt} \qquad \lim_{\delta t \to 0} \frac{\Delta(\text{KE})}{\delta t} = \frac{d(\text{KE})}{dt} \qquad \lim_{\delta t \to 0} \frac{\Delta(\text{PE})}{\delta t} = \frac{d(\text{PE})}{dt}$$

Therefore, the rate equation form of the first law is

$$\dot{Q} = \frac{dU}{dt} + \frac{d(\text{KE})}{dt} + \frac{d(\text{PE})}{dt} + \dot{W} \tag{5.31}$$

We could also write this in the form

$$\dot{Q} = \frac{dE}{dt} + \dot{W} \tag{5.32}$$

EXAMPLE 5.8   During the charging of a storage battery, the current is 20 A and the voltage is 12.8 V. The rate of heat transfer from the battery is 10 W. At what rate is the internal energy increasing?

Solution

Since changes in kinetic and potential energy are insignificant, the first law can be written as a rate equation in the form, Eq. 5.31,

$$\dot{Q} = \frac{dU}{dt} + \dot{W}$$

$$\dot{W} = -\mathscr{E}i = -20 \times 12.8 = -256 \text{ W}$$

Therefore,

$$\frac{dU}{dt} = \dot{Q} - \dot{W} = -10 \text{ W} - (-256) = 246 \text{ J/s}$$

5.9 CONSERVATION OF MASS

In the previous sections we considered the first law of thermodynamics for a control mass undergoing a change of state. A control mass is defined as a fixed quantity of mass. The question now is whether the mass of such a system changes when its energy changes? If it does, our definition of a control mass as a fixed quantity of mass is no longer valid when the energy changes.

We know from relativistic considerations that mass and energy are related by the well-known equation

$$E = mc^2 \tag{5.33}$$

where c = velocity of light and E = energy. We conclude from this equation that the mass of a control mass does change when its energy changes. Let us calculate the magnitude of this change of mass for a typical problem and determine whether this change in mass is significant.

Consider a rigid vessel that contains a 1-kg stoichiometric mixture of a hydrocarbon fuel (such as gasoline) and air. From our knowledge of combustion, we know that after combustion takes place it will be necessary to transfer about 2900 kJ from the system to restore it to its initial temperature. From the first law

$$_1Q_2 = U_2 - U_1 + {}_1W_2$$

we conclude that since $_1W_2 = 0$ and $_1Q_2 = -2900$ kJ, the internal energy of this system decreases by 2900 kJ during the heat transfer process. Let us now calculate the decrease in mass during this process using Eq. 5.33.

The velocity of light, c, is 2.9979×10^8 m/s. Therefore,

$$2900 \text{ kJ} = 2\,900\,000 \text{ J} = m\ (\text{kg}) \times \left(2.9979 \times 10^8 \text{ m/s}\right)^2$$

$$m = 3.23 \times 10^{-11} \text{ kg}$$

Thus, when the energy of the control mass decreases by 2900 kJ, the decrease in mass is 3.23×10^{-11} kg.

A change in mass of this magnitude cannot be detected by even our most accurate chemical balance. Certainly, a fractional change in mass of this magnitude is beyond the accuracy required in essentially all engineering calculations. Therefore, if we use the laws of conservation of mass and conservation of energy as separate laws, we will not introduce significant error into most thermodynamic problems, and our definition of a control mass as having a fixed mass can be used even though the energy changes.

PROBLEMS

5.1 A hydraulic hoist raises a 1750-kg car 1.8 m in an auto repair shop. The hydraulic pump has a constant pressure of 800 kPa on its piston. What is the increase in potential energy of the car and how much volume should the pump displace to deliver that amount of work?

5.2 Airplane takeoff from an aircraft carrier is assisted by a steam driven piston/cylinder with an average pressure of 750 kPa. A 3500 kg airplane should be accelerated from zero to a speed of 30 m/s with 25% of the energy coming from the steam piston. Find the needed piston displacement volume.

5.3 Solve Problem 5.2, but assume the steam pressure in the cylinder starts at 1000 kPa, dropping linearly with volume to reach 100 kPa at the end of the process.

5.4 A piston motion moves a 25-kg hammerhead vertically down 1 m from rest to a velocity of 50 m/s in a stamping machine. What is the change in total energy of the hammerhead?

5.5 A 25-kg piston is above a gas in a long vertical cylinder. Now the piston is released from rest and accelerates up in the cylinder reaching the end 5 m higher at a velocity of 25 m/s. The gas pressure drops during the process so the average is 600 kPa with an outside atmosphere at 100 kPa. Neglect the change in gas kinetic and potential energy, and find the needed change in the gas volume.

5.6 Find the missing properties

    a. H_2O $T = 250°C$, $v = 0.02$ m³/kg $P = ?\ u = ?$

    b. N_2 $T = 277°C$, $P = 0.5$ MPa $x = ?\ h = ?$

    c. H_2O $T = -2°C$, $P = 100$ kPa $u = ?\ v = ?$

    d. R-134a $P = 200$ kPa, $v = 0.12$ m³/kg $u = ?\ T = ?$

    e. NH_3 $T = 65°C$, $P = 600$ kPa $u = ?\ v = ?$

5.7 Find the missing properties and give the phase of the substance

    a. H_2O $u = 2390$ kJ/kg, $T = 90°C$ $h = ?\ v = ?\ x = ?$

    b. H_2O $u = 1200$ kJ/kg, $P = 10$ MPa $T = ?\ x = ?\ v = ?$

    c. R-12 $T = -5°C$, $P = 300$ kPa $h = ?\ x = ?$

    d. R-134a $T = 60°C$, $h = 430$ kJ/kg $v = ?\ x = ?$

    e. NH_3 $T = 20°C$, $P = 100$ kPa $u = ?\ v = ?\ x = ?$

5.8 Find the missing properties and give the phase of the substance

    a. H_2O $T = 120°C$, $v = 0.5$ m³/kg $u = ?\ P = ?\ x = ?$

    b. H_2O $T = 100°C$, $P = 10$ MPa $u = ?\ x = ?\ v = ?$

    c. N_2 $T = 800$ K, $P = 200$ kPa $v = ?\ u = ?$

    d. NH_3 $T = 100°C$, $v = 0.1$ m³/kg $P = ?\ x = ?$

    e. CH_4 $T = 190$ K, $x = 0.75$ $v = ?\ u = ?$

5.9 Find the missing properties among (P, T, v, u, h) together with x if applicable and give the phase of the substance.

    a. R-22 $T = 10°C$, $u = 200$ kJ/kg

    b. H_2O $T = 350°C$, $h = 3150$ kJ/kg

    c. R-12 $P = 600$ kPa, $h = 230$ kJ/kg

    d. R-134a $T = 40°C$, $u = 407$ kJ/kg

    e. NH_3 $T = 20°C$, $v = 0.1$ m³/kg

5.10 A 100-L rigid tank contains nitrogen (N_2) at 900 K, 6 MPa. The tank is now cooled to 100 K. What are the work and heat transfer for this process?

5.11 Water in a 150-L closed, rigid tank is at 100°C, 90% quality. The tank is then cooled to $-10°C$. Calculate the heat transfer during the process.

5.12 A cylinder fitted with a frictionless piston contains 2 kg of superheated refrigerant R-134a vapor at 350 kPa, 100°C. The cylinder is now cooled so the R-134a remains at constant pressure until it reaches a quality of 75%. Calculate the heat transfer in the process.

5.13 A test cylinder with constant volume of 0.1 L contains water at the critical point. It now cools down to room temperature of 20°C. Calculate the heat transfer from the water.

5.14 Ammonia at 0°C, quality 60% is contained in a rigid 200-L tank. The tank and ammonia is now heated to a final pressure of 1 MPa. Determine the heat transfer for the process.

5.15 A 10-L rigid tank contains R-22 at $-10°C$, 80% quality. A 10-A electric current (from a 6-V battery) is passed through a resistor inside the tank for 10 min, after which the R-22 temperature is 40°C. What was the heat transfer to or from the tank during this process?

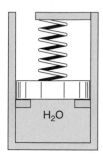

5.16 A piston/cylinder arrangement contains 1 kg of water, shown in Fig. P5.16. The piston is spring loaded and initially rests on some stops. A pressure of 300 kPa will just float the piston and, at a volume of 1.5 m³, a pressure of 500 kPa will balance the piston. The initial state of the water is 100 kPa with a volume of 0.5 m³. Heat is now added until a pressure of 400 kPa is reached.

    a. Find the initial temperature and the final volume.

    b. Find the work and heat transfer in the process and plot the $P-V$ diagram.

5.17 A closed steel bottle contains ammonia at $-20°C$, $x = 20\%$ and the volume is 0.05 m³. It has a safety valve that opens at a pressure of 1.4 MPa. By accident, the bottle is heated until the safety valve opens. Find the temperature and heat transfer when the valve first opens.

5.18 A piston/cylinder arrangement B is connected to a 1-m³ tank A by a line and valve, shown in Fig. P5.18. Initially both contain water, with A at 100 kPa, saturated vapor and B at 400°C, 300 kPa, 1 m³. The valve is now opened, and the water in both A and B comes to a uniform state.

    a. Find the initial mass in A and B.

    b. If the process results in $T_2 = 200°C$, find the heat transfer and the work.

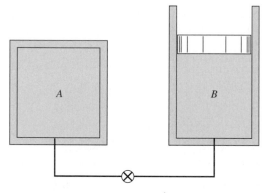

FIGURE P5.18

5.19 Consider the same setup and initial conditions as in the previous problem. Assuming that the process is adiabatic, find the final temperature and the work.

5.20 A vertical cylinder fitted with a piston contains 5 kg of R-22 at 10°C, shown in Fig. P5.20. Heat is transferred to the system, causing the piston to rise until it reaches a set of stops at which point the volume has doubled. Additional heat is transferred until the temperature inside reaches 50°C, at which point the pressure inside the cylinder is 1.3 MPa.

    a. What is the quality at the initial state?

    b. Calculate the heat transfer for the overall process.

5.21 A piston/cylinder contains 50 kg of water at 200 kPa with a volume of 0.1 m³. Stops in the cylinder are placed to restrict the enclosed volume to 0.5 m³ similar to Fig. P5.20. The water is now heated until the piston reaches the stops. Find the necessary heat transfer.

5.22 Ten kilograms of water in a piston/cylinder with constant pressure is at 450°C and a volume of 0.633 m³. It is now cooled to 20°C. Show the $P-v$ diagram and find the work and heat transfer for the process.

5.23 Find the heat transfer in Problem 4.10.

5.24 Find the heat transfer in Problem 4.24.

5.25 A piston/cylinder contains 1 kg of liquid water at 20°C and 300 kPa. There is a linear spring mounted on the piston such that when the water is heated the pressure reaches 3 MPa with a volume of 0.1 m³.

a. Find the final temperature and plot the $P-v$ diagram for the process.

b. Calculate the work and heat transfer for the process.

5.26 An insulated cylinder fitted with a piston contains R-12 at 25°C with a quality of 90% and a volume of 45 L. The piston is allowed to move, and the R-12 expands until it exists as saturated vapor. During this process the R-12 does 7.0 kJ of work against the piston. Determine the final temperature, assuming the process is adiabatic.

5.27 Two kilograms of nitrogen at 100 K, $x = 0.5$ is heated in a constant pressure process to 300 K in a piston/cylinder arrangement. Find the initial and final volumes and the total heat transfer required.

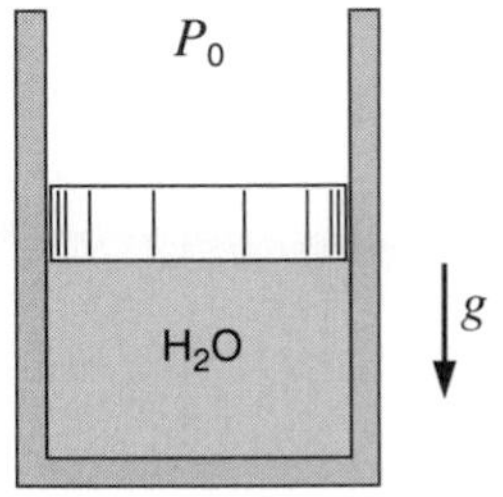

5.28 A piston/cylinder arrangement has the piston loaded with outside atmospheric pressure and the piston mass to a pressure of 150 kPa, shown in Fig. P5.28. It contains water at −2°C, which is then heated until the water becomes saturated vapor. Find the final temperature and specific work and heat transfer for the process.

P_0

H_2O

$\downarrow g$

FIGURE P5.28

5.29 Consider the system shown in Fig. P5.29. Tank A has a volume of 100 L and contains saturated vapor R-134a at 30°C. When the valve is cracked open, R-134a flows slowly into cylinder B. The piston mass requires a pressure of 200 kPa in cylinder B to raise the piston. The process ends when the pressure in tank A has fallen to 200 kPa. During this process heat is exchanged with the surroundings such that the R-134a always remains at 30°C. Calculate the heat transfer for the process.

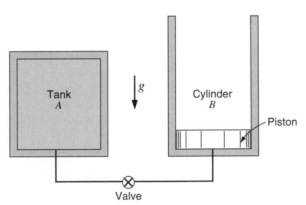

Tank
A

$\downarrow g$

Cylinder
B

Piston

Valve

FIGURE P5.29

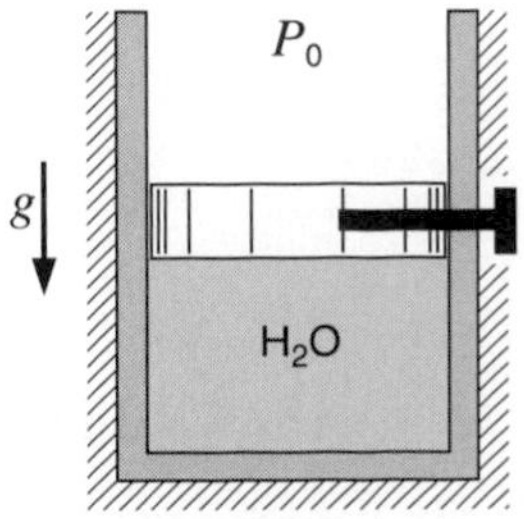

P_0

$g \downarrow$

H_2O

FIGURE P5.31

5.30 A spherical balloon contains 2 kg of R-22 at 0°C, 30% quality. This system is heated until the pressure in the balloon reaches 600 kPa. For this process, it can be assumed that the pressure in the balloon is directly proportional to the balloon diameter. How does pressure vary with volume and what is the heat transfer for the process?

5.31 A piston held by a pin in an insulated cylinder, shown in Fig. P5.31, contains 2 kg water at 100°C, quality 98%. The piston has a mass of 102 kg, with cross-sectional area of 100 cm², and the ambient pressure is 100 kPa. The pin is released, which allows the piston to move. Determine the final state of the water, assuming the process to be adiabatic.

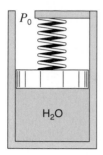

FIGURE P5.32

5.32 A piston/cylinder arrangement has a linear spring and the outside atmosphere acting on the piston shown in Fig. P5.32. It contains water at 3 MPa, 400°C with the volume being 0.1 m³. If the piston is at the bottom, the spring exerts a force such that a pressure of 200 kPa inside is required to balance the forces. The system now cools until the pressure reaches 1 MPa. Find the heat transfer for the process.

5.33 A vertical piston/cylinder has a linear spring mounted as shown in Fig. P5.32. The spring is mounted so at zero cylinder volume a balancing pressure inside is 100 kPa. The cylinder contains 0.5 kg of water at 125°C, 70% quality. Heat is now transferred to the water until the cylinder pressure reaches 300 kPa. How much work is done by the water during this process and what is the heat transfer?

5.34 Two heavily insulated tanks are connected by a valve, as shown in Fig. P5.34. Tank A contains 0.6 kg of water at 300 kPa, 300°C. Tank B has a volume of 300 L and contains water at 600 kPa, 80% quality. The valve is opened, and the two tanks eventually come to a uniform state. Assuming the process to be adiabatic, show the final state (u,v) is two-phase and iterate on final pressure to match required internal energy.

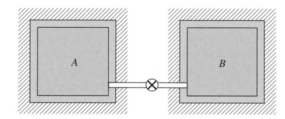

FIGURE P5.34

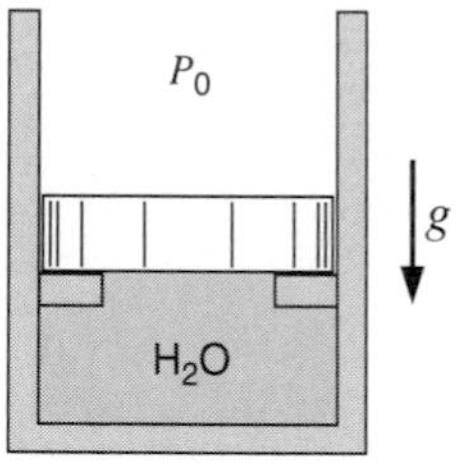

5.35 A piston/cylinder contains 1 kg of ammonia at 20°C with a volume of 0.1 m³, shown in Fig. P5.35. Initially the piston rests on some stops with the top surface open to the atmosphere, P_o, so a pressure of 1400 kPa is required to lift it. To what temperature should the ammonia be heated to lift the piston? If it is heated to saturated vapor find the final temperature, volume, and heat transfer, $_1Q_2$.

5.36 A cylinder/piston arrangement contains 5 kg of water at 100°C with $x = 20\%$ and the piston, $m_p = 75$ kg, resting on some stops, similar to Fig. P5.35. The outside pressure is 100 kPa, and the cylinder area is $A = 24.5$ cm². Heat is now added until the water reaches a saturated vapor state. Find the initial volume, final pressure, work, and heat transfer terms and show the $P-v$ diagram.

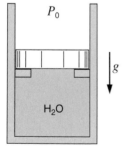

FIGURE P5.35

5.37 A rigid tank is divided into two rooms by a membrane, both containing water, shown in Fig. P5.37. Room A is at 200 kPa, $v = 0.5$ m³/kg, $V_A = 1$ m³, and room B contains 3.5 kg at 0.5 MPa, 400°C. The membrane now ruptures and heat transfer takes place so the water comes to a uniform state at 100°C. Find the heat transfer during the process.

FIGURE P5.36

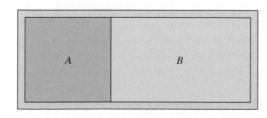

FIGURE P5.37

5.38 Two tanks are connected by a valve and line, as shown in Fig. P5.38. The volumes are both 1 m³ with R-134a at 20°C, quality 15% in A and tank B is evacuated. The valve is opened and saturated vapor flows from A into B until the pressures become equal. The process occurs slowly enough that all temperatures stay at 20°C during the process. Find the total heat transfer to the R-134a during the process.

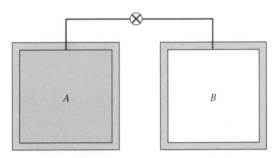

FIGURE P5.38

5.39 Consider the same system as in the previous problem. Let the valve be opened and transfer enough heat to both tanks so that all the liquid disappears. Find the necessary heat transfer.

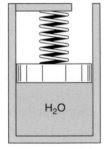

FIGURE P5.40

5.40 A cylinder having a piston restrained by a linear spring contains 0.5 kg of saturated vapor water at 120°C, as shown in Fig. P5.40. Heat is transferred to the water, causing the piston to rise, and with a spring constant of 15 kN/m, piston cross-sectional area 0.05 m², the pressure varies linearly with volume until a final pressure of 500 kPa is reached. Find the final temperature in the cylinder and the heat transfer for the process.

5.41 A water-filled reactor with volume of 1 m³ is at 20 MPa, 360°C and placed inside a containment room, as shown in Fig. P5.41. The room is well insulated and initially evacuated. Due to a failure, the reactor ruptures and the water fills the containment room. Find the minimum room volume so the final pressure does not exceed 200 kPa.

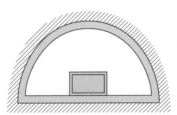

FIGURE P5.41

5.42 Assume the same setup as the previous problem, but the room has a volume of 100 m³. Show that the final state is two-phase and find the final pressure by trial and error.

5.43 Refrigerant-12 is contained in a piston/cylinder arrangement at 2 MPa, 150°C with a massless piston against the stops, at which point V = 0.5 m³. The side above the piston is connected by an open valve to an air line at 10°C, 450 kPa, shown in Fig. P5.43. The whole setup now cools to the surrounding temperature of 10°C. Find the heat transfer and show the process in a P−v diagram.

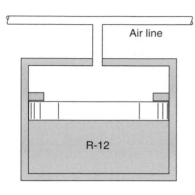

FIGURE P5.43

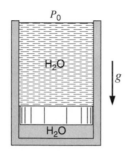

FIGURE P5.44

5.44 A 10-m-high open cylinder, $A_{\mathrm{cyl}} = 0.1$ m^2, contains 20°C water above and 2 kg of 20°C water below a 198.5-kg thin insulated floating piston, shown in Fig. P5.44. Assume standard g, P_o. Now heat is added to the water below the piston so that it expands, pushing the piston up, causing the water on top to spill over the edge. This process continues until the piston reaches the top of the cylinder. Find the final state of the water below the piston (T, P, v) and the heat added during the process.

5.45 A rigid container has two rooms filled with water, each 1 m^3 separated by a wall. Room A has $P = 200$ kPa with a quality $x = 0.80$. Room B has $P = 2$ MPa and $T = 400$°C. The partition wall is removed and the water comes to a uniform state, which after a while due to heat transfer has a temperature of 200°C. Find the final pressure and the heat transfer in the process.

5.46 A piston/cylinder arrangement of initial volume 0.025 m^3 contains saturated water vapor at 180°C. The steam now expands in a polytropic process with exponent $n = 1$ to a final pressure of 200 kPa, while it does work against the piston. Determine the heat transfer in this process.

5.47 Calculate the heat transfer for the process described in Problem 4.23.

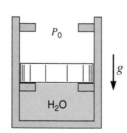

FIGURE P5.48

5.48 Consider the piston/cylinder arrangement shown in Fig. P5.48. A frictionless piston is free to move between two sets of stops. When the piston rests on the lower stops, the enclosed volume is 400 L. When the piston reaches the upper stops, the volume is 600 L. The cylinder initially contains water at 100 kPa, 20% quality. It is heated until the water eventually exists as saturated vapor. The mass of the piston requires 300 kPa pressure to move it against the outside ambient pressure. Determine the final pressure in the cylinder, the heat transfer, and the work for the overall process.

5.49 Calculate the heat transfer for the process described in Problem 4.30.

5.50 A cylinder fitted with a frictionless piston that is restrained by a linear spring contains R-22 at 20°C, quality 60% with a volume of 8 L, shown in Fig. P5.50. The piston cross-sectional area is 0.04 m^2, and the spring constant is 500 kN/m. A total of 62 kJ of heat is now added to the R-22. Verify that the final pressure is around 1600 kPa and find the final temperature of the R-22.

5.51 A 1-L capsule of water at 700 kPa, 150°C is placed in a larger insulated and otherwise evacuated vessel. The capsule breaks and its contents fill the entire volume. If the final pressure should not exceed 125 kPa, what should the vessel volume be?

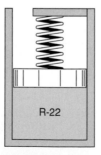

FIGURE P5.50

5.52 A cylinder with a frictionless piston contains steam at 2 MPa, 500°C with a volume of 5 L, shown in Fig. P5.52. The external piston force is proportional to cylin-

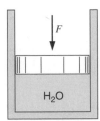

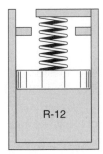

FIGURE P5.52

FIGURE P5.56

der volume cubed. Heat is transferred out of the cylinder, reducing the volume and thus the force until the cylinder pressure has dropped to 500 kPa. Find the work and heat transfer for this process.

5.53 Superheated refrigerant R-134a at 20°C, 0.5 MPa is cooled in a piston/cylinder arrangement at constant temperature to a final two-phase state with quality of 50%. The refrigerant mass is 5 kg, and during this process 500 kJ of heat is removed. Find the initial and final volumes and the necessary work.

5.54 Calculate the heat transfer for the process described in Problem 4.20.

5.55 Calculate the heat transfer for the process described in Problem 4.26.

5.56 A piston/cylinder, shown in Fig. P5.56, contains R-12 at −30°C, $x = 20\%$. The volume is 0.2 m^3. It is known that $V_{stop} = 0.4$ m^3, and if the piston sits at the bottom, the spring force balances the other loads on the piston. It is now heated up to 20°C. Find the mass of the fluid and show the $P{-}v$ diagram. Find the work and heat transfer.

5.57 Ammonia, NH$_3$, is contained in a sealed rigid tank at 0°C, $x = 50\%$ and is then heated to 100°C. Find the final state P_2, u_2 and the specific work and heat transfer.

5.58 A house is being designed to use a thick concrete floor mass as thermal storage material for solar energy heating. The concrete is 30 cm thick and the area exposed to the sun during the daytime is 4 m×6 m. It is expected that this mass will undergo an average temperature rise of about 3°C during the day. How much energy will be available for heating during the nighttime hours?

5.59 A car with mass 1275 kg drives at 60 km/h when the brakes are applied quickly to decrease its speed to 20 km/h. Assume the brake pads are 0.5 kg mass with heat capacity of 1.1 kJ/kg K and the brake discs/drums are 4.0 kg steel where both masses are heated uniformly. Find the temperature increase in the brake assembly.

5.60 A copper block of volume 1 L is heat treated at 500°C and now cooled in a 200-L oil bath initially at 20°C, shown in Fig. P5.60. Assuming no heat transfer with the surroundings, what is the final temperature?

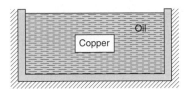

FIGURE P5.60

5.61 Saturated, $x{=}1\%$, water at 25°C is contained in a hollow spherical aluminum vessel with inside diameter of 0.5 m and a 1-cm-thick wall. The vessel is heated until the water inside is saturated vapor. Considering the vessel and water together as a control mass, calculate the heat transfer for the process.

5.62 An ideal gas is heated from 500 to 1500 K. Find the change in enthalpy using constant specific heat from Table A.5 (room temperature value) and discuss the accuracy of the result if the gas is

    a. Argon   c. Carbon dioxide

    b. Oxygen

5.63 A rigid insulated tank is separated into two rooms by a stiff plate. Room A of 0.5 m^3 contains air at 250 kPa, 300 K and room B of 1 m^3 has air at 150 kPa, 1000 K. The plate is removed and the air comes to a uniform state without any heat transfer. Find the final pressure and temperature.

5.64 An insulated cylinder is divided into two parts of 1 m³ each by an initially locked piston, as shown in Fig. P5.64. Side A has air at 200 kPa, 300 K, and side B has air at 1.0 MPa, 1000 K. The piston is now unlocked so that it is free to move, and it conducts heat so that the air comes to a uniform temperature $T_A = T_B$. Find the mass in both A and B, and the final T and P.

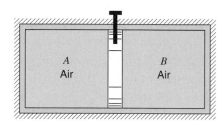

FIGURE P5.64

5.65 A cylinder with a piston restrained by a linear spring contains 2 kg of carbon dioxide at 500 kPa, 400°C. It is cooled to 40°C, at which point the pressure is 300 kPa. Calculate the heat transfer for the process.

5.66 A piston/cylinder in a car contains 0.2 L of air at 90 kPa, 20°C, shown in Fig. P5.66. The air is compressed in a quasi-equilibrium polytropic process with polytropic exponent $n = 1.25$ to a final volume six times smaller. Determine the final pressure, temperature, and the heat transfer for the process.

FIGURE P5.66

5.67 Water at 20°C, 100 kPa, is brought to 200 kPa, 1500°C. Find the change in the specific internal energy, using the water tables and ideal gas table.

5.68 For an application the change in enthalpy of carbon dioxide from 30 to 1500°C at 100 kPa is needed. Consider the following methods and indicate the most accurate one.

a. Constant specific heat, value from Table A.5.

b. Constant specific heat, value at average temperature from the equation in Table A.6.

c. Variable specific heat, integrating the equation in Table A.6.

d. Enthalpy from ideal gas tables in Table A.8.

5.69 Air in a piston/cylinder at 200 kPa, 600 K is expanded in a constant-pressure process to twice the initial volume (state 2), shown in Fig. P5.69. The piston is then locked with a pin and heat is transferred to a final temperature of 600 K. Find P, T, and h for states 2 and 3, and find the work and heat transfer in both processes.

5.70 An insulated floating piston divides a cylinder into two volumes each of 1 m³, as shown in Fig. P5.70. One contains water at 100°C and the other air at −3°C and both pressures are 200 kPa. A line with a safety valve that opens at 400 kPa is attached to the water side of the cylinder. Assume no heat transfer to the water and that the water is incompressible. Show possible air states in a P–v diagram, and

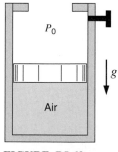

FIGURE P5.69

find the air temperature when the safety valve opens. How much heat transfer is needed to bring the air to 1300 K?

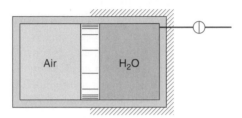

FIGURE P5.70

5.71 Two containers are filled with air, one a rigid tank A, and the other a piston/cylinder B that is connected to A by a line and valve, as shown in Fig. P5.71. The initial conditions are: $m_A = 2$ kg, $T_A = 600$ K, $P_A = 500$ kPa and $V_B = 0.5$ m^3, $T_B = 27°C$, $P_B = 200$ kPa. The piston in B is loaded with the outside atmosphere and the piston mass in the standard gravitational field. The valve is now opened, and the air comes to a uniform condition in both volumes. Assuming no heat transfer, find the initial mass in B, the volume of tank A, the final pressure and temperature, and the work, $_1W_2$.

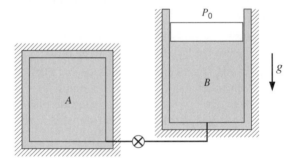

FIGURE P5.71

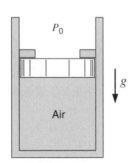

FIGURE P5.73

5.72 A 250-L rigid tank contains methane gas at 500°C, 600 kPa. The tank is cooled to 300 K.

a. Find the final pressure and the heat transfer for the process.

b. What is the percent error in the heat transfer if the specific heat is assumed constant at the room temperature value?

5.73 A piston/cylinder arrangement, shown in Fig. P5.73, contains 5 g of air at 250 kPa, 300°C. The 75-kg piston has a diameter of 0.1 m and initially pushes against the stops. The atmosphere is at 100 kPa and 20°C. The cylinder now cools to 20°C as heat is transferred to the ambient. Calculate the heat transfer.

5.74 Oxygen at 300 kPa, 100°C is in a piston/cylinder arrangement with a volume of 0.1 m^3. It is now compressed in a polytropic process with exponent, $n = 1.2$, to a final temperature of 200°C. Calculate the heat transfer for the process.

5.75 A piston/cylinder contains 2 kg of air at 27°C, 200 kPa, as shown in Fig. P5.75. The piston is loaded with a linear spring, mass, and the atmosphere. Stops are mounted so that $V_{stop} = 3$ m^3, at which point $P = 600$ kPa is required to balance the piston forces. The air is now heated to 400 kPa. Find the final temperature and volume, and the work and heat transfer. Find the work done on the spring.

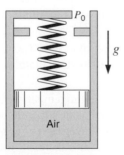

FIGURE P5.75

5.76 A piston/cylinder contains 0.001 m³ air at 300 K, 150 kPa. The air is now compressed in a process in which $P V^{1.25} = C$ to a final pressure of 600 kPa. Find the work performed by the air and the heat transfer.

5.77 An air pistol contains compressed air in a small cylinder, shown in Fig. P5.77. Assume that the volume is 1 cm³, pressure is 1 MPa, and the temperature is 27°C when armed. A bullet, $m = 15$ g, acts as a piston initially held by a pin (trigger); when released, the air expands in an isothermal process ($T =$ constant). If the air pressure is 0.1 MPa in the cylinder as the bullet leaves the gun, find

a. the final volume and the mass of air.

b. the work done by the air and work done on the atmosphere.

c. the work to the bullet and the bullet exit velocity.

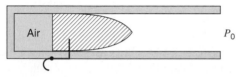

FIGURE P5.77

5.78 A spherical elastic balloon contains nitrogen (N₂) at 20°C, 500 kPa. The initial volume is 0.5 m³. The balloon material is such that the pressure inside is proportional to the balloon diameter. Heat is now transferred to the balloon until its volume reaches 1.0 m³, at which point the process stops.

a. Can the nitrogen be assumed to behave as an ideal gas throughout this process?

b. Calculate the heat transferred to the nitrogen.

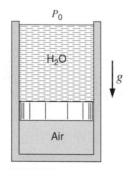

FIGURE P5.79

5.79 A 10-m-high cylinder, cross-sectional area 0.1 m², has a massless piston at the bottom with water at 20°C on top of it, shown in Fig. P5.79. Air at 300 K, volume 0.3 m³, under the piston is heated so that the piston moves up, spilling the water out over the side. Find the total heat transfer to the air when all the water has been pushed out.

5.80 A cylinder fitted with a frictionless piston contains carbon dioxide at 500 kPa, 400 K, at which point the volume is 50 L. The gas is now allowed to expand until the piston reaches a set of fixed stops at 150 L cylinder volume. This process is polytropic, with the polytropic exponent n equal to 1.20. The piston is locked and additional heat is now transferred to the gas, until the final temperature reaches 500 K. Determine

a. the final pressure inside the cylinder.

b. the work and heat transfer for the overall process.

5.81 A cylinder fitted with a frictionless piston contains R-134a at 40°C, 80% quality at which point the volume is 10 L. The external force on the piston is now varied in such a manner that the R-134a slowly expands in a polytropic process to 400 kPa, 20°C. Calculate the work and the heat transfer for this process.

5.82 A piston/cylinder contains argon gas at 140 kPa, 10°C, and the volume is 100 L. The gas is compressed in a polytropic process to 700 kPa, 280°C. Calculate the heat transfer during the process.

5.83 Water at 150°C, 50% quality is contained in a cylinder/piston arrangement with initial volume 0.05 m³. The loading of the piston is such that the inside pressure is linear with the square root of volume as $P = 100 + C V^{0.5}$ kPa. Now heat is transferred to the cylinder to a final pressure of 600 kPa. Find the heat transfer in the process.

5.84 A piston/cylinder has 1 kg propane gas at 700 kPa, 40°C. The piston cross-sectional area is 0.5 m², and the total external force restraining the piston is directly proportional to the cylinder volume squared. Heat is transferred to the propane until its temperature reaches 700°C. Determine the final pressure inside the cylinder, the work done by the propane, and the heat transfer during the process.

5.85 A closed cylinder is divided into two rooms by a frictionless piston held in place by a pin, as shown in Fig. P5.85. Room A has 10 L air at 100 kPa, 30°C, and room B has 300 L saturated water vapor at 30°C. The pin is pulled, releasing the piston, and both rooms come to equilibrium at 30°C, and as the water is compressed it becomes two-phase. Considering a control mass of the air and water, determine the work done by the system and the heat transfer to the cylinder.

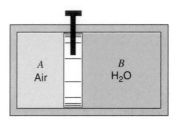

FIGURE P5.85

5.86 A small elevator is being designed for a construction site. It is expected to carry four 75-kg workers to the top of a 100-m-tall building in less than 2 min. The elevator cage will have a counterweight to balance its mass. What is the smallest size (power) electric motor that can drive this unit?

5.87 The rate of heat transfer to the surroundings from a person at rest is about 400 kJ/h. Suppose that the ventilation system fails in an auditorium containing 100 people. Assume the energy goes into the air of volume 1500 m³ initially at 300 K and 101 kPa. Find the rate (degrees per minute) of the air temperature change.

5.88 Consider the 100-L Dewar (a rigid double-walled vessel for storing cryogenic liquids) shown in Fig. P5.88. The Dewar contains nitrogen at 1 atm, 90% liquid, and 10% vapor by volume. The insulation holds heat transfer into the Dewar from the ambient to a very low rate, 5 J/s. The vent valve is accidentally closed so that the pressure inside slowly rises. How long will it take to reach a pressure of 500 kPa?

5.89 A computer in a closed room of volume 200 m³ dissipates energy at a rate of 10 kW. The room has 50 kg wood, 25 kg steel, and air, with all material at 300 K, 100 kPa. Assuming all the mass heats up uniformly, how long will it take to increase the temperature 10°C?

5.90 The heaters in a spacecraft suddenly fail. Heat is lost by radiation at the rate of 100 kJ/h, and the electric instruments generate 75 kJ/h. Initially, the air is at 100 kPa, 25°C with a volume of 10 m³. How long will it take to reach an air temperature of −20°C?

FIGURE P5.88

ADVANCED
PROBLEMS

5.91 A cylinder fitted with a piston restrained by a linear spring has a cross-sectional area of 0.05 m² and initial volume of 20 L, shown in Fig. P5.91. The cylinder contains ammonia at 1 MPa, 60°C. The spring constant is 150 kN/m. Heat is rejected

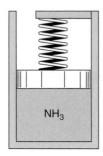

FIGURE P5.91

from the system, and the piston moves until 6.25 kJ of work has been done on the ammonia.

a. Find the final temperature of the ammonia.

b. Calculate the heat transfer for the process.

5.92 A cylinder fitted with a piston contains 2 kg of R-12 at 10°C, 90% quality. The system undergoes a quasi-equilibrium polytropic expansion to 100 kPa, during which the system receives a heat transfer of 52.5 kJ. What is the final temperature of the R-12?

5.93 A spherical balloon initially 150 mm in diameter and containing R-12 at 100 kPa is connected to a 30-L uninsulated, rigid tank containing R-12 at 500 kPa. Everything is at the ambient temperature of 20°C. A valve connecting the tank and balloon is opened slightly and remains so until the pressures equalize. During this process heat is exchanged so the temperature remains constant at 20°C and the pressure inside the balloon is proportional to the diameter at any time. Calculate the final pressure and the work and heat transfer during the process.

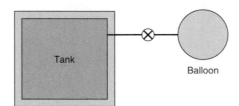

FIGURE P5.93

5.94 Calculate the heat transfer for the process described in Problem 4.44.

5.95 Calculate the heat transfer for the process described in Problem 4.46.

5.96 A cylinder fitted with a frictionless piston contains R-134a at 10°C, quality of 50%, and initial volume of 100 L. The external force on the piston now varies in such a manner that the piston moves, increasing the volume. It is noted that the temperature is 25°C when the last drop of liquid R-134a evaporates. The process continues to a final state of 40°C, 600 kPa. Assume the pressure is piecewise linear in volume and determine the final volume in the cylinder and the work and heat transfer for the overall process.

5.97 A rigid 1-m^3 tank contains butane at 500 K, 100 kPa. The tank is now heated to 1500 K.

a. Is it reasonable to use the specific heat value from Table A.5 to calculate the heat transfer in this process?

b. Calculate the work and the heat transfer for this process.

5.98 A cylinder fitted with a frictionless piston contains 0.2 kg of saturated (both liquid and vapor present) R-12 at −20°C. The external force on the piston is such that the pressure inside the cylinder is related to the volume by the expression

$$P = -47.5 + 4.0 * V^{1.5}, \qquad \text{kPa and L}$$

Heat is now transferred to the cylinder until the pressure inside reaches 250 kPa. Calculate the work and heat transfer.

5.99 A certain elastic balloon will support an internal pressure equal to $P_o = 100$ kPa until the balloon becomes spherical at a diameter of $D_o = 1$ m, beyond which

$$P = P_o + C(1 - x^6)x \ ; \qquad x = D_o/D$$

because of the offsetting effects of balloon curvature and elasticity. This balloon contains helium gas at 250 K, 100 kPa, with a 0.4 m³ volume. The balloon is heated until the volume reaches 2 m³. During the process the maximum pressure inside the balloon is 200 kPa.

a. What is the temperature inside the balloon when pressure is maximum?

b. What are the final pressure and temperature inside the balloon?

c. Determine the work and heat transfer for the overall process.

5.100 A frictionless, thermally conducting piston separates the air and water in the cylinder shown in Fig. P5.100. The initial volumes of A and B are each 500 L, and the initial pressure on each side is 700 kPa. The volume of the liquid in B is 2% of the volume of B at this state. Heat is transferred to both A and B until all the liquid in B evaporates. Notice that $P_A = P_B$ and $T_A = T_B = T_{sat}$ through the process and iterate to find final pressure and then determine the heat transfer.

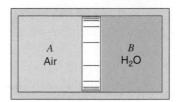

FIGURE P5.100

5.101 A closed, vertical cylinder is divided into two parts A and B by a thermally nonconducting frictionless piston. The upper part A contains air at ambient temperature, 20°C, and the initial volume is 150 L. The lower part B contains R-134a at −15°C, quality of 50%, and initial volume of 50 L. Heat is now transferred from a heat source to part B, causing the piston to move upward until the volume of B reaches 145 L. Neglect the piston mass, such that the pressures in A and B are always equal and assume the temperature in A remains constant during the process.

a. What is the final pressure in A and the final temperature in B?

b. Calculate the work done by the R-134a during the process.

c. Calculate the heat transfer to the R-134a during the process.

d. What is the heat transfer to (or from) the air in A?

5.102E A hydraulic hoist raises a 3650-lbm car 6 ft in an auto repair shop. The hydraulic pump has a constant pressure of 100 lbf/in.² on its piston. What is the increase in potential energy of the car and how much volume should the pump displace to deliver that amount of work?

5.103E A piston motion moves a 50-lbm hammerhead vertically down 3 ft from rest to a velocity of 150 ft/s in a stamping machine. What is the change in total energy of the hammerhead?

5.104E Find the missing properties and give the phase of the substance.

a. H_2O $u = 1000$ Btu/lbm, $T = 270$ F $h = ?$ $v = ?$ $x = ?$

b. H_2O $u = 450$ Btu/lbm, $P = 1500$ lbf/in.² $T = ?$ $x = ?$ $v = ?$

c. R-22 $T = 30$ F, $P = 75$ lbf/in.² $h = ?$ $x = ?$

d. R-134a $T = 140$ F, $h = 185$ Btu/lbm $v = ?$ $x = ?$

e. NH_3 $T = 170$ F, $P = 60$ lbf/in.² $u = ?$ $v = ?$ $x = ?$

5.105E Find the missing properties among (P, T, v, u, h) together with x, if applicable, and give the phase of the substance.

  a. R-22 $T = 50$ F, $u = 85$ Btu/lbm
  b. H_2O $T = 600$ F, $h = 1322$ Btu/lbm
  c. R-22 $P = 150$ lbf/in.^{2}, $h = 115.5$ Btu/lbm
  d. R-134a $T = 100$ F, $u = 175$ Btu/lbm
  e. NH_3 $T = 70$ F, $v = 2$ ft^3/lbm

5.106E Water in a 6-ft^3 closed, rigid tank is at 200 F, 90% quality. The tank is then cooled to 20 F. Calculate the heat transfer during the process.

5.107E A cylinder fitted with a frictionless piston contains 4 lbm of superheated refrigerant R-134a vapor at 400 lbf/in.^{2}, 200 F. The cylinder is now cooled so the R-134a remains at constant pressure until it reaches a quality of 75%. Calculate the heat transfer in the process.

5.108E Ammonia at 30 F, quality 60% is contained in a rigid 8-ft^3 tank. The tank and ammonia are now heated to a final pressure of 140 lbf/in.^{2}. Determine the heat transfer for the process.

5.109E A vertical cylinder fitted with a piston contains 10 lbm of R-22 at 50 F, shown in Fig. P5.20. Heat is transferred to the system, causing the piston to rise until it reaches a set of stops at which point the volume has doubled. Additional heat is transferred until the temperature inside reaches 120 F, at which point the pressure inside the cylinder is 200 lbf/in.^{2}.

  a. What is the quality at the initial state?

  b. Calculate the heat transfer for the overall process.

5.110E A twenty-pound-mass of water in a piston/cylinder with constant pressure is at 1100 F and a volume of 22.6 ft^3. It is now cooled to 100 F. Show the $P-v$ diagram and find the work and heat transfer for the process.

5.111E A piston/cylinder contains 2 lbm of liquid water at 70 F, and 30 lbf/in.^{2}. There is a linear spring mounted on the piston such that when the water is heated the pressure reaches 300 lbf/in.^{2} with a volume of 4 ft^3. Find the final temperature and plot the $P-v$ diagram for the process. Calculate the work and the heat transfer for the process.

5.112E A piston/cylinder arrangement has the piston loaded with outside atmospheric pressure and the piston mass to a pressure of 20 lbf/in.^{2}, shown in Fig P5.28. It contains water at 25 F, which is then heated until the water becomes saturated vapor. Find the final temperature and specific work and heat transfer for the process.

5.113E A piston/cylinder contains 2 lbm of water at 70 F with a volume of 0.1 ft^3, shown in Fig. P5.35. Initially the piston rests on some stops with the top surface open to the atmosphere, P_o, so a pressure of 40 lbf/in.^{2} is required to lift it. To what temperature should the water be heated to lift the piston? If it is heated to saturated vapor, find the final temperature, volume, and the heat transfer.

5.114E Two tanks are connected by a valve and line, as shown in Fig. P5.38. The volumes are both 35 ft^3 with R-134a at 70 F, quality 25% in A, and tank B is evacuated. The valve is opened, and saturated vapor flows from A into B until the pressures become equal. The process occurs slowly enough that all temperatures stay at 70 F during the process. Find the total heat transfer to the R-134a during the process.

5.115E A water-filled reactor with volume of 50 ft^3 is at 2000 lbf/in.^{2}, 560 F and placed inside a containment room, as shown in Fig. P5.41. The room is well insulated and initially evacuated. Due to a failure, the reactor ruptures and the water fills the containment room. Find the minimum room volume so the final pressure does not exceed 30 lbf/in.^{2}.

5.116E A piston/cylinder arrangement of initial volume 0.3 ft^3 contains saturated water vapor at 360 F. The steam now expands in a polytropic process with exponent $n = 1$ to a final pressure of 30 lbf/in.^{2}, while it does work against the piston. Determine the heat transfer in this process.

5.117E Calculate the heat transfer for the process described in Problem 4.72.

5.118E Ammonia, NH_3, is contained in a sealed rigid tank at 30 F, $x = 50\%$ and is then heated to 200 F. Find the final state P_2, u_2 and the specific work and heat transfer.

5.119E A car with mass 3250 lbm drives with 60 mi/h when the brakes are applied to quickly decrease its speed to 20 mi/h. Assume the brake pads are 1 lbm mass with heat capacity of 0.2 Btu/lbm R and the brake discs/drums are 8 lbm steel where both masses are heated uniformly. Find the temperature increase in the brake assembly.

5.120E A copper block of volume 60 in.^{3} is heat treated at 900 F and now cooled in a 3-ft^3 oil bath initially at 70 F. Assuming no heat transfer with the surroundings, what is the final temperature?

5.121E An insulated cylinder is divided into two parts of 10 ft^3 each by an initially locked piston. Side A has air at 2 atm, 600 R, and side B has air at 10 atm, 2000 R, as shown in Fig. P5.64. The piston is now unlocked so it is free to move, and it conducts heat so the air comes to a uniform temperature $T_A = T_B$. Find the mass in both A and B and also the final T and P.

5.122E A cylinder with a piston restrained by a linear spring contains 4 lbm of carbon dioxide at 70 lbf/in.^{2}, 750 F. It is cooled to 75 F, at which point the pressure is 45 lbf/in.^{2}. Calculate the heat transfer for the process.

5.123E A piston/cylinder in a car contains 12 in.^{3} of air at 13 lbf/in.^{2}, 68 F, shown in Fig. P5.66. The air is compressed in a quasi-equilibrium polytropic process with polytropic exponent $n = 1.25$ to a final volume six times smaller. Determine the final pressure, temperature, and heat transfer for the process.

5.124E Water at 70 F, 15 lbf/in.^{2}, is brought to 30 lbf/in.^{2}, 2700 F. Find the change in the specific internal energy, using the water tables and ideal gas table.

5.125E Air in a piston/cylinder at 30 lbf/in.^{2}, 1080 R is shown in Fig. P5.69. It is expanded in a constant-pressure process to twice the initial volume (state 2). The piston is then locked with a pin, and heat is transferred to a final temperature of 1080 R. Find P, T, and h for states 2 and 3, and find the work and heat transfer in both processes.

5.126E Two containers are filled with air, one a rigid tank A, and the other a piston/cylinder B that is connected to A by a line and valve, as shown in Fig. P5.71. The initial conditions are: $m_A = 4$ lbm, $T_A = 1080$ R, $P_A = 75$ lbf/in.^{2} and $V_B = 17$ ft^3, $T_B = 80$ F, $P_B = 30$ lbf/in.^{2}. The piston in B is loaded with the outside atmosphere and the piston mass in the standard gravitational field. The valve is now opened, and the air comes to a uniform condition in both volumes. Assuming no heat transfer, find the initial mass in B, the volume of tank A, the final pressure and temperature, and the work, $_1W_2$.

5.127E Oxygen at 50 lbf/in.^{2}, 200 F is in a piston/cylinder arrangement with a volume of 4 ft^3. It is now compressed in a polytropic process with exponent, $n = 1.2$, to a final temperature of 400 F. Calculate the heat transfer for the process.

5.128E A piston/cylinder contains 4 lbm of air at 100 F, 2 atm, as shown in Fig. P5.75. The piston is loaded with a linear spring, mass, and the atmosphere. Stops are mounted so that $V_{stop} = 100$ ft^3, at which point $P = 6$ atm is required to balance the piston forces. The air is now heated to 60 lbf/in.^{2}. Find the final temperature and volume, and the work and heat transfer. Find the work done on the spring.

5.129E An air pistol contains compressed air in a small cylinder, as shown in Fig. P5.77. Assume that the volume is 1 in.^{3}, pressure is 10 atm, and the temperature is 80 F when armed. A bullet, $m = 0.04$ lbm, acts as a piston initially held by a pin (trigger); when released, the air expands in an isothermal process ($T = $ constant). If the air pressure is 1 atm in the cylinder as the bullet leaves the gun, find

a. the final volume and the mass of air.

b. the work done by the air and work done on the atmosphere.

c. the work to the bullet and the bullet exit velocity.

5.130E A 30-ft-high cylinder, cross-sectional area 1 ft^2, has a massless piston at the bottom with water at 70 F on top of it, as shown in Fig. P5.79. Air at 540 R, volume 10 ft^3 under the piston is heated so that the piston moves up, spilling the water out over the side. Find the total heat transfer to the air when all the water has been pushed out.

5.131E A cylinder fitted with a frictionless piston contains R-134a at 100 F, 80% quality, at which point the volume is 3 gal. The external force on the piston is now varied in such a manner that the R-134a slowly expands in a polytropic process to 50 lbf/in.^{2}, 80 F. Calculate the work and the heat transfer for this process.

5.132E A piston/cylinder contains argon at 20 lbf/in.^{2}, 60 F, and the volume is 4 ft^3. The gas is compressed in a polytropic process to 100 lbf/in.^{2}, 550 F. Calculate the heat transfer during the process.

5.133E Water at 300 F, quality 50% is contained in a cylinder/piston arrangement with initial volume 2 ft^3. The loading of the piston is such that the inside pressure is linear with the square root of volume as $P = 14.7 + CV^{0.5}$ lbf/in.^{2}. Now heat is transferred to the cylinder to a final pressure of 90 lbf/in.^{2}. Find the heat transfer in the process.

5.134E A closed cylinder is divided into two rooms by a frictionless piston held in place by a pin, as shown in Fig. P5.85. Room A has 0.3 ft^3 air at 14.7 lbf/in.^{2}, 90 F, and room B has 10 ft^3 saturated water vapor at 90 F. The pin is pulled, releasing the piston, and both rooms come to equilibrium at 90 F. Considering a control mass of the air and water, determine the work done by the system and the heat transfer to the cylinder.

5.135E A small elevator is being designed for a construction site. It is expected to carry four 150-lbm workers to the top of a 300-ft-tall building in less than 2 min. The elevator cage will have a counterweight to balance its mass. What is the smallest size (power) electric motor that can drive this unit?

5.136E A computer in a closed room of volume 5000 ft^3 dissipates energy at a rate of 10 hp. The room has 100 lbm of wood, 50 lbm of steel, and air, with all material at 540 R, 1 atm. Assuming all the mass heats up uniformly, how long time will it take to increase the temperature 20 F?

5.137 Use the supplied software to track the process in Problem 5.12 in steps of 10°C until the two-phase region is reached, after that step with jumps of 5% in the quality. At each step write out T, x, and the heat transfer to reach that state from the initial state.

5.138 For one of the substances in Table A.6, compare the enthalpy change between any two temperatures, T_1 and T_2, as calculated by integrating the specific heat equation; by assuming constant specific heat at the average temperature; and by assuming constant specific heat at temperature T_1.

5.139 Track the process described in Problem 5.27 so that you can sketch the amount of heat transfer added and the work given out as a function of the volume.

5.140 Using states with given (P,v) and properties from the supplied software, track the process in Problem 5.40. Select five pressures away from the initial toward the final pressure so you can plot the temperature, the heat added, and the work given out as a function of the volume.

5.141 Examine the sensitivity of the final pressure to the containment room volume in Problems 5.41 and 5.42. Solve for the volume for a range of final pressures, 100–250 kPa and sketch the pressure versus volume curve.

5.142 Write a program to solve Problem 5.59 for a range of initial velocities. Let the car mass and final velocity be input variables.

5.143 Write a program for Problem 5.66, where the initial state, the volume ratio, and the polytropic exponent are input variables. To simplify the formulation, use constant specific heat.

5.144 Consider a general version of Problem 5.72 with a substance listed in Table A.6. Write a program where the initial temperature and pressure, and the final temperature are program inputs.

5.145 Examine a process where air at 300 K, 100 kPa is compressed in a piston/cylinder arrangement to 600 kPa. Assume the process is polytropic with exponents in the 1.2–1.6 range. Find the work and heat transfer per unit mass of air. Discuss the different cases and how they may be accomplished by insulating the cylinder or by providing heating or cooling.

5.146 A cylindrical tank of height 2 m with a cross-sectional area of 0.5 m² contains hot water at 80°C, 125 kPa. It is in a room with temperature $T_o = 20°C$, so it slowly loses energy to the room air proportional to the temperature difference as

$$\dot{Q}_{loss} = C\,A(T-T_o)$$

with the tank surface area, A, and C is a constant. For different values of the constant C, estimate the time it takes to bring the water to 50°C. Make enough simplifying assumptions so that you can solve the problem mathematically, that is find a formula for $T(t)$.

6 FIRST LAW ANALYSIS FOR A CONTROL VOLUME

In the preceding chapter we developed the first-law analysis (energy balance) for a control mass going through a process. Many applications in thermodynamics do not readily lend themselves to a control mass approach, but are conveniently handled by the more general control volume technique, as discussed in Chapter 2. The present chapter is concerned with development of the control volume forms of the conservation of mass and energy in situations where there are flows of substance present.

6.1 CONSERVATION OF MASS AND THE CONTROL VOLUME

A control volume is a volume in space in which one has interest for a particular study or analysis. The surface of this control volume is referred to as a control surface and always consists of a closed surface. The size and shape of the control volume are completely arbitrary, and are so defined as to best suit the analysis to be made. The surface may be fixed, or it may move so it expands or contracts. However, the surface must be defined relative to some coordinate system. In some analyses it may be desirable to consider a rotating or moving coordinate system, and to describe the position of the control surface relative to such a coordinate system.

Mass as well as heat and work can cross the control surface and the mass in the control volume, as well as the properties of this mass, can change with time. Figure 6.1 shows a schematic diagram of a control volume that includes heat transfer, shaft work, moving boundary work, accumulation of mass within the control volume and several mass flows. It is important to identify and label each flow of mass and energy and the parts of the control volume that can store (accumulate) mass.

Let us consider the law of the conservation of mass as it relates to the control volume. The physical law concerning mass, recalling Section 5.9, says that we cannot create or destroy mass. We will express this law in a mathematical statement about the mass in the control volume. To do this we must consider all the mass flows into and out of the control volume and the net increase of mass within the control volume. As a somewhat simpler control volume we consider a tank with a cylinder and piston and two pipes attached as shown in Fig. 6.2. The rate of change of mass inside the control volume can be different from zero if we add or take a flow of mass out as

$$\text{Rate of change} = +\text{in} -\text{out}$$

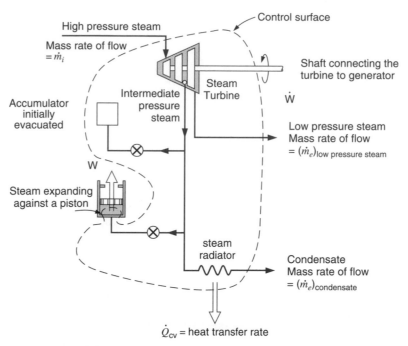

FIGURE 6.1 Schematic diagram of a control volume showing mass and energy transfers and accumulation.

With several possible flows this is written as

$$\frac{dm_{\text{c.v.}}}{dt} = \sum \dot{m}_i - \sum \dot{m}_e \tag{6.1}$$

stating that if the mass inside the control changes with time it is because we add some mass or take some mass out. There are no other means by which the mass inside the control volume could change. Equation 6.1 expressing the conservation of mass is commonly termed the *continuity equation*. While this form of the equation is sufficient for the ma-

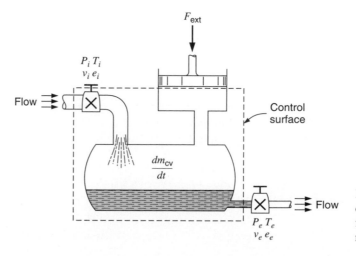

FIGURE 6.2 Schematic diagram of a control volume for the analysis of the continuity equation.

FIGURE 6.3 The flow across a control volume surface, with a flow cross-sectional area of A. Left of valve shown as an average velocity and to the right of valve shown as a distributed flow across area.

jority of applications in thermodynamics, it is frequently rewritten in terms of the local fluid properties in the study of fluid mechanics and heat transfer. In this text we are mainly concerned with the overall mass balance and thus consider Eq. 6.1 as the general expression for the continuity equation.

Since Eq. 6.1 is written for the total mass (lumped form) inside the control volume we may have to consider several contributions to the mass as

$$m_{c.v.} = \int \rho dV = \int (1/v)dV = m_A + m_B + m_C + \cdots$$

Such a summation is needed when the control volume has several accumulation units with different states of the mass.

Let us now consider the mass flow rates across the control volume surface in a little more detail. For simplicity we assume the fluid is flowing in a pipe or duct as illustrated in Fig. 6.3. We wish to relate the total flow rate that appears in Eq. 6.1 to the local properties of the fluid state. The flow across the control volume surface can be indicated with an average velocity shown to the left of the valve or with a distributed velocity over the cross section as shown to the right of the valve.

The volume flow rate is

$$\dot{V} = \mathbf{V}A = \int \mathbf{V}_{local} dA \tag{6.2}$$

so the mass flow rate becomes

$$\dot{m} = \rho_{avg}\dot{V} = \dot{V}/v = \int (\mathbf{V}_{local}/v)dA = \mathbf{V}A/v \tag{6.3}$$

where often the average velocity is used. It should be noted that this result, Eq. 6.3, has been developed for a stationary control surface and we tacitly assumed the flow was normal to the surface. This expression for the mass flow rate applies to any of the various flow streams entering or leaving the control volume, subject to the assumptions mentioned.

EXAMPLE 6.1   Air is flowing in a 0.2-m-diameter pipe at a uniform velocity of 0.1 m/s. The temperature is 25°C and the pressure 150 kPa. Determine the mass flow rate.

Solution

From Eq. 6.3 the mass flow rate is

$$\dot{m} = \mathbf{V}A/v$$

For air, using R from Table A.5, we have

$$v = \frac{RT}{P} = \frac{0.287 \times 298.2}{150} = 0.5705 \text{ m}^3/\text{kg}$$

The cross-sectional area is

$$A = \frac{\pi}{4}(0.2)^2 = 0.0314 \text{ m}^2$$

Therefore,

$$\dot{m} = \mathbf{V}A/v = 0.1 \times 0.0314/0.5705 = 0.0055 \text{ kg/s}$$

6.2 THE FIRST LAW OF THERMODYNAMICS FOR A CONTROL VOLUME

We have already considered the first law of thermodynamics for a control mass, which consists of a fixed quantity of mass, and noted, Eq. 5.5, that it may be written

$$E_2 - E_1 = {}_1Q_2 - {}_1W_2$$

We have also noted that this may be written as an instantaneous rate equation as

$$\frac{dE_{\text{c.m.}}}{dt} = \dot{Q} - \dot{W} \tag{6.4}$$

To write the first law as a rate equation for a control volume, we proceed in a manner analogous to that used in developing a rate equation for the law of conservation of mass. For this purpose a control volume is shown in Fig. 6.4 that involves rate of heat transfer, rates of work and mass flows. The fundamental physical law states that we can not create or destroy energy such that any rate of change of energy must be caused by rates of energy into or out of the control volume. We have already included rates of heat

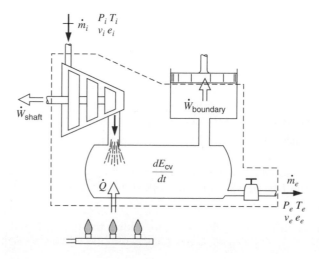

FIGURE 6.4 Schematic diagram to illustrate terms in the energy equation for a general control volume.

transfer and work in Eq. 6.4, so the additional explanation we need are associated with the mass flow rates.

The fluid flowing across the control surface enters or leaves with an amount of energy per unit mass as

$$e = u + \tfrac{1}{2}\mathbf{V}^2 + gZ$$

relating to the state and position of the fluid. Whenever a fluid mass enters a control volume at state i, or exits at state e, there is a boundary movement work associated with that process.

To explain this in more detail consider an amount of mass flowing into the control volume. As it flows in there is a pressure at its back surface so as that surface moves into the control volume it is being pushed by the mass behind it, which is the surroundings. The net effect is that after the mass has entered the control volume the surroundings have pushed it in against the local pressure with a velocity giving it a rate of work in the process. Similarly a fluid exiting the control volume at state e must push the surrounding fluid ahead of it doing work on it, which is work leaving the control volume. The velocity and the area correspond to a certain volume per unit time entering the control volume, enabling us to relate that to the mass flow rate and the specific volume at the state of the mass going in. Now we are able to express the rate of flow work as

$$\dot{W}_{flow} = F\mathbf{V} = \int P\mathbf{V}\,dA = P\dot{V} = Pv\dot{m} \tag{6.5}$$

For the flow that leaves the control volume work is being done by the control volume, $P_e v_e \dot{m}_e$, and for the mass that enters, the surroundings do the rate of work, $P_i v_i \dot{m}_i$. The flow work per unit mass is then Pv, and the total energy associated with the flow of mass is

$$e + Pv = u + Pv + \tfrac{1}{2}\mathbf{V}^2 + gZ = h + \tfrac{1}{2}\mathbf{V}^2 + gZ \tag{6.6}$$

In this equation we have used the definition of the thermodynamic property enthalpy, and it is the appearance of the combination $(u + Pv)$ for the energy in connection with a mass flow that is the primary reason for the definition of the property enthalpy. Its introduction earlier in conjunction with the constant-pressure process was to facilitate use of the tables of thermodynamic properties at that time.

EXAMPLE 6.2

Assume we are standing next to the local city's main water line. The liquid water inside flows at a pressure of say 600 kPa (6 atm) with a temperature of about 10°C. We want to add a smaller amount, 1 kg, of liquid to the line through a side pipe and valve mounted on the main line. How much work will be involved in this process?

If the 1 kg liquid water is in a bucket and we open for the valve to the water main trying to pour it down into the pipe opening we will realize that the water flows the other way. The water will flow from a higher to a lower pressure, i.e., from inside the main line to the atmosphere (from 600 kPa to 101 kPa).

We must take the 1 kg liquid water and put it into a piston cylinder (like a handheld pump) and attach the cylinder to the water pipe. Now we can press on the piston until the water pressure inside is 600 kPa and then open the valve to the main line and slowly squeeze the 1 kg of water in. The work at the piston surface to the water is

$$W = \int P\,dV = P_{water}\,mv = 600 \times 1 \times 0.001 = 0.6\ \text{kJ}$$

and this is the necessary flow work for adding the 1 kg of liquid.

The extension of the first law of thermodynamics from Eq. 6.4 becomes

$$\frac{dE_{\text{c.v.}}}{dt} = \dot{Q}_{\text{c.v.}} - \dot{W}_{\text{c.v.}} + \dot{m}_i e_i - \dot{m}_e e_e + \dot{W}_{\text{flow in}}$$

and the substitution of Eq. 6.5 gives

$$\frac{dE_{\text{c.v.}}}{dt} = \dot{Q}_{\text{c.v.}} - \dot{W}_{\text{c.v.}} + \dot{m}_i\left(e_i + P_i v_i\right) - \dot{m}_e\left(e_e + P_e v_e\right)$$

$$= \dot{Q}_{\text{c.v.}} - \dot{W}_{\text{c.v.}} + \dot{m}_i\left(h_i + \tfrac{1}{2}\mathbf{V}_i^2 + g Z_i\right) - \dot{m}_e\left(h_e + \tfrac{1}{2}\mathbf{V}_e^2 + g Z_e\right)$$

In this form of the energy equation the rate of work term is the sum of all shaft work terms and boundary work terms and any other types of work given out by the control volume; however, the flow work is now listed separately and included with the mass flow rate terms.

For the general control volume we may have several entering or leaving mass flow rates so a summation over those terms is often needed. The final form of the first law of thermodynamics then becomes

$$\frac{dE_{\text{c.v.}}}{dt} = \dot{Q}_{\text{c.v.}} - \dot{W}_{\text{c.v.}} + \sum \dot{m}_i\left(h_i + \tfrac{1}{2}\mathbf{V}_i^2 + g Z_i\right) - \sum \dot{m}_e\left(h_e + \tfrac{1}{2}\mathbf{V}_e^2 + g Z_e\right) \tag{6.7}$$

expressing that the rate of change of energy inside the control volume is due to a net rate of heat transfer, a net rate of work (measured positive out), and the summation of energy fluxes due to mass flows into and out of the control volume. As it was mentioned for the conservation of mass, this equation is written for the total control volume and is therefore in the lumped or integral form, where

$$E_{\text{c.v.}} = \int \rho e \, dV = me = m_A e_A + m_B e_B + m_C e_C + \cdots$$

As the kinetic and potential energy terms per unit mass appear together with the enthalpy in all the flow terms a shorter notation is often used

$$h_{\text{tot}} = h + \tfrac{1}{2}\mathbf{V}^2 + g Z$$

$$h_{\text{stag}} = h + \tfrac{1}{2}\mathbf{V}^2$$

defining the total enthalpy and the stagnation enthalpy (used in fluid mechanics). The shorter equation then becomes

$$\frac{dE_{\text{c.v.}}}{dt} = \dot{Q}_{\text{c.v.}} - \dot{W}_{\text{c.v.}} + \sum \dot{m}_i h_{tot,i} - \sum \dot{m}_e h_{tot,e} \tag{6.8}$$

giving the general energy equation on a rate form. All applications of the energy equation starts with the form in Eq. 6.8, and for special cases this will result in a slightly simpler form as shown in the subsequent sections.

6.3 THE STEADY-STATE, STEADY-FLOW PROCESS (SSSF)

Our first application of the control volume equations will be to develop a suitable analytical model for the long-term steady operation of devices such as turbines, compressors, nozzles, boilers, condensers—a very large class of problems of interest in thermodynamic analysis. This model will not include the short-term transient start-up or shutdown of such devices, but only the steady operating period of time.

Let us consider a certain set of assumptions (beyond those leading to Eqs. 6.1 and 6.7) that lead to a reasonable model for this type of process, which we refer to as the steady-state, steady-flow process. For convenience, we often refer to this process as the SSSF process.

1. The control volume does not move relative to the coordinate frame.
2. The state of the mass at each point in the control volume does not vary with time.
3. As for the mass that flows across the control surface, the mass flux and the state of this mass at each discrete area of flow on the control surface do not vary with time. The rates at which heat and work cross the control surface remain constant.

As an example of a steady-state, steady-flow process consider a centrifugal air compressor that operates with constant mass rate of flow into and out of the compressor, constant properties at each point across the inlet and exit ducts, a constant rate of heat transfer to the surroundings, and a constant power input. At each point in the compressor the properties are constant with time, even though the properties of a given elemental mass of air vary as it flows through the compressor. Usually, such a process is referred to simply as a steady-flow process, since we are concerned primarily with the properties of the fluid entering and leaving the control volume. On the other hand, in the analysis of certain heat transfer problems in which the same assumptions apply, we are primarily interested in the spatial distribution of properties, particularly temperature, and such a process is often referred to as a steady-state process. Since this is an introductory book we use the term steady-state, steady-flow process to emphasize the basic assumptions involved. The student should realize that the terms steady-state process and steady-flow process are both used extensively in the literature.

Let us now consider the significance of each of these assumptions for the SSSF process.

1. The assumption that the control volume does not move relative to the coordinate frame means that all velocities measured relative to the coordinate frame are also velocities relative to the control surface, and there is no work associated with the acceleration of the control volume.
2. The assumption that the state of the mass at each point in the control volume does not vary with time requires that

$$\frac{dm_{c.v.}}{dt} = 0$$

and also

$$\frac{dE_{c.v.}}{dt} = 0$$

Therefore, we conclude that for the SSSF process we can write, from Eqs. 6.1 and 6.7,

Continuity equation:

$$\sum \dot{m}_i = \sum \dot{m}_e \tag{6.9}$$

First law:

$$\dot{Q}_{c.v.} + \sum \dot{m}_i \left(h_i + \frac{\mathbf{V}_i^2}{2} + gZ_i \right) = \sum \dot{m}_e \left(h_e + \frac{\mathbf{V}_e^2}{2} + gZ_e \right) + \dot{W}_{c.v.} \tag{6.10}$$

3. The assumption that the various mass flows, states, and rates at which heat and work cross the control surface remain constant requires that every quantity in Eqs. 6.9 and 6.10 is steady with time. This means that application of Eqs. 6.9 and 6.10 to the operation of some device is independent of time.

Many of the applications of the SSSF model are such that there is only one flow stream entering and one leaving the control volume. For this type of process, we can write

Continuity equation:

$$\dot{m}_i = \dot{m}_e = \dot{m} \tag{6.11}$$

First law:

$$\dot{Q}_{c.v.} + \dot{m} \left(h_i + \frac{\mathbf{V}_i^2}{2} + gZ_i \right) = \dot{m} \left(h_e + \frac{\mathbf{V}_e^2}{2} + gZ_e \right) + \dot{W}_{c.v.} \tag{6.12}$$

Rearranging this equation, we have

$$q + h_i + \frac{\mathbf{V}_i^2}{2} + gZ_i = h_e + \frac{\mathbf{V}_e^2}{2} + gZ_e + w \tag{6.13}$$

where, by definition,

$$q = \frac{\dot{Q}_{c.v.}}{\dot{m}} \quad \text{and} \quad w = \frac{\dot{W}_{c.v.}}{\dot{m}} \tag{6.14}$$

Note that the units for q and w are kJ/kg. From their definition, q and w can be thought of as the heat transfer and work (other than flow work) per unit mass flowing into and out of the control volume for this particular SSSF process.

The symbols q and w are also used for the heat transfer and work per unit mass of a control mass. However, since it is always evident from the context whether it is a control mass (fixed mass) or control volume (involving a flow of mass) with which we are concerned, the significance of the symbols q and w will also be readily evident in each situation.

The SSSF process is often used in the analysis of reciprocating machines, such as reciprocating compressors or engines. In this case the rate of flow, which may actually be pulsating, is considered to be the average rate of flow for an integral number of cycles. A similar assumption is made regarding the properties of the fluid flowing across the control surface, and the heat transfer and work crossing the control surface. It is also assumed that for an integral number of cycles the reciprocating device undergoes, the energy and mass within the control volume do not change.

A number of examples are now given to illustrate the analysis of SSSF processes.

6.4 EXAMPLES OF STEADY-STATE, STEADY-FLOW PROCESSES

In this section, we consider a number of examples of SSSF processes in which there is one fluid stream entering and one leaving the control volume, such that the first law can be written in the form of Eq. 6.13. Some may instead utilize control volumes that include

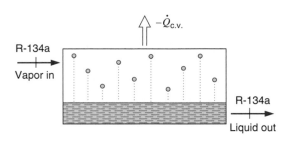

FIGURE 6.5 A refrigeration system condenser.

more than one fluid stream, such that it is necessary to write the first law in the more general form of Eq. 6.10.

Heat Exchanger

A SSSF heat exchanger is a simple fluid flow through a pipe or system of pipes, where heat is transferred to or from the fluid. The fluid may be heated or cooled, and may or may not boil, liquid to vapor, or condense, vapor to liquid. One such example is the condenser in a R-134a refrigeration system, as shown in Fig. 6.5. Superheated vapor enters the condenser, and liquid exits. The process tends to occur at constant pressure, since a fluid flowing in a pipe usually undergoes only a small pressure drop, because of fluid friction at the walls. The pressure drop may or may not be taken into account in a particular analysis. There is no means for doing any work (shaft work, electrical work, etc), and changes in kinetic and potential energies are commonly negligibly small. (One exception may be a boiler tube in which liquid enters and vapor exits at a much larger specific volume. In such a case, it may be necessary to check the exit velocity using Eq. 6.3.) The heat transfer in most heat exchangers is then found from Eq. 6.13 as the change in enthalpy of the fluid. In the condenser shown in Fig. 6.5, the heat transfer out of the condenser then goes to whatever is receiving it, perhaps a stream of air or of cooling water. It is often simpler to write the first law around the entire heat exchanger, including both flow streams, in which case there is little or no heat transfer with the surroundings. Such a situation is the subject of the following example.

EXAMPLE 6.3   Consider a water-cooled condenser in a large refrigeration system in which R-134a is the refrigerant fluid. The refrigerant enters the condenser at 1.0 MPa, 60°C, at the rate of 0.2 kg/s, and exits as a liquid at 0.95 MPa, 35°C. Cooling water enters the condenser at 10°C and exits at 20°C. Determine the rate at which cooling water flows through the condenser.

Control volume: Condenser.

Sketch: Fig. 6.6

Inlet states: R-134a—fixed; water—fixed.

Exit states: R-134a—fixed; water—fixed.

Process: SSSF.

Model: R-134a tables; steam tables.

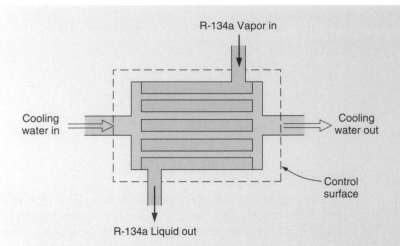

FIGURE 6.6
Schematic diagram of a
R-134a condenser.

Analysis:

With this control volume we have two fluid streams, the R-134a and the water, entering and leaving the control volume. It is reasonable to assume that the kinetic and potential energy changes are negligible. We note that the work is zero, and we make the other reasonable assumption that there is no heat transfer across the control surface. Therefore, the first law, Eq. 6.10, reduces to

$$\sum \dot{m}_i h_i = \sum \dot{m}_e h_e$$

Using the subscript r for refrigerant and w for water,

$$\dot{m}_r(h_i)_r + \dot{m}_w(h_i)_w = \dot{m}_r(h_e)_r + \dot{m}_w(h_e)_w$$

Solution

From the R-134a and steam tables,

$$(h_i)_r = 441.89 \text{ kJ/kg} \qquad (h_i)_w = 42.00 \text{ kJ/kg}$$

$$(h_e)_r = 249.10 \text{ kJ/kg} \qquad (h_e)_w = 83.95 \text{ kJ/kg}$$

Solving the above equation for $\dot{m}_w$, the rate of flow of water,

$$\dot{m}_w = \dot{m}_r \frac{(h_i - h_e)_r}{(h_e - h_i)_w} = 0.2 \text{ kg/s} \frac{(441.89 - 249.10) \text{ kJ/kg}}{(83.95 - 42.00) \text{ kJ/kg}} = 0.919 \text{ kg/s}$$

This problem can also be solved by considering two separate control volumes, one of which has the flow of R-134a across its control surface, and the other having the flow of water across its control surface. Further there is heat transfer from one control volume to the other.

The heat transfer for the control volume involving R-134a is calculated first. In this case the SSSF energy equation, Eq. 6.10, reduces to

$$\dot{Q}_{c.v.} = \dot{m}_r(h_e - h_i)_r$$

$$\dot{Q}_{c.v.} = 0.2 \text{ kg/s} \times (249.10 - 441.89) \text{ kJ/kg} \ -38.558 \text{ kW}$$

This is also the heat transfer to the other control volume, for which $\dot{Q}_{c.v.} = +38.558$ kW.

$$\dot{Q}_{c.v.} = \dot{m}_w (h_e - h_i)_w$$

$$\dot{m}_w = \frac{38.558 \text{ kW}}{(83.95 - 42.00)\text{kJ}/\text{kg}} = 0.919 \text{ kg}/\text{s}$$

Nozzle

A nozzle is a SSSF device whose purpose is to create a high-velocity fluid stream at the expense of its pressure. It is contoured in an appropriate manner to expand a flowing fluid smoothly to a lower pressure, thereby increasing its velocity. There is no means to do any work—there are no moving parts. There is little or no change in potential energy, and usually little or no heat transfer. An exception is the large nozzle on a liquid-propellant rocket, such as was described in Section 1.7, in which the cold propellant is commonly circulated around the outside of the nozzle walls before going to the combustion chamber, in order to keep the nozzle from melting. This case, a nozzle with significant heat transfer, is the exception, and would be noted in such an application. In addition, the kinetic energy of the fluid at the nozzle inlet is usually small, and would be neglected if its value is not known.

EXAMPLE 6.4

Steam at 0.6 MPa, 200°C enters an insulated nozzle with a velocity of 50 m/s. It leaves at a pressure of 0.15 MPa and a velocity of 600 m/s. Determine the final temperature if the steam is superheated in the final state, and the quality if it is saturated.

Control volume: Nozzle.
Inlet state: Fixed (see Fig. 6.7).
Exit state: P_e known.
Process: SSSF.
Model: Steam tables.

Analysis:

$$\dot{Q}_{c.v.} = 0 \text{ (nozzle insulated)}$$

$$\dot{W}_{c.v.} = 0 \qquad \text{PE}_i \approx \text{PE}_e$$

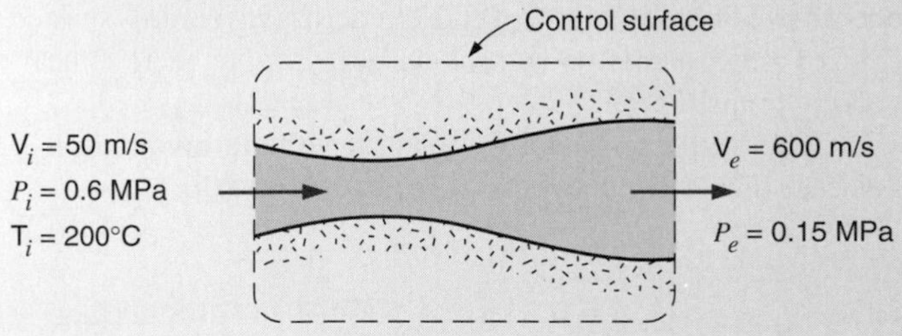

$V_i = 50$ m/s
$P_i = 0.6$ MPa
$T_i = 200$°C

Control surface

$V_e = 600$ m/s

$P_e = 0.15$ MPa

FIGURE 6.7
Illustration for Example 6.4

First law (Eq. 6.13):

$$h_i + \frac{\mathbf{V}_i^2}{2} = h_e + \frac{\mathbf{V}_e^2}{2}$$

Solution

$$h_e = 2850.1 + \frac{(50)^2}{2 \times 1000} - \frac{(600)^2}{2 \times 1000} = 2671.4 \text{ kJ / kg}$$

The two properties of the fluid leaving that we now know are pressure and enthalpy, and therefore the state of this fluid is determined. Since h_e is less than h_g at 0.15 MPa, the quality is calculated.

$$h = h_f + x h_{fg}$$
$$2671.4 = 467.1 + x_e\,2226.5$$
$$x_e = 0.99$$

EXAMPLE 6.4E   Steam at 100 lbf/in.^{2}, 400 F, enters an insulated nozzle with a velocity of 200 ft/s. It leaves at a pressure of 20 lbf/in.^{2} and a velocity of 2000 ft/s. Determine the final temperature if the steam is superheated in the final state, and the quality if it is saturated.

> *Control volume:* Nozzle.
> *Inlet state:* Fixed (see Fig. 6.7E).
> *Exit state:* P_e known.
> *Process:* SSSF.
> *Model:* Steam tables.

Analysis:

$$\dot{Q}_{\text{c.v.}} = 0 \text{ (nozzle insulated)}$$
$$\dot{W}_{\text{c.v.}} = 0 \qquad \text{PE}_i = \text{PE}_e$$

First law (Eq. 6.13):

$$h_i + \frac{\mathbf{V}_i^2}{2} = h_e + \frac{\mathbf{V}_e^2}{2}$$

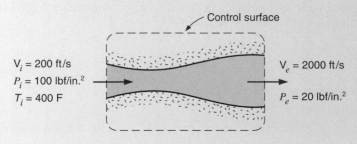

$\text{V}_i = 200 \text{ ft/s}$
$P_i = 100 \text{ lbf/in.}^2$
$T_i = 400 \text{ F}$

Control surface

$\text{V}_e = 2000 \text{ ft/s}$

$P_e = 20 \text{ lbf/in.}^2$

FIGURE 6.7E
Illustration for Example 6.4E.

Solution

$$h_e = 1227.5 + \frac{(200)^2}{2 \times 32.17 \times 778} - \frac{(2000)^2}{2 \times 32.17 \times 778} = 1148.3 \text{ Btu / lbm}$$

The two properties of the fluid leaving that we now know are pressure and enthalpy, and therefore the state of this fluid is determined. Since h_e is less than h_g at 20 lbf/in.^{2}, the quality is calculated.

$$h = h_f + x h_{fg}$$

$$1148.3 = 196.26 + x_e\, 960.1$$

$$x_e = 0.992$$

Diffuser

A SSSF diffuser is a device constructed to decelerate a high-velocity fluid in a manner that results in an increase in pressure of the fluid. In essence, it is the exact opposite of a nozzle, and may be thought of as a fluid flowing in the opposite direction through a nozzle, with the opposite effects. The assumptions are similar to those for a nozzle, with a large kinetic energy at the diffuser inlet, and a small, but usually not negligible, kinetic energy at the exit being the only terms besides the enthalpies remaining in the first law, Eq. 6.13.

Throttle

A throttling process occurs when a fluid flowing in a line suddenly encounters a restriction in the flow passage. This may be a plate with a small hole in it, as shown in Fig. 6.8, it may be a partially closed valve protruding into the flow passage, or it may be a change to a much smaller diameter tube, called a *capillary* tube, which is normally found on a refrigerator. The result of this restriction is an abrupt pressure drop in the fluid, as it is forced to find its way through a suddenly smaller passageway. This process is drastically unlike the smoothly contoured nozzle expansion and area change, which results in a significant velocity increase. There is typically some increase in velocity in a throttle, but both inlet and exit kinetic energies are usually small enough to be neglected. There is no means for doing work, and little or no change in potential energy. Usually, there is not time or opportunity for an appreciable heat transfer, such that the only terms left in the first law, Eq. 6.13, are the inlet and exit enthalpies. We conclude that a SSSF throttling process is approximately a pressure drop at constant enthalpy, and will assume this to be the case unless otherwise noted.

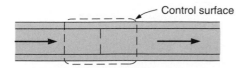

FIGURE 6.8 The throttling process.

Assuming a throttle to be a constant-enthalpy process leads us to define a property called the *Joule-Thomson coefficient, μ_J,* as

$$\mu_J \equiv \left(\frac{\partial T}{\partial P} \right)_h$$

A positive Joule-Thomson means that the fluid temperature drops during throttling, while a negative Joule-Thomson coefficient means that temperature rises during throttling.

EXAMPLE 6.5   Steam at 800 kPa, 300°C is throttled to 200 kPa. Changes in kinetic energy are negligible for this process. Determine the final temperature of the steam, and the average Joule-Thomson coefficient.

Control volume: Throttle valve (or other restriction).

Inlet state: P_i, T_i known; state fixed.

Exit state: P_e known.

Process: SSSF.

Model: Steam tables.

Analysis:

First law, Eq. 6.13: Work equals zero. Neglect heat transfer and changes in kinetic and potential energies. Therefore, the first law reduces to

$$h_i = h_e$$

Solution

Since

$$h_e = h_i = 3056.5 \text{ kJ/kg}$$

and

$$P_e = 200 \text{ kPa}$$

these two properties determine the final state. From the superheat table for steam,

$$T_e = 292.4°C$$

$$\mu_{J(av)} = \left(\frac{\Delta T}{\Delta P} \right)_h = \frac{-7.6}{-600} = 0.0127 \text{ K/kPa}$$

Frequently, a throttling process involves a change in the phase of the fluid. A typical example is the flow through the expansion valve of a vapor-compression refrigeration system. The following example deals with this problem.

EXAMPLE 6.6 Consider the throttling process across the expansion valve or through the capillary tube in a vapor-compression refrigeration cycle. In this process the pressure of the refrigerant drops from the high pressure in the condenser to the low pressure in the evaporator, and during this process some of the liquid flashes into vapor. If we consider this process to be adiabatic, the quality of the refrigerant entering the evaporator can be calculated.

Consider the following process, in which ammonia is the refrigerant. The ammonia enters the expansion valve at a pressure of 1.50 MPa and a temperature of 32°C. Its pressure on leaving the expansion valve is 268 kPa. Calculate the quality of the ammonia leaving the expansion valve.

Control volume: Expansion valve or capillary tube.
Inlet state: P_i, T_i known; state fixed.
Exit state: P_e known.
Process: SSSF.
Model: Ammonia tables.

Analysis:

Standard throttling process analysis and assumptions, as in Example 6.5. The first law reduces to

$$h_i = h_e$$

Solution

From the ammonia tables

$$h_i = 332.6 \text{ kJ/kg}$$

(The enthalpy of a slightly compressed liquid is essentially equal to the enthalpy of saturated liquid at the same temperature.)

$$h_e = h_i = 332.6 = 126.0 + x_e \,(1303.5)$$

$$x_e = 0.1585 = 15.85\%$$

Turbine

A turbine is a rotary SSSF machine whose purpose is the production of shaft work (power, on a rate basis) at the expense of the pressure of the working fluid. Two general classes of turbines are steam (or other working fluid) turbines, in which the steam exiting the turbine passes to a condenser, where it is condensed to liquid, and gas turbines, in which the gas usually exhausts to the atmosphere from the turbine. In either type, the turbine exit pressure is fixed by the environment into which the working fluid exhausts, and the turbine inlet pressure has been reached by previously pumping or compressing the working fluid in another process. Inside the turbine, there are two distinct processes. In the first, the working fluid passes through a set of nozzles, or the equivalent—fixed blade passages contoured to expand the fluid to a lower pressure and to a high velocity. In the second process inside the turbine, this high-velocity fluid stream is directed onto a set of moving (rotating) blades, in which the velocity is reduced before being discharged from the passage. This directed velocity decrease produces a torque on the rotating shaft, resulting in a shaft work output. The low-velocity, low-pressure fluid then exhausts from the turbine.

The first law for this process is either Eq. 6.10 or 6.13. Usually, changes in potential energy are negligible, as is the inlet kinetic energy. Often, the exit kinetic energy is neglected, and any heat rejection from the turbine is undesirable, and is commonly small. We therefore normally assume that a turbine process is adiabatic, and the work output in this case reduces to the decrease in enthalpy from the inlet to exit states. In the following example, however, we include all the terms in the first law and study their relative importance.

EXAMPLE 6.7 The mass rate of flow into a steam turbine is 1.5 kg/s, and the heat transfer from the turbine is 8.5 kW. The following data are known for the steam entering and leaving the turbine.

	Inlet Conditions	Exit Conditions
Pressure	2.0 MPa	0.1 MPa
Temperature	350°C	
Quality		100%
Velocity	50 m/s	100 m/s
Elevation above reference plane	6 m	3 m
$g = 9.8066 \text{ m/s}^2$		

Determine the power output of the turbine.

Control volume: Turbine (Fig. 6.9).

Inlet state: Fixed (above).

Exit state: Fixed (above).

Process: SSSF.

Model: Steam tables.

Analysis:

First law (Eq. 6.12):

$$\dot{Q}_{c.v.} + \dot{m}\left(h_i + \frac{\mathbf{V}_i^2}{2} + gZ_i\right) = \dot{m}\left(h_e + \frac{\mathbf{V}_e^2}{2} + gZ_e\right) + \dot{W}_{c.v.}$$

with

$$\dot{Q}_{c.v.} = -8.5 \text{ kW}$$

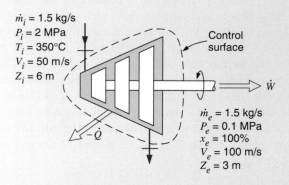

$\dot{m}_i$ = 1.5 kg/s
P_i = 2 MPa
T_i = 350°C
V_i = 50 m/s
Z_i = 6 m

Control
surface

$\dot{W}$

$\dot{m}_e$ = 1.5 kg/s
P_e = 0.1 MPa
x_e = 100%
V_e = 100 m/s
Z_e = 3 m

$-\dot{Q}$

FIGURE 6.9 Illustration for
Example 6.7.

Solution

h_i = 3137.0 kJ/kg (from the steam tables)

$$\frac{V_i^2}{2} = \frac{50 \times 50}{2 \times 1000} = 1.25 \text{ kJ/kg}$$

$$gZ_i = \frac{6 \times 9.8066}{1000} = 0.059 \text{ kJ/kg}$$

Similarly, h_e = 2675.5 kJ/kg

$$\frac{V_e^2}{2} = \frac{100 \times 100}{2 \times 1000} = 5.0 \text{ kJ/kg}$$

$$gZ_e = \frac{3 \times 9.8066}{1000} = 0.029 \text{ kJ/kg}$$

Therefore, substituting into Eq. 6.12,

$$-8.5 + 1.5(3137 + 1.25 + 0.059) = 1.5(2675.5 + 5.0 + 0.029) + \dot{W}_{\text{c.v.}}$$

$$\dot{W}_{\text{c.v.}} = -8.5 + 4707.5 - 4020.8 = 678.2 \text{ kW}$$

If Eq. 6.13 is used, the work per kilogram of fluid flowing is found first.

$$q + h_i + \frac{V_i^2}{2} + gZ_i = h_e + \frac{V_e^2}{2} + gZ_e + w$$

$$q = \frac{-8.5}{1.5} = -5.667 \text{ kJ/kg}$$

Therefore, substituting into Eq. 6.13,

$$-5.667 + 3137 + 1.25 + 0.059 = 2675.5 + 5.0 + 0.029 + w$$

$$w = 452.11 \text{ kJ/kg}$$

$$\dot{W}_{\text{c.v.}} = 1.5 \text{ kg/s} \times 452.11 \text{ kJ/kg} = 678.2 \text{ kW}$$

Two further observations can be made by referring to this example. First, in many engineering problems, potential energy changes are insignificant when compared with the other energy quantities. In the above example the potential energy change did not affect any of the significant figures. In most problems where the change in elevation is small the potential energy terms may be neglected.

Second, if velocities are small, say, under 20 m/s, in many cases the kinetic energy is insignificant compared with other energy quantities. Furthermore, when the velocities entering and leaving the system are essentially the same, the change in kinetic energy is small. Since it is the change in kinetic energy that is important in the SSSF energy equation, the kinetic energy terms can usually be neglected when there is no significant difference between the velocity of the fluid entering and leaving the control volume. Thus, in many thermodynamic problems, one must make judgments as to which quantities may be negligible for a given analysis.

The preceding discussion and example concerned the turbine, which is a rotary work-producing device. There are other nonrotary devices that produce work, which can be called expanders as a general name. In such devices, the first law analysis and assumptions are generally the same as for turbines, except that in a piston/cylinder type expander, there would in most cases be a larger heat loss or rejection during the process.

Compressor/Pump

The purpose of a SSSF compressor (gas) or pump (liquid) is the same, to increase the pressure of a fluid by putting in shaft work (power, on a rate basis). There are two fundamentally different classes of compressors. The most common is a rotary-type compressor (either axial-flow or radial/centrifugal flow), in which the internal processes are essentially the opposite of the two processes occurring inside a turbine. The working fluid enters the compressor at low pressure, moving into a set of rotating blades, from which it exits at high velocity, a result of the shaft work input to the fluid. The fluid then passes through a diffuser section, in which it is decelerated in a manner that results in a pressure increase. The fluid then exits the compressor at high pressure.

The first law for the compressor is either Eq. 6.10 or 6.13. Usually, changes in potential energy are negligible, as is the inlet kinetic energy. Often, the exit kinetic energy is neglected, as well. Heat rejection from the working fluid during compression would be desirable, but is usually small in a rotary compressor, which is a high-volume flowrate machine, and there is not sufficient time to transfer much heat from the working fluid. We therefore normally assume that a rotary compressor process is adiabatic, and the work input in this case reduces to the change in enthalpy from the inlet to exit states.

In a piston/cylinder type compressor, the cylinder usually contains fins to promote heat rejection during compression (or the cylinder may be water-jacketed in a large compressor for even greater cooling rates). In this type of compressor, the heat transfer from the working fluid is significant and is not neglected in the first law. As a general rule, in any example or problem in this text, we will assume that a compressor is adiabatic unless otherwise noted.

EXAMPLE 6.8 The centrifugal air compressor of a gas turbine receives air from the ambient atmosphere where the pressure is 1 bar and the temperature is 300 K. At the discharge of the compressor the pressure is 4 bar, the temperature is 480 K, and the velocity is 100 m/s. The mass rate of flow into the compressor is 15 kg/s. Determine the power required to drive the compressor.

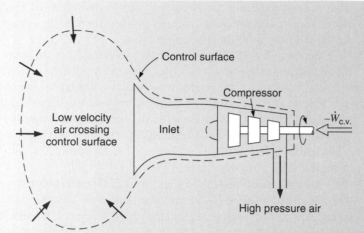

FIGURE 6.10 Sketch for Example 6.8.

Control volume: We consider a control volume around the compressor, and locate the control volume at some distance from the compressor so that the air crossing the control surface has a very low velocity and is essentially at ambient conditions. If we located our control volume directly across the inlet section, it would be necessary to know the temperature and velocity at the compressor inlet.

Sketch: Fig. 6.10.

Inlet and exit states: Both states fixed.

Process: SSSF.

Model: Ideal gas with constant specific heat, value from Table A.5 (300 K).

Analysis:

We assume the compressor to be adiabatic. We also neglect any change in potential energy as well as the inlet kinetic energy. The first law, Eq. 6.13, reduces to

$$h_i = h_e + \frac{\mathbf{V}_e^2}{2} + w$$

Solution

$$-w = h_e - h_i + \frac{\mathbf{V}_e^2}{2} = C_{po}\left(T_e - T_i\right) + \frac{\mathbf{V}_e^2}{2}$$

$$= 1.0035(480 - 300) + \frac{100 \times 100}{2 \times 1000}$$

$$= 180.6 + 5.0 = 185.6 \text{ kJ / kg}$$

$$-\dot{W}_{c.v.} = 15 \times 185.6 = 2784 \text{ kW}$$

A more accurate model for the behavior of the air in this process would be the ideal gas and air tables, A.7. For this model, the solution is

$$h_i = 300.47 \text{ kJ / kg} \qquad h_e = 482.81 \text{ kJ / kg}$$

$$-w = h_e - h_i + \frac{\mathbf{V}_e^2}{2} = 482.81 - 300.47 + \frac{100 \times 100}{2 \times 1000}$$

$$= 182.3 + 5.0 = 187.3 \text{ kJ / kg}$$

$$-\dot{W}_{c.v.} = 15 \times 187.3 = 2810 \text{ kW}$$

EXAMPLE 6.8E   The centrifugal air compressor of a gas turbine receives air from the ambient atmosphere where the pressure is 14.5 lbf/in.^{2} and the temperature is 80F. At the discharge of the compressor the pressure is 54 lbf/in.^{2}, the temperature is 400F, and the velocity is 300 ft/s. The mass rate of flow into the compressor is 2000 lbm/min. Determine the power required to drive the compressor.

Control volume: We consider a control volume around the compressor, and locate the control volume at some distance from the compressor so that the air crossing the control surface has a very low velocity and is essentially at ambient conditions. If we located our control volume directly across the inlet section, it would be necessary to know the temperature and velocity at the compressor inlet.

Sketch: Fig. 6.10

Inlet and exit states: Both states fixed.

Process: SSSF.

Model: Ideal gas with constant specific heat, value from Table C.4 (540 R).

Analysis:

We assume the compressor to be adiabatic. We also neglect any change in potential energy as well as the inlet kinetic energy. The first law, Eq. 6.13, reduces to

$$h_i = h_e + \frac{\mathbf{V}_e^2}{2} + w$$

Solution

$$-w = h_e - h_i + \frac{\mathbf{V}_e^2}{2} = C_{po}(T_e - T_i) + \frac{\mathbf{V}_e^2}{2}$$

$$= 0.24(400 - 80) + \frac{(300)^2}{2 \times 32.17 \times 778}$$

$$= 76.8 + 1.8 = 78.6 \text{ Btu / lbm}$$

$$-\dot{W}_{c.v.} = \frac{78.6(2000)}{2544.4 / 60} = 3708 \text{ hp}$$

A more accurate model for the behavior of the air in this process would be the ideal gas and air tables, C.6. For this model, the solution is

$$h_i = 129.18 \text{ Btu / lbm} \qquad h_e = 206.60 \text{ Btu / lbm}$$

$$-w = h_e - h_i + \frac{\mathbf{V}_e^2}{2}$$

$$= 206.60 - 129.18 + \frac{(300)^2}{2 \times 32.17 \times 778}$$

$$-\dot{W}_{\text{c.v.}} = \frac{79.2(2000)}{2544.4 / 60} = 3736 \text{ hp}$$

Power Plant and Refrigerator

The following examples illustrate the incorporation of several of the devices and machines already discussed in this section into a complete thermodynamic system, which is built for a specific purpose.

EXAMPLE 6.9   Consider the simple steam power plant, as shown in Fig. 6.11. The following data are for such a power plant.

Location	Pressure	Temperature or Quality
Leaving boiler	2.0 MPa	300°C
Entering turbine	1.9 MPa	290°C
Leaving turbine, entering condenser	15 kPa	90%
Leaving condenser, entering pump	14 kPa	45°C
Pump work = 4 kJ/kg		

Determine the following quantities per kilogram flowing through the unit.

1. Heat transfer in line between boiler and turbine.
2. Turbine work.
3. Heat transfer in condenser.
4. Heat transfer in boiler.

There is a certain advantage in assigning a number to various points in the cycle. For this reason the subscripts i and e in the steady-state, steady-flow energy equation are often replaced by appropriate numbers.

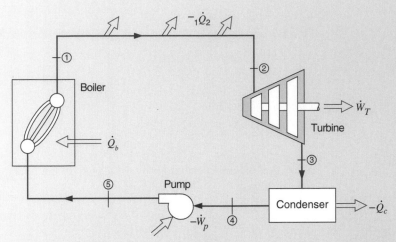

FIGURE 6.11 Simple steam power plant.

Since there are several control volumes to be considered in the solution to this problem, let us consolidate our solution procedure somewhat in this example. Using the notation of Fig. 6.11,

All processes: SSSF.

Model: Steam tables.

From the steam tables:

$$h_1 = 3023.5 \text{ kJ/kg}$$

$$h_2 = 3002.5 \text{ kJ/kg}$$

$$h_3 = 226.0 + 0.9(2373.1) = 2361.8 \text{ kJ/kg}$$

$$h_4 = 188.5 \text{ kJ/kg}$$

All analyses: No changes in kinetic or potential energy will be considered in the solution. In each case, the first law is given by Eq. 6.13.

Now, we proceed to answer the specific questions raised in the problem statement.

1. *Control volume* Pipe line between the boiler and the turbine.

   First law and solution:

   $$_1q_2 + h_1 = h_2$$
   $$_1q_2 = h_2 - h_1 = 3002.5 - 3023.5 = -21.0 \text{ kJ/kg}$$

2. *Control volume* Turbine.

   First law and solution: A turbine is essentially an adiabatic machine. Therefore, it is reasonable to neglect heat transfer, so that

   $$h_2 = h_3 + {_2}w_3$$
   $$_2w_3 = 3002.5 - 2361.8 = 640.7 \text{ kJ/kg}$$

3. *Control volume* Condenser.

First law and solution: There is no work for this control volume. Therefore,

$$_3q_4 + h_3 = h_4$$

$$_3q_4 = 188.5 - 2361.8 = -2173.3 \text{ kJ/kg}$$

4. *Control volume* Boiler.

First law: The work is equal to zero, so that

$$_5q_1 + h_5 = h_1$$

A solution requires a value for h_5, which can be found by taking a control volume around the pump:

$$h_4 = h_5 + _4w_5$$

$$h_5 = 188.5 - (-4) = 192.5 \text{ kJ/kg}$$

Therefore, for the boiler,

$$_5q_1 + h_5 = h_1$$

$$_5q_1 = 3023.5 - 192.5 = 2831 \text{ kJ/kg}$$

EXAMPLE 6.10   The refrigerator shown in Fig. 6.12 uses R-134a as the working fluid. The mass flow rate through each component is 0.1 kg/s, and the power input to the compressor is 5.0 kW. The following state data are known, using the state notation of Fig. 6.12

$$P_1 = 100 \text{ kPa}, \qquad T_1 = -20°\text{C}$$

$$P_2 = 800 \text{ kPa}, \qquad T_2 = 50°\text{C}$$

$$T_3 = 30°\text{C}, \qquad x_3 = 0.0$$

$$T_4 = -25°\text{C}$$

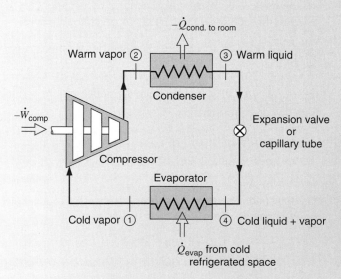

FIGURE 6.12

Determine the following:

1. The quality at the evaporator inlet.
2. The rate of heat transfer to the evaporator.
3. The rate of heat transfer from the compressor.

> *All processes:* SSSF.
>
> *Model:* R-134a.
>
> *All analyses:* No changes in kinetic or potential energy. The first law in each case is given by Eq. 6.10.

Solution

1. *Control volume:* Throttle.
   First law:

$$h_4 = h_3 = 241.8 \text{ kJ/kg}$$

$$h_4 = 241.8 = h_{f4} + x_4 h_{fg4} = 167.4 + x_4 \times 215.6$$

$$x_4 = 0.345$$

2. *Control volume:* Evaporator.
   First law:

$$\dot{Q}_{EVAP} = \dot{m}(h_1 - h_4)$$
$$= 0.1(387.2 - 241.8) = 14.54 \text{ kW}$$

3. *Control volume:* Compressor.
   First law:

$$\dot{Q}_{COMP} = \dot{m}(h_2 - h_1) + \dot{W}_{COMP}$$
$$= 0.1(435.1 - 387.2) - 5.0 = -0.21 \text{ kW}$$

6.5 THE UNIFORM-STATE, UNIFORM-FLOW PROCESS

In Sections 6.3 and 6.4 we considered the steady-state, steady-flow process and several examples of its application. Many processes of interest in thermodynamics involve unsteady flow and do not fit into this category. A certain group of these—for example, filling closed tanks with a gas or liquid, or discharge from closed vessels—can be reasonably represented to a first approximation by another simplified model. We call this process the uniform-state, uniform-flow process, or for convenience, the USUF process. The basic assumptions are as follows:

1. The control volume remains constant relative to the coordinate frame.
2. The state of the mass within the control volume may change with time, but at any instant of time the state is uniform throughout the entire control volume (or over several identifiable regions that make up the entire control volume).

3. The state of the mass crossing each of the areas of flow on the control surface is constant with time although the mass flow rates may be time varying.

Let us examine the consequence of these assumptions and derive an expression for the first law that applies to this process. The assumption that the control volume remains stationary relative to the coordinate frame has already been discussed in Section 6.3. The remaining assumptions lead to the following simplifications for the continuity equation and the first law.

The overall process occurs during time t. At any instant of time during the process, the continuity equation is

$$\frac{dm_{c.v.}}{dt} + \sum \dot{m}_e - \sum \dot{m}_i = 0$$

where the summation is over all areas on the control surface through which flow occurs. Integrating over time t gives the change of mass in the control volume during the overall process.

$$\int_0^t \left(\frac{dm_{c.v.}}{dt} \right) dt = \left(m_2 - m_1 \right)_{c.v.}$$

The total mass leaving the control volume during time t is

$$\int_0^t \left(\sum \dot{m}_e \right) dt = \sum m_e$$

and the total mass entering the control volume during time t is

$$\int_0^t \left(\sum \dot{m}_i \right) dt = \sum m_i$$

Therefore, for this period of time t, we can write the continuity equation for the USUF process as

$$\left(m_2 - m_1 \right)_{c.v.} + \sum m_e - \sum m_i = 0 \tag{6.15}$$

In writing the first law for the USUF process we consider Eq. 6.7, which applies at any instant of time during the process.

$$\dot{Q}_{c.v.} + \sum \dot{m}_i \left(h_i + \frac{\mathbf{V}_i^2}{2} + gZ_i \right) = \frac{dE_{c.v.}}{dt} + \sum \dot{m}_e \left(h_e + \frac{\mathbf{V}_e^2}{2} + gZ_e \right) + \dot{W}_{c.v.}$$

Since at any instant of time the state within the control volume is uniform, the first law for the USUF process becomes

$$\dot{Q}_{c.v.} + \sum \dot{m}_i \left(h_i + \frac{\mathbf{V}_i^2}{2} + gZ_i \right) = \sum \dot{m}_e \left(h_e + \frac{\mathbf{V}_e^2}{2} + gZ_e \right)$$

$$+ \frac{d}{dt} \left[m \left(u + \frac{\mathbf{V}^2}{2} + gZ \right) \right]_{c.v.} + \dot{W}_{c.v.}$$

Let us now integrate this equation over time t, during which time we have

$$\int_0^t \dot{Q}_{c.v.}\,dt = Q_{c.v.}$$

$$\int_0^t \left[\sum \dot{m}_i \left(h_i + \frac{\mathbf{V}_i^2}{2} + gZ_i \right) \right] dt = \sum m_i \left(h_i + \frac{\mathbf{V}_i^2}{2} + gZ_i \right)$$

$$\int_0^t \left[\sum \dot{m}_e \left(h_e + \frac{\mathbf{V}_e^2}{2} + gZ_e \right) \right] dt = \sum m_e \left(h_e + \frac{\mathbf{V}_e^2}{2} + gZ_e \right)$$

$$\int_0^t \dot{W}_{c.v.}\,dt = W_{c.v.}$$

$$\int_0^t \frac{d}{dt}\left[m\left(u + \frac{\mathbf{V}^2}{2} + gZ \right) \right]_{c.v.} dt = \left[m_2\left(u_2 + \frac{\mathbf{V}_2^2}{2} + gZ_2 \right) - m_1\left(u_1 + \frac{\mathbf{V}_1^2}{2} + gZ_1 \right) \right]_{c.v.}$$

Therefore, for this period of time t, we can write the first law for the uniform-state, uniform-flow process as

$$Q_{c.v.} + \sum m_i \left(h_i + \frac{\mathbf{V}_i^2}{2} + gZ_i \right)$$

$$= \sum m_e \left(h_e + \frac{\mathbf{V}_e^2}{2} + gZ_e \right)$$

$$+ \left[m_2\left(u_2 + \frac{\mathbf{V}_2^2}{2} + gZ_2 \right) - m_1\left(u_1 + \frac{\mathbf{V}_1^2}{2} + gZ_1 \right) \right]_{c.v.} + W_{c.v.} \qquad (6.16)$$

As an example of the type of problem for which these assumptions are valid and Eq. 6.16 is appropriate, let us consider the classic problem of flow into an evacuated vessel. This is the subject of Example 6.11.

EXAMPLE 6.11 Steam at a pressure of 1.4 MPa, 300°C is flowing in a pipe, Fig. 6.13. Connected to this pipe through a valve is an evacuated tank. The valve is opened and the tank fills with steam until the pressure is 1.4 MPa, and then the valve is closed. The process takes place adiabatically and kinetic energies and potential energies are negligible. Determine the final temperature of the steam.

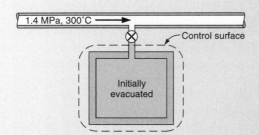

FIGURE 6.13 Flow into an evacuated vessel—control volume analysis.

Control volume: Tank, as shown in Fig. 6.13.

Initial state (in tank): Evacuated, mass $m_1 = 0$.

Final state: P_2 known.

Inlet state: P_i, T_i (in line) known.

Process: USUF.

Model: Steam tables.

Analysis:

First law, Eq. 6.16:

$$Q_{c.v.} + \sum m_i \left(h_i + \frac{\mathbf{V}_i^2}{2} + gZ_i \right)$$

$$= \sum m_e \left(h_e + \frac{\mathbf{V}_e^2}{2} + gZ_e \right)$$

$$+ \left[m_2 \left(u_2 + \frac{\mathbf{V}_2^2}{2} + gZ_2 \right) - m_1 \left(u_1 + \frac{\mathbf{V}_1^2}{2} + gZ_1 \right) \right]_{c.v.} + W_{c.v.}$$

We note that $Q_{c.v.} = 0$, $W_{c.v.} = 0$, $m_e = 0$, and $(m_1)_{c.v.} = 0$. We further assume that changes in kinetic and potential energy are negligible. Therefore, the statement of the first law for this process reduces to

$$m_i h_i = m_2 u_2$$

From the continuity equation for this process, Eq. 6.15, we conclude that

$$m_2 = m_i$$

Therefore, combining the continuity equation with the first law, we have

$$h_i = u_2$$

That is, the final internal energy of the steam in the tank is equal to the enthalpy of the steam entering the tank.

Solution

From the steam tables

$$h_i = u_2 = 3040.4 \text{ kJ/kg}$$

Since the final pressure is given as 1.4 MPa, we know two properties at the final state and therefore the final state is determined. The temperature corresponding to a pressure of 1.4 MPa and an internal energy of 3040.4 kJ/kg is found to be 452°C. Had this problem involved a substance for which internal energies are not listed in the thermodynamic tables, it would have been necessary to calculate a few values for u before an interpolation could be made for the final temperature.

This problem can also be solved by considering the steam that enters the tank and the evacuated space as a control mass, as indicated in Fig. 6.14.

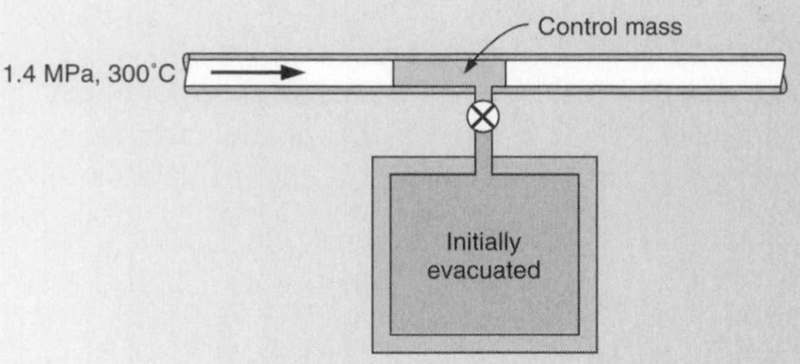

FIGURE 6.14 Flow into an evacuated vessel—control mass.

The process is adiabatic, but we must examine the boundaries for work. If we visualize a piston between the steam that is included in the control mass and the steam that flows behind, we readily recognize that the boundaries move and that the steam in the pipe does work on the steam that comprises the control mass. The amount of this work is

$$-W = P_1 V_1 = m P_1 v_1$$

Writing the first law for the control mass, Eq. 5.11, and noting that kinetic and potential energies can be neglected, we have

$$_1Q_2 = U_2 - U_1 + _1W_2$$

$$0 = U_2 - U_1 - P_1 V_1$$

$$0 = mu_2 - mu_1 - m P_1 v_1 = mu_2 - mh_1$$

Therefore,

$$u_2 = h_1$$

which is the same conclusion that was reached using a control volume analysis.

The two other examples that follow illustrate further the uniform-state, uniform-flow process.

EXAMPLE 6.12

Let the tank of the previous example have a volume of 0.4 m³ and initially contain saturated vapor at 350 kPa. The valve is then opened and steam from the line at 1.4 MPa, 300°C flows into the tank until the pressure is 1.4 MPa.

Calculate the mass of steam that flows into the tank.

Control volume: Tank, as in Fig. 6.13.

Initial state: P_1, saturated vapor; state fixed.

Final state: P_2.

Inlet state: P_i, T_i; state fixed.

Process: USUF.

Model: Steam tables.

Analysis:

Same as in Example 6.11, except that the tank is not evacuated initially. Again we note that $Q_{c.v.} = 0$, $W_{c.v.} = 0$, $m_e = 0$, and we assume that changes in kinetic energy and potential energy are zero. The statement of the first law for this process, Eq. 6.16, reduces to

$$m_i h_i = m_2 u_2 - m_1 u_1$$

The continuity equation, Eq. 6.15, reduces to

$$m_2 - m_1 = m_i$$

Therefore, combining the continuity equation and the first law, we have

$$(m_2 - m_1)h_i = m_2 u_2 - m_1 u_1$$

$$m_2(h_i - u_2) = m_1(h_i - u_1) \qquad (a)$$

There are two unknowns in this equation—m_2 and u_2. However, we have one additional equation:

$$m_2 v_2 = V = 0.4 \text{ m}^3 \qquad (b)$$

Substituting (b) into (a) and rearranging, we have

$$\frac{V}{v_2}(h_i - u_2) - m_1(h_i - u_1) = 0 \qquad (c)$$

in which the only unknowns are v_2 and u_2, both functions of T_2 and P_2. Since T_2 is unknown, it means that there is only one value of T_2 for which Eq. (c) will be satisfied, and we must find it by trial and error.

Solution

$$v_1 = 0.5243 \text{ m}^3/\text{kg} \qquad m_1 = \frac{0.4}{0.5243} = 0.763 \text{ kg}$$

$$h_i = 3040.4 \text{ kJ/kg} \qquad u_1 = 2548.9 \text{ kJ/kg}$$

Assume that

$$T_2 = 300°C$$

For this temperature and the known value of P_2,

$$v_2 = 0.1823 \text{ m}^3/\text{kg} \qquad u_2 = 2785.2 \text{ kJ/kg}$$

Substituting into (c),

$$\frac{0.4}{0.1823}(3040.4 - 2785.2) - 0.763(3040.4 - 2548.9) = +185.0$$

Assuming instead that

$$T_2 = 350°C$$

For this temperature and the known P_2,

$$v_2 = 0.2003 \text{ m}^3/\text{kg} \qquad u_2 = 2869.1 \text{ kJ/kg}$$

Substituting these values into (c),

$$\frac{0.4}{0.2003}(3040.4 - 2869.1) - 0.763(3040.4 - 2548.9) = -32.9$$

and we find that the actual T_2 must be between these two assumed values, in order that (c) be equal to zero. By interpolation,

$$T_2 = 342°C \qquad \text{and} \qquad v_2 = 0.1974 \text{ m}^3/\text{kg}$$

The final mass inside the tank is

$$m_2 = \frac{0.4}{0.1974} = 2.026 \text{ kg}$$

and the mass of steam that flows into the tank is

$$m_i = m_2 - m_1 = 2.026 - 0.763 = 1.263 \text{ kg}$$

EXAMPLE 6.13   A tank of 2 m³ volume contains saturated ammonia at a temperature of 40°C. Initially the tank contains 50% liquid and 50% vapor by volume. Vapor is withdrawn from the top of the tank until the temperature is 10°C. Assuming that only vapor (i.e., no liquid) leaves and that the process is adiabatic, calculate the mass of ammonia that is withdrawn.

> *Control volume:* Tank.
> *Initial state:* T_1, V_{liq}, V_{vap}; state fixed.
> *Final state:* T_2.
> *Exit state:* Saturated vapor (temperature changing).
> *Process:* USUF.
> *Model:* Ammonia tables.

Analysis:

In the first law, Eq. 6.16, we note that $Q_{\text{c.v.}} = 0$, $W_{\text{c.v.}} = 0$, $m_i = 0$, and we assume that changes in kinetic and potential energy are negligible. However, the enthalpy of saturated vapor varies with temperature, and therefore we cannot simply assume that the enthalpy of the vapor leaving the tank remains constant. However, we note that at 40°C, $h_g = 1472.2$ kJ/kg and at 10°C, $h_g = 1453.3$ kJ/kg. Since the change in h_g during this process is small, we may accurately assume that h_e is the average of the two values given above. Therefore,

$$(h_e)_{\text{av}} = 1462.8 \text{ kJ/kg}$$

and the first law reduces to

$$m_e h_e + m_2 u_2 - m_1 u_1 = 0$$

and the continuity equation (from Eq. 6.15)

$$(m_2 - m_1)_{\text{c.v.}} + m_e = 0$$

Combining these two equations we have

$$m_2(h_e - u_2) = m_1 h_e - m_1 u_1$$

Solution

The following values are from the ammonia tables:

$$v_{f1} = 0.001\ 726 \text{ m}^3/\text{kg} \qquad v_{g1} = 0.0833 \text{ m}^3/\text{kg}$$

$$v_{f2} = 0.001\ 601 \qquad v_{fg2} = 0.2040$$

$$u_{f1} = 371.7 - 1554.33 \times 0.001\ 726 = 369.0 \text{ kJ/kg}$$

$$u_{g1} = 1472.2 - 1554.33 \times 0.0833 = 1342.7$$

$$u_{f2} = 227.6 - 614.95 \times 0.001\ 601 = 226.6$$

$$u_{g2} = 1453.3 - 614.95 \times 0.2056 = 1326.9$$

$$u_{fg2} = 1326.9 - 226.6 = 1100.3$$

Calculating first the initial mass, m_1, in the tank, the mass of the liquid initially present, m_{f1}, is

$$m_{f1} = \frac{1.0}{0.001\ 726} = 579.4 \text{ kg}$$

Similarly, the initial mass of vapor, m_{g1}, is

$$m_{g1} = \frac{1.0}{0.0833} = 12.0 \text{ kg}$$

$$m_1 = m_{f1} + m_{g1} = 579.4 + 12.0 = 591.4 \text{ kg}$$

$$m_1 h_e = 591.4 \times 1462.8 = 865\ 100 \text{ kJ}$$

$$m_1 u_1 = (mu)_{f1} + (mu)_{g1} = 579.4 \times 369.0 + 12.0 \times 1342.7$$

$$= 229\ 910 \text{ kJ}$$

Substituting these into the first law,

$$m_2(h_e - u_2) = m_1 h_e - m_1 u_1 = 865\ 100 - 229\ 910 = 635\ 190 \text{ kJ}$$

There are two unknowns, m_2 and u_2, in this equation. However,

$$m_2 = \frac{V}{v_2} = \frac{2.0}{0.001\ 601 + x_2(0.2040)}$$

and

$$u_2 = 226.6 + x_2(1100.3)$$

both are functions only of x_2, the quality at the final state. Consequently,

$$\frac{2.0(1462.8 - 226.6 - 1100.3x_2)}{0.001\,601 + 0.204x_2} = 635\,190$$

Solving,

$$x_2 = 0.01104$$

Therefore,

$$v_2 = 0.001\,601 + 0.011\,04 \times 0.2040 = 0.003\,854 \text{ m}^3/\text{kg}$$

$$m_2 = \frac{2}{0.003\,854} = 518.9 \text{ kg}$$

and the mass of ammonia withdrawn, m_e, is

$$m_e = m_1 - m_2 = 591.4 - 518.9 = 72.5 \text{ kg}$$

EXAMPLE 6.13E   A tank of 50 ft^3 volume contains saturated ammonia at a pressure of 200 lbf/in.^{2}. Initially the tank contains 50% liquid and 50% vapor by volume. Vapor is withdrawn from the top of the tank until the pressure is 100 lbf/in.^{2}. Assuming that only vapor (i.e., no liquid) leaves and that the process is adiabatic, calculate the mass of ammonia that is withdrawn.

> *Control volume:* Tank.
> *Initial state:* T_1, V_{liq}, V_{vap}; state fixed.
> *Final state:* T_2.
> *Exit state:* Saturated vapor (temperature changing).
> *Process:* USUF.
> *Model:* Ammonia tables.

Analysis:

In the first law, Eq. 6.16, we note that $Q_{\text{c.v.}} = 0$, $W_{\text{c.v.}} = 0$, $m_i = 0$, and we assume that changes in kinetic and potential energy are negligible. However, the enthalpy of saturated vapor varies with temperature, and therefore we cannot simply assume that the enthalpy of the vapor leaving the tank remains constant. We note that at 200 lbf/in.^{2}, $h_g = 632.7$ Btu/lbm and at 100 lbf/in.^{2}, $h_g = 626.5$ Btu/lbm. Since the change in h_g during this process is small, we may accurately assume that h_e is the average of the two values given above. Therefore

$$(h_e)_{\text{av}} = 629.6 \text{ Btu/lbm}$$

and the first law reduces to

$$m_e h_e + m_2 u_2 - m_1 u_1 = 0$$

and the continuity equation (from Eq. 6.15)

$$(m_2 - m_1)_{\text{c.v.}} + m_e = 0$$

Combining these two equations we have

$$m_2(h_e - u_2) = m_1 h_e - m_1 u_1$$

The following values are from the ammonia tables:

$$v_{f1} = 0.02732 \text{ ft}^3 / \text{lbm} \qquad v_{g1} = 1.502 \text{ ft}^3 / \text{lbm}$$

$$v_{f2} = 0.02584 \qquad\qquad v_{g2} = 2.952$$

$$u_{f1} = 150.9 - \frac{200 \times 144 \times 0.0273}{778} = 149.9 \text{ Btu} / \text{lbm}$$

$$u_{f2} = 104.7 - \frac{100 \times 144 \times 0.0258}{778} = 104.2$$

$$u_{g1} = 632.7 - \frac{200 \times 144 \times 1.502}{778} = 577.1$$

$$u_{g2} = 626.5 - \frac{100 \times 144 \times 2.952}{778} = 571.9$$

$$u_{fg2} = 571.9 - 104.2 = 467.7$$

Calculating first the initial mass, m_1, in the tank, the mass of the liquid initially present, m_{f1}, is

$$m_{f1} = \frac{25}{0.02732} = 915 \text{ lbm}$$

Similarly, the initial mass of vapor, m_{g1}, is

$$m_{g1} = \frac{25}{1.502} = 16.6 \text{ lbm}$$

$$m_1 = m_{f1} + m_{g1} = 915 + 17 = 932 \text{ lbm}$$

$$m_1 h_e = 932 \times 629.6 = 586\ 800 \text{ Btu}$$

$$m_1 u_1 = (mu)_{f1} + (mu)_{g1} = 915 \times 149.9 + 16.6 \times 577.0 = 146\ 700 \text{ Btu}$$

Substituting these into the first law,

$$m_2(h_e - u_2) = m_1 h_e - m_1 u_1 = 586\ 800 - 146\ 700 = 440\ 100 \text{ Btu}$$

There are two unknowns, m_2 and u_2, in this equation. However,

$$m_2 = \frac{V}{v_2} = \frac{50}{0.0258 + x_2 (2.952 - 0.0258)}$$

and

$$u_2 = 104.2 + x_2 (467.7)$$

both functions only of x_2, the quality at the final state. Consequently,

$$\frac{50(629.6 - 104.2 - 467.7 x_2)}{0.0258 + 2.926 x_2} = 440\ 100$$

Solving, $x_2 = 0.01137$
Therefore,

$$v_2 = 0.0258 + 0.01137(2.926) = 0.0590 \text{ ft}^3 / \text{lbm}$$

$$m_2 = \frac{50}{0.0590} = 847 \text{ lbm}$$

and the mass of ammonia withdrawn, m_e, is

$$m_e = m_1 - m_2 = 932 - 847 = 85 \text{ lbm}$$

PROBLEMS

6.1 Air at 35°C, 105 kPa flows in a 100 mm×150 mm rectangular duct in a heating system. The volumetric flow rate is 0.015 m³/s. What is the velocity of the air flowing in the duct?

6.2 A boiler receives a constant flow of 5000 kg/h liquid water at 5 MPa, 20°C, and it heats the flow such that the exit state is 450°C with a pressure of 4.5 MPa. Determine the necessary minimum pipe flow area in both the inlet and exit pipe(s) if there should be no velocities larger than 20 m/s.

6.3 A natural gas company distributes methane gas in a pipeline flowing at 200 kPa, 275 K. They have carefully measured the average flow velocity to be 5.5 m/s in a 50-cm-diameter pipe. If there is a ±2% uncertainty in the velocity measurement how would you quote the mass flow rate?

6.4 Nitrogen gas flowing in a 50-mm-diameter pipe at 15°C, 200 kPa, at the rate of 0.05 kg/s encounters a partially closed valve. If there is a pressure drop of 30 kPa across the valve and essentially no temperature change, what are the velocities upstream and downstream of the valve?

6.5 Saturated vapor R-134a leaves the evaporator in a heat pump system at 10°C, with a steady mass flow rate of 0.1 kg/s. What is the smallest diameter tubing that can be used at this location if the velocity of the refrigerant is not to exceed 7 m/s?

6.6 Steam at 3 MPa, 400°C enters a turbine with a volume flow rate of 5 m³/s. An extraction of 15% of the inlet mass flow rate exits at 600 kPa, 200°C. The rest exits the turbine at 20 kPa with a quality of 90%, and a velocity of 20 m/s. Determine the volume flow rate of the extraction flow and the diameter of the final exit pipe.

6.7 A pump takes 10°C liquid water in from a river at 95 kPa and pumps it up to an irrigation canal 20 m higher than the river surface. All pipes have diameter of 0.1 m, and the flow rate is 15 kg/s. Assume the pump exit pressure is just enough to carry a water column of the 20 m height with 100 kPa at the top. Find the flow work into and out of the pump and the kinetic energy in the flow.

6.8 A desuperheater mixes superheated water vapor with liquid water in a ratio that produces saturated water vapor as output without any external heat transfer. A flow of 0.5 kg/s superheated vapor at 5 MPa, 400°C and a flow of liquid water at 5 MPa, 40°C enter a desuperheater. If saturated water vapor at 4.5 MPa is produced, determine the flow rate of the liquid water.

6.9 Carbon dioxide enters a steady-state, steady-flow heater at 300 kPa, 15°C, and exits at 275 kPa, 1200°C, as shown in Fig. P6.9. Changes in kinetic and potential energies are negligible. Calculate the required heat transfer per kilogram of carbon dioxide flowing through the heater.

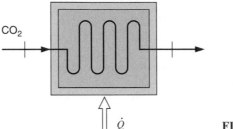

FIGURE P6.9

6.10 Saturated liquid nitrogen at 500 kPa enters a SSSF boiler at a rate of 0.005 kg/s and exits as saturated vapor. It then flows into a superheater also at 500 kPa where it exits at 500 kPa, 275 K. Find the rate of heat transfer in the boiler and the superheater.

6.11 A steam pipe for a 1500-m-tall building receives superheated steam at 200 kPa at ground level. At the top floor the pressure is 125 kPa, and the heat loss in the pipe is 110 kJ/kg. What should the inlet temperature be so that no water will condense inside the pipe?

6.12 In a steam generator, compressed liquid water at 10 MPa, 30°C enters a 30-mm diameter tube at the rate of 3 L/s. Steam at 9 MPa, 400°C exits the tube. Find the rate of heat transfer to the water.

6.13 A heat exchanger, shown in Fig. P6.13, is used to cool an air flow from 800 to 360 K, both states at 1 MPa. The coolant is a water flow at 15°C, 0.1 MPa. If the water leaves as saturated vapor, find the ratio of the flow rates $\dot{m}_{water}/\dot{m}_{air}$.

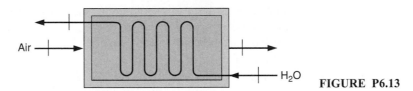

FIGURE P6.13

6.14 A condenser (heat exchanger) brings 1 kg/s water flow at 10 kPa from 300°C to saturated liquid at 10 kPa, as shown in Fig. P6.14. The cooling is done by lake water at 20°C that returns to the lake at 30°C. For an insulated condenser, find the flow rate of cooling water.

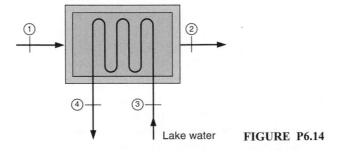

Lake water   **FIGURE P6.14**

6.15 Two kilograms of water at 500 kPa, 20°C are heated in a constant pressure process (SSSF) to 1700°C. Find the best estimate for the heat transfer.

6.16 A mixing chamber with heat transfer receives 2 kg/s of R-22 at 1 MPa, 40°C in one line and 1 kg/s of R-22 at 30°C, quality 50% in a line with a valve. The outgoing flow is at 1 MPa, 60°C. Find the rate of heat transfer to the mixing chamber.

6.17 Compressed liquid R-22 at 1.5 MPa, 10°C is mixed in a steady-state, steady-flow process with saturated vapor R-22 at 1.5 MPa. Both flow rates are 0.1 kg/s, and the exiting flow is at 1.2 MPa and a quality of 85%. Find the rate of heat transfer to the mixing chamber.

6.18 Nitrogen gas flows into a convergent nozzle at 200 kPa, 400 K and very low velocity. It flows out of the nozzle at 100 kPa, 330 K. If the nozzle is insulated, find the exit velocity.

6.19 Superheated vapor ammonia enters an insulated nozzle at 20°C, 800 kPa, shown in Fig. P6.19, with a low velocity and at the steady rate of 0.01 kg/s. The ammonia exits at 300 kPa with a velocity of 450 m/s. Determine the temperature (or quality, if saturated) and the exit area of the nozzle.

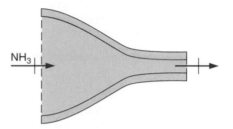

FIGURE P6.19

6.20 A diffuser, shown in Fig. P6.20, has air entering at 100 kPa, 300 K, with a velocity of 200 m/s. The inlet cross-sectional area of the diffuser is 100 mm^2. At the exit, the area is 860 mm^2, and the exit velocity is 20 m/s. Determine the exit pressure and temperature of the air.

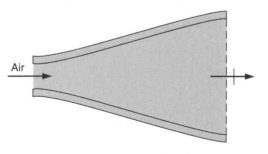

FIGURE P6.20

6.21 A diffuser receives an ideal gas flow at 100 kPa, 300 K with a velocity of 250 m/s, and the exit velocity is 25 m/s. Determine the exit temperature if the gas is argon, helium, or nitrogen.

6.22 The front of a jet engine acts as a diffuser receiving air at 900 km/h, −5°C, 50 kPa, bringing it to 80 m/s relative to the engine before entering the compressor. If the flow area is reduced to 80% of the inlet area, find the temperature and pressure in the compressor inlet.

6.23 Helium is throttled from 1.2 MPa, 20°C to a pressure of 100 kPa. The diameter of the exit pipe is so much larger than the inlet pipe that the inlet and exit velocities are equal. Find the exit temperature of the helium and the ratio of the pipe diameters.

6.24 Water flowing in a line at 400 kPa, saturated vapor is taken out through a valve to 100 kPa. What is the temperature as it leaves the valve, assuming no changes in kinetic energy and no heat transfer?

6.25 Methane at 3 MPa, 300 K is throttled to 100 kPa. Calculate the exit temperature assuming no changes in the kinetic energy and ideal-gas behavior. Repeat the answer for real-gas behavior.

6.26 Water at 1.5 MPa, 150°C, is throttled adiabatically through a valve to 200 kPa. The inlet velocity is 5 m/s, and the inlet and exit pipe diameters are the same. Determine the state and the velocity of the water at the exit.

6.27 An insulated mixing chamber receives 2 kg/s R-134a at 1 MPa, 100°C in a line with low velocity. Another line with R-134a as saturated liquid, 60°C flows through a valve to the mixing chamber at 1 MPa after the valve. The exit flow is saturated vapor at 1 MPa flowing at 20 m/s. Find the flow rate for the second line.

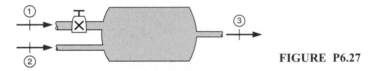

FIGURE P6.27

6.28 A mixing chamber receives 2 kg/s R-134a at 1 MPa, 100°C in a line with low velocity and 1 kg/s from a line with R-134a as saturated liquid, 60°C flows through a valve to the mixing chamber at 1 MPa after the valve. There is heat transfer so the exit flow is saturated vapor at 1 MPa flowing at 20 m/s. Find the rate of heat transfer and the exit pipe diameter.

6.29 A steam turbine receives water at 15 MPa, 600°C at a rate of 100 kg/s, shown in Fig. P6.29. In the middle section 20 kg/s is withdrawn at 2 MPa, 350°C, and the rest exits the turbine at 75 kPa, 95% quality. Assuming no heat transfer and no changes in kinetic energy, find the total turbine power output.

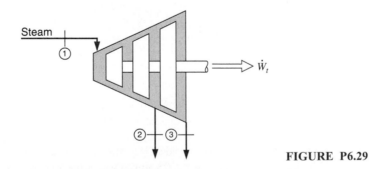

FIGURE P6.29

6.30 A small, high-speed turbine operating on compressed air produces a power output of 100W. The inlet state is 400 kPa, 50°C, and the exit state is 150 kPa, −30°C. Assuming the velocities to be low and the process to be adiabatic, find the required mass flow rate of air through the turbine.

6.31 A steam turbine receives steam from two boilers. One flow is 5 kg/s at 3 MPa, 700°C and the other flow is 15 kg/s at 800 kPa, 500°C. The exit state is 10 kPa, with a quality of 96%. Find the total power out of the adiabatic turbine.

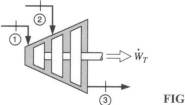

FIGURE P6.31

6.32 A small turbine, shown in Fig. P6.32, is operated at part load by throttling a 0.25-kg/s steam supply at 1.4 MPa, 250°C down to 1.1 MPa before it enters the turbine and the exhaust is at 10 kPa. If the turbine produces 110 kW, find the exhaust temperature (and quality if saturated).

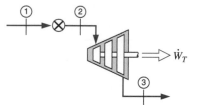

FIGURE P6.32

6.33 Hoover Dam across the Colorado River dams up Lake Mead 200 m higher than the river downstream. The electric generators driven by water-powered turbines deliver 1300 MW of power. If the water is 17.5°C, find the minimum amount of water running through the turbines.

6.34 A large SSSF expansion engine has two low-velocity flows of water entering. High-pressure steam enters at point 1 with 2.0 kg/s at 2 MPa, 500°C and 0.5 kg/s cooling water at 120 kPa, 30°C enters at point 2. A single flow exits at point 3 with 150 kPa, 80% quality, through a 0.15-m diameter exhaust pipe. There is a heat loss of 300 kW. Find the exhaust velocity and the power output of the engine.

6.35 A small water pump is used in an irrigation system. The pump takes water in from a river at 10°C, 100 kPa at a rate of 5 kg/s. The exit line enters a pipe that goes up to an elevation 20 m above the pump and river, where the water runs into an open channel. Assume the process is adiabatic and that the water stays at 10°C. Find the required pump work.

6.36 The compressor of a large gas turbine receives air from the ambient at 95 kPa, 20°C, with a low velocity. At the compressor discharge, air exits at 1.52 MPa, 430°C, with velocity of 90 m/s. The power input to the compressor is 5000 kW. Determine the mass flow rate of air through the unit.

6.37 Two steady flows of air enter a control volume, shown in Fig. P6.37. One is 0.025 kg/s flow at 350 kPa, 150°C, state 1, and the other enters at 350 kPa, 15°C, both flowing with low velocity. A single flow of air exits at 100 kPa, −40°C through a 25-mm diameter pipe, state 3. The control volume rejects 1.2 kW heat to the surroundings and produces 4.5 kW of power. Determine the flow rate of air at the inlet at state 2.

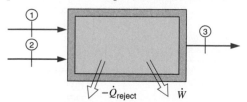

FIGURE P6.37

6.38 An air compressor takes in air at 100 kPa, 17°C and delivers it at 1 MPa, 600 K to a constant-pressure cooler, which it exits at 300 K. Find the specific compressor work and the specific heat transfer.

6.39 The following data are for a simple steam power plant, as shown in Fig. P6.39.

State	1	2	3	4	5	6	7
P MPa	6.2	6.1	5.9	5.7	5.5	0.01	0.009
T °C		45	175	500	490		40

State 6 has $x_6 = 0.92$, and velocity of 200 m/s. The rate of steam flow is 25 kg/s, with 300 kW power input to the pump. Piping diameters are 200 mm from steam generator to the turbine and 75 mm from the condenser to the steam generator. Determine the power output of the turbine and the heat transfer rate in the condenser.

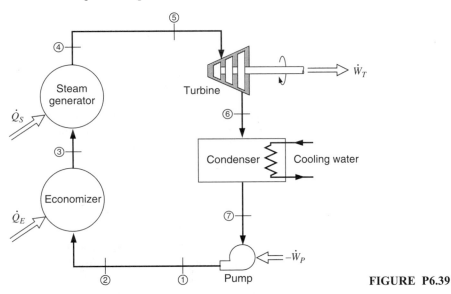

FIGURE P6.39

6.40 For the same steam power plant as shown in Fig. P6.39 and Problem 6.39 determine the rate of heat transfer in the economizer, which is a low temperature heat exchanger, and in the steam generator. Determine also the flow rate of cooling water through the condenser, if the cooling water increases from 15° to 25°C in the condenser.

6.41 Cogeneration is often used where a steam supply is needed for industrial process energy. Assume a supply of 5 kg/s steam at 0.5 MPa is needed. Rather than generating this from a pump and boiler, the setup in Fig. P6.41 is used so the supply is ex-

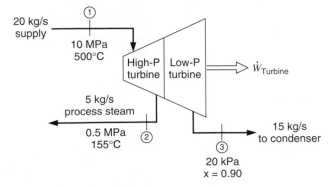

FIGURE P6.41

tracted from the high-pressure turbine. Find the power the turbine now cogenerates in this process.

6.42 A somewhat simplified flow diagram for a nuclear power plant shown in Fig. 1.4 is given in Fig. P6.42. Mass flow rates and the various states in the cycle are shown in the accompanying table.

Point	$\dot{m}$, kg/s	P, kPa	T, °C	h, kJ/kg
1	75.6	7240	sat vap	
2	75.6	6900		2765
3	62.874	345		2517
4		310		
5		7		2279
6	75.6	7	33	
7		415		140
8	2.772	35		2459
9	4.662	310		558
10		35	34	
11	75.6	380	68	
12	8.064	345		2517
13	75.6	330		
14				349
15	4.662	965	139	584
16	75.6	7930		565
17	4.662	965		2593
18	75.6	7580		688
19	1386	7240	277	
20	1386	7410		1221
21	1386	7310		

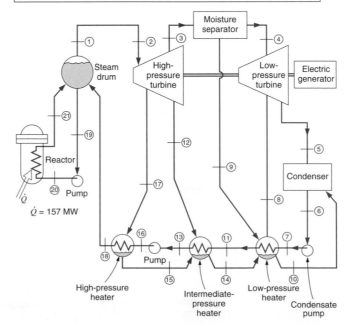

FIGURE P6.42

The cycle includes a number of heaters in which heat is transferred from steam, taken out of the turbine at some intermediate pressure, to liquid water pumped from the condenser on its way to the steam drum. The heat exchanger in the reactor supplies 157 MW, and it may be assumed that there is no heat transfer in the turbines.

a. Assuming the moisture separator has no heat transfer between the two turbine sections, determine the enthalpy and quality (h_4, x_4).

b. Determine the power output of the low-pressure turbine.

c. Determine the power output of the high-pressure turbine.

d. Find the ratio of the total power output of the two turbines to the total power delivered by the reactor.

6.43 Consider the powerplant as described in the previous problem.

a. Determine the quality of the steam leaving the reactor.

b. What is the power to the pump that feeds water to the reactor?

6.44 Consider the powerplant as described in Problem 6.42.

a. Determine the temperature of the water leaving the intermediate pressure heater, T_{13}, assuming no heat transfer to the surroundings.

b. Determine the pump work, between states 13 and 16.

6.45 Consider the powerplant as described in Problem 6.42.

a. Find the power removed in the condenser by the cooling water (not shown).

b. Find the power to the condensate pump.

c. Do the energy terms balance for the low-pressure heater or is there a heat transfer not shown?

6.46 A proposal is made to use a geothermal supply of hot water to operate a steam turbine, as shown in Fig. P6.46. The high-pressure water at 1.5 MPa, 180°C is throttled into a flash evaporator chamber, which forms liquid and vapor at a lower pressure of 400 kPa. The liquid is discarded while the saturated vapor feeds the turbine and exits at 10 kPa, 90% quality. If the turbine should produce 1 MW, find the required mass flow rate of hot geothermal water in kilograms per hour.

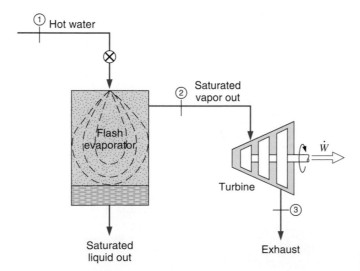

FIGURE P6.46

6.47 An R-12 heat pump cycle shown in Fig. P6.47 has an R-12 flow rate of 0.05 kg/s with 4 kW into the compressor. The following data are given:

State	1	2	3	4	5	6
P kPa	1250	1230	1200	320	300	290
T °C	120	110	45		0	5

Calculate the heat transfer from the compressor, the heat transfer from the R-12 in the condenser and the heat transfer to the R-12 in the evaporator.

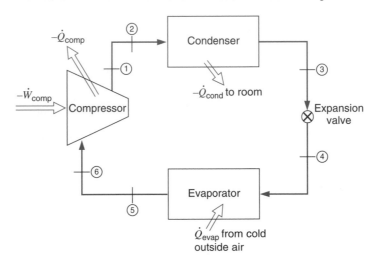

FIGURE P6.47

6.48 A rigid 100-L tank contains air at 1 MPa, 200°C. A valve on the tank is now opened, and air flows out until the pressure drops to 100 kPa. During this process, heat is transferred from a heat source at 200°C, such that when the valve is closed, the temperature inside the tank is 50°C. What is the heat transfer?

6.49 A 25-L tank, shown in Fig. P6.49, that is initially evacuated is connected by a valve to an air supply line flowing air at 20°C, 800 kPa. The valve is opened, and air flows into the tank until the pressure reaches 600 kPa. Determine the final temperature and mass inside the tank, assuming the process is adiabatic. Develop an expression for the relation between the line temperature and the final temperature using constant specific heats.

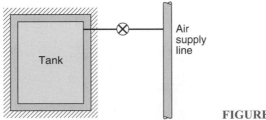

FIGURE P6.49

6.50 A 100-L rigid tank contains carbon dioxide gas at 1 MPa, 300 K. A valve is cracked open, and carbon dioxide escapes slowly until the tank pressure has dropped to 500 kPa. At this point the valve is closed. The gas remaining inside the tank may be assumed to have undergone a polytropic expansion, with polytropic exponent $n = 1.15$. Find the final mass inside and the heat transferred to the tank during the process.

6.51 A 1-m³ tank contains ammonia at 150 kPa, 25°C. The tank is attached to a line flowing ammonia at 1200 kPa, 60°C. The valve is opened, and mass flows in until the tank is half full of liquid, by volume at 25°C. Calculate the heat transferred from the tank during this process.

6.52 A nitrogen line, 300 K and 0.5 MPa, shown in Fig. P6.52, is connected to a turbine that exhausts to a closed initially empty tank of 50 m³. The turbine operates to a tank pressure of 0.5 MPa, at which point the temperature is 250 K. Assuming the entire process is adiabatic, determine the turbine work.

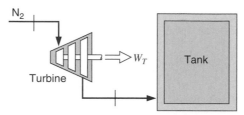

FIGURE P6.52

6.53 An evacuated 150-L tank is connected to a line flowing air at room temperature, 25°C, and 8 MPa pressure. The valve is opened, allowing air to flow into the tank until the pressure inside is 6 MPa. At this point the valve is closed. This filling process occurs rapidly and is essentially adiabatic. The tank is then placed in storage where it eventually returns to room temperature. What is the final pressure?

6.54 A 0.5-m diameter balloon containing air at 200 kPa, 300 K is attached by a valve to an air line flowing air at 400 kPa, 400 K. The valve is now opened, allowing air to flow into the balloon until the pressure inside reaches 300 kPa, at which point the valve is closed. The final temperature inside the balloon is 350 K. The pressure is directly proportional to the diameter of the balloon. Find the work and heat transfer during the process.

6.55 A 500-L insulated tank contains air at 40°C, 2 MPa. A valve on the tank is opened, and air escapes until half the original mass is gone, at which point the valve is closed. What is the pressure inside then?

6.56 A steam engine based on a turbine is shown in Fig. P6.56. The boiler tank has a volume of 100 L and initially contains saturated liquid with a very small amount of vapor at 100 kPa. Heat is now added by the burner, and the pressure regulator does not open before the boiler pressure reaches 700 kPa, which it keeps constant. The saturated vapor enters the turbine at 700 kPa and is discharged to the atmosphere as saturated vapor at 100 kPa. The burner is turned off when no more liquid is present in the boiler. Find the total turbine work and the total heat transfer to the boiler for this process.

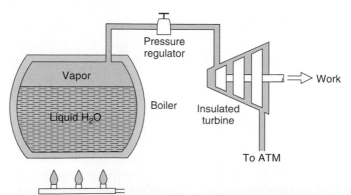

FIGURE P6.56

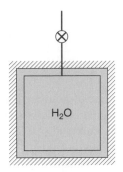

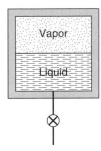

FIGURE P6.57

6.57 A 2-m^3 insulated vessel, shown in Fig. P6.57, contains saturated vapor steam at 4 MPa. A valve on the top of the tank is opened, and steam is allowed to escape. During the process any liquid formed collects at the bottom of the vessel, so that only saturated vapor exits. Calculate the total mass that has escaped when the pressure inside reaches 1 MPa.

6.58 A 1-m^3 insulated, 40-kg rigid steel tank contains air at 500 kPa, and both tank and air are at 20°C. The tank is connected to a line flowing air at 2 MPa, 20°C. The valve is opened, allowing air to flow into the tank until the pressure reaches 1.5 MPa and is then closed. Assume the air and tank are always at the same temperature and find the final temperature.

6.59 A 750-L rigid tank, shown in Fig. P6.59, initially contains water at 250°C, 50% liquid and 50% vapor, by volume. A valve at the bottom of the tank is opened, and liquid is slowly withdrawn. Heat transfer takes place such that the temperature remains constant. Find the amount of heat transfer required to reach the state where half the initial mass is withdrawn.

6.60 An initially empty bottle, $V = 0.25$ m^3, is filled with water from a line at 0.8 MPa, 350°C. Assume no heat transfer and that the bottle is closed when the pressure reaches line pressure. Find the final temperature and mass in the bottle.

FIGURE P6.59

6.61 A supply line of ammonia at 0°C, 450 kPa is used to fill a 0.05-m^3 container initially storing ammonia at 20°C, 100 kPa. The supply line valve is closed when the pressure inside reaches 290.9 kPa. Find the final mass and temperature in the container.

6.62 An insulated spring-loaded piston/cylinder, shown in Fig. P6.62, is connected to an air line flowing air at 600 kPa, 700 K by a valve. Initially the cylinder is empty and the spring force is zero. The valve is then opened until the cylinder pressure reaches 300 kPa. By noting that $u_2 = u_{line} + C_v(T_2 - T_{line})$ and $h_{line} - u_{line} = RT_{line}$ find an expression for T_2 as a function of P_2, P_0, T_{line}. With $P_0 = 100$ kPa, find T_2.

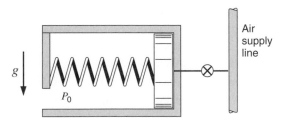

FIGURE P6.62

6.63 A mass-loaded piston/cylinder, shown in Fig. P6.63, containing air is at 300 kPa, 17°C with a volume of 0.25 m^3, while at the stops $V = 1$ m^3. An air line, 500 kPa, 600 K, is connected by a valve that is then opened until a final inside pressure of 400 kPa is reached, at which point $T = 350$ K. Find the air mass that enters, the work, and heat transfer.

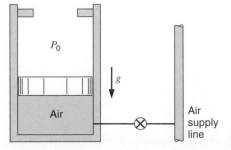

FIGURE P6.63

6.64 An elastic balloon behaves such that pressure is proportional to diameter and the balloon contains 0.5 kg air at 200 kPa, 30°C. The balloon is momentarily connected to an air line at 400 kPa, 100°C. Air is let in until the volume doubles, during which process there is a heat transfer of 50 kJ out of the balloon. Find the final temperature and the mass of air that enters the balloon.

6.65 A 2-m³ storage tank contains 95% liquid and 5% vapor by volume of liquified natural gas (LNG) at 160 K, as shown in Fig. P6.65. It may be assumed that LNG has the same properties as pure methane. Heat is transferred to the tank and saturated vapor at 160 K flows into the steady flow heater which it leaves at 300 K. The process continues until all the liquid in the storage tank is gone. Calculate the total amount of heat transfer to the tank and the total amount of heat transferred to the heater.

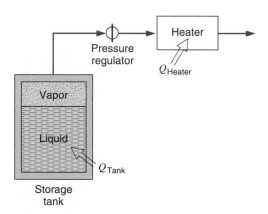

FIGURE P6.65

6.66 A spherical balloon is constructed of a material such that the pressure inside is proportional to the balloon diameter to the power 1.5. The balloon contains argon gas at 1200 kPa, 700°C, at a diameter of 2.0 m. A valve is now opened, allowing gas to flow out until the diameter reaches 1.8 m, at which point the temperature inside is 600°C. The balloon then continues to cool until the diameter is 1.4 m.

a. How much mass was lost from the balloon?

b. What is the final temperature inside?

c. Calculate the heat transferred from the balloon during the overall process.

6.67 A rigid tank initially contains 100 L of saturated-liquid R-12 and 100 L of saturated-vapor R-12 at 0°C. A valve on the bottom of the tank is connected to a line flowing R-12 at 10°C, 900 kPa. A pressure-relief valve on the top of the tank is set

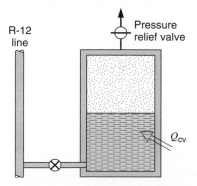

FIGURE P6.67

at 745 kPa (when tank pressure reaches that value, mass escapes such that the tank pressure cannot exceed 745 kPa). The line valve is now opened, allowing 10 kg of R-12 to flow in from the line, and then this valve is closed. Heat is transferred slowly to the tank, until the final mass inside is 100 kg, at which point the process is stopped.

a. How much mass exits the pressure-relief valve during the overall process?

b. How much heat is transferred to the tank?

6.68 A cylinder with a constant load on the piston contains water at 500 kPa, 20°C, and volume of 1 L. The bottom of the cylinder is connected with a line and valve to a steam supply line carrying steam at 1 MPa, 200°C. The valve is now opened for a short time to let steam in to a final volume of 10 L. The final uniform state is two-phase, and there is no heat transfer in the process. What is the final mass inside the cylinder?

ADVANCED PROBLEMS

6.69 A 2-m^3 insulated tank containing ammonia at -20°C, 80% quality, is connected by a valve to a line flowing ammonia at 2 MPa, 60°C. The valve is opened, allowing ammonia to flow into the tank. At what pressure should the valve be closed if the manufacturer wishes to have 15 kg of ammonia inside at the final state?

6.70 Air is contained in the insulated cylinder shown in Fig. P6.70. At this point the air is at 140 kPa, 25°C, and the cylinder volume is 15 L. The piston cross-sectional area is 0.045 m^2, and the spring is linear with spring constant 35 kN/m. The valve is opened, and air from the line at 700 kPa, 25°C flows into the cylinder until the pressure reaches 700 kPa, and then the valve is closed. Find the final temperature.

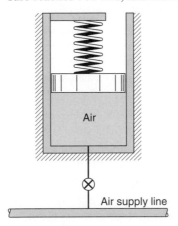

Air

Air supply line

FIGURE P6.70

6.71 An inflatable bag, initially flat and empty, is connected to a supply line of saturated vapor R-22 at ambient temperature of 10°C. The valve is opened, and the bag slowly inflates at constant temperature to a final diameter of 2 m. The bag is inflated at constant pressure, $P_o = 100$ kPa, until it becomes spherical at $D_o = 1$ m. After this the pressure and diameter are related according to

$$P = P_o + C\left[1 - \left(\frac{D_o}{D}\right)^6\right]\frac{D_o}{D}$$

A maximum pressure of 500 kPa is recorded for the whole process. Find the heat transfer to the bag during the inflation process.

6.72 A cylinder, shown in Fig. P6.72, fitted with a piston restrained by a linear spring contains 1 kg of R-12 at 100°C, 800 kPa. The spring constant is 50 kN/m, and the piston cross-sectional area is 0.05 m². A valve on the cylinder is opened, and R-12 flows out until half the initial mass is left. Heat is transferred so the final temperature of the R-12 is 10°C. Find the final state of the R-12, (P_2, x_2), and the heat transfer to the cylinder.

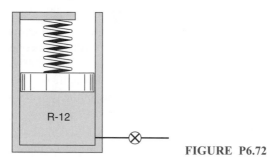

FIGURE P6.72

ENGLISH UNIT PROBLEMS

6.73E Air at 95 F, 16 lbf/in.² flows in a 4 in.×6 in. rectangular duct in a heating system. The volumetric flow rate is 30 cfm (ft³/min). What is the velocity of the air flowing in the duct?

6.74E Saturated vapor R-134a leaves the evaporator in a heat pump at 50 F, with a steady mass flow rate of 0.2 lbm/s. What is the smallest diameter tubing that can be used at this location if the velocity of the refrigerant is not to exceed 20 ft/s?

6.75E A pump takes 40 F liquid water from a river at 14 lbf/in.² and pumps it up to an irrigation canal 60 ft higher than the river surface. All pipes have diameter of 4 in., and the flow rate is 35 lbm/s. Assume the pump exit pressure is just enough to carry a water column of the 60 ft height with 15 lbf/in.² at the top. Find the flow work into and out of the pump and the kinetic energy in the flow.

6.76E Carbon dioxide gas enters a steady-state, steady-flow heater at 45 lbf/in.² 60 F and exits at 40 lbf/in.², 1800 F. It is shown in Fig. P6.9, where changes in kinetic and potential energies are negligible. Calculate the required heat transfer per lbm of carbon dioxide flowing through the heater.

6.77E In a steam generator, compressed liquid water at 1500 lbf/in.², 100 F enters a 1-in.-diameter tube at the rate of 5 ft³/min. Steam at 1250 lbf/in.², 750 F exits the tube. Find the rate of heat transfer to the water.

6.78E A heat exchanger is used to cool an air flow from 1400 to 680 R, both states at 150 lbf/in.². The coolant is a water flow at 60 F, 15 lbf/in.², and it is shown in Fig. P6.13. If the water leaves as saturated vapor, find the ratio of the flow rates $\dot{m}_{water}/\dot{m}_{air}$.

6.79E A condenser, as the heat exchanger shown in Fig. P6.14, brings 1 lbm/s water flow at 1 lbf/in.² from 500 F to saturated liquid at 1 lbf/in.². The cooling is done by lake water at 70 F that returns to the lake at 90 F. For an insulated condenser, find the flow rate of cooling water.

6.80E Four pound-mass of water at 80 lbf/in.², 70 F is heated in a constant pressure process (SSSF) to 2600 F. Find the best estimate for the heat transfer.

6.81E Nitrogen gas flows into a convergent nozzle at 30 lbf/in.^{2}, 600 R and very low velocity. It flows out of the nozzle at 15 lbf/in.^{2}, 500 R. If the nozzle is insulated, find the exit velocity.

6.82E A diffuser shown in Fig. P6.20 has air entering at 14.7 lbf/in.^{2}, 540 R, with a velocity of 600 ft/s. The inlet cross-sectional area of the diffuser is 0.2 in.^{2}. At the exit, the area is 1.75 in.^{2}, and the exit velocity is 60 ft/s. Determine the exit pressure and temperature of the air.

6.83E Helium is throttled from 175 lbf/in.^{2}, 70 F, to a pressure of 15 lbf/in.^{2}. The diameter of the exit pipe is so much larger than the inlet pipe that the inlet and exit velocities are equal. Find the exit temperature of the helium and the ratio of the pipe diameters.

6.84E Water flowing in a line at 60 lbf/in.^{2}, saturated vapor is taken out through a valve to 14.7 lbf/in.^{2}. What is the temperature as it leaves the valve assuming no changes in kinetic energy and no heat transfer?

6.85E An insulated mixing chamber receives 4 lbm/s R-134a at 150 lbf/in.^{2}, 220 F in a line with low velocity. Another line with R-134a as saturated liquid, 130 F flows through a valve to the mixing chamber at 150 lbf/in.^{2} after the valve. The exit flow is saturated vapor at 150 lbf/in.^{2} flowing at 60 ft/s. Find the mass flow rate for the second line.

6.86E A steam turbine receives water at 2000 lbf/in.^{2}, 1200 F at a rate of 200 lbm/s, as shown in Fig. P6.29. In the middle section 40 lbm/s is withdrawn at 300 lbf/in.^{2}, 650 F and the rest exits the turbine at 10 lbf/in.^{2}, 95% quality. Assuming no heat transfer and no changes in kinetic energy, find the total turbine power output.

6.87E A small, high-speed turbine operating on compressed air produces a power output of 0.1 hp. The inlet state is 60 lbf/in.^{2}, 120 F, and the exit state is 14.7 lbf/in.^{2}, −20 F. Assuming the velocities to be low and the process to be adiabatic, find the required mass flow rate of air through the turbine.

6.88E Hoover Dam across the Colorado River dams up Lake Mead 600 ft higher than the river downstream. The electric generators driven by water-powered turbines deliver 1.2×10^6 Btu/s. If the water is 65 F, find the minimum amount of water running through the turbines.

6.89E A small water pump is used in an irrigation system. The pump takes water in from a river at 50 F, 1 atm at a rate of 10 lbm/s. The exit line enters a pipe that goes up to an elevation 60 ft above the pump and river, where the water runs into an open channel. Assume the process is adiabatic and that the water stays at 50 F. Find the required pump work.

6.90E An air compressor takes in air at 14 lbf/in.^{2}, 60 F and delivers it at 140 lbf/in.^{2}, 1080 R to a constant-pressure cooler, which it exits at 560 R. Find the specific compressor work and the specific heat transfer.

6.91E The following data are for a simple steam power plant as shown in Fig. P6.39.

State	1	2	3	4	5	6	7
P lbf/in^2	900	890	860	830	800	1.5	1.4
T F		115	350	920	900		110

State 6 has $x_6 = 0.92$ and velocity of 600 ft/s. The rate of steam flow is 200 000 lbm/h, with 400 hp input to the pump. Piping diameters are 8 in. from steam generator to the turbine and 3 in. from the condenser to the steam generator. Determine the power output of the turbine and the heat transfer rate in the condenser.

6.92E For the same steam power plant as shown in Fig. P6.39 and Problem 6.91 determine the rate of heat transfer in the economizer which is a low temperature heat exchanger and the steam generator. Determine also the flow rate of cooling water through the condenser, if the cooling water increases from 55 to 75 F in the condenser.

6.93E A proposal is made to use a geothermal supply of hot water to operate a steam turbine, as shown in Fig. P6.46. The high pressure water at 200 lbf/in.^{2}, 350 F is throttled into a flash evaporator chamber, which forms liquid and vapor at a lower pressure of 60 lbf/in.^{2}. The liquid is discarded while the saturated vapor feeds the turbine and exits at 1 lbf/in.^{2}, 90% quality. If the turbine should produce 1000 hp, find the required mass flow rate of hot geothermal water in pound-mass per hour.

6.94E A 1-ft^3 tank, shown in Fig. P6.49, that is initially evacuated is connected by a valve to an air supply line flowing air at 70 F, 120 lbf/in.^{2}. The valve is opened, and air flows into the tank until the pressure reaches 90 lbf/in.^{2}. Determine the final temperature and mass inside the tank, assuming the process is adiabatic. Develop an expression for the relation between the line temperature and the final temperature using constant specific heats.

6.95E A 20-ft^3 tank contains ammonia at 20 lbf/in.^{2}, 80 F. The tank is attached to a line flowing ammonia at 180 lbf/in.^{2}, 140 F. The valve is opened, and mass flows in until the tank is half full of liquid, by volume at 80 F. Calculate the heat transferred from the tank during this process.

6.96E A 18-ft^3 insulated tank contains air at 100 F, 300 lbf/in.^{2}. A valve on the tank is opened, and air escapes until half the original mass is gone, at which point the valve is closed. What is the pressure inside then?

6.97E Air is contained in the insulated cylinder shown in Fig. P6.70. At this point the air is at 20 lbf/in.^{2}, 80 F, and the cylinder volume is 0.5 ft^3. The piston cross-sectional area is 0.5 ft^2, and the spring is linear with spring constant 200 lbf/in. The valve is opened, and air from the line at 100 lbf/in.^{2}, 80 F flows into the cylinder until the pressure reaches 100 lbf/in.^{2}, and then the valve is closed. Find the final temperature.

6.98E A 35-ft^3 insulated, 90-lbm rigid steel tank contains air at 75 lbf/in.^{2}, and both tank and air are at 70 F. The tank is connected to a line flowing air at 300 lbf/in.^{2}, 70 F. The valve is opened, allowing air to flow into the tank until the pressure reaches 250 lbf/in.^{2} and is then closed. Assume the air and tank are always at the same temperature and find the final temperature.

6.99E A cylinder fitted with a piston restrained by a linear spring contains 2 lbm of R-22 at 220 F, 125 lbf/in.^{2}. The system is shown in Fig. P6.72 where the spring constant is 285 lbf/in., and the piston cross-sectional area is 75 in.^{2}. A valve on the cylinder is opened and R-22 flows out until half the initial mass is left. Heat is transferred so the final temperature of the R-22 is 30 F. Find the final state of the R-22, (P_2, x_2) and the heat transfer to the cylinder.

6.100E An initially empty bottle, $V = 10$ ft^3, is filled with water from a line at 120 lbf/in.^{2}, 500 F. Assume no heat transfer and that the bottle is closed when the pressure reaches line pressure. Find the final temperature and mass in the bottle.

6.101E A mass-loaded piston/cylinder containing air is at 45 lbf/in.^{2}, 60 F with a volume of 9 ft^3, while at the stops $V = 36$ ft^3. An air line, 75 lbf/in.^{2}, 1100 R is connected by a valve, as shown in Fig. P6.63. The valve is then opened until a final inside pressure of 60 lbf/in.^{2} is reached, at which point $T = 630R$. Find the air mass that enters, the work, and heat transfer.

6.102E A nitrogen line, 540 R, 75 lbf/in.^{2} is connected to a turbine that exhausts to a closed initially empty tank of 2000 ft^3, as shown in Fig. P6.52. The turbine operates to a tank pressure of 75 lbf/in.^{2}, at which point the temperature is 450 R. Assuming the entire process is adiabatic, determine the turbine work.

COMPUTER, DESIGN, AND OPEN-ENDED PROBLEMS

6.103 Fit a polynomial expression of degree n in the temperature for ideal gas specific heat. Use the ideal gas enthalpy values for one of the substances listed in Table A.8 as data. The accuracy of the correlation should be studied as a function of the temperature range of the fit, as well as of the polynomial degree n.

6.104 An insulated tank of volume V contains a specified ideal gas (with constant specific heat) at P_1, T_1. A valve is opened, allowing the gas to flow out until the pressure inside drops to P_2. Determine T_2 and m_2 using a stepwise solution in increments of pressure between P_1 and P_2; the number of increments is variable.

6.105 We wish to solve Problem 6.57, using a stepwise solution, whereby the process is subdivided into several parts to minimize the effects of a linear average enthalpy approximation. Divide the process into two or three steps so you can get a better estimate for the mass times enthalpy leaving the tank.

6.106 Examine a process where air at 300 K, 100 kPa is compressed in a piston/cylinder arrangement to 600 kPa. Assume the process is polytropic with exponents in the 1.2–1.6 range. Find the work and heat transfer per unit mass of air. Discuss the different cases and how they may be accomplished by insulating the cylinder or by providing heating or cooling.

6.107 A cylindrical tank of height 2 m with a cross-sectional area of 0.5 m^2 contains hot water at 80°C, 125 kPa. It is in a room with temperature $T_0 = 20$°C so it slowly loses energy to the room air proportional to the temperature difference as

$$\dot{Q}_{loss} = CA(T - T_0)$$

with the tank surface area, A, and C is a constant. For different values of the constant C, estimate the time it takes to bring the water to 50°C. Make enough simplifying assumptions so you can solve the problem mathematically (i.e., find a formula for $T(t)$).

6.108 The air–water counterflowing heat exchanger given in Problem 6.13 has an air exit temperature of 360 K. Suppose the air exit temperature is listed as 300 K, then a ratio of the mass flow rates is found from the energy equation to be 5. Show that this is an impossible process by looking at air and water temperatures at several locations inside the heat exchanger. Discuss how this puts a limit on the energy that can be extracted from the air.

6.109 A coflowing heat exchanger receives air at 800 K, 1 MPa and liquid water at 15°C, 100 kPa, as shown in Fig. P6.109. The air line heats the water so at the exit the air temperature is 20°C above the water temperature. Investigate the limits for the air and water exit temperatures as a function of the ratio of the two mass flow rates. Plot the temperatures of the air and water inside the heat exchanger along the flow path.

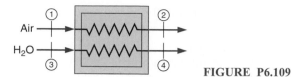

FIGURE P6.109

6.110 Consider the geothermal supply used to feed a turbine in Problem 6.46. Investigate the possibility for two flash evaporators, as shown in Fig. P6.110. Assume the turbine exit state is as listed and examine if there is an optimal choice for P_2 that will give the most turbine work per unit mass of the hot water supply.

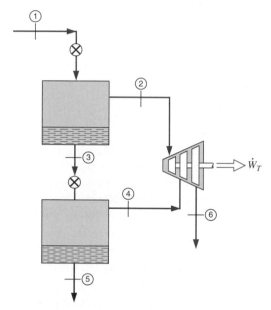

FIGURE P6.110

6.111 Consider the R-12 heat pump described in Problem 6.47. The compressor had a heat loss since the temperature is significantly higher than the surrounding temperature, 20°C. Investigate the effect of the heat loss on the work required to drive the compressor. Assume the same inlet state and exit pressure as given and vary the exit temperature. To calculate the specific work term, assume the process is polytropic and make a listing of corresponding work, heat transfer, and exit temperatures. Should the compressor be insulated?

THE SECOND LAW OF THERMODYNAMICS 7

The first law of thermodynamics states that during any cycle that a system undergoes, the cyclic integral of the heat is equal to the cyclic integral of the work. The first law, however, places no restrictions on the direction of flow of heat and work. A cycle in which a given amount of heat is transferred from the system and an equal amount of work is done on the system satisfies the first law just as well as a cycle in which the flows of heat and work are reversed. However, we know from our experience that because a proposed cycle does not violate the first law does not ensure that the cycle will actually occur. It is this kind of experimental evidence that led to the formulation of the second law of thermodynamics. Thus, a cycle will occur only if both the first and second laws of thermodynamics are satisfied.

In its broader significance the second law acknowledges that processes proceed in a certain direction but not in the opposite direction. A hot cup of coffee cools by virtue of heat transfer to the surroundings, but heat will not flow from the cooler surroundings to the hotter cup of coffee. Gasoline is used as a car drives up a hill, but the fuel level in the gasoline tank cannot be restored to its original level when the car coasts down the hill. Such familiar observations as these, and a host of others, are evidence of the validity of the second law of thermodynamics.

In this chapter, we consider the second law for a system undergoing a cycle, and in the next two chapters we extend the principles to a system undergoing a change of state and then to a control volume.

7.1 HEAT ENGINES AND REFRIGERATORS

Consider the system and the surroundings previously cited in the development of the first law, as shown in Fig. 7.1. Let the gas constitute the system and, as in our discussion of the first law, let this system undergo a cycle in which work is first done on the system by the paddle wheel as the weight is lowered. Then let the cycle be completed by transferring heat to the surroundings.

We know from our experience that we cannot reverse this cycle. That is, if we transfer heat to the gas, as shown by the dotted arrow, the temperature of the gas will increase, but the paddle wheel will not turn and raise the weight. With the given surroundings (the container, the paddle wheel, and the weight) this system can operate in a cycle in which the heat transfer and work are both negative, but it cannot operate in a cycle in which both the heat transfer and work are positive, even though this would not violate the first law.

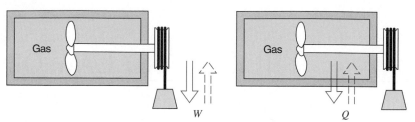

FIGURE 7.1 A system that undergoes a cycle involving work and heat.

Consider another cycle, which we know from our experience is impossible actually to complete. Let two systems, one at a high temperature and the other at a low temperature, undergo a process in which a quantity of heat is transferred from the high-temperature system to the low-temperature system. We know that this process can take place. We also know that the reverse process, in which heat is transferred from the low-temperature system to the high-temperature system, does not occur, and that it is impossible to complete the cycle by heat transfer only. This impossibility is illustrated in Fig. 7.2.

These two examples lead us to a consideration of the heat engine and the refrigerator, which is also referred to as a heat pump. With the heat engine we can have a system that operates in a cycle and performs a net positive work and a net positive heat transfer. With the heat pump we can have a system that operates in a cycle and has heat transferred to it from a low-temperature body and heat transferred from it to a high-temperature body, though work is required to do this. Three simple heat engines and two simple refrigerators will be considered.

The first heat engine is shown in Fig. 7.3. It consists of a cylinder fitted with appropriate stops and a piston. Let the gas in the cylinder constitute the system. Initially the piston rests on the lower stops, with a weight on the platform. Let the system now undergo a process in which heat is transferred from some high-temperature body to the gas, causing it to expand and raise the piston to the upper stops. At this point the weight is removed. Now let the system be restored to its initial state by transferring heat from the gas to a low-temperature body, thus completing the cycle. Since the weight was raised during the cycle, it is evident that work was done by the gas during the cycle. From the first law we conclude that the net heat transfer was positive and equal to the work done during the cycle.

Such a device is called a heat engine, and the substance to which and from which heat is transferred is called the working substance or working fluid. A heat engine may be defined as a device that operates in a thermodynamic cycle and does a certain amount of net positive work through the transfer of heat from a high-temperature body to a low-temperature body. Often the term heat engine is used in a broader sense to include all devices that produce work, either through heat transfer or through combustion, even though the device does not operate in a thermodynamic cycle. The internal combustion engine and

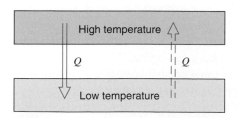

FIGURE 7.2 An example showing the impossibility of completing a cycle by transferring heat from a low-temperature body to a high-temperature body.

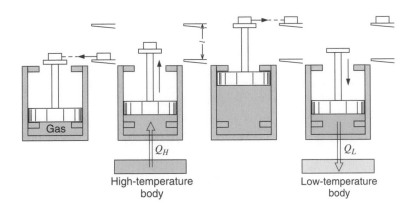

FIGURE 7.3 A simple heat engine.

the gas turbine are examples of such devices, and calling them heat engines is an acceptable use of the term. In this chapter, however, we are concerned with the more restricted form of heat engine, as just defined, one which operates on a thermodynamic cycle.

A simple steam power plant is an example of a heat engine in this restricted sense. Each component in this plant may be analyzed individually as a steady-state, steady-flow process, but as a whole it may be considered a heat engine (Fig. 7.4) in which water (steam) is the working fluid. An amount of heat, Q_H, is transferred from a high-temperature body, which may be the products of combustion in a furnace, a reactor, or a secondary fluid that in turn has been heated in a reactor. In Fig. 7.4 the turbine is shown schematically as driving the pump. What is significant, however, is the net work that is delivered during the cycle. The quantity of heat Q_L is rejected to a low-temperature body, which is usually the cooling water in a condenser. Thus, the simple steam power plant is a heat engine in the restricted sense, for it has a working fluid, to which and from which heat is transferred, and which does a certain amount of work as it undergoes a cycle.

Another example of a heat engine is the thermoelectric power generation device that was discussed in Chapter 1 and shown schematically in Fig. 1.10. Heat is transferred from a high-temperature body to the hot junction (Q_H), and heat is transferred from the cold junction to the surroundings (Q_L). Work is done in the form of electrical energy. Since there is no working fluid, we do not usually think of this as a device that operates in a cycle.

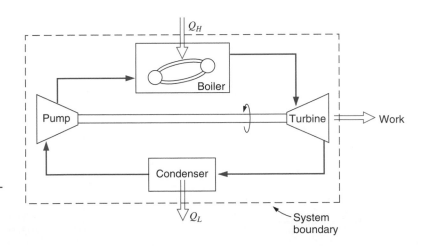

FIGURE 7.4 A heat engine involving steady-state, steady-flow processes.

However, if we adopt a microscopic point of view, we could regard a cycle as the flow of electrons. Furthermore, as with the steam power plant, the state at each point in the thermoelectric power generator does not change with time under steady-state conditions.

Thus, by means of a heat engine, we are able to have a system operate in a cycle and have both the net work and the net heat transfer positive, which we were not able to do with the system and surroundings of Fig. 7.1.

We note that in using the symbols Q_H and Q_L, we have departed from our sign connotation for heat, because for a heat engine Q_L is negative when the working fluid is considered as the system. In this chapter it will be advantageous to use the symbol Q_H to represent the heat transfer to or from the high-temperature body, and Q_L to represent the heat transfer to or from the low-temperature body. The direction of the heat transfer will be evident from the context.

At this point it is appropriate to introduce the concept of thermal efficiency of a heat engine. In general, we say that efficiency is the ratio of output, the energy sought, to input, the energy that costs, but these output and input must be clearly defined. At the risk of oversimplification, we may say that in a heat engine the energy sought is the work, and the energy that costs money is the heat from the high-temperature source (indirectly, the cost of the fuel). Thermal efficiency is defined as

$$\eta_{\text{thermal}} = \frac{W\left(\text{energy sought}\right)}{Q_H\left(\text{energy that costs}\right)} = \frac{Q_H - Q_L}{Q_H} = 1 - \frac{Q_L}{Q_H} \qquad (7.1)$$

The second cycle that we were not able to complete was the one indicating the impossibility of transferring heat directly from a low-temperature body to a high-temperature body. This can of course be done with a refrigerator or heat pump. A vapor-compression refrigerator cycle, which was introduced in Chapter 1 and shown in Fig. 1.7, is shown again in Fig. 7.5. The working fluid is the refrigerant, such as R-134a or ammonia, which goes through a thermodynamic cycle. Heat is transferred to the refrigerant in the evaporator, where its pressure and temperature are low. Work is done on the refrigerant in the compressor and heat is transferred from it in the condenser, where its pressure and temperature are high. The pressure drops as the refrigerant flows through the throttle valve or capillary tube.

Thus, in a refrigerator or heat pump, we have a device that operates in a cycle, that requires work, and that accomplishes the objective of transferring heat from a low-temperature body to a high-temperature body.

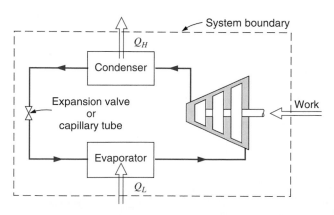

FIGURE 7.5 A simple refrigeration cycle.

The thermoelectric refrigerator, which was discussed in Chapter 1 and is shown schematically in Fig. 1.9, is another example of a device that meets our definition of a refrigerator. The work input to the thermoelectric refrigerator is in the form of electrical energy, and heat is transferred from the refrigerated space to the cold junction (Q_L) and from the hot junction to the surroundings (Q_H).

The "efficiency" of a refrigerator is expressed in terms of the coefficient of performance, which we designate with the symbol β. For a refrigerator the objective, that is, the energy sought, is Q_L, the heat transferred from the refrigerated space. The energy that costs is the work W. Thus, the coefficient of performance, β,[1] is

$$\beta = \frac{Q_L(\text{energy sought})}{W(\text{energy that costs})} = \frac{Q_L}{Q_H - Q_L} = \frac{1}{Q_H / Q_L - 1} \tag{7.2}$$

Before we state the second law, the concept of a thermal reservoir should be introduced. A thermal reservoir is a body to which and from which heat can be transferred indefinitely without change in the temperature of the reservoir. Thus, a thermal reservoir always remains at constant temperature. The ocean and the atmosphere approach this definition very closely. Frequently it will be useful to designate a high-temperature reservoir and a low-temperature reservoir. Sometimes a reservoir from which heat is transferred is called a source, and a reservoir to which heat is transferred is called a sink.

7.2 THE SECOND LAW OF THERMODYNAMICS

On the basis of the matter considered in the previous section, we are now ready to state the second law of thermodynamics. There are two classical statements of the second law, known as the Kelvin–Planck statement and the Clausius statement.

The Kelvin–Planck statement: It is impossible to construct a device that will operate in a cycle and produce no effect other than the raising of a weight and the exchange of heat with a single reservoir. See Fig. 7.6.

This statement ties in with our discussion of the heat engine. In effect, it states that it is impossible to construct a heat engine that operates in a cycle, receives a given amount of heat from a high-temperature body, and does an equal amount of work. The

---

[1]It should be noted that a refrigeration or heat pump cycle can be used with either of two objectives. It can be used as a refrigerator, in which case the primary objective is Q_L, the heat transferred to the refrigerant from the refrigerated space. It can also be used as a heating system (in which case it is usually referred to as a heat pump), the objective being Q_H, the heat transferred from the refrigerant to the high-temperature body, which is the space to be heated. Q_L is transferred to the refrigerant from the ground, the atmospheric air, or well water. The coefficient of performance for this case, β' is

$$\beta' = \frac{Q_H(\text{energy sought})}{W(\text{energy that costs})} = \frac{Q_H}{Q_H - Q_L} = \frac{1}{1 - Q_L / Q_H}$$

It also follows that for a given cycle,

$$\beta' - \beta = 1$$

Unless otherwise specified, the term coefficient of performance will always refer to a refrigerator as defined by Eq. 7.2.

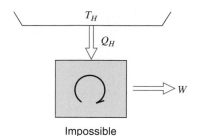

Impossible

FIGURE 7.6 The Kelvin-Planck statement.

only alternative is that some heat must be transferred from the working fluid at a lower temperature to a low-temperature body. Thus, work can be done by the transfer of heat only if there are two temperature levels, and heat is transferred from the high-temperature body to the heat engine and also from the heat engine to the low-temperature body. This implies that it is impossible to build a heat engine that has a thermal efficiency of 100%.

The Clausius statement: It is impossible to construct a device that operates in a cycle and produces no effect other than the transfer of heat from a cooler body to a hotter body. See Fig. 7.7

This statement is related to the refrigerator or heat pump. In effect, it states that it is impossible to construct a refrigerator that operates without an input of work. This also implies that the coefficient of performance is always less than infinity.

Three observations should be made about these two statements. The first observation is that both are negative statements. It is of course impossible to "prove" a negative statement. However, we can say that the second law of thermodynamics (like every other law of nature) rests on experimental evidence. Every relevant experiment that has been conducted either directly or indirectly verifies the second law, and no experiment has ever been conducted that contradicts the second law. The basis of the second law is therefore experimental evidence.

A second observation is that these two statements of the second law are equivalent. Two statements are equivalent if the truth of each statement implies the truth of the other, or if the violation of each statement implies the violation of the other. That a violation of the Clausius statement implies a violation of the Kelvin–Planck statement may be shown. The device at the left in Fig. 7.8 is a refrigerator that requires no work and thus violates the Clausius statement. Let an amount of heat Q_L be transferred from the low-temperature reservoir to this refrigerator, and let the same amount of heat Q_L be transferred to the

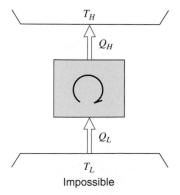

Impossible

FIGURE 7.7 The Clausius statement.

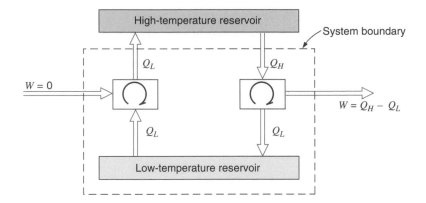

FIGURE 7.8
Demonstration of the equivalence of the two statements of the second law.

high-temperature reservoir. Let an amount of heat Q_H that is greater than Q_L be transferred from the high-temperature reservoir to the heat engine, and let the engine reject the amount of heat Q_L as it does an amount of work W, which equals $Q_H - Q_L$. Because there is no net heat transfer to the low-temperature reservoir, the low-temperature reservoir, the heat engine, and the refrigerator can be considered together as a device that operates in a cycle and produces no effect other than the raising of a weight (work) and the exchange of heat with a single reservoir. Thus, a violation of the Clausius statement implies a violation of the Kelvin–Planck statement. The complete equivalence of these two statements is established when it is also shown that a violation of the Kelvin–Planck statement implies a violation of the Clausius statement. This is left as an exercise for the student.

The third observation is that frequently the second law of thermodynamics has been stated as the impossibility of constructing a perpetual-motion machine of the second kind. A perpetual-motion machine of the first kind would create work from nothing or create mass or energy, thus violating the first law. A perpetual-motion machine of the second kind would extract heat from a source and then convert this heat completely into other forms of energy, thus violating the second law. A perpetual-motion machine of the third kind would have no friction, and thus would run indefinitely but produce no work.

A heat engine that violated the second law could be made into a perpetual-motion machine of the second kind by taking the following steps. Consider Fig. 7.9, which might be the power plant of a ship. An amount of heat Q_L is transferred from the ocean to a high-temperature body by means of a heat pump. The work required is W', and the heat

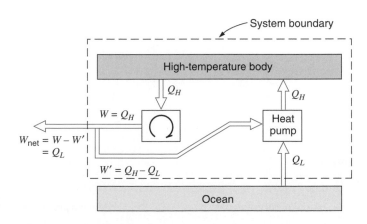

FIGURE 7.9 A perpetual-motion machine of the second kind.

transferred to the high-temperature body is Q_H. Let the same amount of heat be transferred to a heat engine that violates the Kelvin–Planck statement of the second law and does an amount of work $W = Q_H$. Of this work an amount $Q_H - Q_L$ is required to drive the heat pump, leaving the net work ($W_{net} = Q_L$) available for driving the ship. Thus, we have a perpetual-motion machine in the sense that work is done by utilizing freely available sources of energy such as the ocean or atmosphere.

7.3 THE REVERSIBLE PROCESS

The question that can now logically be posed is this: If it is impossible to have a heat engine of 100% efficiency, what is the maximum efficiency one can have? The first step in the answer to this question is to define an ideal process, which is called a reversible process.

A reversible process for a system is defined as a process that once having taken place can be reversed and in so doing leave no change in either system or surroundings.

Let us illustrate the significance of this definition for a gas contained in a cylinder that is fitted with a piston. Consider first Fig. 7.10, in which a gas, which we define as the system, is restrained at high pressure by a piston that is secured by a pin. When the pin is removed, the piston is raised and forced abruptly against the stops. Some work is done by the system, since the piston has been raised a certain amount. Suppose we wish to restore the system to its initial state. One way of doing this would be to exert a force on the piston and thus compress the gas until the pin can be reinserted in the piston. Since the pressure on the face of the piston is greater on the return stroke than on the initial stroke, the work done on the gas in this reverse process is greater than the work done by the gas in the initial process. An amount of heat must be transferred from the gas during the reverse stroke so that the system has the same internal energy as it had originally. Thus, the system is restored to its initial state, but the surroundings have changed by virtue of the fact that work was required to force the piston down and heat was transferred to the surroundings. The initial process therefore is an irreversible one because it could not be reversed without leaving a change in the surroundings.

In Fig. 7.11 let the gas in the cylinder comprise the system, and let the piston be loaded with a number of weights. Let the weights be slid off horizontally one at a time, allowing the gas to expand and do work in raising the weights that remain on the piston. As the size of the weights is made smaller and their number is increased, we approach a process that can be reversed, for at each level of the piston during the reverse process there will be a small weight that is exactly at the level of the platform and thus can be placed on the platform without requiring work. In the limit, therefore, as the weights be-

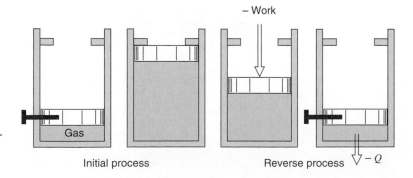

FIGURE 7.10 An example of an irreversible process.

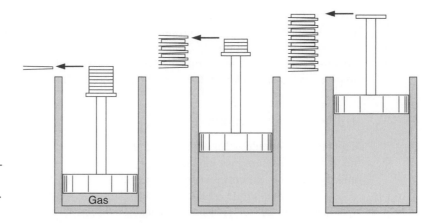

FIGURE 7.11 An example of a process that approaches being reversible.

come very small, the reverse process can be accomplished in such a manner that both the system and surroundings are in exactly the same state they were initially. Such a process is a reversible process.

7.4 FACTORS THAT RENDER PROCESSES IRREVERSIBLE

There are many factors that make processes irreversible. Four of those factors—friction, unrestrained expansion, heat transfer through a finite temperature difference, and mixing of two different substances—are considered in this section.

Friction

It is readily evident that friction makes a process irreversible, but a brief illustration may amplify the point. Let a block and an inclined plane make up a system, as in Fig. 7.12, and let the block be pulled up the inclined plane by weights that are lowered. A certain amount of work is needed to do this. Some of this work is required to overcome the friction between the block and the plane, and some is required to increase the potential energy of the block. The block can be restored to its initial position by removing some of the weights and thus allowing the block to slide back down the plane. Some heat transfer

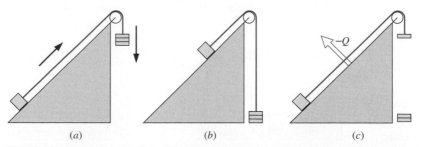

FIGURE 7.12 Demonstration of the fact that friction makes processes irreversible.

from the system to the surroundings will no doubt be required to restore the block to its initial temperature. Since the surroundings are not restored to their initial state at the conclusion of the reverse process, we conclude that friction has rendered the process irreversible. Another type of frictional effect is that associated with the flow of viscous fluids in pipes and passages and in the movement of bodies through viscous fluids.

Unrestrained Expansion

The classic example of an unrestrained expansion, as shown in Fig. 7.13, is a gas separated from a vacuum by a membrane. Consider what happens when the membrane breaks and the gas fills the entire vessel. It can be shown that this is an irreversible process by considering what would be necessary to restore the system to its original state. The gas would have to be compressed and heat transferred from the gas until its initial state is reached. Since the work and heat transfer involve a change in the surroundings, the surroundings are not restored to their initial state, indicating that the unrestrained expansion was an irreversible process. The process described in Fig. 7.10 is also an example of an unrestrained expansion.

In the reversible expansion of a gas there must be only an infinitesimal difference between the force exerted by the gas and the restraining force, so that the rate at which the boundary moves will be infinitesimal. In accordance with our previous definition, this is a quasi-equilibrium process. However, actual systems have a finite difference in forces, which causes a finite rate of movement of the boundary, and thus the processes are irreversible in some degree.

Heat Transfer Through a Finite Temperature Difference

Consider as a system a high-temperature body and a low-temperature body, and let heat be transferred from the high-temperature body to the low-temperature body. The only way in which the system can be restored to its initial state is to provide refrigeration, which requires work from the surroundings, and some heat transfer to the surroundings will also be necessary. Because of the heat transfer and the work, the surroundings are not restored to their original state, indicating that the process was irreversible.

An interesting question is now posed. Heat is defined as energy that is transferred through a temperature difference. We have just shown that heat transfer through a temperature difference is an irreversible process. Therefore, how can we have a reversible heat-transfer process? A heat-transfer process approaches a reversible process as the temperature difference between the two bodies approaches zero. Therefore, we define a reversible heat-transfer process as one in which the heat is transferred through an infinitesimal temperature difference. We realize of course that to transfer a finite amount of heat through an infinitesimal temperature difference would require an infinite amount of time

FIGURE 7.13
Demonstration of the fact that unrestrained expansion makes processes irreversible.

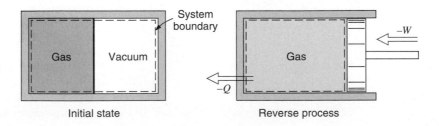

Initial state   Reverse process

or infinite area. Therefore, all actual heat transfers are through a finite temperature difference and hence are irreversible, the greater the temperature difference, the greater the irreversibility. We will find, however, that the concept of reversible heat transfer is very useful in describing ideal processes.

Mixing of Two Different Substances

Figure 7.14 illustrates the process of mixing two different gases separated by a membrane. When the membrane is broken, a homogeneous mixture of oxygen and nitrogen fills the entire volume. This process will be considered in some detail in Chapter 12. We can say here that this may be considered a special case of an unrestrained expansion, for each gas undergoes an unrestrained expansion as it fills the entire volume. A certain amount of work is necessary to separate these gases. Thus, an air separation plant such as described in Chapter 1 requires an input of work to accomplish the separation.

Other Factors

There are a number of other factors that make processes irreversible, but they will not be considered in detail here. Hysteresis effects and the i^2R loss encountered in electrical circuits are both factors that make processes irreversible. Ordinary combustion is also an irreversible process.

It is frequently advantageous to distinguish between internal and external irreversibility. Figure 7.15 shows two identical systems to which heat is transferred. Assuming each system to be a pure substance, the temperature remains constant during the heat-transfer process. In one system the heat is transferred from a reservoir at a temperature $T + dT$, and in the other the reservoir is at a much higher temperature, $T + \Delta T$, than the system. The first is a reversible heat-transfer process and the second is an irreversible heat-transfer process. However, as far as the system itself is concerned, it passes through exactly the same states in both processes, which we assume are reversible. Thus, we can say for the second system that the process is internally reversible but externally irreversible because the irreversibility occurs outside the system.

We should also note the general interrelation of reversibility, equilibrium, and time. In a reversible process, the deviation from equilibrium is infinitesimal, and therefore it occurs at an infinitesimal rate. Since it is desirable that actual processes proceed at a finite rate, the deviation from equilibrium must be finite, and therefore the actual process is irreversible in some degree. The greater the deviation from equilibrium, the greater the irreversibility, and the more rapidly the process will occur. It should also be noted that the quasi-equilibrium process, which was described in Chapter 2, is a reversible process, and hereafter the term reversible process will be used.

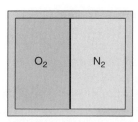 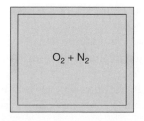

FIGURE 7.14 Demonstration of the fact that the mixing of two different substances is an irreversible process.

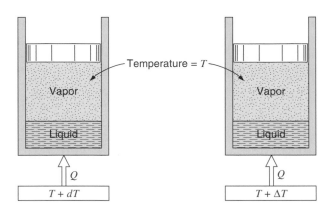

FIGURE 7.15 Illustration of the difference between an internally and externally reversible process.

7.5 THE CARNOT CYCLE

Having defined the reversible process and considered some factors that make processes irreversible, let us again pose the question raised in Section 7.3. If the efficiency of all heat engines is less than 100%, what is the most efficient cycle we can have? Let us answer this question for a heat engine that receives heat from a high-temperature reservoir and rejects heat to a low-temperature reservoir. Since we are dealing with reservoirs, we recognize that both the high temperature and the low temperature of the reservoirs are constant and remain constant regardless of the amount of heat transferred.

Let us assume that this heat engine, which operates between the given high-temperature and low-temperature reservoirs, does so in a cycle in which every process is reversible. If every process is reversible, the cycle is also reversible; and if the cycle is reversed, the heat engine becomes a refrigerator. In the next section we will show that this is the most efficient cycle that can operate between two constant-temperature reservoirs. It is called the Carnot cycle and is named after a French engineer, Nicolas Leonard Sadi Carnot (1796–1832), who expressed the foundations of the second law of thermodynamics in 1824.

We now turn our attention to the Carnot cycle. Figure 7.16 shows a power plant that is similar in many respects to a simple steam power plant and, we assume, operates on the Carnot cycle. Consider the working fluid to be a pure substance, such as steam. Heat is transferred from the high-temperature reservoir to the water (steam) in the boiler. For this process to be a reversible heat transfer, the temperature of the water (steam) must be only infinitesimally lower than the temperature of the reservoir. This result also implies, since the temperature of the reservoir remains constant, that the temperature of the water must remain constant. Therefore, the first process in the Carnot cycle is a reversible isothermal process in which heat is transferred from the high-temperature reservoir to the working fluid. A change of phase from liquid to vapor at constant pressure is of course an isothermal process for a pure substance.

The next process occurs in the turbine without heat transfer and is therefore adiabatic. Since all processes in the Carnot cycle are reversible, this must be a reversible adiabatic process, during which the temperature of the working fluid decreases from the temperature of the high-temperature reservoir to the temperature of the low-temperature reservoir.

In the next process heat is rejected from the working fluid to the low-temperature reservoir. This must be a reversible isothermal process in which the temperature of the

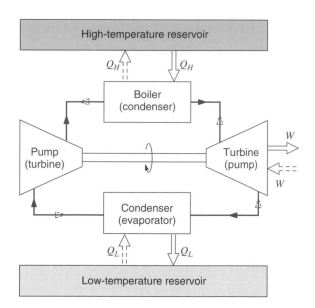

FIGURE 7.16 Example of a heat engine that operates on a Carnot cycle.

working fluid is infinitesimally higher than that of the low-temperature reservoir. During this isothermal process some of the steam is condensed.

The final process, which completes the cycle, is a reversible adiabatic process in which the temperature of the working fluid increases from the low temperature to the high temperature. If this were to be done with water (steam) as the working fluid, a mixture of liquid and vapor would have to be taken from the condenser and compressed. (This would be very inconvenient in practice, and therefore in all power plants the working fluid is completely condensed in the condenser. The pump handles only the liquid phase.)

Since the Carnot heat engine cycle is reversible, every process could be reversed, in which case it would become a refrigerator. The refrigerator is shown by the dotted lines and parentheses in Fig. 7.16. The temperature of the working fluid in the evaporator would be infinitesimally less that the temperature of the low-temperature reservoir, and in the condenser it is infinitesimally higher than that of the high-temperature reservoir.

It should be emphasized that the Carnot cycle can be executed in many different ways. Many different working substances can be used, such as a gas or a thermoelectric device such as described in Chapter 1. There are also various possible arrangements of machinery. For example, a Carnot cycle can be devised that takes place entirely within a cylinder, using a gas as a working substance, as shown in Fig. 7.17.

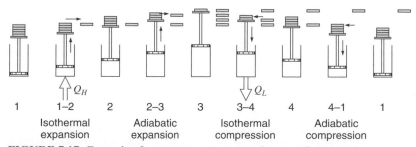

FIGURE 7.17 Example of a gaseous system operating on a Carnot cycle.

The important point to be made here is that the Carnot cycle, regardless of what the working substance may be, always has the same four basic processes. These processes are

1. A reversible isothermal process in which heat is transferred to or from the high-temperature reservoir.
2. A reversible adiabatic process in which the temperature of the working fluid decreases from the high temperature to the low temperature.
3. A reversible isothermal process in which heat is transferred to or from the low-temperature reservoir.
4. A reversible adiabatic process in which the temperature of the working fluid increases from the low temperature to the high temperature.

7.6 TWO PROPOSITIONS REGARDING THE EFFICIENCY OF A CARNOT CYCLE

There are two important propositions regarding the efficiency of a Carnot cycle.

First Proposition

It is impossible to construct an engine that operates between two given reservoirs and is more efficient than a reversible engine operating between the same two reservoirs.

The proof of this statement is accomplished through a "thought experiment." An initial assumption is made, and it is then shown that this assumption leads to impossible conclusions. The only possible conclusion is that the initial assumption was incorrect.

Let us assume that there is an irreversible engine operating between two given reservoirs that has a greater efficiency than a reversible engine operating between the same two reservoirs. Let the heat transfer to the irreversible engine be Q_H, the heat rejected be Q_L', and the work be W_{IE} (which equals $Q_H - Q_L'$) as shown in Fig. 7.18. Let the reversible engine operate as a refrigerator (this is possible since it is reversible). Finally,

FIGURE 7.18
Demonstration of the fact that the Carnot cycle is the most efficient cycle operating between two fixed-temperature reservoirs.

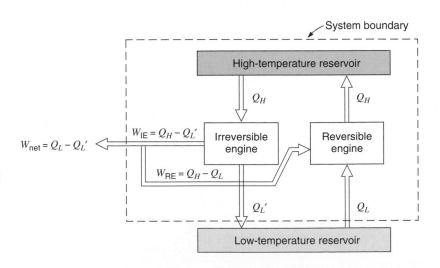

let the heat transfer with the low-temperature reservoir be Q_L, the heat transfer with the high-temperature reservoir be Q_H, and the work required be W_{RE} (which equals $Q_H - Q_L$).

Since the initial assumption was that the irreversible engine is more efficient, it follows (because Q_H is the same for both engines) that $Q'_L < Q_L$ and $W_{IE} > W_{RE}$. Now the irreversible engine can drive the reversible engine and still deliver the net work W_{net}, which equals $W_{IE} - W_{RE} = Q_L - Q'_L$. If we consider the two engines and the high-temperature reservoir as a system, as indicated in Fig. 7.18 it is a system that operates in a cycle, exchanges heat with a single reservoir, and does a certain amount of work. However, this would constitute a violation of the second law, and we conclude that our initial assumption (that the irreversible engine is more efficient than a reversible engine) is incorrect. Therefore, we cannot have an irreversible engine that is more efficient than a reversible engine operating between the same two reservoirs.

Second Proposition

All engines that operate on the Carnot cycle between two given constant-temperature reservoirs have the same efficiency. The proof of this proposition is similar to the proof just outlined, which assumes that there is one Carnot cycle that is more efficient than another Carnot cycle operating between the same temperature reservoirs. Let the Carnot cycle with the higher efficiency replace the irreversible cycle of the previous argument, and the Carnot cycle with the lower efficiency operate as the refrigerator. The proof proceeds with the same line of reasoning as in the first proposition. The details are left as an exercise for the student.

7.7 THE THERMODYNAMIC TEMPERATURE SCALE

In discussing the matter of temperature in Chapter 2, we pointed out that the zeroth law of thermodynamics provides a basis for temperature measurement, but that a temperature scale must be defined in terms of a particular thermometer substance and device. A temperature scale that is independent of any particular substance, which might be called an absolute temperature scale, would be most desirable. In the last paragraph we noted that the efficiency of a Carnot cycle is independent of the working substance and depends only on the temperature. This fact provides the basis for such an absolute temperature scale, which we call the thermodynamic scale.

The concept of this temperature scale may be developed with the help of Fig. 7.19, which shows three reservoirs and three engines that operate on the Carnot cycle. T_1 is the highest temperature, T_3 is the lowest temperature, and T_2 is an intermediate temperature, and the engines operate between the various reservoirs as indicated. Q_1 is the same for both A and C and, since we are dealing with reversible cycles, Q_3 is the same for B and C.

Since the efficiency of a Carnot cycle is a function only of the temperature, we can write

$$\eta_{thermal} = 1 - \frac{Q_L}{Q_H} = 1 - \psi(T_L, T_H) \tag{7.3}$$

where ψ designates a functional relation.

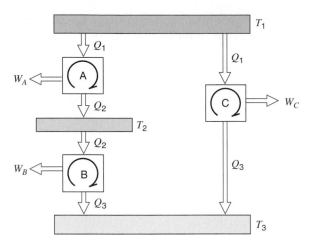

FIGURE 7.19 Arrangement of heat engines to demonstrate the thermodynamic temperature scale.

Let us apply this functional relation to the three Carnot cycles of Fig. 7.19.

$$\frac{Q_1}{Q_2} = \psi(T_1, T_2)$$

$$\frac{Q_2}{Q_3} = \psi(T_2, T_3)$$

$$\frac{Q_1}{Q_3} = \psi(T_1, T_3)$$

Since

$$\frac{Q_1}{Q_3} = \frac{Q_1 Q_2}{Q_2 Q_3}$$

it follows that

$$\psi(T_1, T_3) = \psi(T_1, T_2) \times \psi(T_2, T_3) \tag{7.4}$$

Note that the left side is a function of T_1 and T_3 (and not of T_2), and therefore the right side of this equation must also be a function of T_1 and T_3 (and not of T_2). From this fact we can conclude that the form of the function ψ must be such that

$$\psi(T_1, T_2) = \frac{f(T_1)}{f(T_2)}$$

$$\psi(T_2, T_3) = \frac{f(T_2)}{f(T_3)}$$

for in this way $f(T_2)$ will cancel from the product of $\psi(T_1, T_2) \times \psi(T_2, T_3)$. Therefore, we conclude that

$$\frac{Q_1}{Q_3} = \psi(T_1, T_3) = \frac{f(T_1)}{f(T_3)} \tag{7.5}$$

In general terms,

$$\frac{Q_H}{Q_L} = \frac{f(T_H)}{f(T_L)} \tag{7.6}$$

Now there are several functional relations that will satisfy this equation. For the thermodynamic scale of temperature, which was originally proposed by Lord Kelvin, the selected relation is

$$\frac{Q_H}{Q_L} = \frac{(T_H)}{(T_L)} \tag{7.7}$$

With absolute temperatures so defined, the efficiency of a Carnot cycle may be expressed in terms of the absolute temperatures.[2]

$$\eta_{\text{thermal}} = 1 - \frac{Q_L}{Q_H} = 1 - \frac{T_L}{T_H} \tag{7.8}$$

This means that if the thermal efficiency of a Carnot cycle operating between two given constant-temperature reservoirs is known, the ratio of the two absolute temperatures is also known.

It should be noted that Eq. 7.7 gives us a ratio of absolute temperatures, but it does not give us information about the magnitude of the degree. Let us first consider a qualitative approach to this matter and then a more rigorous statement.

Suppose we had a heat engine operating on the Carnot cycle that received heat at the temperature of the steam point and rejected heat at the temperature of the ice point. (Because a Carnot cycle involves only reversible processes, it is impossible to construct such a heat engine and perform the proposed experiment. However, we can follow the reasoning as a "thought experiment" and gain additional understanding of the thermodynamic temperature scale.) If the efficiency of such an engine could be measured, we would find it to be 26.80%. Therefore, from Eq. 7.8,

$$\eta_{th} = 1 - \frac{T_L}{T_H} = 1 - \frac{T_{\text{ice point}}}{T_{\text{steam point}}} = 0.2680$$

$$\frac{T_{\text{ice point}}}{T_{\text{steam point}}} = 0.7320$$

---

[2]Lord Kelvin also proposed a logarithmic scale of the form

$$\frac{Q_H}{Q_L} = \frac{e^{"T_H"}}{e^{"T_L"}}$$

where $"T_H"$ and $"T_L"$ designate the absolute temperatures on this proposed logarithmic scale. This relation can also be written

$$\ln\frac{Q_H}{Q_L} = "T_H" - "T_L"$$

The form Kelvin actually proposed was

$$\log_{10}\frac{Q_H}{Q_L} = "T_H" - "T_L"$$

Thus, the relation between the scale in use and the proposed logarithmic scale is

$$"T" = \log_{10} T + L$$

where L is a constant that determines the level of temperature that corresponds to zero on the logarithmic scale. On this logarithmic scale temperatures range from $-\infty$ to $+\infty$, whereas on the thermodynamic scale in use they vary from 0 to $+\infty$ for ordinary systems.

This gives us one equation concerning the two unknowns T_H and T_L. The second equation comes from an arbitrary decision regarding the magnitude of the degree on the thermodynamic temperature scale. If we wish to have the magnitude of the degree on the absolute scale correspond to the magnitude of the degree on the Celsius scale, we can write

$$T_{\text{steam point}} - T_{\text{ice point}} = 100$$

Solving these two equations simultaneously, we find

$$T_{\text{steam point}} = 373.15 \text{ K}, \qquad T_{\text{ice point}} = 273.15 \text{ K}$$

It follows that

$$T(^\circ\text{C}) + 273.15 = T(\text{K})$$

The absolute scale related to the Fahrenheit scale is the Rankine scale, designated by R. On both these scales there are 180 degrees between the ice point and the steam point. Therefore, for a Carnot cycle heat engine operating between the steam point and the ice point, we would have the two relations

$$T_{\text{steam point}} - T_{\text{ice point}} = 180$$
$$\frac{T_{\text{ice point}}}{T_{\text{steam point}}} = 0.7320$$

Solving these two equations simultaneously, we find

$$T_{\text{steam point}} = 671.67 \text{ R}, \qquad T_{\text{ice point}} = 491.67 \text{ R}$$

It follows that temperatures on the Fahrenheit and Rankine scales are related as follows:

$$T(\text{F}) + 459.67 = T(\text{R})$$

As already noted, the measurement of efficiencies of Carnot cycles is, however, not a practical way to approach the problem of temperature measurement on the thermodynamic scale of temperature. The actual approach is based on the ideal-gas thermometer and an assigned value for the triple point of water. At the Tenth Conference on Weights and Measures, which was held in 1954, the temperature of the triple point of water was assigned the value 273.16 K. [The triple point of water is approximately 0.01°C above the ice point. The ice point is defined as the temperature of a mixture of ice and water at a pressure of 1 atm (101.3 kPa) of air that is saturated with water vapor.] The ideal-gas thermometer is discussed in the following section.

EXAMPLE 7.1 Let us consider the heat engine, schematically shown in Fig. 7.20, that receives a heat transfer rate of 1 MW at a high temperature of 550°C and rejects energy to the ambient at 300 K. Work is produced at a rate of 450 kW. We would like to know how much energy is discarded to the ambient, the engine efficiency and compare both of these to a Carnot heat engine operating between the same two reservoirs.

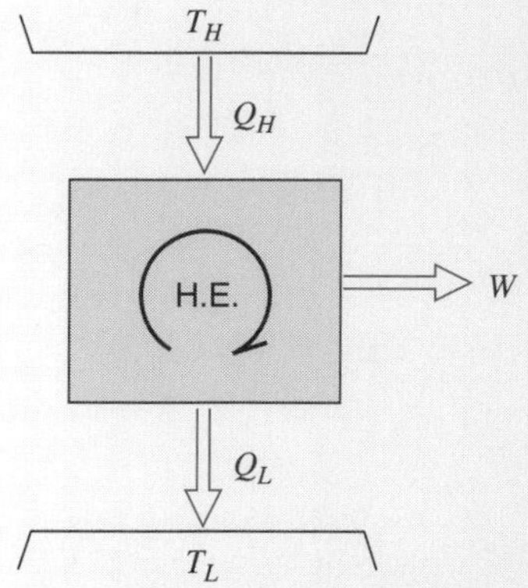

FIGURE 7.20 A heat engine operating between two constant temperature energy reservoirs for Ex. 7.1.

Solution

Taking the heat engine as a control volume, the energy equation gives

$$\dot{Q}_L = \dot{Q}_H - \dot{W} = 1000 - 450 = 550 \text{ kW}$$

and from the definition of the efficiency

$$\eta_{thermal} = \dot{W}/\dot{Q}_H = 450/1000 = 0.45$$

For the Carnot heat engine, the efficiency is given by the temperature of the reservoirs

$$\eta_{\text{Carnot}} = 1 - \frac{T_L}{T_H} = 1 - \frac{300}{550 + 273} = 0.635$$

The rates of work and heat rejection become

$$\dot{W} = \eta_{\text{Carnot}}\dot{Q}_H = 0.635 \times 1000 = 635 \text{ kW}$$
$$\dot{Q}_L = \dot{Q}_H - \dot{W} = 1000 - 635 = 365 \text{ kW}$$

The actual heat engine thus has a lower efficiency than the Carnot (ideal) heat engine, with a value of 45% typical for a modern steam power plant. This also implies that the actual engine rejects a larger amount of energy to the ambient (55%) compared with the Carnot heat engine (36%).

EXAMPLE 7.2

As one mode of operation of an air-conditioner is the cooling of a room on a hot day, it works as a refrigerator, shown in Fig. 7.21. A total of 4 kW should be removed from a room at 24°C to the outside atmosphere at 35°C. We would like to estimate the magnitude of the required work. To do this we will not analyze the processes inside the refrigerator, which is deferred to Chapter 11, but we can give a lower limit for the rate of work assuming it is a Carnot cycle refrigerator.

Solution

The coefficient of performance (COP) is

$$\beta = \frac{\dot{Q}_L}{\dot{W}} = \frac{\dot{Q}_L}{\dot{Q}_H - \dot{Q}_L} = \frac{T_L}{T_H - T_L} = \frac{273 + 24}{35 - 24} = 27$$

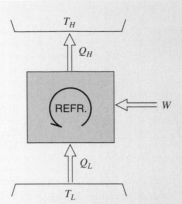

FIGURE 7.21 An air-conditioner in cooling mode where T_L is the room, Example 7.2.

so the rate of work or power input will be

$$\dot{W} = \dot{Q}_L / \beta = 4/27 = 0.15 \text{ kW}$$

Since the power was estimated assuming a Carnot refrigerator, it is the smallest amount possible. Recall also the expressions for heat transfer rates in Chapter 4. If the refrigerator should push 4.15 kW out to the atmosphere at 35°C the high temperature side of it should be at a higher temperature, maybe 45°C, to have a reasonably small sized heat exchanger. As it cools the room, a flow of air of less than say 18°C would be needed. Redoing the COP with a high of 45°C and a low of 18°C gives 10.8 which is more realistic. A real refrigerator would operate with a COP of the order of 5.

7.8 THE IDEAL-GAS TEMPERATURE SCALE

In this section we consider the ideal-gas scale of temperature. This scale is based on the observation that as the pressure of a gas approaches zero, its equation of state approaches that of an ideal gas:

$$Pv = RT$$

Consider how an ideal gas might be used to measure temperature in a constant-volume gas thermometer, which is shown schematically in Fig. 7.22. Let the gas bulb be placed in the location where the temperature is to be measured, and let the mercury column be adjusted so that the level of mercury stands at the reference mark A. Thus, the volume of the gas remains constant. Assume that the gas in the capillary tube is at the same temperature as the gas in the bulb. Then the pressure of the gas, which is indicated by the height L of the mercury column, is a measure of the temperature.

Let the pressure that is associated with the temperature of the triple point of water (273.16 K) also be measured, and let us designate this pressure $P_{\text{t.p.}}$. Then, from the definition of an ideal gas, any other temperature T could be determined from a pressure measurement P by the relation

$$T = 273.16 \left(\frac{P}{P_{t.p.}} \right)$$

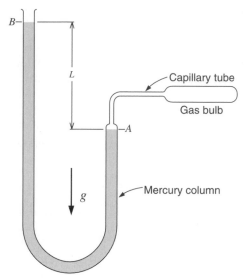

FIGURE 7.22 Schematic diagram of a constant-volume gas thermometer.

The temperature so measured is referred to as the ideal-gas temperature, and it can be shown that the temperature so measured is exactly equal to the thermodynamic temperature.

From a practical point of view, we have the problem that no gas behaves exactly like an ideal gas. However, we do know that as the pressure approaches zero, the behavior of all gases approaches that of an ideal gas. Suppose then that a series of measurements is made with varying amounts of gas in the gas bulb. This means that the pressure measured at the triple point, and also the pressure at any other temperature, will vary. If the indicated temperature T_i (obtained by assuming that the gas is ideal) is plotted against the pressure of gas with the bulb at the triple point of water, a curve like the one shown in Fig. 7.23 is obtained. When this curve is extrapolated to zero pressure, the correct ideal-gas temperature is obtained. Different curves might result from different gases, but they would all indicate the same temperature at zero pressure.

We have outlined only the general features and principles for measuring temperature on the ideal-gas scale of temperatures. Precision work in this field is difficult and laborious, and there are only a few laboratories in the world where this precision work is carried on. The International Temperature Scale, which was mentioned in Chapter 2,

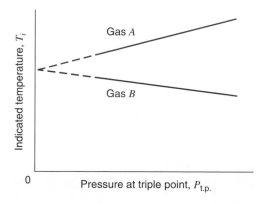

FIGURE 7.23 Sketch showing how the ideal-gas temperature is determined.

closely approximates the thermodynamic temperature scale and is much easier to work with in actual temperature measurement.

We now demonstrate that the ideal-gas temperature scale discussed above is, in fact, identical to the thermodynamic temperature scale, which was defined in the discussion of the Carnot cycle and the second law. Our objective can be achieved by using an ideal gas as the working fluid for a Carnot-cycle heat engine and analyzing the four processes that make up the cycle. The four state points, 1, 2, 3, 4, and the four processes are as shown in Fig. 7.24. For convenience, let us consider a unit mass of gas inside the cylinder. Now for each of the four processes, the reversible work done at the moving boundary is given by Eq. 4.2.

$$\delta w = P\,dv$$

Similarly, for each process the gas behavior is, from the ideal-gas relation, Eq. 3.2,

$$Pv = RT$$

and the internal energy change, from Eq. 5.20, is

$$du = C_{v0}dT$$

Assuming no changes in kinetic or potential energies, the first law is, from Eq. 5.7 at unit mass,

$$\delta q = du + \delta w$$

Substituting the three previous expressions into this equation, we have for each of the four processes

$$\delta q = C_{v0}\,dT + \frac{RT}{v}\,dv \tag{7.9}$$

The shape of the two isothermal processes shown in Fig. 7.24 is known, since Pv is constant in each case. The process 1–2 is an expansion at T_H, such that v_2 is larger than v_1. Similarly, the processes 3–4 is a compression at a lower temperature, T_L, and v_4 is smaller than v_3. The adiabatic process 2–3 is an expansion from T_H to T_L, with an in-

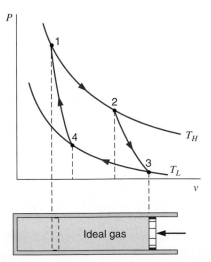

FIGURE 7.24 The ideal gas Carnot cycle.

crease in specific volume, while the adiabatic process 4–1 is a compression from T_L to T_H, with a decrease in specific volume. The area below each process line represents the work for that process, as given by Eq. 4.2

We now proceed to integrate Eq. 7.9 for each of the four processes that make up the Carnot cycle. For the isothermal heat addition process 1–2, we have

$$q_H = {}_1q_2 = 0 + RT_H \ln\frac{v_2}{v_1}$$
(7.10)

For the adiabatic expansion process 2–3,

$$0 = \int_{T_H}^{T_L} \frac{C_{v0}}{T}\, dT + R\ln\frac{v_3}{v_2}$$
(7.11)

For the isothermal heat rejection process 3–4,

$$q_L = -{}_3q_4 = -0 - RT_L \ln\frac{v_4}{v_3}$$

$$= +RT_L \ln\frac{v_3}{v_4}$$
(7.12)

and for the adiabatic compression process 4–1,

$$0 = \int_{T_L}^{T_H} \frac{C_{v0}}{T}\, dT + R\ln\frac{v_1}{v_4}$$
(7.13)

From Eqs. 7.11 and 7.13,

$$\int_{T_L}^{T_H} \frac{C_{v0}}{T}\, dT = R\ln\frac{v_3}{v_2} = -R\ln\frac{v_1}{v_4}$$

Therefore,

$$\frac{v_3}{v_2} = \frac{v_4}{v_1}, \quad \text{or} \quad \frac{v_3}{v_4} = \frac{v_2}{v_1}$$
(7.14)

Thus, from Eqs. 7.10 and 7.12 and substituting Eq. 7.14, we find that

$$\frac{q_H}{q_L} = \frac{RT_H \ln\dfrac{v_2}{v_1}}{RT_L \ln\dfrac{v_3}{v_4}} = \frac{T_H}{T_L}$$

which is Eq. 7.7, the definition of the thermodynamic temperature scale in connection with the second law.

A final point needs to be made about the significance of absolute-zero temperature in connection with the second law and the thermodynamic temperature scale. Consider a Carnot-cycle heat engine that receives a given amount of heat from a given high-temperature reservoir. As the temperature at which heat is rejected from the cycle is lowered, the net work output increases and the amount of heat rejected decreases. In the limit, the heat rejected is zero, and the temperature of the reservoir corresponding to this limit is absolute zero.

Similarly, for a Carnot-cycle refrigerator, the amount of work required to produce a given amount of refrigeration increases as the temperature of the refrigerated space decreases. Absolute zero represents the limiting temperature that can be achieved, and the amount of work required to produce a finite amount of refrigeration approaches infinity as the temperature at which refrigeration is provided approaches zero.

PROBLEMS

7.1 Calculate the thermal efficiency of the steam power plant cycle described in Problem 6.39.

7.2 Calculate the coefficient of performance of the R-12 heat pump cycle described in Problem 6.47.

7.3 Prove that a cyclic device that violates the Kelvin–Planck statement of the second law also violates the Clausius statement of the second law.

7.4 Discuss the factors that would make the power plant cycle described in Problem 6.39 an irreversible cycle.

7.5 Discuss the factors that would make the heat pump described in Problem 6.47 an irreversible cycle.

7.6 Calculate the thermal efficiency of a Carnot-cycle heat engine operating between reservoirs at 500°C and 40°C. Compare the result with that of Problem 7.1.

7.7 Calculate the coefficient of performance of a Carnot-cycle heat pump operating between reservoirs at 0°C and 45°C. Compare the result with that of Problem 7.2.

7.8 A car engine burns 5 kg fuel (equivalent to addition of Q_H) at 1500 K and rejects energy to the radiator and the exhaust at an average temperature of 750 K. If the fuel provides 40 000 kJ/kg, what is the maximum amount of work the engine can provide?

7.9 In a steam power plant 1 MW is added at 700°C in the boiler, 0.58 MW is taken out at 40°C in the condenser, and the pump work is 0.02 MW. Find the plant thermal efficiency. Assuming the same pump work and heat transfer to the boiler is given, how much turbine power could be produced if the plant were running in a Carnot cycle?

7.10 At certain locations geothermal energy in undergound water is available and used as the energy source for a power plant. Consider a supply of saturated liquid water at 150°C. What is the maximum possible thermal efficiency of a cyclic heat engine using this source of energy with the ambient at 20°C? Would it be better to locate a source of saturated vapor at 150°C than to use the saturated liquid at 150°C?

7.11 Find the maximum coefficient of performance for the refrigerator in your kitchen, assuming it runs in a Carnot cycle.

7.12 An air-conditioner provides 1 kg/s of air at 15°C cooled from outside atmospheric air at 35°C. Estimate the amount of power needed to operate the air-conditioner. Clearly state all assumptions made.

7.13 A salesperson selling refrigerators and deep freezers will guarantee a minimum coefficient of performance of 4.5 year round. How would you evaluate that? Are they all the same?

7.14 A car engine operates with a thermal efficiency of 35%. Assume the air-conditioner has a coefficient of performance that is one third of the theoretical maximum and it is mechanically pulled by the engine. How much fuel energy should you spend extra to remove 1 kJ at 15°C when the ambient is at 35°C?

7.15 We propose to heat a house in the winter with a heat pump. The house is to be maintained at 20°C at all times. When the ambient temperature outside drops to −10°C, the rate at which heat is lost from the house is estimated to be 25 kW. What is the minimum electrical power required to drive the heat pump?

7.16 Electric solar cells can produce power with 15% efficiency. Assume a heat engine with a low temperature heat rejection at 30°C driving an electric generator with 80% efficiency. What should the effective high temperature in the heat engine be to have the same overall efficiency as the solar cells?

7.17 A cyclic machine, shown in Fig. P7.17, receives 325 kJ from a 1000 K energy reservoir. It rejects 125 kJ to a 400 K energy reservoir and the cycle produces 200 kJ of work as output. Is this cycle reversible, irreversible, or impossible?

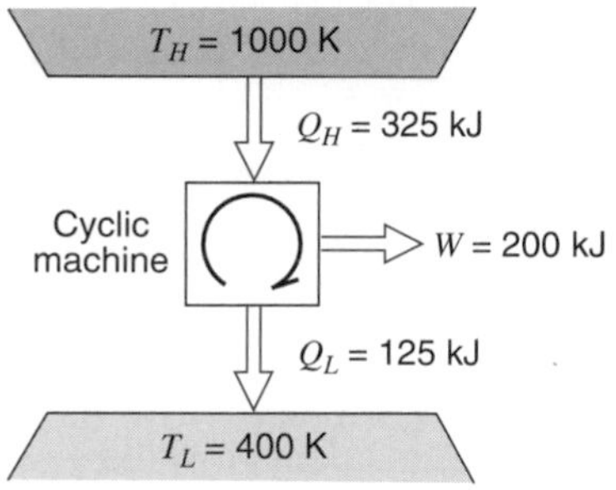

FIGURE P7.17

7.18 A household freezer operates in a room at 20°C. Heat must be transferred from the cold space at a rate of 2 kW to maintain its temperature at −30°C. What is the theoretically smallest (power) motor required to operate this freezer?

7.19 A heat pump has a coefficient of performance that is 50% of the theoretical maximum. It maintains a house at 20°C, which leaks energy of 0.6 kW per degree temperature difference to the ambient. For a maximum of 1.0 kW power input find the minimum outside temperature for which the heat pump is a sufficient heat source.

7.20 A heat pump cools a house at 20°C with a maximum of 1.2 kW power input. The house gains 0.6 kW per degree temperature difference to the ambient and the heat pump coefficient of performance is 60% of the theoretical maximum. Find the maximum outside temperature for which the heat pump provides sufficient cooling.

7.21 Differences in surface water and deep water temperature can be utilized for power generation. It is proposed to construct a cyclic heat engine that will operate near Hawaii, where the ocean temperature is 20°C near the surface and 5°C at some depth. What is the possible thermal efficiency of such a heat engine?

7.22 A thermal storage is made with a rock (granite) bed of 2 m³ which is heated to 400 K using solar energy. A heat engine receives a Q_H from the bed and rejects heat to the ambient at 290 K. The rock bed therefore cools down and as it reaches

290 K the process stops. Find the energy the rock bed can give out. What is the heat engine efficiency at the beginning of the process and what is it at the end of the process?

7.23 An inventor has developed a refrigeration unit that maintains the cold space at −10°C, while operating in a 25°C room. A coefficient of performance of 8.5 is claimed. How do you evaluate this?

7.24 A steel bottle $V = 0.1$ m^3 contains R-134a at 20°C, 200 kPa. It is placed in a deep freezer where it is cooled to –20°C. The deep freezer sits in a room with ambient temperature of 20°C and has an inside temperature of –20°C. Find the amount of energy the freezer must remove from the R-134a and the extra amount of work input to the freezer to do the process.

7.25 A certain solar-energy collector produces a maximum temperature of 100°C. The energy is used in a cyclic heat engine that operates in a 10°C environment. What is the maximum thermal efficiency? What is it if the collector is redesigned to focus the incoming light to produce a maximum temperature of 300°C?

7.26 Liquid sodium leaves a nuclear reactor at 800°C and is used as the energy source in a steam power plant. The condenser cooling water comes from a cooling tower at 15°C. Determine the maximum thermal efficiency of the power plant. Is it misleading to use the temperatures given to calculate this value?

7.27 A 4L jug of milk at 25°C is placed in your refrigerator where it is cooled down to 5°C. The high temperature in the Carnot refrigeration cycle is 45°C and the properties of milk are the same as for liquid water. Find the amount of energy that must be removed from the milk and the additional work needed to drive the refrigerator.

7.28 A house is heated by a heat pump driven by an electric motor using the outside as the low-temperature reservoir. The house loses energy in direct proportion to the temperature difference as $\dot{Q}_{loss} = K(T_H - T_L)$. Determine the minimum electric power required to drive the heat pump as a function of the two temperatures.

7.29 A house is heated by an electric heat pump using the outside as the low-temperature reservoir. For several different winter outdoor temperatures, estimate the percent savings in electricity if the house is kept at 20°C instead of 24°C. Assume that the house is losing energy to the outside as described in the previous problem.

7.30 An air-conditioner with a power input of 1.2 kW is working as a refrigerator ($\beta = 3$) or as a heat pump ($\beta' = 4$). It maintains an office at 20°C year round which exchanges 0.5 kW per degree temperature difference with the atmosphere. Find the maximum and minimum outside temperature for which this unit is sufficient.

7.31 A house is cooled by an electric heat pump using the outside as the high-temperature reservoir. For several different summer outdoor temperatures, estimate the percent savings in electricity if the house is kept at 25°C instead of 20°C. Assume that the house is gaining energy from the outside in direct proportion to the temperature difference.

7.32 Helium has the lowest normal boiling point of any of the elements at 4.2 K. At this temperature the enthalpy of evaporation is 83.3 kJ/kmol. A Carnot refrigeration cycle is analyzed for the production of 1 kmol of liquid helium at 4.2 K from saturated vapor at the same temperature. What is the work input to the refrigerator and the coefficient of performance for the cycle with an ambient at 300 K?

7.33 We wish to produce refrigeration at −30°C. A reservoir, shown in Fig. P7.33, is available at 200°C and the ambient temperature is 30°C. Thus, work can be done by a cyclic heat engine operating between the 200°C reservoir and the ambient. This work is used to drive the refrigerator. Determine the ratio of the heat transferred from the 200°C reservoir to the heat transferred from the −30°C reservoir, assuming all processes are reversible.

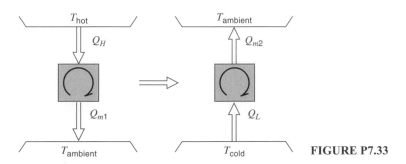

FIGURE P7.33

7.34 A combination of a heat engine driving a heat pump (similar to Fig. P7.33) takes waste energy at 50°C as a source Q_{w1} to the heat engine rejecting heat at 30°C. The remainder Q_{w2} goes into the heat pump that delivers a Q_H at 150°C. If the total waste energy is 5 MW, find the rate of energy delivered at the high temperature.

7.35 A temperature of about 0.01 K can be achieved by magnetic cooling. In this process a strong magnetic field is imposed on a paramagnetic salt, maintained at 1 K by transfer of energy to liquid helium boiling at low pressure. The salt is then thermally isolated from the helium, the magnetic field is removed, and the salt temperature drops. Assume that 1 mJ is removed at an average temperature of 0.1 K to the helium by a Carnot-cycle heat pump. Find the work input to the heat pump and the coefficient of performance with an ambient at 300 K.

7.36 The lowest temperature that has been achieved is about 1×10^{-6} K. To achieve this an additional stage of cooling is required beyond that described in the previous problem, namely nuclear cooling. This process is similar to magnetic cooling, but it involves the magnetic moment associated with the nucleus rather than that associated with certain ions in the paramagnetic salt. Suppose that 10 μJ is to be removed from a specimen at an average temperature of 10^{-5} K (10 microjoules is about the potential energy loss of a pin dropping 3 mm). Find the work input to a Carnot heat pump and its coefficient of performance to do this assuming the ambient is at 300 K.

7.37 A heat pump heats a house in the winter and then reverses to cool it in the summer. The interior temperature should be 20°C in the winter and 25°C in the summer. Heat transfer through the walls and ceilings is estimated to be 2400 kJ per hour per degree temperature difference between the inside and outside.

a. If the winter outside temperature is 0°C, what is the minimum power required to drive the heat pump?

b. For the same power as in part (a), what is the maximum outside summer temperature for which the house can be maintained at 25°C?

7.38 It is proposed to build a 1000-MW electric power plant with steam as the working fluid. The condensers are to be cooled with river water (see Fig. P7.38). The maximum steam temperature is 550°C, and the pressure in the condensers will be 10 kPa. Estimate the temperature rise of the river downstream from the power plant.

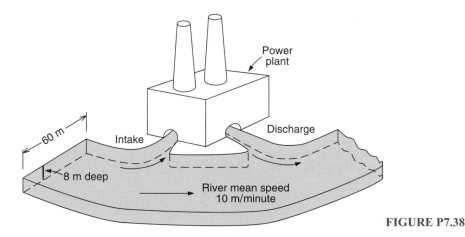

FIGURE P7.38

7.39 Two different fuels can be used in a heat engine, operating between the fuel-burning temperature and a low temperature of 350 K. Fuel A burns at 2500 K, delivering 52 000 kJ/kg and costs $1.75/kg. Fuel B burns at 1700 K, delivering 40000 kJ/kg and cost $1.50/kg. Which fuel would you buy and why?

7.40 A refrigerator uses a power input of 2.5 kW to cool a 5°C space with the high temperature in the cycle as 50°C. The Q_H is pushed to the ambient air at 35°C in a heat exchanger where the transfer coefficient is 50 W/m²K. Find the required minimum heat transfer area.

7.41 Refrigerant-12 at 95°C, $x = 0.1$ flowing at 2 kg/s is brought to saturated vapor in a constant-pressure heat exchanger. The energy is supplied by a heat pump with a low temperature of 10°C. Find the required power input to the heat pump.

7.42 A furnace, shown in Fig. P7.42, can deliver heat, Q_{H1} at T_{H1} and it is proposed to use this to drive a heat engine with a rejection at T_{atm} instead of direct room heating. The heat engine drives a heat pump that delivers Q_{H2} at T_{room} using the atmosphere as the cold reservoir. Find the ratio Q_{H2}/Q_{H1} as a function of the temperatures. Is this a better set-up than direct room heating from the furnace?

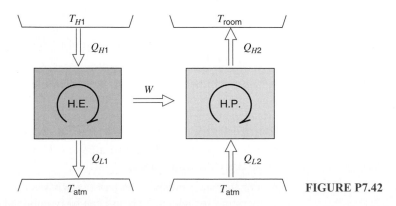

FIGURE P7.42

7.43 A heat engine has a solar collector receiving 0.2 kW/m^2 inside which a transfer media is heated to 450 K. The collected energy powers a heat engine which rejects heat at 40°C. If the heat engine should deliver 2.5 kW, what is the minimum size (area) solar collector?

7.44 In a cryogenic experiment you need to keep a container at −125°C although it gains 100 W due to heat tranfer. What is the smallest motor you would need for a heat pump absorbing heat from the container and rejecting heat to the room at 20°C?

7.45 Sixty kilograms per hour of water runs through a heat exchanger, entering as saturated liquid at 200 kPa and leaving as saturated vapor. The heat is supplied by a Carnot heat pump operating from a low-temperature reservoir at 16°C. Find the rate of work into the heat pump.

7.46 Air in a rigid 1 m^3 box is at 300 K, 200 kPa. It is heated to 600 K by heat transfer from a reversible heat pump that receives energy from the ambient at 300 K besides the work input. Use constant specific heat at 300 K. Since the coefficient of performance changes write $\delta Q = m_{\text{air}} C_v dT$ and find δW. Integrate δW with temperature to find the required heat pump work.

7.47 Consider the rock bed thermal storage in Problem 7.22. Use the specific heat so you can write δQ_H in terms of dT_{rock} and find the expression for δW out of the heat engine. Integrate this expression over temperature and find the total heat engine work output.

7.48 A Carnot heat engine, shown in Fig. P7.48, receives energy from a reservoir at T_{res} through a heat exchanger where the heat transferred is proportional to the temperature difference as $\dot{Q}_H = K(T_{\text{res}} - T_H)$. It rejects heat at a given low temperature T_L. To design the heat engine for maximum work output, show that the high temperature, T_H, in the cycle should be selected as $T_H = \sqrt{(T_L T_{\text{res}})}$.

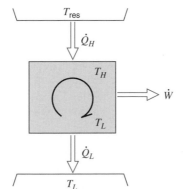

FIGURE P7.48

7.49 A 10-m^3 tank of air at 500 kPa, 600 K acts as the high-temperature reservoir for a Carnot heat engine that rejects heat at 300 K. A temperature difference of 25°C between the air tank and the Carnot cycle high temperature is needed to transfer the heat. The heat engine runs until the air temperature has dropped to 400 K and then stops. Assume constant specific heat capacities for air and find how much work is given out by the heat engine.

7.50 Consider a Carnot cycle heat engine operating in outer space. Heat can be rejected from this engine only by thermal radiation, which is proportional to the radiator area and the fourth power of absolute temperature, $\dot{Q}_{rad} \sim KAT^4$. Show that for a given engine work output and given T_H, the radiator area will be minimum when the ratio $T_L/T_H = \text{fl}$.

7.51 Air in a piston/cylinder goes through a Carnot cycle with the P-v diagram shown in Fig. 7.24. The high and low temperatures are 600 K and 300 K, respectively. The heat added at the high temperature is 250 kJ/kg, and the lowest pressure in the cycle is 75 kPa. Find the specific volume and pressure at all four states in the cycle assuming constant specific heats at 300 K.

7.52 Hydrogen gas is used in a Carnot cycle having an efficiency of 60% with a low temperature of 300 K. During the heat rejection the pressure changes from 90 kPa to 120 kPa. Find the high and low temperature heat transfer and the net cycle work per unit mass hydrogen.

7.53 Obtain information from manufacturers of heat pumps for domestic use. Make a listing of the coefficient of performance and compare those to corresponding Carnot cycle devices operating between the same temperature reservoirs.

ENGLISH UNIT PROBLEMS

7.54E Calculate the thermal efficiency of the steam power plant cycle described in Problem 6.91.

7.55E Calculate the thermal efficiency of a Carnot-cycle heat engine operating between reservoirs at 920 F and 110 F. Compare the result with that of Problem 7.54.

7.56E A car engine burns 10 lbm of fuel (equivalent to addition of Q_H) at 2600 R and rejects energy to the radiator and the exhaust at an average temperature of 1300 R. If the fuel provides 17 200 Btu/lbm, what is the maximum amount of work the engine can provide?

7.57E In a steam power plant 1000 Btu/s is added at 1200 F in the boiler, 580 Btu/s is taken out at 100 F in the condenser, and the pump work is 20 Btu/s. Find the plant thermal efficiency. Assuming the same pump work and heat transfer to the boiler as given, how much turbine power could be produced if the plant were running in a Carnot cycle?

7.58E An air-conditioner provides 1 lbm/s of air at 60 F cooled from outside atmospheric air at 95 F. Estimate the amount of power needed to operate the air-conditioner. Clearly state all assumptions made.

7.59E A car engine operates with a thermal efficiency of 35%. Assume the air-conditioner has a coefficient of performance that is one-third the theoretical maximum and it is mechanically pulled by the engine. How much fuel energy should you spend extra to remove 1 Btu at 60 F when the ambient is at 95 F?

7.60E We propose to heat a house in the winter with a heat pump. The house is to be maintained at 68 F at all times. When the ambient temperature outside drops to 15 F, the rate at which heat is lost from the house is estimated to be 80 000 Btu/h. What is the minimum electrical power required to drive the heat pump?

7.61E A heat pump cools a house at 70 F with a maximum of 4000 Btu/h power input. The house gains 2000 Btu/h per degree temperature difference to the ambient and the refrigerator coefficient of performance is 60% of the theoretical maximum. Find the maximum outside temperature for which the heat pump provides sufficient cooling.

7.62E A thermal storage is made with a rock (granite) bed of 70 ft^3, which is heated to 720 R using solar energy. A heat engine receives a Q_H from the bed and rejects heat to the ambient at 520 R. The rock bed therefore cools down, and as it reaches 520 R the process stops. Find the energy the rock bed can give out. What is the heat engine efficiency at the beginning of the process and what is it at the end of the process?

7.63E An inventor has developed a refrigeration unit that maintains the cold space at 14 F, while operating in a 77 F room. A coefficient of performance of 8.5 is claimed. How do you evaluate this?

7.64E Liquid sodium leaves a nuclear reactor at 1500 F and is used as the energy source in a steam power plant. The condenser cooling water comes from a cooling tower at 60 F. Determine the maximum thermal efficiency of the power plant. Is it misleading to use the temperatures given to calculate this value?

7.65E A house is heated by an electric heat pump using the outside as the low-temperature reservoir. For several different winter outdoor temperatures, estimate the percent savings in electricity if the house is kept at 68 F instead of 75 F. Assume that the house is losing energy to the outside directly proportional to the temperature difference as $\dot{Q}_{loss} = K(T_H - T_L)$.

7.66E A house is cooled by an electric heat pump using the outside as the high-temperature reservoir. For several different summer outdoor temperatures estimate the percent savings in electricity if the house is kept at 77 F instead of 68 F. Assume that the house is gaining energy from the outside directly proportional to the temperature difference.

7.67E We wish to produce refrigeration at −20 F. A reservoir is available at 400 F and the ambient temperature is 80 F, as shown in Fig. P7.33. Thus, work can be done by a cyclic heat engine operating between the 400 F reservoir and the ambient. This work is used to drive the refrigerator. Determine the ratio of the heat transferred from the 400 F reservoir to the heat transferred from the −20 F reservoir, assuming all processes are reversible.

7.68E Refrigerant-22 at 180 F, $x = 0.1$ flowing at 4 lbm/s is brought to saturated vapor in a constant-pressure heat exchanger. The energy is supplied by a heat pump with a low temperature of 50 F. Find the required power input to the heat pump.

7.69E A heat engine has a solar collector receiving 600 Btu/h per square foot inside which a transfer media is heated to 800 R. The collected energy powers a heat engine that rejects heat at 100 F. If the heat engine should deliver 8500 Btu/h, what is the minimum size (area) solar collector?

7.70E Six-hundred pound-mass per hour of water runs through a heat exchanger, entering as saturated liquid at 30 lbf/in.^{2} and leaving as saturated vapor. The heat is supplied by a Carnot heat pump operating from a low-temperature reservoir at 60 F. Find the rate of work into the heat pump.

7.71E Air in a rigid 40 ft^3 box is at 540 R, 30 lbf/in.^{2}. It is heated to 1100 R by heat transfer from a reversible heat pump that receives energy from the ambient at 540 R besides the work input. Use constant specific heat at 540 R. Since the coefficient of performance changes, write $\delta Q = m_{air} C_v dT$ and find δW. Integrate δW with temperature to find the required heat pump work.

7.72E A 350 ft^3 tank of air at 80 lbf/in.^{2}, 1080 R acts as the high-temperature reservoir for a Carnot heat engine that rejects heat at 540 R. A temperature difference of 45 F between the air tank and the Carnot cycle high temperature is needed to transfer the heat. The heat engine runs until the air temperature has dropped to 700 R and then stops. Assume constant specific heat capacities for air and find how much work is given out by the heat engine.

7.73E Air in a piston/cylinder goes through a Carnot cycle with the *P-v* diagram shown in Fig. 7.24. The high and low temperatures are 1200 R and 600 R, respectively. The heat added at the high temperature is 100 Btu/lbm, and the lowest pressure in the cycle is 10 lbf/in.^{2}. Find the specific volume and pressure at all four states in the cycle assuming constant specific heats at 80 F.

ENTROPY 8

Up to this point in our consideration of the second law of thermodynamics, we have dealt only with thermodynamic cycles. Although this is a very important and useful approach, we are often concerned with processes rather than cycles. Thus, we might be interested in the second-law analysis of processes we encounter daily, such as the combustion process in an automobile engine, the cooling of a cup of coffee, or the chemical processes that take place in our bodies. It would also be beneficial to be able to deal with the second law quantitatively as well as qualitatively.

In our consideration of the first law, we initially stated the law in terms of a cycle, but then defined a property, the internal energy, which enabled us to use the first law quantitatively for processes. Similarly, we have stated the second law for a cycle, and we now find that the second law leads to a property, entropy, which enables us to treat the second law quantitatively for processes. Energy and entropy are both abstract concepts that help to describe certain observations. As we noted in Chapter 2, thermodynamics can be described as the science of energy and entropy. The significance of this statement will become increasingly evident.

8.1 THE INEQUALITY OF CLAUSIUS

The first step in our consideration of the property we call entropy is to establish the inequality of Clausius, which is

$$\oint \frac{\delta Q}{T} \leq 0$$

The inequality of Clausius is a corollary or a consequence of the second law of thermodynamics. It will be demonstrated to be valid for all possible cycles, including both reversible and irreversible heat engines and refrigerators. Since any reversible cycle can be represented by a series of Carnot cycles, in this analysis we need consider only a Carnot cycle that leads to the inequality of Clausius.

Consider first a reversible (Carnot) heat engine cycle operating between reservoirs at temperatures T_H and T_L, as shown in Fig. 8.1. For this cycle, the cyclic integral of the heat transfer, $\oint \delta Q$, is greater than zero.

$$\oint \delta Q = Q_H - Q_L > 0$$

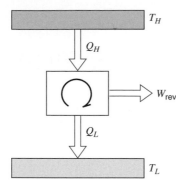

FIGURE 8.1 Reversible heat engine cycle for demonstration of the inequality of Clausius.

Since T_H and T_L are constant, from the definition of the absolute temperature scale and from the fact this is a reversible cycle, it follows that

$$\oint \frac{\delta Q}{T} = \frac{Q_H}{T_H} - \frac{Q_L}{T_L} = 0$$

If $\oint \delta Q$, the cyclic integral of δQ, approaches zero (by making T_H approach T_L) and the cycle remains reversible, the cyclic integral of $\delta Q/T$ remains zero. Thus, we conclude that for all reversible heat engine cycles

$$\oint \delta Q \geq 0$$

and

$$\oint \frac{\delta Q}{T} = 0$$

Now consider an irreversible cyclic heat engine operating between the same T_H and T_L as the reversible engine of Fig. 8.1 and receiving the same quantity of heat Q_H. Comparing the irreversible cycle with the reversible one, we conclude from the second law that

$$W_{irr} < W_{rev}$$

Since $Q_H - Q_L = W$ for both the reversible and irreversible cycles, we conclude that

$$Q_H - Q_{L\,irr} < Q_H - Q_{L\,rev}$$

and therefore

$$Q_{L\,irr} > Q_{L\,rev}$$

Consequently, for the irreversible cyclic engine,

$$\oint \delta Q = Q_H - Q_{L\,irr} > 0$$

$$\oint \frac{\delta Q}{T} = \frac{Q_H}{T_H} - \frac{Q_{L\,irr}}{T_L} < 0$$

Suppose that we cause the engine to become more and more irreversible, but keep Q_H, T_H, and T_L fixed. The cyclic integral of δQ then approaches zero, and that for $\delta Q/T$

becomes a progressively larger negative value. In the limit, as the work output goes to zero,

$$\oint \delta Q = 0$$

$$\oint \frac{\delta Q}{T} < 0$$

Thus, we conclude that for all irreversible heat engine cycles

$$\oint \delta Q \geq 0$$

$$\oint \frac{\delta Q}{T} < 0$$

To complete the demonstration of the inequality of Clausius, we must perform similar analyses for both reversible and irreversible refrigeration cycles. For the reversible refrigeration cycle shown in Fig. 8.2,

$$\oint \delta Q = -Q_H + Q_L < 0$$

and

$$\oint \frac{\delta Q}{T} = -\frac{Q_H}{T_H} + \frac{Q_L}{T_L} = 0$$

As the cyclic integral of δQ approaches zero reversibly (T_H approaches T_L), the cyclic integral of $\delta Q/T$ remains at zero. In the limit,

$$\oint \delta Q = 0$$

$$\oint \frac{\delta Q}{T} = 0$$

Thus, for all reversible refrigeration cycles,

$$\oint \delta Q \leq 0$$

$$\oint \frac{\delta Q}{T} = 0$$

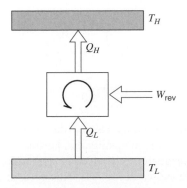

FIGURE 8.2 Reversible refrigeration cycle for demonstration of the inequality of Clausius.

Finally, let an irreversible cyclic refrigerator operate between temperatures T_H and T_L and receive the same amount of heat Q_L as the reversible refrigerator of Fig. 8.2. From the second law, we conclude that the work input required will be greater for the irreversible refrigerator, or

$$W_{\text{irr}} > W_{\text{rev}}$$

Since $Q_H - Q_L = W$ for each cycle, it follows that

$$Q_{H\,\text{irr}} - Q_L > Q_{H\,\text{rev}} - Q_L$$

and therefore,

$$Q_{H\,\text{irr}} > Q_{H\,\text{rev}}$$

That is, the heat rejected by the irreversible refrigerator to the high-temperature reservoir is greater than the heat rejected by the reversible refrigerator. Therefore, for the irreversible refrigerator,

$$\oint \delta Q = -Q_{H\,\text{irr}} + Q_L < 0$$

$$\oint \frac{\delta Q}{T} = -\frac{Q_{H\,\text{irr}}}{T_H} + \frac{Q_L}{T_L} < 0$$

As we make this machine progressively more irreversible, but keep Q_L, T_H, and T_L constant, the cyclic integrals of δQ and $\delta Q/T$ both become larger in the negative direction. Consequently, a limiting case as the cyclic integral of δQ approaches zero does not exist for the irreversible refrigerator.

Thus, for all irreversible refrigeration cycles,

$$\oint \delta Q < 0$$

$$\oint \frac{\delta Q}{T} < 0$$

Summarizing, we note that, in regard to the sign of $\oint \delta Q$, we have considered all possible reversible cycles (i.e., $\oint \delta Q \gtrless 0$), and for each of these reversible cycles

$$\oint \frac{\delta Q}{T} = 0$$

We have also considered all possible irreversible cycles for the sign of $\oint \delta Q$ (that is, $\oint \delta Q \gtrless 0$), and for all these irreversible cycles

$$\oint \frac{\delta Q}{T} < 0$$

Thus, for all cycles we can write

$$\oint \frac{\delta Q}{T} \leq 0 \tag{8.1}$$

where the equality holds for reversible cycles and the inequality for irreversible cycles. This relation, Eq. 8.1, is known as the inequality of Clausius.

The significance of the inequality of Clausius may be illustrated by considering the simple steam power plant cycle shown in Fig. 8.3. This cycle is slightly different from the

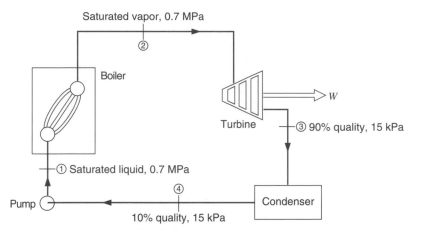

FIGURE 8.3 A simple steam power plant that demonstrates the inequality of Clausius.

usual cycle for steam power plants in that the pump handles a mixture of liquid and vapor in such proportions that saturated liquid leaves the pump and enters the boiler. Suppose that someone reports that the pressure and quality at various points in the cycle are as given in Fig. 8.3. Does this cycle satisfy the inequality of Clausius?

Heat is transferred in two places, the boiler and the condenser. Therefore,

$$\oint \frac{\delta Q}{T} = \int \left(\frac{\delta Q}{T} \right)_{boiler} + \int \left(\frac{\delta Q}{T} \right)_{condenser}$$

Since the temperature remains constant in both the boiler and condenser, this may be integrated as follows:

$$\oint \frac{\delta Q}{T} = \frac{1}{T_1} \int_1^2 \delta Q + \frac{1}{T_3} \int_3^4 \delta Q = \frac{_1Q_2}{T_1} + \frac{_3Q_4}{T_3}$$

Let us consider a 1-kg mass as the working fluid.

$$_1q_2 = h_2 - h_1 = 2066.3 \text{ kJ / kg}, \qquad T_1 = 164.97°\text{C}$$

$$_3q_4 = h_4 - h_3 = 463.4 - 2361.8 = -1898.4 \text{ kJ / kg} \qquad T_3 = 53.97°\text{C}$$

Therefore,

$$\oint \frac{\delta Q}{T} = \frac{2066.3}{164.97 + 273.15} - \frac{1898.4}{53.97 + 273.15} = -1.087 \text{ kJ / kg K}$$

Thus, this cycle satisfies the inequality of Clausius, which is equivalent to saying that it does not violate the second law of thermodynamics.

8.2 ENTROPY—A PROPERTY OF A SYSTEM

By applying Eq. 8.1 and Fig. 8.4 we can demonstrate that the second law of thermodynamics leads to a property of a system that we call entropy. Let a system (control mass) undergo a reversible process from state 1 to state 2 along a path A, and let the cycle be completed along path B, which is also reversible.

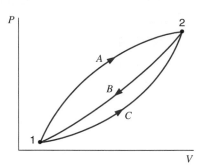

FIGURE 8.4 Two reversible cycles demonstrating the fact that entropy is a property of a substance.

Because this is a reversible cycle, we can write

$$\oint \frac{\delta Q}{T} = 0 = \int_1^2 \left(\frac{\delta Q}{T} \right)_A + \int_2^1 \left(\frac{\delta Q}{T} \right)_B$$

Now consider another reversible cycle, which proceeds first along path C and is then completed along path B. For this cycle we can write

$$\oint \frac{\delta Q}{T} = 0 = \int_1^2 \left(\frac{\delta Q}{T} \right)_C + \int_2^1 \left(\frac{\delta Q}{T} \right)_B$$

Subtracting the second equation from the first, we have

$$\int_1^2 \left(\frac{\delta Q}{T} \right)_A = \int_1^2 \left(\frac{\delta Q}{T} \right)_C$$

Since the $\int \delta Q/T$ is the same for all reversible paths between states 1 and 2, we conclude that this quantity is independent of the path and it is a function of the end states only; it is therefore a property. This property is called entropy and is designated S. It follows that entropy may be defined as a property of a substance in accordance with the relation

$$dS \equiv \left(\frac{\delta Q}{T} \right)_{\text{rev}} \tag{8.2}$$

Entropy is an extensive property, and the entropy per unit mass is designated s. It is important to note that entropy is defined here in terms of a reversible process.

The change in the entropy of a system as it undergoes a change of state may be found by integrating Eq. 8.2. Thus,

$$S_2 - S_1 = \int_1^2 \left(\frac{\delta Q}{T} \right)_{\text{rev}} \tag{8.3}$$

To perform this integration, we must know the relation between T and Q, and illustrations will be given subsequently. The important point is that since entropy is a property, the change in the entropy of a substance in going from one state to another is the same for all processes, both reversible and irreversible, between these two states. Equation 8.3 enables us to find the change in entropy only along a reversible path. However, once the change has been evaluated, this value is the magnitude of the entropy change for all processes between these two states.

Equation 8.3 enables us to determine changes of entropy, but it tells us nothing about absolute values of entropy. However, from the third law of thermodynamics, which is discussed in Chapter 14, it follows that the entropy of all pure substances can be assigned the value of zero at the absolute zero of temperature. This value allows in turn the assignment of absolute values of entropy and is particularly important when chemical reactions are involved.

However, when there is no change of composition, it is quite adequate to give values of entropy relative to some arbitrarily selected reference state. This is the procedure followed in most tables of thermodynamic properties, such as the steam tables and ammonia tables. Therefore, until absolute entropy is introduced in Chapter 14, values of entropy will always be given relative to some arbitrary reference state.

A word should be added here regarding the role of T as an integrating factor. We noted in Chapter 4 that Q is a path function, and therefore δQ is an inexact differential. However, since $(\delta Q/T)_{rev}$ is a thermodynamic property, it is an exact differential. From a mathematical perspective, we note that an inexact differential may be converted to an exact differential by the introduction of an integrating factor. Therefore, $1/T$ serves as the integrating factor in converting the inexact differential δQ to the exact differential $\delta Q/T$ for a reversible process.

8.3 THE ENTROPY OF A PURE SUBSTANCE

Entropy is an extensive property of a system. Values of specific entropy (entropy per unit mass) are given in tables of thermodynamic properties in the same manner as specific volume and specific enthalpy. The units of specific entropy in the steam tables, refrigerant tables, and ammonia tables are kJ/kg K, and the values are given relative to an arbitrary reference state. In the steam tables the entropy of saturated liquid at 0.01°C is given the value of zero. For many refrigerants, the entropy of saturated liquid at −40°C is assigned the value of zero.

In general, we use the term "entropy" to refer to both total entropy and entropy per unit mass, since the context or appropriate symbol will clearly indicate the precise meaning of the term.

In the saturation region the entropy may be calculated using the quality. The relations are similar to those for specific volume and enthalpy.

$$s = (1-x)s_f + xs_g$$

$$s = s_f + xs_{fg} \tag{8.4}$$

The entropy of a compressed liquid is tabulated in the same manner as the other properties. These properties are primarily a function of the temperature and are not greatly different from those for saturated liquid at the same temperature. Table 4 of the steam tables, which is summarized in Table B.1.4, gives the entropy of compressed-liquid water in the same manner as for other properties.

The thermodynamic properties of a substance are often shown on a temperature–entropy diagram and on an enthalpy–entropy diagram, which is also called a Mollier diagram, after Richard Mollier (1863–1935) of Germany. Figures 8.5 and 8.6 show the essential elements of temperature–entropy and enthalpy–entropy diagrams for steam. The general features of such diagrams are the same for all pure substances. A more complete temperature–entropy diagram for steam is shown in Fig F.1.

These diagrams are valuable both because they present thermodynamic data and because they enable us to visualize the changes of state that occur in various processes. As our study progresses, the student should acquire facility in visualizing thermodynamic

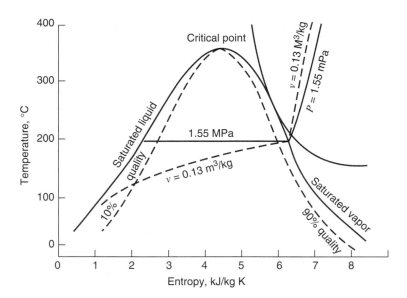

FIGURE 8.5
Temperature–entropy
diagram for steam.

processes on these diagrams. The temperature–entropy diagram is particularly useful for this purpose.

One more observation should be made here concerning the compressed-liquid lines on the temperature–entropy diagram for water. Reference to Table 4 of the steam tables (Table B.1.4) indicates that the entropy of the compressed liquid is less than that of saturated liquid at the same temperature for all the temperatures listed except 0°C. It can be shown that each constant-pressure line crosses the saturated-liquid line at the point of maximum density, which is about 4°C for water (the exact temperature at which maximum density occurs varies with pressure). For temperatures less than the temperature at which the density is a maximum, the entropy of compressed liquid is greater than that of the saturated liquid. Thus, lines of constant pressure in the compressed-liquid region appear (on a magnified scale) as shown in Fig. 8.7*a*. It is important to understand the general shape of these lines when showing the pumping process for liquids.

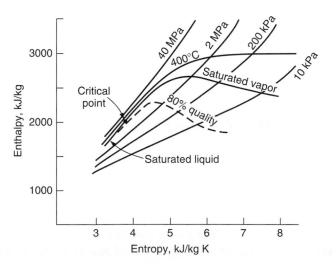

FIGURE 8.6 Enthalpy–
entropy diagram for steam.

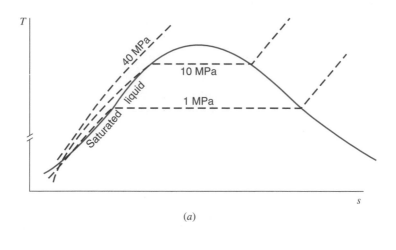

(a)

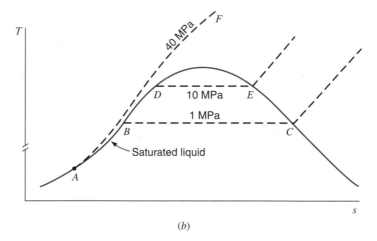

(b)

FIGURE 8.7
Temperature–entropy diagram to show properties of a compressed liquid, water.

Having made this observation for water, we should state that for most substances the difference in the entropy of a compressed liquid and a saturated liquid at the same temperature is so small that a process in which liquid is heated at constant pressure nearly coincides with the saturated-liquid line until the saturation temperature is reached (Fig. 8.7b). Thus, if water at 10 MPa is heated from 0°C to the saturation temperature, it would be shown by line *ABD*, which coincides with the saturated-liquid line.

8.4 ENTROPY CHANGE IN REVERSIBLE PROCESSES

Having established that entropy is a thermodynamic property of a system, we now consider its significance in various processes. In this section we will limit ourselves to systems that undergo reversible processes and consider the Carnot cycle, reversible heat-transfer processes, and reversible adiabatic processes.

Let the working fluid of a heat engine operating on the Carnot cycle make up the system. The first process is the isothermal transfer of heat to the working fluid from the high-temperature reservoir. For this process we can write

$$S_2 - S_1 = \int_1^2 \left(\frac{\delta Q}{T}\right)_{rev}$$

Since this is a reversible process in which the temperature of the working fluid remains constant, the equation can be integrated to give

$$S_2 - S_1 = \frac{1}{T_H}\int_1^2 \delta Q = \frac{{}_1 Q_2}{T_H}$$

This process is shown in Fig. 8.8a, and the area under line 1–2, area 1–2–b–a–1, represents the heat transferred to the working fluid during the process.

The second process of a Carnot cycle is a reversible adiabatic one. From the definition of entropy,

$$dS = \left(\frac{\delta Q}{T}\right)_{rev}$$

it is evident that the entropy remains constant in a reversible adiabatic process. A constant-entropy process is called an isentropic process. Line 2–3 represents this process, and this process is concluded at state 3 when the temperature of the working fluid reaches T_L.

The third process is the reversible isothermal process in which heat is transferred from the working fluid to the low-temperature reservoir. For this process we can write

$$S_4 - S_3 = \int_3^4 \left(\frac{\delta Q}{T}\right)_{rev} = \frac{{}_3 Q_4}{T_L}$$

Because during this process the heat transfer is negative (in regard to the working fluid), the entropy of the working fluid decreases. Moreover, because the final process 4–1, which completes the cycle, is a reversible adiabatic process (and therefore isentropic), it is evident that the entropy decrease in process 3–4 must exactly equal the entropy increase in process 1–2. The area under line 3–4, area 3–4–a–b–3, represents the heat transferred from the working fluid to the low-temperature reservoir.

Since the net work of the cycle is equal to the net heat transfer, it is evident that area 1–2–3–4–1 represents the net work of the cycle. The efficiency of the cycle may also be expressed in terms of areas.

$$\eta_{th} = \frac{W_{net}}{Q_H} = \frac{\text{area } 1-2-3-4-1}{\text{area } 1-2-b-a-1}$$

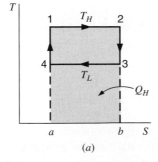
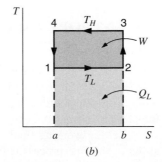

FIGURE 8.8 The Carnot cycle on the temperature–entropy diagram.

Some statements made earlier about efficiencies may now be understood graphically. For example, increasing T_H while T_L remains constant increases the efficiency. Decreasing T_L as T_H remains constant increases the efficiency. It is also evident that the efficiency approaches 100% as the absolute temperature at which heat is rejected approaches zero.

If the cycle is reversed, we have a refrigerator or heat pump. The Carnot cycle for a refrigerator is shown in Fig. 8.8b. Notice that the entropy of the working fluid increases at T_L, since heat is transferred to the working fluid at T_L. The entropy decreases at T_H because of heat transfer from the working fluid.

Let us next consider reversible heat-transfer processes. Actually, we are concerned here with processes that are internally reversible, that is, processes that have no irreversibilities within the boundary of the system. For such processes the heat transfer to or from a system can be shown as an area on a temperature–entropy diagram. For example, consider the change of state from saturated liquid to saturated vapor at constant pressure. This process would correspond to the process 1–2 on the T–s diagram of Fig. 8.9 (note that absolute temperature is required here), and the area 1–2–b–a–1 represents the heat transfer. Since this is a constant-pressure process, the heat transfer per unit mass is equal to h_{fg}. Thus,

$$s_2 - s_1 = s_{fg} = \frac{1}{m} \int_1^2 \left(\frac{\delta Q}{T} \right)_{rev} = \frac{1}{mT} \int_1^2 \delta Q = \frac{{}_1q_2}{T} = \frac{h_{fg}}{T}$$

This relation gives a clue about how s_{fg} is calculated for tabulation in tables of thermodynamic properties. For example, consider steam at 10 MPa. From the steam tables we have

$$h_{fg} = 1317.1 \, \text{kJ} / \text{kg}$$

$$T = 311.06 + 273.15 = 584.21 \, \text{K}$$

Therefore,

$$s_{fg} = \frac{h_{fg}}{T} = \frac{1317.1}{584.21} = 2.2544 \, \text{kJ} / \text{kg K}$$

This is the value listed for s_{fg} in the steam tables.

If heat is transferred to the saturated vapor at constant pressure, the steam is superheated along line 2–3. For this process we can write

$$\eta_{th} = \frac{W_{net}}{Q_H} = \frac{\text{area } 1\text{-}2\text{-}3\text{-}}{\text{area } 1\text{-}2\text{-}b\text{-}}$$

Since T is not constant, this equation cannot be integrated unless we know a relation between temperature and entropy. However, we do realize that the area under line 2–3, area

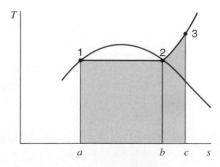

FIGURE 8.9 A temperature–entropy diagram to show areas that represent heat transfer for an internally reversible process.

2–3–*c*–*b*–2, represents $\int_2^3 T\,ds$, and therefore represents the heat transferred during this reversible process.

The important conclusion to draw here is that for processes that are internally reversible, the area underneath the process line on a temperature–entropy diagram represents the quantity of heat transferred. This is not true for irreversible processes, as will be demonstrated later.

There are many situations in which essentially adiabatic processes take place. We have already noted that in such cases the ideal process, which is a reversible adiabatic process, is isentropic. We will consider here an example of this ideal process for a control mass and then consider later in the chapter the reversible adiabatic process for a control volume. We will also note in a later section of this chapter that by comparing an actual process with the ideal or isentropic process, we have a basis for defining the efficiency of certain classes of machines.

EXAMPLE 8.1 Consider a cylinder fitted with a piston that contains saturated R-134a vapor at −5°C. Let this vapor be compressed in a reversible adiabatic process until the pressure is 1.0 MPa. Determine the work per kilogram of R-134a for this process.

> *Control mass:* R-134a
>
> *Initial state:* T_1, saturated vapor; state fixed.
>
> *Final state:* P_2 known.
>
> *Process:* Reversible and adiabatic.
>
> *Model:* R-134a tables.

Analysis:

First law, adiabatic:

$$_1q_2 = u_2 - u_1 + {}_1w_2 = 0$$
$$_1w_2 = u_1 - u_2$$

Second law, reversible and adiabatic:

$$s_1 = s_2$$

Therefore, we know entropy and pressure in the final state, which is sufficient to specify the final state, since we are dealing with a pure substance.

Solution

From the R-134a tables,

$$h_1 = 395.34 \text{ kJ/kg} \qquad s_1 = 1.7288 \text{ kJ/kg K}$$
$$v_1 = 0.082\,58 \text{ m}^3/\text{kg} \qquad P_1 = P_g = 245 \text{ kPa}$$

Then,

$$u_1 = 395.34 - 245 \times 0.082\,58 = 375.11 \text{ kJ/kg}$$

Since

$$P_2 = 1.0 \text{ MPa}, \ s_2 = 1.7288 \text{ kJ / kg K},$$

from the superheat tables for R-134a, we find

$$T_2 = 44.1°C$$

$$h_2 = 424.70 \text{ kJ / kg}, \quad v_2 = 0.021\ 03 \text{ m}^3 / \text{kg}$$

Therefore,

$$u_2 = 424.70 - 1000 \times 0.021\ 03 = 403.67 \text{ kJ / kg}$$

$$_1w_2 = u_1 - u_2 = 375.11 - 403.67 = -28.56 \text{ kJ / kg}$$

8.5 THE THERMODYNAMIC PROPERTY RELATION

At this point we derive two important thermodynamic relations for a simple compressible substance. These relations are

$$T \, dS = dU + P \, dV$$

$$T \, dS = dH - V \, dP$$

The first of these relations can be derived by considering a simple compressible substance in the absence of motion or gravitational effects. The first law for a change of state under these conditions can be written

$$\delta Q = dU + \delta W$$

The equations we are deriving here deal first with the changes of state in which the state of the substance can be identified at all times. Thus, we must consider a quasi-equilibrium process or, to use the term introduced in the last chapter, a reversible process. For a reversible process of a simple compressible substance, we can write

$$\delta Q = T \, dS \quad \text{and} \quad \delta W = P \, dV$$

Substituting these relations into the first-law equation, we have

$$T \, dS = dU + P \, dV \tag{8.5}$$

which is the first equation we set out to derive. Note that this equation was derived by assuming a reversible process, and this equation can therefore be integrated for any reversible process, for during such a process the state of the substance can be identified at any point during the process. We also note that Eq. 8.5 deals only with properties. Suppose we have an irreversible process taking place between the given initial and final states. The properties of a substance depend only on the state, and therefore the change in the properties during a given change of state are the same for an irreversible process as for a reversible process. Therefore, Eq. 8.5 is often applied to an irreversible process between two given states, but the integration of Eq. 8.5 is performed along a reversible path between the same two states.

Since enthalpy is defined as

$$H = U + PV$$

it follows that

$$dH = dU + P\,dV + V\,dP$$

Substituting this relation into Eq. 8.5, we have

$$T\,dS = dH - V\,dP \tag{8.6}$$

which is the second relation that we set out to derive. These two expressions, Eqs. 8.5 and 8.6, are two forms of the thermodynamic property relation, and are frequently called Gibbs equations.

These equations can also be written for a unit mass,

$$T\,ds = du + P\,dv$$

$$T\,ds = dh - v\,dP \tag{8.7}$$

or on a mole basis,

$$T\,d\bar{s} = d\bar{u} + P\,d\bar{v}$$

$$T\,d\bar{s} = d\bar{h} - \bar{v}\,dP \tag{8.8}$$

The Gibbs equations will be used extensively in certain subsequent sections of this book.

If we consider substances of fixed composition other than a simple compressible substance, we can write "$T\,dS$" equations other than those just given for a simple compressible substance. In Eq. 4.16 we noted that for a reversible process we can write the following expression for work:

$$\delta W = P\,dV - \mathscr{T}\,dL - \mathscr{S}\,d\mathscr{A} - \mathscr{E}dZ + \cdots$$

It follows that a more general expression for the thermodynamic property relation would be

$$T\,dS = dU + P\,dV - \mathscr{T}\,dL - \mathscr{S}\,d\mathscr{A} - \mathscr{E}\,dZ + \cdots \tag{8.9}$$

8.6 ENTROPY CHANGE OF A CONTROL MASS DURING AN IRREVERSIBLE PROCESS

Consider a control mass that undergoes the cycles shown in Fig. 8.10. The cycle made up of the reversible processes A and B is a reversible cycle. Therefore we can write

$$\oint \frac{\delta Q}{T} = \int_1^2 \left(\frac{\delta Q}{T}\right)_A + \int_2^1 \left(\frac{\delta Q}{T}\right)_B = 0$$

The cycle made of the irreversible process C and the reversible process B is an irreversible cycle. Therefore, for this cycle the inequality of Clausius may be applied, giving the result

$$\oint \frac{\delta Q}{T} = \int_1^2 \left(\frac{\delta Q}{T}\right)_C + \int_2^1 \left(\frac{\delta Q}{T}\right)_B < 0$$

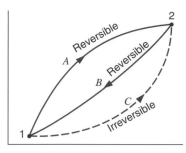

FIGURE 8.10 Entropy change of a control mass during an irreversible process.

Subtracting the second equation from the first and rearranging, we have

$$\int_1^2 \left(\frac{\delta Q}{T} \right)_A > \int_1^2 \left(\frac{\delta Q}{T} \right)_C$$

Since path A is reversible, and since entropy is a property

$$\int_1^2 \left(\frac{\delta Q}{T} \right)_A = \int_1^2 dS_A = \int_1^2 dS_C$$

Therefore,

$$\int_1^2 dS_C > \int_1^2 \left(\frac{\delta Q}{T} \right)_C$$

As path C was arbitrary, the general result is

$$dS \geq \frac{\delta Q}{T}$$

$$S_2 - S_1 \geq \int_1^2 \frac{\delta Q}{T} \tag{8.10}$$

In these equations the equality holds for a reversible process and the inequality for an irreversible process.

This is one of the most important equations of thermodynamics. It is used to develop a number of concepts and definitions. In essence, this equation states the influence of irreversibility on the entropy of a control mass. Thus, if an amount of heat δQ is transferred to a control mass at temperature T in a reversible process, the change of entropy is given by the relation

$$dS = \left(\frac{\delta Q}{T} \right)_{rev}$$

If any irreversible effects occur while the amount of heat δQ is transferred to the control mass at temperature T, however, the change of entropy will be greater than for the reversible process. We would then write

$$dS > \left(\frac{\delta Q}{T} \right)_{irr}$$

Equation 8.10 holds when $\delta Q = 0$, when $\delta Q < 0$, and when $\delta Q > 0$. If δQ is negative, the entropy will tend to decrease as a result of the heat transfer. However, the influence of irreversibilities is still to increase the entropy of the mass and from the absolute numerical perspective we can still write for δQ:

$$dS \geq \frac{\delta Q}{T}$$

8.7 ENTROPY GENERATION

The conclusion from the previous considerations is that the entropy change for an irreversible process is larger than the change in a reversible process for the same δQ and T. This can be written out in a common form as an equality

$$dS = \frac{\delta Q}{T} + \delta S_{gen} \tag{8.11}$$

provided the last term is positive

$$\delta S_{gen} \geq 0 \tag{8.12}$$

The amount of entropy, δS_{gen}, is the entropy generation in the process due to irreversibilities occuring inside the system, a control mass for now, but later extended to the more general control volume. This internal generation can be caused by the processes mentioned in Section 7.4 such as friction, unrestrained expansions, internal transfer of energy (redistribution) over a finite temperature difference. In addition to this internal entropy generation, external irreversibilities are possible by heat transfer over finite temperature differences as the δQ is transferred from a reservoir or by the mechanical transfer of work.

Equation 8.12 is then valid with the equal sign for a reversible process and the greater than sign for an irreversible process. Since the entropy generation is always positive and the smallest in a reversible process, namely zero, we may deduce some limits for the heat transfer and work terms.

Consider a reversible process, for which the entropy generation is zero, and the heat transfer and work terms therefore are

$$\delta Q = T\, dS \qquad \text{and} \qquad \delta W = P\, dV$$

For an irreversible process with a nonzero entropy generation, the heat transfer from Eq. 8.11 becomes

$$\delta Q_{irr} = T\, dS - T\delta S_{gen}$$

and thus smaller than the reversible case for the same change of state, dS. We also note that for the irreversible process, the work is no longer equal to $P\, dV$, but is smaller. Further, since the first law is

$$\delta Q_{irr} = dU + \delta W_{irr}$$

and the property relation is valid,

$$T\, dS = dU + P\, dV$$

it is found that

$$\delta W_{irr} = P\, dV - T\delta S_{gen} \tag{8.13}$$

showing that the work is reduced by an amount proportional to the entropy generation. For this reason the term $T\,\delta S_{gen}$ is often called "lost work," although it is not a real work or energy quantity lost, but a lost opportunity to extract work.

Equation 8.11 can be integrated between initial and final states, to

$$S_2 - S_1 = \int_1^2 dS = \int_1^2 \frac{\delta Q}{T} +_1 S_{2\,gen} \tag{8.14}$$

Thus we have an expression for the change of entropy for an irreversible process as an equality, whereas in the last section we had an inequality. In the limit of a reversible process, with a zero entropy generation, the change in S expressed in Eq. 8.14 becomes identical to Eq. 8.10 as the equal sign applies and the work term becomes $\int P\, dV$.

Some important conclusions can now be drawn from Eqs. 8.11, 8.12, and 8.13. First, there are two ways in which the entropy of a system can be increased—by transferring heat to it and by having an irreversible process. Since the entropy generation cannot be less than zero, there is only one way in which the entropy of a system can be decreased, and that is to transfer heat from the system.

Second, as we have already noted for an adiabatic process, $\delta Q = 0$, and therefore the increase in entropy is always associated with the irreversibilities.

Finally, the presence of irreversibilities will cause the work to be smaller than the reversible work. This means less work out in an expansion process and more work into the control mass ($\delta W < 0$) in a compression process.

One other point concerning the representation of irreversible processes on P–V and T–S diagrams should be made. The work for an irreversible process is not equal to $\int P\, dV$, and the heat transfer is not equal to $\int T\, dS$. Therefore, the area underneath the path does not represent work and heat on the P–V and T–S diagrams, respectively. In fact, in many situations we are not certain of the exact state through which a system passes when it undergoes an irreversible process. For this reason it is advantageous to show irreversible processes as dashed lines and reversible processes as solid lines. Thus, the area underneath the dashed line will never represent work or heat. Figure 8.11a shows an irreversible process and, because the heat transfer and work for this process are zero, the area underneath the dashed line has no significance. Figure 8.11b shows the reversible process, and area 1–2–b–a–1 represents the work on the P–V diagram and the heat transfer on the T–S diagram.

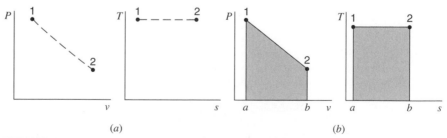

FIGURE 8.11 Reversible and irreversible processes on pressure-volume and temperature–entropy diagrams.

8.8 PRINCIPLE OF THE INCREASE OF ENTROPY

In the previous section, we considered irreversible processes in which the irreversibilities occurred inside the system or control mass. We also found that the entropy change of a control mass could be either positive or negative, since entropy can be increased by internal entropy generation and either increased or decreased by heat transfer, depending on the direction of that transfer. In this section, we examine the effect of heat transfer on the change of state in the surroundings, as well as the control mass itself.

Consider the process shown in Fig. 8.12 in which a quantity of heat δQ is transferred from the surroundings at temperature T_0 to the control mass at temperature T. Let the work done during this process be δW. For this process we can apply Eq. 8.10 to the control mass and write

$$dS_{c.m.} \geq \frac{\delta Q}{T}$$

For the surrounding at T_0, δQ is negative and we assume a reversible heat extraction so

$$dS_{surr} = \frac{-\delta Q}{T_0}$$

The total net change of entropy is therefore

$$dS_{net} = dS_{c.m.} + dS_{surr} \geq \frac{\delta Q}{T} - \frac{\delta Q}{T_0}$$

$$\geq \delta Q \left(\frac{1}{T} - \frac{1}{T_0} \right) \qquad (8.15)$$

Since $T_0 > T$, the quantity $[(1/T) - (1/T_0)]$ is positive and we conclude that

$$dS_{net} = dS_{c.m.} + dS_{surr} \geq 0$$

If $T > T_0$, the heat transfer is from the control mass to the surroundings, and both δQ and the quantity $[(1/T) - (1/T_0)]$ are negative, thus yielding the same result.

It should be noted that the right-hand side of Eq. 8.15 represents an external entropy generation due to heat transfer through a finite temperature difference. To amplify this point, take as a control mass the system that connects the surroundings at T_0 with the previous control mass at T, which typically is the walls. This mass does not experience any change of state, yet it has fluxes of entropy flowing in and out due to the heat transfer, and it is an irreversible process. For this mass, Eq. 8.11 gives

$$dS_{c.m.2} = 0 = \frac{\delta Q}{T_0} - \frac{\delta Q}{T} + \delta S_{gen,2}$$

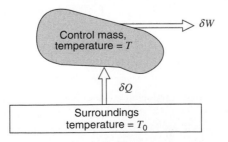

FIGURE 8.12 Entropy change for the control mass plus surroundings.

and it is realized that the difference in the two $\delta Q/T$ terms (fluxes of S) is the entropy generated in this control mass

$$\delta S_{gen,2} = \delta Q\left(\frac{1}{T} - \frac{1}{T_0}\right)$$

This is also precisely the location in space, where the heat transfer takes place over the finite temperature difference $T_0 - T$. This term is always positive (or zero for an adiabatic process), but as the temperature difference is made to approach zero, this term also approaches zero.

There could also be additional entropy generation terms in the surroundings of the types discussed in the previous section, if those factors were also present in the surroundings, and those will be positive, as well. Thus we conclude that the net entropy change is the sum of a number of terms, each of which is positive, due to a specific cause of irreversible entropy generation, such that the net entropy change could also be termed the total entropy generation.

$$dS_{net} = dS_{c.m.} + dS_{surr} = \sum \delta S_{gen} \geq 0 \tag{8.16}$$

where the equality holds for reversible processes and the inequality for irreversible processes. This is a very important equation, not only for thermodynamics but also for philosophical thought. This equation is referred to as the principle of the increase of entropy. The great significance is that the only processses that can take place are those in which the net change in entropy of the control mass plus its surroundings increase (or in the limit, remains constant). The reverse process, in which both the control mass and surroundings are returned to their original state, can never be made to occur. In other words, Eq. 8.16 dictates the single direction in which any process can proceed. Thus, the principle of the increase of entropy can be considered a quantitative general statement of the second law from the macroscopic point of view and applies to the combustion of fuel in our automobile engines, the cooling of our coffee, and the processes that take place in our body.

Sometimes this principle of the increase of entropy is stated in terms of an isolated system, one in which there is no interaction between the the system and its surroundings. Then there is no change in the surroundings, and we then conclude that

$$dS_{isolated\ system} = \delta S_{gen,\ system} \geq 0 \tag{8.17}$$

That is, in an isolated system, the only processes that can occur are those that have an associated increase in entropy.

The development of Eq. 8.16 as the principle of the increase of entropy was made for an infinitesimal change of state. When we wish to test a claimed process to see whether it satisfies the second law of thermodynamics, it will necessarily be for a finite change of state. Consider a control mass undergoing a process from initial state 1 to final state 2, with an associated heat transfer $_1Q_2$, which may be known or calculated from the first law. The heat transfer is to or from a reservoir at temperature T_0. For this process,

$$\Delta S_{c.m.} = S_2 - S_1, \ \Delta S_{surr} = -\frac{_1Q_2}{T_0}$$

$$\Delta S_{net} = \Delta S_{c.m.} + \Delta S_{surr} \tag{8.18}$$

and the net entropy change as calculated from Eq. 8.18 must be greater than zero (irreversible process), or in the limit be equal to zero (completely reversible process, internally and externally). This type of calculation is illustrated in the following example.

EXAMPLE 8.2   Suppose that 1 kg of saturated water vapor at 100°C is condensed to a saturated liquid at 100°C in a constant-pressure process by heat transfer to the surrounding air, which is at 25°C. What is the net increase in entropy of the water plus surroundings?

Solution

For the control mass (water), from the steam tables,

$$\Delta S_{c.m.} = -ms_{fg} = -1 \times 6.0480 = -6.0480 \text{ kJ / K}$$

Concerning the surroundings, we have

$$Q_{\text{to surroundings}} = mh_{fg} = 1 \times 2257.0 = 2257 \text{ kJ} \backslash \text{K}$$

$$\Delta S_{surr} = \frac{Q}{T_0} = \frac{2257}{298.15} = 7.5700 \text{ kJ / K}$$

$$\Delta S_{net} = \Delta S_{c.m.} + \Delta S_{surr} = -6.0480 + 7.5700 = 1.5220 \text{ kJ / K}$$

This increase in entropy is in accordance with the principle of the increase of entropy, and tells us, as does our experience, that this process can take place.

It is interesting to note how this heat transfer from the water to the surroundings might have taken place reversibly. Suppose that an engine operating on the Carnot cycle received heat from the water and rejected heat to the surroundings, as shown in Fig. 8.13. The decrease in the entropy of the water is equal to the increase in the entropy of the surroundings.

$$\Delta S_{c.m.} = -6.0480 \text{ kJ / K}$$

$$\Delta S_{surr} = 6.0480 \text{ kJ / K}$$

$$Q_{\text{to surroundings}} = T_0 \Delta S = 298.15(6.0480) = 1803.2 \text{ kJ}$$

$$W = Q_H - Q_L = 2257 - 1803.2 = 453.8 \text{ kJ}$$

Since this is a reversible cycle, the engine could be reversed and operated as a heat pump. For this cycle the work input to the heat pump would be 453.8 kJ.

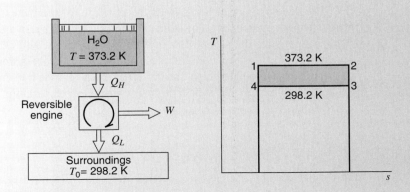

FIGURE 8.13
Reversible heat transfer with the surroundings.

EXAMPLE 8.2E Suppose that 1 lbm of saturated water vapor at 212 F is condensed to a saturated liquid at 212 F in a constant-pressure process by heat transfer to the surrounding air, which is at 80 F. What is the net increase in entropy of the water plus surroundings?

Solution

For the control mass (water), from the steam tables,

$$\Delta S_{system} = -s_{fg} = -1.4446 \text{ Btu / 1bm R}$$

Considering the surroundings, we have

$$Q_{\text{to surroundings}} = h_{fg} = 970.3 \text{ Btu / 1bm}$$

$$\Delta S_{surr} = \frac{Q}{T_0} = \frac{970.3}{540} = 1.7980 \text{ Btu / 1bm R}$$

$$\Delta S_{system} + \Delta S_{surr} = -1.4446 + 1.7980 = 0.3534 \text{ Btu / 1bm R}$$

This increase in entropy is in accordance with the principle of the increase of entropy, and tells us, as does our experience, that this process can take place.

It is interesting to note how this heat transfer from the water to the surroundings might have taken place reversibly. Suppose that an engine operating on the Carnot cycle received heat from the water and rejected heat to the surroundings, as shown in Fig. 8.13E. The decrease in the entropy of the water is equal to the increase in the entropy of the surroundings.

$$\Delta S_{c.m.} = -1.4446 \text{ Btu / 1bm R}$$

$$\Delta S_{surr} = 1.4446 \text{ Btu / 1bm R}$$

$$Q_{\text{to surroundings}} = T_0 \Delta S = 540(1.4446) = 780.1 \text{ Btu / 1bm}$$

$$W = Q_H - Q_L = 970.3 - 780.1 = 190.2 \text{ Btu / 1bm}$$

Since this is a reversible cycle, the engine could be reversed and operated as a heat pump. For this cycle the work input to the heat pump would be 190.2 Btu/lbm.

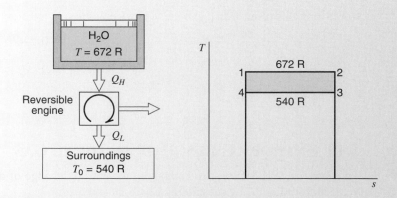

FIGURE 8.13E
Reversible heat transfer with the surroundings.

8.9 ENTROPY CHANGE OF A SOLID OR LIQUID

In Section 5.6 we considered the calculation of the internal energy and enthalpy changes with temperature for solids and liquids and found that, in general, it is possible to express both in terms of the specific heat, in the simple manner of Eq. 5.17, and in most instances in the integrated form of Eq. 5.18. We can now use this result and the thermodynamic property relation, Eq. 8.7, to calculate the entropy change for a solid or liquid. Note that for such a phase the specific volume term in Eq. 8.7 is very small, so that substituting Eq. 5.17 yields

$$ds \simeq \frac{du}{T} \simeq \frac{C}{T} dT \tag{8.19}$$

Now, as was mentioned in Section 5.6, for many processes involving a solid or liquid, we may assume that the specific heat remains constant, in which case Eq. 8.19 can be integrated. The result is

$$s_2 - s_1 \simeq C \ln \frac{T_2}{T_1} \tag{8.20}$$

If the specific heat is not constant, then commonly C is known as a function of T, in which case Eq. 8.19 can also be integrated to find the entropy change.

EXAMPLE 8.3   One kilogram of liquid water is heated from 20°C to 90°C. Calculate the entropy change, assuming constant specific heat, and compare the result with that found when using the steam tables.

   Control mass: Water.

   Initial and final states: Known.

   Model: Constant specific heat, value at room temperature.

Solution

For constant specific heat, from Eq. 8.20,

$$s_2 - s_1 = 4.184 \ln\left(\frac{363.2}{293.2}\right) = 0.8958 \text{ kJ / kg K}$$

Comparing this result with that obtained by using the steam tables, we have

$$s_2 - s_1 = s_{f\,90°C} - s_{f\,20°C} = 1.1925 - 0.2966$$

$$= 0.8959 \text{ kJ / kg K}$$

8.10 ENTROPY CHANGE OF AN IDEAL GAS

Two very useful equations for computing the entropy change of an ideal gas can be developed from Eq. 8.7 by substituting Eqs. 5.20 and 5.24:

$$T\,ds = du + P\,dv$$

For an ideal gas

$$du = C_{v0}dT \qquad \text{and} \qquad \frac{P}{T} = \frac{R}{v}$$

Therefore,

$$ds = C_{v0}\frac{dT}{T} + \frac{R\,dv}{v} \tag{8.21}$$

$$s_2 - s_1 = \int_1^2 C_{v0}\frac{dT}{T} + R\ln\frac{v_2}{v_1} \tag{8.22}$$

Similarly,

$$T\,ds = dh - v\,dP$$

For an ideal gas

$$dh = C_{p0}\,dT \qquad \text{and} \qquad \frac{v}{T} = \frac{R}{P}$$

Therefore,

$$ds = C_{p0}\frac{dT}{T} - R\frac{dp}{P} \tag{8.23}$$

$$s_2 - s_1 = \int_1^2 C_{p0}\frac{dT}{T} - R\ln\frac{P_2}{P_1} \tag{8.24}$$

To integrate Eqs. 8.22 and 8.24, we must know the temperature dependence of the specific heats. However, if we recall that their difference is always constant as expressed by Eq. 5.27, we realize that we need to examine the temperature dependence of only one of the specific heats.

As in Section 5.7, let us consider the specific heat C_{p0}. Again, there are three possibilities to examine, the simplest of which is the assumption of constant specific heat. In this instance it is possible to integrate Eq. 8.24 directly, to

$$s_2 - s_1 = C_{p0}\ln\frac{T_2}{T_1} - R\ln\frac{P_2}{P_1} \tag{8.25}$$

Similarly, integrating Eq. 8.22 for constant specific heat, we have

$$s_2 - s_1 = C_{v0}\ln\frac{T_2}{T_1} + R\ln\frac{v_2}{v_1} \tag{8.26}$$

The second possibility for the specific heat is to use an analytical equation for C_{p0} as a function of temperature, for example, one of those listed in Table A.6. The third possibility is to integrate the results of the calculations of statistical thermodynamics from reference temperature T_0 to any other temperature T and define a function

$$s_T^0 = \int_{T_0}^T \frac{C_{p0}}{T}dT \tag{8.27}$$

This function can then be tabulated in the single-entry (temperature) ideal-gas table, as for air in Table A.7 or for other gases in Table A.8. The entropy change between any two states 1 and 2 is then given by

$$s_2 - s_1 = \left(s_{T2}^0 - s_{T1}^0\right) - R\ln\frac{P_2}{P_1} \tag{8.28}$$

As with the energy functions discussed in Section 5.7, the ideal-gas tables, Tables A.7 and A.8, would give the most accurate results, and the equations listed in Table A.6 would give a close empirical approximation. Constant specific heat would be less accurate, except for monatomic gases and for other gases below room temperature. Again, it should be remembered that all these results are part of the ideal-gas model, which may or may not be appropriate in any particular problem.

EXAMPLE 8.4

Consider Example 5.6, in which oxygen is heated from 300 to 1500 K. Assume that during this process the pressure dropped from 200 to 150 kPa. Calculate the change in entropy per kilogram.

Solution

The most accurate answer for the entropy change, assuming ideal-gas behavior, would be from the ideal-gas tables, Table A.8. This result is, using Eq. 8.28,

$$\bar{s}_2 - \bar{s}_1 = (258.068 - 205.329) - 8.3145 \ln\left(\frac{150}{200}\right)$$

$$= 52.739 + 2.392 = 55.131 \text{ kJ / kmol K}$$

$$s_2 - s_1 = \frac{55.131}{32} = 1.7228 \text{ kJ / kg K}$$

The empirical equation from Table A.6 should give a good approximation to this result. Integrating Eq. 8.24, we have

$$\bar{s}_2 - \bar{s}_1 = \int_{T_1}^{T_2} \bar{C}_{p0} \frac{dT}{T} - \bar{R} \ln \frac{P_2}{P_1}$$

$$\bar{s}_2 - \bar{s}_1 = \left(37.432 \ln \theta + \frac{0.020\,102}{1.5} \theta^{1.5}\right.$$

$$\left. + \frac{178.57}{1.5} \theta^{-1.5} - \frac{236.88}{2} \theta^{-2}\right)\Bigg|_{\theta_1=3}^{\theta_2=15} - 8.3145 \ln\left(\frac{150}{200}\right)$$

$$= 52.726 + 2.392 = 55.118 \text{ kJ / kmol K}$$

$$s_2 - s_1 = \frac{\bar{s}_2 - \bar{s}_1}{M} = \frac{55.118}{32} = 1.7224 \text{ kJ / kg K}$$

which is within 0.1% of the previous value. For constant specific heat, using the value at 300 K from Table A.5, we have

$$s_2 - s_1 = 0.9216 \ln\left(\frac{1500}{300}\right) - 0.259\,83 \ln\left(\frac{150}{200}\right)$$

$$= 1.4833 + 0.0747 = 1.558 \text{ kJ / kg K}$$

which is too low by 9.6%. If, on the other hand, we assume that the specific heat is constant at its value at 900 K, the average temperature, as in Example 5.6, then

$$s_2 - s_1 = 1.0714 \ln\left(\frac{1500}{300}\right) + 0.0747 = 1.7991 \text{ kJ} / \text{kg K}$$

which is high by 4.4%.

EXAMPLE 8.5 Calculate the change in entropy per kilogram as air is heated from 300 to 600 K while pressure drops from 400 to 300 kPa. Assume:

1. Constant specific heat.
2. Variable specific heat.

Solution

1. From Table A.5 for air at 300 K,

$$C_{p0} = 1.0035 \text{ kJ} / \text{kg K}$$

Therefore, using Eq. 8.25, we have

$$s_2 - s_1 = 1.0035 \ln\left(\frac{600}{300}\right) - 0.287 \ln\left(\frac{300}{400}\right) = 0.7781 \text{ kJ} / \text{kg K}$$

2. From Table A.7,

$$s_{T1}^0 = 6.8693 \text{ kJ} / \text{kg K}, \qquad s_{T2}^0 = 7.5764 \text{ kJ} / \text{kg K}$$

Using Eq. 8.28 gives

$$s_2 - s_1 = 7.5764 - 6.8693 - 0.287 \ln\left(\frac{300}{400}\right) = 0.7897 \text{ kJ} / \text{kg K}$$

EXAMPLE 8.5E Calculate the change in entropy per pound as air is heated from 540 R to 1200 R while pressure drops from 50 lbf/in.^{2} to 40 lbf/in.^{2}. Assume:

1. Constant specific heat.
2. Variable specific heat.

Solution

1. From Table C.4 for air at 80 F,

$$C_{p0} = 0.24 \text{ Btu} / \text{lbm R}$$

Therefore, using Eq. 8.25, we have

$$s_2 - s_1 = 0.24 \ln\left(\frac{1200}{540}\right) - \frac{53.34}{778} \ln\left(\frac{40}{50}\right) = 0.2068 \text{ Btu} / \text{lbm R}$$

2. From Table C.6

$$s^0_{T_1} = 0.6008 \text{ Btu / lbm R} \qquad s^0_{T_2} = 0.7963 \text{ Btu / lbm R}$$

Using Eq. 8.28 gives

$$s_2 - s_1 = 0.7963 - 0.6008 - \frac{53.34}{778} \ln \frac{40}{50} = 0.2108 \text{ Btu / lbm R}$$

The air tables can be used for reversible adiabatic processes by employing the relative pressure P_r and relative specific volume v_r. The definition of these terms and the derivation follow.

For the reversible adiabatic process

$$T \, ds = dh - v \, dP = 0$$

Therefore,

$$dh = C_{p0} dT = v \, dP = RT \frac{dP}{P}$$

$$\frac{dP}{P} = \frac{C_{p0}}{R} \frac{dT}{T}$$

Let this equation be integrated between a reference state having a temperature T_0 and a pressure P_0, and a given arbitrary state having a temperature T and a pressure P. Then

$$\ln \frac{P}{P_0} = \frac{1}{R} \int_{T_0}^{T} C_{p0} \frac{dT}{T}$$

The right side of this equation is a function of temperature only. The relative pressure P_r is defined as

$$\ln P_r \equiv \ln \frac{P}{P_0} = \frac{1}{R} \int_{T_0}^{T} C_{p0} \frac{dT}{T} = \frac{s^0_T}{R} \tag{8.29}$$

Thus, a value of P_r can be tabulated as a function of temperature.

If we consider two states, 1 and 2, along a constant-entropy line, it follows from Eq. 8.29 that

$$\frac{P_1}{P_2} = \left(\frac{P_{r1}}{P_{r2}} \right)_{s=\text{constant}} \tag{8.30}$$

This equation states that the ratio of the relative pressures for two states having the same entropy is equal to the ratio of the absolute pressures.

The development of the relative specific volume is similar, and the ratio of the relative specific volumes v_r, in an isentropic process is equal to the ratio of the specific volumes. That is,

$$\frac{v_1}{v_2} = \left(\frac{v_{r1}}{v_{r2}} \right)_{s=\text{constant}} \tag{8.31}$$

The isentropic functions P_r and v_r are tabulated for air, Table A.7, but are not listed in the other gas tables in the Appendix, Table A.8. In analyzing an isentropic process for these gases, we need to use Eq. 8.28 with the left side of the equation equal to zero and the standard-state entropies from Table A.8.

EXAMPLE 8.6

One kilogram of air is contained in a cylinder fitted with a piston at a pressure of 400 kPa and a temperature of 600 K. The air is expanded to 150 kPa in a reversible, adiabatic process. Calculate the work done by the air.

Control mass: Air.

Initial state: P_1, T_1; state 1 fixed.

Final state: P_2.

Process: Reversible and adiabatic.

Model: Ideal-gas and air tables, Table A.7.

Analysis:

First law:

$$0 = u_2 - u_1 + w$$

Second law:

$$s_2 = s_1$$

Solution

From Table A.7,

$$u_1 = 435.10 \text{ kJ / kg} \qquad P_{r1} = 13.0923$$

From Eq. 8.30,

$$P_{r2} = P_{r1} \times \frac{P_2}{P_1} = 13.0923 \times \frac{150}{400} = 4.9096$$

From Table A.7,

$$T_2 = 457 \text{ K} \qquad u_2 = 328.14$$

Therefore,

$$w = 435.10 - 328.14 = 106.96 \text{ kJ / kg}$$

At this point, it is advantageous to introduce the specific heat ratio k, which is defined as the ratio of the constant-pressure specific heat to the constant-volume specific heat at zero pressure.

$$k = \frac{C_{p0}}{C_{v0}} \tag{8.32}$$

Because the difference between C_{p0} and C_{v0} is a constant, Eq. 5.27, and because C_{p0} and C_{v0} are functions of temperature, it follows that k is also a function of temperature. However, when we consider the specific heat to be constant, k is also constant.

From the definition of k and Eq. 5.27, it follows that

$$C_{v0} = \frac{R}{k-1} \qquad C_{p0} = \frac{kR}{k-1} \tag{8.33}$$

Some very useful and simple relations for the reversible adiabatic process can be developed when the specific heats are assumed to be constant.

For the reversible adiabatic process, $ds = 0$. Therefore,

$$T\, ds = du + P\, dv = C_{v0} dT + P\, dv = 0$$

From the equation of state for an ideal gas,

$$dT = \frac{1}{R}\left(P\, dv + v\, dP\right)$$

Therefore,

$$\frac{C_{v0}}{R}\left(P\, dv + v\, dP\right) + P\, dv = 0$$

Substituting Eq. 8.33 into the expression and rearranging gives

$$\frac{1}{k-1}\left(P\, dv + v\, dP\right) + P\, dv = 0$$

$$v\, dP + kP\, dv = 0$$

$$\frac{dP}{P} + k\frac{dv}{v} = 0$$

Because k is constant when the specific heat is constant, this equation can be integrated under these conditions to give

$$Pv^k = \text{constant} \tag{8.34}$$

Equation 8.4 holds for all reversible adiabatic processes that involve an ideal gas with constant specific heat. It is usually advantageous to express this constant in terms of the initial and final states.

$$Pv^k = P_1 v_1^k = P_2 v_2^k = \text{constant} \tag{8.35}$$

From this equation and the ideal-gas equation of state, the following expressions relating the initial and final states of an isentropic process can be derived.

$$\frac{P_2}{P_1} = \left(\frac{v_1}{v_2}\right)^k = \left(\frac{V_1}{V_2}\right)^k \tag{8.36}$$

$$\frac{T_2}{T_1} = \left(\frac{P_2}{P_1}\right)^{(k-1)/k} = \left(\frac{v_1}{v_2}\right)^{k-1} \tag{8.37}$$

With the assumption of constant specific heat, some convenient equations can be derived for the work done by an ideal gas during an adiabatic process. Consider a control mass consisting of an ideal gas that undergoes a process in which work is done only at the moving boundary.

$$_1Q_2 = m(u_2 - u_1) + {}_1W_2 = 0$$

$$_1W_2 = -m(u_2 - u_1) = -mC_{v0}(T_2 - T_1)$$

$$= \frac{mR}{1-k}(T_2 - T_1) = \frac{P_2V_2 - P_1V_1}{1-k} \qquad (8.38)$$

It should be noted that Eq. 8.38 applies to adiabatic processes only. Since no assumption was made regarding reversibility, it applies to both reversible and irreversible processes. Equation 8.38 is frequently derived for reversible processes by starting with the relation

$$_1W_2 = \int_1^2 P\, dV$$

for the control mass.

8.11 THE REVERSIBLE POLYTROPIC PROCESS FOR AN IDEAL GAS

When a gas undergoes a reversible process in which there is heat transfer, the process frequently takes place in such a manner that a plot of log P versus log V is a straight line, as shown in Fig. 8.14. For such a process PV^n is a constant.

A process having this relation between pressure and volume is called a polytropic process. An example is the expansion of the combustion gases in the cylinder of a water-cooled reciprocating engine. If the pressure and volume are measured during the expansion stroke of a polytropic process, as might be done with an engine indicator, and the logarithms of the pressure and volume are plotted, the result would be similar to the straight line in Fig. 8.14. From this figure it follows that

$$\frac{d \ln P}{d \ln V} = -n$$

$$d \ln P + nd \ln V = 0$$

If n is a constant (which implies a straight line on the log P versus log V plot), this equation can be integrated to give the following relation:

$$PV^n = \text{ constant} = P_1V_1^n = P_2V_2^n \qquad (8.39)$$

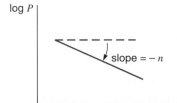

FIGURE 8.14 Example of a polytropic process.

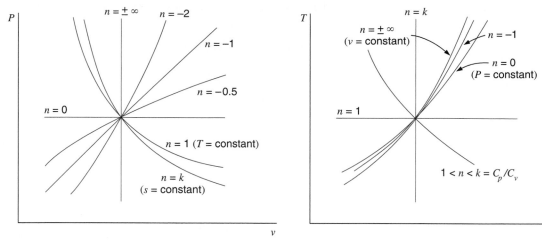

FIGURE 8.15 Polytropic processes on P–v and T–s diagrams.

From this equation it is evident that the following relations can be written for a polytropic process.

$$\frac{P_2}{P_1} = \left(\frac{V_1}{V_2}\right)^n$$

$$\frac{T_2}{T_1} = \left(\frac{P_2}{P_1}\right)^{(n-1)/n} = \left(\frac{V_1}{V_2}\right)^{n-1} \tag{8.40}$$

For a control mass consisting of an ideal gas, the work done at the moving boundary during a reversible polytropic process can be derived from the relations

$$_1W_2 = \int_1^2 P\,dV \qquad \text{and} \qquad PV^n = \text{constant}$$

$$_1W_2 = \int_1^2 P\,dV = \text{constant} \int_1^2 \frac{dV}{V^n}$$

$$= \frac{P_2V_2 - P_1V_1}{1-n} = \frac{mR(T_2 - T_1)}{1-n} \tag{8.41}$$

for any value of n except $n = 1$.

The polytropic processes for various values of n are shown in Fig. 8.15 on P–v and T–s diagrams. The values of n for some familiar processes are

Isobaric process	$n = 0$,	$P = $ constant
Isothermal process	$n = 1$,	$T = $ constant
Isentropic process	$n = k$,	$s = $ constant
Isochoric process	$n = \infty$,	$v = $ constant

EXAMPLE 8.7   In a reversible process, nitrogen is compressed in a cylinder from 100 kPa, 20°C to 500 kPa. During this compression process the relation between pressure and volume is $PV^{1.3}$

= constant. Calculate the work and heat transfer per kilogram, and show this process on *P–V* and *T–S* diagrams.

> *Control mass:* Nitrogen.
>
> *Initial state:* P_1, T_1; state 1 known.
>
> *Final state:* P_2.
>
> *Process:* Reversible, polytropic with exponent $n < k$
>
> *Diagram:* Fig. 8.16.
>
> *Model:* Ideal gas, constant specific heat—value at 300 K.

Analysis:

Boundary movement work. From Eq. 8.41,

$$_1W_2 = \int_1^2 P\,dV = \frac{P_2V_2 - P_1V_1}{1-n} = \frac{mR(T_2 - T_1)}{1-n}$$

First law:

$$_1q_2 = u_2 - u_1 + {}_1w_2 = C_{v0}(T_2 - T_1) + {}_1w_2$$

Solution

From Eq. 8.40,

$$\frac{T_2}{T_1} = \left(\frac{P_2}{P_1}\right)^{(n-1)/n} = \left(\frac{500}{100}\right)^{(1.3-1)/1.3} = 1.4498$$

$$T_2 = 293.2 \times 1.4498 = 425\ \text{K}$$

Then

$$_1w_2 = \frac{R(T_2 - T_1)}{1-n} = \frac{0.2968(425 - 293.2)}{(1-1.3)} = -130.4\ \text{kJ/kg}$$

and from the first law,

$$_1q_2 = C_{v0}(T_2 - T_1) + {}_1w_2$$

$$= 0.7448(425 - 293.2) - 130.4 = -32.2\ \text{kJ/kg}$$

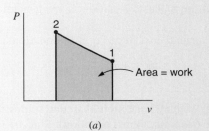
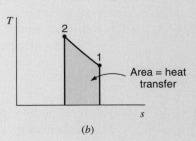

FIGURE 8.16 Diagram for Example 8.7.

The reversible isothermal process for an ideal gas is of particular interest. In this process

$$PV = \text{constant} = P_1V_1 = P_2V_2 \tag{8.42}$$

The work done at the boundary of a simple compressible mass during a reversible isothermal process can be found by integrating the equation

$$_1W_2 = \int_1^2 P\,dV$$

The integration is

$$_1W_2 = \int_1^2 P\,dV = \text{constant} \int_1^2 \frac{dV}{V} = P_1V_1 \ln\frac{V_2}{V_1} = P_1V_1 \ln\frac{P_1}{P_2} \tag{8.43}$$

or

$$_1W_2 = mRT \ln\frac{V_2}{V_1} = mRT \ln\frac{P_1}{P_2} \tag{8.44}$$

Because there is no change in internal energy or enthalpy in an isothermal process, the heat transfer is equal to the work (neglecting changes in kinetic and potential energy). Therefore, we could have derived Eq. 8.43 by calculating the heat transfer.

For example, using Eq. 8.7, we have

$$\int_1^2 T\,ds = {}_1q_2 = \int_1^2 du + \int_1^2 P\,dv$$

But $du = 0$ and $Pv = \text{constant} = P_1v_1 = P_2v_2$, such that

$$_1q_2 = \int_1^2 P\,dv = P_1v_1 \ln\frac{v_2}{v_1}$$

which yields the same result as Eq. 8.43.

8.12 ENTROPY AS A RATE EQUATION

The second law of thermodynamics was used to write the balance of entropy in Eq. 8.11 for a variation and in Eq. 8.14 for a finite change. In some cases the equation is needed in a rate form so a given process can be tracked in time. The rate form is also the basis for the development of the entropy balance equation in the general control volume analysis for an unsteady situation.

Take the incremental change in S from Eq. 8.11 and divide by δt. We get

$$\frac{dS}{\delta t} = \frac{1}{T}\frac{\delta Q}{\delta t} + \frac{\delta S_{\text{gen}}}{\delta t} \tag{8.45}$$

For a given control volume we may have more than one source of heat transfer, each at a certain surface temperature (semidistributed situation). Since we did not have to consider the temperature at which the heat transfer crossed the control surface for the energy equation, all the terms were added into a net heat transfer in a rate form in Eq. 5.32. Using this and a dot to indicate a rate, the final form for the entropy equation in the limit is

$$\frac{dS_{\text{c.m.}}}{dt} = \sum \frac{1}{T}\dot{Q} + \dot{S}_{\text{gen}} \tag{8.46}$$

expressing the rate of entropy change as due to the flux of entropy into the control mass from heat transfer and an increase due to irreversible processes inside the control mass. If only reversible processes take place inside the control volume, the rate of change of entropy is determined by the rate of heat transfer divided by the temperature terms alone.

EXAMPLE 8.8   Consider an electric space heater that converts 1 kW of electric power into a heat flux of 1 kW delivered at 600 K from the hot wire surface. Let us look at the process of the energy conversion from electricity to heat transfer and find the rate of total entropy generation.

Solution

Control mass: The electric heater wire.
State: Constant wire temperature 600 K.

The first and the second law of thermodynamics in rate form become

$$\frac{dE_{\text{c.m.}}}{dt} = \frac{dU_{\text{c.m.}}}{dt} = 0 = \dot{W}_{\text{el,in}} - \dot{Q}_{\text{out}}$$

$$\frac{dS_{\text{c.m.}}}{dt} = \frac{dU_{\text{c.m.}}}{dt} / T = 0 = -\dot{Q}_{\text{out}} / T_{\text{surface}} + \dot{S}_{\text{gen}}$$

Notice we neglected kinetic and potential energy changes to go from rate of E to rate of U, then the left hand side is zero since it is steady state and the right-hand side of the energy equation is electric work in minus the heat transfer out. For the entropy equation the left-hand side is zero because of steady state and the right-hand side has a flux of entropy out due to heat transfer and entropy is generated in the wire.
    We now get the entropy generation as

$$\dot{S}_{\text{gen}} = \dot{Q}_{\text{out}} / T = 1/600 = 0.00167 \quad \text{kW} / \text{K}$$

PROBLEMS

8.1 Consider the steam power plant in Problem 7.9 and the heat engine in Problem 7.17. Show whether these cycles satisfy the inequality of Clausius.

8.2 Find the missing properties and give the phase of the substance

a. H_2O	$s = 7.70$ kJ/kg K, $P = 25$ kPa	$h = ?$ $T = ?$ $x = ?$
b. H_2O	$u = 3400$ kJ/kg, $P = 10$ MPa	$T = ?$ $x = ?$ $s = ?$
c. R-12	$T = 0°C$, $P = 250$ kPa	$s = ?$ $x = ?$
d. R-134a	$T = -10°C$, $x = 0.45$	$v = ?$ $s = ?$
e. NH_3	$T = 20°C$, $s = 5.50$ kJ/kg K	$u = ?$ $x = ?$

8.3 Consider a Carnot-cycle heat engine with water as the working fluid. The heat transfer to the water occurs at 300°C, during which process the water changes from saturated liquid to saturated vapor. The heat is rejected from the water at 40°C. Show the cycle on a T–s diagram and find the quality of the water at the beginning and end of the heat rejection process. Determine the net work output per kilogram of water and the cycle thermal efficiency.

8.4 In a Carnot engine with water as the working fluid, the high temperature is 250°C and as Q_H is received, the water changes from saturated liquid to saturated vapor. The water pressure at the low temperature is 100 kPa. Find T_L, the cycle thermal efficiency, the heat added per kilogram, and the entropy, s, at the beginning of the heat rejection process.

8.5 Water is used as the working fluid in a Carnot-cycle heat engine, where it changes from saturated liquid to saturated vapor at 200°C as heat is added. Heat is rejected in a constant pressure process (also constant T) at 20 kPa. The heat engine powers a Carnot-cycle refrigerator that operates between −15°C and +20°C. Find the heat added to the water per kg water. How much heat should be added to the water in the heat engine so the refrigerator can remove 1 kJ from the cold space?

8.6 Consider a Carnot-cycle heat pump with R-22 as the working fluid. Heat is rejected from the R-22 at 40°C, during which process the R-22 changes from saturated vapor to saturated liquid. The heat is transferred to the R-22 at 0°C.

  a. Show the cycle on a T–s diagram.

  b. Find the quality of the R-22 at the beginning and end of the isothermal heat addition process at 0°C.

  c. Determine the coefficient of performance for the cycle.

8.7 Do Problem 8.6 using refrigerant R-134a instead of R-22.

8.8 Water at 200 kPa, $x = 1.0$ is compressed in a piston/cylinder to 1 MPa, 250°C in a reversible process. Find the sign for the work and the sign for the heat transfer.

8.9 One kilogram of ammonia in a piston/cylinder at 50°C, 1000 kPa is expanded in a reversible isothermal process to 100 kPa. Find the work and heat transfer for this process.

8.10 One kilogram of ammonia in a piston/cylinder at 50°C, 1000 kPa is expanded in a reversible isobaric process to 140°C. Find the work and heat transfer for this process.

8.11 One kilogram of ammonia in a piston/cylinder at 50°C, 1000 kPa is expanded in a reversible adiabatic process to 100 kPa. Find the work and heat transfer for this process.

8.12 A cylinder fitted with a piston contains ammonia at 50°C, 20% quality with a volume of 1 L. The ammonia expands slowly, and during this process heat is transferred to maintain a constant temperature. The process continues until all the liquid is gone. Determine the work and heat transfer for this process.

8.13 An insulated cylinder fitted with a piston contains 0.1 kg of water at 100°C, 90% quality. The piston is moved, compressing the water until it reaches a pressure of 1.2 MPa. How much work is required in the process?

8.14 A cylinder fitted with a frictionless piston contains water, as shown in Fig. P8.14. A constant hydraulic pressure on the back face of the piston maintains a cylinder pressure of 10 MPa. Initially, the water is at 700°C, and the volume is 100 L. The water is now cooled and condensed to saturated liquid. The heat released during this process is the Q supply to a cyclic heat engine that in turn rejects heat to the ambient at 30°C. If the overall process is reversible, what is the net work output of the heat engine?

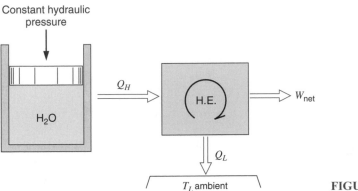

FIGURE P8.14

8.15 One kilogram of water at 300°C expands against a piston in a cylinder until it reaches ambient pressure, 100 kPa, at which point the water has a quality of 90%. It may be assumed that the expansion is reversible and adiabatic. What was the initial pressure in the cylinder and how much work is done by the water?

8.16 A piston/cylinder has 2 kg ammonia at 50°C, 100 kPa, which is compressed to 1000 kPa. The process happens so slowly that the temperature is constant. Find the heat transfer and work for the process assuming it to be reversible.

8.17 A heavily insulated cylinder/piston contains ammonia at 1200 kPa, 60°C. The piston is moved, expanding the ammonia in a reversible process until the temperature is −20°C. During the process 600 kJ of work is given out by the ammonia. What was the initial volume of the cylinder?

8.18 A closed tank, $V = 10$ L, containing 5 kg of water initially at 25°C, is heated to 175°C by a heat pump that is receiving heat from the surroundings at 25°C. Assume that this process is reversible. Find the heat transfer to the water and the work input to the heat pump.

8.19 A rigid, insulated vessel contains superheated vapor steam at 3 MPa, 400°C. A valve on the vessel is opened, allowing steam to escape. The overall process is irreversible, but the steam remaining inside the vessel goes through a reversible adiabatic expansion. Determine the fraction of steam that has escaped when the final state inside is saturated vapor.

8.20 A cylinder containing R-134a at 10°C, 150 kPa, has an initial volume of 20 L. A piston compresses the R-134a in a reversible, isothermal process until it reaches the saturated vapor state. Calculate the required work and heat transfer to accomplish this process.

8.21 An insulated cylinder fitted with a piston contains 0.1 kg of superheated vapor steam. The steam expands to ambient pressure, 100 kPa, at which point the steam inside the cylinder is at 150°C. The steam does 50 kJ of work against the piston during the expansion. Verify that the initial pressure is 1.19 MPa and find the initial temperature.

8.22 A heavily insulated cylinder fitted with a frictionless piston, as shown in Fig. P8.22, contains ammonia at 6°C, 90% quality, at which point the volume is 200 L. The external force on the piston is now increased slowly, compressing the ammonia until its temperature reaches 50°C. How much work is done on the ammonia during this process?

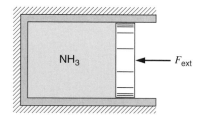

FIGURE P8.22

8.23 A piston/cylinder with constant loading of piston contains 1 L water at 400 kPa, quality 15%. It has some stops mounted so the maximum possible volume is 11 L. A reversible heat pump extracting heat from the ambient at 300 K, 100 kPa heats the water to 300°C. Find the total work and heat transfer for the water and the work input to the heat pump.

8.24 A piston/cylinder contains 2 kg water at 200°C, 10 MPa. The piston is slowly moved to expand the water in an isothermal process to a pressure of 200 kPa. Any heat transfer takes place with an ambient at 200°C and the whole process may be assumed reversible. Sketch the process in a P-V diagram and calculate both the heat transfer and the total work.

8.25 An insulated cylinder/piston has an initial volume of 0.15 m^3 and contains steam at 400 kPa, 200°C. The steam is expanded adiabatically, and the work output is measured very carefully to be 30 kJ. It is claimed that the final state of the water is in the two-phase (liquid and vapor) region. What is your evaluation of the claim?

8.26 An amount of energy, say 1000 kJ, comes from a furnace at 800°C going into water vapor at 400°C, from which it goes to a solid metal at 200°C and then into some air at 70°C. For each location calculate the flux of s through a surface as (Q/T). What makes the flux larger and larger?

8.27 An insulated cylinder/piston contains R-134a at 1 MPa, 50°C, with a volume of 100 L. The R-134a expands, moving the piston until the pressure in the cylinder has dropped to 100 kPa. It is claimed that the R-134a does 190 kJ of work against the piston during the process. Is that possible?

8.28 A piece of hot metal should be cooled rapidly (quenched) to 25°C, which requires removal of 1000 kJ from the metal. The cold space that absorbs the energy could be one of three possibilities: (1) Submerge the metal into a bath of liquid water and ice, thus melting the ice. (2) Let saturated liquid R-22 at −20°C absorb the energy so that it becomes saturated vapor. (3) Absorb the energy by vaporizing liquid nitrogen at 101.3 kPa pressure.

  a. Calculate the change of entropy of the cooling media for each of the three cases.

  b. Discuss the significance of the results.

8.29 A mass- and atmosphere-loaded piston/cylinder contains 2 kg of water at 5 MPa, 100°C. Heat is added from a reservoir at 700°C to the water until it reaches 700°C. Find the work, heat transfer, and total entropy production for the system and surroundings.

8.30 A cylinder fitted with a movable piston contains water at 3 MPa, 50% quality, at which point the volume is 20 L. The water now expands to 1.2 MPa as a result of receiving 600 kJ of heat from a large source at 300°C. It is claimed that the water does 124 kJ of work during this process. Is this possible?

8.31 A 4 L jug of milk at 25°C is placed in your refrigerator where it is cooled down to the refrigerator's inside constant temperature of 5°C. Assume the milk has the properties of liquid water and find the entropy generated in the cooling process.

8.32 A piston/cylinder contains 1 kg water at 150 kPa, 20°C. The piston is loaded so pressure is linear in volume. Heat is added from a 600°C source until the water is at 1 MPa, 500°C. Find the heat transfer and the total change in entropy.

8.33 Water in a piston/cylinder shown in Fig. P8.33 is at 1 MPa, 500°C. There are two stops, a lower one at which $V_{min} = 1$ m³ and an upper one at $V_{max} = 3$ m³. The piston is loaded with a mass and outside atmosphere such that it floats when the pressure is 500 kPa. This setup is now cooled to 100°C by rejecting heat to the surroundings at 20°C. Find the total entropy generated in the process.

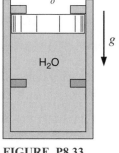

FIGURE P8.33

8.34 Two tanks contain steam, and they are both connected to a piston/cylinder, as shown in Fig. P8.34. Initially the piston is at the bottom and the mass of the piston is such that a pressure of 1.4 MPa below it will be able to lift it. Steam in A is 4 kg at 7 MPa, 700°C and B has 2 kg at 3 MPa, 350°C. The two valves are opened, and the water comes to a uniform state. Find the final temperature and the total entropy generation, assuming no heat transfer.

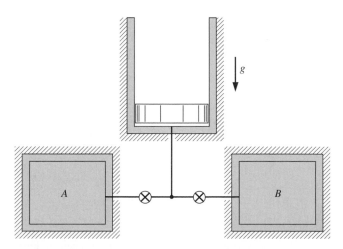

FIGURE P8.34

8.35 A cylinder/piston contains 3 kg of water at 500 kPa, 600°C. The piston has a cross-sectional area of 0.1 m² and is restrained by a linear spring with spring constant 10 kN/m. The setup is allowed to cool down to room temperature due to heat transfer to the room at 20°C. Calculate the total (water and surroundings) change in entropy for the process.

8.36 A cylinder/piston contains water at 200 kPa, 200°C with a volume of 20 L. The piston is moved slowly, compressing the water to a pressure of 800 kPa. The loading on the piston is such that the product PV is a constant. Assuming that the room temperature is 20°C, show that this process does not violate the second law.

8.37 One kilogram of ammonia (NH_3) is contained in a spring-loaded piston/cylinder as saturated liquid at −20°C. Heat is added from a reservoir at 100°C until a final condition of 800 kPa, 70°C is reached. Find the work, heat transfer, and entropy generation, assuming the process is internally reversible.

8.38 A piston/cylinder has a piston loaded so pressure is linear with volume, and it contains 2 kg water at 100°C, quality 10%. Heat is added from a 700°C energy reservoir so a final state of 500°C, 1 MPa is reached. Find the specific work and heat transfer for the water and the total entropy generation for the process.

8.39 An insulated cylinder fitted with a frictionless piston contains saturated vapor R-12 at ambient temperature, 20°C. The initial volume is 10 L. The R-12 is now expanded to a temperature of −30°C. The insulation is then removed from the cylinder, allowing it to warm at constant pressure to ambient temperature. Calculate the net work and the net entropy change for the overall process.

8.40 A foundry form box with 25 kg of 200°C hot sand is dumped into a bucket with 50 L water at 15°C. Assuming no heat transfer with the surroundings and no boiling away of liquid water, calculate the net entropy change for the process.

8.41 A large slab of concrete, $5 \times 8 \times 0.3$ m, is used as a thermal storage mass in a solar-heated house. If the slab cools overnight from 23°C to 18°C in an 18°C house, what is the net entropy change associated with this process?

8.42 Find the total work the heat engine can give out as it receives energy from the rock bed as described in Problem 7.22 (see Fig. P8.42). *Hint:* write the entropy balance equation for the control volume that is the combination of the rock bed and the heat engine.

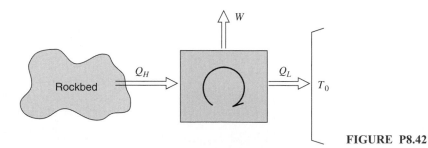

FIGURE P8.42

8.43 Liquid lead initially at 500°C is poured into a form so that it holds 2 kg. It then cools at constant pressure down to room temperature of 20°C as heat is transferred to the room. The melting point of lead is 327°C and the enthalpy change between the phases, h_{if}, is 24.6 kJ/kg. The specific heat is 0.138 kJ/kg K for the solid and 0.155 kJ/kg K for the liquid. Calculate the net entropy change for this process.

8.44 A hollow steel sphere with a 0.5-m inside diameter and a 2-mm thick wall contains water at 2 MPa, 250°C. The system (steel plus water) cools to the ambient temperature, 30°C. Calculate the net entropy change of the system and surroundings for this process.

8.45 A mass of 1 kg of air contained in a cylinder at 1.5 MPa, 1000 K, expands in a reversible isothermal process to a volume 10 times larger. Calculate the heat transfer during the process and the change of entropy of the air.

8.46 A mass of 1 kg of air contained in a cylinder at 1.5 MPa, 1000 K, expands in a reversible adiabatic process to 100 kPa. Calcuate the final temperature and the work done during the process, using

a. Constant specific heat, value from Table A.5

b. The ideal gas tables, Table A.7

8.47 Consider a Carnot-cycle heat pump having 1 kg of nitrogen gas in a cylinder/piston arrangement. This heat pump operates between reservoirs at 300 K and 400 K. At the beginning of the low-temperature heat addition, the pressure is 1 MPa. During this processs the volume triples. Analyze each of the four processes in the cycle and determine

a. The pressure, volume, and temperature at each point

b. The work and heat transfer for each process

8.48 A rigid tank contains 2 kg of air at 200 kPa and ambient temperature, 20°C. An electric current now passes through a resistor inside the tank. After a total of 100 kJ of electrical work has crossed the boundary, the air temperature inside is 80°C. Is this possible?

8.49 A handheld pump for a bicycle has a volume of 25 cm^3 when fully extended. You now press the plunger (piston) in while holding your thumb over the exit hole so that an air pressure of 300 kPa is obtained. The outside atmosphere is at P_0, T_0. Consider two cases: (1) it is done quickly (~1 s), and (2) it is done very slowly (~1 h).

a. State assumptions about the process for each case.

b. Find the final volume and temperature for both cases.

8.50 An insulated cylinder/piston contains carbon dioxide gas at 120 kPa, 400 K. The gas is compressed to 2.5 MPa in a reversible adiabatic process. Calculate the final temperature and the work per unit mass, assuming

a. Variable specific heat, Table A.8

b. Constant specific heat, value from Table A.5

c. Constant specific heat, value at an intermediate temperature from Table A.6

8.51 Consider a small air pistol with a cylinder volume of 1 cm^3 at 250 kPa, 27°C. The bullet acts as a piston initially held by a trigger. The bullet is released so the air expands in an adiabatic process. If the pressure should be 100 kPa as the bullet leaves the cylinder find the final volume and the work done by the air.

8.52 A rigid storage tank of 1.5 m^3 contains 1 kg argon at 30°C. Heat is then transferred to the argon from a furnace operating at 1300°C until the specific entropy of the argon has increased by 0.343 kJ/kg K. Find the total heat transfer and the entropy generated in the process.

8.53 A piston/cylinder, shown in Fig. P8.53, contains air at 1380 K, 15 MPa, with $V_1 =$ 10 cm^3, $A_{cyl} = 5$ cm^2. The piston is released, and just before the piston exits the end of the cylinder the pressure inside is 200 kPa. If the cylinder is insulated, what is its length? How much work is done by the air inside?

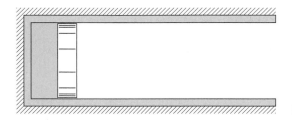

FIGURE P8.53

8.54 Two rigid tanks shown in Fig. P8.54 each contain 10 kg N$_2$ gas at 1000 K, 500 kPa. They are now thermally connected to a reversible heat pump, which heats one and cools the other with no heat transfer to the surroundings. When one tank is heated to 1500 K the process stops. Find the final (P, T) in both tanks and the work input to the heat pump, assuming constant heat capacities.

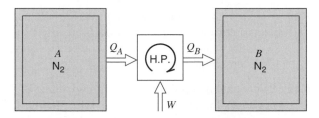

FIGURE P8.54

8.55 Repeat the previous problem, but with variable heat capacities.

8.56 We wish to obtain a supply of cold helium gas by applying the following technique. Helium contained in a cylinder at ambient conditions, 100 kPa, 20°C, is compressed in a reversible isothermal process to 600 kPa, after which the gas is expanded back to 100 kPa in a reversible adiabatic process.

  a. Show the process on a T–s diagram.

  b. Calculate the final temperature and the net work per kilogram of helium.

  c. If a diatomic gas, such as nitrogen or oxygen, is used instead, would the final temperature be higher, lower, or the same?

8.57 A 1-m^3 insulated, rigid tank contains air at 800 kPa, 25°C. A valve on the tank is opened, and the pressure inside quickly drops to 150 kPa, at which point the valve is closed. Assuming that the air remaining inside has undergone a reversible adiabatic expansion, calculate the mass withdrawn during the process.

8.58 An uninsulated cylinder fitted with a piston contains air at 500 kPa, 200°C, at which point the volume is 10 L. The external force on the piston is now varied in such a manner that the air expands to 150 kPa, 25 L volume. It is claimed that in this process the air produces 70% of the work that would have resulted from a reversible, adiabatic expansion from the same initial pressure and temperature to the same final pressure. Room temperature is 20°C.

a. What is the amount of work claimed?

b. Is this claim possible?

8.59 A rigid container with volume 200 L is divided into two equal volumes by a partition. Both sides contain nitrogen, one side is at 2 MPa, 200°C, and the other at 200 kPa, 100°C. The partition ruptures, and the nitrogen comes to a uniform state at 70°C. Assume the temperature of the surroundings is 20°C, determine the work done and the net entropy change for the process.

8.60 Nitrogen at 600 kPa, 127°C is in a 0.5 m^3 insulated tank connected to a pipe with a valve to a second insulated initially empty tank of volume 0.5 m^3. The valve is opened and the nitrogen fills both tanks. Find the final pressure and temperature and the entropy generation this process causes. Why is the process irreversible?

8.61 Neon at 400 kPa, 20°C is brought to 100°C in a polytropic process with $n = 1.4$. Give the sign for the heat transfer and work terms and explain.

8.62 A cylinder/piston contains carbon dioxide at 1 MPa, 300°C with a volume of 200 L. The total external force acting on the piston is proportional to V^3. This system is allowed to cool to room temperature, 20°C. What is the total entropy generation for the process?

8.63 A cylinder/piston contains 1 kg methane gas at 100 kPa, 20°C. The gas is compressed reversibly to a pressure of 800 kPa. Calculate the work required if the process is

a. Adiabatic

b. Isothermal

c. Polytropic, with exponent $n = 1.15$

8.64 The power stroke in an internal combustion engine can be approximated with a polytropic expansion. Consider air in a cylinder volume of 0.2 L at 7 MPa, 1800 K. It now expands in a reversible polytropic process with exponent, $n = 1.5$, through a volume ratio of 8:1. Show this process on P–v and T–s diagrams, and calculate the work and heat transfer for the process.

8.65 Helium in a piston/cylinder at 20°C, 100 kPa is brought to 400 K in a reversible polytropic process with exponent $n = 1.25$. You may assume helium is an ideal gas with constant specific heat. Find the final pressure and both the specific heat-transfer and specific work.

8.66 A cylinder/piston contains air at ambient conditions, 100 kPa and 20°C with a volume of 0.3 m^3. The air is compressed to 800 kPa in a reversible polytropic process with exponent, $n = 1.2$, after which it is expanded back to 100 kPa in a reversible adiabatic process.

a. Show the two processes in P–v and T–s diagrams.

b. Determine the final temperature and the net work.

c. What is the potential refrigeration capacity (in kilojoules) of the air at the final state?

8.67 An ideal gas having a constant specific heat undergoes a reversible polytropic expansion with exponent, $n = 1.4$. If the gas is carbon dioxide will the heat transfer for this process be positive, negative, or zero?

8.68 A cylinder fitted with a piston contains 0.5 kg of R-134a at 60°C, with a quality of 50 percent. The R-134a now expands in an internally reversible polytropic process to ambient temperature, 20°C at which point the quality is 100%. Any heat transfer is with a constant-temperature source, which is at 60°C. Find the polytropic exponent n and show that this process satisfies the second law of thermodynamics.

8.69 A cylinder/piston contains 100 L of air at 110 kPa, 25°C. The air is compressed in a reversible polytropic process to a final state of 800 kPa, 200°C. Assume the heat transfer is with the ambient at 25°C and determine the polytropic exponent n and the final volume of the air. Find the work done by the air, the heat transfer and the total entropy generation for the process.

8.70 A mass of 2 kg ethane gas at 500 kPa, 100°C, undergoes a reversible polytropic expansion with exponent, $n = 1.3$, to a final temperature of the ambient, 20°C. Calculate the total entropy generation for the process if the heat is exchanged with the ambient.

8.71 A cylinder/piston contains saturated vapor R-22 at 10°C; the volume is 10 L. The R-22 is compressed to 2 MPa, 60°C in a reversible (internally) polytropic process. If all the heat transfer during the process is with the ambient at 10°C, calculate the net entropy change.

8.72 A closed, partly insulated cylinder divided by an insulated piston contains air in one side and water on the other, as shown in Fig. P8.72. There is no insulation on the end containing water. Each volume is initially 100 L, with the air at 40°C and the water at 90°C, quality 10%. Heat is slowly transferred to the water, until a final pressure of 500 kPa. Calculate the amount of heat transferred.

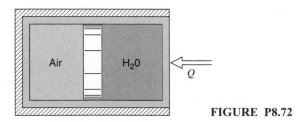

FIGURE P8.72

8.73 A spring-loaded piston/cylinder, shown in Fig. P8.73, contains water at 100 kPa with $v = 0.07237$ m³/kg. The water is now heated to a pressure of 3 MPa by a reversible heat pump extracting Q from a reservoir at 300 K. It is known that the water will pass through saturated vapor at 1.5 MPa and that pressure varies linearly with volume. Find the final temperature, the specific heat transfer to the water, and the work input to the heat pump.

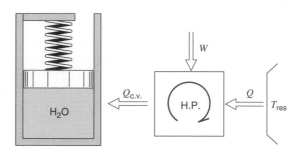

FIGURE P8.73

8.74 A cylinder with a linear spring-loaded piston contains carbon dioxide gas at 2 MPa with a volume of 50 L. The device is of aluminum and has a mass of 4 kg. Everything (Al and gas) is initially at 200°C. By heat transfer the whole system cools to the ambient temperature of 25°C, at which point the gas pressure is 1.5 MPa. Find the total entropy generation for the process.

8.75 A cylinder fitted with a piston contains air at 400 K, 1.0 MPa, at which point the volume is 100 L. The air now expands to a final state at 300 K, 200 kPa, and during the process the cylinder receives heat transfer from a heat source at 400 K. The work done by the air is 70% of what the work would have been for a reversible polytropic process between the same initial and final states. Calculate the heat transfer and the net entropy change for the process.

<table>
<tr><td>

ADVANCED
PROBLEMS

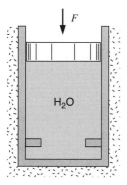

FIGURE P8.76

</td><td>

8.76 An insulated cylinder with a frictionless piston, shown in Fig. P8.76, contains water at ambient pressure, 100 kPa, a quality of 0.8 and the volume is 8 L. A force is now applied, slowly compressing the water until it reaches a set of stops, at which point the cylinder volume is 1 L. The insulation is then removed from the cylinder walls, and the water cools to ambient temperature, 20°C. Calculate the work and the heat transfer for the overall process.

8.77 Consider the process shown in Fig. P8.77. The insulated tank A has a volume of 600 L, and contains steam at 1.4 MPa, 300°C. The uninsulated tank B has a volume of 300 L and contains steam at 200 kPa, 200°C. A valve connecting the two tanks is opened, and steam flows from A to B until the temperature in A reaches 250°C. The valve is closed. During the process heat is transferred from B to the surroundings at 25°C, such that the temperature in B remains at 200°C. It may be assumed that the steam remaining in A has undergone a reversible adiabatic expansion. Determine the final pressure in tank A, the final pressure and mass in tank B, and the net entropy change, system plus surroundings, for the process

</td></tr>
</table>

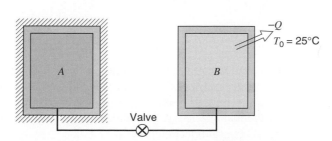

FIGURE P8.77

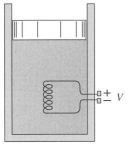

FIGURE P8.78

8.78 A vertical cylinder/piston contains R-22 at $-20°C$, 70% quality, and the volume is 50 L, shown in Fig. P8.78. This cylinder is brought into a $20°C$ room, and an electric current of 10 A is passed through a resistor inside the cylinder. The voltage drop across the resistor is 12 V. It is claimed that after 30 min the temperature inside the cylinder is $40°C$. Is this possible?

8.79 Redo Problem 8.57, but calculate the mass withdrawn by a first-law, control-volume analysis. Compare the result to that obtained in Problem 8.57. Show from a differential step of mass out that the first law leads to the same result. (Find the relation between dP and dT.)

8.80 A vertical cylinder is fitted with a frictionless piston that is initially resting on stops. The cylinder contains carbon dioxide gas at 200 kPa, 300 K, and at this point the volume is 50 L. A cylinder pressure of 400 kPa is required to make the piston rise from the stops. Heat is now transferred to the gas from an aluminum cubic block, 0.1 m on each side. The block is initially at 700 K.

   a. What is the temperature of the aluminum block when the piston first begins to rise?

   b. The process continues until the gas and block reach a common final temperature. What is this temperature?

   c. Calculate the net entropy change for the overall process.

8.81 A piston/cylinder contains 2 kg water at 5 MPa, $800°C$. The piston is loaded so pressure is proportional to volume, $P = CV$. It is now cooled by an external reservoir at $0°C$ to a final state of saturated vapor. Find the final pressure, work, heat transfer, and the entropy generation for the process.

8.82 A gas in a rigid vessel is at ambient temperature and at a pressure, P_1, slightly higher than ambient pressure, P_0. A valve on the vessel is opened, so gas escapes and the pressure drops quickly to ambient pressure. The valve is closed and after a long time the remaining gas returns to ambient temperature at which point the pressure is P_2. Develop an expression that allows a determination of the ratio of specific heats, k, in terms of the pressures.

ENGLISH UNIT PROBLEMS

8.83E Consider the steam power plant in Problem 7.57 and show that this cycle satisfies the inequality of Clausius.

8.84E Find the missing properties and give the phase of the substance.

   a. H_2O $s = 1.75$ Btu/lbm R, $P = 4$ lbf/in.^{2} $h = ? \ T = ? \ x = ?$

   b. H_2O $u = 1350$ Btu/lbm, $P = 1500$ lbf/in.^{2} $T = ? \ x = ? \ s = ?$

   c. R-22 $T = 30$ F, $P = 60$ lbf/in.^{2} $s = ? \ x = ?$

   d. R-134a $T = 10$ F, $x = 0.45$ $v = ? \ s = ?$

   e. NH_3 $T = 60$ F, $s = 1.35$ Btu/lbm R $u = ? \ x = ?$

8.85E In a Carnot engine with water as the working fluid, the high temperature is 450 F and as Q_H is received, the water changes from saturated liquid to saturated vapor. The water pressure at the low temperature is 14.7 lbf/in.^{2}. Find T_L, cycle thermal efficiency, heat added per pound-mass, and entropy, s, at the beginning of the heat rejection process.

8.86E Consider a Carnot-cycle heat pump with R-22 as the working fluid. Heat is rejected from the R-22 at 100 F, during which process the R-22 changes from saturated vapor to saturated liquid. The heat is transferred to the R-22 at 30 F.

a. Show the cycle on a T–s diagram.

b. Find the quality of the R-22 at the beginning and end of the isothermal heat addition process at 30 F.

c. Determine the coefficient of performance for the cycle.

8.87E Do Problem 8.86 using refrigerant R-134a instead of R-22.

8.88E Water at 30 lbf/in.^{2}, $x = 1.0$ is compressed in a piston/cylinder to 140 lbf/in.^{2}, 600 F in a reversible process. Find the sign for the work and the sign for the heat transfer.

8.89E Two pound-mass of ammonia in a piston/cylinder at 120 F, 150 lbf/in.^{2} is expanded in a reversible adiabatic process to 15 lbf/in.^{2}. Find the work and heat transfer for this process.

8.90E A cylinder fitted with a piston contains ammonia at 120 F, 20% quality with a volume of 60 in.^{3}. The ammonia expands slowly, and during this process heat is transferred to maintain a constant temperature. The process continues until all the liquid is gone. Determine the work and heat transfer for this process.

8.91E One pound-mass of water at 600 F expands against a piston in a cylinder until it reaches ambient pressure, 14.7 lbf/in.^{2}, at which point the water has a quality of 90%. It may be assumed that the expansion is reversible and adiabatic.

a. What was the initial pressure in the cylinder?

b. How much work is done by the water?

8.92E A closed tank, $V = 0.35$ ft^3, containing 10 lbm of water initially at 77 F is heated to 350 F by a heat pump that is receiving heat from the surroundings at 77 F. Assume that this process is reversible. Find the heat transfer to the water and the work input to the heat pump.

8.93E A cylinder containing R-134a at 50 F, 20 lbf/in.^{2} has an initial volume of 1 ft^3. A piston compresses the R-134a in a reversible, isothermal process until it reaches the saturated vapor state. Calculate the required work and heat transfer to accomplish this process.

8.94E A rigid, insulated vessel contains superheated vapor steam at 450 lbf/in.^{2}, 700 F. A valve on the vessel is opened, allowing steam to escape. It may be assumed that the steam remaining inside the vessel goes through a reversible adiabatic expansion. Determine the fraction of steam that has escaped, when the final state inside is saturated vapor.

8.95E A cylinder/piston contains 5 lbm of water at 80 lbf/in.^{2}, 1000 F. The piston has cross-sectional area of 1 ft^2 and is restrained by a linear spring with spring constant 60 lbf/in. The setup is allowed to cool down to room temperature due to heat transfer to the room at 70 F. Calculate the total (water and surroundings) change in entropy for the process.

8.96E An insulated cylinder/piston contains R-134a at 150 lbf/in.^{2}, 120 F, with a volume of 3.5 ft^3. The R-134a expands, moving the piston until the pressure in the cylinder has dropped to 15 lbf/in.^{2}. It is claimed that the R-134a does 180 Btu of work against the piston during the process. Is that possible?

8.97E A mass- and atmosphere-loaded piston/cylinder contains 4 lbm of water at 500 lbf/in.^{2}, 200 F. Heat is added from a reservoir at 1200 F to the water until it reaches 1200 F. Find the work, heat transfer, and total entropy production for the system and surroundings.

8.98E A 1-gallon jug of milk at 75 F is placed in your refrigerator where it is cooled down to the refrigerator's inside temperature of 40 F. Assume the milk has the properties of liquid water and find the entropy generated in the cooling process.

8.99E Water in a piston/cylinder is at 150 lbf/in.^{2}, 900 F, as shown in Fig. P8.33. There are two stops, a lower one at which $V_{min} = 35$ ft^3 and an upper one at $V_{max} = 105$ ft^3. The piston is loaded with a mass and outside atmosphere such that it floats when the pressure is 75 lbf/in.^{2}. This setup is now cooled to 210 F by rejecting heat to the surroundings at 70 F. Find the total entropy generated in the process.

8.100E A cylinder/piston contains water at 30 lbf/in.^{2}, 400 F with a volume of 1 ft^3. The piston is moved slowly, compressing the water to a pressure of 120 lbf/in.^{2}. The loading on the piston is such that the product PV is a constant. Assuming that the room temperature is 70 F, show that this process does not violate the second law.

8.101E One pound mass of ammonia (NH_3) is contained in a linear spring-loaded piston/cylinder as saturated liquid at 0 F. Heat is added from a reservoir at 225 F until a final condition of 125 lbf/in.^{2}, 160 F is reached. Find the work, heat transfer, and entropy generation, assuming the process is internally reversible.

8.102E A foundry form box with 50 lbm of 400 F hot sand is dumped into a bucket with 2 ft^3 water at 60 F. Assuming no heat transfer with the surroundings and no boiling away of liquid water, calculate the net entropy change for the process.

8.103E A hollow steel sphere with a 2-ft inside diameter and a 0.1-in. thick wall contains water at 300 lbf/in.^{2}, 500 F. The system (steel plus water) cools to the ambient temperature, 90 F. Calculate the net entropy change of the system and surroundings for this process.

8.104E A handheld pump for a bicycle has a volume of 2 in.^{3} when fully extended. You now press the plunger (piston) in while holding your thumb over the exit hole so an air pressure of 45 lbf/in.^{2} is obtained. The outside atmosphere is at P_0, T_0. Consider two cases: (1) it is done quickly (~1 s), and (2) it is done very slowly (~1 h).

a. State assumptions about the process for each case.

b. Find the final volume and temperature for both cases.

8.105E A piston/cylinder contains air at 2500 R, 2200 lbf/in.^{2}, with $V_1 = 1$ in.^{3}, $A_{cyl} = 1$ in.^{2}, as shown in Fig. P7.42. The piston is released and just before the piston exits the end of the cylinder the pressure inside is 30 lbf/in.^{2}. If the cylinder is insulated, what is its length? How much work is done by the air inside?

8.106E A 25-ft^3 insulated, rigid tank contains air at 110 lbf/in.^{2}, 75 F. A valve on the tank is opened, and the pressure inside quickly drops to 15 lbf/in.^{2}, at which point the valve is closed. Assuming that the air remaining inside has undergone a reversible adiabatic expansion, calculate the mass withdrawn during the process.

8.107E A rigid container with volume 7 ft^3 is divided into two equal volumes by a partition. Both sides contain nitrogen, one side is at 300 lbf/in.^{2}, 400 F, and the other at 30 lbf/in.^{2}, 200 F. The partition ruptures, and the nitrogen comes to a uniform state at 160 F. Assuming the temperature of the surroundings is 68 F, determine the work done and the net entropy change for the process.

8.108E Nitrogen at 90 lbf/in.^{2}, 260 F is in a 20 ft^3 insulated tank connected to a pipe with a valve to a second insulated initially empty tank of volume 20 ft^3. The valve is opened and the nitrogen fills both tanks. Find the final pressure and temperature and the entropy generation this process causes. Why is the process irreversible?

8.109E A cylinder/piston contains carbon dioxide at 150 lbf/in.^{2}, 600 F, with a volume of 7 ft^3. The total external force acting on the piston is proportional to V^3. This system is allowed to cool to room temperature, 70 F. What is the total entropy generation for the process?

8.110E Helium in a piston/cylinder at 70 F, 15 lbf/in.^{2} is brought to 720 R in a reversible polytropic process with exponent $n = 1.25$. You may assume helium is an ideal gas with constant specific heat. Find the final pressure and both the specific heat transfer and specific work.

8.111E A cylinder/piston contains air at ambient conditions, 14.7 lbf/in.^{2} and 70 F, with a volume of 10 ft^3. The air is compressed to 100 lbf/in.^{2} in a reversible polytropic process with exponent, $n = 1.2$, after which it is expanded back to 14.7 lbf/in.^{2} in a reversible adiabatic process.

a. Show the two processes in P–v and T–s diagrams.

b. Determine the final temperature and the net work.

c. What is the potential refrigeration capacity (in British thermal units) of the air at the final state?

8.112E A cylinder/piston contains 4 ft^3 of air at 16 lbf/in.^{2}, 77 F. The air is compressed in a reversible polytropic process to a final state of 120 lbf/in.^{2}, 400 F. Assume the heat transfer is with the ambient at 77 F and determine the polytropic exponent n and the final volume of the air. Find the work done by the air, the heat transfer, and the total entropy generation for the process.

8.113E A cylinder with a linear spring-loaded piston contains carbon dioxide gas at 300 lbf/in.^{2} with a volume of 2 ft^3. The device is of aluminum and has a mass of 8 lbm. Everything (Al and gas) is initially at 400 F. By heat transfer the whole system cools to the ambient temperature of 77 F, at which point the gas pressure is 220 lbf/in.^{2}. Find the total entropy generation for the process.

COMPUTER, DESIGN, AND OPEN-ENDED PROBLEMS

8.114 Use the menu-driven software to get the properties for the calculation of the isentropic efficiency of the pump in the steam power plant of Problem 6.39.

8.115 Write a computer program to solve Problem 8.40 using constant specific heat for both the sand and the liquid water. Let the amount and the initial temperatures be input variables.

8.116 Write a program to solve Problem 8.42 with the thermal storage rock bed in Problem 7.22. Let the size and temperatures be input variables so the heat engine work output can be studied as a function of the system parameters.

8.117 Write a program to solve the following problem. One of the gases listed in Table A.6 undergoes a reversible adiabatic process in a cylinder from P_1, T_1 to P_2. We wish to calculate the final temperature and the work for the process by three methods:

a. Integrating the specific heat equation.

b. Assuming constant specific heat at temperature, T_1.

c. Assuming constant specific heat at the average temperature (by iteration).

8.118 Write a program to solve Problem 8.45. Let the initial state and the expansion ratio be input variables.

8.119 Write a program to solve a problem similar to 8.46, but instead of the ideal gas tables use the formula for the specific heat as a function of temperature in Table A.6.

8.120 Write a program to study a general polytropic process in an ideal gas with constant specific heat. Take Problem 8.61 as an example.

8.121 Write a program to solve the general case of Problem 8.64, in which the initial state and the expansion ratio are input variables.

8.122 A piston/cylinder maintaining constant pressure contains 0.5 kg of water at room temperature 20°C and 100 kPa. An electric heater of 500 W heats the water up to 500°C. Assume no heat losses to the ambient and plot the temperature and total accumulated entropy production as a function of time. Investigate the first part of the process, namely bringing the water to the boiling point, by measuring it in your kitchen and knowing the rate of power added.

8.123 Air in a piston/cylinder is used as a small air-spring that should support a steady load of 200 N. Assume that the load can vary with ±10% over a period of 1 s and that the displacement should be limited to ±0.01 m. For some choice of sizes show the spring displacement, x, as a function of load and compare that to an elastic linear coil spring designed for the same conditions.

8.124 Consider a piston/cylinder arrangement with ammonia at −10°C, 50 kPa that is compressed to 200 kPa. Examine the effect of heat transfer to/from the ambient at 15°C on the process and the required work. Some limiting processes are a reversible adiabatic compression giving an exit temperature of about 90°C and as mentioned in the text an isothermal compression. Evaluate the work and heat transfer for both cases and for cases in between assuming a polytropic process. Which processes are actually possible and how would they proceed?

8.125 Consider a compression or expansion process for a control mass that is an ideal gas. Assume the heat transfer rate is proportional to the temperature difference, $T - T_0$, so the energy equation on a rate form becomes

$$\frac{dE}{dt} = \dot{Q} - \dot{W}$$

$$mC_v\dot{T} = -C_H\left(T - T_o\right) - Pm\dot{v}$$

This can be written as a rate equation for the temperature, T, as

$$\dot{T}/T = -C_1\left(1 - T_o/T\right) - \frac{R}{C_v}\frac{\dot{v}}{v}$$

with the factor, $C_1 = C_H/mC_v$, units 1/s. Investigate different processes (vary C_1) with expansion or compression at a constant relative volume change, $\dot{v}/v$ and different initial conditions. Plot various properties along the process path and include total entropy generation.

9 SECOND-LAW ANALYSIS FOR A CONTROL VOLUME

In the preceeding two chapters we discussed the second law of thermodynamics and the thermodynamic property entropy. As was done with the first law-analysis, we now consider the more general application of these concepts, the control volume analysis, and a number of cases of special interest. We will also discuss usual definitions of thermodynamic efficiencies.

9.1 THE SECOND LAW OF THERMODYNAMICS FOR A CONTROL VOLUME

The second law of thermodynamics can be applied to a control volume by a procedure similar to that used in Section 6.1, where the first law was developed for a control volume. We start with the second law expressed as a change of the entropy for a control mass in a rate form from Eq. 8.46,

$$\frac{dS_{\text{c.m.}}}{dt} = \sum \frac{\dot{Q}}{T} + \dot{S}_{\text{gen}} \tag{9.1}$$

to which we now will add the contributions from the mass flow rates in and out of the control volume. A simple example of such a situation is illustrated in Fig. 9.1. The flow of mass does carry an amount of entropy, s, per unit mass flowing, but it does not give rise to any other contributions. As a process may take place in the flow, entropy can be generated, but this is attributed to the space it belongs to, i.e., either inside or outside of the control volume.

The balance of entropy as an equation then states that the rate of change in total entropy inside the control volume is equal to the net sum of fluxes across the control surface plus the generation rate

$$\frac{dS_{\text{c.v.}}}{dt} = \sum \dot{m}_i s_i - \sum \dot{m}_e s_e + \sum \frac{\dot{Q}_{\text{c.v.}}}{T} + \dot{S}_{\text{gen}} \tag{9.2}$$

These fluxes are mass flow rates carrying a level of entropy and the rate of heat transfer that takes place at a certain temperature (the temperature right at the control surface). The accumulation and generations terms cover the total control volume and are expressed in the lumped (integral form) so that

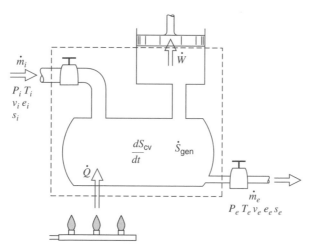

FIGURE 9.1 The entropy balance for a conrol volume on a rate form.

$$S_{c.v.} = \int \rho s\, dV = m_{c.v.}s = m_A s_A + m_B s_B + m_C s_C + \cdots$$

$$\dot{S}_{gen} = \int \rho \dot{s}_{gen}\, dV = \dot{S}_{gen,A} + \dot{S}_{gen,B} + \dot{S}_{gen,C} + \cdots \tag{9.3}$$

If the control volume has several different accumulation units with different fluid states and processes occuring in them we may have to sum the various contributions over the different domains. If the heat transfer is distributed over the control surface that also has to be done as an integral over the total surface area using the local temperature and rate of heat transfer per unit area, $\dot{Q}/A$, as

$$\sum \frac{\dot{Q}_{c.v.}}{T} = \int \frac{d\dot{Q}}{T} = \int_{surface} \frac{(\dot{Q}/A)}{T}\, dA \tag{9.4}$$

These distributed cases typically require a much more detailed analysis, which is beyond the scope of the current presentation of the second law.

The generation term(s) in Eq. 9.2 from a summation of individual positive internal-irreversibility entropy generation terms in Eq. 9.3 is necessarily positive (or zero), such that an inequality is often written as

$$\frac{dS_{c.v.}}{dt} \geq \sum \dot{m}_i s_i - \sum \dot{m}_e s_e + \sum \frac{\dot{Q}_{c.v.}}{T} \tag{9.5}$$

Now the equality applies to internally reversible processes and the inequality to internally irreversible processes. The form of the second law in Eq. 9.2 or 9.5 is general, such that any particular case results in a form that is a subset (simplification) of this form. Examples of various classes of problems are illustrated in the following sections.

If there is no mass flow into or out of the control volume it simplifies to a control mass and the equation for the total entropy reverts back to Eq. 8.45. Since that version of the second law has been covered in Chapter 8 we will consider here the remaining cases that were done for the first law of thermodynamics in Chapter 6.

9.2 THE STEADY-STATE, STEADY-FLOW PROCESS AND THE UNIFORM-STATE, UNIFORM-FLOW PROCESS

We now consider the application of the second-law control volume equation, Eq. 9.2 or 9.5, to the two control volume model processes developed in Chapter 6.

For the steady-state, steady-flow process, which has been defined in Section 6.3, we conclude that there is no change with time of the entropy per unit mass at any point within the control volume, and therefore the first term of Eq. 9.2 equals zero. That is,

$$\frac{dS_{c.v.}}{dt} = 0 \tag{9.6}$$

so that, for the SSSF process,

$$\sum \dot{m}_e s_e - \sum \dot{m}_i s_i = \sum_{c.v.} \frac{\dot{Q}_{c.v.}}{T} + \dot{S}_{gen} \tag{9.7}$$

in which the various mass flows, heat transfer and entropy generation, rates and states are all constant with time.

If in a steady-state, steady-flow process there is only one area over which mass enters the control volume at a uniform rate and only one area over which mass leaves the control volume at a uniform rate, we can write

$$\dot{m}(s_e - s_i) = \sum_{c.v.} \frac{\dot{Q}_{c.v.}}{T} + \dot{S}_{gen} \tag{9.8}$$

For an adiabatic process with these assumptions, it follows that

$$s_e \geq s_i \tag{9.9}$$

where the equality holds for a reversible adiabatic process.

EXAMPLE 9.1 Steam enters a steam turbine at a pressure of 1 MPa, a temperature of 300°C, and a velocity of 50 m/s. The steam leaves the turbine at a pressure of 150 kPa and a velocity of 200 m/s. Determine the work per kilogram of steam flowing through the turbine, assuming the process to be reversible and adiabatic.

Control volume: Turbine.

Sketch: Fig. 9.2.

Inlet state: Fixed (Fig. 9.2).

Exit state: P_e, $\mathbf{V}_e$ known.

Process: SSSF.

Model: Steam tables.

Analysis:

Continuity equation:

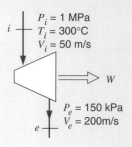

$P_i = 1$ MPa
$T_i = 300°C$
$V_i = 50$ m/s

W

$P_e = 150$ kPa
$V_e = 200$m/s

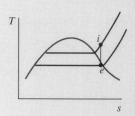

FIGURE 9.2 Sketch for Example 9.1.

$$\dot{m}_e = \dot{m}_i = \dot{m}$$

First law:

$$h_i + \frac{\mathbf{V}_i^2}{2} = h_e + \frac{\mathbf{V}_e^2}{2} + w$$

Second law:

$$s_e = s_i$$

Solution

From the steam tables,

$$h_i = 3051.2 \text{ kJ / kg} \qquad s_i = 7.1229 \text{ kJ / kg K}$$

The two properties known in the final state are pressure and entropy.

$$P_e = 0.15 \text{ MPa} \qquad s_e = s_i = 7.1229 \text{ kJ / kg K}$$

The quality and enthalpy of the steam leaving the turbine can be determined.

$$s_e = 7.1229 = s_f + x_e s_{fg} = 1.4336 + x_e 5.7897$$

$$x_e = 0.9827$$

$$h_e = h_f + x_e h_{fg} = 467.1 + 0.9827(2226.5)$$

$$= 2655.0 \text{ kJ / kg}$$

Therefore, the work per kilogram of steam for this isentropic process may be found using the equation for the first law.

$$w = 3051.2 + \frac{50 \times 50}{2 \times 1000} - 2655.0 - \frac{200 \times 200}{2 \times 1000} = 377.5 \text{ kJ / kg}$$

EXAMPLE 9.2 Consider the reversible adiabatic flow of steam through a nozzle. Steam enters the nozzle at 1 MPa, 300°C, with a velocity of 30 m/s. The pressure of the steam at the nozzle exit is 0.3 MPa. Determine the exit velocity of the steam from the nozzle, assuming a reversible, adiabatic, steady-state, steady-flow process.

Control volume: Nozzle.

Sketch: Fig. 9.3.

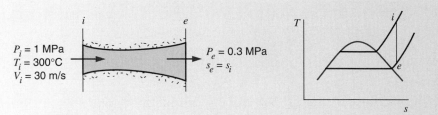

FIGURE 9.3 Sketch for Example 9.2.

Inlet state: Fixed (Fig. 9.3).

Exit state: P_e known.

Process: SSSF.

Model: Steam tables.

Analysis:

Because this is a steady-state, steady-flow process in which the work, the heat transfer, and the changes in potential energy are zero, we can write

Continuity equation:

$$\dot{m}_e = \dot{m}_i = \dot{m}$$

First law:

$$h_i + \frac{\mathbf{V}_i^2}{2} = h_e + \frac{\mathbf{V}_e^2}{2}$$

Second law:

$$s_e = s_i$$

Solution

From the steam tables,

$$h_i = 3051.2 \text{ kJ / kg} \qquad s_i = 7.1229 \text{ kJ / kg K}$$

The two properties that we know in the final state are entropy and pressure.

$$s_e = s_i = 7.1229 \text{ kJ / kg K} \qquad P_e = 0.3 \text{ MPa}$$

Therefore,

$$T_e = 159.1° \text{C} \qquad h_e = 2780.2 \text{ kJ / kg}$$

Substituting into the equation for the first law, we have

$$\frac{\mathbf{V}_e^2}{2} = h_i - h_e + \frac{\mathbf{V}_i^2}{2}$$

$$= 3051.2 - 2780.2 + \frac{30 \times 30}{2 \times 1000} = 271.5 \text{ kJ / kg}$$

$$\mathbf{V}_e = 737 \text{ m/s}$$

EXAMPLE 9.2E Consider the reversible adiabatic flow of steam through a nozzle. Steam enters the nozzle at 100 lbf/in.^{2}, 500 F, with a velocity of 100 ft/s. The pressure of the steam at the nozzle exit is 40 lbf/in.^{2}. Determine the exit velocity of the steam from the nozzle, assuming a reversible adiabatic, steady-state, steady-flow process.

> *Control volume:* Nozzle.
>
> *Sketch:* Fig. 9.3E.
>
> *Inlet state:* Fixed (Fig. 9.3E).
>
> *Exit state:* P_e known.
>
> *Process:* SSSF.
>
> *Model:* Steam tables.

Analysis:

Because this is a steady-state, steady-flow process in which the work, the heat transfer, and the changes in potential energy are zero, we can write
Continuity equation:

$$\dot{m}_e = \dot{m}_i = \dot{m}$$

First law:

$$h_i + \frac{\mathbf{V}_i^2}{2} = h_e + \frac{\mathbf{V}_e^2}{2}$$

Second law:

$$s_e = s_i$$

Solution

From the steam tables,

$$h_i = 1279.1 \text{ Btu / lbm} \qquad s_i = 1.7085 \text{ Btu / lbm R}$$

The two properties that we know in the final state are entropy and pressure.

$$s_e = s_i = 1.7085 \text{ Btu / lbm R}, \ P_e = 40 \text{ lbf / in.}^2$$

Therefore,

$$T_e = 314.2 \text{ F} \qquad h_e = 1193.9 \text{ Btu / lbm}$$

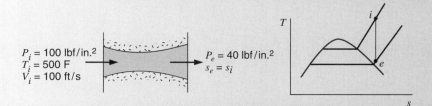

FIGURE 9.3E Sketch for Example 9.2E.

Substituting into the equation for the first law, we have

$$\frac{\mathbf{V}_e^2}{2} = h_i - h_e + \frac{\mathbf{V}_i^2}{2}$$

$$= 1279.1 - 1193.9 + \frac{100 \times 100}{2 \times 32.17 \times 778} = 85.4 \text{ Btu / lbm}$$

$$\mathbf{V}_e = \sqrt{2 \times 32.17 \times 778 \times 85.4} = 2070 \text{ ft / s}$$

EXAMPLE 9.3

An inventor reports that she has a refrigeration compressor that receives saturated R-134a vapor at −20°C and delivers the vapor at 1 MPa, 40°C. The compression process is adiabatic. Does the process described violate the second law?

Control volume: Compressor.

Inlet state: Fixed (saturated vapor at T_i).

Exit state: Fixed (P_e, T_e known).

Process: SSSF, adiabatic.

Model: R-134a tables.

Analysis:

Because this is a steady-state, steady-flow, adiabatic process, we can write
Second law:

$$s_e \geq s_i$$

Solution

From the R-134a tables,

$$s_e = 1.71479 \text{ kJ / kg K} \qquad s_i = 1.7395 \text{ kJ / kg K}$$

Therefore, $s_e < s_i$, whereas for this process the second law requires that $s_e \geq s_i$. The process described involves a violation of the second law and thus would be impossible.

EXAMPLE 9.4

Air is compressed in a centrifugal compressor from ambient conditions, 100 kPa and 300 K, to a pressure of 450 kPa. Assume the process to be reversible and adiabatic with negligible changes in kinetic and potential energy. Calculate the work input per kilogram of air flowing through the compressor.

Control volume: Compressor.

Inlet state: P_i, T_i known; state fixed.

Exit state: P_e known.

Process: SSSF.

Model: Ideal-gas and air tables, A.7.

Analysis:

Because this is a steady-state, steady-flow reversible process, we can write
Continuity equation:

$$\dot{m}_e = \dot{m}_i = \dot{m}$$

First law:

$$h_i = h_e + w$$

Second law:

$$s_e = s_i$$

Solution

From Table A.7,

$$h_i = 300.47 \text{ kJ / kg} \qquad P_{ri} = 1.1146$$

From Eq. 8.29,

$$P_{re} = P_{ri} \frac{P_e}{P_i} = 1.1146 \frac{450}{100} = 5.0157$$

Therefore,

$$T_e = 460 \text{ K} \qquad h_e = 462.14$$

and from the first law,

$$w = h_i - h_e = 300.47 - 462.14 = -161.67 \text{ kJ / kg}$$

EXAMPLE 9.4E   Air is compressed in a centrifugal compressor from ambient conditions, 14.7 lbf/in.^{2} and 520 R, to a pressure of 65 lbf/in.^{2}. Assume the process to be reversible and adiabatic with negligible changes in kinetic and potential energy. Calculate the work input per pound mass of air flowing through the compressor.

     Control volume: Compressor.

     Inlet state: P_i, T_i known; state fixed.

     Exit state: P_e known.

     Process: SSSF.

     Model: Ideal-gas and air tables, C.6.

Analysis:

Since this is a steady-state, steady-flow reversible process we can write
Continuity equation:

$$\dot{m}_e = \dot{m}_i = \dot{m}$$

First law:

$$h_i = h_e + w$$

Second law:

$$s_e = s_i$$

Solution

From Table C.6,

$$h_i = 124.38 \text{ Btu / lbm} \qquad P_{ri} = 0.9767$$

From Eq. 8.29,

$$P_{re} = P_{ri}\left(\frac{P_e}{P_i}\right) = 0.9767\frac{65}{14.7} = 4.3187$$

Therefore,

$$T_e = 794 \text{ R} \qquad h_e = 190.37$$

and from the first law,

$$w = h_i - h_e = 124.38 - 190.37 = -66.0 \text{ kJ / kg}$$

For the uniform-state, uniform-flow process, which was described in Section 6.5, the second law for a control volume, Eq. 9.5, can be written in the following form.

$$\frac{d}{dt}(ms)_{\text{c.v.}} + \sum \dot{m}_e s_e - \sum \dot{m}_i s_i \geq \sum_{\text{c.v.}} \frac{\dot{Q}_{\text{c.v.}}}{T} \qquad (9.10)$$

If this is integrated over the time interval t, we have

$$\int_0^t \frac{d}{dt}(ms)_{\text{c.v.}} \, dt = (m_2 s_2 - m_1 s_1)_{\text{c.v.}}$$

$$\int_0^t \left(\sum \dot{m}_e s_e\right) dt = \sum m_e s_e \qquad \int_0^t \left(\sum \dot{m}_i s_i\right) dt = \sum m_i s_i$$

Therefore, for this period of time t, we can write the second law for the uniform-state, uniform-flow process as

$$(m_2 s_2 - m_1 s_1)_{\text{c.v.}} + \sum m_e s_e - \sum m_i s_i \geq \int_0^t \sum_{\text{c.v.}} \frac{\dot{Q}_{\text{c.v.}}}{T} dt \qquad (9.11)$$

Since in this process the temperature is uniform throughout the control volume at any instant of time, the integral on the right reduces to

$$\int_0^t \sum_{\text{c.v.}} \frac{\dot{Q}_{\text{c.v.}}}{T} dt = \int_0^t \frac{1}{T} \sum_{\text{c.v.}} \dot{Q}_{\text{c.v.}} dt = \int_0^t \frac{\dot{Q}_{\text{c.v.}}}{T} dt$$

and therefore the second law for the uniform-state, uniform-flow process can be written

$$(m_2 s_2 - m_1 s_1)_{\text{c.v.}} + \sum m_e s_e - \sum m_i s_i \geq \int_0^t \frac{\dot{Q}_{\text{c.v.}}}{T} dt \qquad (9.12)$$

By introducing the rate of internal entropy generation, we can write this as an equality. Integrating over the time interval t, we have the total amount of internal entropy generation during the process, $_1S_{2\text{gen}}$. Therefore,

$$\left(m_2 s_2 - m_1 s_1\right)_{\text{c.v.}} + \Sigma m_e s_e - \Sigma m_i s_i = \int_0^t \frac{\dot{Q}_{\text{c.v.}}}{T}\,dt +\ _1 S_{2\,\text{gen}} \qquad (9.13)$$

9.3 THE REVERSIBLE STEADY-STATE, STEADY-FLOW PROCESS

An expression can be derived for the work in a reversible adiabatic, steady-state, steady-flow process that is of great help in understanding its significant variables. We have noted that when a steady-state, steady-flow process involves a single flow of fluid into and out of the control volume, the first law, Eq. 6.13, can be written,

$$q + h_i + \frac{\mathbf{V}_i^2}{2} + gZ_i = h_e + \frac{\mathbf{V}_e^2}{2} + gZ_e + w$$

and the second law, Eq. 9.8 is

$$\dot{m}\left(s_e - s_i\right) = \sum_{\text{c.v.}} \frac{\dot{Q}_{\text{c.v.}}}{T} + \dot{S}_{\text{gen}}$$

Let us now consider two types of flow, a reversible adiabatic process and a reversible isothermal process.

If the process is reversible and adiabatic, the second-law equation reduces to

$$s_e = s_i$$

It follows from the property relation

$$T\,ds = dh - v\,dP$$

that

$$h_e - h_i = \int_i^e v\,dP \qquad (9.14)$$

Substituting these relations into Eq. 6.13 and noting that $q = 0$, we have for the reversible, adiabatic process

$$w = \left(h_i - h_e\right) + \frac{\mathbf{V}_i^2 - \mathbf{V}_e^2}{2} + g\left(Z_i - Z_e\right)$$

$$= -\int_i^e v\,dP + \frac{\mathbf{V}_i^2 - \mathbf{V}_e^2}{2} + g\left(Z_i - Z_e\right) \qquad (9.15)$$

If, instead, the process is reversible and isothermal, the second law reduces to

$$\dot{m}\left(s_e - s_i\right) = \frac{1}{T}\sum_{\text{c.v.}} \dot{Q}_{\text{c.v.}} = \frac{\dot{Q}_{\text{c.v.}}}{T} \qquad (9.16)$$

or

$$T\left(s_e - s_i\right) = \frac{\dot{Q}_{\text{c.v.}}}{\dot{m}} = q$$

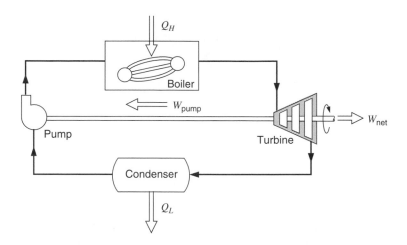

FIGURE 9.4 Simple
steam power plant.

and the property relation can be integrated to give

$$T(s_e - s_i) = (h_e - h_i) - \int_i^e v \, dP \tag{9.17}$$

Substituting Eqs. 9.16 and 9.17 into the first law, Eq. 6.13, gives us the same expression as for the reversible adiabatic process, Eq. 9.15. We further note that any other reversible process can be constructed, in the limit, from a series of alternate adiabatic and isothermal processes. Thus, we may conclude that Eq. 9.15 is valid for any reversible, steady-state, steady-flow process without the restriction that it be either adiabatic or isothermal.

This expression has a wide range of application. If we consider a reversible steady-state, steady-flow process in which the work is zero (such as flow through a nozzle) and the fluid is incompressible (v = constant), Eq. 9.15 can be integrated to give

$$v(P_e - P_i) + \frac{\mathbf{V}_e^2 - \mathbf{V}_i^2}{2} + g(Z_e - Z_i) = 0 \tag{9.18}$$

Known as the Bernoulli equation (after Daniel Bernoulli), this is a very important equation in fluid mechanics.

Equation 9.15 is also frequently applied to the large class of flow processes involving work (such as turbines and compressors) in which changes in kinetic and potential energies of the working fluid are small. The model process for these machines is then a reversible, SSSF process with no change in kinetic or potential energy (and commonly, although not necessarily, adiabatic as well). For this process Eq. 9.15 reduces to the form

$$w = -\int_i^e v \, dP \tag{9.19}$$

From this result, we conclude that the shaft work associated with this type of process is closely related to the specific volume of the fluid during the process. To amplify this point further, consider the simple steam power plant shown in Fig. 9.4. Suppose this is an ideal power plant with no pressure drop in the piping, the boiler, or the condenser. Thus, the pressure increase in the pump is equal to the pressure decrease in the turbine. Neglecting kinetic and potential energy changes, the work done in each of these processes is given by Eq. 9.19. Since the pump handles liquid, which has a very small specific volume compared to that of the vapor that flows through the turbine, the power input to the pump is much less than the power output of the turbine. The difference is the net power output of the power plant.

This same line of reasoning can be qualitatively applied to actual devices that involve steady-state, steady-flow processes, even though the processes are not exactly reversible and adiabatic.

EXAMPLE 9.5   Calculate the work per kilogram to pump water isentropically from 100 kPa, 30°C, to 5 MPa.

> *Control volume:*   Pump.
>
> *Inlet state:* P_i, T_i known; state fixed.
>
> *Exit state:* P_e known.
>
> *Process:*   SSSF.
>
> *Model:*   Steam tables.

Analysis:

Since the process is SSSF, reversible, and adiabatic, and changes in kinetic and potential energies can be neglected,

First law:

$$h_i = h_e + w$$

Second law:

$$s_e - s_i = 0$$

Solution

Since P_e and s_e are known, state e is fixed and therefore h_e is known and w can be found from the first law. However, the process is reversible, SSSF, with negligible changes in kinetic and potential energies, so that Eq. 9.19 is also valid. Furthermore, since a liquid is being pumped, the specific volume will change very little during the process.

From the steam tables, $v_i = 0.001\,004$ m³/kg. Assuming that the specific volume remains constant and using Eq. 9.19, we have

$$-w = \int_1^2 v\,dP = v\left(P_2 - P_1\right) = 0.001\,004\left(5000 - 100\right) = 4.92 \text{ kJ/kg}$$

As a final application of Eq. 9.15, we recall the reversible polytropic process for an ideal gas, discussed in Section 8.11 for a control mass process. For the SSSF process with no change in kinetic and potential energies, from the relations,

$$w = -\int_i^e v\,dP \qquad \text{and} \qquad Pv^n = \text{constant} = C^n$$

$$w = -\int_i^e v\,dP = -C\int_i^e \frac{dP}{P^{1/n}}$$

$$= -\frac{n}{n-1}\left(P_e v_e - P_i v_i\right) = -\frac{nR}{n-1}\left(T_e - T_i\right) \tag{9.20}$$

If the process is isothermal, then $n = 1$ and the integral becomes

$$w = -\int_i^e v \, dP = -\text{constant} \int_i^e \frac{dP}{P} = -P_i v_i \ln \frac{P_e}{P_i} \tag{9.21}$$

These evaluations of the integral

$$\int_i^e v \, dP$$

may also be used in conjunction with Eq. 9.15 for instances in which kinetic and potential energy changes are not negligibly small.

9.4 PRINCIPLE OF THE INCREASE OF ENTROPY

The principle of the increase of entropy for a control mass analysis was discussed in Section 8.8. The same general conclusion is reached for a control volume analysis. To demonstrate this, consider a control volume, Fig. 9.5, that exchanges both mass and heat with its surroundings. At the point in the surroundings where the heat transfer occurs, the temperature is T_0. From Eq. 9.5, the second law for this process is

$$\frac{dS_{\text{c.v.}}}{dt} + \sum \dot{m}_e s_e - \sum \dot{m}_i s_i \geq \sum_{\text{c.v.}} \frac{\dot{Q}_{\text{c.v.}}}{T}$$

We recall that the first term represents the rate of change of entropy within the control volume, and the next terms the net entropy flow out of the control volume resulting from the mass flow. Therefore, for the surroundings, we can write

$$\frac{dS_{\text{surr}}}{dt} = \sum \dot{m}_e s_e - \sum \dot{m}_i s_i - \frac{\dot{Q}_{\text{c.v.}}}{T_0} \tag{9.22}$$

Adding Eqs. 9.5 and 9.22, we have

$$\frac{dS_{\text{net}}}{dt} = \frac{dS_{\text{c.v.}}}{dt} + \frac{dS_{\text{surr}}}{dt} \geq \sum_{\text{c.v.}} \frac{\dot{Q}_{\text{c.v.}}}{T} - \frac{\dot{Q}_{\text{c.v.}}}{T_0} \tag{9.23}$$

Because $\dot{Q}_{\text{c.v.}} > 0$ when $T_0 > T$ and $\dot{Q}_{\text{c.v.}} < 0$ when $T_0 < T$, it follows that

$$\frac{dS_{\text{net}}}{dt} = \frac{dS_{\text{c.v.}}}{dt} + \frac{dS_{\text{surr}}}{dt} = \sum \dot{S}_{\text{gen}} \geq 0 \tag{9.24}$$

which can be termed the general statement of the principle of the increase of entropy.

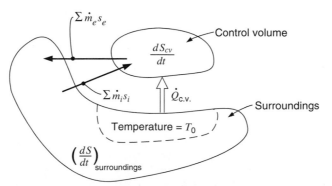

FIGURE 9.5 Entropy change for a control volume plus surroundings.

When we use Eq. 9.24 to check any particular process for a possible violation of the second law, it will be in connection with one of our model processes. For example, in a steady-state, steady-flow process, as we consider the two terms in Eq. 9.24, we realize that, in accordance with Eq. 9.6, the first term is zero. As a result, all the entropy change that is due to irreversibilities in this type of process is observed in the surroundings. This term may then be evaluated using Eq. 9.22. On the other hand, for the uniform-state, uniform-flow process, there are both control volume and surroundings terms to evaluate. Each term is integrated over the time t of the process, as was done in Section 9.2. Thus, Eq. 9.24 is integrated to

$$\Delta S_{\text{net}} = \Delta S_{\text{c.v.}} + \Delta S_{\text{surr}} \tag{9.25}$$

in which the control volume term is

$$\Delta S_{\text{c.v.}} = \left(m_2 s_2 - m_1 s_1 \right)_{\text{c.v.}} \tag{9.26}$$

The term for the surroundings is, after applying Eq. 9.13 to the surroundings and integrating,

$$\Delta S_{\text{surr}} = \frac{-Q_{\text{c.v.}}}{T_0} + \sum m_e s_e - \sum m_i s_i \tag{9.27}$$

9.5 EFFICIENCY

In Chapter 7 we noted that the second law of thermodynamics led to the concept of thermal efficiency for a heat engine cycle, namely

$$\eta_{\text{th}} = \frac{W_{\text{net}}}{Q_H}$$

where W_{net} is the net work of the cycle and Q_H is the heat transfer from the high-temperature body.

In this chapter we have extended our consideration of the second law to control volume processes, which leads us now to consider the efficiency of a process. For example, we might be interested in the efficiency of a turbine in a steam power plant or the compressor in a gas turbine engine.

In general, we can say that to determine the efficiency of a machine in which a process takes place, we compare the actual performance of the machine under given conditions and the performance that would have been achieved in an ideal process. It is in the definition of this ideal process that the second law becomes a major consideration. For example, a steam turbine is intended to be an adiabatic machine. The only heat transfer is the unavoidable heat transfer that takes place between the given turbine and the surroundings. We also note that for a given steam turbine operating in a steady-state, steady-flow manner, the state of the steam entering the turbine and the exhaust pressure are fixed. Therefore, the ideal process is a reversible adiabatic process, which is an isentropic process, between the inlet state, and the turbine exhaust pressure. In other words, the variables P_i, T_i, and P_e are the design variables, the first two because the working fluid has been prepared in prior processes to be at these conditions at the turbine inlet, while the exit pressure is fixed by the environment into which the turbine exhausts. Thus, the ideal turbine process would go from state i to state es, as shown in Fig. 9.6, while the real tur-

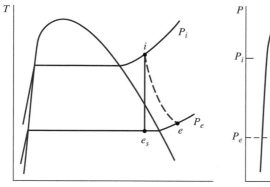
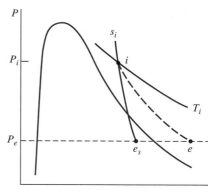

FIGURE 9.6 The process in a reversible adiabatic steam turbine and an actual turbine.

bine process is irreversible, with the exhaust at a larger entropy at the real exit state e. Figure 9.6 shows typical states for a steam turbine, where state es is in the two-phase region, and state e may be as well, or may be in the superheated vapor region, depending upon the extent of irreversibility of the real process. Denoting the work done in the real process from i to e as w, and that done in the ideal, isentropic process from the same P_i, T_i to the same P_e as w_s, we define the efficiency of the turbine as

$$\eta_{\text{turbine}} = \frac{w}{w_s} \qquad (9.28)$$

The same definition applies to a gas turbine, where all states are in the gaseous phase. Typical turbine efficiencies are 0.70–0.85, with large turbines usually having higher efficiencies than small ones.

EXAMPLE 9.6 A steam turbine receives steam at a pressure of 1 MPa, 300°C. The steam leaves the turbine at a pressure of 15 kPa. The work output of the turbine is measured and is found to be 600 kJ/kg of steam flowing through the turbine. Determine the efficiency of the turbine.

> *Control volume:* Turbine.
>
> *Inlet state:* P_i, T_i known; state fixed.
>
> *Exit state:* P_e known.
>
> *Process:* SSSF.
>
> *Model:* Steam tables.

Analysis:

The efficiency of the turbine is given by Eq. 9.28.

$$\eta_{\text{turbine}} = \frac{w_a}{w_s}$$

Thus, to determine the turbine efficiency, we calculate the work that would be done in an isentropic process between the given inlet state and final pressure. For this isentropic process,

Continuity equation:

$$\dot{m}_i = \dot{m}_e = \dot{m}$$

First law:

$$h_i = h_{es} + w_s$$

Second law:

$$s_i = s_{es}$$

Solution

From the steam tables,

$$h_i = 3051.2 \text{ kJ / kg} \qquad s_i = 7.1229 \text{ kJ / kg K}$$

Therefore, at $P_e = 15$ kPa,

$$s_{es} = s_i = 7.1229 = 0.7549 + x_{es} 7.2536$$

$$x_{es} = 0.8779$$

$$h_{es} = 225.9 + 0.8779(2373.1) = 2309.3 \text{ kJ / kg}$$

From the first law for the isentropic process,

$$w_s = h_i - h_{es} = 3051.2 - 2309.3 = 741.9 \text{ kJ / kg}$$

But, since

$$w_a = 600 \text{ kJ / kg}$$

we find that

$$\eta_{\text{turbine}} = \frac{w_a}{w_s} = \frac{600}{741.9} = 0.809 = 80.9\%$$

In connection with this example, it should be noted that to find the actual state e of the steam exiting the turbine, we need to analyze the real process taking place. For the real process

$$\dot{m}_i = \dot{m}_e = \dot{m}$$

$$h_i = h_e + w_a$$

$$s_e > s_i$$

Therefore, from the first law for the real process, we have

$$h_e = 3051.2 - 600 = 2451.2 \text{ kJ / kg}$$

$$2451.2 = 225.9 + x_e 2373.1$$

$$x_e = 0.9377$$

It is important to keep in mind that the turbine efficiency is defined in terms of an ideal, isentropic process from P_i, T_i, to P_e, even when one or more of these variables is unknown. This is illustrated in the following example.

EXAMPLE 9.7   Air enters a gas turbine at 1600 K, and exits at 100 kPa, 830 K. The turbine efficiency is estimated to be 85%. What is the turbine inlet pressure?

Control volume: Turbine.

Inlet state: T_i known.

Exit state: P_e, T_e known; state fixed.

Process: SSSF.

Model: Air tables, A.7.

Analysis:

The efficiency, which is 85%, is given by Eq. 9.28,

$$\eta_{\text{turbine}} = \frac{w}{w_s}$$

The first law for the real, irreversible process is

$$h_i = h_e + w$$

For the ideal, isentropic process from P_i, T_i to P_e, the first law is

$$h_i = h_{es} + w_s$$

and the second law is, from Eq. 8.29,

$$\frac{P_i}{P_e} = \frac{P_{ri}}{P_{res}}$$

Solution

From the air tables, A.7, at 1600 K,

$$h_i = 1757.3 \text{ kJ} / \text{kg}, \quad P_{ri} = 634.97$$

From the air tables at 830 K, (the actual turbine exit temperature),

$$h_e = 855.3 \text{ kJ} / \text{kg}$$

Therefore, from the first law for the real process,

$$w = 1757.3 - 855.3 = 902.0 \text{ kJ} / \text{kg}$$

Using the definition of turbine efficiency,

$$w_s = 902.0 / 0.85 = 1061.2 \text{ kJ} / \text{kg}$$

From the first law for the isentropic process,

$$h_{es} = 1757.3 - 1061.2 = 696.1 \text{ kJ} / \text{kg}$$

so that, from the air tables,

$$T_{es} = 684\,\text{K}, \quad P_{res} = 21.211$$

and the turbine inlet pressure is

$$P_i = 0.1 \times \frac{634.97}{21.211} = 3.0\,\text{MPa}$$

It is important to note that the second law relation, Eq. 8.29, applies to the process from state i to state es, not to the real state e.

As was discussed in Section 6.4, unless specifically noted to the contrary, we normally assume compressors or pumps to be adiabatic. In this case the fluid enters the compressor at P_i and T_i, the condition at which it exists, and exits at the desired value of P_e, the reason for building the compressor. Thus, the ideal process between the given inlet state i and the exit pressure would be an isentropic process between state i and state es, as shown in Fig. 9.7, with a work input of w_s. The real process, however, is irreversible, and the fluid exits at the real state e with a larger entropy, and a larger amount of work input w is required. The compressor (or pump, in the case of a liquid) efficiency is defined as

$$\eta_{\text{comp}} = \frac{w_s}{w} \tag{9.29}$$

Typical compressor efficiencies are in the range 0.70–0.85, with large compressors usually having higher efficiencies than small ones.

If an effort is made to cool a gas during compression by using a water jacket or fins, the ideal process is considered a reversible isothermal process, the work input for which is w_T, compared to the larger required work w for the real compressor. The efficiency of the cooled compressor is then

$$\eta_{\text{cooled comp}} = \frac{w_T}{w} \tag{9.30}$$

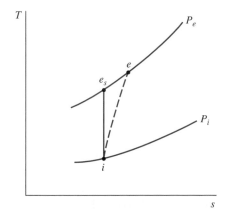 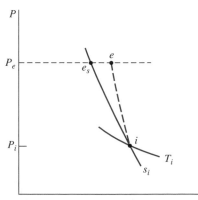

FIGURE 9.7 The compression process in an ideal and actual adiabatic compressor.

EXAMPLE 9.8   Air enters an automotive supercharger at 100 kPa, 300 K, and is compressed to 150 kPa. The efficiency is 70 percent. What is the required work input per kg of air? What is the exit temperature?

> *Control volume:*   Supercharger (compressor).
>
> *Inlet state:* P_i, T_i known; state fixed.
>
> *Exit state:* P_e known.
>
> *Process:*   SSSF.
>
> *Model:*   Ideal gas, 300 K specific heat, Table A.5.

Analysis:

The efficiency, which is 70%, is given by Eq. 9.29,

$$\eta_{\text{comp}} = \frac{w_s}{w}$$

The first law for the real, irreversible process is

$$h_i = h_e + w, \quad w = C_{P0}(T_i - T_e)$$

For the ideal, isentropic process from P_i, T_i to P_e, the first law is

$$h_i = h_{es} + w_s, \quad w_s = C_{P0}(T_i - T_{es})$$

and the second law is, from Eq. 8.36,

$$\frac{T_{es}}{T_i} = \left(\frac{P_e}{P_i}\right)^{(k-1)/k}$$

Solution

Using C_{P0} and k from Table A.5, from the second law,

$$T_{es} = 300\left(\frac{150}{100}\right)^{0.286} = 336.9 \text{ K}$$

From the first law for the isentropic process;

$$h_{es} = 1.0035(300 - 336.9) = -37.1 \text{ kJ / kg}$$

so that, from the efficiency, the real work input is

$$w = -37.1 / 0.70 = -53.0 \text{ kJ / kg}$$

and from the first law for the real process, the temperature at the supercharger exit is

$$T_e = 300 - \frac{-53.0}{1.0035} = 352.8 \text{ K}$$

Our final example is that of nozzle efficiency. As discussed in Section 6.4, the purpose of a nozzle is to produce a high-velocity fluid stream, or in terms of energy, a

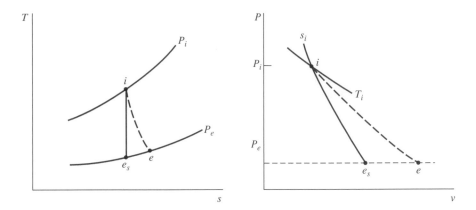

FIGURE 9.8 The ideal and actual process in an adiabatic nozzle.

large kinetic energy, at the expense of the fluid pressure. The design variables are the same as for a turbine, P_i, T_i, and P_e. A nozzle is usually assumed to be adiabatic, such that the ideal process is an isentropic process from state i to state es, as shown in Fig. 9.8, with the production of velocity $\mathbf{V}_{es}$. The real process is irreversible, with the exit state e having a larger entropy, and a smaller exit velocity $\mathbf{V}_e$. The nozzle efficiency is defined in terms of the corresponding kinetic energies,

$$\eta_{\text{nozz}} = \frac{\mathbf{V}_e^2 / 2}{\mathbf{V}_{es}^2 / 2}$$

Nozzles are simple devices with no moving parts. As a result, nozzle efficiencies may be very high, typically 0.90–0.97.

In summary, to determine the efficiency of a device that carries out a process (rather than a cycle), we compare the actual performance to what would be achieved in a related, but well-defined ideal process.

9.6 SOME GENERAL COMMENTS REGARDING ENTROPY

It is quite possible at this point that a student may have a good grasp of the material that has been covered, and yet may have only a vague understanding of the significance of entropy. In fact, the question "What is entropy?" is frequently raised by students with the implication that no one really knows! This section has been included in an attempt to give insight into the qualitative and philosophical aspects of the concept of entropy, and to illustrate the broad application of entropy to many different disciplines.

First, we recall that the concept of energy rises from the first law of thermodynamics and the concept of entropy from the second law of thermodynamics. Actually it is just as difficult to answer the question "What is energy?" as it is to answer the question "What is entropy?" However, since we regularly use the term energy and are able to relate this term to phenomena that we observe every day, the word energy has a definite meaning to us and thus serves as an effective vehicle for thought and communication. The word entropy could serve in the same capacity. If, when we observed a highly irreversible process (such as cooling coffee by placing an ice cube in it), we said, "That surely increases the entropy," we would soon be as familiar with the word *entropy* as we are with the word *energy*. In many cases when we speak about a higher efficiency we are actually speaking about accomplishing a given objective with a smaller total increase in entropy.

A second point to be made regarding entropy is that in statistical thermodynamics, the property entropy is defined in terms of probability. Although this topic will not be examined in detail in this text, a few brief remarks regarding entropy and probability may prove helpful. From this point of view the net increase in entropy that occurs during an irreversible process can be associated with a change of state from a less probable state to a more probable state. For instance, to use a previous example, one is more likely to find gas on both sides of ruptured membrane in Fig. 7.11 than to find a gas on one side and a vacuum on the other. Thus, when the membrane ruptures, the direction of the process is from a less probable state to a more probable state and associated with this process is an increase in entropy. Similarly, the more probable state is that a cup of coffee will be at the same temperature as its surroundings than at a higher (or lower) temperature. Therefore, as the coffee cools as the result of a transferring of heat to the surroundings, there is a change from a less probable to a more probable state, and associated with this is an increase in entropy.

PROBLEMS

9.1 Steam enters a turbine at 3 MPa, 450°C, expands in a reversible adiabatic process and exhausts at 10 kPa. Changes in kinetic and potential energies between the inlet and the exit of the turbine are small. The power output of the turbine is 800 kW. What is the mass flow rate of steam through the turbine?

9.2 In a heat pump that uses R-134a as the working fluid, the R-134a enters the compressor at 150 kPa, −10°C at a rate of 0.1 kg/s. In the compressor the R-134a is compressed in an adiabatic process to 1 MPa. Calculate the power input required to the compressor, assuming the process to be reversible.

9.3 Consider the design of a nozzle in which nitrogen gas flowing in a pipe at 500 kPa, 200°C, and at a velocity of 10 m/s, is to be expanded to produce a velocity of 300 m/s. Determine the exit pressure and cross-sectional area of the nozzle if the mass flow rate is 0.15 kg/s, and the expansion is reversible and adiabatic.

9.4 A compressor is surrounded by cold R-134a so it works as an isothermal compressor. The inlet state is 0°C, 100 kPa, and the exit state is saturated vapor. Find the specific heat transfer and specific work.

9.5 Air at 100 kPa, 17°C is compressed to 400 kPa after which it is expanded through a nozzle back to the atmosphere. The compressor and the nozzle are both reversible and adiabatic and kinetic energy in/out of the compressor can be neglected. Find the compressor work and its exit temperature, and find the nozzle exit velocity.

9.6 A small turbine delivers 150 kW and is supplied with steam at 700°C, 2 MPa. The exhaust passes through a heat exchanger where the pressure is 10 kPa and exits as saturated liquid. The turbine is reversible and adiabatic. Find the specific turbine work, and the heat transfer in the heat exchanger.

9.7 A counterflowing heat exchanger, shown in Fig. P9.7, is used to cool air at 540 K, 400 kPa to 360 K by using a 0.05 kg/s supply of water at 20°C, 200 kPa.

The air flow is 0.5 kg/s in a 10-cm diameter pipe. Find the air inlet velocity, the water exit temperature, and total entropy generation in the process.

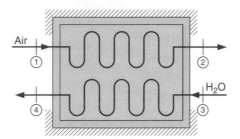

FIGURE P9.7

9.8 Analyze the steam turbine described in Problem 6.29. Is it possible?

9.9 A coflowing heat exchanger has one line with 2 kg/s saturated water vapor at 100 kPa entering. The other line is 1 kg/s air at 200 kPa, 1200 K. The heat exchanger is very long so the two flows exit at the same temperature. Find the exit temperature by trial and error. Calculate the rate of entropy generation.

9.10 Atmospheric air at –45°C, 60 kPa enters the front diffuser of a jet engine with a velocity of 900 km/h and frontal area of 1 m². After the adiabatic diffuser the velocity is 20 m/s. Find the diffuser exit temperature and the maximum pressure possible.

9.11 A Hilch tube has an air inlet flow at 20°C, 200 kPa and two exit flows of 100 kPa, one at 0°C and the other at 40°C. The tube has no external heat transfer and no work and all the flows are SSSF and have negligible kinetic energy. Find the fraction of the inlet flow that comes out at 0°C. Is this setup possible?

9.12 Two flowstreams of water, one at 0.6 MPa, saturated vapor, and the other at 0.6 MPa, 600°C, mix adiabatically in a SSSF process to produce a single flow out at 0.6 MPa, 400°C. Find the total entropy generation for this process.

9.13 In a heat-driven refrigerator with ammonia as the working fluid, a turbine with inlet conditions of 2.0 MPa, 70°C is used to drive a compressor with inlet saturated vapor at −20°C. The exhausts, both at 1.2 MPa, are then mixed together. The ratio of the mass flow rate to the turbine to the total exit flow was measured to be 0.62. Can this be true?

9.14 A diffuser is a steady-state, steady-flow device in which a fluid flowing at high velocity is decelerated such that the pressure increases in the process. Air at 120 kPa, 30°C enters a diffuser with velocity 200 m/s and exits with a velocity of 20 m/s. Assuming the process is reversible and adiabatic, what are the exit pressure and temperature of the air?

9.15 A reversible SSSF device receives a flow of 1 kg/s air at 400 K, 450 kPa and the air leaves at 600 K, 100 kPa. Heat transfer of 800 kW is added from a 1000 K reservoir, 100 kW is rejected at 350 K, and some heat transfer takes place at 500 K. Find the heat transferred at 500 K and the rate of work produced.

9.16 One technique for operating a steam turbine in part-load power output is to throttle the steam to a lower pressure before it enters the turbine, as shown in Fig. P9.16. The steamline conditions are 2 MPa, 400°C, and the turbine exhaust pressure is fixed at 10 kPa. Assuming the expansion inside the turbine to be reversible and adiabatic, determine

a. The full-load specific work output of the turbine.

b. The pressure the steam must be throttled to for 80% of full-load output.

c. Show both processes in a *T–s* diagram.

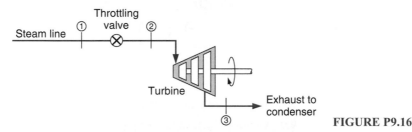

FIGURE P9.16

9.17 Carbon dioxide at 300 K, 200 kPa is brought through a SSSF device where it is heated to 500 K by a 600 K reservoir in a constant pressure process. Find the specific work, heat transfer, and entropy generation.

9.18 One type of feedwater heater for preheating the water before entering a boiler operates on the principle of mixing the water with steam that has been bled from the turbine. For the states as shown in Fig. P9.18, calculate the rate of net entropy increase for the process, assuming the process to be steady flow and adiabatic.

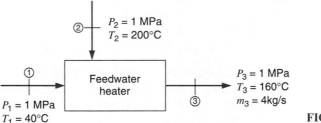

FIGURE P9.18

9.19 Air at 327°C, 400 kPa with a volume flow 1 m³/s runs through an adiabatic turbine with exhaust pressure of 100 kPa. Neglect kinetic energies and use constant specific heats. Find the lowest and highest possible exit temperature. For each case find also the rate of work and the rate of entropy generation.

9.20 A certain industrial process requires a steady supply of saturated vapor steam at 200 kPa, at a rate of 0.5 kg/s. Also required is a steady supply of compressed air at

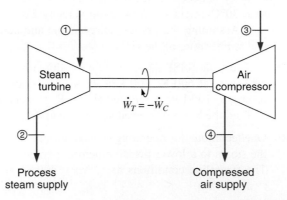

FIGURE P9.20

500 kPa, at a rate of 0.1 kg/s. Both are to be supplied by the process shown in Fig. P9.20. Steam is expanded in a turbine to supply the power needed to drive the air compressor, and the exhaust steam exits the turbine at the desired state. Air into the compressor is at the ambient conditions, 100 kPa, 20°C. Give the required steam inlet pressure and temperature, assuming that both the turbine and the compressor are reversible and adiabatic.

9.21 Air enters a turbine at 800 kPa, 1200 K, and expands in a reversible adiabatic process to 100 kPa. Calculate the exit temperature and the work output per kilogram of air, using

a. The ideal gas tables, Table A.7.

b. Constant specific heat, value at 300 K from Table A.5.

c. Constant specific heat, value at an intermediate temperature from Fig. 5.10.

Discuss why the method of part (b) gives a poor value for the exit temperature and yet a relatively good value for the work output.

9.22 Consider a steam turbine power plant operating at supercritical pressure, as shown in Fig. P9.22. As a first approximation, it may be assumed that the turbine and the pump processes are reversible and adiabatic. Neglecting any changes in kinetic and potential energies, calculate

a. The specific turbine work output and the turbine exit state.

b. The pump work input and enthalpy at the pump exit state.

c. The thermal efficiency of the cycle.

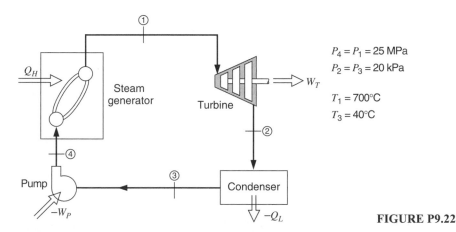

$P_4 = P_1 = 25$ MPa
$P_2 = P_3 = 20$ kPa

$T_1 = 700°C$
$T_3 = 40°C$

FIGURE P9.22

9.23 A supply of 5 kg/s ammonia at 500 kPa, 20°C is needed. Two sources are available: one is saturated liquid at 20°C and the other is at 500 kPa, 140°C. Flows from the two sources are fed through valves to an insulated SSSF mixing chamber, which then produces the desired output state. Find the two source mass flow rates and the total rate of entropy generation by this setup.

9.24 A turbo charger boosts the inlet air pressure to an automobile engine. It consists of an exhaust gas driven turbine directly connected to an air compressor, as shown in Fig. P9.24. For a certain engine load the conditions are given in the figure. Assume that both the turbine and the compressor are reversible and adiabatic having also the same mass flow rate. Calculate the turbine exit temperature and power output. Find also the compressor exit pressure and temperature.

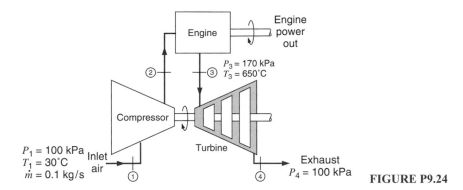

FIGURE P9.24

9.25 A stream of ammonia enters a steady flow device at 100 kPa, 50°C, at the rate of 1 kg/s. Two streams exit the device at equal mass flow rates; one is at 200 kPa, 50°C, and the other as saturated liquid at 10°C. It is claimed that the device operates in a room at 25°C on an electrical power input of 250 kW. Is this possible?

9.26 An initially empty 0.1 m³ cannister is filled with R-12 from a line flowing saturated liquid at −5°C. This is done quickly such that the process is adiabatic. Find the final mass, liquid and vapor volumes, if any, in the cannister. Is the process reversible?

9.27 Air from a line at 12 MPa, 15°C, flows into a 500-L rigid tank that initially contained air at ambient conditions, 100 kPa, 15°C. The process occurs rapidly and is essentially adiabatic. The valve is closed when the pressure inside reaches some value, P_2. The tank eventually cools to room temperature, at which time the pressure inside is 5 MPa. What is the pressure P_2? What is the net entropy change for the overall process?

9.28 An initially empty spring-loaded piston/cylinder requires 100 kPa to float the piston. A compressor with a line and valve now charges the cylinder with water to a final pressure of 1.4 MPa at which point the volume is 0.6 m³, state 2. The inlet condition to the reversible adiabatic compressor is saturated vapor at 100 kPa. After charging, the valve is closed and the water eventually cools to room temperature, 20°C, state 3. Find the final mass of water, the piston work from 1 to 2, the required compressor work, and the final pressure, P_3.

9.29 An initially empty cannister of volume 0.2 m³ is filled with carbon dioxide from a line at 1000 kPa, 500 K. Assume the process is adiabatic and the flow continues until it stops by itself. Find the final mass and temperature of the carbon dioxide in the cannister and the total entropy generated by the process.

9.30 A 1-m³ rigid tank contains 100 kg R-22 at ambient temperature, 15°C. A valve on top of the tank is opened, and saturated vapor is throttled to ambient pressure, 100 kPa, and flows to a collector system. During the process the temperature inside the tank remains at 15°C. The valve is closed when no more liquid remains inside. Calculate the heat transfer to the tank and the total entropy generation in the process.

9.31 An old abandoned saltmine, 100 000 m³ in volume, contains air at 290 K, 100 kPa. The mine is used for energy storage so the local power plant pumps it up to 2.1 MPa using outside air at 290 K, 100 kPa. Assume the pump is ideal and the process is adiabatic. Find the final mass and temperature of the air and the required pump work. Overnight, the air in the mine cools down to 400 K. Find the final pressure and heat transfer.

9.32 A rigid steel bottle, $V = 0.25$ m^3, contains air at 100 kPa, 300 K. The bottle is now charged with air from a line at 260 K, 6 MPa to a bottle pressure of 5 MPa, state 2, and the valve is closed. Assume that the process is adiabatic, and the charge always is uniform. In storage, the bottle slowly returns to room temperature at 300 K, state 3. Find the final mass, the temperature T_2, the final pressure P_3, the heat transfer $_1Q_3$, and the total entropy generation.

9.33 An insulated 2 m^3 tank is to be charged with R-134a from a line flowing the refrigerant at 3 MPa. The tank is initially evacuated, and the valve is closed when the pressure inside the tank reaches 3 MPa. The line is supplied by an insulated compressor that takes in R-134a at 5°C, quality of 96.5%, and compresses it to 3 MPa in a reversible process. Calculate the total work input to the compressor to charge the tank.

9.34 A horizontal, insulated cylinder has a frictionless piston held against stops by an external force of 500 kN, as shown in Fig. P9.34. The piston cross-sectional area is 0.5 m^2, and the initial volume is 0.25 m^3. Argon gas in the cylinder is at 200 kPa, 100°C. A valve is now opened to a line flowing argon at 1.2 MPa, 200°C, and gas flows in until the cylinder pressure just balances the external force, at which point the valve is closed. The external force is now slowly reduced so the gas expands, moving the piston to a final pressure of 100 kPa. Find the final temperature of the argon and the work done during the overall process.

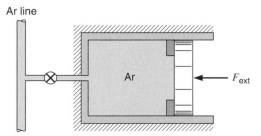

FIGURE P9.34

9.35 A rigid 1.0 m^3 tank contains water initially at 120°C, with 50% liquid and 50% vapor, by volume. A pressure-relief valve on the top of the tank is set to 1.0 MPa (the tank pressure cannot exceed 1.0 MPa—water will be discharged instead). Heat is now transferred to the tank from a 200°C heat source until the tank contains saturated vapor at 1.0 MPa. Calculate the heat transfer to the tank and show that this process does not violate the second law.

9.36 A frictionless piston/cylinder is loaded with a linear spring, spring constant 100 kN/m and the piston cross-sectional area is 0.1 m^2. The cylinder initial volume of 20 L contains air at 200 kPa and ambient temperature, 10°C. The cylinder has a set of stops that prevents its volume from exceeding 50 L. A valve connects to a line flowing air at 800 kPa, 50°C, as shown in Fig. P9.36. The valve is now opened, allowing air to flow in until the cylinder pressure reaches 800 kPa, at which point the temperature inside the cylinder is 80°C. The valve is then closed and the process ends.

a. Is the piston at the stops at the final state?

b. Taking the inside of the cylinder as a control volume, calculate the heat transfer during the process.

c. Calculate the net entropy change for this process.

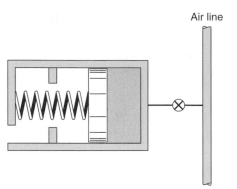

FIGURE P9.36

9.37 An insulated piston/cylinder contains R-22 at 20°C, 85% quality, at a cylinder volume of 50 L. A valve at the closed end of the cylinder is connected to a line flowing R-22 at 2 MPa, 60°C. The valve is now opened, allowing R-22 to flow in, and at the same time the external force on the piston is decreased, and the piston moves. When the valve is closed, the cylinder contents are at 800 kPa, 20°C, and a positive work of 50 kJ has been done against the external force. What is the final volume of the cylinder? Does this process violate the second law of thermodynamics?

9.38 Liquid water at ambient conditions, 100 kPa, 25°C, enters a pump at the rate of 0.5 kg/s. Power input to the pump is 3 kW. Assuming the pump process to be reversible, determine the pump exit pressure and temperature.

9.39 A firefighter on a ladder 25 m above ground should be able to spray water an additional 10 m up with the hose nozzle of exit diameter 2.5 cm. Assume a water pump on the ground and a reversible flow (hose, nozzle included) and find the minimum required power.

9.40 A large storage tank contains liquefied natural gas (LNG), which may be assumed to be pure methane. The tank contains saturated liquid at ambient pressure, 100 kPa; it is to be pumped to 500 kPa and fed to a pipeline at the rate of 0.5 kg/s. How much power input is required for the pump, assuming it to be reversible?

9.41 A small dam has a pipe carrying liquid water at 150 kPa, 20°C with a flow rate of 2000 kg/s in a 0.5-m diameter pipe. The pipe runs to the bottom of the dam 15 m lower into a turbine with pipe diameter 0.35 m. Assume no friction or heat transfer in the pipe and find the pressure of the turbine inlet. If the turbine exhausts to 100 kPa with negligible kinetic energy, what is the rate of work?

9.42 A small pump is driven by a 2 kW motor with liquid water at 150 kPa, 10°C entering. Find the maximum water flow rate you can get with an exit pressure of 1 MPa and negligible kinetic energies. The exit flow goes through a small hole in a spray nozzle out to the atmosphere at 100 kPa. Find the spray velocity.

9.43 Saturated R-134a at −10°C is pumped/compressed to a pressure of 1.0 MPa at the rate of 0.5 kg/s in a reversible adiabatic SSSF process. Calculate the power required and the exit temperature for the two cases of inlet state of the R-134a:

a. Quality of 100%.   b. Quality of 0%.

9.44 A small water pump on ground level has an inlet pipe down into a well at a depth *H* with the water at 100 kPa, 15°C. The pump delivers water at 400 kPa to a building. The absolute pressure of the water must be at least twice the saturation pressure to avoid cavitation. What is the maximum depth this setup will allow?

9.45 Atmospheric air at 100 kPa, 17°C blows at 60 km/h toward the side of a building. Assuming the air is nearly incompressible, find the pressure and the temperature at the stagnation point (zero velocity) on the wall.

9.46 A small pump takes in water at 20°C, 100 kPa, and pumps it to 2.5 MPa at a flow rate of 100 kg/min. Find the required pump power input.

9.47 Helium gas enters a steady-flow expander at 800 kPa, 300°C, and exits at 120 kPa. The mass flow rate is 0.2 kg/s, and the expansion process can be considered as a reversible polytropic process with exponent, $n = 1.3$. Calculate the power output of the expander.

9.48 A pump/compressor pumps a substance from 100 kPa, 10°C to 1 MPa in a reversible adiabatic SSSF process. The exit pipe has a small crack, so that a small amount leaks to the atmosphere at 100 kPa. If the substance is (a) water, (b) R-12, find the temperature after compression and the temperature of the leak flow as it enters the atmosphere, neglecting kinetic energies.

9.49 A certain industrial process requires a steady 0.5 kg/s of air at 200 m/s, at the condition of 150 kPa, 300 K, as shown in Fig. P9.49. This air is to be the exhaust from a specially designed turbine whose inlet pressure is 400 kPa. The turbine process may be assumed to be reversible and polytropic, with polytropic exponent $n = 1.20$.
a. What is the turbine inlet temperature?
b. What are the power output and heat transfer rate for the turbine?
c. Calculate the rate of net entropy increase, if the heat transfer comes from a source at a temperature 100°C higher than the turbine inlet temperature.

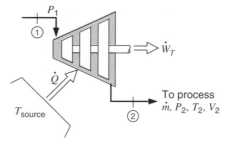

FIGURE P9.49

9.50 A mixing chamber receives 5 kg/min ammonia as saturated liquid at −20°C from one line and ammonia at 40°C, 250 kPa from another line through a valve. The chamber also receives 325 kJ/min energy as heat transferred from a 40°C reservoir. This should produce saturated ammonia vapor at −20°C in the exit line. What is the mass flow rate in the second line and what is the total entropy generation in the process?

9.51 A compressor is used to bring saturated water vapor at 1 MPa up to 17.5 MPa, where the actual exit temperature is 650°C. Find the isentropic compressor efficiency and the entropy generation.

9.52 Liquid water enters a pump at 15°C, 100 kPa, and exits at a pressure of 5 MPa. If the isentropic efficiency of the pump is 75%, determine the enthalpy (steam table reference) of the water at the pump exit.

9.53 A centrifugal compressor takes in ambient air at 100 kPa, 15°C, and discharges it at 450 kPa. The compressor has an isentropic efficiency of 80%. What is your best estimate for the discharge temperature?

9.54 Repeat Problem 9.20 assuming the steam turbine and the air compressor each have an isentropic efficiency of 80%.

9.55 A small air turbine with an isentropic efficiency of 80% should produce 270 kJ/kg of work. The inlet temperature is 1000 K and it exhausts to the atmosphere. Find the required inlet pressure and the exhaust temperature.

9.56 Carbon dioxide, CO_2, enters an adiabatic compressor at 100 kPa, 300 K, and exits at 1000 kPa, 520 K. Find the compressor efficiency and the entropy generation for the process.

9.57 Repeat Problem 9.22 assuming the turbine and the pump each have an isentropic efficiency of 85%.

9.58 Air enters an insulated compressor at ambient conditions, 100 kPa, 20°C, at the rate of 0.1 kg/s and exits at 200°C. The isentropic efficiency of the compressor is 70%. What is the exit pressure? How much power is required to drive the compressor?

9.59 Steam enters a turbine at 300°C and exhausts at 20 kPa. It is estimated that the isentropic efficiency of the turbine is 70%. What is the maximum turbine inlet pressure if the exhaust is not to be in the two-phase region?

9.60 A nozzle is required to produce a flow of air at 200 m/s at 20°C, 100 kPa. It is estimated that the nozzle has an isentropic efficiency of 92%. What nozzle inlet pressure and temperature is required assuming the inlet kinetic energy is negligible?

9.61 A turbine receives air at 1500 K, 1000 kPa, and expands it to 100 kPa. The turbine has an isentropic efficiency of 85%. Find the actual turbine exit air temperature and the specific entropy increase in the actual turbine.

9.62 Assume both the compressor and the nozzle in Problem 9.5 have an isentropic efficiency of 90%, the rest being unchanged. Find the actual compressor work and its exit temperature and find the actual nozzle exit velocity.

9.63 The small turbine in Problem 9.6 was ideal. Assume instead that the isentropic turbine efficiency is 88%. Find the actual specific turbine work, the entropy generated in the turbine and the heat transfer in the heat exchanger.

9.64 A geothermal supply of hot water at 500 kPa, 150°C is fed to an insulated flash evaporator at the rate of 1.5 kg/s. A stream of saturated liquid at 200 kPa is drained from the bottom of the chamber, and a stream of saturated vapor at 200 kPa is drawn from the top and fed to a turbine. The turbine has an isentropic efficiency of 70% and an exit pressure of 15 kPa. Evaluate the second law for a control volume that includes the flash evaporator and the turbine.

9.65 Redo Problem 9.39 if the water pump has an isentropic efficiency of 85% (hose, nozzle included).

9.66 A flow of 20 kg/s steam at 10 MPa, 550°C enters a two-stage turbine. The exit of the first stage is at 2 MPa where 4 kg/s is taken out for process steam and the rest continues through the second stage, which has an exit at 50 kPa. Assuming both stages have an isentropic efficiency of 85%, find the total actual turbine work and the entropy generation.

9.67 Air flows into an insulated nozzle at 1 MPa, 1200 K with 15 m/s and mass flow rate of 2 kg/s. It expands to 650 kPa, and exit temperature is 1100 K. Find the exit velocity and the nozzle efficiency.

9.68 A nozzle is required to produce a steady stream of R–134a at 240 m/s at ambient conditions, 100 kPa, 20°C. The isentropic efficiency may be assumed to be 90%. Find by trial and error or verify that the inlet pressure is 375 kPa. What is the required inlet temperature in the line upstream of the nozzle?

9.69 Calculate the isentropic efficiency for each of the stages in the steam turbine shown in Problem 6.41. Find also the total entropy generated in the turbine.

9.70 A two-stage compressor having an interstage cooler takes in air, 300 K, 100 kPa, and compresses it to 2 MPa, as shown in Fig. P9.70. The cooler then cools the air to 340 K, after which it enters the second stage, which has an exit pressure of 15.74 MPa. The isentropic efficiency of stage one is 90% and the air exits the second stage at 630 K. Both stages are adiabatic, and the cooler dumps Q to reservoir at T_0. Find Q in the cooler, the efficiency of the second stage, and the total entropy generated in this process.

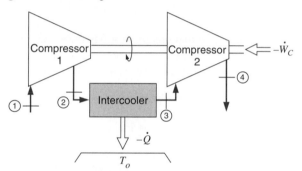

FIGURE P9.70

9.71 A two-stage turbine receives air at 1160 K, 5.0 MPa. The first stage exit is at 1 MPa, and the air then enters stage 2, which has an exit pressure of 200 kPa. Each stage has an isentropic efficiency of 85%. Find the specific work in each stage, the overall isentropic efficiency, and the total entropy generation.

9.72 A paper mill, shown in Fig. P9.72, has two steam generators, one at 4.5 MPa, 300°C and one at 8 MPa, 500°C. Each generator feeds a turbine, both of which have an exhaust pressure of 1.2 MPa and isentropic efficiency of 87%, such that their combined power output is 20 MW. The two exhaust flows are mixed adiabatically to produce saturated vapor at 1.2 MPa. Find the two mass flow rates and the entropy produced in each turbine and in the mixing chamber.

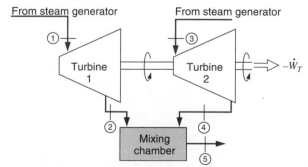

FIGURE P9.72

9.73 A heat-powered portable air compressor consists of three components: (a) an adiabatic compressor; (b) a constant pressure heater (heat supplied from an outside source); and (c) an adiabatic turbine (see Fig. P9.73). The compressor and the turbine each have an isentropic efficiency of 85%. Ambient air enters the compressor at 100 kPa, 300 K, and is compressed to 600 kPa. All of the power from the turbine goes into the compressor, and the turbine exhaust is the supply of compressed air. If this pressure is required to be 200 kPa, what must the temperature be at the exit of the heater?

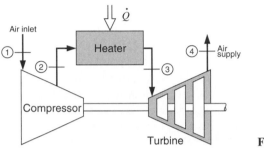

FIGURE P9.73

9.74 Assume an actual compressor has the same exit pressure and specific heat transfer as the ideal isothermal compressor in Problem 9.4 with an isothermal efficiency of 80%. Find the specific work and exit temperature for the actual compressor.

9.75 A watercooled air compressor takes air in at 20°C, 90 kPa and compresses it to 500 kPa. The isothermal efficiency is 80%, and the actual compressor has the same heat transfer as the ideal one. Find the specific compressor work and the exit temperature.

9.76 Repeat Problem 9.33 when the compressor has an isentropic efficiency of 80%.

9.77 Saturated vapor R-22 enters an insulated compressor with an isentropic efficiency of 75%, and the R-22 exits at 3.5 MPa, 120°C. Find the compressor inlet temperature by trial and error.

9.78 Air enters an insulated turbine at 50°C and exits the turbine at –30°C, 100 kPa. The isentropic turbine efficiency is 70%, and the inlet volumetric flow rate is 20 L/s. What is the turbine inlet pressure and the turbine power output?

9.79 Repeat Problem 9.43 for a pump/compressor isentropic efficiency of 70%.

9.80 A certain industrial process requires a steady 0.5 kg/s supply of compressed air at 500 kPa, at a maximum temperature of 30°C, as shown in Fig. P9.80. This air is to be supplied by installing a compressor and aftercooler. Local ambient conditions

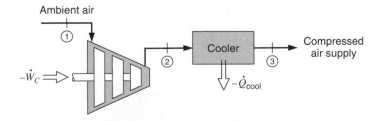

FIGURE P9.80

are 100 kPa, 20°C. Using an isentropic compressor efficiency of 80%, determine the power required to drive the compressor and the rate of heat rejection in the aftercooler.

9.81 The turbo charger in Problem 9.24 has isentropic efficiencies of 70% for both the compressor and the turbine. Repeat the questions when the actual compressor has the same flow rate as the ideal but a lower exit pressure.

9.82 In a heat-powered refrigerator, a turbine is used to drive the compressor using the same working fluid. Consider the combination shown in Fig. P9.82 where the turbine produces just enough power to drive the compressor and the two exit flows are mixed together. List any assumptions made and find the ratio of mass flow rates $\dot{m}_3/\dot{m}_1$ and T_5 (x_5 if in two-phase region) if

a. The turbine and the compressor are reversible and adiabatic.

b. The turbine and the compressor both have an isentropic efficiency of 70%.

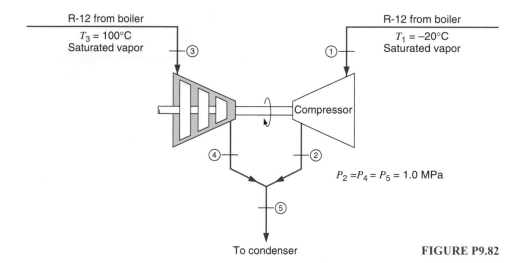

FIGURE P9.82

ADVANCED PROBLEMS

9.83 An air turbine with inlet conditions 1200 K, 1 MPa and exhaust pressure of 100 kPa pulls a sledge over a leveled plane surface, $T = 20°C$. The turbine work overcomes the friction between the sledge and the surface. Find the total entropy generation per kilogram of air through the turbine.

9.84 Consider the scheme shown in Fig. P9.84 for producing fresh water from salt water. The conditions are as shown in the figure. Assume that the properties of salt water are the same as for pure water, and that the pump is reversible and adiabatic.

a. Determine the ratio $(\dot{m}_7/\dot{m}_1)$, the fraction of salt water purified.

b. Determine the input quantities, w_p and q_H.

c. Make a second law analysis of the overall system.

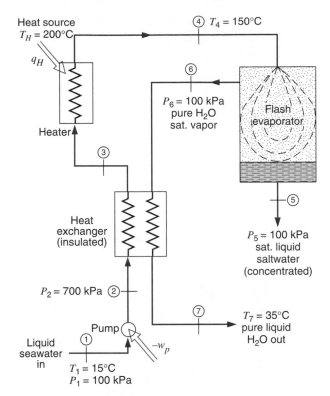

Heat source
$T_H = 200°C$

q_H

④ $T_4 = 150°C$

⑥
$P_6 = 100$ kPa
pure H_2O
sat. vapor

Flash
evaporator

Heater

③

Heat
exchanger
(insulated)

$P_5 = 100$ kPa
sat. liquid
saltwater
(concentrated)

⑤

$P_2 = 700$ kPa ②

⑦ $T_7 = 35°C$
pure liquid
H_2O out

Pump

Liquid
seawater
in
① $-w_p$

$T_1 = 15°C$
$P_1 = 100$ kPa

FIGURE P9.84

9.85 A cylinder/piston containing 2 kg of ammonia at $-10°C$, 90% quality is brought into a 20°C room and attached to a line flowing ammonia at 800 kPa, 40°C. The total restraining force on the piston is proportional to the cylinder volume squared. The valve is opened, and ammonia flows into the cylinder until the mass inside is twice the initial mass and the valve is closed. An electrical current of 15 A is passed through a 2-Ω resistor inside the cylinder for 20 min. It is claimed that the final pressure in the cylinder is 600 kPa. Is this possible?

9.86 A certain industrial process requires a steady stream of saturated vapor water at 200 kPa at a rate of 2 kg/s. There are two alternatives for supplying this steam from ambient liquid water at 20°C, 100 kPa. Assume pump efficiency of 80%.

1. Pump the water to 200 kPa and feed it to a steam generator (heater).

2. Pump the water to 5 MPa, feed it to a steam generator and heat to 450°C, then expand it through a turbine from which the steam exhausts at the desired state.

a. Compare these two alternatives in terms of heat transfer and work. Is the turbine isentropic efficiency reasonable?

b. What is the total entropy generation for each alternative?

9.87 Ammonia enters a nozzle at 800 kPa, 50°C, at a velocity of 10 m/s and at the rate of 0.1 kg/s. The nozzle expansion is assumed to be a reversible, polytropic SSSF process. Ammonia exits the nozzle at 200 kPa; the rate of heat transfer to the nozzle is 8.2 kW. Verify that the exit temperature is close to $-10°C$. What is the velocity of the ammonia exiting the nozzle?

9.88 A cylinder fitted with a spring-loaded piston serves as the supply of steam for a steam turbine (see Fig. P9.88). Initially, the cylinder pressure is 2 MPa and the volume is 1.0 m³. The force exerted by the spring is zero at zero cylinder volume, and the top of the piston is open to the ambient. The cylinder temperature is maintained at a constant 300°C by heat transfer from a source at that temperature. A pressure regulator between the cylinder and turbine maintains a steady 500 kPa, 300°C at the turbine inlet, such that when the cylinder pressure drops to 500 kPa, the process stops. The turbine process is reversible and adiabatic, and the exhaust is to a condenser at 50 kPa.

a. What is the total work output of the turbine during the process?

b. What is the turbine exhaust temperature (or quality)?

c. What is the total heat transfer to the cylinder during the process?

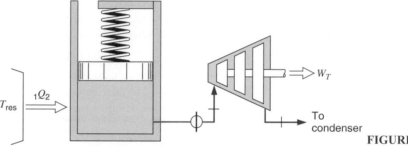

FIGURE P9.88

9.89 Supercharging of an engine is used to increase the inlet air density so that more fuel can be added, the result of which is an increased power output. Assume that ambient air, 100 kPa and 27°C, enters the supercharger at a rate of 250 L/s. The supercharger (compressor) has an isentropic efficiency of 75%, and uses 20 kW of power input. Assume that the ideal and actual compressor have the same exit pressure. Find the ideal specific work and verify that the exit pressure is 175 kPa. Find the percent increase in air density entering the engine due to the supercharger and the entropy generation.

9.90 A jet-ejector pump, shown schematically in Fig. P9.90, is a device in which a low-pressure (secondary) fluid is compressed by entrainment in a high-velocity (primary) fluid stream. The compression results from the deceleration in a diffuser. For purposes of analysis this can be considered as equivalent to the turbine-compressor unit shown in Fig. P9.82 with the states 1, 3, and 5 corresponding to those in Fig. P9.90. Consider a steam jet-pump with state 1 as saturated vapor at 35 kPa; state 3 is 300 kPa, 150°C; and the discharge pressure, P_5, is 100 kPa.

a. Calculate the ideal mass flow ratio, $\dot{m}_1 / \dot{m}_3$.

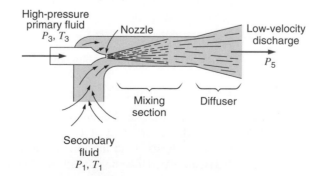

FIGURE P9.90

b. The efficiency of a jet pump is defined as

$$\eta_{\text{jet pump}} = \frac{(\dot{m}_1 / \dot{m}_3)_{\text{actual}}}{(\dot{m}_1 / \dot{m}_3)_{\text{ideal}}}$$

for the same inlet conditions and discharge pressure. Determine the discharge temperature of the jet pump if its efficiency is 10%.

ENGLISH UNIT PROBLEMS

9.91E Steam enters a turbine at 450 lbf/in.^{2}, 900 F, expands in a reversible adiabatic process and exhausts at 2 lbf/in.^{2}. Changes in kinetic and potential energies between the inlet and the exit of the turbine are small. The power output of the turbine is 800 Btu/s. What is the mass flow rate of steam through the turbine?

9.92E In a heat pump that uses R-134a as the working fluid, the R-134a enters the compressor at 30 lbf/in.^{2}, 20 F at a rate of 0.1 lbm/s. In the compressor the R-134a is compressed in an adiabatic process to 150 lbf/in.^{2}. Calculate the power input required to the compressor, assuming the process to be reversible.

9.93E Air at 1 atm, 60 F is compressed to 4 atm, after which it is expanded through a nozzle back to the atmosphere. The compressor and the nozzle are both reversible and adiabatic, and kinetic energy in/out of the compressor can be neglected. Find the compressor work and its exit temperature and find the nozzle exit velocity.

9.94E Analyse the steam turbine described in Problem 6.86. Is it possible?

9.95E Two flowstreams of water, one at 100 lbf/in.^{2}, saturated vapor, and the other at 100 lbf/in.^{2}, 1000 F, mix adiabatically in a SSSF process to produce a single flow out at 100 lbf/in.^{2}, 600 F. Find the total entropy generation for this process.

9.96E A diffuser is a steady-state, steady-flow device in which a fluid flowing at high velocity is decelerated such that the pressure increases in the process. Air at 18 lbf/in.^{2}, 90 F enters a diffuser with velocity 600 ft/s and exits with a velocity of 60 ft/s. Assuming the process is reversible and adiabatic, what are the exit pressure and temperature of the air?

9.97E One technique for operating a steam turbine in part-load power output is to throttle the steam to a lower pressure before it enters the turbine, as shown in Fig. P9.16. The steamline conditions are 200 lbf/in.^{2}, 600 F, and the turbine exhaust pressure is fixed at 1 lbf/in.^{2}. Assuming the expansion inside the turbine to be reversible and adiabatic,

a. Determine the full-load specific work output of the turbine.

b. Determine the pressure the steam must be throttled to for 80% of full-load output.

c. Show both processes in a *T–s* diagram.

9.98E Air at 540 F, 60 lbf/in.^{2} with a volume flow 40 ft^3/s runs through an adiabatic turbine with exhaust pressure of 15 lbf/in.^{2}. Neglect kinetic energies and use constant specific heats. Find the lowest and highest possible exit temperature. For each case find also the rate of work and the rate of entropy generation.

9.99E A supply of 10 lbm/s ammonia at 80 lbf/in.^{2}, 80 F is needed. Two sources are available: one is saturated liquid at 80 F and the other is at 80 lbf/in.^{2}, 260 F. Flows from the two sources are fed through valves to an insulated SSSF mixing chamber, which then produces the desired output state. Find the two source mass flow rates and the total rate of entropy generation by this setup.

9.100E Air from a line at 1800 lbf/in.^{2}, 60 F, flows into a 20-ft^3 rigid tank that initially contained air at ambient conditions, 14.7 lbf/in.^{2}, 60 F. The process occurs rapidly and is essentially adiabatic. The valve is closed when the pressure inside reaches some value, P_2. The tank eventually cools to room temperature, at which time the pressure inside is 750 lbf/in.^{2}. What is the pressure P_2? What is the net entropy change for the overall process?

9.101E An old abandoned saltmine, 3.5×10^6 ft^3 in volume, contains air at 520 R, 14.7 lbf/in.^{2}. The mine is used for energy storage so the local power plant pumps it up to 310 lbf/in.^{2} using outside air at 520 R, 14.7 lbf/in.^{2}. Assume the pump is ideal and the process is adiabatic. Find the final mass and temperature of the air and the required pump work. Overnight, the air in the mine cools down to 720 R. Find the final pressure and heat transfer.

9.102E A rigid 35 ft^3 tank contains water initially at 250 F, with 50% liquid and 50% vapor, by volume. A pressure-relief valve on the top of the tank is set to 140 lbf/in.^{2} (the tank pressure cannot exceed 140 lbf/in.^{2} - water will be discharged instead). Heat is now transferred to the tank from a 400 F heat source until the tank contains saturated vapor at 140 lbf/in.^{2}. Calculate the heat transfer to the tank and show that this process does not violate the second law.

9.103E Liquid water at ambient conditions, 14.7 lbf/in.^{2}, 75 F, enters a pump at the rate of 1 lbm/s. Power input to the pump is 3 Btu/s. Assuming the pump process to be reversible, determine the pump exit pressure and temperature.

9.104E A fireman on a ladder 80 ft above ground should be able to spray water an additional 30 ft up with the hose nozzle of exit diameter 1 in. Assume a water pump on the ground and a reversible flow (hose, nozzle included) and find the minimum required power.

9.105E Saturated R-134a at 10 F is pumped/compressed to a pressure of 150 lbf/in.^{2} at the rate of 1.0 lbm/s in a reversible adiabatic SSSF process. Calculate the power required and the exit temperature for the two cases of inlet state of the R-134a:

    a. Quality of 100%.   b. Quality of 0%.

9.106E A small pump takes in water at 70 F, 14.7 lbf/in.^{2} and pumps it to 250 lbf/in.^{2} at a flow rate of 200 lbm/min. Find the required pump power input.

9.107E Helium gas enters a steady-flow expander at 120 lbf/in.^{2}, 500 F, and exits at 18 lbf/in.^{2}. The mass flow rate is 0.4 lbm/s, and the expansion process can be considered as a reversible polytropic process with exponent, $n = 1.3$. Calculate the power output of the expander.

9.108E A mixing chamber receives 10 lbm/min ammonia as saturated liquid at 0 F from one line and ammonia at 100 F, 40 lbf/in.^{2} from another line through a valve. The chamber also receives 340 Btu/min energy as heat transferred from a 100-F reservoir. This should produce saturated ammonia vapor at 0 F in the exit line. What is the mass flow rate at state 2, and what is the total entropy generation in the process?

9.109E A compressor is used to bring saturated water vapor at 150 lbf/in.^{2} up to 2500 lbf/in.^{2}, where the actual exit temperature is 1200 F. Find the isentropic compressor efficiency and the entropy generation.

9.110E A small air turbine with an isentropic efficiency of 80% should produce 120 Btu/lbm of work. The inlet temperature is 1800 R, and it exhausts to the atmosphere. Find the required inlet pressure and the exhaust temperature.

9.111E Air enters an insulated compressor at ambient conditions, 14.7 lbf/in.^{2}, 70 F, at the rate of 0.1 lbm/s and exits at 400 F. The isentropic efficiency of the compressor is 70%. What is the exit pressure? How much power is required to drive the compressor?

9.112E Air at 1 atm, 60 F is compressed to 4 atm, after which it is expanded through a nozzle back to the atmosphere. The compressor and the nozzle both have efficiency of 90%, and kinetic energy in/out of the compressor can be neglected. Find the actual compressor work and its exit temperature, and find the actual nozzle exit velocity.

9.113E A geothermal supply of hot water at 80 lbf/in.^{2}, 300 F is fed to an insulated flash evaporator at the rate of 10 000 lbm/h. A stream of saturated liquid at 30 lbf/in.^{2} is drained from the bottom of the chamber, and a stream of saturated vapor at 30 lbf/in.^{2} is drawn from the top and fed to a turbine. The turbine has an isentropic efficiency of 70% and an exit pressure of 2 lbf/in.^{2}. Evaluate the second law for a control volume that includes the flash evaporator and the turbine.

9.114E Redo Problem 9.104 if the water pump has an isentropic efficiency of 85% (hose, nozzle included).

9.115E A nozzle is required to produce a steady stream of R-134a at 790 ft/s at ambient conditions, 14.7 lbf/in.^{2}, 70 F. The isentropic efficiency may be assumed to be 90%. What pressure and temperature are required in the line upstream of the nozzle?

9.116E A two-stage turbine receives air at 2100 R, 750 lbf/in.^{2}. The first stage exit is at 150 lbf in.^{2} and the air then enters stage 2, which has an exit pressure of 30 lbf/in.^{2}. Each stage has an isentropic efficiency of 85%. Find the specific work in each stage, the overall isentropic efficiency, and the total entropy generation.

9.117E A watercooled air compressor takes air in at 70 F, 14 lbf/in.^{2} and compresses it to 80 lbf/in.^{2}. The isothermal efficiency is 80%, and the actual compressor has the same heat transfer as the ideal one. Find the specific compressor work and the exit temperature.

9.118E Repeat Problem 9.105 for a pump/compressor isentropic efficiency of 70%.

9.119E A paper mill has two steam generators, one at 600 lbf/in.^{2}, 550 F and one at 1250 lbf/in.^{2}, 900 F. The setup is shown in Fig. P9.72. Each generator feeds a turbine, both of which have an exhaust pressure of 160 lbf/in.^{2} and isentropic efficiency of 87%, such that their combined power output is 20 000 Btu/s. The two exhaust flows are mixed adiabatically to produce saturated vapor at 160 lbf/in.^{2}. Find the two mass flow rates and the entropy produced in each turbine and in the mixing chamber.

COMPUTER, DESIGN, AND OPEN-ENDED PROBLEMS

9.120 Use the menu-driven software to get the properties for the calculation of the isentropic efficiency of the pump in the steam power plant of Problem 6.39.

9.121 Write a program to solve the general case of Problem 9.10, in which the states, velocities, and area are input variables. Use a constant specific heat and find diffuser exit area, temperature, and pressure.

9.122 Write a program to solve Problem 9.11 in which the inlet and exit flow states are input variables. Use a constant specific heat and let the program calculate the split of the mass flow and the overall entropy generation.

9.123 Write a program to solve the general version of Problem 9.46. Initial state, flow rate, and final pressure are input variables. Compute the required pump power from the assumption of constant specific volume equal to the inlet state value.

9.124 Write a program to solve Problem 9.32 with the final bottle pressure as an input variable. Print out the temperature right after charging and the temperature, pressure and heat transfer after state 3 is reached.

9.125 Consider a small air compressor taking atmospheric air in and compressing it to 1 MPa in a SSSF process. For a maximum flowrate of 0.1 kg/s, discuss the necessary sizes for the piping and the motor to drive the unit.

9.126 Small gasoline engine or electric motor-driven air compressors are used to supply compressed air to power tools, machine shops, etc. The compressor charges air into a tank that acts as a storage buffer. Find examples of these and discuss their sizes in terms of tank volume, charging pressure, engine, or motor power. Also find the time it will take to charge the system from start up and its continuous supply capacity.

9.127 A coflowing heat exchanger receives air at 800 K, 15 MPa and water at 15°C, 100 kPa. The two flows exchange energy as they flow alongside each other to the exit, where the air should be cooled to 350 K. Investigate the range of water flows necessary per kilogram per second air flow and the possible water exit temperatures with the restriction that the minimum temperature difference between the water and air should be 25°C. Include an estimation for the overall entropy generation in the process per kilogram of air flow.

9.128 Consider a geothermal supply of hot water available as saturated liquid at $P_1 = 1.5$ MPa. The liquid is to be flashed (throttled) to some lower pressure, P_2. The saturated liquid and saturated vapor at this pressure are separated, and the vapor is expanded through a reversible adiabatic turbine to the exhaust pressure, $P_3 = 10$ kPa. Study the turbine power output per unit initial mass, m_1 as a function of the pressure, P_2.

9.129 A reversible adiabatic compressor receives air at the state of the surroundings, 20°C, 100 kPa. It should compress the air to a pressure of 1.2 MPa in two stages with a constant pressure intercooler between the two stages. Investigate the work input as a function of the pressure between the two stages assuming the intercooler brings the air down to 50°C.

9.130 (Adv.) Investigate the optimal pressure, P_2, for a constant pressure intercooler between two stages in a compressor. Assume the compression process in each stage follows a polytropic process and that the intercooler brings the substance to the original inlet temperature, T_1. Show that the minimal work for the combined stages arises when

$$P_2 = (P_1 P_3)^{1/2}$$

where P_3 is the final exit pressure.

9.131 (Adv.) Reexamine the previous problem when the intercooler cools the substance to a temperature, $T_2 > T_1$, due to finite heat transfer rates. What is the effect of having isentropic efficiencies for the compressor stages less than 100% on the total work and selection of P_2?

9.132 Investigate the sizes of turbochargers and superchargers available for automobiles. Look at their boost pressures and check if they also have intercoolers mounted. Analyze an example with respect to the power input and the air it can deliver to the engine and estimate its isentropic efficiency if enough data are found.

9.133 (Adv.) Write a program to solve the following problem. An ideal gas with constant specific heat enters a nozzle at a known inlet state P_1, T_1 with velocity $\mathbf{V}_1$ and area A_1, and expands in a reversible adiabatic process. We wish to map out the nozzle contour in terms of the cross-sectional area A as a function of the local pressure P, where P is any value less than the inlet pressure. This problem should be studied over a range of conditions for given values of the constants R and k.

IRREVERSIBILITY AND AVAILABILITY 10

We now turn our attention to irreversibility and availability, two additional concepts that have found increasing use in recent years. These concepts are particularly applicable in the analysis of complex thermodynamic systems, for with the aid of a digital computer, irreversibility and availability are very powerful tools in design and optimization studies of such systems.

10.1 AVAILABLE ENERGY, REVERSIBLE WORK, AND IRREVERSIBILITY

In the previous chapter, we introduced the concept of the efficiency of a device, such as a turbine, nozzle, or compressor (perhaps more correctly termed a first-law efficiency, since it is given as the ratio of two energy terms). We proceed now to develop concepts that include more meaningful second-law analysis. Our ultimate goal is to use this analysis to manage our natural resources and environment better.

We first focus our attention on the potential for producing useful work from some source or supply of energy. Consider the simple situation shown in Fig. 10.1a, in which there is an energy source Q in the form of heat transfer from a very large and, therefore, constant-temperature reservoir at temperature T. What is the ultimate potential for producing work?

To answer this question, we imagine that a cyclic heat engine is available, as shown in Fig. 10.1b. To convert the maximum fraction of Q to work requires that the engine be completely reversible, that is, a Carnot cycle, and that the lower-temperature reservoir be at the lowest temperature possible, often, but not necessarily, at the ambient temperature. From the first and second laws for the Carnot cycle and the usual consideration of all the Q's as positive quantities, we find

$$W_{\text{rev H.E.}} = Q - Q_0$$

$$\frac{Q}{T} = \frac{Q_0}{T_0}$$

so that

$$W_{\text{rev H.E.}} = Q\left(1 - \frac{T_0}{T}\right) \tag{10.1}$$

We might say that the fraction of Q given by the right side of Eq. 10.1 is the available portion of the total energy quantity Q. To carry this thought one step farther, consider

311

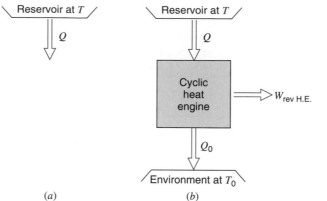

(a)   (b)

FIGURE 10.1 Constant-temperature energy source.

the situation shown on the T–S diagram in Fig. 10.2. The total shaded area is Q. The portion of Q that is below T_0, the environment temperature, cannot be converted into work by the heat engine and must instead be thrown away. This portion is therefore the unavailable portion of total energy Q, and the portion lying between the two temperatures T and T_0 is the available energy.

Let us next consider the same situation, except that the heat transfer Q is available from a constant-pressure source, for example, a simple heat exchanger as shown in Fig. 10.3a. The Carnot cycle must now be replaced by a sequence of such engines, with the result shown in Fig. 10.3b. The only difference between the first and second examples is that the second includes an integral, which corresponds to ΔS.

$$\Delta S = \int \frac{\delta Q_{rev}}{T} = \frac{Q_0}{T_0} \tag{10.2}$$

Substituting into the first law, we have

$$W_{rev\,H.E.} = Q - T_0\,\Delta S \tag{10.3}$$

Note that this ΔS quantity does not include the standard sign convention. It corresponds to the amount of change of entropy shown in Fig. 10.3b. Equation 10.2 specifies the available portion of the quantity Q. The portion unavailable for producing work in this circumstance lies below T_0 in Fig. 10.3b.

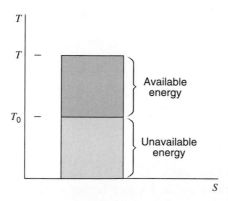

FIGURE 10.2 T–S diagram for constant-temperature energy source.

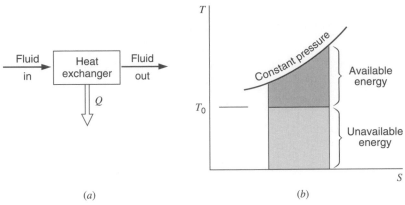

FIGURE 10.3 Changing-temperature energy source.

The Control Mass Process

Consider the real process shown in Fig. 10.4, in wihch a control mass receives an amount of heat $_1Q_2$ from a reservoir at temperature T_H, undergoes a change of state from 1–2, and does an amount of work $_1W_2$. The first law is, assuming no changes in kinetic or potential energies,

$$_1Q_2 = (U_2 - U_1) + {_1W_2} \tag{10.4}$$

This real process is irreversible, such that

$$\Delta S_{\text{net/real}} = \Delta S_{\text{c.m.}} + \Delta S_{\text{surr}}$$

$$= (S_2 - S_1) - \frac{_1Q_2}{T_H} > 0 \tag{10.5}$$

We wish to establish a quantitative measure in energy terms of the extent or degree to which any particular real process is irreversible. This can be accomplished by comparison with a control mass undergoing the same change of state, with the same amount of heat transferred from the reservoir, with everything being reversible. That is, the control mass change of state is a reversible change from U_1, S_1, to U_2, S_2, and heat transfer $_1Q_2$ leaves the reservoir at T_H. If the entire process is reversible, then the net entropy change must equal zero. Noting that all the terms in Eq. 10.5 are the same as for the real irreversible process, we conclude that there must be an additional negative term in the second law for the reversible case. This can only be a heat transfer from the surroundings, divided by the temperature at which it leaves the surroundings. In order to make this heat transfer as small as possible, it should come from the lowest possible temperature, usually the ambi-

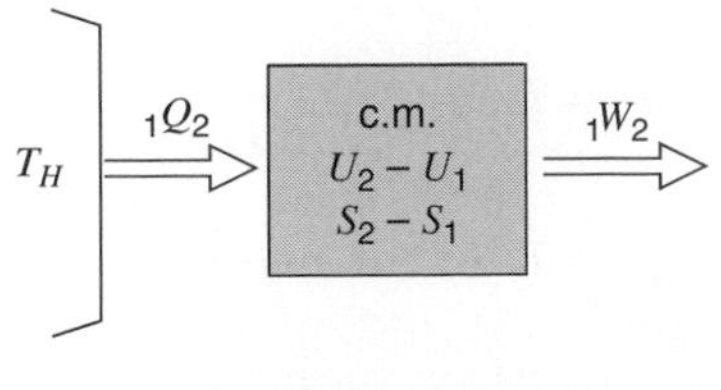

T_H $_1Q_2$ c.m. $U_2 - U_1$ $S_2 - S_1$ $_1W_2$

FIGURE 10.4 A real irreversible process

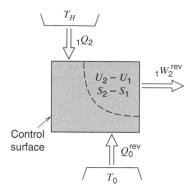

FIGURE 10.5 An ideal reversible process

ent temperature T_0. This is the situation shown in Fig. 10.5, such that the second law for the reversible process is

$$\Delta S_{\text{net/rev}} = \left(S_2 - S_1\right) - \frac{{}_1Q_2}{T_H} - \frac{Q_0^{\text{rev}}}{T_0} = 0 \qquad (10.6)$$

Since any reversible heat transfer must occur over only an infinitesimal temperature difference, we recognize that both heat transfers shown in Fig. 105 must be transferred through reversible heat engines or heat pumps. These are located inside the system boundary, which includes the original control mass plus any necessary heat engines and pumps. Only the net work and the two heat transfers cross this extended boundary. Equation 10.6 can be rewritten as

$$Q_0^{\text{rev}} = T_0\left(S_2 - S_1\right) - {}_1Q_2 \frac{T_0}{T_H} \qquad (10.7)$$

The first law for the reversible process of Fig. 10.5 is

$${}_1W_2^{\text{rev}} = {}_1Q_2 + Q_0^{\text{rev}} - \left(U_2 - U_1\right) \qquad (10.8)$$

and substituting Eq. 10.7,

$${}_1W_2^{\text{rev}} = T_0\left(S_2 - S_1\right) - \left(U_2 - U_1\right) + {}_1Q_2\left(1 - \frac{T_0}{T_H}\right) \qquad (10.9)$$

This expression establishes the theoretical upper limit for the work that could be produced by a control mass undergoing the change of state 1–2 in which heat ${}_1Q_2$ is transferred from a reservoir at T_H, all occurring in the environment T_0. The difference between this quantity and the work actually done in the real process of Eqs. 10.4 and 10.5 is a measure of the extent of the irreversibility I of the real process, or

$${}_1I_2 = {}_1W_2^{\text{rev}} - {}_1W_2 \qquad (10.10)$$

The irreversibility ${}_1I_2$ of the real process can also be expressed in another form, by substituting Eqs. 10.4 and 10.9 into 10.10, which results in

$$\begin{aligned} {}_1I_2 &= T_0\left(S_2 - S_1\right) - \frac{T_0}{T_H} {}_1Q_2 \\ &= T_0\left[\left(S_2 - S_1\right) - \frac{{}_1Q_2}{T_H}\right] \\ &= T_0\left[\Delta S_{\text{net/real}}\right] \qquad (10.11) \end{aligned}$$

and we note that the irreversibility of a real process is another way of expressing the second law for that process, in energy units instead of entropy units.

EXAMPLE 10.1 Consider the reversible process shown in Fig. 10.5. Analyze the individual control mass, heat pump, and heat engine processes occurring inside the extended system boundary of Fig. 10.5, and show that the net reversible work is that given by Eq. 10.9.

Figure: Fig. 10.6

Analysis and solution

For the overall process,

$$_1W_2^{\text{rev}} = {}_1W_{2\,\text{c.m.}}^{\text{rev}} + W_{\text{H.E.}} - W_{\text{H.P. in}} \tag{a}$$

The first law for the control mass reversible process is

$$_1W_{2\,\text{c.m.}}^{\text{rev}} = {}_1Q_2^{\text{rev}} - (U_2 - U_1) \tag{b}$$

and for the reversible heat pump,

$$W_{\text{H.P. in}} = {}_1Q_2^{\text{rev}} - Q_0 \tag{c}$$

The second law for the control mass and heat pump is

$$\Delta S_{\text{net}} = \Delta S_{\text{c.m.}} + \Delta S_{\text{surr}}$$

$$= (S_2 - S_1) - \frac{Q_0}{T_0} = 0 \tag{d}$$

For the reversible heat engine receiving $_1Q_2$ from reservoir T_H, from Eq. 10.1,

$$W_{\text{H.E.}} = {}_1Q_2 - Q_0' = {}_1Q_2\left(1 - \frac{T_0}{T_H}\right) \tag{e}$$

Substituting Eqs. (b), (c), (d), and (e) into (a) results in

$$_1W_2^{\text{rev}} = T_0(S_2 - S_1) - (U_2 - U_1) + {}_1Q_2\left(1 - \frac{T_0}{T_H}\right)$$

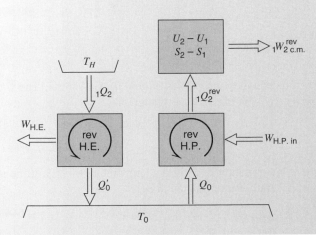

FIGURE 10.6 The reversible process of Example 10.1

which is the same as Eq. 10.9. Note that the reversible Q_0^{rev} of Fig. 10.5 and Eq. 10.7 is the difference between the Q_0 and Q_0' of Fig. 10.6

EXAMPLE 10.2

An insulated rigid tank is divided into two parts A and B by a diaphragm. Each part has a volume of 1 m³. Initially, part A contains water at room temperature, 20°C, with a quality of 50%, while part B is evacuated. The diaphragm then ruptures and the water fills the total volume. Determine the reversible work for this change of state, and the irreversibility of the process.

Control mass: Water.

Initial state: T_1, x_1 known; state fixed.

Final state: V_2 known.

Process: Adiabatic, no change in kinetic or potential energy.

Model: Steam tables.

Analysis:

There is a boundary movement for the water, but since it occurs against no resistance, there is no work done. Therefore, the first law reduces to

$$m(u_2 - u_1) = 0$$

From Eq. 10.9 with no change in internal energy and no heat transfer,

$$_1W_2^{\text{rev}} = T_0(S_2 - S_1) = T_0 m(s_2 - s_1)$$

From Eq. 10.10,

$$_1I_2 = {_1W_2^{\text{rev}}} - {_1W_2} = {_1W_2^{\text{rev}}}$$

Solution

From the steam tables at state 1,

$$u_1 = 1243.5 \quad v_1 = 28.895 \quad s_1 = 4.4819$$

Therefore,

$$v_2 = V_2/m = 2 \times v_1 = 57.79 \quad u_2 = u_1 = 1243.5$$

These two independent properties, v_2 and u_2, fix state 2. The final temperature T_2 must be found by trial and error in the steam tables.

For $\quad T_2 = 5°C$ and $v_2 \Rightarrow x = 0.3928, u = 948.5 \text{ kJ/kg}$

For $\quad T_2 = 10°C$ and $v_2 \Rightarrow x = 0.5433, u = 1317 \text{ kJ/kg}$

so the final interpolation in u gives a temperature of 9°C. If the menu-driven software is used the final state is interpolated to be

$$T_2 = 9.1°C \quad x_2 = 0.513 \quad s_2 = 4.644$$

with the given u and v. Since the actual work is zero we have

$$_1I_2 = {_1}W_2^{\text{rev}} = T_0 m (s_2 - s_1)$$
$$= 293.2 (1 / 28.895)(4.644 - 4.4819) = 1.645 \text{ kJ}$$

For processes in which kinetic and potential energy changes are significant, the development of the expressions for work, reversible work, and irreversibility are all the same, substituting $E = U + KE + PE$ for U in any equation involving energy.

The Steady-State Steady-Flow Process

The preceding development of the expressions for reversible work, Eq. 10.9, and irreversibility, Eq. 10.10 or 10.11, were made for a control mass undergoing a given change of state. We can follow an analogous procedure for a control volume undergoing a SSSF process, one stream in and one stream out, such that the first and second laws can both be written per unit mass of flow through the control volume, inlet state i to exit state e, with heat transfer per unit mass q with reservoir T_H, all in the environment T_0. Neglecting kinetic and potential energy terms, this results in the expression for reversible work per unit mass

$$w^{\text{rev}} = T_0 (s_e - s_i) - (h_e - h_i) + q \left(1 - \frac{T_0}{T_H} \right) \tag{10.12}$$

with the irreversibility of the real process for the same states i and e and the same q with reservoir T_H and environment T_0 being

$$i = w^{\text{rev}} - w \tag{10.13}$$

The irreversibility of the real process can also be expressed as

$$i = T_0 (s_e - s_i) - \frac{T_0}{T_H} q$$
$$= T_0 \left[(s_e - s_i) - \frac{q}{T_H} \right]$$
$$= T_0 \left[\frac{1}{\dot{m}} \frac{dS_{\text{net/real}}}{dt} \right] \tag{10.14}$$

EXAMPLE 10.3   Consider an air compressor that receives ambient air at 100 kPa, 25°C. It compresses the air to a pressure of 1 MPa, where it exits at a temperature of 540 K. Since the air and compressor housing are hotter than the ambient, it loses 50 kJ per kilogram air flowing through the compressor. Find the reversible work, reversible heat transfer, and irreversibility in the process.

Control volume: The air compressor.

Inlet state: P_i, T_i known; state fixed.

Exit state: P_e, T_e known; state fixed.

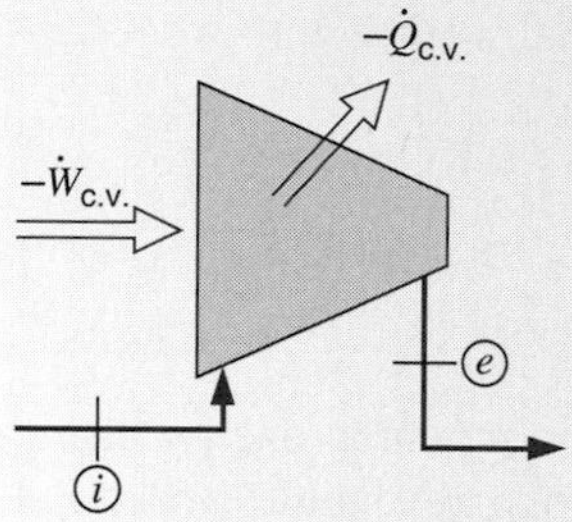

FIGURE 10.7 Illustration for Example 10.3.

Process: Nonadiabatic compression with no change in kinetic or potential energy.

Model: Ideal gas.

Analysis:

This SSSF process has a single inlet and exit flow so all quantities are done on a mass basis as specific quantities. From the ideal gas air tables:

$$h_i = 298.62 \text{ kJ} / \text{kg} \qquad s^0_{T_i} = 6.8629 \text{ kJ} / \text{kg K}$$
$$h_e = 544.69 \text{ kJ} / \text{kg} \qquad s^0_{T_e} = 7.4664 \text{ kJ} / \text{kg K}$$

so the energy equation for the actual compressor gives the work as

$$q = -50 \text{ kJ} / \text{kg}$$
$$w = h_i - h_e + q = 298.62 - 544.69 - 50 = -296.07 \text{ kJ} / \text{kg}$$

The reversible work for the given change of state is, from Eq. 10.12,

$$w^{\text{rev}} = T_0\left(s_e - s_i\right) - \left(h_e - h_i\right) + q\left(1 - \frac{T_0}{T_H}\right)$$
$$= 298.2\left(7.4664 - 6.8629 - 0.287 \ln 10\right) - \left(544.69 - 298.62\right) + 0$$
$$= -17.10 - 246.07 = -263.17 \text{ kJ} / \text{kg}$$

From Eq. 10.13,

$$i = w^{\text{rev}} - w$$
$$= -263.17 - \left(-296.07\right) = 32.9 \text{ kJ} / \text{kg}$$

EXAMPLE 10.3E Consider an air compressor that receives ambient air at 14.7 lbf/in.^{2}, 80 F. It compresses the air to a pressure of 150 lbf/in.^{2}, where it exits at a temperature of 960 R. Since the air and compressor housing are hotter than the ambient, it loses 22 Btu/lbm air flowing through the compressor. Find the reversible work, reversible heat transfer, and irreversibility in the process.

Control volume: The air compressor.

Inlet state: P_i, T_i known; state fixed.

Exit state: P_e, T_e known; state fixed.

Process: Nonadiabatic compression with no change in kinetic or potential energy.

Model: Ideal gas.

Analysis:

This SSSF process has a single inlet and exit flow so all quantities are done on a mass basis as specific quantities. From the ideal gas air tables

$$h_i = 129.18 \text{ Btu / lbm} \qquad s^0_{T_i} = 1.6405 \text{ Btu / lbm R}$$
$$h_e = 231.20 \text{ Btu / lbm} \qquad s^0_{T_e} = 1.7803 \text{ Btu / lbm R}$$

so the energy equation for the actual compressor gives the work as

$$q = -22 \text{ Btu / lbm}$$
$$w = h_i + h_e + q = 129.18 - 231.20 - 22 = -124.02 \text{ Btu / lbm}$$

The reversible work for the given change of state is, from Eq. 10.12,

$$w^{\text{rev}} = T_0\left(s_e - s_i\right) - \left(h_e - h_i\right) + q\left(1 - \frac{T_0}{T_H}\right)$$
$$= 539.7\left(1.7803 - 1.6405 - 0.06855 \ln 10.2\right) - \left(231.20 - 129.18\right)$$
$$= -10.47 - 192.02 = -112.49 \text{ Btu / lbm}$$

From Eq. 10.13,

$$i = w^{\text{rev}} - w$$
$$= -112.49 - \left(-124.02\right) = 11.53 \text{ Btu / lbm}$$

EXAMPLE 10.4   A feedwater heater has 5 kg/s water at 5 MPa, 40°C flowing through it, being heated from two sources as shown in Fig. 10.8. One source adds 900 kW from a 100°C reservoir and the other source adds heat transfer from a 200°C reservoir such that the water exit condition is 5 MPa, 180°C. Find the reversible work and the irreversibility.

Control volume: Feedwater heater extending out to the two reservoirs.

Inlet state: P_i, T_i known; state fixed.

Exit state: P_e, T_e known; state fixed.

Process: Constant pressure heat addition with no change in kinetic or potential energy.

Model: Steam tables.

Analysis:

This control volume has a single inlet and exit flow with two heat transfer rates coming from reservoirs different from the ambient. There is no actual work or actual heat transfer with the surroundings at 25°C. For the actual feedwater heater, the energy equation becomes

$$h_i + q_1 + q_2 = h_e$$

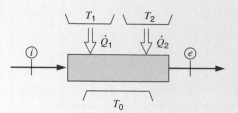

FIGURE 10.8 The feedwater heater for Example 10.4.

The reversible work for the given change of state is, from Eq. 10.12, with heat transfer q_1 from reservoir T_1 and heat transfer q_2 from reservoir T_2,

$$w^{rev} = T_0 \left(s_e - s_i \right) - \left(h_e - h_i \right) + q_1 \left(1 - \frac{T_0}{T_1} \right) + q_2 \left(1 - \frac{T_0}{T_2} \right)$$

From Eq. 10.13, since the actual work is zero,

$$i = w^{rev} - w = w^{rev}$$

Solution

From the steam tables the inlet and exit state properties are

$$h_i = 171.97 \qquad s_i = 0.5705$$

$$h_e = 765.25 \qquad s_e = 2.1341$$

The second heat transfer is found from the energy equation as

$$q_2 = h_e - h_i - q_1 = 765.25 - 171.97 - 900 / 5 = 413.28 \text{ kJ} / \text{kg}$$

The reversible work is

$$w^{rev} = T_0 \left(s_e - s_i \right) - \left(h_e - h_i \right) + q_1 \left(1 - \frac{T_0}{T_1} \right) + q_2 \left(1 - \frac{T_0}{T_2} \right)$$

$$= 298.2 \left(2.1341 - 0.5705 \right) - \left(765.25 - 171.97 \right)$$

$$+ 180 \left(1 - \frac{298.2}{373.2} \right) + 413.28 \left(1 - \frac{298.2}{473.2} \right)$$

$$= 466.27 - 593.28 + 36.17 + 152.84 = 62.0 \text{ kJ} / \text{kg}$$

The irreversibility is

$$i = w^{rev} = 62.0 \text{ kJ} / \text{kg}$$

The expression for reversible work for the SSSF process, Eq. 10.12, was derived without including kinetic and potential energy terms. Whenever necessary, especially in nozzles and diffusers where the kinetic energy change is the reason for building the device, these terms can be included along with the enthalpy terms of the fluid stream in and out of the control volume. Another way of including these terms would be to say that the enthalpies in Eq. 10.12 are the total enthalpies, as used in Eq. 6.8. There are also SSSF processes involving more than one fluid stream entering or exiting the control volume. In such cases, it is necessary to rewrite Eq. 10.12 on a rate basis, including the mass flow rates of the different streams involved in the process.

The Uniform-State Uniform-Flow Process

The uniform-state uniform-flow process has a control volume change from state 1 to state 2 with possible mass flow in at state i and/or flow out at state e. The procedure for developing an expression for reversible work for this process is analogous to the original example followed for the control mass. In this case, assuming no kinetic or potential energy terms are included, the equation for reversible work will contain control volume entropy and energy terms of the same form as in Eq. 10.9 (but recognizing that the masses at states 1 and 2 are different), and will also contain entropy and enthalpy flow terms the same as in Eq. 10.12 (each one including the appropriate mass flow). The result is

$$W_{\text{c.v.}}^{\text{rev}} = T_0\left(m_2 s_2 - m_1 s_1\right) - \left(m_2 u_2 - m_1 u_1\right)$$
$$+ T_0\left(m_e s_e - m_i s_i\right) - \left(m_e h_e - m_i h_i\right)$$
$$+ Q_{\text{c.v.}}\left(1 - \frac{T_0}{T_H}\right) \tag{10.15}$$

This expression can also be grouped as

$$W_{\text{c.v.}}^{\text{rev}} = m_i\left(h_i - T_0 s_i\right) - m_e\left(h_e - T_0 s_e\right)$$
$$+ m_1\left(u_1 - T_0 s_1\right) - m_2\left(u_2 - T_0 s_2\right) + Q_{\text{c.v.}}\left(1 - \frac{T_0}{T_H}\right) \tag{10.16}$$

As in the previous developments, h_i and h_e can be replaced by $h_{\text{TOT } i}$ and $h_{\text{TOT } e}$, and u_1 and u_2 can be replaced by e_1 and e_2, whenever kinetic and potential energies are significant. Also, summations can be added to the flow terms in cases where there is more than one flow stream in or out of the control volume.

The irreversibility for a USUF process is found from the general definition,

$$I_{\text{c.v.}} = W_{\text{c.v.}}^{\text{rev}} - W_{\text{c.v.}} \tag{10.17}$$

which, by substitution of Eq. 10.15 and also the first law, can also be expressed as

$$I_{\text{c.v.}} = T_0\left[\left(m_2 s_2 - m_1 s_1\right) + m_e s_e - m_i s_i - \frac{Q_{\text{c.v.}}}{T_H}\right]$$
$$= T_0\left[\Delta S_{\text{c.v.}} + \Delta S_{\text{surr}}\right] = T_0\left[\Delta S_{\text{net/real}}\right] \tag{10.18}$$

EXAMPLE 10.5 A 1-m³ rigid tank contains ammonia at 200 kPa and the ambient temperature 20°C. The tank is connected with a valve to a line flowing saturated liquid ammonia at −10°C. The valve is opened and the tank is charged quickly until the flow stops and the valve is closed. As the process happens very quickly there is no heat transfer. Determine the final mass in the tank and the irreversibility in the process.

Control volume: The tank and the valve.

Initial state: T_1, P_1 known; state fixed.

Inlet state: T_i, x_i known; state fixed.

Final state: $P_2 = P_{\text{line}}$ known.

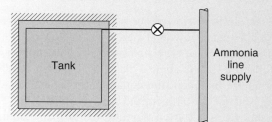

FIGURE 10.9 Ammonia tank and line for Example 10.5.

Process: Adiabatic, no kinetic or potential energy change.

Model: Ammonia tables.

Analysis:

Since the line pressure is higher than the initial pressure inside the tank flow is going into the tank and the flow stops when the tank pressure has increased to the line pressure. The continuity, energy, and entropy equations are

$$m_2 - m_1 = m_i$$

$$m_2 u_2 - m_1 u_1 = m_i h_i = (m_2 - m_1) h_i$$

$$m_2 s_2 - m_1 s_1 = m_i s_i +_1 S_{2,\text{gen}}$$

where kinetic and potential energies are zero for the initial and final states and neglected for the inlet flow.

Solution

From the ammonia tables the initial and line state properties are

$$v_1 = 0.6995 \text{ m}^3 \qquad u_1 = h_1 - P_1 v_1 = 1369.5 \text{ kJ} / \text{kg} \qquad s_1 = 5.927 \text{ kJ} / \text{kg K}$$

$$h_i = 134.41 \text{ kJ} / \text{kg} \qquad s_i = 0.5408 \text{ kJ} / \text{kg K}$$

The initial mass is therefore

$$m_1 = V / v_1 = 1 / 0.6995 = 1.4296 \text{ kg}$$

It is observed that only the final pressure is known so one property is needed. The unknowns are the final mass and final internal energy in the energy equation. Since only one property is unknown, the two quantities are not independent. From the energy equation we have

$$m_2 (u_2 - h_i) = m_1 (u_1 - h_i)$$

from which it is seen that $u_2 > h_i$ and the state therefore is two-phase or superheated vapor. Assume that the state is two phase, then

$$m_2 = V / v_2 = 1 / (0.001534 + x_2 \times 0.41684)$$

$$u_2 = 133.964 + x_2 \times 1175.257$$

so the energy equation is

$$\frac{133.964 + x_2 \times 1175.257 - 134.41}{0.001534 + x_2 \times 0.041684} = 1.4296(1369.5 - 134.41) = 1765.67 \text{ kJ}$$

This equation is solved for the quality and the rest of the properties to give

$$x_2 = 0.007182 \qquad v_2 = 0.0045276 \text{ m}^3/\text{kg} \qquad s_2 = 0.5762 \text{ kJ/kg}$$

Now the final mass and the irreversibility are found

$$m_2 = V/v_2 = 1/0.0045276 = 220.87 \text{ kg}$$

$$\Delta S_{\text{net}} = m_2 s_2 - m_1 s_1 - m_i s_i = 127.265 - 8.473 - 118.673 = 0.119 \text{ kJ/K}$$

$$I_{\text{c.v.}} = T_0 \ \Delta S_{\text{net}} = 293.15 \times 0.119 = 34.885 \text{ kJ}$$

10.2 AVAILABILITY AND SECOND-LAW EFFICIENCY

What is the maximum reversible work that can be done by a given mass in a given state? In the previous section, we developed expressions for the reversible work for a given change of state for a control mass and control volume undergoing specific types of processes. For any given case, what final state will give the maximum reversible work?

The answer to this question is that, for any type of process, when the mass comes into equilibrium with the environment, no spontaneous change of state will occur and the mass will be incapable of doing any work. Therefore, if a mass in a given state undergoes a completely reversible process until it reaches a state in which it is in equilibrium with the environment, the maximum reversible work will have been done by the mass. In this sense, we refer to the availability at the original state in terms of the potential for achieving the maximum possible work by the mass.

If a control mass is in equilibrium with the surroundings, it must certainly be in pressure and temperature equilibrium with the surroundings, that is, at pressure P_0 and temperature T_0. It must also be in chemical equilibrium with the surroundings, which implies that no further chemical reaction will take place. Equilibrium with the surroundings also requires that the system have zero velocity and minimum potential energy. Similar requirements could be set forth regarding electrical and surface effects if these are relevant to a given problem.

The same general remarks can be made about a quantity of mass that undergoes a steady-state steady-flow process. With a given state for the mass entering the control volume, the reversible work will be a maximum when this mass leaves the control volume in equilibrium with the surroundings. This means that as the mass leaves the control volume, it must be at the pressure and temperature of the surroundings, in chemical equilibrium with the surroundings, and have minimum potential energy and zero velocity. (The mass leaving the control volume must of necessity have some velocity but it can be made to approach zero.)

Let us first consider the availability associated with a steady-state steady-flow process. For a control volume with a single-flow stream, the reversible work is given by

Eq. 10.12. Including kinetic and potential energies, this expression is rewritten in the form,

$$w^{\text{rev}} = \left(h_{\text{TOT}\,i} - T_0 s_i\right) - \left(h_{\text{TOT}\,e} - T_0 s_e\right) + q\left(1 - \frac{T_0}{T_H}\right)$$

From the discussion of the heat engine that led to Eq. 10.1, it is clear that the last term in the expression for the reversible work is the contribution to the net reversible work from the heat transfers. These can be viewed as transfer of availability associated with q, which gives a potential to do work as in a heat engine. Such contributions are separate from the availability in the flow itself. The SSSF flow reversible work will be maximum, relative to the surroundings, when the mass leaving the control volume is in equilibrium with the surroundings. The state in which the fluid is in equilibrium with the surroundings is designated with subscript 0, and the reversible work will be maximum when $h_e = h_0$, $s_e = s_0$, $V_e = 0$, and $Z_e = Z_0$. This maximum reversible work per unit mass flow without the additional heat transfers is the flow availability or exergy and assigned the symbol ψ

$$\psi = \left(h - T_0 s + \frac{1}{2}V^2 + gZ\right) - \left(h_0 - T_0 s_0 + gZ_0\right) \tag{10.19}$$

It is written without subscript for the inlet state to indicate that this is the flow availability associated with a substance in any state as it enters the control volume in a steady-state steady-flow process. The reversible work is therefore seen to be equal to the decrease in flow availability plus the reversible work that can be extracted from heat engines operating with the heat transfer at T_H and the ambient.

Irreversibility is as usual defined as the difference between the reversible work and the actual work. If we write this as a general expression on a rate basis, including summations to account for the possibility of more than one flow stream and also more than one heat transfer, the result is

$$\dot{I}_{\text{c.v.}} = \left(\sum \dot{m}_i \psi_i - \sum \dot{m}_e \psi_e\right) + \sum \left(1 - \frac{T_0}{T_j}\right)\dot{Q}_{\text{c.v.},j} - \dot{W}_{\text{c.v.}} \tag{10.20}$$

In this form the irreversibility is equal to the decrease in the availability of the mass flows plus the decrease of availability of each heat transfer rate j at reservoir T_j minus the increase in availability of the surroundings that receive the actual work. The rate of irreversibility is thus seen to be the rate of destruction of availability, which is also directly proportional to the net rate of entropy increase, as noted in Eq. 10.14.

For a control mass, a similar consideration of the maximum reversible work will lead to a nonflow availability concept. In this case the volume may change and some work is exchanged with the ambient, which is not available as useful work. The reversible work for a control mass is given by Eq. 10.9. Including kinetic and potential energies (e instead of u), this expression is rewritten as

$$_1 w_2^{\text{rev}} = \left(e_1 - T_0 s_1\right) - \left(e_2 - T_0 s_2\right) + {_1}q_2\left(1 - \frac{T_0}{T_H}\right)$$

which is the maximum between the two given states. This work is available if the final state is in equilibrium with the surroundings, for which we must have $e_2 = e_0 = u_0 + gZ_0$, the kinetic energy being zero, and $s_2 = s_0$. The work done against the surroundings, w_{surr}, is

$$w_{\text{surr}} = P_0\left(v_0 - v_1\right) = -P_0\left(v_1 - v_0\right)$$

such that the maximum available work is

$$w_{\text{avail}}^{\text{max}} = w_{\text{max}}^{\text{rev}} - w_{\text{surr}}$$

$$= (e - T_0 s) - (e_0 - T_0 s_0) + P_0 (v - v_0) + {}_1 q_2 \left(1 - \frac{T_0}{T_H}\right) \qquad (10.21)$$

The subscript is again dropped to indicate that this is the maximum available work at a given state having also $_1 q_2$ available from a source at T_H. The nonflow availability is defined as the maximum available work from a state without the heat transfers included as

$$\phi = (e - T_0 s) - (e_o - T_o s_o) + P_o (v - v_0)$$

$$\phi = (e + P_0 v - T_0 s) - (e_0 + P_0 v_0 - T_0 s_0) \qquad (10.22)$$

Sometimes the definition excludes the kinetic and potential energies, in which case u should be used instead of e.

The irreversibility may again be related to the changes in availability through the difference between the reversible work and the actual work. The reversible work from above is expressed with the availability from which the actual work is subtracted to give

$$_1 I_2 = m(\phi_1 - \phi_2) + \sum \left(1 - \frac{T_0}{T_j}\right) Q_j - \left({}_1 W_2^{\text{ac}} - P_0 (V_2 - V_1)\right) \qquad (10.23)$$

including the possibility of having heat transfer with more than one reservoir. The irreversibility is then equal to the decrease in availability of the control mass plus the decrease in availability of the heat transfers at reservoirs T_j minus the increase in availability of the surroundings that received the actual work. It is noted again that the irreversibility expresses the net destruction of availability of the control mass and surroundings, which is proportional to the net entropy increase, as given in Eq. 10.11.

The use of availability and irreversibility in an actual thermodynamic problem is shown in Fig. 10.10. A theoretical analysis was made of a reciprocating internal-combustion automotive engine to see what happened to the availability of the air–fuel mixture that entered the engine and where the irreversibility occurred during the process. The abscissa on Fig. 10.10 is the crank angle, the left side representing bottom dead center when the cylinder is assumed to be filled with an air–fuel mixture having the availability indicated on the ordinate. As the compression process takes place, this mixture becomes more available through the work done in compressing the mixture. When the piston passes top dead center, the expansion process takes place. The beginning and end of combustion are also indicated on the diagram. During the combustion and expansion process irreversibilities occur. Those associated with the combustion process itself and those associated with heat transfer to the cooling water or surroundings are both indicated. The work done during the expansion process and the availability at the end of the expansion are indicated on the right ordinate. The availability that remains in the cylinder at the end of the expansion stroke is exhausted to the atmosphere.

The less the irreversibility associated with a given change of state, the greater the amount of work that will be done (or the smaller the amount of work that will be required). This relation is significant in at least two regards. The first is that availability is one of our natural resources. This availability is found in such forms as oil reserves, coal reserves, and uranium reserves. Suppose we wish to accomplish a given objective that requires a certain amount of work. If this work is produced reversibly while drawing on one of the availability reserves, the decrease in availability is exactly equal to the reversible work. However, since there are irreversibilities in producing this required amount of

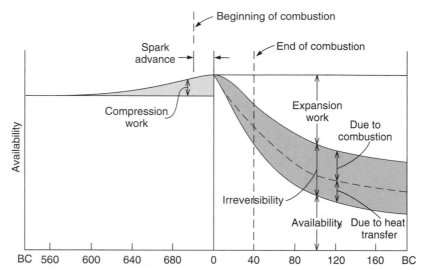

FIGURE 10.10 Availability versus crank angle of the charge in a spark-ignited internal combustion engine. (From D. J. Patterson and G. J. Van Wylen, "A Digitial Computer Simulation for Spark Ignited Engine Cycles," *SAE Progress in Technology Series,* 7, p. 88. Published by SAE Inc., New York, 1964.)

work, the actual work will be less than the reversible work, and the decrease in availability will be greater (by the amount of the irreversibility) than if this work had been produced reversibly. Thus the more irreversibilities we have in all our processes, the greater will be the decrease in our availability reserves.[1] The conservation and effective use of these availability reserves is an important responsibility for all of us.

The second reason that it is desirable to accomplish a given objective with the smallest irreversibility is an economic one. Work costs money, and in many cases a given objective can be accomplished at less cost when the irreversibility is less. It should be noted, however, that many factors enter into the total cost of accomplishing a given objective, and an optimization process that considers many factors is often necessary to arrive at the most economical design. For example, in a heat-transfer process, the smaller the temperature difference across which the heat is transferred, the less the irreversibility. However, for a given rate of heat transfer, a smaller temperature difference will require a larger (and therefore more expensive) heat exchanger. These various factors must all be considered in developing the optimum and most economical design.

In many engineering decisions other factors, such as the impact on the environment (for example, air pollution and water pollution) and the impact on society must be considered in developing the optimum design.

Along with the increased use of availability analysis in recent years, a term called the "second-law efficiency" has come into more common use. This term refers to comparison of the desired output of a process with the cost, or input, in terms of the thermodynamic availability. Thus, the isentropic turbine efficiency defined by Eq. 7.72 as the actual work output divided by the work for a hypothetical isentropic expansion from the

---

[1]In many popular talks reference is made to our energy reserves. From a thermodynamic point of view, availability reserves would be a much more acceptable term. There is much energy in the atmosphere and the ocean, but relatively little availability.

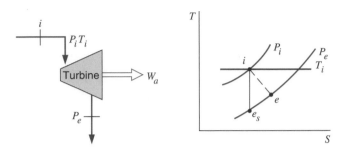

FIGURE 10.11 Irreversible turbine.

same inlet state to the same exit pressure might well be called a "first-law efficiency," in that it is a comparison of two energy quantities. The "second-law efficiency," as just described, would be the actual work output of the turbine divided by the decrease in availability from the same inlet state to the same exit state. For the turbine shown in Fig. 10.11, the second-law efficiency is

$$n_{\text{2nd law}} = \frac{w_a}{\psi_i - \psi_e} \tag{10.24}$$

In this sense, this concept provides a rating or measure of the real process in terms of the actual change of state, and is simply another convenient way of utilizing the concept of thermodynamic availability. In a similar manner, the second-law efficiency of a pump or compressor is the ratio of the increase in availability to the work input to the device.

EXAMPLE 10.6   An insulated steam turbine, Fig. 10.12, receives 30 kg of steam per second at 3 MPa, 350°C. At the point in the turbine where the pressure is 0.5 MPa, steam is bled off for processing equipment at the rate of 5 kg/s. The temperature of this steam is 200°C. The balance of the steam leaves the turbine at 15 kPa, 90% quality. Determine the availability per kilogram of the steam entering and at both points at which steam leaves the turbine, the isentropic efficiency and the second-law efficiency for this process.

Control volume: Turbine.

Inlet state: P_1, T_1 known; state fixed.

Exit state: P_2, T_2 known; P_3, x_3 known; both states fixed.

Process: SSSF.

Model: Steam tables.

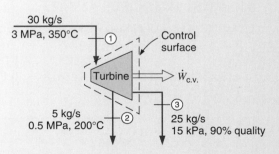

FIGURE 10.12 Sketch for Example 10.6.

Analysis:

The availability at any point for the steam entering or leaving the turbine is given by Eq. 10.19,

$$\psi = (h - h_0) - T_0(s - s_0) + \frac{\mathbf{V}^2}{2} + g(Z - Z_0)$$

Since there are no changes in kinetic and potential energy in this problem, this equation reduces to

$$\psi = (h - h_0) - T_0(s - s_0)$$

For the ideal isentropic turbine,

$$\dot{W}_s = \dot{m}_1 h_1 - \dot{m}_2 h_{2s} - \dot{m}_3 h_{3s}$$

For the actual turbine,

$$\dot{W} = \dot{m}_1 h_1 - \dot{m}_2 h_2 - \dot{m}_3 h_3$$

Solution

At the pressure and temperature of the surroundings, 0.1 MPa, 25°C, the water is a slightly compressed liquid, and the properties of the water are essentially equal to those for saturated liquid at 25°C.

$$h_0 = 104.9 \text{ kJ / kg} \qquad s_0 = 0.3674 \text{ kJ / kg K}$$

From Eq. 10.19

$$\psi_1 = (3115.3 - 104.9) - 298.15(6.7428 - 0.3674) = 1109.6 \text{ kJ / kg}$$

$$\psi_2 = (2855.4 - 104.9) - 298.15(7.0592 - 0.3674) = 755.3 \text{ kJ / kg}$$

$$\psi_3 = (2361.8 - 104.9) - 298.15(7.2831 - 0.3674) = 195.0 \text{ kJ / kg}$$

$$\dot{m}_1 \psi_1 - \dot{m}_2 \psi_2 - \dot{m}_3 \psi_3 = 30(1109.6) - 5(755.3) - 25(195.0) = 24\ 637 \text{ kW}$$

For the ideal isentropic turbine,

$$s_{2s} = 6.7428 = 1.8606 + x_{2s} \times 4.9606, \qquad x_{2s} = 0.9842$$
$$h_{2s} = 640.2 + 0.9842 \times 2108.5 = 2715.4$$
$$s_{3s} = 6.7428 = 0.7549 + x_{3s} \times 7.2536, \qquad x_{3s} = 0.8255$$
$$h_{3s} = 225.9 + 0.8255 \times 2373.1 = 2184.9$$
$$\dot{W}_s = 30(3115.3) - 5(2715.4) - 25(2184.9) = 25\ 260 \text{ kW}$$

For the actual turbine,

$$\dot{W} = 30(3115.3) - 5(2855.4) - 25(2361.8) = 20\ 137 \text{ kW}$$

The isentropic efficiency is

$$\eta_s = \frac{20\ 137}{25\ 260} = 0.797$$

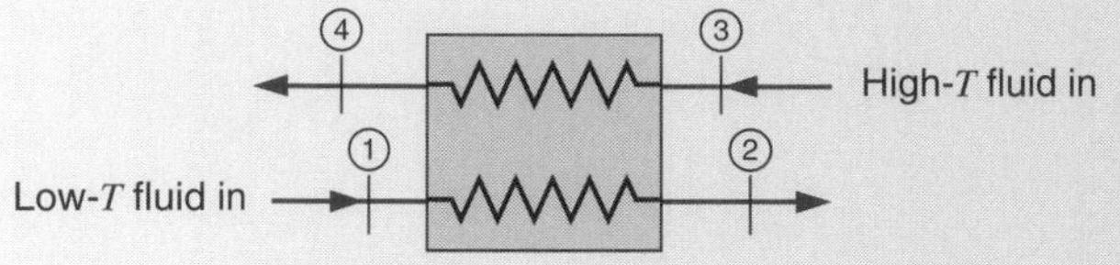

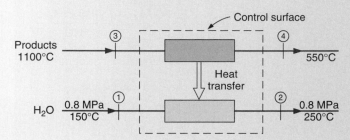

FIGURE 10.13 A two-fluid heat exchanger.

and the second-law efficiency is

$$\eta_{2\text{nd law}} = \frac{20\ 137}{24\ 637} = 0.817$$

For a device that does not involve the production or the input of work, the definition of second-law efficiency refers to the accomplishment of the goal of the process relative to the process input, in terms of availability changes or transfers. For example, in a heat exchanger, where energy is transferred from a high-temperature fluid stream to a low-temperature fluid stream, as shown in Fig. 10.13, in which case the second-law efficiency is defined as

$$\eta_{2\text{nd law}} = \frac{\dot{m}_1(\psi_2 - \psi_1)}{\dot{m}_3(\psi_3 - \psi_4)} \tag{10.25}$$

EXAMPLE 10.7   In a boiler, heat is transferred from the products of combustion to the steam. The temperature of the products of combustion decreases from 1100°C to 550°C while the pressure remains constant at 0.1 MPa. The average constant-pressure specific heat of the products of combustion is 1.09 kJ/kg K. The water enters at 0.8 MPa, 150°C and leaves at 0.8 MPa, 250°C. Determine the second-law efficiency for this process and the irreversibility per kilogram of water evaporated.

Control volume: Overall heat exchanger.

Sketch: Fig. 10.14.

Inlet states: Both known, given in Fig. 10.14.

Exit states: Both known, given in Fig. 10.14.

Process: Overall, adiabatic.

Diagram: Fig. 10.15.

Model: Products—ideal gas, constant specific heat. Water—steam tables.

FIGURE 10.14 Sketch for Example 10.7.

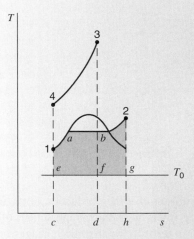

FIGURE 10.15 Temperature-entropy diagram for Example 10.7.

Analysis:

For the products, the entropy change for this constant-pressure process is

$$\left(s_e - s_i\right)_{\text{prod}} = C_{po} \ln \frac{T_e}{T_i}$$

For this control volume we can write the following governing equations:
Continuity equation:

$$\left(\dot{m}_i\right)_{H_2O} = \left(\dot{m}_e\right)_{H_2O} \qquad \text{(a)}$$

$$\left(\dot{m}_i\right)_{\text{prod}} = \left(\dot{m}_e\right)_{\text{prod}} \qquad \text{(b)}$$

First law (a steady-state steady-flow process):

$$\left(\dot{m}_i h_i\right)_{H_2O} + \left(\dot{m}_i h_i\right)_{\text{prod}} = \left(\dot{m}_e h_e\right)_{H_2O} + \left(\dot{m}_e h_e\right)_{\text{prod}} \qquad \text{(c)}$$

Second law (the process is adiabatic for the control volume shown):

$$\left(\dot{m}_e s_e\right)_{H_2O} + \left(\dot{m}_e s_e\right)_{\text{prod}} \geq \left(\dot{m}_i s_i\right)_{H_2O} + \left(\dot{m}_i s_i\right)_{\text{prod}}$$

Solution

From Eqs. a, b, and c, we can calculate the ratio of the mass flow of products to the mass flow of water.

$$\dot{m}_{\text{prod}}\left(h_i - h_e\right)_{\text{prod}} = \dot{m}_{H_2O}\left(h_e - h_i\right)_{H_2O}$$

$$\frac{\dot{m}_{\text{prod}}}{\dot{m}_{H_2O}} = \frac{\left(h_e - h_i\right)_{H_2O}}{\left(h_i - h_e\right)_{\text{prod}}} = \frac{2950 - 632.2}{1.09(1100 - 550)} = 3.866$$

The increase in availability of the water is, per kilogram of water,

$$\psi_2 - \psi_1 = \left(h_2 - h_1\right) - T_0\left(s_2 - s_1\right)$$

$$= (2950 - 632.2) - 298.15(7.0384 - 1.8418)$$

$$= 768.4 \text{ kJ / kg } H_2O$$

The decrease in availability of the products, per kilogram of water, is

$$\frac{\dot{m}_{prod}}{\dot{m}_{H_2O}}\left(\psi_3 - \psi_4\right) = \frac{\dot{m}_{prod}}{\dot{m}_{H_2O}}\left[\left(h_3 - h_4\right) - T_0\left(s_3 - s_4\right)\right]$$

$$= 3.866\left[1.09(1100 - 550) - 298.15\left(1.09\ln\frac{1373.15}{823.15}\right)\right]$$

$$= 1674.7 \text{ kJ / kg } H_2O$$

Therefore, the second-law efficiency is, from Eq. 10.25,

$$\eta_{2nd\ law} = \frac{768.4}{1674.7} = 0.459$$

From Eq. 10.20, the process irreversibility per kilogram of water is

$$\frac{\dot{I}}{\dot{m}_{H_2O}} = \sum_i \frac{\dot{m}_i}{\dot{m}_{H_2O}}\psi_i - \sum_2 \frac{\dot{m}_e}{\dot{m}_{H_2O}}\psi_e$$

$$= \left(\psi_1 - \psi_2\right) + \frac{\dot{m}_{prod}}{\dot{m}_{H_2O}}\left(\psi_3 - \psi_4\right)$$

$$= \left(-768.4 + 1674.7\right) = 906.3 \text{ kJ / kg } H_2O$$

It is also of interest to determine the net change of entropy. The change in the entropy of the water is

$$\left(s_2 - s_1\right)_{H_2O} = 7.0384 - 1.8418 = 5.1966 \text{ kJ / kg } H_2O \text{ K}$$

The change in the entropy of the products is

$$\frac{\dot{m}_{prod}}{\dot{m}_{H_2O}}\left(s_4 - s_3\right)_{prod} = -3.866\left(1.09\ln\frac{1373.15}{823.15}\right) = -2.1564 \text{ kJ / kg } H_2O \text{ K}$$

Thus there is a net increase in entropy during the process. The irreversibility could also have been calculated from Eq. 10.14:

$$\dot{I} = \sum \dot{m}_e T_0 s_e - \sum \dot{m}_i T_0 s_i - \dot{Q}_{c.v.}$$

For the control volume selected, $\dot{Q}_{c.v.} = 0$, and therefore

$$\frac{\dot{I}}{\dot{m}_{H_2O}} = T_0\left(s_2 - s_1\right)_{H_2O} + \frac{\dot{m}_{prod}}{\dot{m}_{H_2O}} T_0\left(s_4 - s_3\right)_{prod}$$

$$= 298.15(5.1966) + 298.15(-2.1564)$$

$$= 906.3 \text{ kJ / kg } H_2O$$

These two processes are shown on the T–s diagram of Fig. 10.15. Line 3–4 represents the process for the 3.866 kg of products. Area 3–4–c–d–3 represents the heat transferred from the 3.866 kg of products of combustion, and area 3–4–e–f–3 represents the decrease in availability of these products. Area 1–a–b–2–h–c–1 represents the heat transferred to the water, and this is equal to area 3–4–c–d–3, which represents the heat transferred from the products of combustion. Area 1–a–b–2–g–e–1 represents the increase in availability of the water. The difference between area 3–4–e–f–3 and area 1–a–b–2–g–e–1 represents the net decrease in availability. It is readily shown that this net change is equal to area f–g–h–d–f, or $T_0(\Delta s)_{\text{net}}$. Since the actual work is zero, this area also represents the irreversibility, which agrees with our calculation above.

It is essential to note that when the change of state involving heat transfer takes place reversibly, the net change in entropy is zero, and therefore the decrease in the entropy of the body from which heat is transferred must be equal to the increase in entropy of the body to which heat is transferred. This is best demonstrated by considering an example similar to Example 10.6.

EXAMPLE 10.8

Consider the process in Example 10.7, but assume that the heat transfer could be made to be entirely reversible, that is, that the heat transfer takes place through a reversible engine.

Sketch: Fig. 10.16.

Analysis:

Schematically this would involve heat transfer from the products of combustion to reversible engines that reject heat to the water, as shown in Fig. 10.16.
We again write the governing equations:

Continuity equation:

$$\left(\dot{m}_i\right)_{\text{H}_2\text{O}} = \left(\dot{m}_e\right)_{\text{H}_2\text{O}}$$

$$\left(\dot{m}_i\right)_{\text{prod}} = \left(\dot{m}_e\right)_{\text{prod}}$$

First law:

$$\left(\dot{m}_i h_i\right)_{\text{H}_2\text{O}} + \left(\dot{m}_i h_i\right)_{\text{prod}} = \left(\dot{m}_e h_e\right)_{\text{H}_2\text{O}} + \left(\dot{m}_e h_e\right)_{\text{prod}} + \dot{W}_{\text{c.v.}}$$

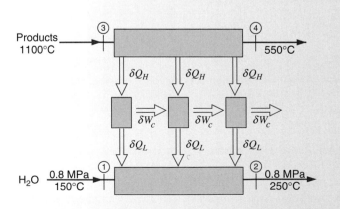

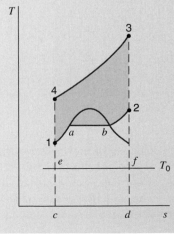

FIGURE 10.16
Diagram for
Example 10.8.

Second law: Since this is a reversible adiabatic process,

$$\left(\dot{m}_e s_e\right)_{H_2O} + \left(\dot{m}_e s_e\right)_{prod} = \left(\dot{m}_i s_i\right)_{H_2O} + \left(\dot{m}_i s_i\right)_{prod}$$

Solution

From the continuity equation and second law we can determine flow of products per unit flow of water.

$$\left(s_2 - s_1\right)_{H_2O} + \frac{\dot{m}_{prod}}{\dot{m}_{H_2O}}\left(s_4 - s_3\right)_{prod} = 0$$

$$\frac{\dot{m}_{prod}}{\dot{m}_{H_2O}}\left(s_3 - s_4\right)_{prod} = \left(7.0384 - 1.8418\right)$$

$$\frac{\dot{m}_{prod}}{\dot{m}_{H_2O}}\left(1.09 \ln \frac{1373.15}{823.15}\right) = 5.1966$$

$$\frac{\dot{m}_{prod}}{\dot{m}_{H_2O}} = 9.317$$

We now calculate the decrease in availability of the water and the products:

$$\left(\psi_1 - \psi_2\right) = \left(h_1 - h_2\right) - T_0\left(s_1 - s_2\right)$$

$$\frac{\dot{m}_{prod}}{\dot{m}_{H_2O}}\left(\psi_3 - \psi_4\right) = \left[\left(h_3 - h_4\right) - T_0\left(s_3 - s_4\right)\right]\frac{\dot{m}_{prod}}{\dot{m}_{H_2O}}$$

But, since

$$\left(s_2 - s_1\right)_{H_2O} = \frac{\dot{m}_{prod}}{\dot{m}_{H_2O}}\left(s_3 - s_4\right)_{prod}$$

it follows that

$$\left(\psi_1 - \psi_2\right) + \frac{\dot{m}_{prod}}{\dot{m}_{H_2O}}\left(\psi_3 - \psi_4\right) = \left(h_1 - h_2\right) + \frac{\dot{m}_{prod}}{\dot{m}_{H_2O}}\left(h_3 - h_4\right)$$

We note that this net decrease in availability is exactly equal to the work that would be determined from the first law. We would expect this, of course, since the process is completely reversible.

$$w = w^{rev} = \left(632.2 - 2950\right) + 9.317 \times 1.09\left(1100 - 550\right)$$
$$= 3267.7 \text{ kJ / kg } H_2O$$

Since the net change in entropy is zero, the T–s diagram is as shown in Fig. 10.16. Area 3–4–c–d–3 represents the heat transferred from the products to the engines, and area 1–a–b–2–d–c–1 represents the heat received by the water. Area 3–4–1–a–b–2–3 represents the work done by the heat engines, which is equal to the reversible work for this process.

PROBLEMS

10.1 Calculate the reversible work and irreversibility for the process described in Problem 5.18, assuming that the heat transfer is with the surroundings at 20°C.

10.2 Calculate the reversible work and irreversibility for the process described in Problem 5.65, assuming that the heat transfer is with the surroundings at 20°C.

10.3 The compressor in a refrigerator takes refrigerant R-134a in at 100 kPa, −20°C and compresses it to 1 MPa, 40°C. With the room at 20°C find the minimum compressor work.

10.4 Calculate the reversible work out of the two-stage turbine shown in Problem 6.41, assuming the ambient is at 25°C. Compare this to the actual work which was found to be 18.08 MW.

10.5 A household refrigerator has a freezer at T_F and a cold space at T_C from which energy is removed and rejected to the ambient at T_A as shown in Fig. P10.5. Assuming that the rate of heat transfer from the cold space, $\dot{Q}_C$, is the same as from the freezer, $\dot{Q}_F$, find an expression for the minimum power into the heat pump. Evaluate this power when $T_A = 20°C$, $T_C = 5°C$, $T_F = -10°C$, and $\dot{Q}_F = 3$ kW.

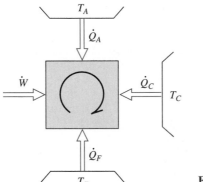

FIGURE P10.5

10.6 An air compressor takes air in at the state of the surroundings, 100 kPa, 300 K. The air exits at 400 kPa, 200°C at the rate of 2 kg/s. Determine the minimum compressor work input.

10.7 A supply of steam at 100 kPa, 150°C is needed in a hospital for cleaning purposes at a rate of 15 kg/s. A supply of steam at 150 kPa, 250°C is available from a boiler, and tap water at 100 kPa, 15°C is also available. The two sources are then mixed in a SSSF mixing chamber to generate the desired state as output. Determine the rate of irreversibility of the mixing process.

10.8 Two flows of air both at 200 kPa of equal flow rates mix in an insulated mixing chamber. One flow is at 1500 K, and the other is at 300 K. Find the irreversibility in the process per kilogram of air flowing out.

10.9 A steam turbine receives steam at 6 MPa, 800°C. It has a heat loss of 49.7 kJ/kg and an isentropic efficiency of 90%. For an exit pressure of 15 kPa and surroundings at 20°C, find the actual work and the reversible work between the inlet and the exit.

10.10 A 2-kg piece of iron is heated from room temperature 25°C to 400°C by a heat source at 600°C. What is the irreversibility in the process?

10.11 A 2-kg/s flow of steam at 1 MPa, 700°C should be brought to 500°C by spraying in liquid water at 1 MPa, 20°C in a SSSF setup. Find the rate of irreversibility, assuming that surroundings are at 20°C.

10.12 Fresh water can be produced from saltwater by evaporation and subsequent condensation. An example is shown in Fig. P10.12 where 150-kg/s saltwater, state 1, comes from the condenser in a large power plant. The water is throttled to the saturated pressure in the flash evaporator, and the vapor, state 2, is then condensed by cooling with sea water. As the evaporation takes place below atmospheric pressure, pumps must bring the liquid water flows back up to P_0. Assume that the saltwater has the same properties as pure water, the ambient is at 20°C, and there are no external heat transfers. With the states as shown in the table below find the irreversibility in the throttling valve and in the condenser.

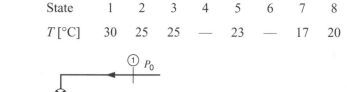

State	1	2	3	4	5	6	7	8
T [°C]	30	25	25	—	23	—	17	20

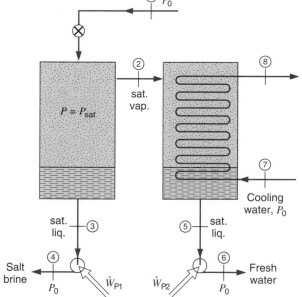

FIGURE P10.12

10.13 An air compressor receives atmospheric air at $T_0 = 17$°C, 100 kPa, and compresses it up to 1400 kPa. The compressor has an isentropic efficiency of 88%, and it loses energy by heat transfer to the atmosphere as 10% of the isentropic work. Find the actual exit temperature and the reversible work.

10.14 A piston/cylinder has forces on the piston so it keeps constant pressure. It contains 2 kg of ammonia at 1 MPa, 40°C and is now heated to 100°C by a reversible heat engine that receives heat from a 200°C source. Find the work out of the heat engine.

10.15 Air flows through a constant pressure heating device, shown in Fig. P10.15. It is heated up in a reversible process with a work input of 200 kJ/kg air flowing. The device exchanges heat with the ambient at 300 K. The air enters at 300 K, 400 kPa. Assuming constant specific heat, develop an expression for the exit temperature and solve for it.

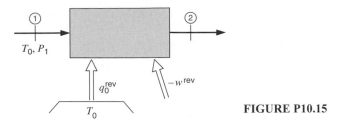

FIGURE P10.15

10.16 Air enters the turbocharger compressor (see Fig. P10.16), of an automotive engine at 100 kPa, 30°C, and exits at 170 kPa. The air is cooled by 50°C in an intercooler before entering the engine. The isentropic efficiency of the compressor is 75%. Determine the temperature of the air entering the engine and the irreversibility of the compression-cooling process.

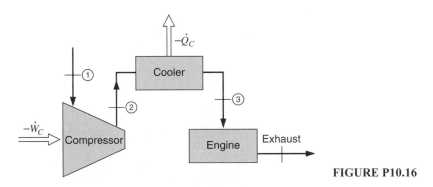

FIGURE P10.16

10.17 A car air-conditioning unit has a 0.5-kg aluminum storage cylinder that is sealed with a valve, and it contains 2 L of refrigerant R-134a at 500 kPa and both are at room temperature 20°C. It is now installed in a car sitting outside where the whole system cools down to ambient temperature at −10°C. What is the irreversibility of this process?

10.18 A steady combustion of natural gas yields 0.15 kg/s of products (having approximately the same properties as air) at 1100°C, 100 kPa. The products are passed through a heat exchanger and exit at 550°C. What is the maximum theoretical power output from a cyclic heat engine operating on the heat rejected from the combustion products, assuming that the ambient temperature is 20°C?

10.19 A counterflowing heat exchanger cools air at 600 K, 400 kPa to 320 K using a supply of water at 20°C, 200 kPa. The water flow rate is 0.1 kg/s, and the air flow rate is 1 kg/s. Assume this can be done in a reversible process by the use of heat engines and neglect kinetic energy changes. Find the water exit temperature and the power out of the heat engine(s).

10.20 Water as saturated liquid at 200 kPa goes through a constant pressure heat exchanger as shown in Fig. P10.20. The heat input is supplied from a reversible heat pump extracting heat from the surroundings at 17°C. The water flow rate is 2 kg/min and the whole process is reversible, that is, there is no overall net entropy change. If the heat pump receives 40 kW of work, find the water exit state and the increase in availability of the water.

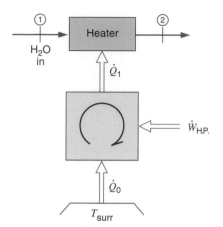

FIGURE P10.20

10.21 Calculate the irreversibility for the process described in Problem 6.63, assuming that heat transfer is with the surroundings at 17°C.

10.22 The high-temperature heat source for a cyclic heat engine is a SSSF heat exchanger where R-134a enters at 80°C, saturated vapor, and exits at 80°C, saturated liquid at a flow rate of 5 kg/s. Heat is rejected from the heat engine to a SSSF heat exchanger where air enters at 150 kPa and ambient temperature 20°C, and exits at 125 kPa, 70°C. The rate of irreversibility for the overall process is 175 kW. Calculate the mass flow rate of the air and the thermal efficiency of the heat engine.

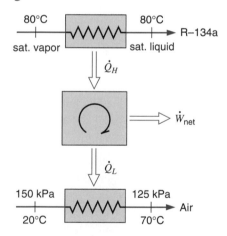

FIGURE P10.22

10.23 A control mass gives out 10 kJ of energy in the form of

   a. Electrical work from a battery   c. Heat transfer at 500°C

   b. Mechanical work from a spring

   Find the change in availability of the control mass for each of the three cases.

10.24 Calculate the availability of the water at the initial and final states of Problem 8.32, and the irreversibility of the process.

10.25 A steady stream of R-22 at ambient temperature, 10°C, and at 750 kPa enters a solar collector. The stream exits at 80°C, 700 kPa. Calculate the change in availability of the R-22 between these two states.

10.26 Nitrogen flows in a pipe with velocity 300 m/s at 500 kPa, 300°C. What is its availability with respect to an ambient at 100 kPa, 20°C?

10.27 A 10-kg iron disk brake on a car is initially at 10°C. Suddenly the brake pad hangs up, increasing the brake temperature by friction to 110°C while the car maintains constant speed. Find the change in availability of the disk and the energy depletion of the car's gas tank due to this process alone. Assume that the engine has a thermal efficiency of 35%.

10.28 A 1-kg block of copper at 350°C is quenched in a 10-kg oil bath initially at ambient temperature of 20°C. Calculate the final uniform temperature (no heat transfer to/from ambient) and the change of availability of the system (copper and oil).

10.29 Calculate the availability of the system (aluminum plus gas) at the initial and final states of Problem 8.74, and also the process irreversibility.

10.30 Consider the springtime melting of ice in the mountains, which gives cold water running in a river at 2°C while the air temperature is 20°C. What is the availability of the water (SSSF) relative to the temperature of the ambient?

10.31 Refrigerant R-12 at 30°C, 0.75 MPa enters a SSSF device and exits at 30°C, 100 kPa. Assume the process is isothermal and reversible. Find the change in availability of the refrigerant.

10.32 A geothermal source provides 10 kg/s of hot water at 500 kPa, 150°C flowing into a flash evaporator that separates vapor and liquid at 200 kPa. Find the three fluxes of availability (inlet and two outlets) and the irreversibility rate.

10.33 Air flows at 1500 K, 100 kPa through a constant pressure heat exchanger giving energy to a heat engine and comes out at 500 K. At what constant temperature should the same heat transfer be delivered to provide the same availability?

10.34 A wooden bucket (2 kg) with 10 kg hot liquid water, both at 85°C, is lowered 400 m down into a mineshaft. What is the availability of the bucket and water with respect to the surface ambient at 20°C?

10.35 A rigid container with volume 200 L is divided into two equal volumes by a partition. Both sides contain nitrogen, one side is at 2 MPa, 300°C, and the other at 1 MPa, 50°C. The partition ruptures, and the nitrogen comes to a uniform state at 100°C. Assuming the surroundings are at 25°C, find the actual heat transfer and the irreversibility in the process.

10.36 An air compressor is used to charge an initially empty 200-L tank with air up to 5 MPa. The air inlet to the compressor is at 100 kPa, 17°C, and the compressor isentropic efficiency is 80%. Find the total compressor work and the change in availability of the air.

10.37 Air enters a compressor at ambient conditions, 100 kPa, 300 K, and exits at 800 kPa. If the isentropic compressor efficiency is 85%, what is the second-law efficiency of the compressor process?

10.38 Steam enters a turbine at 25 MPa, 550°C and exits at 5 MPa, 325°C at a flow rate of 70 kg/s. Determine the total power output of the turbine, its isentropic efficiency, and the second-law efficiency.

10.39 A compressor is used to bring saturated water vapor at 1 MPa up to 17.5 MPa, where the actual exit temperature is 650°C. Find the irreversibility and the second-law efficiency.

10.40 A flow of steam at 10 MPa, 550°C goes through a two-stage turbine. The pressure between the stages is 2 MPa, and the second stage has an exit at 50 kPa. Assume both stages have an isentropic efficiency of 85%. Find the second-law efficiencies for both stages of the turbine.

10.41 Consider the two-stage turbine in the previous problem as a single turbine from inlet to final actual exit and find its second-law efficiency.

10.42 The simple steam power plant shown in Problem 6.39 has a turbine with given inlet and exit states. Find the availability at the turbine exit, state 6. Find the second-law efficiency for the turbine, neglecting kinetic energy at state 5.

10.43 A compressor takes in saturated vapor R-134a at −20°C and delivers it at 30°C, 0.4 MPa. Assuming that the compression is adiabatic, find the isentropic efficiency and the second-law efficiency.

10.44 Steam is supplied in a line at 3 MPa, 700°C. A turbine with an isentropic efficiency of 85% is connected to the line by a valve, and it exhausts to the atmosphere at 100 kPa. If the steam is throttled down to 2 MPa before entering the turbine, find the **a**ctual turbine specific work. Find the change in availability through the valve and the second-law efficiency of the turbine.

10.45 The condenser in a refrigerator receives R-134a at 700 kPa, 50°C, and it exits as saturated liquid at 25°C. The flowrate is 0.1 kg/s, and the condenser has air flowing in at ambient 15°C and leaving at 35°C. Find the minimum flow rate of air and the heat exchanger second-law efficiency.

10.46 A piston/cylinder arrangement has a load on the piston so it maintains constant pressure. It contains 1 kg of steam at 500 kPa, 50% quality. Heat from a reservoir at 700°C brings the steam to 600°C. Find the second-law efficiency for this process. Note that no formula is given for this particular case, so determine a reasonable expression for it.

10.47 Air flows into a heat engine at ambient conditions 100 kPa, 300 K, as shown in Fig. P10.47. Energy is supplied as 1200 kJ/kg air from a 1500-K source, and in some part of the process a heat transfer loss of 300 kJ/kg air happens at 750 K. The air leaves the engine at 100 kPa, 800 K. Find the first- and the second-law efficiencies.

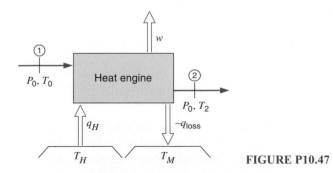

FIGURE P10.47

10.48 Consider the high-pressure closed feedwater heater in the nuclear power plant described in Problem 6.42. Determine its second-law efficiency.

10.49 Consider a gasoline engine for a car as a SSSF device where air and fuel enter at the surrounding conditions 25°C, 100 kPa and leave the engine exhaust manifold at 1000 K, 100 kPa as products assumed to be air. The engine cooling system removes 750 kJ/kg air through the engine to the ambient. For the analysis, take the fuel as air where the extra energy of 2200 kJ/kg of air released in the combustion process is added as heat transfer from a 1800-K reservoir. Find the work out of the engine, the irreversibility per kilogram of air, and the first- and second-law efficiencies.

10.50 Air enters a steady-flow turbine at 1600 K and exhausts to the atmosphere at 1000 K. The second-law efficiency is 85%. What is the turbine inlet pressure?

10.51 Consider the two turbines in Problem 9.72. What is the second-law efficiency for the combined system?

10.52 Air in a piston/cylinder arrangement is at 110 kPa, 25°C, with a volume of 50 L. It goes through a reversible polytropic process to a final state of 700 kPa, 500 K, and exchanges heat with the ambient at 25°C through a reversible device. Find the total work (including the external device) and the heat transfer from the ambient.

ADVANCED PROBLEMS

10.53 Refrigerant-22 is flowing in a pipeline at 10°C, 600 kPa, with a velocity of 200 m/s, at a steady flowrate of 0.1 kg/s. It is desired to decelerate the fluid and increase its pressure by installing a diffuser in the line (a diffuser is basically the opposite of a nozzle in this respect). The R-22 exits the diffuser at 30°C, with a velocity of 100 m/s. It may be assumed that the diffuser process is SSSF, polytropic, and internally reversible. Determine the diffuser exit pressure and the rate of irreversibility for the process.

10.54 A piston/cylinder contains ammonia at −20°C, quality 80%, and a volume of 10 L. A force is now applied to the piston so it compresses the ammonia in an adiabatic process to a volume of 5 L, where the piston is locked. Now heat transfer with the ambient takes place so the ammonia reaches the temperature of the ambient at 20°C. Find the work and heat transfer. If it is done in a reversible process, how much work and heat transfer would be involved?

10.55 Consider the irreversible process in Problem 8.34. Assume that the process could be done reversibly by adding heat engines/pumps between tanks A and B and the cylinder. The total system is insulated, so there is no heat transfer to or from the ambient. Find the final state, the work given out to the piston, and the total work to or from the heat engines/pumps.

10.56 Water in a piston/cylinder is at 100 kPa, 34°C, shown in Fig. P10.56. The cylinder has stops mounted so $V_{min} = 0.01$ m^3 and $V_{max} = 0.5$ m^3. The piston is loaded with a mass and outside P_0, so a pressure inside of 5 MPa will float it. Heat of 15 000 kJ from a 400°C source is added. Find the total change in availability of the water and the total irreversibility.

10.57 A rock bed consists of 6000 kg granite and is at 70°C. A small house with lumped mass of 12 000 kg wood and 1000 kg iron is at 15°C. They are now brought to a uniform final temperature with no external heat transfer.

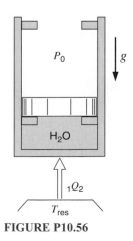

FIGURE P10.56

a. For a reversible process, find the final temperature and the work done in the process.

b. If they are connected thermally by circulating water between the rock bed and the house, find the final temperature and the irreversibility of the process, assuming that surroundings are at 15°C.

10.58 Consider the heat engine in Problem 10.47. The exit temperature was given as 800 K, but what are the theoretical limits for this temperature? Find the lowest and the highest, assuming that the heat transfers are as given. For an exit temperature that is the average of the highest and lowest possible, find the first- and second-law efficiencies for the heat engine.

10.59 Air in a piston/cylinder arrangement, shown in Fig. P10.59, is at 200 kPa, 300 K with a volume of 0.5 m^3. If the piston is at the stops, the volume is 1 m^3 and a pressure of 400 kPa is required. The air is then heated from the initial state to 1500 K by a 1900 K reservoir. Find the total irreversibility in the process, assuming surroundings are at 20°C.

10.60 Consider two rigid containers each of volume 1m^3 containing air at 100 kPa, 400 K. An internally reversible Carnot heat pump is then thermally connected between them so it heats one up and cools the other down. In order to transfer heat at a reasonable rate, the temperature difference between the working substance inside the heat pump and the air in the containers is set to 20°C. The process stops when the air in the coldest tank reaches 300 K. Find the final temperature of the air that is heated up, the work input to the heat pump, and the overall second-law efficiency.

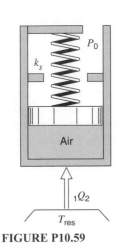

FIGURE P10.59

ENGLISH UNIT PROBLEMS

10.61E Calculate the reversible work and irreversibility for the process described in Problem 5.122, assuming that the heat transfer is with the surroundings at 68 F.

10.62E The compressor in a refrigerator takes refrigerant R-134a in at 15 lbf/in.^{2}, 0 F and compresses it to 125 lbf/in.^{2}, 100 F. With the room at 70 F find the reversible heat transfer and the minimum compressor work.

10.63E A supply of steam at 14.7 lbf/in.^{2}, 320 F is needed in a hospital for cleaning purposes at a rate of 30 lbm/s. A supply of steam at 20 lbf/in.^{2}, 500 F is available from a boiler, and tap water at 14.7 lbf/in.^{2}, 60 F is also available. The two sources are then mixed in a SSSF mixing chamber to generate the desired state as output. Determine the rate of irreversibility of the mixing process.

10.64E A 4-lbm piece of iron is heated from room temperature 77 F to 750 F by a heat source at 1100 F. What is the irreversibility in the process?

10.65E Fresh water can be produced from saltwater by evaporation and subsequent condensation. An example is shown in Fig. P10.12 where 300-lbm/s saltwater, state 1, comes from the condenser in a large power plant. The water is throttled to the saturated pressure in the flash evaporator, and the vapor, state 2, is then condensed by cooling with sea water. As the evaporation takes place below atmospheric pressure, pumps must bring the liquid water flows back up to P_0. Assume that the saltwater has the same properties as pure water, the ambient is at 68 F, and that there are no external heat transfers. With the states as shown in the table below find the irreversibility in the throttling valve and in the condenser.

State	1	2	3	4	5	6	7	8
T [F]	86	77	77	—	74	—	63	68

10.66E Air flows through a constant pressure heating device as shown in Fig. P10.15. It is heated up in a reversible process with a work input of 85 Btu/lbm air flowing. The device exchanges heat with the ambient at 540 R. The air enters at 540 R, 60 lbf/in.^{2}. Assuming constant specific heat, develop an expression for the exit temperature and solve for it.

10.67E Air enters the turbocharger compressor of an automotive engine at 14.7 lbf/in.^{2}, 90 F, and exits at 25 lbf/in.^{2}, as shown in Fig. P10.16. The air is cooled by 90 F in an intercooler before entering the engine. The isentropic efficiency of the compressor is 75%. Determine the temperature of the air entering the engine and the irreversibility of the compression-cooling process.

10.68E Calculate the irreversibility for the process described in Problem 6.101, assuming that the heat transfer is with the surroundings at 61 F.

10.69E A control mass gives out 1000 Btu of energy in the form of
  a. Electrical work from a battery   c. Heat transfer at 700 F
  b. Mechanical work from a spring

Find the change in availability of the control mass for each of the three cases.

10.70E A steady stream of R-22 at ambient temperature, 50 F, and at 110 lbf/in.^{2} enters a solar collector. The stream exits at 180 F, 100 lbf/in.^{2}. Calculate the change in availability of the R-22 between these two states.

10.71E A 20-lbm iron disk brake on a car is at 50 F. Suddenly the brake pad hangs up, increasing the brake temperature by friction to 230 F while the car maintains constant speed. Find the change in availability of the disk and the energy depletion of the car's gas tank due to this process alone. Assume that the engine has a thermal efficiency of 35%.

10.72E Calculate the availability of the system (aluminum plus gas) at the initial and final states of Problem 8.113, and also the irreversibility.

10.73E Consider the springtime melting of ice in the mountains, which gives cold water running in a river at 34 F while the air temperature is 68 F. What is the availability of the water (SSSF) relative to the temperature of the ambient?

10.74E A geothermal source provides 20 lbm/s of hot water at 80 lbf/in.^{2}, 300 F flowing into a flash evaporator that separates vapor and liquid at 30 lbf/in.^{2}. Find the three fluxes of availability (inlet and two outlets) and the irreversibility rate.

10.75E A wood bucket (4 lbm) with 20 lbm hot liquid water, both at 180 F, is lowered 1300 ft down into a mineshaft. What is the availability of the bucket and water with respect to the surface ambient at 70 F?

10.76E An air compressor is used to charge an initially empty 7-ft^3 tank with air up to 750 lbf/in.^{2}. The air inlet to the compressor is at 14.7 lbf/in.^{2}, 60 F, and the compressor isentropic efficiency is 80%. Find the total compressor work and the change in energy of the air.

10.77E A compressor is used to bring saturated water vapor at 150 lbf/in.^{2} up to 2500 lbf/in.^{2}, where the actual exit temperature is 1200 F. Find the irreversibility and the second-law efficiency.

10.78E The simple steam power plant in Problem 6.91, shown in Fig P6.39 has a turbine with given inlet and exit states. Find the availability at the turbine exit, state 6. Find the second-law efficiency for the turbine, neglecting kinetic energy at state 5.

10.79E Steam is supplied in a line at 450 lbf/in.^{2}, 1200 F. A turbine with an isentropic efficiency of 85% is connected to the line by a valve, and it exhausts to the atmosphere at 14.7 lbf/in.^{2}. If the steam is throttled down to 300 lbf/in.^{2} before entering the turbine, find the actual turbine specific work. Find the change in availability through the valve and the second law efficiency of the turbine.

10.80E A piston/cylinder arrangement has a load on the piston so it maintains constant pressure. It contains 1 lbm of steam at 80 lbf/in.^{2}, 50% quality. Heat from a reservoir at 1300 F brings the steam to 1000 F. Find the second-law efficiency for this process. Note that no formula is given for this particular case, so determine a reasonable expression for it.

10.81E Air flows into a heat engine at ambient conditions 14.7 lbf/in.^{2}, 540 R, as shown in Fig. P10.47. Energy is supplied as 540 Btu per lbm air from a 2700 R source, and in some part of the process a heat transfer loss of 135 Btu per lbm air happens at 1350 R. The air leaves the engine at 14.7 lbf/in.^{2}, 1440 R. Find the first- and the second-law efficiencies.

10.82E Consider a gasoline engine for a car as a SSSF device where air and fuel enters at the surrounding conditions 77 F, 14.7 lbf/in.^{2} and leaves the engine exhaust manifold at 1800 R, 14.7 lbf/in.^{2} as products assumed to be air. The engine cooling system removes 320 Btu/lbm air through the engine to the ambient. For the analysis, take the fuel as air where the extra energy of 950 Btu/lbm of air released in the combustion process is added as heat transfer from a 3240 R reservoir. Find the work out of the engine, the irreversibility per pound-mass of air, and the first- and second-law efficiencies.

10.83E Consider the two turbines in Problem 9.119, shown in Fig. P9.72. What is the second-law efficiency of the combined system?

10.84E (Adv.) Refrigerant-22 is flowing in a pipeline at 40 F, 80 lbf/in.^{2}, with a velocity of 650 ft/s, at a steady flowrate of 0.2 lbm/s. It is desired to decelerate the fluid and increase its pressure by installing a diffuser in the line (a diffuser is basically the opposite of a nozzle in this respect). The R-22 exits the diffuser at 80 F, with a velocity of 160 ft/s. It may be assumed that the diffuser process is SSSF, polytropic, and internally reversible. Determine the diffuser exit pressure and the rate of irreversibility for the process.

10.85E (Adv.) Water in a piston/cylinder is at 14.7 lbf/in.^{2}, 90 F, as shown in Fig. P10.56. The cylinder has stops mounted so that $V_{min} = 0.36$ ft^3 and $V_{max} = 18$ ft^3. The piston is loaded with a mass and outside P_0, so a pressure inside of 700 lbf/in.^{2} will float it. Heat of 14 000 Btu from a 750 F source is added. Find the total change in availability of the water and the total irreversibility.

10.86E A rock bed consists of 12 000 lbm granite and is at 160 F. A small house with lumped mass of 24 000 lbm wood and 2000 lbm iron is at 60 F. They are now brought to a uniform final temperature with no external heat transfer.

a. For a reversible process, find the final temperature and the work done in the process.

b. If they are connected thermally by circulating water between the rock bed and the house, find the final temperature and the irreversibility of the process assuming that surroundings are at 60 F.

10.87E Air in a piston/cylinder arrangement, shown in Fig. P10.59, is at 30 lbf/in.², 540 R with a volume of 20 ft³. If the piston is at the stops, the volume is 40 ft³ and a pressure of 60 lbf/in.² is required. The air is then heated from the initial state to 2700 R by a 3400 R reservoir. Find the total irreversibility in the process, assuming surroundings are at 70 F.

COMPUTER, DESIGN, AND OPEN-ENDED PROBLEMS

10.88 Use the menu-driven software to get the properties of water as needed for consideration of the moisture separator in Problem 6.42. Steam comes in at state 3 and leaves as liquid, state 9, with the rest at state 4 going to the low-pressure turbine. Assume no heat transfer to the surroundings at 20°C and find the total entropy generation and irreversibility in the process.

10.89 Use the menu-driven software to get the properties of water as needed and calculate the second-law efficiency of the low-pressure turbine in Problem 6.42.

10.90 Write a program to solve the general case of Problem 10.10. The initial state is to be the program input variable, and the output should also include the change in availability for both the iron and the source.

10.91 Write a program to solve Problem 10.15. Use constant specific heat and let the work input be a program input variable.

10.92 The maximum power a windmill can possibly extract from the wind is

$$\dot{W} = \frac{16}{27}\rho A \mathbf{V}\frac{1}{2}\mathbf{V}^2 = \frac{16}{27}\dot{m}_{air} \times \text{KE}$$

Water flowing through Hoover Dam, see Problem 5.88, produces $\dot{W} = 0.8\dot{m}_{water}$ gh. Burning 1 kg of coal gives 24 000 kJ delivered at 900 K to a heat engine. Find other examples in the literature and from problems in the previous chapters with steam and gases into turbines. Make a list of the availability (exergy) for a flow of 1 kg/s of substance with the above examples. Use a reasonable choice for the values of the parameters and do the necessary analysis.

10.93 Consider the condenser in the simple steam power plant described in Problem 6.39. The cooling water is lake water at 20°C, and it should not be heated more than 5°C as it goes back to the lake. Assume the heat transfer rate inside the condenser is 350 W/m²K so $\dot{Q} = 350 \times A\Delta T$ in watts. Estimate the flow rate of the cooling water and the needed interface area inside the condenser, A. Find the change in the availability of the cooling water and the steam inside the condenser and compare. Discuss your estimates and the size of the pump for the cooling water.

10.94 Consider the nuclear power plant shown in Problem 6.42. Select one feedwater heater and one pump and make an analysis of their performance. Check the energy balances and do the second-law analysis. Determine the change of availability in all the flows and discuss measures of performance for both the pump and the feedwater heater.

10.95 Reconsider the use of the geothermal energy as discussed in Problem 6.46. The analysis that was done and the original problem statement specified the turbine exit state as 10 kPa, 90% quality. Reconsider this problem with an adiabatic turbine having an isentropic efficiency of 85% and an exit pressure of 10 kPa. Include a second-law analysis and discuss the changes in availability. Describe another way of using the geothermal energy and make appropriate calculations.

10.96 Reconsider the dual flash chamber version of the use of geothermal hot water as described in Problem 6.110. With your knowledge of the second law, repeat the problem assuming that the turbine exit pressure is 10 kPa and disregard the listed exit quality. Include a comparison of the change in availability of the geothermal water with the turbine work.

10.97 An air gun should shoot a harpoon of mass 5 kg out so that it has a velocity of 75 m/s as it leaves the gun. The harpoon acts as the piston in a cylinder, and air is trapped below the piston (end of harpoon) that can be initially locked. The air is charged so the initial state is at high pressure and temperature. Determine sizes for the cylinder diameter, cylinder length, air mass, and initial (P,T) of the air. Make reasonable assumptions about the process and include a determination of the state of the air during the process.

10.98 Energy can be stored in many different forms. Thermal energy can be stored as internal energy in a mass like a rock bed, water, metals, etc. Mechanical energy (potential or kinetic) can be stored in springs, rotating flywheels, elevated masses, etc. A tank with a compressed gas that can drive a turbine is used. Batteries are used in cars. Make a list with at least five different ways of storing 1000 MJ of energy and size the systems. Note how the energy is taken out and find the availability for each case. Discuss the various alternatives.

10.99 Find from the literature the amount of energy that must be stored in a car to start the engine. Size three different systems to provide that energy and compare those to an ordinary car battery. Discuss the feasibility and cost.

11 POWER AND REFRIGERATION SYSTEMS

Some power plants, such as the simple steam power plant, which we have considered several times, operate in a cycle. That is, the working fluid undergoes a series of processes and finally returns to the initial state. In other power plants, such as the internal-combustion engine and the gas turbine, the working fluid does not go through a thermodynamic cycle, even though the engine itself may operate in a mechanical cycle. In this instance the working fluid has a different composition or is in a different state at the conclusion of the process than it had or was at the beginning. Such equipment is sometimes said to operate on the open cycle (the word cycle is really a misnomer), whereas the steam power plant operates on a closed cycle. The same distinction between open and closed cycles can be made regarding refrigeration devices. For both the open- and closed-cycle apparatus, however, it is advantageous to analyze the performance of an idealized closed cycle similar to the actual cycle. Such a procedure is particularly advantageous for determining the influence of certain variables on performance. For example, the spark-ignition internal-combustion engine is usually approximated by the Otto cycle. From an analysis of the Otto cycle we conclude that increasing the compression ratio increases the efficiency. This is also true for the actual engine, even though the Otto-cycle efficiencies may deviate significantly from the actual efficiencies.

This chapter is concerned with these idealized cycles for both power and refrigeration apparatus. Both vapors and ideal gases are considered as working fluids. An attempt will be made to point out how the processes in actual apparatus deviate from the ideal. Consideration is also given to certain modifications of the basic cycles that are intended to improve performance. These modifications include the use of devices such as regenerators, multistage compressors and expanders, and intercoolers. Various combinations of these types of systems and also special applications, such as cogeneration of electrical power and energy, combined cycles, topping and bottoming cycles, and binary cycle systems, are also discussed in this chapter.

11.1 INTRODUCTION TO POWER SYSTEMS

In introducing the second law of thermodynamics in Chapter 7, we considered cyclic heat engines consisting of four separate processes. We noted there that it is possible to have these engines operate as steady-state, steady-flow devices involving shaft work, as shown in Fig. 7.16, or instead as cylinder/piston devices involving boundary-movement work, as shown in Fig. 7.17. The former may have a working fluid that changes phase

346

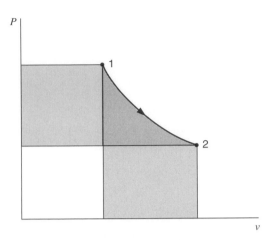

FIGURE 11.1 Comparison of shaft work and boundary-movement work.

during the processes in the cycle, or may have a single-phase working fluid throughout. The latter type would normally have a gaseous working fluid throughout the cycle.

For a reversible SSSF process involving negligible kinetic and potential energy changes, the shaft work per unit mass is given by Eq. 9.19,

$$w = -\int v\,dP$$

For a reversible process involving a simple compressible substance, the boundary movement work per unit mass is given by Eq. 4.3,

$$w = \int P\,dv$$

The areas represented by these two integrals are shown in Fig. 11.1. It is of interest to note that, in the former case, there is no work involved in a constant-pressure process, while in the latter case, there is no work involved in a constant-volume process.

Let us now consider a power system consisting of four SSSF processes, as in Fig. 7.16. We assume that each process is internally reversible and has negligible changes in kinetic and potential energies, which results in the work for each process being given by Eq. 9.19. For convenience of operation, we will make the two heat-transfer processes (boiler and condenser) constant-pressure processes, such that those are simple heat exchangers involving no work. Let us also assume that the turbine and pump processes are both adiabatic, such that they are therefore isentropic processes. Thus, the four processes comprising the cycle are as shown in Fig. 11.2. Note that if the entire cycle takes place inside the two-phase liquid-vapor dome, the resulting cycle is the Carnot cycle, since the two constant-pressure processes are also isothermal. Otherwise, this cycle is not a Carnot cycle. In either case, we find that the net work output for this power system is given by

$$w_{\text{net}} = -\int_1^2 v\,dP + 0 - \int_3^4 v\,dP + 0 = -\int_1^2 v\,dP + \int_4^3 v\,dP$$

and, since $P_2 = P_3$ and also $P_1 = P_4$, we find that the system produces a net work output because the specific volume is larger during the expansion from 3 to 4 than it is during the compression from 1 to 2. This result is also evident from the areas $-\int v\,dP$ in Fig. 11.2. We conclude that it would be advantageous to have this difference in specific volume be as large as possible, as, for example, the difference between a vapor and a liquid.

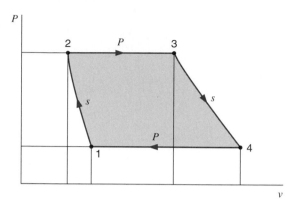

FIGURE 11.2 Four-process power cycle.

If the four-process cycle shown in Fig. 11.2 were accomplished in a cylinder/piston system involving boundary movement work, then the net work output for this power system is given by

$$w_{net} = \int_1^2 P \, dv + \int_2^3 P \, dv + \int_3^4 P \, dv + \int_4^1 P \, dv$$

and from these four areas on Fig. 11.2, we note that the pressure is higher during any given change in volume in the two expansion processes than in the two compression processes, resulting in a net positive area and a net work output.

For either of the two cases just analyzed, it is noted from Fig. 11.2 that the net work output of the cycle is equal to the area enclosed by the process lines 1–2–3–4–1, and is equal for both, even though the work terms for the four individual processes are different for the two cases.

In the next several sections, we consider the Rankine cycle, which is the ideal, four-SSSF process cycle shown in Fig. 11.2, utilizing a phase change between vapor and liquid in order to maximize the difference in specific volume during expansion and compression. This is the idealized model for a steam powerplant system.

11.2 THE RANKINE CYCLE

We now consider the idealized four-SSSF-process cycle shown in Fig. 11.2, in which state 1 is saturated liquid and state 3 either saturated vapor or superheated vapor. This system is termed the Rankine cycle and is the model for the simple steam powerplant. It is convenient to show the states and processes on a T–s diagram, as given in Fig. 11.3. The four processes are

1–2: Reversible adiabatic pumping process in the pump

2–3: Constant-pressure transfer of heat in the boiler

3–4: Reversible adiabatic expansion in the turbine (or other prime mover such as a steam engine)

4–1: Constant-pressure transfer of heat in the condenser

As mentioned above, the Rankine cycle also includes the possibility of superheating the vapor, as cycle 1–2–3′–4′–1.

If changes of kinetic and potential energy are neglected, heat transfer and work may be represented by various areas on the T–s diagram. The heat transferred to the working

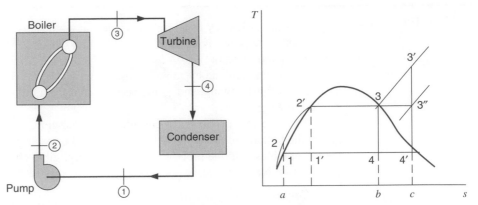

FIGURE 11.3 Simple steam power plant which operates on the Rankine cycle.

fluid is represented by area a–2–$2'$–3–b–a, and the heat transferred from the working fluid by area a–1–4–b–a. From the first law we conclude that the area representing the work is the difference between these two areas—area 1–2–$2'$–3–4–1. The thermal efficiency is defined by the relation

$$\eta_{th} = \frac{w_{net}}{q_H} = \frac{\text{area } 1-2-2'-3-4-1}{\text{area } a-2-2'-3-b-a} \tag{11.1}$$

For analyzing the Rankine cycle, it is helpful to think of efficiency as depending on the average temperature at which heat is supplied and the average temperature at which heat is rejected. Any changes that increase the average temperature at which heat is supplied or decrease the average temperature heat is rejected will increase the Rankine-cycle efficiency.

It should be stated that in analyzing the ideal cycles in this chapter, the changes in kinetic and potential energies from one point in the cycle to another are neglected. In general, this is a reasonable assumption for the actual cycles.

It is readily evident that the Rankine cycle has a lower efficiency than a Carnot cycle with the same maximum and minimum temperatures as a Rankine cycle, because the average temperature between 2 and $2'$ is less than the temperature during evaporation. We might well ask, why choose the Rankine cycle as the ideal cycle? Why not select the Carnot cycle $1'$–$2'$–3–4–$1'$? At least two reasons can be given. The first reason concerns the pumping process. State $1'$ is a mixture of liquid and vapor. Great difficulties are encountered in building a pump that will handle the mixture of liquid and vapor at $1'$ and deliver saturated liquid at $2'$. It is much easier to condense the vapor completely and handle only liquid in the pump: the Rankine cycle is based on this fact. The second reason concerns superheating the vapor. In the Rankine cycle the vapor is superheated at constant pressure, process 3–$3'$. In the Carnot cycle all the heat transfer is at constant temperature, and therefore the vapor is superheated in process 3–$3''$. Note, however, that during this process the pressure is dropping, which means that the heat must be transferred to the vapor as it undergoes an expansion process in which work is done. This heat transfer is also very difficult to achieve in practice. Thus, the Rankine cycle is the ideal cycle that can be approximated in practice. In the following sections we will consider some variations on the Rankine cycle that enable it to approach more closely the efficiency of the Carnot cycle.

Before we discuss the influence of certain variables on the performance of the Rankine cycle, an example is given.

EXAMPLE 11.1

Determine the efficiency of a Rankine cycle using steam as the working fluid in which the condenser pressure is 10 kPa. The boiler pressure is 2 MPa. The steam leaves the boiler as saturated vapor.

In solving Rankine-cycle problems, we let w_p denote the work into the pump per kilogram of fluid flowing, and q_L the heat rejected from the working fluid per kilogram of fluid flowing.

To solve this problem we consider, in succession, a control surface around the pump, the boiler, the turbine, and the condenser. For each the thermodynamic model is the steam tables, and the process is SSSF with negligible changes in kinetic and potential energies. Now, in turn,

Control volume: Pump.

Inlet state: P_1 known, saturated liquid; state fixed.

Exit state: P_2 known.

Analysis:

First law:

$$w_p = h_2 - h_1$$

Second law:

$$s_2 = s_1$$

Because

$$s_2 = s_1, \qquad h_2 - h_1 = \int_1^2 v\, dP$$

Solution

Assuming the liquid to be incompressible, we have

$$w_p = v(P_2 - P_1) = (0.001\,01)(2000 - 10) = 2.0 \text{ kJ / kg}$$
$$h_2 = h_1 + w_p = 191.8 + 2.0 = 193.8$$

Control volume: Boiler.

Inlet state: P_2, h_2 known; state fixed.

Exit state: P_3 known, saturated vapor; state fixed.

Analysis:

First law:

$$q_H = h_3 - h_2$$

Solution

$$q_H = h_3 - h_2 = 2799.5 - 193.8 = 2605.7 \text{ kJ / kg}$$

Control volume: Turbine.

Inlet state: State 3 known (above).

Exit state: P_4 known.

Analysis:

First law:

$$w_t = h_3 - h_4$$

Second law:

$$s_3 = s_4$$

Solution

We can determine the quality at state 4 as follows:

$$s_3 = s_4 = 6.3409 = 0.6493 + x_4 \, 7.5009 \qquad x_4 = 0.7588$$

$$h_4 = 191.8 + 0.7588(2392.8) = 2007.5$$

$$w_t = 2799.5 - 2007.5 = 792.0 \text{ kJ} / \text{kg}$$

Control volume: Condenser.

Inlet state: State 4 known (as given).

Exit state: State 1 known (as given).

Analysis:

First law:

$$q_L = h_4 - h_1$$

Solution

$$q_L = h_4 - h_1 = 2007.5 - 191.8 = 1815.7 \text{ kJ} / \text{kg}$$

We can now calculate the thermal efficiency:

$$\eta_{\text{th}} = \frac{w_{\text{net}}}{q_H} = \frac{q_H - q_L}{q_H} = \frac{w_t - w_p}{q_H} = \frac{792.0 - 2.0}{2605.7} = 30.3\%$$

We could also write an expression for thermal efficiency in terms of properties at various points in the cycle.

$$\eta_{\text{th}} = \frac{(h_3 - h_2) - (h_4 - h_1)}{h_3 - h_2} = \frac{(h_3 - h_4) - (h_2 - h_1)}{h_3 - h_2}$$

$$= \frac{2605.7 - 1815.7}{2605.7} = \frac{792.0 - 2.0}{2605.7} = 30.3\%$$

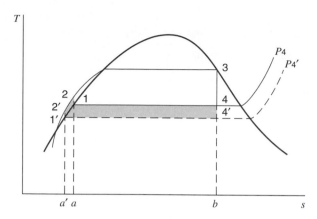

FIGURE 11.4 Effect of exhaust pressure on Rankine-cycle efficiency.

11.3 EFFECT OF PRESSURE AND TEMPERATURE ON THE RANKINE CYCLE

Let us first consider the effect of exhaust pressure and temperature on the Rankine cycle. This effect is shown on the T-s diagram of Fig. 11.4. Let the exhaust pressure drop from P_4 to P_4', with the corresponding decrease in temperature at which heat is rejected. The net work is increased by area 1–4–4′–1′–2′–2–1 (shown by the cross-hatching). The heat transferred to the steam is increased by area a'–2′–2–a–a'. Since these two areas are approximately equal, the net result is an increase in cycle efficiency. This is also evident from the fact that the average temperature at which heat is rejected is decreased. Note, however, that lowering the back pressure causes the moisture content of the steam leaving the turbine to increase. This is a significant factor because if the moisture in the low-pressure stages of the turbine exceeds about 10%, not only is there a decrease in turbine efficiency, but erosion of the turbine blades may also be a very serious problem.

Next, consider the effect of superheating the steam in the boiler, as shown in Fig. 11.5. It is readily evident that the work is increased by area 3–3′–4′–4–3, and the heat transferred in the boiler is increased by area 3–3′–b'–b–3. Since the ratio of these two areas is greater than the ratio of net work to heat supplied for the rest of the cycle, it is evident that for given pressures, superheating the steam increases the Rankine-cycle efficiency. This increase in efficiency would also follow from the fact that the average

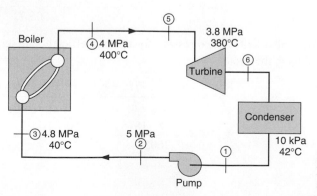

FIGURE 11.5 Effect of superheating on Rankine-cycle efficiency.

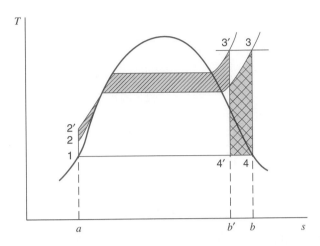

FIGURE 11.6 Effect of boiler pressure on Rankine-cycle efficiency.

temperature at which heat is transferred to the steam is increased. Note also that when the steam is superheated, the quality of the steam leaving the turbine increases.

Finally, the influence of the maximum pressure of the steam must be considered, and this is shown in Fig. 11.6. In this analysis the maximum temperature of the steam, as well as the exhaust pressure, is held constant. The heat rejected decreases by area b'–$4'$–4–b–b'. The net work increases by the amount of the single cross-hatching and decreases by the amount of the double cross-hatching. Therefore, the net work tends to remain the same, but the heat rejected decreases, and hence the Rankine-cycle efficiency increases with an increase in maximum pressure. Note that in this instance too the average temperature at which heat is supplied increases with an increase in pressure. The quality of the steam leaving the turbine decreases as the maximum pressure increases.

To summarize this section, we can say that the efficiency of the Rankine cycle can be increased by lowering the exhaust pressure, by increasing the pressure during heat addition, and by superheating the steam. The quality of the steam leaving the turbine is increased by superheating the steam and decreased by lowering the exhaust pressure and by increasing the pressure during heat addition.

EXAMPLE 11.2   In a Rankine cycle steam leaves the boiler and enters the turbine at 4 MPa, 400°C. The condenser pressure is 10 kPa. Determine the cycle efficiency.

To determine the cycle efficiency, we must calculate the turbine work, the pump work, and the heat transfer to the steam in the boiler. We do this by considering a control surface around each of these components in turn. In each case the thermodynamic model is the steam tables, and the process is SSSF with negligible changes in kinetic and potential energies.

Control volume: Pump.

Inlet state: P_1 known, saturated liquid; state fixed.

Exit state: P_2 known.

Analysis:

First law:

$$w_p = h_2 - h_1$$

Second law:

$$s_2 = s_1$$

Since $s_2 = s_1$,

$$h_2 - h_1 = \int_1^2 v \, dP = v(P_2 - P_1)$$

Solution

$$w_p = v(P_2 - P_1) = (0.001\,01)(4000 - 10) = 4.0 \text{ kJ / kg}$$

$$h_1 = 191.8$$

$$h_2 = 191.8 + 4.0 = 195.8$$

Control volume: Turbine.
Inlet state: P_3, T_3 known; state fixed.
Exit state: P_4 known.

Analysis:

First law:

$$w_t = h_3 - h_4$$

Second law:

$$s_4 = s_3$$

Solution

$$h_3 = 3213.6, \qquad s_3 = 6.7690$$

$$s_3 = s_4 = 6.7690 = 0.6493 + x_4 7.5009, \qquad x_4 = 0.8159$$

$$h_4 = 191.8 + 0.8159(2392.8) = 2144.1$$

$$w_t = h_3 - h_4 = 3213.6 - 2144.1 = 1069.5 \text{ kJ / kg}$$

$$\text{w}_{\text{net}} = w_t - w_p = 1069.5 - 4.0 = 1065.5 \text{ kJ / kg}$$

Control volume: Boiler.
Inlet state: P_2, h_2 known; state fixed.
Exit state: State 3 fixed (as given).

Analysis:

First law:

$$q_H = h_3 - h_2$$

Solution

$$q_H = h_3 - h_2 = 3213.6 - 195.8 = 3017.8 \text{ kJ / kg}$$

$$\eta_{\text{th}} = \frac{w_{\text{net}}}{q_H} = \frac{1065.5}{3017.8} = 35.3\%$$

The net work could also be determined by calculating the heat rejected in the condenser, q_L, and noting, from the first law, that the net work for the cycle is equal to the net heat transfer. Considering a control surface around the condenser, we have

$$q_L = h_4 - h_1 = 2144.1 - 191.8 = 1952.3 \text{ kJ / kg}$$

Therefore,

$$w_{\text{net}} = q_H - q_L = 3017.8 - 1952.3 = 1065.5 \text{ kJ / kg}$$

EXAMPLE 11.2E   In a Rankine cycle, steam leaves the boiler and enters the turbine at 600 lbf/in.², 800 F. The condenser pressure is 1 lbf/in.² Determine the cycle efficiency.

To determine the cycle efficiency, we must calculate the turbine work, the pump work, and the heat transfer to the steam in the boiler. We do this by considering a control surface around each of these components in turn. In each case the thermodynamic model is the steam tables, and the process is SSSF with negligible changes in kinetic and potential energies.

Control volume: Pump.

Inlet state: P_1 known, saturated liquid; state fixed.

Exit state: P_2 known.

Analysis:

First law:

$$w_P = h_2 - h_1$$

Second law:

$$s_2 = s_1$$

Since $s_2 = s_1$,

$$h_2 - h_1 = \int_1^2 v \, dP = v \left(P_2 - P_1 \right)$$

Solution

$$w_p = v \left(P_2 - P_1 \right) = 0.01614 \left(600 - 1 \right) \times \tfrac{144}{778} = 1.8 \text{ Btu / 1bm}$$

$$h_1 = 69.70$$

$$h_2 = 69.7 + 1.8 = 71.5 \text{ Btu / 1bm}$$

Control volume: Turbine.

Inlet state: P_3, T_3 known; state fixed.

Exit state: P_4 known.

Analysis:

First law:

$$w_t = h_3 - h_4$$

Second law:

$$s_4 = s_3$$

Solution

$$h_3 = 1407.6 \qquad s_3 = 1.6343$$
$$s_3 = s_4 = 1.6343 = 1.9779 - (1-x)_4 \, 1.8453$$
$$(1-x)_4 = 0.1861$$
$$h_4 = 1105.8 - 0.1861(1036.0) = 913.0$$
$$w_t = h_3 - h_4 = 1407.6 - 913.0 = 494.6 \text{ Btu} / \text{lbm}$$
$$\mathrm{W}_{\text{net}} = w_t - w_p = 494.6 - 1.8 = 492.8 \text{ Btu} / \text{lbm}$$

Control volume: Boiler.

Inlet state: P_2, h_2 known; state fixed.

Exit state: State 3 fixed (above).

Analysis:

First law:

$$q_H = h_3 - h_2$$

Solution

$$q_H = h_3 - h_2 = 1407.6 - 71.5 = 1336.1 \text{ Btu} / \text{lbm}$$
$$\eta_{\text{th}} = \frac{w_{\text{net}}}{q_H} = \frac{492.8}{1336.1} = 36.9\%$$

The net work could also be determined by calculating the heat rejected in the condenser, q_L, and noting, from the first law, that the net work for the cycle is equal to the net heat transfer. Considering a control surface around the condenser, we have

$$q_L = h_4 - h_1 = 913.0 - 69.7 = 843.3 \text{ Btu} / \text{lbm}$$

Therefore,

$$w_{\text{net}} = q_H - q_L = 1336.1 - 843.3 = 492.8 \text{ Btu} / \text{lbm}$$

11.4 THE REHEAT CYCLE

In the last section we noted that the efficiency of the Rankine cycle could be increased by increasing the pressure during the addition of heat. However, the increase in pressure also increases the moisture content of the steam in the low-pressure end of the turbine. The reheat cycle has been developed to take advantage of the increased efficiency with higher pressures, and yet avoid excessive moisture in the low-pressure stages of the turbine. This cycle is shown schematically and on a *T–s* diagram in Fig. 11.7. The unique feature of

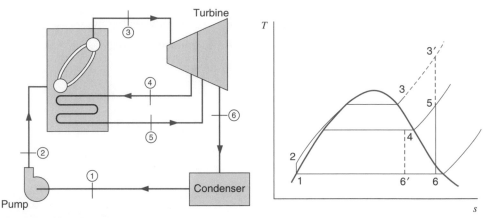

FIGURE 11.7 The ideal reheat cycle.

this cycle is that the steam is expanded to some intermediate pressure in the turbine and is then reheated in the boiler, after which it expands in the turbine to the exhaust pressure. It is evident from the T–s diagram that there is very little gain in efficiency from reheating the steam, because the average temperature at which heat is supplied is not greatly changed. The chief advantage is in decreasing to a safe value the moisture content in the low-pressure stages of the turbine. If metals could be found that would enable us to superheat the steam to 3′, the simple Rankine cycle would be more efficient than the reheat cycle, and there would be no need for the reheat cycle.

EXAMPLE 11.3   Consider a reheat cycle utilizing steam. Steam leaves the boiler and enters the turbine at 4 MPa, 400°C. After expansion in the turbine to 400 kPa, the steam is reheated to 400°C and then expanded in the low-pressure turbine to 10 kPa. Determine the cycle efficiency.

For each control volume analyzed, the thermodynamic model is the steam tables, the process is SSSF, and changes in kinetic and potential energies are negligible.

> *Control volume:*   High-pressure turbine.
>
> *Inlet state:* P_3, T_3 known; state fixed.
>
> *Exit state:* P_4 known.

Analysis:

First law:

$$w_{h\text{-}p} = h_3 - h_4$$

Second law:

$$s_3 = s_4$$

Solution

$$h_3 = 3213.6, \qquad s_3 = 6.7690$$
$$s_4 = s_3 = 6.7690 = 1.7766 + x_4 5.1193, \qquad x_4 = 0.9752$$
$$h_4 = 604.7 + 0.9752(2133.8) = 2685.6$$

Control volume: Low-pressure turbine.

Inlet state: P_5, T_5 known; state fixed.

Exit state: P_6 known.

Analysis:

First law:

$$w_{1-p} = h_5 - h_6$$

Second law:

$$s_5 = s_6$$

Solution

$$h_5 = 3273.4 \qquad s_5 = 7.8985$$
$$s_6 = s_5 = 7.8985 = 0.6493 + x_6 7.5009, \qquad x_6 = 0.9664$$
$$h_6 = 191.8 + 0.9664(2392.8) = 2504.3$$

For the overall turbine, the total work output w_t is the sum of w_{h-p} and w_{1-p}, so that

$$w_t = (h_3 - h_4) + (h_5 - h_6)$$
$$= (3213.6 - 2685.6) + (3273.4 - 2504.3)$$
$$= 1297.1 \text{ kJ} / \text{kg}$$

Control volume: Pump.

Inlet state: P_1 known, saturated liquid; state fixed.

Exit state: P_2 known.

Analysis:

First law:

$$w_p = h_2 - h_1$$

Second law:

$$s_2 = s_1$$

Since $s_2 = s_1$,

$$h_2 - h_1 = \int_1^2 v \, dP = v(P_2 - P_1)$$

Solution

$$w_p = v(P_2 - P_1) = (0.001\,01)(4000 - 10) = 4.0 \text{ kJ} / \text{kg}$$
$$h_2 = 191.8 + 4.0 = 195.8$$

Control volume: Boiler.

Inlet states: States 2 and 4 both known (above).

Exit states: States 3 and 5 both known (as given).

Analysis:

First law:

$$q_H = \left(h_3 - h_2\right) + \left(h_5 - h_4\right)$$

Solution

$$q_H = \left(h_3 - h_2\right) + \left(h_5 - h_4\right)$$
$$= \left(3213.6 - 195.8\right) + \left(3273.4 - 2685.6\right) = 3605.6 \text{ kJ / kg}$$

Therefore,

$$w_{\text{net}} = w_t - w_p = 1297.1 - 4.0 = 1293.1 \text{ kJ / kg}$$

$$\eta_{\text{th}} = \frac{w_{\text{net}}}{q_H} = \frac{1293.1}{3605.6} = 35.9\%$$

By comparing this example with Example 11.2, we find that through reheating the gain in efficiency is relatively small, but the moisture content of the vapor leaving the turbine is decreased from 18.4 to 3.4%.

11.5 THE REGENERATIVE CYCLE

Another important variation from the Rankine cycle is the regenerative cycle, which uses feedwater heaters. The basic concepts of this cycle can be demonstrated by considering the Rankine cycle without superheat as shown in Fig. 11.8. During the process between states

FIGURE 11.8
Temperature-entropy diagram showing the relationship between Carnot-cycle efficiency and Rankine-cycle efficiency.

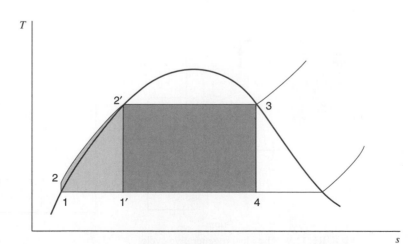

2 and 2′, the working fluid is heated while in the liquid phase, and the average temperature of the working fluid is much lower than during the vaporization process 2′–3. The process between states 2 and 2′ causes the average temperature at which heat is supplied in the Rankine cycle to be lower than in the Carnot cycle 1′–2′–3–4–1′. Consequently, the efficiency of the Rankine cycle is lower than that of the corresponding Carnot cycle. In the regenerative cycle the working fluid enters the boiler at some state between 2 and 2′, and consequently the average temperature at which heat is supplied is higher.

Consider first an idealized regenerative cycle, as shown in Fig. 11.9. The unique feature of this cycle compared to the Rankine cycle is that after leaving the pump, the liquid circulates around the turbine casing, counterflow to the direction of vapor flow in the turbine. Thus, it is possible to transfer to the liquid flowing around the turbine the heat from the vapor as it flows through the turbine. Let us assume for the moment that this is a reversible heat transfer, that is, at each point the temperature of the vapor is only infinitesimally higher than the temperature of the liquid. In this instance line 4–5 on the *T–s* diagram of Fig. 11.9, which represents the states of the vapor flowing through the turbine, is exactly parallel to line 1–2–3, which represents the pumping process (1–2) and the states of the liquid flowing around the turbine. Consequently, areas 2–3–*b*–*a*–2 and 5–4–*d*–*c*–5 are not only equal but congruous, and these areas, respectively, represent the heat transferred to the liquid and from the vapor. Heat is also transferred to the working fluid at constant temperature in process 3–4, and area 3–4–*d*–*b*–3 represents this heat transfer. Heat is transferred from the working fluid in process 5–1, and area 1–5–*c*–*a*–1 represents this heat transfer. This area is exactly equal to area 1′–5′–*d*–*b*–1′, which is the heat rejected in the related Carnot cycle 1′–3–4–5′–1′. Thus, the efficiency of this idealized regenerative cycle is exactly equal to the efficiency of the Carnot cycle with the same heat supply and heat rejection temperatures.

Quite obviously this idealized regenerative cycle is impractical. First, it would be impossible to effect the necessary heat transfer from the vapor in the turbine to the liquid feedwater. Furthermore, the moisture content of the vapor leaving the turbine increases considerably as a result of the heat transfer. The disadvantage of this has been noted previously. The practical regenerative cycle extracts some of the vapor after it has partially expanded in the turbine and uses feedwater heaters, as shown in Fig. 11.10.

Steam enters the turbine at state 5. After expansion to state 6, some of the steam is extracted and enters the feedwater heater. The steam that is not extracted is expanded in the turbine to state 7 and is then condensed in the condenser. This condensate is pumped

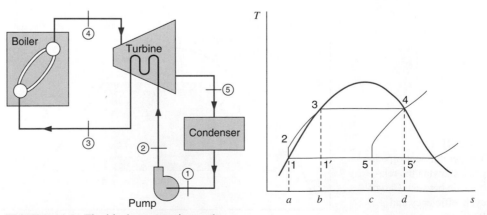

FIGURE 11.9 The ideal regenerative cycle.

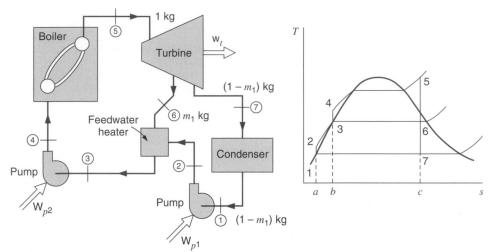

FIGURE 11.10 Regenerative cycle with open feedwater heater.

into the feedwater heater where it mixes with the steam extracted from the turbine. The proportion of steam extracted is just sufficient to cause the liquid leaving the feedwater heater to be saturated at state 3. Note that the liquid has not been pumped to the boiler pressure, but only to the intermediate pressure corresponding to state 6. Another pump is required to pump the liquid leaving the feedwater heater to boiler pressure. The significant point is that the average temperature at which heat is supplied has been increased.

This cycle is somewhat difficult to show on a T–s diagram because the masses of steam flowing through the various components vary. The T–s diagram of Fig. 11.10 simply shows the state of the fluid at the various points.

Area 4–5–c–b–4 in Fig. 11.10 represents the heat transferred per kilogram of working fluid. Process 7–1 is the heat rejection process, but since not all the steam passes through the condenser, area 1–7–c–a–1 represents the heat transfer per kilogram flowing through the condenser, which does not represent the heat transfer per kilogram of working fluid entering the turbine. Between states 6 and 7 only part of the steam is flowing through the turbine. The example that follows illustrates the calculations for the regenerative cycle.

EXAMPLE 11.4   Consider a regenerative cycle using steam as the working fluid. Steam leaves the boiler and enters the turbine at 4 MPa, 400°C. After expansion to 400 kPa, some of the steam is extracted from the turbine for the purpose of heating the feedwater in an open feedwater heater. The pressure in the feedwater heater is 400 kPa and the water leaving it is saturated liquid at 400 kPa. The steam not extracted expands to 10 kPa. Determine the cycle efficiency.

The line diagram and T–s diagram for this cycle are shown in Fig. 11.10.

As in previous examples, the model for each control volume is the steam tables, the process is SSSF, and kinetic and potential energy changes are negligible.

From Examples 11.2 and 11.3 we have the following properties:

$$h_5 = 3213.6 \qquad h_6 = 2685.6$$

$$h_7 = 2144.1 \qquad h_1 = 191.8$$

Control volume: Low-pressure pump.

Inlet state: P_1 known, saturated liquid; state fixed.

Exit state: P_2 known.

Analysis:

First law:

$$w_{p1} = h_2 - h_1$$

Second law:

$$s_2 = s_1$$

Therefore,

$$h_2 - h_1 = \int_1^2 v\,dP = v\left(P_2 - P_1\right)$$

Solution

$$w_{p1} = v\left(P_2 - P_1\right) = \left(0.001\,01\right)\left(400 - 10\right) = 0.4 \text{ kJ / kg}$$

$$h_2 = h_1 + w_p = 191.8 + 0.4 = 192.2$$

Control volume: Turbine.

Inlet state: P_5, T_5 known; state fixed.

Exit state: P_6 known; P_7 known.

Analysis:

First law:

$$w_t = \left(h_5 - h_6\right) + \left(1 - m_1\right)\left(h_6 - h_7\right)$$

Second law:

$$s_5 = s_6 = s_7$$

Solution

From the second law, the values for h_6 and h_7 given previously were calculated in Examples 11.2 and 11.3.

Control volume: Feedwater heater.

Inlet states: States 2 and 6 both known (as given).

Exit state: P_3 known, saturated liquid; state fixed.

Analysis:

First law:

$$m_1\left(h_6\right) + \left(1 - m_1\right)h_2 = h_3$$

Solution

$$m_1(2685.6) + (1 - m_1)(192.2) = 604.7$$

$$m_1 = 0.1654$$

We can now calculate the turbine work.

$$
\begin{aligned}
w_t &= (h_5 - h_6) + (1 - m_1)(h_6 - h_7) \\
&= (3213.6 - 2685.6) + (1 - 0.1654)(2685.6 - 2144.1) \\
&= 979.9 \text{ kJ / kg}
\end{aligned}
$$

Control volume: High-pressure pump.

Inlet state: State 3 known (as given).

Exit state: P_4 known.

Analysis:

First law:

$$w_{p2} = h_4 - h_3$$

Second law:

$$s_4 = s_3$$

Solution

$$w_{p2} = v(P_4 - P_3) = (0.001\,084)(4000 - 400) = 3.9 \text{ kJ / kg}$$

$$h_4 = h_3 + w_{p2} = 604.7 + 3.9 = 608.6$$

Therefore,

$$
\begin{aligned}
w_{\text{net}} &= w_t - (1 - m_1)w_{p1} - w_{p2} \\
&= 979.9 - (1 - 0.1654)(0.4) - 3.9 = 975.7 \text{ kJ / kg}
\end{aligned}
$$

Control volume: Boiler.

Inlet state: P_4, h_4 known (as given); state fixed.

Exit state: State 5 known (as given).

Analysis:

First law:

$$q_H = h_5 - h_4$$

Solution

$$q_H = h_5 - h_4 = 3213.6 - 608.6 = 2605.0 \text{ kJ / kg}$$

$$\eta_{\text{th}} = \frac{w_{\text{net}}}{q_H} = \frac{975.7}{2605.0} = 37.5\%$$

Note the increase in efficiency over the efficiency of the Rankine cycle of Example 11.2.

Up to this point the discussion and examples have tacitly assumed that the extraction steam and feedwater are mixed in the feedwater heater. Another much-used type of feedwater heater, known as a closed heater, is one in which the steam and feedwater do not mix; rather heat is transferred from the extracted steam as it condenses on the outside of tubes while the feedwater flows through the tubes. In a closed heater, a schematic sketch of which is shown in Fig. 11.11, the steam and feedwater may be at considerably different pressures. The condensate may be pumped into the feedwater line, or it may be removed through a trap to a lower-pressure heater or to the condenser. (A trap is a device that permits liquid but not vapor to flow to a region of lower pressure.)

Open feedwater heaters have the advantage of being less expensive and having better heat-transfer characteristics compared to closed feedwater heaters. They have the disadvantage of requiring a pump to handle the feedwater between each heater.

In many power plants a number of stages of extraction are used, though only rarely more than five. The number is, of course, determined by economics. It is evident that using a very large number of extraction stages and feedwater heaters allows the cycle efficiency to approach that of the idealized regenerative cycle of Fig. 11.9, where the feedwater enters the boiler as saturated liquid at the maximum pressure. In practice, however, this could not be economically justified because the savings effected by the increase in efficiency would be more than offset by the cost of additional equipment (feedwater heaters, piping, and so forth).

A typical arrangement of the main components in an actual power plant is shown in Fig. 11.12. Note that one open feedwater heater is a deaerating feedwater heater; this heater has the dual purpose of heating and removing the air from the feedwater. Unless the air is removed, excessive corrosion occurs in the boiler. Note also that the condensate from the high-pressure heater drains (through a trap) to the intermediate heater, and the intermediate heater drains to the deaerating feedwater heater. The low-pressure heater drains to the condenser.

Many actual power plants combine one reheat stage with a number of extraction stages. The principles already considered are readily applied to such a cycle.

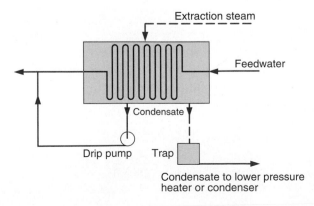

FIGURE 11.11 Schematic arrangement for a closed feedwater heater.

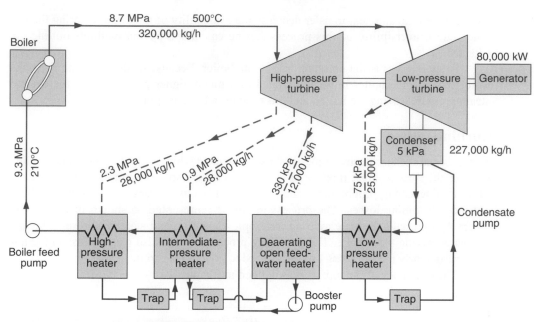

FIGURE 11.12 Arrangement of heaters in an actual power plant utilizing regenerative feedwater heaters.

11.6 DEVIATION OF ACTUAL CYCLES FROM IDEAL CYCLES

Before we leave the matter of vapor power cycles, a few comments are in order regarding the ways in which an actual cycle deviates from an ideal cycle. The losses associated with the combustion process are considered in a later chapter. The most important of these losses are the following.

Piping Losses

Pressure drops caused by frictional effects and heat transfer to the surroundings are the most important piping losses. Consider, for example, the pipe connecting the turbine to the boiler. If only frictional effects occur, states a and b in Fig. 11.13 would represent the states of the steam leaving the boiler and entering the turbine, respectively. Note that the frictional effects cause an increase in entropy. Heat transferred to the surroundings at constant pressure can be represented by process bc. This effect decreases entropy. Both

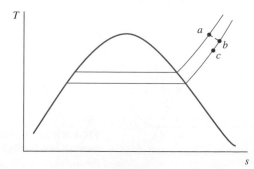

FIGURE 11.13 Temperature–entropy diagram showing effect of losses between boiler and turbine.

the pressure drop and heat transfer decrease the availability of the steam entering the turbine. The irreversibility of this process can be calculated by the methods outlined in Chapter 10.

A similar loss is the pressure drop in the boiler. Because of this pressure drop, the water entering the boiler must be pumped to a much higher pressure than the desired steam pressure leaving the boiler, and this requires additional pump work.

Turbine Losses

The losses in the turbine are primarily those associated with the flow of the working fluid through the turbine. Heat transfer to the surroundings also represents a loss, but this is usually of secondary importance. The effects of these two losses are the same as those outlined for piping losses. The process might be represented as shown in Fig. 11.14, where 4_s represents the state after an isentropic expansion and state 4 represents the actual state leaving the turbine. The governing procedures may also cause a loss in the turbine, particularly if a throttling process is used to govern the turbine.

The efficiency of the turbine was defined in Chapter 9 as

$$\eta_t = \frac{w_t}{h_3 - h_{4s}}$$

where the states are as designated in Fig. 11.14.

Pump Losses

The losses in the pump are similar to those of the turbine and are primarily due to the irreversibilities associated with the fluid flow. Heat transfer is usually a minor loss.

The pump efficiency is defined as

$$\eta_p = \frac{h_{2s} - h_1}{w_p}$$

where the states are as shown in Fig. 11.14, and w_p is the actual work input per kilogram of fluid.

Condenser Losses

The losses in the condenser are relatively small. One of these minor losses is the cooling below the saturation temperature of the liquid leaving the condenser. This represents a

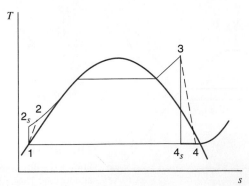

FIGURE 11.14 Temperature–entropy diagram showing effect of turbine and pump inefficiencies on cycle performance.

loss because additional heat transfer is necessary to bring the water to its saturation temperature.

The influence of these losses on the cycle is illustrated in the following example, which should be compared to Example 11.2.

EXAMPLE 11.5 A steam power plant operates on a cycle with pressures and temperatures as designated in Fig. 11.15. The efficiency of the turbine is 86% and the efficiency of the pump is 80%. Determine the thermal efficiency of this cycle.

As in previous examples, for each control volume the model used is the steam tables, and each process is SSSF with no changes in kinetic or potential energy. This cycle is shown on the T–s diagram of Fig. 11.16.

Control volume: Turbine.

Inlet state: P_5, T_5 known; state fixed.

Exit state: P_6 known.

Analysis:

First law:

$$w_t = h_5 - h_6$$

Second law:

$$s_{6s} = s_5$$

$$\eta_t = \frac{w_t}{h_5 - h_{6s}} = \frac{h_5 - h_6}{h_5 - h_{6s}}$$

Solution

From the steam tables.

$$h_5 = 3169.1, \qquad s_5 = 6.7235$$
$$s_{6s} = s_5 = 6.7235 = 0.6493 + x_{6s} 7.5009, \qquad x_{6s} = 0.8098$$
$$h_{6s} = 191.8 + 0.8098(2392.8) = 2129.5$$
$$w_t = \eta_t (h_5 - h_{6s}) = 0.86(3169.1 - 2129.5) = 894.1 \text{ kJ} / \text{kg}$$

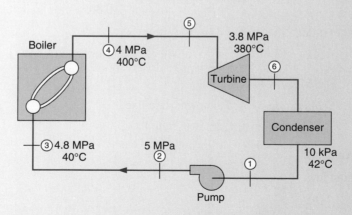

FIGURE 11.15 Schematic diagram for Example 11.5.

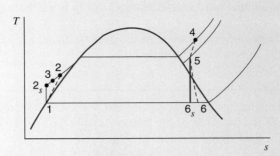

FIGURE 11.16 Temperature–entropy diagram for Example 11.5.

Control volume: Pump.

Inlet state: P_1, T_1 known; state fixed.

Exit state: P_2 known.

Analysis:

First law:

$$w_p = h_2 - h_1$$

Second law:

$$s_{2s} = s_1$$

$$\eta_p = \frac{h_{2s} - h_1}{w_p} = \frac{h_{2s} - h_1}{h_2 - h_1}$$

Since $s_{2s} = s_1$,

$$h_{2s} - h_1 = v(P_2 - P_1)$$

Therefore,

$$w_p = \frac{h_{2s} - h_1}{\eta_p} = \frac{v(P_2 - P_1)}{\eta_p}$$

Solution

$$w_p = \frac{v(P_2 - P_1)}{\eta_p} = \frac{(0.001\ 009)(5000 - 10)}{0.80} = 6.3\ \text{kJ}\,/\,\text{kg}$$

Therefore,

$$w_{\text{net}} = w_t - w_p = 894.1 - 6.3 = 887.8\ \text{kJ}\,/\,\text{kg}$$

Control volume: Boiler.

Inlet state: P_3, T_3 known; state fixed.

Exit state: P_4, T_4 known; state fixed.

Analysis:

First law:

$$q_H = h_4 - h_3$$

Solution

$$q_H = h_4 - h_3 = 3213.6 - 171.8 = 3041.8 \text{ kJ} / \text{kg}$$

$$\eta_{th} = \frac{887.8}{3041.8} = 29.2\%$$

This result compares to the Rankine efficiency of 35.3% for the similar cycle of Example 11.2.

EXAMPLE 11.5E   A steam power plant operates on a cycle with pressure and temperatures as designated in Fig. 11.15E. The efficiency of the turbine is 86% and the efficiency of the pump is 80%. Determine the thermal efficiency of this cycle.

As in previous examples, for each control volume the model used is the steam tables, and each process is SSSF with no changes in kinetic or potential energy. This cycle is shown on the T–s diagram of Fig. 11.16.

Control volume: Turbine.

Inlet state: P_5, T_5 known; state fixed.

Exit state: P_6 known.

Analysis:

First law:

$$w_t = h_5 - h_6$$

Second law:

$$s_{6s} = s_5$$

$$\eta_t = \frac{w_t}{h_5 - h_{6s}} = \frac{h_5 - h_6}{h_5 - h_{6s}}$$

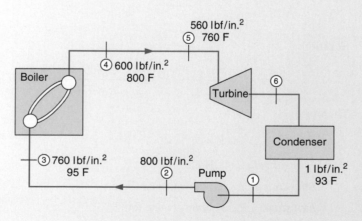

FIGURE 11.15E
Schematic diagram for
Example 11.5E.

Solution

From the steam tables,

$$h_5 = 1386.8 \quad s_5 = 1.6248$$

$$s_{6s} = s_5 = 1.6248 = 1.9779 - (1-x)_{6s} 1.8453$$

$$(1-x)_{6s} = \frac{0.3531}{1.8453} = 0.1912$$

$$h_{6s} = 1105.8 - 0.1912(1036.0) = 907.6$$

$$w_t = \eta_t(h_5 - h_{6s}) = 0.86(1386.8 - 907.6)$$

$$= 0.86(479.2) = 412.1 \text{ Btu/lbm}$$

Control volume: Pump.

Inlet state: P_1, T_1 known; state fixed.

Exit state: P_2 known.

Analysis:

First law:

$$w_p = h_2 - h_1$$

Second law:

$$s_{2s} = s_1$$

$$\eta_p = \frac{h_{2s} - h_1}{w_p} = \frac{h_{2s} - h_1}{h_2 - h_1}$$

Since $s_{2s} = s_1$,

$$h_{2s} - h_1 = v(P_2 - P_1)$$

Therefore,

$$w_p = \frac{h_{2s} - h_1}{\eta_p} = \frac{v(P_2 - P_1)}{\eta_p}$$

Solution

$$w_p = \frac{v(P_2 - P_1)}{\eta_p} = \frac{0.016\ 15(800-1)144}{0.8 \times 778} = 3.0 \text{ Btu/lbm}$$

Therefore,

$$w_{net} = w_t - w_p = 412.1 - 3.0 = 409.1 \text{ Btu/lbm}$$

Control volume: Boiler.

Inlet state: P_3, T_3 known; state fixed.

Exit state: P_4, T_4 known; state fixed.

Analysis:

First law:

$$q_H = h_4 - h_3$$

Solution

$$q_H = h_4 - h_3 = 1407.6 - 65.1 = 1342.5 \text{ Btu / lbm}$$

$$\eta_{th} = \frac{409.1}{1342.5} = 30.4\%$$

This compares to an efficiency of 36.9% for the Rankine efficiency of the similar cycle of Example 11.2E.

11.7 COGENERATION

There are many occasions in industrial settings where the need arises for a specific source or supply of energy within the environment in which a steam powerplant is being used to generate electricity. In such cases, it is appropriate to consider supplying this source of energy in the form of steam that has already been expanded through the high-pressure section of the turbine in the powerplant cycle, thereby eliminating the construction and use of a second boiler or other energy source. Such an arrangement is shown in Fig. 11.17, in which the turbine is tapped at some intermediate pressure to furnish the necessary amount of process steam required for the particular energy need—perhaps to operate a special process in the plant, or in many cases simply for the purpose of space heating the facilities. This type of application is termed cogeneration, and if the system is designed as a package with both the electrical and the process steam requirements in mind,

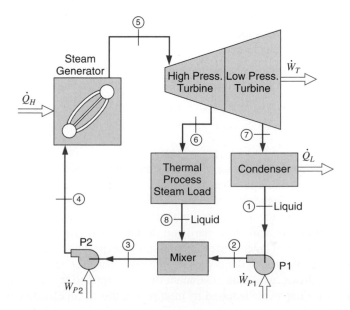

FIGURE 11.17 Example of a cogeneration system.

it is possible to achieve a substantial savings in capital cost of equipment and also in the operating cost, through careful consideration of all the requirements and optimization of the various parameters involved. Specific examples of cogeneration systems are considered in the problems at the end of the chapter.

11.8 Air-Standard Power Cycles

In Section 11.1 we considered idealized four-process cycles, including both SSSF-process and cylinder/piston boundary-movement cycles. The question of phase-change cycles and single-phase cycles was also mentioned. We then proceeded to examine the Rankine powerplant cycle in detail, the idealized model of a phase-change power cycle. However, many work-producing devices (engines) utilize a working fluid that is always a gas. The spark-ignition automotive engine is a familiar example, and so are the Diesel engine and the conventional gas turbine. In all these engines there is a change in the composition of the working fluid, because during combustion it changes from air and fuel to combustion products. For this reason these engines are called internal-combustion engines. In contrast, the steam power plant may be called an external combustion engine, because heat is transferred from the products of combustion to the working fluid. External-combustion engines using a gaseous working fluid (usually air) have been built. To date they have had only limited application, but the use of the gas-turbine cycle in conjunction with a nuclear reactor has been investigated extensively. Other external-combustion engines are currently receiving serious attention in an effort to combat air pollution.

Because the working fluid does not go through a complete thermodynamic cycle in the engine (even though the engine operates in a mechanical cycle) the internal-combustion engine operates on the so-called open cycle. However, for analyzing internal-combustion engines, it is advantageous to devise closed cycles that closely approximate the open cycles. One such approach is the air-standard cycle, which is based on the following assumptions.

1. A fixed mass of air is the working fluid throughout the entire cycle, and the air is always an ideal gas. Thus, there is no inlet process or exhaust process.
2. The combustion process is replaced by a process transferring heat from an external source.
3. The cycle is completed by heat transfer to the surroundings (in contrast to the exhaust and intake process of an actual engine).
4. All processes are internally reversible.
5. An additional assumption is often made, that air has a constant specific heat.

The principal value of the air-standard cycle is to enable us to examine qualitatively the influence of a number of variables on performance. The quantitative results obtained from the air-standard cycle, such as efficiency and mean effective pressure, will differ from those of the actual engine. Our emphasis, therefore, in our consideration of the air-standard cycle, will be primarily on the qualitative aspects.

The term "mean effective pressure" (mep), which is used in conjunction with reciprocating engines, is defined as the pressure that, if it acted on the piston during the entire power stroke, would do an amount of work equal to that actually done on the piston. The work for one cycle is found by multiplying this mean effective pressure by the area of the piston (minus the area of the rod on the crank end of a double-acting engine) and by the stroke.

11.9 THE BRAYTON CYCLE

In discussing idealized four-SSSF-process power cycles in Section 11.1, a cycle involving two constant-pressure and two isentropic processes was examined, and the results shown in Fig. 11.2. This cycle used with a condensing working fluid is the Rankine cycle, but when used with a single-phase, gaseous working fluid it is termed the Brayton cycle. The air-standard Brayton cycle is the ideal cycle for the simple gas turbine. The simple open-cycle gas turbine utilizing an internal-combustion process and the simple closed-cycle gas turbine, which utilizes heat-transfer processes, are both shown schematically in Fig. 11.18. The air-standard Brayton cycle is shown on the P–v and T–s diagrams of Fig. 11.19.

The efficiency of the air-standard Brayton cycle is found as follows:

$$\eta_{th} = 1 - \frac{Q_L}{Q_H} = 1 - \frac{C_p(T_4 - T_1)}{C_p(T_3 - T_2)} = 1 - \frac{T_1(T_4/T_1 - 1)}{T_2(T_3/T_2 - 1)}$$

We note, however, that

$$\frac{P_3}{P_4} = \frac{P_2}{P_1}$$

$$\frac{P_2}{P_1} = \left(\frac{T_2}{T_1}\right)^{k/(k-1)} = \frac{P_3}{P_4} = \left(\frac{T_3}{T_4}\right)^{k/(k-1)}$$

$$\frac{T_3}{T_4} = \frac{T_2}{T_1} \quad \therefore \frac{T_3}{T_2} = \frac{T_4}{T_1} \quad \text{and} \quad \frac{T_3}{T_2} - 1 = \frac{T_4}{T_1} - 1$$

$$\eta_{th} = 1 - \frac{T_1}{T_2} = 1 - \frac{1}{(P_2/P_1)^{(k-1)/k}} \tag{11.2}$$

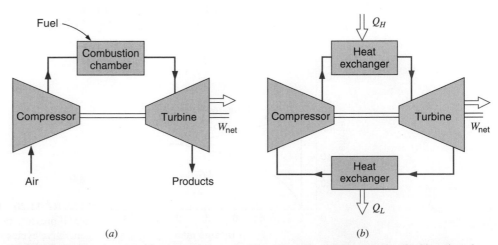

(a)	(b)

FIGURE 11.18 A gas turbine operating on the Brayton cycle. (a) Open cycle. (b) Closed cycle.

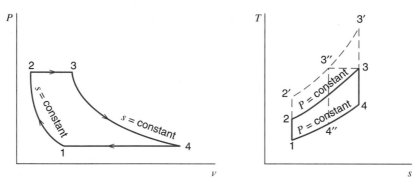

FIGURE 11.19   The air-standard Brayton cycle.

The efficiency of the air-standard Brayton cycle is therefore a function of the isentropic pressure ratio; Fig. 11.20 shows a plot of efficiency versus pressure ratio. The fact that efficiency increases with pressure ratio is evident from the $T–s$ diagram of Fig. 11.19 because increasing the pressure ratio changes the cycle from 1–2–3–4–1 to 1–2′–3′–4–1. The latter cycle has a greater heat supply and the same heat rejected as the original cycle; therefore, it has a greater efficiency. Note that the latter cycle has a higher maximum temperature, $T_3′$, than the original cycle, T_3. In the actual gas turbine the maximum temperature of the gas entering the turbine is fixed by material considerations. Therefore, if we fix the temperature T_3 and increase the pressure ratio, the resulting cycle is 1–2′–3″–4″–1. This cycle would have a higher efficiency than the original cycle, but the work per kilogram of working fluid is thereby changed.

With the advent of nuclear reactors, the closed-cycle gas turbine has become more important. Heat is transferred, either directly or via a second fluid, from the fuel in the nuclear reactor to the working fluid in the gas turbine. Heat is rejected from the working fluid to the surroundings.

The actual gas-turbine engine differs from the ideal cycle primarily because of irreversibilities in the compressor and turbine, and because of pressure drop in the flow passages and combustion chamber (or in the heat exchanger of a closed-cycle turbine). Thus, the state points in a simple open-cycle gas turbine might be as shown in Fig. 11.21.

The efficiencies of the compressor and turbine are defined in relation to isentropic processes. With the states designated as in Fig. 11.21, the definitions of compressor and

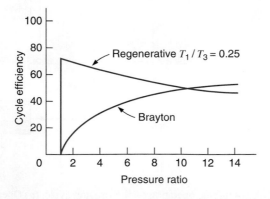

FIGURE 11.20   Cycle efficiency as a function of pressure ratio for the Brayton and regenerative cycles.

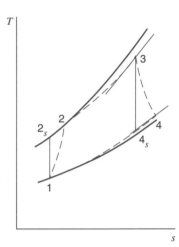

FIGURE 11.21 Effect of inefficiencies on the gas-turbine cycle.

turbine efficiencies are

$$\eta_{comp} = \frac{h_{2s} - h_1}{h_2 - h_1} \tag{11.3}$$

$$\eta_{turb} = \frac{h_3 - h_4}{h_3 - h_{4s}} \tag{11.4}$$

Another important feature of the Brayton cycle is the large amount of compressor work (also called back work) compared to turbine work. Thus, the compressor might require from 40 to 80% of the output of the turbine. This is particularly important when the actual cycle is considered, because the effect of the losses is to require a larger amount of compression work from a smaller amount of turbine work, and thus, the overall efficiency drops very rapidly with a decrease in the efficiencies of the compressor and turbine. In fact, if these efficiencies drop below about 60%, all the work of the turbine will be required to drive the compressor, and the overall efficiency will be zero. This is in sharp contrast to the Rankine cycle, where only 1 or 2% of the turbine work is required to drive the pump. This demonstrates the inherent advantage of the cycle utilizing a condensing working fluid, such that a much larger difference in specific volume between the expansion and compression processes is utilized effectively.

EXAMPLE 11.6   In an air-standard Brayton cycle the air enters the compressor at 0.1 MPa, 15°C. The pressure leaving the compressor is 1.0 MPa, and the maximum temperature in the cycle is 1100°C. Determine

1. The pressure and temperature at each point in the cycle
2. The compressor work, turbine work, and cycle efficiency

For each of the control volumes analyzed, the model is ideal gas with constant specific heat, value at 300 K, and each process is SSSF with no kinetic or potential energy changes. The diagram for this example is Fig. 11.19.

Control volume: Compressor.
Inlet state: P_1, T_1 known; state fixed.
Exit state: P_2 known.

Analysis:

First law:

$$w_c = h_2 - h_1$$

(Note that the compressor work w_c is here defined as work input to the compressor.)

Second law:

$$s_2 = s_1$$

so that

$$\frac{T_2}{T_1} = \left(\frac{P_2}{P_1}\right)^{(k-1)/k}$$

Solution

$$\left(\frac{P_2}{P_1}\right)^{(k-1)/k} = 10^{0.286} = 1.932 \qquad T_2 = 556.8 \text{ K}$$

$$w_c = h_2 - h_1 = C_p\left(T_2 - T_1\right)$$

$$= 1.0035\left(556.8 - 288.2\right) = 269.5 \text{ kJ/kg}$$

Control volume: Turbine.
Inlet state: P_3 ($= P_2$) known, T_3 known; state fixed.
Exit state: P_4 ($= P_1$) known.

Analysis:

First law:

$$w_t = h_3 - h_4$$

Second law:

$$s_3 = s_4$$

so that

$$\frac{T_3}{T_4} = \left(\frac{P_3}{P_4}\right)^{(k-1)/k}$$

Solution

$$\left(\frac{P_3}{P_4}\right)^{(k-1)/k} = 10^{0.286} = 1.932 \qquad T_4 = 710.8 \text{ K}$$

$$w_t = h_3 - h_4 = C_p(T_3 - T_4)$$

$$= 1.0035(1373.2 - 710.8) = 664.7 \text{ kJ/kg}$$

$$w_{\text{net}} = w_t - w_c = 664.7 - 269.5 = 395.2 \text{ kJ/kg}$$

Control volume: High-temperature heat exchanger.

Inlet state: State 2 fixed (as given).

Exit state: State 3 fixed (as given).

Analysis:

First law:

$$q_H = h_3 - h_2 = C_p(T_3 - T_2)$$

Solution

$$q_H = h_3 - h_2 = C_p(T_3 - T_2) = 1.0035(1373.2 - 556.8) = 819.3 \text{ kJ/kg}$$

Control volume: Low-temperature heat exchanger.

Inlet state: State 4 fixed (above).

Exit state: State 1 fixed (above).

Analysis:

First law:

$$q_L = h_4 - h_1 = C_p(T_4 - T_1)$$

Solution

$$q_L = h_4 - h_1 = C_p(T_4 - T_1) = 1.0035(710.8 - 288.2) = 424.1 \text{ kJ/kg}$$

Therefore,

$$\eta_{\text{th}} = \frac{w_{\text{net}}}{q_H} = \frac{395.2}{819.3} = 48.2\%$$

This may be checked by using Eq. 11.2.

$$\eta_{\text{th}} = 1 - \frac{1}{(P_2/P_1)^{(k-1)/k}} = 1 - \frac{1}{10^{0.286}} = 48.2\%$$

EXAMPLE 11.7   Consider a gas turbine with air entering the compressor under the same conditions as in Example 11.6 and leaving at a pressure of 1.0 MPa. The maximum temperature is 1100°C. Assume a compressor efficiency of 80%, a turbine efficiency of 85%, and a pressure drop between the compressor and turbine of 15 kPa. Determine the compressor work, turbine work, and cycle efficiency.

As in the previous example, for each control volume the model is ideal gas, constant specific heat, value at 300 K, and each process is SSSF with no kinetic or potential energy changes. In this example the diagram is Fig. 11.21.

Control volume: Compressor.

Inlet state: P_1, T_1 known; state fixed.

Exit state: P_2 known.

Analysis:

First law, real process:

$$w_c = h_2 - h_1$$

Second law, ideal process:

$$s_{2s} = s_1$$

so that

$$\frac{T_{2s}}{T_1} = \left(\frac{P_2}{P_1}\right)^{(k-1)/k}$$

In addition,

$$\eta_c = \frac{h_{2s} - h_1}{h_2 - h_1} = \frac{T_{2s} - T_1}{T_2 - T_1}$$

Solution

$$\left(\frac{P_2}{P_1}\right)^{(k-1)/k} = \frac{T_{2s}}{T_1} = 10^{0.286} = 1.932 \qquad T_{2s} = 556.8 \text{ K}$$

$$\eta_c = \frac{h_{2s} - h_1}{h_2 - h_1} = \frac{T_{2s} - T_1}{T_2 - T_1} = \frac{556.8 - 288.2}{T_2 - T_1} = 0.80$$

$$T_2 - T_1 = \frac{556.8 - 288.2}{0.80} = 335.8 \qquad T_2 = 624.0 \text{ K}$$

$$w_c = h_2 - h_1 = C_p\left(T_2 - T_1\right)$$

$$= 1.0035(624.0 - 288.2) = 337.0 \text{ kJ} / \text{kg}$$

Control volume: Turbine.

Inlet state: P_3 $(P_2 - \text{drop})$ known, T_3 known; state fixed.

Exit state: P_4 known.

Analysis:

First law, real process:

$$w_t = h_3 - h_4$$

Second law, ideal process:

$$s_{4s} = s_3$$

So that

$$\frac{T_3}{T_{4s}} = \left(\frac{P_3}{P_4}\right)^{(k-1)/k}$$

In addition,

$$\eta_t = \frac{h_3 - h_4}{h_3 - h_{4s}} = \frac{T_3 - T_4}{T_3 - T_{4s}}$$

Solution

$$P_3 = P_2 - \text{pressure drop} = 1.0 - 0.015 = 0.985 \text{ MPa}$$

$$\left(\frac{P_3}{P_4}\right)^{(k-1)/k} = \frac{T_3}{T_{4s}} = 9.85^{0.286} = 1.9236 \qquad T_{4s} = 713.9 \text{ K}$$

$$\eta_t = \frac{h_3 - h_4}{h_3 - h_{4s}} = \frac{T_3 - T_4}{T_3 - T_{4s}} = 0.85$$

$$T_3 - T_4 = 0.85(1373.2 - 713.9) = 560.4$$

$$T_4 = 812.8 \text{ K}$$

$$w_t = h_3 - h_4 = C_p(T_3 - T_4)$$

$$= 1.0035(1373.2 - 812.8) = 562.4 \text{ kJ / kg}$$

$$w_{\text{net}} = w_t - w_c = 562.4 - 337.0 = 225.4 \text{ kJ / kg}$$

Control volume: High-temperature heat exchanger.
Inlet state: State 2 fixed (as given).
Exit state: State 3 fixed (as given).

Analysis:

First law:

$$q_H = h_3 - h_2$$

Solution

$$q_H = h_3 - h_2 = C_p(T_3 - T_2)$$

$$= 1.0035 (1373.2 - 624.0) = 751.8 \text{ kJ / 1}$$

so that

$$\eta_{\text{th}} = \frac{w_{\text{net}}}{q_H} = \frac{225.4}{751.8} = 30.0\%$$

The following comparisons can be made between Examples 11.6 and 11.7.

	w_c	w_t	w_{net}	q_H	η_{th}
Example 11.6 (Ideal)	269.5	664.7	395.2	819.3	48.2
Example 11.7 (Actual)	337.0	562.4	225.4	751.8	30.0

As stated previously, the irreversibilities decrease the turbine work and increase the compressor work. Since the net work is the difference between these two, it decreases very rapidly as compressor and turbine efficiencies decrease. The development of compressors and turbines of high efficiency is therefore an important aspect of the development of gas turbines.

Note that in the ideal cycle (Example 11.6) about 41% of the turbine work is required to drive the compressor and 59% is delivered as net work. In the actual turbine (Example 11.7) 60% of the turbine work is required to drive the compressor and 40% is delivered as net work. Thus, if the net power of this unit is to be 10 000 kW, a 25 000-kW turbine and a 15 000-kW compressor are required. This result demonstrates that a gas turbine has a high back-work ratio.

11.10 THE SIMPLE GAS-TURBINE CYCLE WITH A REGENERATOR

The efficiency of the gas-turbine cycle may be improved by introducing a regenerator. The simple open-cycle gas-turbine cycle with a regenerator is shown in Fig. 11.22, and the corresponding ideal air-standard cycle with a regenerator is shown on the P–v and T–s diagrams. In cycle 1–2–x–3–4–y–1, the temperature of the exhaust gas leaving the turbine in state 4 is higher than the temperature of the gas leaving the compressor. Therefore,

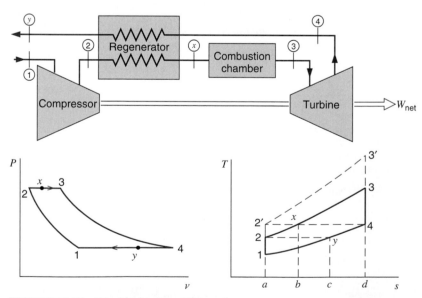

FIGURE 11.22 The ideal regenerative cycle.

heat can be transferred from the exhaust gases to the high-pressure gases leaving the compressor. If this is done in a counterflow heat exchanger, which is known as a regenerator, the temperature of the high-pressure gas leaving the regenerator, T_x, may, in the ideal case, have a temperature equal to T_4, the temperature of the gas leaving the turbine. Heat transfer from the external source is necessary only to increase the temperature from T_x to T_3. Area x–3–d–b–x represents the heat transferred, and area y–1–a–c–y represents the heat rejected.

The influence of pressure ratio on the simple gas-turbine cycle with a regenerator is shown by considering cycle 1–$2'$–$3'$–4–1. In this cycle the temperature of the exhaust gas leaving the turbine is just equal to the temperature of the gas leaving the compressor; therefore, there is no possibility of utilizing a regenerator. This can be shown more exactly by determining the efficiency of the ideal gas-turbine cycle with a regenerator.

The efficiency of this cycle with regeneration is found as follows, where the states are as given in Fig. 11.22.

$$\eta_{th} = \frac{w_{net}}{q_H} = \frac{w_t - w_c}{q_H}$$

$$q_H = C_p\left(T_3 - T_x\right)$$

$$w_t = C_p\left(T_3 - T_4\right)$$

But for an ideal regenerator, $T_4 = T_x$, and therefore $q_H = w_t$. Consequently,

$$\eta_{th} = 1 - \frac{w_c}{w_t} = 1 - \frac{C_p\left(T_2 - T_1\right)}{C_p\left(T_3 - T_4\right)}$$

$$= 1 - \frac{T_1\left(T_2 / T_1 - 1\right)}{T_3\left(1 - T_4 / T_3\right)} = 1 - \frac{T_1}{T_3}\frac{\left[\left(P_2 / P_1\right)^{(k-1)/k} - 1\right]}{\left[1 - \left(P_1 / P_2\right)^{(k-1)/k}\right]}$$

$$\eta_{th} = 1 - \frac{T_1}{T_3}\left(\frac{P_2}{P_1}\right)^{(k-1)/k}$$

Thus, for the ideal cycle with regeneration the thermal efficiency depends not only on the pressure ratio but also on the ratio of the minimum to maximum temperature. We note that, in contrast to the Brayton cycle, the efficiency decreases with an increase in pressure ratio. The thermal efficiency versus pressure ratio for this cycle is plotted in Fig. 11.20 for a value of

$$\frac{T_1}{T_3} = 0.25$$

The effectiveness or efficiency of a regenerator is given by the regenerator efficiency, which can best be defined by reference to Fig. 11.23. State x represents the high-pressure gas leaving the regenerator. In the ideal regenerator there would be only an infinitesimal temperature difference between the two streams, and the high-pressure gas would leave the regenerator at temperature T'_x, and $T'_x = T_4$. In an actual regenerator, which must operate with a finite temperature difference T_x, the actual temperature leaving

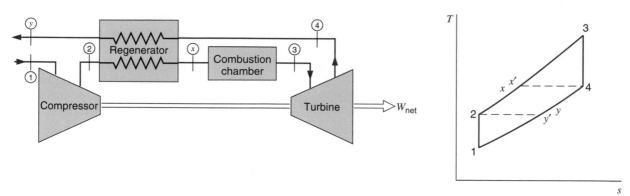

FIGURE11.23 Temperature–entropy diagram to illustrate the definition of regenerator efficiency.

the regenerator is therefore less than T'_x. The regenerator efficiency is defined by

$$\eta_{\text{reg}} = \frac{h_x - h_2}{h'_x - h_2}$$

If the specific heat is assumed to be constant, the regenerator efficiency is also given by the relation

$$\eta_{\text{reg}} = \frac{T_x - T_2}{T'_x - T_2}$$

It should be pointed out that a higher efficiency can be achieved by using a regenerator with a greater heat-transfer area. However, this also increases the pressure drop, which represents a loss, and both the pressure drop and the regenerator efficiency must be considered in determining which regenerator gives maximum thermal efficiency for the cycle. From an economic point of view, the cost of the regenerator must be weighed against the saving that can be effected by its use.

EXAMPLE 11.8   If an ideal regenerator is incorporated into the cycle of Example 11.6, determine the thermal efficiency of the cycle.

The diagram for this example is Fig. 11.23. Values are from Example 11.6. Therefore, for the analysis of the high-temperature heat exchanger (combustion chamber), from the first law,

$$q_H = h_3 - h_x$$

so that the solution is

$$T_x = T_4 = 710.8 \text{ K}$$

$$q_H = h_3 - h_x = C_p(T_3 - T_x) = 1.0035(1373.2 - 710.8) = 664.7 \text{ kJ/kg}$$

$$w_{\text{net}} = 395.2 \text{ kJ/kg} \quad \text{(from Example 11.6)}$$

$$\eta_{\text{th}} = \frac{395.2}{664.7} = 59.5\%$$

11.11 GAS-TURBINE POWER CYCLE CONFIGURATIONS

The Brayton cycle, as the idealized model for the gas turbine powerplant, has a reversible, adiabatic compressor and a reversible, adiabatic turbine. In the following example, we consider the effect of replacing these components with reversible, isothermal processes.

EXAMPLE 11.9   An air-standard power cycle has the same states as given in Example 11.6. In this cycle, however, the compressor and turbine are both reversible, isothermal processes. Calculate the compressor work and the turbine work, and compare with the results of Example 11.6.

Control volumes: Compressor, turbine.

Analysis:

For each reversible, isothermal process, from Eq. 9.21,

$$w = -\int_i^e v\, dP = -P_i v_i \ln\frac{P_e}{P_i} = -RT_i \ln\frac{P_e}{P_i}$$

Solution

For the compressor,

$$w = -0.287 \times 288.2 \times \ln 10 = -190.5 \text{ kJ / kg}$$

compared with −269.5 kJ/kg in the adiabatic compressor.

For the turbine,

$$w = -0.287 \times 1373.2 \times \ln 0.1 = +907.5 \text{ kJ / kg}$$

compared with + 664.7 kJ/kg in the adiabatic turbine.

It is found that the isothermal process would be preferable to the adiabatic process in both the compressor and turbine. The resulting cycle, called the Ericsson cycle, consists of two reversible, constant-pressure processes and two reversible, constant-temperature processes. The reason that the actual gas turbine does not attempt to emulate this cycle rather than the Brayton cycle, is that the compressor and turbine processes are both high-flow rate processes involving work-related devices in which it is not practical to attempt to transfer large quantities of heat. As a consequence, the processes tend to be essentially adiabatic, so that this becomes the process in the model cycle.

There is a modification of the Brayton/gas turbine cycle that tends to change its performance in the direction of the Ericsson cycle. This modification is to use multiple stages of compression with intercooling, and also multiple stages of expansion with reheat. Such a cycle with two stages of compression and expansion, and also incorporating a regenerator, is shown in Fig. 11.24. The air-standard cycle is given on the corresponding *T–s* diagram. It may be shown that for this cycle the maximum efficiency is obtained

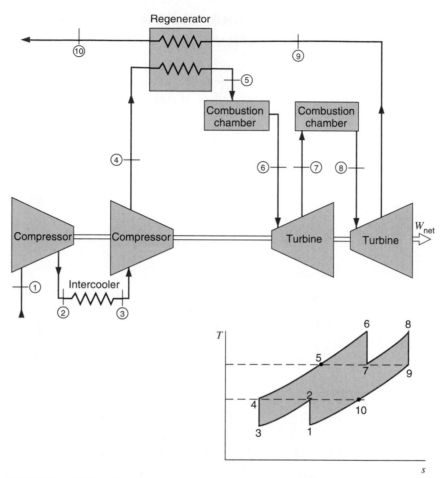

FIGURE 11.24 The ideal gas-turbine cycle utilizing intercooling, reheat, and a regenerator.

if equal pressure ratios are maintained across the two compressors and the two turbines. In this ideal cycle it is assumed that the temperature of the air leaving the intercooler, T_3, is equal to the temperature of the air entering the first stage of compression, T_1, and that the temperature after reheating, T_8, is equal to the temperature entering the first turbine, T_6. Furthermore, in the ideal cycle it is assumed that the temperature of the high-pressure air leaving the regenerator, T_5, is equal to the temperature of the low-pressure air leaving the turbine, T_9.

If a large number of stages of compression and expansion are used, it is evident that the Ericsson cycle is approached. This is shown in Fig. 11.25. In practice, the economical limit to the number of stages is usually two or three. The turbine and compressor losses and pressure drops that have already been discussed would be involved in any actual unit employing this cycle.

There are a variety of ways in which the turbines and the compressors using this cycle can be utilized. Two possible arrangements for closed cycles are shown in Fig. 11.26. One advantage frequently sought in a given arrangement is ease of control of the unit under various loads. Detailed discussion of this point, however, is beyond the scope of this text.

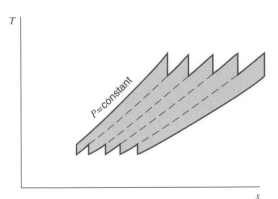

FIGURE 11.25 Temperature–entropy diagram that shows how the gas-turbine cycle with many stages approaches the Ericsson cycle.

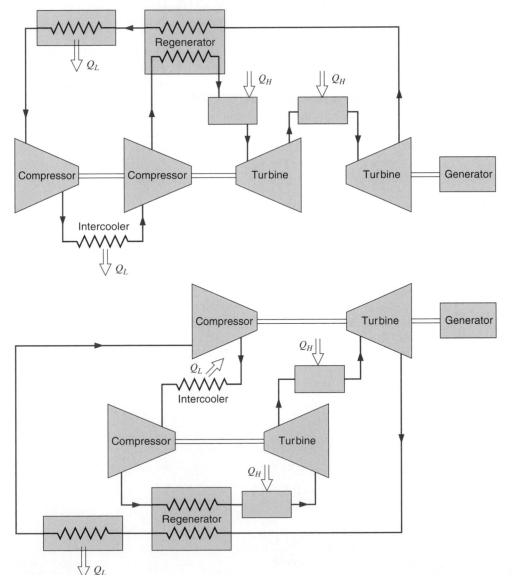

FIGURE 11.26 Some arrangements of components that may be utilized in stationary gas-turbine power plants.

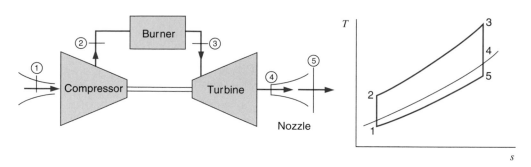

FIGURE 11.27 The ideal gas-turbine cycle for a jet engine.

11.12 THE AIR-STANDARD CYCLE FOR JET PROPULSION

The next air-standard power cycle we consider is utilized in jet propulsion. In this cycle the work done by the turbine is just sufficient to drive the compressor. The gases are expanded in the turbine to a pressure for which the turbine work is just equal to the compressor work. The exhaust pressure of the turbine will then be greater than that of the surroundings, and the gas can be expanded in a nozzle to the pressure of the surroundings. Since the gases leave at a high velocity, the change in momentum that the gases undergo gives a thrust to the aircraft in which the engine is installed. The air-standard cycle for this situation is shown in Fig. 11.27. The principles governing this cycle follow from the analysis of the Brayton cycle plus that for a reversible, adiabatic nozzle.

EXAMPLE 11.10 Consider an ideal jet propulsion cycle in which air enters the compressor at 0.1 MPa, 15°C. The pressure leaving the compressor is 1.0 MPa, and the maximum temperature is 1100°C. The air expands in the turbine to a pressure at which the turbine work is just equal to the compressor work. On leaving the turbine, the air expands in a nozzle to 0.1 MPa. The process is reversible and adiabatic. Determine the velocity of the air leaving the nozzle.

The model used is ideal gas, constant specific heat, value at 300 K, and each process is SSSF with no potential energy change. The only kinetic energy change occurs in the nozzle. The diagram is shown in Fig. 11.27.

The compressor analysis is the same as in Example 11.6. From the results of that solution,

$$P_1 = 0.1 \text{ MPa} \qquad T_1 = 288.2 \text{ K}$$

$$P_2 = 1.0 \text{ MPa} \qquad T_2 = 556.8 \text{ K}$$

$$w_c = 269.5 \text{ kJ / kg}$$

The turbine analysis is also the same as in Example 11.6. Here, however,

$$P_3 = 1.0 \text{ MPa} \qquad T_3 = 1373.2 \text{ K}$$

$$w_c = w_t = C_p(T_3 - T_4) = 269.5 \text{ kJ / kg}$$

$$T_3 - T_4 = \frac{269.5}{1.0035} = 268.6 \qquad T_4 = 1104.6 \text{ K}$$

so that

$$\frac{T_3}{T_4} = \left(\frac{P_3}{P_4}\right)^{(k-1)/k} = \frac{1373.2}{1104.6} = 1.2432$$

$$\frac{P_3}{P_4} = 2.142 \qquad P_4 = 0.4668 \text{ MPa}$$

Control volume: Nozzle.

Inlet state: State 4 fixed (above).

Exit state: P_5 known.

Analysis:

First law:

$$h_4 = h_5 + \frac{\mathbf{V}_5^2}{2}$$

Second law:

$$s_4 = s_5$$

Solution

Since P_5 is 0.1 MPa, from the second law we find that $T_5 = 710.8$ K.

$$\mathbf{V}_5^2 = 2C_{p0}(T_4 - T_5)$$
$$\mathbf{V}_5^2 = 2 \times 1000 \times 1.0035(1104.6 - 710.8)$$
$$\mathbf{V}_5 = 889 \text{ m/s}$$

11.13 THE OTTO CYCLE

In Section 11.1, we discussed power cycles incorporating either SSSF processes or cylinder/piston boundary work processes. In that section, it was noted that for the SSSF-process, there is no work in a constant-pressure process. Each of the SSSF power cycles presented in subsequent sections of this chapter incorporated two constant-pressure heat transfer processes. It was further noted in Section 11.1 that in a boundary movement work process, there is no work in a constant-volume process. In the next three sections, we will present ideal air-standard power cycles for cylinder/piston boundary movement work processes, each example of which includes either one or two constant-volume heat transfer processes.

The air-standard Otto cycle is an ideal cycle that approximates a spark-ignition internal-combustion engine. This cycle is shown on the *P–v* and *T–s* diagrams of Fig. 11.28. Process 1–2 is an isentropic compression of the air as the piston moves from crank-end dead center to head-end dead center. Heat is then added at constant volume while the piston is momentarily at rest at head-end dead center. (This process corresponds to the ignition of the fuel–air mixture by the spark and the subsequent burning in the ac-

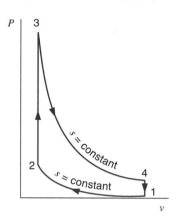 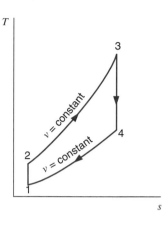

FIGURE 11.28 The air-standard Otto cycle.

tual engine.) Process 3–4 is an isentropic expansion, and process 4–1 is the rejection of heat from the air while the piston is at crank-end dead center.

The thermal efficiency of this cycle is found as follows, assuming constant specific heat of air.

$$\eta_{th} = \frac{Q_H - Q_L}{Q_H} = 1 - \frac{Q_L}{Q_H} = 1 - \frac{mC_v(T_4 - T_1)}{mC_v(T_3 - T_2)}$$

$$= 1 - \frac{T_1(T_4/T_1 - 1)}{T_2(T_3/T_2 - 1)}$$

We note further that

$$\frac{T_2}{T_1} = \left(\frac{V_1}{V_2}\right)^{k-1} = \left(\frac{V_4}{V_3}\right)^{k-1} = \frac{T_3}{T_4}$$

Therefore,

$$\frac{T_3}{T_2} = \frac{T_4}{T_1}$$

and

$$\eta_{th} = 1 - \frac{T_1}{T_2} = 1 - (r_v)^{1-k} = 1 - \frac{1}{r_v^{k-1}} \tag{11.5}$$

where

$$r_v = \text{compression ratio} = \frac{V_1}{V_2} = \frac{V_4}{V_3}$$

The important thing to note is that the efficiency of the air-standard Otto cycle is a function only of the compression ratio, and that the efficiency is increased by increasing the compression ratio. Figure 11.29 shows a plot of the air-standard cycle thermal efficiency versus compression ratio. It is also true of an actual spark-ignition engine that the efficiency can be increased by increasing the compression ratio. The trend toward higher compression ratios is prompted by the effort to obtain higher thermal efficiency. In the actual engine there is an increased tendency for the fuel to detonate as the compression ratio is increased. After detonation the fuel burns rapidly, and strong pressure waves pre-

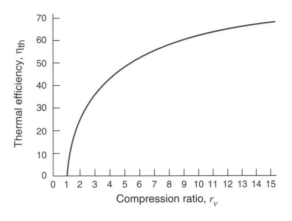

FIGURE 11.29 Thermal efficiency of the Otto cycle as a function of compression ratio.

sent in the engine cylinder give rise to the so-called spark knock. Therefore, the maximum compression ratio that can be used is fixed by the fact that detonation must be avoided. The advance in compression ratios over the years in the actual engine was originally made possible by developing fuels with better antiknock characteristics, primarily through the addition of tetraethyl lead. More recently, however, nonleaded gasolines with good antiknock characteristics have been developed in an effort to reduce atmospheric contamination.

Some of the most important ways in which the actual open-cycle spark-ignition engine deviates from the air-standard cycle are as follows.

1. The specific heats of the actual gases increase with an increase in temperature.

2. The combustion process replaces the heat-transfer process at high temperature, and combustion may be incomplete.

3. Each mechanical cycle of the engine involves an inlet and an exhaust process and, because of the pressure drop through the valves, a certain amount of work is required to charge the cylinder with air and exhaust the products of combustion.

4. There will be considerable heat transfer between the gases in the cylinder and the cylinder walls.

5. There will be irreversibilities associated with pressure and temperature gradients.

EXAMPLE 11.11 The compression ratio in an air-standard Otto cycle is 8. At the beginning of the compression stroke the pressure is 0.1 MPa and the temperature is 15°C. The heat transfer to the air per cycle is 1800 kJ/kg air. Determine

1. The pressure and temperature at the end of each process of the cycle

2. The thermal efficiency

3. The mean effective pressure

Control mass: Air inside cylinder.
Diagram: Fig. 11.28.

State information: $P_1 = 0.1$ MPa, $T_1 = 288.2$ K.

Process information: Four processes known (Fig. 11.28). Also $r_v = 8$ and $q_H = 1800$ kJ/kg.

Model: Ideal gas, constant specific heat, value at 300 K.

Analysis:

Second law for compression process 1–2:

$$s_2 = s_1$$

so that

$$\frac{T_2}{T_1} = \left(\frac{V_1}{V_2}\right)^{k-1}$$

$$\frac{P_2}{P_1} = \left(\frac{V_1}{V_2}\right)^{k}$$

First law for heat addition process 2–3:

$$q_H = {}_2q_3 = u_3 - u_2 = C_v\left(T_3 - T_2\right)$$

Second law for expansion process 3–4:

$$s_4 = s_3$$

so that

$$\frac{T_3}{T_4} = \left(\frac{V_4}{V_3}\right)^{k-1}$$

$$\frac{P_3}{P_4} = \left(\frac{V_4}{V_3}\right)^{k}$$

In addition,

$$\eta_{th} = 1 - \frac{1}{r_v^{k-1}} \qquad \mathrm{mep} = \frac{w_{net}}{v_1 - v_2}$$

Solution

$$v_1 = \frac{0.287 \times 288.2}{100} = 0.827 \text{ m}^3/\text{kg}$$

$$\frac{T_2}{T_1} = \left(\frac{V_1}{V_2}\right)^{k-1} = 8^{0.4} = 2.3 \qquad T_2 = 662 \text{ K}$$

$$\frac{P_2}{P_1} = \left(\frac{V_1}{V_2}\right)^{k} = 8^{1.4} = 18.38 \qquad P_2 = 1.838 \text{ MPa}$$

$$v_2 = \frac{0.827}{8} = 0.1034 \text{ m}^3/\text{kg}$$

$${}_2q_3 = C_v\left(T_3 - T_2\right) = 1800 \text{ kJ/kg}$$

$$T_3 - T_2 = \frac{1800}{0.7165} = 2512 \qquad T_3 = 3174 \text{ K}$$

$$\frac{T_3}{T_2} = \frac{P_3}{P_2} = \frac{3174}{662} = 4.795 \qquad P_3 = 8.813 \text{ MPa}$$

$$\frac{T_3}{T_4} = \left(\frac{V_4}{V_3}\right)^{k-1} = 8^{0.4} = 2.3 \qquad T_4 = 1380 \text{ K}$$

$$\frac{P_3}{P_4} = \left(\frac{V_4}{V_3}\right)^{k} = 8^{1.4} = 18.38 \qquad P_4 = 0.4795 \text{ MPa}$$

$$\eta_{\text{th}} = 1 - \frac{1}{r_v^{k-1}} = 1 - \frac{1}{8^{0.4}} = 1 - \frac{1}{2.3} = 1 - 0.435 = 0.565 = 56.5\%$$

This can be checked by finding the heat rejected.

$$_4q_1 = C_v(T_1 - T_4) = 0.7165(288.2 - 1380) = -782.3 \text{ kJ/kg}$$

$$\eta_{\text{th}} = 1 - \frac{782.3}{1800} = 1 - 0.435 = 0.565 = 56.5\%$$

$$w_{\text{net}} = 1800 - 782.3 = 1017.7 \text{ kJ/kg} = (v_1 - v_2)\text{mep}$$

$$\text{mep} = \frac{1017.7}{(0.827 - 0.1034)} = 1406 \text{ kPa}$$

This is a high value for mean effective pressure, largely because the two constant-volume heat transfer processes keep the total volume change to a minimum (compared with a Brayton cycle, for example). Thus, the Otto cycle is a good model to emulate in the cylinder/piston internal combustion engine. At the other extreme, a low mean effective pressure means a large piston displacement for a given power output, which in turn means high frictional losses in an actual engine.

EXAMPLE 11.11E   The compression ratio in an air-standard Otto cycle is 8. At the beginning of the compression stroke the pressure is 14.7 lbf/in.² and the temperature is 60 F. The heat transfer to the air per cycle is 800 Btu/lbm air. Determine:

1. The pressure and temperature at the end of each process of the cycle

2. The thermal efficiency

3. The mean effective pressure

Control mass: Air inside cylinder.

Diagram: Fig. 11.28.

State information: $P_1 = 14.7$ lbf/in.² $T_1 = 520$ R.

Process of information: Four processes known (Fig. 11.28). Also, $r_v = 8$ and $q_H = 800$ Btu/lbm.

Model: Ideal gas, constant specific heat, value at 80 F.

Analysis:

Second law for compression process 1–2:

$$s_2 = s_1$$

so that

$$\frac{T_2}{T_1} = \left(\frac{V_1}{V_2}\right)^{k-1}$$

$$\frac{P_2}{P_1} = \left(\frac{V_1}{V_2}\right)^{k}$$

First law for heat addition process 2–3:

$$q_H = {}_2q_3 = u_3 - u_2 = C_v\left(T_3 - T_2\right)$$

Second law for expansion process 3–4:

$$s_4 = s_3$$

so that

$$\frac{T_3}{T_4} = \left(\frac{V_4}{V_3}\right)^{k-1}$$

$$\frac{P_3}{P_4} = \left(\frac{V_4}{V_3}\right)^{k}$$

In addition,

$$\eta_{\text{th}} = 1 - \frac{1}{r_v^{k-1}} \qquad \text{mep} = \frac{w_{\text{net}}}{\left(v_1 - v_2\right)}$$

Solution

$$v_1 = \frac{53.34 \times 520}{14.7 \times 144} = 13.08 \text{ ft}^3/\text{lbm}$$

$$\frac{T_2}{T_1} = \left(\frac{V_1}{V_2}\right)^{k-1} = 8^{0.4} = 2.3 \qquad T_2 = 2.3(520) = 1197 \text{ R}$$

$$\frac{P_2}{P_1} = \left(\frac{V_1}{V_2}\right)^{k} = 8^{1.4} = 18.4 \qquad P_2 = 18.4(14.7) = 270.3 \text{ lbf/in.}^2$$

$$v_2 = \frac{13.08}{8} = 1.637 \text{ ft}^3/\text{lbm}$$

$${}_2q_3 = C_v\left(T_3 - T_2\right) = 800 \text{ Btu/lbm}$$

$$T_3 - T_2 = \frac{800}{0.171} = 4690 \text{ R} \qquad T_3 = 1197 + 4690 = 5887 \text{ R}$$

$$\frac{T_3}{T_2} = \frac{P_3}{P_2} = \frac{5887}{1197} = 4.92 \qquad P_3 = 4.92(270.3) = 1331 \text{ lbf/in.}^2$$

$$\frac{T_3}{T_4} = \left(\frac{V_4}{V_3}\right)^{k-1} = 8^{0.4} = 2.3 \qquad T_4 = \frac{5887}{2.3} = 2558 \text{ R}$$

$$\frac{P_3}{P_4} = \left(\frac{V_4}{V_3}\right)^{k} = 8^{1.4} = 18.4 \qquad P_4 = \frac{1331}{18.4} = 73.0 \text{ lbf/in.}^2$$

$$\eta_{\text{th}} = 1 - \frac{1}{r_v^{k-1}} = 1 - \frac{1}{8^{0.4}} = 1 - \frac{1}{2.3} = 1 - 0.435 = 0.565$$

This can be checked by finding the heat rejected.

$$_4q_1 = C_v(T_1 - T_4) = 0.171(520 - 2558) = -348 \text{ Btu/lbm}$$

$$\eta_{\text{th}} = 1 - \frac{348}{800} = 1 - 0.435 = 0.565$$

$$w_{\text{net}} = 800 - 348 = 452 \text{ Btu/lbm} = (v_1 - v_2)\text{mep}$$

$$\text{mep} = \frac{452 \times 778}{(13.08 - 1.637)144} = 213.5 \text{ lbf/in.}^2$$

This is a high value for mean effective pressure, largely because the two constant-volume heat transfer processes keeps the total volume change to a minimum (compared with a Brayton cycle, for example). Thus, the Otto cycle is a good model to emulate in the cylinder/piston internal combustion engine. At the other extreme, a low mean effective pressure means a large piston displacement for a given power output, which in turn means high frictional losses in an actual engine.

11.14 THE DIESEL CYCLE

The air-standard Diesel cycle is shown in Fig. 11.30. This is the ideal cycle for the Diesel engine, which is also called the compression-ignition engine.

In this cycle the heat is transferred to the working fluid at constant pressure. This process corresponds to the injection and burning of the fuel in the actual engine. Since the gas is expanding during the heat addition in the air-standard cycle, the heat transfer must be just sufficient to maintain constant pressure. When state 3 is reached the heat addition ceases and the gas undergoes an isentropic expansion, process 3–4, until the piston reaches crank-end dead center. As in the air-standard Otto cycle, a constant-volume rejection of heat at crank-end dead center replaces the exhaust and intake processes of the actual engine.

The efficiency of the Diesel cycle is given by the relation

$$\eta_{\text{th}} = 1 - \frac{Q_L}{Q_H} = 1 - \frac{C_v(T_4 - T_1)}{C_p(T_3 - T_2)} = 1 - \frac{T_1(T_4/T_1 - 1)}{kT_2(T_3/T_2 - 1)} \qquad (11.6)$$

It is important to note that the isentropic compression ratio is greater than the isentropic expansion ratio in the Diesel cycle. In addition, for a given state before compres-

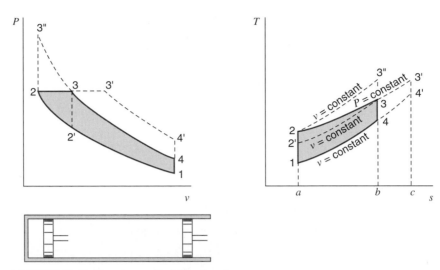

FIGURE 11.30 The air-standard diesel cycle.

sion and a given compression ratio (that is, given states 1 and 2), the cycle efficiency decreases as the maximum temperature increases. This is evident from the *T–s* diagram, because the constant-pressure and constant-volume lines converge, and increasing the temperature from 3 to 3′ requires a large addition of heat (area 3–3′–*c*–*b*–3) and results in a relatively small increase in work (area 3–3′–4′–4–3).

There are a number of comparisons between the Otto cycle and the Diesel cycle, but here we will note only two. Consider Otto cycle 1–2–3″–4–1 and Diesel cycle 1–2–3–4–1, which have the same state at the beginning of the compression stroke and the same piston displacement and compression ratio. It is evident from the *T–s* diagram that the Otto cycle has the higher efficiency. In practice, however, the Diesel engine can operate on a higher compression ratio than the spark-ignition engine. The reason is that in the spark-ignition engine an air–fuel mixture is compressed, and detonation (spark knock) becomes a serious problem if too high a compression ratio is used. This problem does not exist in the Diesel engine because only air is compressed during the compression stroke.

Therefore, we might compare an Otto cycle with a Diesel cycle and in each case select a compression ratio that might be achieved in practice. Such a comparison can be made by considering Otto cycle 1–2′–3–4–1 and Diesel cycle 1–2–3–4–1. The maximum pressure and temperature are the same for both cycles, which means that the Otto cycle has a lower compression ratio than the Diesel cycle. It is evident from the *T–s* diagram that in this case the Diesel cycle has the higher efficiency. Thus, the conclusions drawn from a comparison of these two cycles must always be related to the basis on which the comparison has been made.

The actual compression-ignition open cycle differs from the air-standard Diesel cycle in much the same way that the spark-ignition open cycle differs from the air-standard Otto cycle.

EXAMPLE 11.12 An air-standard Diesel cycle has a compression ratio of 18, and the heat transferred to the working fluid per cycle is 1800 kJ/kg. At the beginning of the compression process the pressure is 0.1 MPa and the temperature is 15°C. Determine

1. The pressure and temperature at each point in the cycle
2. The thermal efficiency
3. The mean effective pressure

Control mass: Air inside cylinder.

Diagram: Figure 11.30.

State information: $P_1 = 0.1$ MPa, $T_1 = 288.2$ K.

Process information: Four processes known (Fig. 11.30). Also $r_v = 18$ and $q_H = 1800$ kJ/kg.

Model: Ideal gas, constant specific heat, value at 300 K.

Analysis:

Second law for compression process 1–2:

$$s_2 = s_1$$

so that

$$\frac{T_2}{T_1} = \left(\frac{V_1}{V_2}\right)^{k-1}$$

$$\frac{P_2}{P_1} = \left(\frac{V_1}{V_2}\right)^{k}$$

First law for heat addition process 2–3:

$$q_H = {}_2q_3 = C_p\left(T_3 - T_2\right)$$

Second law for expansion process 3–4:

$$s_4 = s_3$$

so that

$$\frac{T_3}{T_4} = \left(\frac{V_4}{V_3}\right)^{k-1}$$

In addition,

$$\eta_{\text{th}} = \frac{w_{\text{net}}}{q_H} \qquad \text{mep} = \frac{w_{\text{net}}}{v_1 - v_2}$$

Solution

$$v_1 = \frac{0.287 \times 288.2}{100} = 0.827 \text{ m}^3/\text{kg}$$

$$v_2 = \frac{v_1}{18} = \frac{0.827}{18} = 0.045\,95 \text{ m}^3/\text{kg}$$

$$\frac{T_2}{T_1} = \left(\frac{V_1}{V_2}\right)^{k-1} = 18^{0.4} = 3.1777 \qquad T_2 = 915.8 \text{ K}$$

$$\frac{P_2}{P_1} = \left(\frac{V_1}{V_2}\right)^{k} = 18^{1.4} = 57.2 \qquad P_2 = 5.72 \text{ MPa}$$

$$q_H = {}_2q_3 = C_p(T_3 - T_2) = 1800 \text{ kJ/kg}$$

$$T_3 - T_2 = \frac{1800}{1.0035} = 1794 \qquad T_3 = 2710 \text{ K}$$

$$\frac{V_3}{V_2} = \frac{T_3}{T_2} = \frac{2710}{915.8} = 2.959 \qquad v_3 = 0.135\,98 \text{ m}^3/\text{kg}$$

$$\frac{T_3}{T_4} = \left(\frac{V_4}{V_3}\right)^{k-1} = \left(\frac{0.827}{0.135\,98}\right)^{0.4} = 2.0588 \qquad T_4 = 1316 \text{ K}$$

$$q_L = {}_4q_1 = C_v(T_1 - T_4) = 0.7165(288.2 - 1316) = -736.6 \text{ kJ/kg}$$

$$w_{net} = 1800 - 736.6 = 1063.4 \text{ kJ/kg}$$

$$\eta_{th} = \frac{w_{net}}{q_H} = \frac{1063.4}{1800} = 59.1\%$$

$$\text{mep} = \frac{w_{net}}{v_1 - v_2} = \frac{1063.4}{0.827 - 0.045\,95} = 1362 \text{ kPa}$$

11.15 The Stirling Cycle

The final air-standard power cycle to be discussed is the Stirling cycle, which is shown on the *P–v* and *T–s* diagrams of Fig. 11.31. Heat is transferred to the working fluid during the constant-volume process 2–3 and also during the isothermal expansion process 3–4. Heat is rejected during the constant-volume process 4–1 and also during the isothermal compression process 1–2. Thus, this cycle is the same as the Otto cycle with the adiabatic

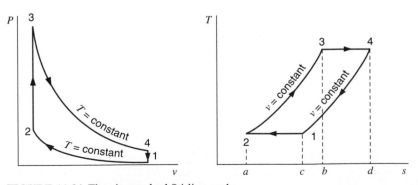

FIGURE 11.31 The air-standard Stirling cycle.

processes of that cycle replaced with isothermal processes. Since the Stirling cycle includes two constant-volume heat transfer processes, keeping the total volume change during the cycle to a minimum, it is a good candidate for a cylinder/piston boundary work application; it should have a high mean effective pressure.

Stirling-cycle engines have been developed in recent years as external-combustion engines with regeneration. The significance of regeneration is noted from the ideal case shown in Fig. 11.31. Note that the heat transfer to the gas between states 2 and 3, area 2–3–*b*–*a*–2, is exactly equal to the heat transfer from the gas between states 4 and 1, area 1–4–*d*–*c*–1. Thus, in the ideal cycle, all external heat supplied Q_H takes place in the isothermal expansion process 3–4, and all external heat rejection Q_L takes place in the isothermal compression process 1–2. Since all heat is supplied and rejected isothermally, the efficiency of this cycle equals the efficiency of a Carnot cycle operating between the same temperatures. The same conclusions would be drawn in the case of an Ericsson cycle, which was discussed briefly in Section 11.11, if that cycle were to include a regenerator as well.

11.16 INTRODUCTION TO REFRIGERATION SYSTEMS

In Section 11.1, we discussed cyclic heat engines consisting of four separate processes, either SSSF or cylinder/piston boundary movement work devices. We further allowed for a working fluid that changes phase or for one that is single-phase throughout the cycle. We then considered a power system comprised of four reversible SSSF processes, two of which were constant-pressure heat transfer processes, for simplicity of equipment requirements, since these two processes involve no work. It was further assumed that the other two work-involved processes were adiabatic and therefore isentropic. The resulting power cycle appeared as in Fig. 11.2.

We now consider the basic ideal refrigeration system cycle in exactly the same terms as those described above, except that each process is the reverse of that in the power cycle. The result is the ideal cycle shown in Fig. 11.32. Note that if the entire cycle takes place inside the two-phase liquid-vapor dome, the resulting cycle is, as with the power cycle, the Carnot cycle, since the two constant-pressure processes are also isothermal. Otherwise, this cycle is not a Carnot cycle. It is also noted, as before, that the net

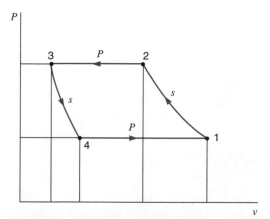

FIGURE 11.32 Four-process refrigeration cycle.

work input to the cycle is equal to the area enclosed by the process lines 1–2–3–4–1, independently of whether the individual processes are SSSF or cylinder/piston boundary movement.

In the next section, we make one modification to this idealized basic refrigeration system cycle in presenting and applying the model of refrigeration and heat pump systems.

11.17 THE VAPOR-COMPRESSION REFRIGERATION CYCLE

In this section, we consider the ideal refrigeration cycle for a working substance that changes phase during the cycle, in a manner equivalent to that done with the Rankine power cycle in Section 11.2. In doing so, we note that state 3 in Fig. 11.32 is saturated liquid at the condenser temperature and state 1 is saturated vapor at the evaporator temperature. This means that the isentropic expansion process from 3–4 will be in the two-phase region, and mostly liquid. As a consequence, there will be very little work output from this process, such that it is not worth the cost of including this piece of equipment in the system. We therefore replace the turbine with a throttling device, usually a valve or a length of small-diameter tubing, by which the working fluid is throttled from the high-pressure to the low-pressure side. The resulting cycle becomes the ideal model for a vapor-compression refrigeration system, which is shown in Fig. 11.33. Saturated vapor at low pressure enters the compressor and undergoes a reversible adiabatic compression, process 1–2. Heat is then rejected at constant pressure in process 2–3, and the working fluid exits the condenser as saturated liquid. An adiabatic throttling process, 3–4, follows, and the working fluid is then evaporated at constant pressure, process 4–1, to complete the cycle.

The similarity of this cycle to the reverse of the Rankine cycle has already been noted. We also note the difference between this cycle and the ideal Carnot cycle, in which the working fluid always remains inside the two-phase region, 1′–2′–3–4′–1′. It is much more expedient to have a compressor handle only vapor than a mixture of liquid and vapor, as would be required in process 1′–2′ of the Carnot cycle. It is virtually impossible to compress, at a reasonable rate, a mixture such as that represented by state 1′ and still maintain equilibrium between liquid and vapor. The other difference, that of replacing the turbine by the throttling process, has already been discussed.

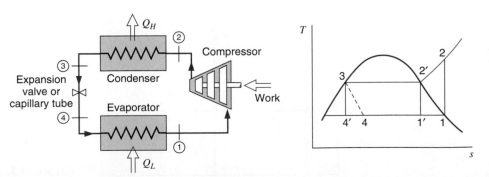

FIGURE 11.33 The ideal vapor-compression refrigeration cycle.

It should be pointed out that the system described in Fig. 11.33 can be used for either of two purposes. The first use is as a refrigeration system, in which case it is desired to maintain a space at a low temperature T_1 relative to the ambient temperature T_3. (In a real system, it would be necessary to allow a finite temperature difference in both the evaporator and condenser to provide a finite rate of heat transfer in each.) Thus, the reason for building the system in this case is the quantity q_L. The measure of performance of a refrigeration system is given in terms of the coefficient of performance, β, which was defined in Chapter 7 as

$$\beta = \frac{q_L}{w_c} \tag{11.7}$$

The second use of the system described in Fig. 11.33 is as a heat pump system, in which case it is desired to maintain a space at a temperature T_3 above that of the ambient (or other source) T_1. In this case, the reason for building the system is the quantity q_H, and the coefficient of performance for the heat pump, β', is now

$$\beta' = \frac{q_H}{w_c} \tag{11.8}$$

Refrigeration systems and heat pump systems are, of course, different in terms of design variables, but the analysis of the two is the same. When we discuss refrigerators in this and the following two sections, it should be kept in mind that the same comments generally apply to heat pump systems, as well.

EXAMPLE 11.13   Consider an ideal refrigeration cycle which uses R-12 as the working fluid. The temperature of the refrigerant in the evaporator is −20°C and in the condenser it is 40°C. The refrigerant is circulated at the rate of 0.03 kg/s. Determine the coefficient of performance and the capacity of the plant in rate of refrigeration.

The diagram for this example is as shown in Fig. 11.33. For each control volume analyzed, the thermodynamic model is the R-12 tables. Each process is SSSF with no changes in kinetic or potential energy.

    Control volume: Compressor.

    Inlet state: T_1 known, saturated vapor; state fixed.

    Exit state: P_2 known (saturation presssure at T_3).

Analysis:

First law:

$$w_c = h_2 - h_1$$

Second law:

$$s_2 = s_1$$

Solution

At $T_3 = 40°C$,

$$P_g = P_2 = 0.9607 \text{ MPa}$$

From the R-12 tables,

$$h_1 = 178.61 \qquad s_1 = 0.7082$$

Therefore,

$$s_2 = s_1 = 0.7082$$

so that

$$T_2 = 50.8°C \quad \text{and} \quad h_2 = 211.38$$

$$w_c = h_2 - h_1 = 211.38 - 178.61 = 32.77 \text{ kJ / kg}$$

Control volume: Expansion valve.
Inlet state: T_3 known, saturated liquid; state fixed.
Exit state: T_4 known.

Analysis:

First law:

$$h_3 = h_4$$

Solution

$$h_4 = h_3 = 74.53$$

Control volume: Evaporator.
Inlet state: State 4 known (as given).
Exit state: State 1 known (as given).

Analysis:

First law:

$$q_L = h_1 - h_4$$

Solution

$$q_L = h_1 - h_4 = 178.61 - 74.53 = 104.08 \text{ kJ / kg}$$

Therefore,

$$\beta = \frac{q_L}{w_c} = \frac{104.08}{32.77} = 3.18$$

$$\text{Capacity} = 104.08 \times 0.03 = 3.12 \text{ kW}$$

11.18 WORKING FLUIDS FOR VAPOR-COMPRESSION REFRIGERATION SYSTEMS

A much larger number of different working fluids (refrigerants) are utilized in vapor-compression refrigeration systems than in vapor power cycles. Ammonia and sulfur dioxide were important in the early days of vapor-compression refrigeration, but both are highly toxic and therefore dangerous substances. For many years now, the principal refrigerants have been the halogenated hydrocarbons, which are marketed under the trade names of Freon and Genatron. For example, dichlorodifluoromethane (CCl_2F_2) is known as Freon-12 and Genatron-12, and therefore as refrigerant-12 or R-12. This group of substances, known commonly as chlorofluorocarbons or CFCs, are chemically very stable at ambient temperature, especially those lacking any hydrogen atoms. This characteristic is necessary for a refrigerant working fluid. This same characteristic, however, has devastating consequences if the gas, having leaked from an appliance into the atmosphere, spends many years slowly diffusing upward into the stratosphere. There it is broken down, releasing chlorine, which destroys the protective ozone layer of the stratosphere. It is therefore of overwhelming importance to us all to eliminate completely the widely used but life-threatening CFCs, particularly R-11 and R-12, and to develop suitable and acceptable replacements. The CFCs containing hydrogen (often termed HCFCs), such as R-22, have shorter atmospheric lifetimes, and therefore are not as likely to reach the stratosphere before being broken up and rendered harmless. The most desirable fluids, called HFCs, contain no chlorine atoms at all.

There are two important considerations when selecting refrigerant working fluids: the temperature at which refrigeration is needed and the type of equipment to be used.

As the refrigerant undergoes a change of phase during the heat transfer process, the pressure of the refrigerant will be the saturation pressure during the heat supply and heat rejection processes. Low pressures mean large specific volumes and correspondingly large equipment. High pressures mean smaller equipment, but it must be designed to withstand higher pressure. In particular, the pressures should be well below the critical pressure. For extremely low temperature applications a binary fluid system may be used by cascading two separate systems.

The type of compressor used has a particular bearing on the refrigerant. Reciprocating compressors are best adapted to low specific volumes, which means higher pressures, whereas centrifugal compressors are most suitable for low pressures and high specific volumes.

It is also important that the refrigerants used in domestic appliances be nontoxic. Other beneficial characteristics, in addition to being environmentally acceptable, are miscibility with compressor oil, dielectric strength, stability, and low cost. Refrigerants, however, have an unfortunate tendency to cause corrosion. For given temperatures during evaporation and condensation, not all refrigerants have the same coefficient of performance for the ideal cycle. It is, of course, desirable to use the refrigerant with the highest coefficient of performance, other factors permitting.

11.19 DEVIATION OF THE ACTUAL VAPOR-COMPRESSION REFRIGERATION CYCLE FROM THE IDEAL CYCLE

The actual refrigeration cycle deviates from the ideal cycle primarily because of pressure drops associated with fluid flow and heat transfer to or from the surroundings. The actual cycle might approach the one shown in Fig. 11.34.

The vapor entering the compressor will probably be superheated. During the compression process there are irreversibilities and heat transfer either to or from the surroundings, depending on the temperature of the refrigerant and the surroundings. Therefore, the

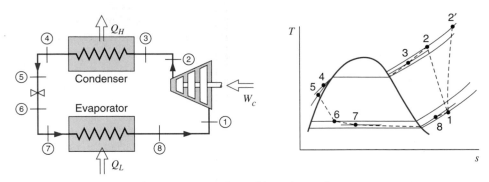

FIGURE 11.34 The actual vapor-compression refrigeration cycle.

entropy might increase or decrease during this process, for the irreversibility and the heat transferred to the refrigerant cause an increase in entropy, and the heat transferred from the refrigerant causes a decrease in entropy. These possibilities are represented by the two dashed lines 1–2 and 1–2′. The pressure of the liquid leaving the condenser will be less than the pressure of the vapor entering, and the temperature of the refrigerant in the condenser will be somewhat higher than that of the surroundings to which heat is being transferred. Usually the temperature of the liquid leaving the condenser is lower than the saturation temperature. It might drop somewhat more in the piping between the condenser and expansion valve. This represents a gain, however, because as a result of this heat transfer the refrigerant enters the evaporator with a lower enthalpy, which permits more heat to be transferred to the refrigerant in the evaporator.

There is some drop in pressure as the refrigerant flows through the evaporator. It may be slightly superheated as it leaves the evaporator, and through heat transferred from the surroundings its temperature will increase in the piping between the evaporator and the compressor. This heat transfer represents a loss, because it increases the work of the compressor, since the fluid entering it has an increased specific volume.

EXAMPLE 11.14   A refrigeration cycle utilizes R-12 as the working fluid. Following are the properties at various points of the cycle designated in Fig. 11.34.

$$P_1 = 125 \text{ kPa} \qquad T_1 = -10°\text{C}$$

$$P_2 = 1.2 \text{ MPa} \qquad T_2 = 100°\text{C}$$

$$P_3 = 1.19 \text{ MPa} \qquad T_3 = 80°\text{C}$$

$$P_4 = 1.16 \text{ MPa} \qquad T_4 = 45°\text{C}$$

$$P_5 = 1.15 \text{ MPa} \qquad T_5 = 40°\text{C}$$

$$P_6 = P_7 = 140 \text{ kPa} \qquad x_6 = x_7$$

$$P_8 = 130 \text{ kPa} \qquad T_8 = -20°\text{C}$$

The heat transfer from R-12 during the compression process is 4 kJ/kg. Determine the coefficient of performance of this cycle.

For each control volume, the model is the R-12 tables. Each process is SSSF with no changes in kinetic or potential energy.

Control volume: Compressor.
Inlet state: P_1, T_1 known; state fixed.
Exit state: P_2, T_2 known; state fixed.

Analysis:

First law:

$$q + h_1 = h_2 + w$$
$$w_c = -w = h_2 - h_1 - q$$

Solution

From the R-12 tables,

$$h_1 = 185.16 \qquad h_2 = 245.52$$

Therefore,

$$w_c = 245.52 - 185.16 - (-4) = 64.36 \text{ kJ} / \text{kg}$$

Control volume: Throttling valve plus line.
Inlet state: P_5, T_5 known; state fixed.
Exit state: $P_7 = P_6$ known, $x_7 = x_6$.

Analysis:

First law:

$$h_5 = h_6$$

Since $x_7 = x_6$, it follows that $h_7 = h_6$.

Solution

$$h_5 = h_6 = h_7 = 74.53$$

Control volume: Evaporator.
Inlet state: P_7, h_7 known (above).
Exit state: P_8, T_8 known; state fixed.

Analysis:

First law:

$$q_L = h_8 - h_7$$

Solution

$$q_L = h_8 - h_7 = 179.12 - 74.53 = 104.59 \text{kJ} / \text{kg}$$

Therefore,

$$\beta = \frac{q_L}{w_c} = \frac{104.59}{64.36} = 1.625$$

11.20 THE AMMONIA ABSORPTION REFRIGERATION CYCLE

The ammonia absorption refrigeration cycle differs from the vapor-compression cycle in the manner in which compression is achieved. In the absorption cycle the low-pressure ammonia vapor is absorbed in water and the liquid solution is pumped to a high pressure by a liquid pump. Figure 11.35 shows a schematic arrangement of the essential elements of such a system.

The low-pressure ammonia vapor leaving the evaporator enters the absorber where it is absorbed in the weak ammonia solution. This process takes place at a temperature slightly higher than that of the surroundings. Heat must be transferred to the surroundings during this process. The strong ammonia solution is then pumped through a heat exchanger to the generator where a higher pressure and temperature are maintained. Under these conditions ammonia vapor is driven from the solution as heat is transferred from a high-temperature source. The ammonia vapor goes to the condenser

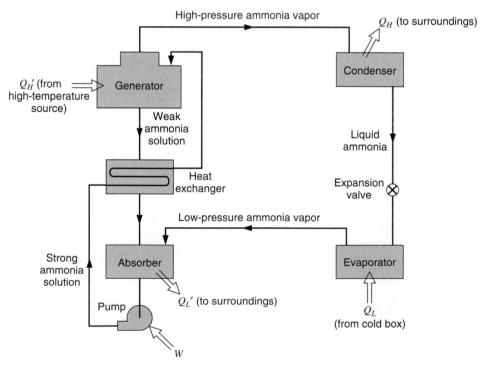

FIGURE 11.35 The ammonia-absorption refrigeration cycle.

where it is condensed, as in a vapor-compression system, and then to the expansion valve and evaporator. The weak ammonia solution is returned to the absorber through the heat exchanger.

The distinctive feature of the absorption system is that very little work input is required because the pumping process involves a liquid. This follows from the fact that for a reversible steady-flow process with negligible changes in kinetic and potential energy, the work is equal to $-\int v \, dP$ and the specific volume of the liquid is much less than the specific volume of the vapor. On the other hand, a relatively high-temperature source of heat must be available ($100°$ to $200°C$). There is more equipment in an absorption system than in a vapor-compression system, and it can usually be economically justified only when a suitable source of heat is available that would otherwise be wasted. In recent years, the absorption cycle has been given increased attention in connection with alternate energy sources, for example, solar energy or supplies of geothermal energy.

This cycle brings out the important principle that since the shaft work in a reversible steady-flow process with negligible changes in kinetic and potential energy is $-\int v \, dP$, a compression process should take place with the smallest possible specific volume.

11.21 THE AIR-STANDARD REFRIGERATION CYCLE

If we consider the original ideal four-process refrigeration cycle of Fig. 11.32 with a non-condensing (gaseous) working fluid, then the work output during the isentropic expansion process is not negligibly small, as was the case with a condensing working fluid. Therefore, we retain the turbine in the four-SSSF process ideal air-standard refrigeration cycle shown in Fig. 11.36. This cycle is seen to be the reverse Brayton cycle, and is used in practice in the liquefaction of air and other gases and also in certain special situations that require refrigeration, such as aircraft cooling systems. After compression from states

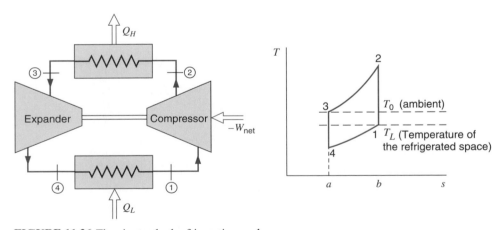

FIGURE 11.36 The air-standard refrigeration cycle.

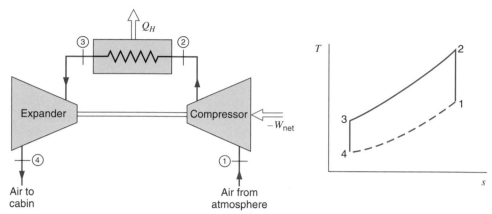

FIGURE 11.37 An air refrigeration cycle that might be utilized for aircraft cooling.

1 to 2, the air is cooled as heat is transferred to the surroundings at temperature T_0. The air is then expanded in process 3–4 to the pressure entering the compressor, and the temperature drops to T_4 in the expander. Heat may then be transferred to the air until temperature T_L is reached. The work for this cycle is represented by area 1–2–3–4–1, and the refrigeration effect is represented by area 4–1–*b*–*a*–4. The coefficient of performance is the ratio of these two areas.

In practice, this cycle has been used to cool aircraft in an open cycle. A simplified form is shown in Fig. 11.37. Upon leaving the expander, the cool air is blown directly into the cabin, thus providing the cooling effect where needed.

When counterflow heat exchangers are incorporated, very low temperatures can be obtained. This is essentially the cycle used in low-pressure air liquefaction plants and in other liquefaction devices such as the Collins helium liquefier. The ideal cycle is as shown in Fig. 11.38. It is evident that the expander operates at very low temperature, which presents unique problems to the designer in providing lubrication and choosing materials.

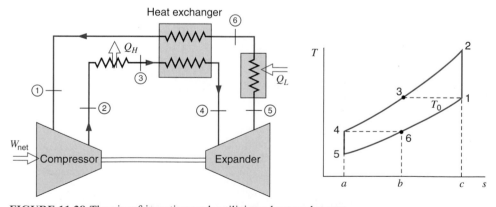

FIGURE 11.38 The air refrigeration cycle utilizing a heat exchanger.

EXAMPLE 11.15   Consider the simple air-standard refrigeration cycle of Fig. 11.36. Air enters the compressor at 0.1 MPa, −20°C, and leaves at 0.5 MPa. Air enters the expander at 15°C. Determine

1. The coefficient of performance for this cycle
2. The rate at which air must enter the compressor to provide 1 kW of refrigeration

For each control volume in this example, the model is ideal gas with constant specific heat, value at 300 K, and each process is SSSF with no kinetic or potential energy changes. The diagram for this example is Fig. 11.36.

> *Control volume:*   Compressor.
>
> *Inlet state:* P_1, T_1 known; state fixed.
>
> *Exit state:* P_2 known.

Analysis:

First law:

$$w_c = h_2 - h_1$$

(Here w_c designates work into the compressor.)
Second law:

$$s_1 = s_2$$

so that

$$\frac{T_2}{T_1} = \left(\frac{P_2}{P_1}\right)^{(k-1)/k}$$

Solution

$$\frac{T_2}{T_1} = \left(\frac{P_2}{P_1}\right)^{(k-1)/k} = 5^{0.286} = 1.5845 \qquad T_2 = 401.2 \text{ K}$$

$$w_c = h_2 - h_1 = C_p\left(T_2 - T_1\right)$$

$$= 1.0035\left(401.2 - 253.2\right) = 148.5 \text{ kJ/kg}$$

> *Control volume:*   Expander.
>
> *Inlet state:* $P_3 (= P_2)$ known, T_3 known; state fixed.
>
> *Exit state:* $P_4 (= P_1)$ known.

Analysis:

First law:

$$w_t = h_3 - h_4$$

Second law:

$$s_3 = s_4$$

so that

$$\frac{T_3}{T_4} = \left(\frac{P_3}{P_4}\right)^{(k-1)/k}$$

Solution

$$\frac{T_3}{T_4} = \left(\frac{P_3}{P_4}\right)^{(k-1)/k} = 5^{0.286} = 1.5845 \qquad T_4 = 181.9 \text{ K}$$

$$w_t = h_3 - h_4 = 1.0035(288.2 - 181.9) = 106.7 \text{ kJ/kg}$$

Control volume: High-temperature heat exchanger.

Inlet state: State 2 known (as given).

Exit state: State 3 known (as given).

Analysis:

First law:

$$q_H = h_2 - h_3 \quad \text{(heat rejected)}$$

Solution

$$q_H = h_2 - h_3 = C_p(T_2 - T_3) = 1.0035(401.2 - 288.2) = 113.4 \text{ kJ/kg}$$

Control volume: Low-temperature heat exchanger.

Inlet state: State 4 known (as given).

Exit state: State 1 known (as given).

Analysis:

First law:

$$q_L = h_1 - h_4$$

Solution

$$q_L = h_1 - h_4 = C_p(T_1 - T_4) = 1.0035(253.2 - 181.9) = 71.6 \text{ kJ/kg}$$

Therefore,

$$w_{net} = w_c - w_t = 148.5 - 106.7 = 41.8 \text{ kJ/kg}$$

$$\beta = \frac{q_L}{w_{net}} = \frac{71.6}{41.8} = 1.713$$

To provide 1 kW of refrigeration capacity, we have

$$\dot{m} = \frac{\dot{Q}_L}{q_L} = \frac{1}{71.6} = 0.014 \text{ kg/s}$$

EXAMPLE 11.15E Consider the simple air-standard refrigeration cycle of Fig. 11.36. Air enters the compressor at 14.7 lbf/in.^{2}, 0 F, and leaves at 80 lbf/in.^{2} Air enters the expander at 60 F. Determine

1. The coefficient of performance for this cycle
2. The rate at which air must enter the compressor in order to provide 1 kW of refrigeration

For each control volume in this example, the model is ideal gas with constant specific heat, value at 80 F, and each process is SSSF with no kinetic or potential energy changes. The diagram for this example is Fig. 11.36.

Control volume: Compressor.

Inlet state: P_1, T_1 known; state fixed.

Exit state: P_2 known.

Analysis

First law:

$$w_c = h_2 - h_1$$

(w_c designates work into the compressor)

Second law:

$$s_1 = s_2$$

so that

$$\frac{T_2}{T_1} = \left(\frac{P_2}{P_1}\right)^{(k-1)/k}$$

Solution

$$\frac{T_2}{T_1} = \left(\frac{P_2}{P_1}\right)^{(k-1)/k} = \left(\frac{80}{14.7}\right)^{0.286} = 1.624 \qquad T_2 = 747 \text{ R}$$

$$w_c = h_2 - h_1 = C_p(T_2 - T_1) = 0.24(747 - 460) = 68.9 \text{ Btu/lbm}$$

Control volume: Expander.

Inlet state: $P_3 (= P_2)$ known, T_3 known; state fixed.

Exit state: $P_4 (= P_1)$ known.

Analysis:

First law:

$$w_t = h_3 - h_4$$

Second law:

$$s_3 = s_4$$

so that

$$\frac{T_3}{T_4} = \left(\frac{P_3}{P_4}\right)^{(k-1)/k}$$

Solution

$$\frac{T_3}{T_4} = \left(\frac{P_3}{P_4}\right)^{(k-1)/k} = \left(\frac{80}{14.7}\right)^{0.286} = 1.624 \qquad T_4 = 320 \text{ R}$$

$$w_t = h_3 - h_4 = 0.24(520 - 320) = 48.0 \text{ Btu/lbm}$$

Control volume: High-temperature heat exchanger.

Inlet state: State 2 known (as given).

Exit state: State 3 known (as given).

Analysis:

First law:

$$q_H = h_2 - h_3 \qquad \text{(heat rejected)}$$

Solution

$$q_H = h_2 - h_3 = C_p(T_2 - T_3) = 0.24(747 - 520) = 54.5 \text{ Btu/lbm}$$

Control volume: Low-temperature heat exchanger.

Inlet state: State 4 known (as given).

Exit state: State 1 known (as given).

Analysis:

First law:

$$q_L = h_1 - h_4$$

Solution

$$q_L = h_1 - h_4 = C_p(T_1 - T_4) = 0.24(460 - 320) = 33.6 \text{ Btu/lbm}$$

Therefore,

$$w_{net} = w_c - w_t = 68.9 - 48.0 = 20.9 \text{ Btu / lbm}$$

$$\beta = \frac{q_L}{w_{net}} = \frac{33.6}{20.9} = 1.61$$

In order to provide 1 kW of refrigeration capacity (3412 Btu/h)

$$\dot{m} = \frac{\dot{Q}_L}{q_L} = \frac{3412}{33.6} = 101.5 \text{ lbm / h}$$

11.22 COMBINED-CYCLE POWER AND REFRIGERATION SYSTEMS

There are many situations in which it is desirable to combine two cycles in series, either power systems or refrigeration systems, in order to take advantage of a very wide temperature range or to utilize what would otherwise be waste heat to improve efficiency. One combined power cycle, shown in Fig. 11.39 as a simple steam cycle with a mercury-topping cycle, is often referred to as a binary cycle. The advantage of this combined system is that mercury has a very low vapor pressure relative to that for water; therefore it is possible for an isothermal boiling process in the mercury to take place at a high temperature, much higher than the critical temperature of water, but still at a moderate pressure. The mercury condenser then provides an isothermal heat source as input to the steam boiler, such that the two cycles can be closely matched by proper selection of the cycle variables, with the resulting combined cycle then having a high thermal efficiency. Saturation pressures and temperatures for a typical mercury/water binary cycle are shown in the T–s diagram of Fig. 11.39.

A different type of combined cycle that has seen considerable attention is to use the "waste heat" exhaust from a Brayton cycle gas turbine engine (or another combustion en-

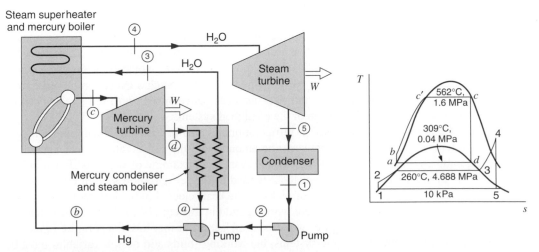

FIGURE 11.39 Mercury/water binary power system.

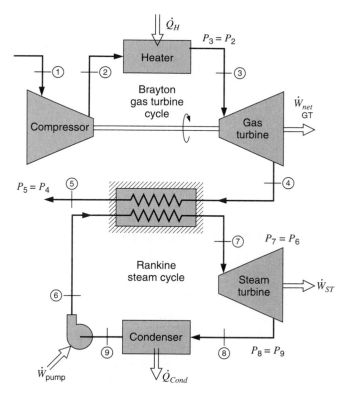

FIGURE 11.40 Combined Brayton/Rankine cycle power system.

gine such as a Diesel engine) as the heat source for a steam or other vapor power cycle, in which case the vapor cycle acts as a bottoming cycle for the gas engine, in order to improve the overall thermal efficiency of the combined power system. Such a system, utilizing a gas turbine and a steam Rankine cycle, is shown in Fig. 11.40. In such a combination, there is a natural mismatch using the cooling of a noncondensing gas as the energy source to effect an isothermal boiling process plus superheating the vapor, and careful design is required in order to avoid a pinch point, a condition at which the gas has cooled to the vapor boiling temperature without having provided sufficient energy to complete the boiling process.

One way to take advantage of the cooling exhaust gas in the Brayton cycle portion of the combined system is to utilize a mixture as the working fluid in the Rankine cycle. An example of this type of application is the Kalina cycle, which uses ammonia–water mixtures as the working fluid in the Rankine-type cycle. Such a cycle can be made very efficient, inasmuch as the temperature differences between the two fluid streams can be controlled through careful design of the combined system.

Combined cycles are used in refrigeration systems in cases in which there is a very large temperature difference between the ambient and the refrigerated space. Such a refrigeration system is often called a cascade system, an example of which is shown in Fig. 11.41. In this case, the refrigerant R-22 is used in the refrigeration system rejecting heat to the ambient, while its evaporator picks up the heat rejected in the low-temperature system condenser, the low temperature working fluid in this case being R-23, whose thermodynamic properties are suited to work as a refrigerant in this low-temperature range. As with the other combined-cycle systems, the working fluids and design variables must be considered very carefully, in order to optimize the performance of each unit.

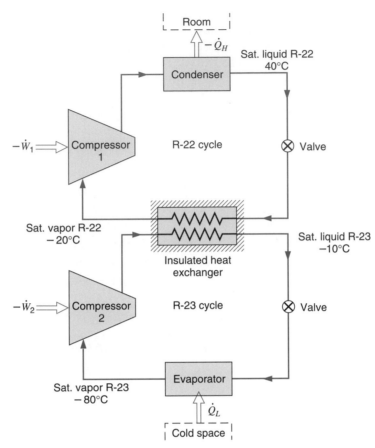

FIGURE 11.41 Combined-cycle cascade refrigeration system.

We have described only a few combined-cycle systems here, as examples of the types of applications that can be dealt with, and the resulting improvement in overall performance that can occur. Obviously, there are many other combinations of power and refrigeration systems, and some of these are discussed in the problems at the end of the chapter.

PROBLEMS

11.1 A steam power plant, as shown in Fig. 11.3, operating in a Rankine cycle has saturated vapor at 3.5 MPa leaving the boiler. The turbine exhausts to the condenser operating at 10 kPa. Find the specific work and heat transfer in each of the ideal components and the cycle efficiency.

11.2 Consider a solar-energy-powered ideal Rankine cycle that uses water as the working fluid. Saturated vapor leaves the solar collector at 175°C, and the condenser pressure is 10 kPa. Determine the thermal efficiency of this cycle.

11.3 A utility runs a Rankine cycle with a water boiler at 3.5 MPa and the cycle has the highest and lowest temperatures of 450°C and 45°C, respectively. Find the plant efficiency and the efficiency of a Carnot cycle with the same temperatures.

11.4 A steam power plant operating in an ideal Rankine cycle has a high pressure of 5 MPa and a low pressure of 15 kPa. The turbine exhaust state should have a quality of at least 95% and the turbine power generated should be 7.5 MW. Find the necessary boiler exit temperature and the total mass flow rate.

11.5 A supply of geothermal hot water is to be used as the energy source in an ideal Rankine cycle, with R-134a as the cycle working fluid. Saturated vapor R-134a leaves the boiler at a temperature of 85°C, and the condenser temperature is 40°C. Calculate the thermal efficiency of this cycle.

11.6 Do Problem 11.5 with R-22 as the working fluid.

11.7 Do Problem 11.5 with ammonia as the working fluid.

11.8 Consider the ammonia Rankine-cycle power plant shown in Fig. P11.8, a plant that was designed to operate in a location where the ocean water temperature is 25°C near the surface and 5°C at some greater depth.

   a. Determine the turbine power output and the pump power input for the cycle.

   b. Determine the mass flow rate of water through each heat exchanger.

   c. What is the thermal efficiency of this power plant?

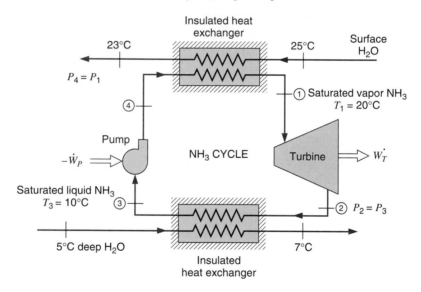

$$\eta_{s\,Pump} = 0.80; \ \eta_{s\,Turbine} = 0.80; \ \dot{m}_{NH_3} = 1000 \text{ kg/s}$$

FIGURE P11.8

11.9 Do Problem 11.8 with R-134a as the working fluid in the Rankine cycle.

11.10 Consider the boiler in Problem 11.5 where the geothermal hot water brings the R-134a to saturated vapor. Assume a counter flowing heat exchanger arrangement. The geothermal water temperature should be equal to or greater than the R-134a temperature at any location inside the heat exchanger. The point with the smallest temperature difference between the source and the working fluid is called the pinch point. If 2 kg/s of geothermal water is available at 95°C, what is the maximum power output of this cycle for R-134a as the working fluid? (Hint: split the heat exchanger C.V. into two so the pinch point with $\Delta T = 0$, $T = 85°C$ appears.)

11.11 Do the previous problem with R-22 as the working fluid.

11.12 The power plant in Problem 11.1 is modified to have a superheater section following the boiler so the steam leaves the super heater at 3.5 MPa, 400°C. Find the specific work and heat transfer in each of the ideal components and the cycle efficiency.

11.13 A steam power plant has a steam generator exit at 4 MPa, 500°C and a condenser exit temperature of 45°C. Assume all components are ideal and find the cycle efficiency and the specific work and heat transfer in the components.

11.14 Consider an ideal Rankine cycle using water with a high-pressure side of the cycle at a supercritical pressure. Such a cycle has a potential advantage of minimizing local temperature differences between the fluids in the steam generator, such as the instance in which the high-temperature energy source is the hot exhaust gas from a gas-turbine engine. Calculate the thermal efficiency of the cycle if the state entering the turbine is 25 MPa, 500°C, and the condenser pressure is 5 kPa. What is the steam quality at the turbine exit?

11.15 Steam enters the turbine of a power plant at 5 MPa and 400°C, and exhausts to the condenser at 10 kPa. The turbine produces a power output of 20 000 kW with an isentropic efficiency of 85%. What is the mass flow rate of steam around the cycle and the rate of heat rejection in the condenser? Find the thermal efficiency of the power plant. How does this compare with a Carnot cycle?

11.16 Consider an ideal steam reheat cycle where steam enters the high-pressure turbine at 3.5 MPa, 400°C, and then expands to 0.8 MPa. It is then reheated to 400°C and expands to 10 kPa in the low-pressure turbine. Calculate the cycle thermal efficiency and the moisture content of the steam leaving the low-pressure turbine.

11.17 The reheat pressure affects the operating variables and thus turbine performance. Repeat Problem 11.16 twice, using 0.6 and 1.0 MPa for the reheat pressure.

11.18 The effect of a number of reheat stages on the ideal steam reheat cycle is to be studied. Repeat Problem 11.16 using two reheat stages, one stage at 1.2 MPa and the second at 0.2 MPa, instead of the single reheat stage at 0.8 MPa.

11.19 A closed feedwater heater in a regenerative steam power cycle heats 20 kg/s of water from 100°C, 20 MPa to 250°C, 20 MPa. The extraction steam from the turbine enters the heater at 4 MPa, 275°C, and leaves as saturated liquid. What is the required mass flow rate of the extraction steam?

11.20 An open feedwater heater in a regenerative steam power cycle receives 20 kg/s of water at 100°C, 2 MPa. The extraction steam from the turbine enters the heater at 2 MPa, 275°C, and all the feedwater leaves as saturated liquid. What is the required mass flow rate of the extraction steam?

11.21 A power plant with one closed feedwater heater has a condenser temperature of 45°C, a maximum pressure of 5 MPa, and boiler exit temperature of 900°C. Extraction steam at 1 MPa to the feedwater heater condenses and is pumped up to the 5 MPa feedwater line where all the water goes to the boiler at 200°C. Find the fraction of extraction steam flow and the two specific pump work inputs.

11.22 A power plant with one open feedwater heater has a condenser temperature of 45°C, a maximum pressure of 5 MPa, and boiler exit temperature of 900°C.

Extraction steam at 1 MPa to the feedwater heater is mixed with the feedwater line so the exit is saturated liquid into the second pump. Find the fraction of extraction steam flow and the two specific pump work inputs.

11.23 A steam power plant operates with a boiler output of 20 kg/s steam at 2 MPa, 600°C. The condenser operates at 50°C, dumping energy to a river that has an average temperature of 20°C. There is one open feedwater heater with extraction from the turbine at 600 kPa and its exit is saturated liquid. Find the mass flow rate of the extraction flow. If the river water should not be heated more than 5°C, how much water should be pumped from the river to the heat exchanger (condenser)?

11.24 Consider an ideal steam regenerative cycle in which steam enters the turbine at 3.5 MPa, 400°C, and exhausts to the condenser at 10 kPa. Steam is extracted from the turbine at 0.8 MPa for an open feedwater heater. The feedwater leaves the heater as saturated liquid. The appropriate pumps are used for the water leaving the condenser and the feedwater heater. Calculate the thermal efficiency of the cycle and the net work per kilogram of steam.

11.25 Repeat Problem 11.24, but assume a closed instead of an open feedwater heater. A single pump is used to pump the water leaving the condenser up to the boiler pressure of 3.5 MPa. Condensate from the feedwater heater is drained through a trap to the condenser.

11.26 A steam power plant has high and low pressures of 25 MPa and 10 kPa, and one open feedwater heater operating at 1 MPa with the exit as saturated liquid. The maximum temperature is 800°C and the turbine has a total power output of 5 MW. Find the fraction of the flow for extraction to the feedwater and the total condenser heat transfer rate.

11.27 Do Problem 11.26 with a closed feedwater heater instead of an open heater and a drip pump to add the extraction flow to the feed water line at 25 MPa. Assume the temperature is 175°C after the drip pump flow is added to the line. One main pump brings the water to 25 MPa from the condenser.

11.28 Consider an ideal combined reheat and regenerative cycle in which steam enters the high-pressure turbine at 3.5 MPa, 400°C, and is extracted to an open feedwater heater at 0.8 MPa with exit as saturated liquid. The remainder of the steam is reheated to 400°C at this pressure, 0.8 MPa, and is fed to the low-pressure turbine. The condenser pressure is 10 kPa. Calculate the thermal efficiency of the cycle and the net work per kilogram of steam.

11.29 An ideal steam power plant is designed to operate on the combined reheat and regenerative cycle and to produce a net power output of 10 MW. Steam enters the high-pressure turbine at 8 MPa, 550°C, and is expanded to 0.6 MPa, at which pressure some of the steam is fed to an open feedwater heater, and the remainder is reheated to 550°C. The reheated steam is then expanded in the low-pressure turbine to 10 kPa. Determine the steam flow rate to the high-pressure turbine and the power required to drive each pump.

11.30 The low pressure turbine in a reheat and regenerative cycle receives 10 kg/s steam at 600 kPa, 550°C. The turbine exhausts to a condenser operating at 10 kPa. The condenser cooling water temperature is restricted to a maximum of 10°C increase, so what is the needed flow rate of the cooling water? The steam velocity in the turbine-condenser connecting pipe is restricted to a maximum of 100 m/s. What is the diameter of the connecting pipe?

11.31 A steam power plant has a high pressure of 5 MPa and maintains 50°C in the condenser. The boiler exit temperature is 600°C. All the components are ideal except the turbine which has an actual exit state of saturated vapor at 50°C. Find the cycle efficiency with the actual turbine and the turbine isentropic efficiency.

11.32 A steam power cycle has a high pressure of 3.5 MPa and a condenser exit temperature of 45°C. The turbine efficiency is 85%, and other cycle components are ideal. If the boiler superheats to 800°C, find the cycle thermal efficiency.

11.33 A steam power plant operates with a high pressure of 5 MPa and has a boiler exit temperature of 600°C receiving heat from a 700°C source. The ambient at 20°C provides cooling for the condenser so it can maintain 45°C inside. All the components are ideal except for the turbine, which has an exit state with a quality of 97%. Find the work and heat transfer in all components per kg water and the turbine isentropic efficiency. Find the rate of entropy generation per kg water in the boiler/heat source setup.

11.34 Repeat Problem 11.25 assuming the turbine has an isentropic efficiency of 85%.

11.35 Steam leaves a power plant steam generator at 3.5 MPa, 400°C, and enters the turbine at 3.4 MPa, 375°C. The isentropic turbine efficiency is 88%, and the turbine exhaust pressure is 10 kPa. Condensate leaves the condenser and enters the pump at 35°C, 10 kPa. The isentropic pump efficiency is 80%, and the discharge pressure is 3.7 MPa. The feedwater enters the steam generator at 3.6 MPa, 30°C. Calculate the thermal efficiency of the cycle and the entropy generation for the process in the line between the steam generator exit and the turbine inlet, assuming an ambient temperature of 25°C.

11.36 For the steam power plant described in Problem 11.1, assume the isentropic efficiencies of the turbine and pump are 85% and 80%, respectively. Find the component specific work and heat transfers and the cycle efficiency.

11.37 A small steam power plant has a boiler exit of 3 MPa, 400°C while it maintains 50 kPa in the condenser. All the components are ideal except the turbine, which has an isentropic efficiency of 80% and it should deliver a shaft power of 9.0 MW to an electric generator. Find the specific turbine work, the needed flow rate of steam, and the cycle efficiency.

11.38 In a particular reheat-cycle power plant, steam enters the high-pressure turbine at 5 MPa, 450°C and expands to 0.5 MPa, after which it is reheated to 450°C. The steam is then expanded through the low-pressure turbine to 7.5 kPa. Liquid water leaves the condenser at 30°C, is pumped to 5 MPa, and then returned to the steam generator. Each turbine is adiabatic with an isentropic efficiency of 87% and a pump efficiency of 82%. If the total power output of the turbines is 10 MW, determine the mass flow rate of steam, the pump power input, and the thermal efficiency of the power plant.

11.39 A supercritical steam power plant has a high pressure of 30 MPa and an exit condenser temperature of 50°C. The maximum temperature in the boiler is 1000°C, and the turbine exhaust is saturated vapor. There is one open feedwater heater receiving extraction from the turbine at 1 MPa, and its exit is saturated liquid flowing to pump 2. The isentropic efficiency for the first section and the overall turbine are both 88.5%. Find the ratio of the extraction mass flow to total flow into turbine. What is the boiler inlet temperature with and without the feedwater heater?

11.40 In one type of nuclear power plant, heat is transferred in the nuclear reactor to liquid sodium. The liquid sodium is then pumped through a heat exchanger where heat is transferred to boiling water. Saturated vapor steam at 5 MPa exits this heat exchanger and is then superheated to 600°C in an external gas-fired superheater. The steam enters the turbine, which has one (open-type) feedwater extraction at 0.4 MPa. The isentropic turbine efficiency is 87%, and the condenser pressure is 7.5 kPa. Determine the heat transfer in the reactor and in the superheater to produce a net power output of 1 MW.

11.41 A cogenerating steam power plant, as in Fig. 11.17, operates with a boiler output of 25 kg/s steam at 7 MPa, 500°C. The condenser operates at 7.5 kPa, and the process heat is extracted as 5 kg/s from the turbine at 500 kPa, state 6, and after use is returned as saturated liquid at 100 kPa, state 8. Assume all components are ideal and find the temperature after pump 1, the total turbine output and the total process heat transfer.

11.42 A 10 kg/s steady supply of saturated-vapor steam at 500 kPa is required for drying a wood pulp slurry in a paper mill (see Fig. P11.42). It is decided to supply this steam by cogeneration, that is, the steam supply will be the exhaust from a steam turbine. Water at 20°C, 100 kPa, is pumped to a pressure of 5 MPa and then fed to a steam generator. It may be assumed that the isentropic efficiency of the pump is 75%, and that of the turbine is 85%. What is the steam temperature exiting the steam generator? What is the additional heat transfer rate to the steam generator beyond what would have been required to produce only the desired steam supply? What is the difference in net power?

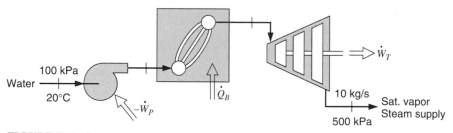

FIGURE P11.42

11.43 An industrial application has the following steam requirement: one 10-kg/s stream at a pressure of 0.5 MPa and one 5-kg/s stream at 1.4 MPa (both saturated or slightly superheated vapor). It is obtained by cogeneration, whereby a high-pressure boiler supplies steam at 10 MPa, 500°C to a turbine. The required amount is withdrawn at 1.4 MPa, and the remainder is expanded in the low-pressure end of the turbine to 0.5 MPa providing the second required steam flow. Assuming both turbine sections have an isentropic efficiency of 85%, determine the following.

a. The power output of the turbine and the heat transfer rate in the boiler.

b. Compute the rates needed if the steam were generated in a low-pressure boiler without cogeneration. Assume that for each, 20°C liquid water is pumped to the required pressure and fed to a boiler.

11.44 In a cogenerating steam power plant the turbine receives steam from a high-pressure steam drum and a low-pressure steam drum, as shown in Fig. P11.44. The condenser is made as two closed heat exchangers used to heat water running in a separate loop for district heating. The high-temperature heater adds 30 MW and the low-temperature heaters adds 31 MW to the district heating water flow. Find the power cogenerated by the turbine and the temperature in the return line to the deaerator.

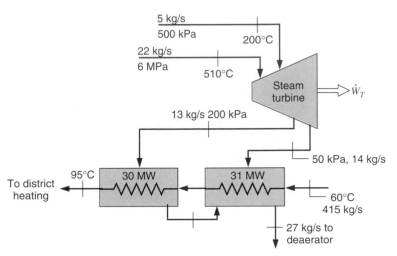

FIGURE P11.44

11.45 A boiler delivers steam at 10 MPa, 550°C to a two-stage turbine as shown in Fig. 11.17. After the first stage, 25% of the steam is extracted at 1.4 MPa for a process application and returned at 1 MPa, 90°C to the feedwater line. The remainder of the steam continues through the low-pressure turbine stage, which exhausts to the condenser at 10 kPa. One pump brings the feedwater to 1 MPa, and a second pump brings it to 10 MPa. Assume the first and second stages in the steam turbine have isentropic efficiencies of 85% and 80% and that both pumps are ideal. If the process application requires 5 MW of power, how much power can then be cogenerated by the turbine?

11.46 Consider an ideal air-standard Brayton cycle in which the air into the compressor is at 100 kPa, 20°C, and the pressure ratio across the compressor is 12:1. The maximum temperature in the cycle is 1100°C, and the air flow rate is 10 kg/s. Assume constant specific heat for the air, value from Table A.5. Determine the compressor work, the turbine work, and the thermal efficiency of the cycle.

11.47 Repeat Problem 11.46, but assume variable specific heat for the air, Table A.7.

11.48 An ideal regenerator is incorporated into the ideal air-standard Brayton cycle of Problem 11.46. Find the thermal efficiency of the cycle with this modification.

11.49 A Brayton cycle inlet is at 300 K, 100 kPa, and the combustion adds 670 kJ/kg. The maximum temperature is 1200 K due to material considerations. What is the maximum allowed compression ratio? For this ratio calculate the net work and cycle efficiency assuming variable specific heat for the air, Table A.7.

11.50 A large stationary Brayton cycle gas-turbine power plant delivers a power output of 100 MW to an electric generator. The minimum temperature in the cycle is

300 K, and the maximum temperature is 1600 K. The minimum pressure in the cycle is 100 kPa, and the compressor pressure ratio is 14 to 1. Calculate the power output of the turbine. What fraction of the turbine output is required to drive the compressor? What is the thermal efficiency of the cycle?

11.51 Repeat Problem 11.50, but assume that the compressor has an isentropic efficiency of 85% and the turbine an isentropic efficiency of 88%.

11.52 Repeat Problem 11.51, but include a regenerator with 75% efficiency in the cycle.

11.53 A gas turbine with air as the working fluid has two ideal turbine sections, as shown in Fig. P11.53, the first of which drives the ideal compressor, with the second producing the power output. The compressor input is at 290 K, 100 kPa, and the exit is at 450 kPa. A fraction of flow, x, bypasses the burner and the rest $(1 - x)$ goes through the burner where 1200 kJ/kg is added by combustion. The two flows then mix before entering the first turbine and continue through the second turbine, with exhaust at 100 kPa. If the mixing should result in a temperature of 1000 K into the first turbine find the fraction x. Find the required pressure and temperature into the second turbine and its specific power output.

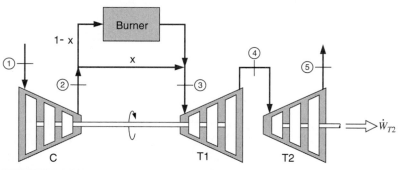

FIGURE P11.53

11.54 The gas-turbine cycle shown in Fig. P11.54 is used as an automotive engine. In the first turbine, the gas expands to pressure P_5, just low enough for this turbine to drive the compressor. The gas is then expanded through the second turbine connected to the drive wheels. The data for the engine are shown in the figure and assume that all processes are ideal. Determine the intermediate pressure P_5, the net specific work output of the engine, and the mass flow rate through the engine. Find also the air temperature entering the burner T_3, and the thermal efficiency of the engine.

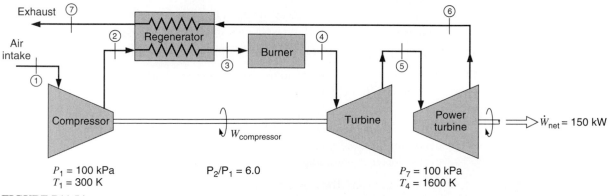

FIGURE P11.54

11.55 Repeat Problem 11.54, but assume that the compressor has an efficiency of 82%, that both turbines have efficiencies of 87%, and that the regenerator efficiency is 70%.

11.56 Repeat the questions in Problem 11.54 when we assume that friction causes pressure drops in the burner and on both sides of the regenerator. In each case, the pressure drop is estimated to be 2% of the inlet pressure to that component of the system, so $P_3 = 588$ kPa, $P_4 = 0.98\ P_3$, and $P_6 = 102$ kPa.

11.57 Consider an ideal gas-turbine cycle with two stages of compression and two stages of expansion. The pressure ratio across each compressor stage and each turbine stage is 8 to 1. The pressure at the entrance to the first compressor is 100 kPa, the temperature entering each compressor is 20°C, and the temperature entering each turbine is 1100°C. An ideal regenerator is also incorporated into the cycle. Determine the compressor work, the turbine work, and the thermal efficiency of the cycle.

11.58 Repeat Problem 11.57, but assume that each compressor stage and each turbine stage has an isentropic efficiency of 85%. Also assume that the regenerator has an efficiency of 70%.

11.59 A gas turbine cycle has two stages of compression, with an intercooler between the stages. Air enters the first stage at 100 kPa, 300 K. The pressure ratio across each compressor stage is 5 to 1, and each stage has an isentropic efficiency of 82%. Air exits the intercooler at 330 K. The maximum cycle temperature is 1500 K, and the cycle has a single turbine stage with an isentropic efficiency of 86%. The cycle also includes a regenerator with an efficiency of 80%. Calculate the temperature at the exit of each compressor stage, the second-law efficiency of the turbine, and the cycle thermal efficiency.

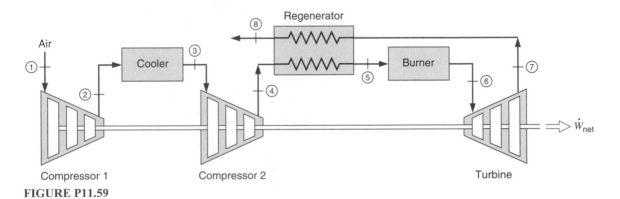

FIGURE P11.59

11.60 A two-stage air compressor has an intercooler between the two stages, as shown in Fig. P11.60. The inlet state is 100 kPa, 290 K, and the final exit pressure is 1.6 MPa. Assume that the constant pressure intercooler cools the air to the inlet temperature, $T_3 = T_1$. It can be shown, see Problem 9.130, that the optimal pressure, $P_2 = (P_1 P_4)^{1/2}$, for minimum total compressor work. Find the specific compressor works and the intercooler heat transfer for the optimal P_2.

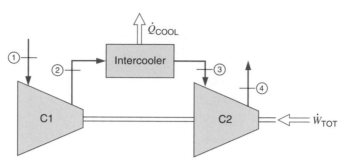

FIGURE P11.60

11.61 Repeat Problem 11.60 when the intercooler brings the air to $T_3 = 320$ K. The corrected formula for the optimal pressure is $P_2 =[\ P_1 P_4\ (T_3/T_1)^{n/(n-1)}\ ]^{1/2}$, see Problem 9.131, where n is the exponent in the assumed polytropic process.

11.62 Consider an ideal air-standard Ericsson cycle that has an ideal regenerator, as shown in Fig. P11.62. The high pressure is 1 MPa and the cycle efficiency is 70%. Heat is rejected in the cycle at a temperature of 300 K, and the cycle pressure at the beginning of the isothermal compression process is 100 kPa. Determine the high temperature, the compressor work, and the turbine work per kilogram of air.

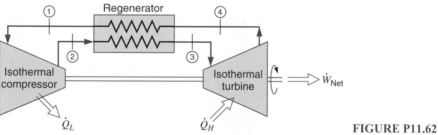

FIGURE P11.62

11.63 An air-standard Ericsson cycle has an ideal regenerator. Heat is supplied at 1000°C and heat is rejected at 20°C. Pressure at the beginning of the isothermal compression process is 70 kPa. The heat added is 600 kJ/kg. Find the compressor work, the turbine work, and the cycle efficiency.

11.64 Consider an ideal air-standard cycle for a gas-turbine, jet propulsion unit, such as that shown in Fig. 11.27. The pressure and temperature entering the compressor are 90 kPa, 290 K. The pressure ratio across the compressor is 14 to 1, and the turbine inlet temperature is 1500 K. When the air leaves the turbine, it enters the nozzle and expands to 90 kPa. Determine the pressure at the nozzle inlet and the velocity of the air leaving the nozzle.

11.65 The turbine in a jet engine receives air at 1250 K, 1.5 MPa. It exhausts to a nozzle at 250 kPa, which in turn exhausts to the atmosphere at 100 kPa. The isentropic efficiency of the turbine is 85%, and the nozzle efficiency is 95%. Find the nozzle inlet temperature and the nozzle exit velocity. Assume negligible kinetic energy out of the turbine.

11.66 Repeat Problem 11.64, but assume that the isentropic compressor efficiency is 87%, the isentropic turbine efficiency is 89%, and the isentropic nozzle efficiency is 96%.

11.67 Consider an air standard jet engine cycle operating in a 280 K, 100 kPa environment. The compressor requires a shaft power input of 4000 kW. Air enters the turbine state 3 at 1600 K, 2 MPa, at the rate of 9 kg/s, and the isentropic efficiency of the turbine is 85%. Determine the pressure and temperature entering the nozzle at state 4. If the nozzle efficiency is 95%, determine the temperature and velocity exiting the nozzle at state 5.

11.68 A jet aircraft is flying at an altitude of 4900 m, where the ambient pressure is approximately 55 kPa and the ambient temperature is −18°C. The velocity of the aircraft is 280 m/s, the pressure ratio across the compressor is 14:1, and the cycle maximum temperature is 1450 K. Assume the inlet flow goes through a diffuser to zero relative velocity at state 1. Find the temperature and pressure at state 1 and the velocity (relative to the aircraft) of the air leaving the engine at 55 kPa.

11.69 Air flows into a gasoline engine at 95 kPa, 300 K. The air is then compressed with a volumetric compression ratio of 8:1. In the combustion process 1300 kJ/kg of energy is released as the fuel burns. Find the temperature and pressure after combustion.

11.70 A gasoline engine has a volumetric compression ratio of 9. The state before compression is 290 K, 90 kPa, and the peak cycle temperature is 1800 K. Find the pressure after expansion, the cycle net work, and the cycle efficiency using properties from Table A.7.

11.71 To approximate an actual spark-ignition engine, consider an air-standard Otto cycle that has a heat addition of 1800 kJ/kg of air, a compression ratio of 7, and a pressure and temperature at the beginning of the compression process of 90 kPa, 10°C. Assuming constant specific heat, with the value from Table A.5, determine the maximum pressure and temperature of the cycle, the thermal efficiency of the cycle, and the mean effective pressure.

11.72 Repeat Problem 11.71, but assume variable specific heat. The ideal gas air tables, Table A.7, are recommended for this calculation (and the specific heat from Fig. 5.10 at high temperature).

11.73 A gasoline engine takes air in at 290 K, 90 kPa and then compresses it. The combustion adds 1000 kJ/kg to the air after which the temperature is 2050 K. Use the cold air properties (i.e., constant heat capacities at 300 K) and find the compression ratio, the compression specific work, and the highest pressure in the cycle.

11.74 Answer the same three questions for the previous problem, but use variable heat capacities (use Table A.7).

11.75 When methanol produced from coal is considered as an alternative fuel to gasoline for automotive engines, it is recognized that the engine can be designed with a higher compression ratio, say 10 instead of 7, but that the energy release with combustion for a stoichiometric mixture with air is slightly smaller, about 1700 kJ/kg. Repeat Problem 11.71 using these values.

11.76 It is found experimentally that the power stroke expansion in an internal combustion engine can be approximated with a polytropic process with a value of the polytropic exponent n somewhat larger than the specific heat ratio k. Repeat Problem 11.71 but assume that the expansion process is reversible and polytropic (instead of the isentropic expansion in the Otto cycle) with n equal to 1.50.

11.77 In the Otto cycle, all the heat transfer q_H occurs at constant volume. It is more realistic to assume that part of q_H occurs after the piston has started its downward motion in the expansion stroke. Therefore, consider a cycle identical to the Otto cycle, except that the first two-thirds of the total q_H occurs at constant volume and the last one-third occurs at constant pressure. Assume that the total q_H is 2100 kJ/kg, that the state at the beginning of the compression process is 90 kPa, 20°C, and that the compression ratio is 9. Calculate the maximum pressure and temperature and the thermal efficiency of this cycle. Compare the results with those of a conventional Otto cycle having the same given variables.

11.78 A diesel engine has a compression ratio of 20:1 with an inlet of 95 kPa, 290 K, state 1, with volume 0.5 L. The maximum cycle temperature is 1800 K. Find the maximum pressure, the net specific work, and the thermal efficiency.

11.79 A diesel engine has a bore of 0.1 m, a stroke of 0.11 m, and a compression ratio of 19:1 running at 2000 RPM (revolutions per minute). Each cycle takes two revolutions and has a mean effective pressure of 1400 kPa. With a total of 6 cylinders, find the engine power in kW and horsepower, hp.

11.80 At the beginning of compression in a diesel cycle $T = 300$ K, $P = 200$ kPa and after combustion (heat addition) is complete $T = 1500$ K and $P = 7.0$ MPa. Find the compression ratio, the thermal efficiency, and the mean effective pressure.

11.81 Consider an ideal air-standard diesel cycle in which the state before the compression process is 95 kPa, 290 K, and the compression ratio is 20. Find the maximum temperature (by iteration) in the cycle to have a thermal efficiency of 60%.

11.82 Consider an ideal Stirling-cycle engine in which the state at the beginning of the isothermal compression process is 100 kPa, 25°C, the compression ratio is 6, and the maximum temperature in the cycle is 1100°C. Calculate the maximum cycle pressure and the thermal efficiency of the cycle with and without regenerators.

11.83 An air-standard Stirling cycle uses helium as the working fluid. The isothermal compression brings helium from 100 kPa, 37°C to 600 kPa. The expansion takes place at 1200 K and there is no regenerator. Find the work and heat transfer in all of the four processes per kg helium and the thermal cycle efficiency.

11.84 Consider an ideal air-standard Stirling cycle with an ideal regenerator. The minimum pressure and temperature in the cycle are 100 kPa, 25°C, the compression ratio is 10, and the maximum temperature in the cycle is 1000°C. Analyze each of the four processes in this cycle for work and heat transfer, and determine the overall performance of the engine.

11.85 The air-standard Carnot cycle was not shown in the text; show the $T–s$ diagram for this cycle. In an air-standard Carnot cycle, the low temperature is 280 K and the efficiency is 60%. If the pressure before compression and after heat rejection is 100 kPa, find the high temperature and the pressure just before heat addition.

11.86 Air in a piston/cylinder goes through a Carnot cycle in which $T_L = 26.8$°C and the total cycle efficiency is $\eta = 2/3$. Find T_H, the specific work and volume ratio in the adiabatic expansion for constant C_p, C_v. Repeat the calculation for variable heat capacities.

11.87 Consider an ideal refrigeration cycle that has a condenser temperature of 45°C and an evaporator temperature of –15°C. Determine the coefficient of performance of this refrigerator for the working fluids R-12 and R-22.

11.88 The environmentally safe refrigerant R-134a is one of the replacements for R-12 in refrigeration systems. Repeat Problem 11.87 using R-134a and compare the result with that for R-12.

11.89 A refrigerator using R-22 is powered by a small natural–gas fired heat engine with a thermal efficiency of 25%, as shown in Fig. P11.89. The R-22 condenses at 40°C, it evaporates at –20°C and the cycle is standard. Find the two specific heat transfers in the refrigeration cycle. What is the overall coefficient of performance as Q_L/Q_1?

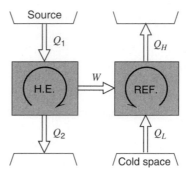

FIGURE P11.89

11.90 A refrigerator with R-12 as the working fluid has a minimum temperature of –10°C and a maximum pressure of 1 MPa. Assume an ideal refrigeration cycle as in Fig. 11.32. Find the specific heat transfer from the cold space and that to the hot space, and the coefficient of performance.

11.91 A refrigerator in a meat warehouse must keep a low temperature of –15°C and the outside temperature is 20°C. It uses R-12 as the refrigerant, which must remove 5 kW from the cold space. Find the flow rate of the R-12 needed, assuming a standard vapor compression refrigeration cycle with a condenser at 20°C.

11.92 A refrigerator with R-12 as the working fluid has a minimum temperature of –10°C and a maximum pressure of 1 MPa. The actual adiabatic compressor exit temperature is 60°C. Assume no pressure loss in the heat exchangers. Find the specific heat transfer from the cold space and that to the hot space, the coefficient of performance, and the isentropic efficiency of the compressor.

11.93 Consider an ideal heat pump that has a condenser temperature of 50°C and an evaporator temperature of 0°C. Determine the coefficient of performance of this heat pump for the working fluids R-12, R-22, and ammonia.

11.94 The air conditioner in a car uses R-134a, and the compressor power input is 1.5 kW, bringing the R-134a from 201.7 kPa to 1200 kPa by compression. The cold space is a heat exchanger that cools atmospheric air from the outside down to 10°C and blows it into the car. What is the mass flow rate of the R-134a and what is the low temperature heat transfer rate? How much is the mass flow rate of air at 10°C?

11.95 A refrigerator using R-134a is located in a 20°C room. Consider the cycle to be ideal, except that the compressor is neither adiabatic nor reversible. Saturated vapor at −20°C enters the compressor, and the R-134a exits the compressor at 50°C. The condenser temperature is 40°C. The mass flow rate of refrigerant around the cycle is 0.2 kg/s, and the coefficient of performance is measured and found to be 2.3. Find the power input to the compressor and the rate of entropy generation in the compressor process.

11.96 A small heat pump unit is used to heat water for a hot-water supply. Assume that the unit uses R-22 and operates on the ideal refrigeration cycle. The evaporator temperature is 15°C, and the condenser temperature is 60°C. If the amount of hot water needed is 0.1 kg/s, determine the amount of energy saved by using the heat pump instead of directly heating the water from 15 to 60°C.

11.97 The refrigerant R-22 is used as the working fluid in a conventional heat pump cycle. Saturated vapor enters the compressor of this unit at 10°C; its exit temperature from the compressor is measured and found to be 85°C. If the isentropic efficiency of the compressor is estimated to be 70%, what is the coefficient of performance of the heat pump?

11.98 In an actual refrigeration cycle using R-12 as the working fluid, the refrigerant flow rate is 0.05 kg/s. Vapor enters the compressor at 150 kPa, −10°C, and leaves at 1.2 MPa, 75°C. The power input to the non-adiabatic compressor is measured and found to be 2.4 kW. The refrigerant enters the expansion valve at 1.15 MPa, 40°C, and leaves the evaporator at 175 kPa, −15°C. Determine the entropy generation in the compression process, the refrigeration capacity, and the coefficient of performance for this cycle.

11.99 Consider a small ammonia absorption refrigeration cycle that is powered by solar energy and is to be used as an air conditioner. Saturated vapor ammonia leaves the generator at 50°C, and saturated vapor leaves the evaporator at 10°C. If 7000 kJ of heat is required in the generator (solar collector) per kilogram of ammonia vapor generated, determine the overall performance of this system.

11.100 The performance of an ammonia absorption cycle refrigerator is to be compared with that of a similar vapor-compression system. Consider an absorption system having an evaporator temperature of −10°C and a condenser temperature of 50°C. The generator temperature in this system is 150°C. In this cycle 0.42 kJ is transferred to the ammonia in the evaporator for each kilojoule transferred from the high-temperature source to the ammonia solution in the generator. To make the comparison, assume that a reservoir is available at 150°C, and that heat is transferred from this reservoir to a reversible engine that rejects heat to the surroundings at 25°C. This work is then used to drive an ideal vapor-compression system with ammonia as the refrigerant. Compare the amount of refrigeration that can be achieved per kilojoule from the high-temperature source with the 0.42 kJ that can be achieved in the absorption system.

11.101 A heat exchanger is incorporated into an ideal air-standard refrigeration cycle, as shown in Fig. P11.101. It may be assumed that both the compression and the expansion are reversible adiabatic processes in this ideal case. Determine the coefficient of performance for the cycle.

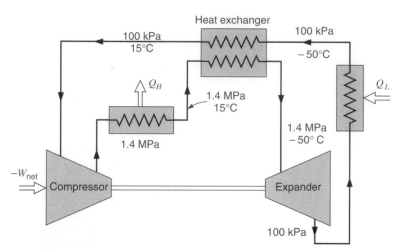

FIGURE P11.101

11.102 Repeat Problem 11.101, but assume an isentropic efficiency of 75% for both the compressor and the expander.

11.103 Repeat Problems 11.101 and 11.102, but assume that helium is the cycle working fluid instead of air. Discuss the significance of the results.

11.104 A binary system power plant uses mercury for the high-temperature cycle and water for the low-temperature cycle, as shown in Fig. 11.39. The temperatures and pressures are shown in the corresponding T–s diagram. The maximum temperature in the steam cycle is where the steam leaves the superheater at point 4 where it is 500°C. Determine the ratio of the mass flow rate of mercury to the mass flow rate of water in the heat exchanger that condenses mercury and boils the water and the thermal efficiency of this ideal cycle.

The following saturation properties for mercury are known:

P, MPa	T_g, °C	h_f, kJ/kg	h_g, kJ/kg	s_f, kJ/kgK	s_g, kJ/kgK
0.04	309	42.21	335.64	0.1034	0.6073
1.60	562	75.37	364.04	0.1498	0.4954

11.105 A Rankine steam power plant should operate with a high pressure of 3 MPa, a low pressure of 10 kPa, and the boiler exit temperature should be 500°C. The available high-temperature source is the exhaust of 175 kg/s air at 600°C from a gas turbine. If the boiler operates as a counterflowing heat exchanger where the temperature difference at the pinch point is 20°C, find the maximum water mass flow rate possible and the air exit temperature.

11.106 A simple Rankine cycle with R-22 as the working fluid is to be used as a bottoming cycle for an electrical generating facility driven by the exhaust gas from a Diesel engine as the high temperature energy source in the R-22 boiler. Diesel inlet conditions are 100 kPa, 20°C, the compression ratio is 20, and the maximum temperature in the cycle is 2800°C. Saturated vapor R-22 leaves the bottoming cycle boiler at 110°C, and the condenser temperature is 30°C. The power output of the Diesel engine is 1 MW. Assuming ideal cycles throughout, determine

a. The flow rate required in the diesel engine.

b. The power output of the bottoming cycle, assuming that the diesel exhaust is cooled to 200°C in the R-22 boiler.

11.107 For a cryogenic experiment, heat should be removed from a space at 75 K to a reservoir at 180 K. A heat pump is designed to use nitrogen and methane in a cascade arrangement (see Fig. 11.41), where the high temperature of the nitrogen condensation is at 10 K higher than the low-temperature evaporation of the methane. The two other phase changes take place at the listed reservoir temperatures. Find the saturation temperatures in the heat exchanger between the two cycles that give the best coefficient of performance for the overall system.

11.108 A cascade system is composed of two ideal refrigeration cycles, as shown in Fig. 11.41. The high-temperature cycle uses R-22. Saturated liquid leaves the condenser at 40°C, and saturated vapor leaves the heat exchanger at −20°C. The low-temperature cycle uses a different refrigerant, R-23 (Fig. F.3 or the software). Saturated vapor leaves the evaporator at −80°C, and saturated liquid leaves the heat exchanger at −10°C. Calculate the ratio of the mass flow rates through the two cycles and the coefficient of performance of the system.

11.109 Consider an ideal dual-loop heat-powered refrigeration cycle using R-12 as the working fluid, as shown in Fig. P11.109. Saturated vapor at 105°C leaves the boiler and expands in the turbine to the condenser pressure. Saturated vapor at −15°C leaves the evaporator and is compressed to the condenser pressure. The ratio of the flows through the two loops is such that the turbine produces just enough power to drive the compressor. The two exiting streams mix together and enter the condenser. Saturated liquid leaving the condenser at 45°C is then separated into two streams in the necessary proportions. Determine the ratio of mass flow rate through the power loop to that through the refrigeration loop. Find also the performance of the cycle, in terms of the ratio Q_L/Q_H.

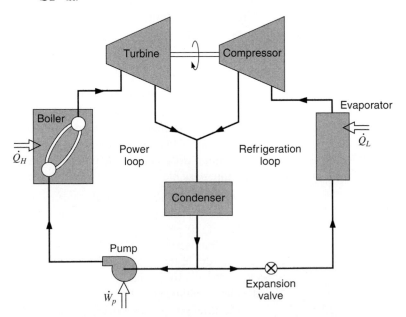

FIGURE P11.109

ADVANCED PROBLEMS

11.110 Find the availability of the water at all four states in the Rankine cycle described in Problem 11.12. Assume that the high-temperature source is 500°C and the low-temperature reservoir is at 25°C. Determine the flow of availability in or out of the reservoirs per kilogram of steam flowing in the cycle. What is the overall cycle second law efficiency?

11.111 The effect of a number of open feedwater heaters on the thermal efficiency of an ideal cycle is to be studied. Steam leaves the steam generator at 20 MPa, 600°C, and the cycle has a condenser pressure of 10 kPa. Determine the thermal efficiency for each of the following cases. **A:** No feedwater heater. **B:** One feedwater heater operating at 1 MPa. **C:** Two feedwater heaters, one operating at 3 MPa and the other at 0.2 MPa.

11.112 Find the availability of the water at all the states in the steam power plant described in Problem 11.36. Assume the heat source in the boiler is at 600°C and the low-temperature reservoir is at 25°C. Give the second law efficiency of all the components.

11.113 The power plant shown in Fig. 11.40 combines a gas-turbine cycle and a steam-turbine cycle. The following data are known for the gas-turbine cycle. Air enters the compressor at 100 kPa, 25°C, the compressor pressure ratio is 14, and the isentropic compressor efficiency is 87%; the heater input rate is 60 MW; the turbine inlet temperature is 1250°C, the exhaust pressure is 100 kPa, and the isentropic turbine efficiency is 87%; the cycle exhaust temperature from the heat exchanger is 200°C. The following data are known for the steam-turbine cycle. The pump inlet state is saturated liquid at 10 kPa, the pump exit pressure is 12.5 MPa, and the isentropic pump efficiency is 85%; turbine inlet temperature is 500°C, and the isentropic turbine efficiency is 87%. Determine

a. The mass flow rate of air in the gas-turbine cycle.

b. The mass flow rate of water in the steam cycle.

c. The overall thermal efficiency of the combined cycle.

11.114 For Problem 11.105, determine the change of availability of the water flow and that of the air flow. Use these to determine a second law efficiency for the boiler heat exchanger.

11.115 One means of improving the performance of a refrigeration system that operates over a wide temperature range is to use a two-stage compressor. Consider an ideal refrigeration system of this type that uses R-12 as the working fluid, as shown in Fig. P11.115. Saturated liquid leaves the condenser at 40°C and is throttled to −20°C. The liquid and vapor at this temperature are separated, and the liquid is throttled to the evaporator temperature, −70°C. Vapor leaving the evaporator is compressed to the saturation pressure corresponding to −20°C, after which it is mixed with the vapor leaving the flash chamber. It may be assumed that both the flash chamber and the mixing chamber are well insulated to prevent heat transfer from the ambient. Vapor leaving the mixing chamber is compressed in the second stage of the compressor to the saturation pressure corresponding to the condenser temperature, 40°C. Determine the following:

a. The coefficient of performance of the system.

b. The coefficient of performance of a simple ideal refrigeration cycle operating over the same condenser and evaporator ranges as those of the two-stage compressor unit studied in this problem.

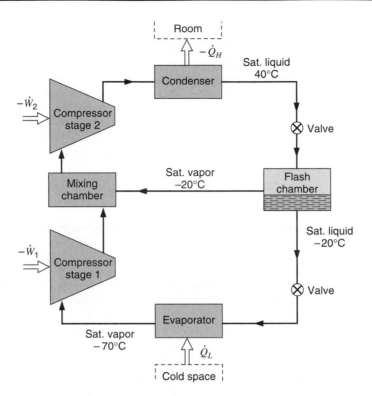

FIGURE P11.115

11.116 A jet ejector, a device with no moving parts, functions as the equivalent of a coupled turbine-compressor unit (see Problems 9.82 and 9.90). Thus, the turbine-compressor in the dual-loop cycle of Fig. P11.109 could be replaced by a jet

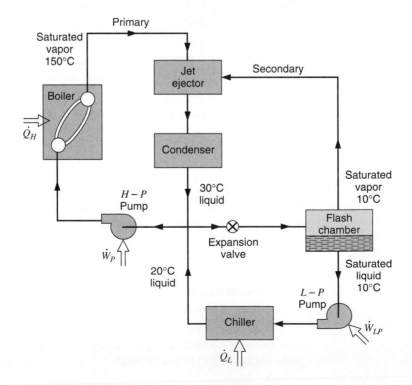

FIGURE P11.116

ejector. The primary stream of the jet ejector enters from the boiler, the secondary stream enters from the evaporator, and the discharge flows to the condenser. Alternatively, a jet ejector may be used with water as the working fluid. The purpose of the device is to chill water, usually for an air-conditioning system. In this application the physical setup is as shown in Fig. P11.116. Using the data given on the diagram, evaluate the performance of this cycle in terms of the ratio Q_L/Q_H.

a. Assume an ideal cycle.

b. Assume an ejector efficiency of 20% (see Problem 9.90).

ENGLISH UNIT PROBLEMS

11.117E A steam power plant, as shown in Fig. 11.3, operating in a Rankine cycle has saturated vapor at 600 lbf/in.^{2} leaving the boiler. The turbine exhausts to the condenser operating at 2 lbf/in.^{2}. Find the specific work and heat transfer in each of the ideal components and the cycle efficiency.

11.118E Consider a solar-energy-powered ideal Rankine cycle that uses water as the working fluid. Saturated vapor leaves the solar collector at 350 F, and the condenser pressure is 1 lbf/in.^{2}. Determine the thermal efficiency of this cycle.

11.119E A supply of geothermal hot water is to be used as the energy source in an ideal Rankine cycle, with R-134a as the cycle working fluid. Saturated vapor R-134a leaves the boiler at a temperature of 180 F, and the condenser temperature is 100 F. Calculate the thermal efficiency of this cycle.

11.120E Do Problem 11.119 with R-22 as the working fluid.

11.121E The power plant in Problem 11.117 is modified to have a superheater section following the boiler so the steam leaves the superheater at 600 lbf/in.^{2}, 700 F. Find the specific work and heat transfer in each of the ideal components and the cycle efficiency.

11.122E Consider a simple ideal Rankine cycle using water at a supercritical pressure. Such a cycle has a potential advantage of minimizing local temperature differences between the fluids in the steam generator, such as the instance in which the high-temperature energy source is the hot exhaust gas from a gas-turbine engine. Calculate the thermal efficiency of the cycle if the state entering the turbine is 3500 lbf/in.^{2}, 1100 F, and the condenser pressure is 1 lbf/in.^{2}. What is the steam quality at the turbine exit?

11.123E Consider an ideal steam reheat cycle in which the steam enters the high-pressure turbine at 600 lbf/in.^{2}, 700 F, and then expands to 120 lbf/in.^{2}. It is then reheated to 700 F and expands to 2 lbf/in.^{2} in the low-pressure turbine. Calculate the thermal efficiency of the cycle and the moisture content of the steam leaving the low-pressure turbine.

11.124E A closed feedwater heater in a regenerative steam power cycle heats 40 lbm/s of water from 200 F, 2000 lbf/in.^{2} to 450 F, 2000 lbf/in.^{2}. The extraction steam from the turbine enters the heater at 500 lbf/in.^{2}, 550 F and leaves as saturated liquid. What is the required mass flow rate of the extraction steam?

11.125E Consider an ideal steam regenerative cycle in which steam enters the turbine at 600 lbf/in.^{2}, 700 F, and exhausts to the condenser at 2 lbf/in.^{2}. Steam is extracted from the turbine at 120 lbf/in.^{2} for an open feedwater heater. The feedwater leaves the heater as saturated liquid. The appropriate pumps are used for the water leaving the condenser and the feedwater heater. Calculate the thermal efficiency of the cycle and the net work per pound-mass of steam.

11.126E Consider an ideal combined reheat and regenerative cycle in which steam enters the high-pressure turbine at 500 lbf/in.^{2}, 700 F, and is extracted to an open feedwater heater at 120 lbf/in.^{2} with exit as saturated liquid. The remainder of the steam is reheated to 700 F at this pressure, 120 lbf/in.^{2}, and is fed to the low-pressure turbine. The condenser pressure is 2 lbf/in.^{2}. Calculate the thermal efficiency of the cycle and the net work per pound-mass of steam.

11.127E A steam power cycle has a high pressure of 500 lbf/in.^{2} and a condenser exit temperature of 110 F. The turbine efficiency is 85%, and other cycle components are ideal. If the boiler superheats to 1400 F, find the cycle thermal efficiency.

11.128E The steam power cycle in Problem 11.117 has an isentropic efficiency of the turbine of 85% and that for the pump of 80%. Find the cycle efficiency and the specific work and heat transfer in the components.

11.129E Steam leaves a power plant steam generator at 500 lbf/in.^{2}, 650 F, and enters the turbine at 490 lbf/in.^{2}, 625 F. The isentropic turbine efficiency is 88%, and the turbine exhaust pressure is 1.7 lbf/in.^{2}. Condensate leaves the condenser and enters the pump at 110 F, 1.7 lbf/in.^{2}. The isentropic pump efficiency is 80%, and the discharge pressure is 520 lbf/in.^{2}. The feedwater enters the steam generator at 510 lbf/in.^{2}, 100 F. Calculate the thermal efficiency of the cycle and the entropy generation of the flow in the line between the steam generator exit and the turbine inlet, assuming an ambient temperature of 77 F.

11.130E In one type of nuclear power plant, heat is transferred in the nuclear reactor to liquid sodium. The liquid sodium is then pumped through a heat exchanger where heat is transferred to boiling water. Saturated vapor steam at 700 lbf/in.^{2} exits this heat exchanger and is then superheated to 1100 F in an external gas-fired superheater. The steam enters the turbine, which has one (open-type) feedwater extraction at 60 lbf/in.^{2}. The isentropic turbine efficiency is 87%, and the condenser pressure is 1 lbf/in.^{2}. Determine the heat transfer in the reactor and in the superheater to produce a net power output of 1000 Btu/s.

11.131E A boiler delivers steam at 1500 lbf/in.^{2}, 1000 F to a two-stage turbine, as shown in Fig. 11.17. After the first stage, 25% of the steam is extracted at 200 lbf/in.^{2} for a process application and returned at 150 lbf/in.^{2}, 190 F to the feedwater line. The remainder of the steam continues through the low-pressure turbine stage, which exhausts to the condenser at 2 lbf/in.^{2}. One pump brings the feedwater to 150 lbf/in.^{2} and a second pump brings it to 1500 lbf/in.^{2}. Assume the first and second stages in the steam turbine have isentropic efficiencies of 85% and 80% and that both pumps are ideal. If the process application requires 5000 Btu/s of power, how much power can then be cogenerated by the turbine?

11.132E A large stationary Brayton cycle gas-turbine power plant delivers a power output of 100 000 hp to an electric generator. The minimum temperature in the cycle is 540 R, and the maximum temperature is 2900 R. The minimum pressure in the cycle is 1 atm, and the compressor pressure ratio is 14 to 1. Calculate the power output of the turbine, the fraction of the turbine output required to drive the compressor, and the thermal efficiency of the cycle.

11.133E An ideal regenerator is incorporated into the ideal air-standard Brayton cycle of Problem 11.132. Calculate the cycle thermal efficiency with this modification.

11.134E Consider an ideal gas-turbine cycle with two stages of compression and two stages of expansion. The pressure ratio across each compressor stage and each turbine stage is 8 to 1. The pressure at the entrance to the first compressor is 14 lbf/in.^{2}, the temperature entering each compressor is 70 F, and the temperature entering each turbine is 2000 F. An ideal regenerator is also incorporated into the cycle. Determine the compressor work, the turbine work, and the thermal efficiency of the cycle.

11.135E Repeat Problem 11.134, but assume that each compressor stage and each turbine stage has an isentropic efficiency of 85%. Also assume that the regenerator has an efficiency of 70%.

11.136E An air-standard Ericsson cycle has an ideal regenerator, as shown in Fig. P11.62. Heat is supplied at 1800 F, and heat is rejected at 68 F. Pressure at the beginning of the isothermal compression process is 10 lbf/in.^{2}. The heat added is 275 Btu/lbm. Find the compressor work, the turbine work, and the cycle efficiency.

11.137E The turbine in a jet engine receives air at 2200 R, 220 lbf/in.^{2}. It exhausts to a nozzle at 35 lbf/in.^{2}, which in turn exhausts to the atmosphere at 14.7 lbf/in.^{2}. The isentropic efficiency of the turbine is 85%, and the nozzle efficiency is 95%. Find the nozzle inlet temperature and the nozzle exit velocity. Assume negligible kinetic energy out of the turbine.

11.138E Air flows into a gasoline engine at 14 lbf/in.^{2}, 540 R. The air is then compressed with a volumetric compression ratio of 8 : 1. In the combustion process 560 Btu/lbm of energy is released as the fuel burns. Find the temperature and pressure after combustion.

11.139E To approximate an actual spark-ignition engine consider an air-standard Otto cycle that has a heat addition of 800 Btu/lbm of air, a compression ratio of 7, and a pressure and temperature at the beginning of the compression process of 13 lbf/in.^{2}, 50 F. Assuming constant specific heat, with the value from Table C.4, determine the maximum pressure and temperature of the cycle, the thermal efficiency of the cycle, and the mean effective pressure.

11.140E In the Otto cycle, all the heat transfer q_H occurs at constant volume. It is more realistic to assume that part of q_H occurs after the piston has started its downwards motion in the expansion stroke. Therefore, consider a cycle identical to the Otto cycle, except that the first two-thirds of the total q_H occurs at constant volume and the last one-third occurs at constant pressure. Assume the total q_H is 700 Btu/lbm, that the state at the beginning of the compression process is 13 lbf/in.^{2}, 68 F, and that the compression ratio is 9. Calculate the maximum pressure and temperature and the thermal efficiency of this cycle. Compare the results with those of a conventional Otto cycle having the same given variables.

11.141E It is found experimentally that the power stroke expansion in an internal combustion engine can be approximated with a polytropic process with a value of the polytropic exponent n somewhat larger than the specific heat ratio k. Repeat Problem 11.139 but assume the expansion process is reversible and polytropic (instead of the isentropic expansion in the Otto cycle) with n equal to 1.50.

11.142E A diesel engine has a bore of 4 in., a stroke of 4.3 in., and a compression ratio of 19:1 running at 2000 RPM (revolutions per minute). Each cycle takes two revolutions and has a mean effective pressure of 200 $lbf/in.^2$. With a total of six cylinders find the engine power in Btu/s and horsepower, hp.

11.143E At the beginning of compression in a diesel cycle $T = 540$ R, $P = 30$ $lbf/in.^2$ and the state after combustion (heat addition) is 2600 R and 1000 $lbf/in.^2$. Find the compression ratio, the thermal efficiency, and the mean effective pressure.

11.144E Consider an ideal air-standard diesel cycle where the state before the compression process is 14 $lbf/in.^2$, 63 F and the compression ratio is 20. Find the maximum temperature (by iteration) in the cycle to have a thermal efficiency of 60%.

11.145E Consider an ideal Stirling-cycle engine in which the pressure and temperature at the beginning of the isothermal compression process are 14.7 $lbf/in.^2$, 80 F, the compression ratio is 6, and the maximum temperature in the cycle is 2000 F. Calculate the maximum pressure in the cycle and the thermal efficiency of the cycle with and without regenerators.

11.146E An ideal air-standard Stirling cycle uses helium as working fluid. The isothermal compression brings the helium from 15 $lbf/in.^2$, 70 F to 90 $lbf/in.^2$. The expansion takes place at 2100 R, and there is no regenerator. Find the work and heat transfer in all four processes per lbm helium and the cycle efficiency.

11.147E The air-standard Carnot cycle was not shown in the text; show the T–s diagram for this cycle. In an air-standard Carnot cycle the low temperature is 500 R and the efficiency is 60%. If the pressure before compression and after heat rejection is 14.7 $lbf/in.^2$, find the high temperature and the pressure just before heat addition.

11.148E Air in a piston/cylinder goes through a Carnot cycle in which $T_L = 80.3$ F and the total cycle efficiency is $\eta = 2/3$. Find T_H, the specific work and volume ratio in the adiabatic expansion for constant C_p, C_v. Repeat the calculation for variable heat capacities.

11.149E Consider an ideal refrigeration cycle that has a condenser temperature of 110 F and an evaporator temperature of 5 F. Determine the coefficient of performance of this refrigerator for the working fluids R-12 and R-22.

11.150E The environmentally safe refrigerant R-134a is one of the replacements for R-12 in refrigeration systems. Repeat Problem 11.149 using R-134a and compare the result with that for R-12.

11.151E Consider an ideal heat pump that has a condenser temperature of 120 F and an evaporator temperature of 30 F. Determine the coefficient of performance of this heat pump for the working fluids R-12, R-22, and ammonia.

11.152E The refrigerant R-22 is used as the working fluid in a conventional heat pump cycle. Saturated vapor enters the compressor of this unit at 50 F; its exit temperature from the compressor is measured and found to be 185 F. If the isentropic efficiency of the compressor is estimated to be 70%, what is the coefficient of performance of the heat pump?

11.153E Consider a small ammonia absorption refrigeration cycle that is powered by solar energy and is to be used as an air conditioner. Saturated vapor ammonia leaves the generator at 120 F, and saturated vapor leaves the evaporator at 50 F. If 3000 Btu of heat is required in the generator (solar collector) per pound-mass of ammonia vapor generated, determine the overall performance of this system.

11.154E Consider an ideal dual-loop heat-powered refrigeration cycle using R-12 as the working fluid, as shown in Fig. P11.109. Saturated vapor at 220 F leaves the boiler and expands in the turbine to the condenser pressure. Saturated vapor at 0 F leaves the evaporator and is compressed to the condenser pressure. The ratio of the flows through the two loops is such that the turbine produces just enough power to drive the compressor. The two exiting streams mix together and enter the condenser. Saturated liquid leaving the condenser at 110 F is then separated into two streams in the necessary proportions. Determine the ratio of mass flow rate through the power loop to that through the refrigeration loop. Find also the performance of the cycle, in terms of the ratio Q_L/Q_H.

11.155E (Adv.) Find the availability of the water at all four states in the Rankine cycle described in Problem 11.121. Assume the high-temperature source is 900 F and the low-temperature reservoir is at 65 F. Determine the flow of availability in or out of the reservoirs per pound-mass of steam flowing in the cycle. What is the overall cycle second-law efficiency?

COMPUTER, DESIGN, AND OPEN-ENDED PROBLEMS

11.156 The effect of turbine exhaust pressure on the performance of the ideal steam Rankine cycle given in Problem 11.12 is to be studied. Calculate the thermal efficiency of the cycle and the moisture content of the steam leaving the turbine for turbine exhaust pressures of 5, 10, 50, and 100 kPa. Plot the thermal efficiency versus turbine exhaust pressure for the specified turbine inlet pressure and temperature.

11.157 The effect of turbine inlet pressure on the performance of the ideal steam Rankine cycle given in Problem 11.12 is to be studied. Calculate the thermal efficiency of the cycle and the moisture content of the steam leaving the turbine for turbine inlet pressures of 1, 3.5, 6, and 10 MPa. Plot the thermal efficiency versus turbine inlet pressure for the specified turbine inlet temperature and exhaust pressure.

11.158 The effect of turbine inlet temperature on the performance of the ideal steam Rankine cycle given in Problem 11.12 is to be studied. Calculate the thermal efficiency of the cycle and the moisture content of the steam leaving the turbine for turbine inlet temperatures of 400°, 500°, 800°C, and saturated vapor (at 3.5 MPa). Plot the thermal efficiency versus turbine inlet temperature for the specified turbine inlet pressure and exhaust pressure.

11.159 Write a program to solve the following problem. The effects of varying parameters on the performance of an air-standard Brayton cycle are to be determined. Consider a compressor inlet condition of 100 kPa, 20°C, and assume constant specific heat. The thermal efficiency of the cycle and the net specific work output should be determined for the combinations of the following variables.

a. Compressor pressure ratio of 6, 9, 12, and 15.

b. Maximum cycle temperature of 900, 1100, 1300, and 1500°C.

c. Compressor and turbine isentropic efficiencies each 100, 90, 80, and 70%.

11.160 The effect of variable specific heat on the result of the previous problem is to be studied. Write a modified program that uses the heat capacity curvefits from Table A.6 that can be integrated mathematically to give formulas for the enthalpy and standard entropy (see Section 8.10).

11.161 The effect of adding a regenerator to the gas-turbine cycle in the previous two problems is to be studied. Repeat one of these problems by including a regenerator with various values of the regenerator efficiency.

11.162 Write a program to simulate the Otto cycle using nitrogen as the working fluid. Use the variable specific heat as given in Table A.6. The beginning of compression has a state of 100 kPa, 20°C. Determine the net specific work output and the cycle thermal efficiency for various combinations of compression ratio and maximum cycle temperature. Compare the result with those found when constant specific heat is assumed.

11.163 Write a program to compare the performance of the Otto and Diesel cycles. Repeat the previous problem and add the Diesel cycle also using nitrogen as the working fluid. Use the specific heat equation given in Table A.6.

11.164 Write a program to study the performance of the air-standard refrigeration cycle, incorporating a heat exchanger as shown in Fig. 11.38. The compressor inlet condition is 100 kPa, 15°C, and the pressure ratio across the compressor and expander is a variable. Include isentropic efficiencies for the compressor and expander.

11.165 A power plant is built to provide district heating of buildings that requires 90°C liquid water at 150 kPa. The district heating water is returned at 50°C, 100 kPa, in a closed loop in an amount such that 20 MW of power is delivered. This hot water is produced from a steam power cycle with a boiler making steam at 5 MPa, 600°C delivered to the steam turbine. The steam cycle could have its condenser operate at 90°C providing the power to the district heating. It could also be done with extraction of steam from the turbine. Suggest a system and evaluate its performance in terms of the cogenerated amount of turbine work.

11.166 The effect of evaporator temperature on the coefficient of performance of a heat pump is to be studied. Consider an ideal cycle with R-22 as the working fluid and a condenser temperature of 40°C. Plot a curve for the coefficient of performance versus the evaporator temperature for temperatures from +15 to −25°C.

11.167 A hospital requires 2 kg/s steam at 200°C, 125 kPa for sterilization purposes, and space heating requires 15 kg/s hot water at 90°C, 100 kPa. Both of these requirements are provided by the hospital's steam power plant. Discuss some arrangement that will accomplish this.

11.168 Use the menu-driven software to get properties for the different refrigerants to solve the following problem. A heat pump should be designed to operate in the ideal refrigeration cycle where the high-temperature condensation is at $T_H +$

10°C, and the low-temperature evaporation is at $T_L - 10$°C. Suggest a suitable refrigerant (from among those included in the software) to use for the cases where the temperatures are as follows:

a. High temperature is 30°C, low temperature is −10°C.

b. High temperature is −20°C, low temperature is −50°C.

11.169 Investigate the maximum power out of a steam power plant with operating conditions as in Problem 11.12. The energy source is 100 kg/s combustion products (air) at 125 kPa, 1200 K. Make sure the air temperature is higher than the water temperature throughout the boiler.

11.170 In Problem 11.105, a steam cycle was powered by the exhaust from a gas turbine. With a single water flow and air flow heat exchanger, the air is leaving with a relatively high temperature. Analyze how some more of the energy in the air can be used before the air is flowing out to the chimney. Can it be used in a feedwater heater?

12 GAS MIXTURES

Up to this point in our development of thermodynamics we have considered primarily pure substances. A large number of thermodynamic problems involve mixtures of different pure substances. Sometimes these mixtures are referred to as solutions, particularly in the liquid and solid phases.

In this chapter we shall turn our attention to various thermodynamic considerations of gas mixtures. We begin with a consideration of a rather simple problem: mixtures of ideal gases. This leads to a consideration of a simplified but very useful model of certain mixtures, such as air and water vapor, which may involve a condensed (solid or liquid) phase of one of the components.

An understanding of the materials in this chapter is a necessary foundation for the consideration of chemical reactions and chemical and phase equilibrium. These topics are covered in subsequent chapters.

12.1 GENERAL CONSIDERATIONS AND MIXTURES OF IDEAL GASES

Let us consider a general mixture of N components, each a pure substance, so the total mass and the total number of moles are

$$m_{\text{tot}} = m_1 + m_2 + \cdots + m_N = \sum m_i$$

$$n_{\text{tot}} = n_1 + n_2 + \cdots + n_N = \sum n_i$$

The mixture is usually described by a mass fraction (concentration)

$$c_i = \frac{m_i}{m_{\text{tot}}} \tag{12.1}$$

or a mole fraction for each component as

$$y_i = \frac{n_i}{n_{\text{tot}}} \tag{12.2}$$

which are related through the molecular weight, M_i, as $m_i = n_i M_i$. We may then convert from a mole basis to a mass basis as

$$c_i = \frac{m_i}{m_{\text{tot}}} = \frac{n_i M_i}{\sum n_j M_j} = \frac{n_i M_i / n_{\text{tot}}}{\sum n_j M_j / n_{\text{tot}}} = \frac{y_i M_i}{\sum y_j M_j} \tag{12.3}$$

438

and from a mass basis to a mole basis as

$$y_i = \frac{n_i}{n_{\text{tot}}} = \frac{m_i / M_i}{\sum m_j / M_j} = \frac{m_i / (M_i m_{\text{tot}})}{\sum m_j / (M_j m_{\text{tot}})} = \frac{c_i / M_i}{\sum c_j / M_j} \tag{12.4}$$

The molecular weight for the mixture becomes

$$M_{\text{mix}} = \frac{m_{\text{tot}}}{n_{\text{tot}}} = \frac{\sum n_i M_i}{n_{\text{tot}}} = \sum y_i M_i \tag{12.5}$$

Consider a mixture of two gases (not necessarily ideal gases) such as shown in Fig. 12.1. What properties can we experimentally measure for such a mixture? Certainly we can measure the pressure, temperature, volume, and mass of the mixture. We can also experimentally measure the composition of the mixture, and thus determine the mole and mass fractions.

Suppose that this mixture undergoes a process or a chemical reaction and we wish to perform a thermodynamic analysis of this process or reaction. What type of thermodynamic data would we use in performing such an analysis? One possibility would be to have tables of thermodynamic properties of mixtures. However, the number of different mixtures that is possible, both as regards the substances involved and the relative amounts of each, is such that we would need a library full of tables of thermodynamic properties to handle all possible situations. It would be much simpler if we could determine the thermodynamic properties of a mixture from the properties of the pure components. This is in essence the approach that is used in dealing with ideal gases and certain other simplified models of mixtures.

One exception to this procedure is the case where a particular mixture is encountered very frequently, the most familiar being air. Tables and charts of the thermodynamic properties of air are available. However, even in this case it is necessary to define the composition of the "air" for which the tables are given, because the composition of the atmosphere varies with altitude, with the number of pollutants, and with other variables at a given location. The composition of air on which air tables are usually based is as follows:

Component	% on Mole Basis
Nitrogen	78.10
Oxygen	20.95
Argon	0.92
CO_2 & trace elements	0.03

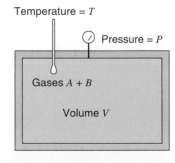

FIGURE 12.1 A mixture of two gases.

Consider again the mixture of Fig. 12.1. For the general case the properties of the mixture are defined in terms of the partial molal properties of the individual components. A partial molal property is defined as the value of a property, such as internal energy, for a given component as it exists in the mixture. With this definition, the internal energy of the mixture of Fig. 12.1 would be

$$U_{\text{mix}} = n_A \overline{U}_A + n_B \overline{U}_B \tag{12.6}$$

where $\overline{U}$ designates the partial molal internal energy. Similar equations can be written for other properties.

In this chapter we focus on mixtures of ideal gases. We assume that each component is uninfluenced by the presence of the other components, and that each component can be treated as an ideal gas. In an actual case of a gaseous mixture at high pressure this assumption would probably not be true because of the nature of the interaction between the molecules of the different components.

Two models are used in analyzing the mixtures of gases, namely, the Dalton model and the Amagat model.

Dalton Model

For the Dalton model, the properties of each component are considered as though each component existed separately at the volume and temperature of the mixture, as shown in Fig. 12.2.

Consider this model for the special case in which both the mixture and the separated components can be considered an ideal gas.

For the mixture:
$$PV = n\overline{R}T$$

$$n = n_A + n_B \tag{12.7}$$

For the components:
$$P_A V = n_A \overline{R}T$$
$$P_B V = n_B \overline{R}T \tag{12.8}$$

On substituting, we have

$$n = n_A + n_B$$

$$\frac{PV}{\overline{R}T} = \frac{P_A V}{\overline{R}T} + \frac{P_B V}{\overline{R}T} \tag{12.9}$$

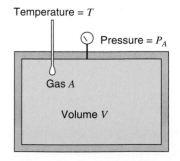
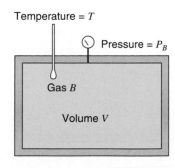

FIGURE 12.2 The Dalton model.

or

$$P = P_A + P_B$$

where P_A and P_B are referred to as partial pressures.

Thus for a mixture of ideal gases, the pressure is the sum of the partial pressures of the individual components.

It should be stressed that the term *partial pressure* is relevant only for ideal gases. This concept assumes that the molecules of each component are uninfluenced by the other components, and that the total pressure is the sum of partial pressures of the individual components. It should also be noted that partial pressure is not a partial molal property in the sense as defined by Eq. 12.3, since partial molal properties relate only to extensive properties.

Amagat Model

In the Amagat model the properties of each component are considered as though each component existed separately at the pressure and temperature of the mixture, as shown in Fig. 12.3. The volumes of A and B under these conditions are V_A and V_B, respectively.

In the general case, the sum of the volumes when separated, namely, $V_A + V_B$, need not be equal to the volume of the mixture. However, let us consider the special case in which both the separated components and the mixture are considered to be ideal gases. In this case we can write:

For the mixture:
$$PV = n\overline{R}T$$
$$n = n_A + n_B \tag{12.10}$$

For the components:
$$PV_A = n_A\overline{R}T$$
$$PV_B = n_B\overline{R}T \tag{12.11}$$

On substituting we have

$$n = n_A + n_B$$
$$\frac{PV}{\overline{R}T} = \frac{PV_A}{\overline{R}T} + \frac{PV_B}{\overline{R}T}$$

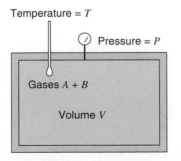
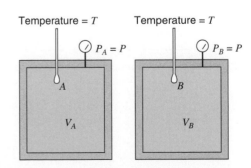

FIGURE 12.3 The Amagat model.

Therefore

$$V_A + V_B = V$$

or

$$\frac{V_A}{V} + \frac{V_B}{V} = 1 \tag{12.12}$$

where V_A/V and V_B/V are referred to as the volume fractions.

Thus for ideal gases, Amagat's model leads to the conclusion that the sum of the volume fractions is unity, and that there would be no volume change if the components were mixed while holding the temperature and pressure constant.

From Eqs. 12.7, 12.8, 12.10, and 12.11 it is evident that

$$\frac{V_A}{V} = \frac{n_A}{n} = \frac{P_A}{P}$$

$$\frac{V_A}{V} = y_A = \frac{P_A}{P} \tag{12.13}$$

That is, for each component of a mixture of ideal gases, the volume fraction, the mole fraction, and the ratio of the partial pressure to the total pressure are equal.

In determining the internal energy, enthalpy, and entropy of a mixture of ideal gases, the Dalton model proves useful because the assumption is made that each constituent behaves as though it occupies the entire volume by itself. Thus, the internal energy, enthalpy, and entropy can be evaluated as the sum of the respective properties of the constituent gases at the condition at which the component exists in the mixture. Since for ideal gases the internal energy and enthalpy are functions only of temperature, it follows that

$$U = n\bar{u} = n_A \bar{u}_A + n_B \bar{u}_B \tag{12.14}$$

$$H = n\bar{h} = n_A \bar{h}_A + n_B \bar{h}_B \tag{12.15}$$

where $\bar{u}_A$ and $\bar{h}_A$ are the internal energy and enthalpy per mole for pure A and $\bar{u}_B$ and $\bar{h}_B$ are the same quantities for pure B, all at the temperature of the mixture.

The entropy of an ideal gas is a function of pressure as well as temperature. Since each component exists in the mixture at its partial pressure,

$$S = n\bar{s} = n_A \bar{s}_A + n_B \bar{s}_B \tag{12.16}$$

where $\bar{s}_A$ is the entropy per mole for pure A at T and P_A (the partial pressure of A), and $\bar{s}_B$ is the entropy per mole for pure B at T and P_B.

EXAMPLE 12.1   A volumetric analysis of a gaseous mixture yields the following results:

CO_2	12.0%
O_2	4.0
N_2	82.0
CO	2.0

TABLE 12.1

Constituent	Percent by Volume	Mole Fraction		Molecular Weight		Mass kg per kmol of Mixture	Analysis on Mass Basis, Percent
CO_2	12	0.12	×	44.0	=	5.28	$\dfrac{5.28}{30.08} = 17.55$
O_2	4	0.04	×	32.0	=	1.28	$\dfrac{1.28}{30.08} = 4.26$
N_2	82	0.82	×	28.0	=	22.96	$\dfrac{22.96}{30.08} = 76.33$
CO	2	0.02	×	28.0	=	$\dfrac{0.56}{30.08}$	$\dfrac{0.56}{30.08} = \dfrac{1.86}{100.00}$

Determine the analysis on a mass basis, and the molecular weight and the gas constant on a mass basis for the mixture. Assume ideal gas behavior.

Control mass: Gas mixture.

State: Composition known.

Solution

It is convenient to set up and solve the problem as shown below in Table 12.1 From this table, we note that

$$\text{Molecular weight of mixture} = 30.08$$

$$R \text{ for mixture} = \frac{\overline{R}}{M} = \frac{8.3145}{30.08} = 0.2764 \text{ kJ / kg K}$$

If the analysis has been given on a mass basis, and the mole fraction or volumetric analysis is desired, the procedure shown in Table 12.2 can be used.

$$M = \frac{1}{\text{kmol / kg mixture}} = \frac{1}{0.033\,24} = 30.08$$

$$R = \frac{\overline{R}}{M} = \frac{8.3145}{30.08} = 0.2764 \text{ kJ / kg K}$$

TABLE 12.2

Constituent	Mass Fraction	Molecular Weight		kmol per kg of Mixture	Mole Fraction	Volumetric Analysis, Percent
CO_2	0.1755	÷	44.0	= 0.003 99	0.120	12.0
O_2	0.0426	÷	32.0	= 0.001 33	0.040	4.0
N_2	0.7633	÷	28.0	= 0.027 26	0.820	82.0
CO	0.0186	÷	28.0	= 0.000 66	0.020	2.0
				0.033 24	1.000	100.0

EXAMPLE 12.2 Let n_A moles of gas A at a given pressure and temperature be mixed with n_B moles of gas B at the same pressure and temperature in an adiabatic constant-volume process, as shown in Fig. 12.4. Determine the increase in entropy for this process.

> *Control mass:* All gas (A and B).
>
> *Initial states:* P, T known for A and B.
>
> *Final state:* P, T of mixture known.
>
> *Sketch:* Figure 12.4.

Analysis and Solution

The final partial pressure of gas A is P_A and for gas B it is P_B. Since there is no change in temperature, Eq. 8.23 reduces to

$$\left(S_2 - S_1\right)_A = -n_A \overline{R} \ln \frac{P_A}{P} = -n_A \overline{R} \ln y_A$$

$$\left(S_2 - S_1\right)_B = -n_B \overline{R} \ln \frac{P_B}{P} = -n_B \overline{R} \ln y_B$$

The total change in entropy is the sum of the entropy changes for gases A and B.

$$S_2 - S_1 = -\overline{R}\left(n_A \ln y_A + n_B \ln y_B\right)$$

The result of Example 12.2 can readily be generalized to account for the mixing of any number of components at the same temperature and pressure. The result is

$$S_2 - S_1 = -\overline{R} \sum_k n_k \ln y_k \tag{12.17}$$

The interesting thing about this equation is that the increase in entropy depends only on the number of moles of component gases, and is independent of the composition of the gas. For example, when 1 mol of oxygen and 1 mol of nitrogen are mixed, the increase in entropy is the same as when 1 mol of hydrogen and 1 mol of nitrogen are mixed. But we also know that if 1 mol of nitrogen is "mixed" with another mole of nitrogen there is no increase in entropy. The question that arises is how dissimilar must the gases be in order to have an increase in entropy? The answer lies in our ability to distinguish between the two gases. The entropy increases whenever we can distinguish between the gases being mixed. When we cannot distinguish between the gases, there is no increase in entropy.

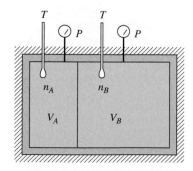
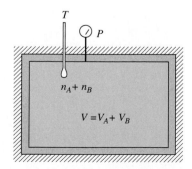

FIGURE 12.4 Sketch for Example 12.2.

12.2 A SIMPLIFIED MODEL OF A MIXTURE INVOLVING GASES AND A VAPOR

Let us now consider a simplification, which is often a reasonable one, of the problem involving a mixture of ideal gases that is in contact with a solid or liquid phase of one of the components. The most familiar example is a mixture of air and water vapor in contact with liquid water or ice, such as encountered in air conditioning or in drying. We are all familiar with the condensation of water from the atmosphere when it cools on a summer day.

This problem and a number of similar problems can be analyzed quite simply and with considerable accuracy if the following assumptions are made:

1. The solid or liquid phase contains no dissolved gases.
2. The gaseous phase can be treated as a mixture of ideal gases.
3. When the mixture and the condensed phase are at a given pressure and temperature, the equilibrium between the condensed phase and its vapor is not influenced by the presence of the other component. This means that when equilibrium is achieved, the partial pressure of the vapor will be equal to the saturation pressure corresponding to the temperature of the mixture.

Since this approach is used extensively and with considerable accuracy, let us give some attention to the terms that have been defined and the type of problems for which this approach is valid and relevant. In our discussion we will refer to this as a gas–vapor mixture.

The dew point of a gas–vapor mixture is the temperature at which the vapor condenses or solidifies when it is cooled at constant pressure. This is shown on the T–s diagram for the vapor shown in Fig. 12.5. Suppose that the temperature of the gaseous mixture and the partial pressure of the vapor in the mixture are such that the vapor is initially superheated at state 1. If the mixture is cooled at constant pressure, the partial pressure of the vapor remains constant until point 2 is reached, and then condensation begins. The temperature at state 2 is the dew-point temperature. Lines 1–3 on the diagram indicate that if the mixture is cooled at constant volume the condensation begins at point 3, which is slightly lower than the dew-point temperature.

If the vapor is at the saturation pressure and temperature, the mixture is referred to as a saturated mixture, and for an air–water vapor mixture, the term *saturated air* is used.

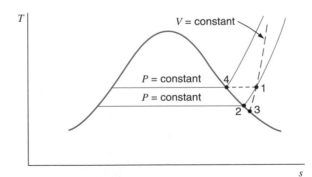

FIGURE 12.5 Temperature–entropy diagram to show definition of the dew point.

The relative humidity ϕ is defined as the ratio of the mole fraction of the vapor in the mixture to the mole fraction of vapor in a saturated mixture at the same temperature and total pressure. Since the vapor is considered an ideal gas, the definition reduces to the ratio of the partial pressure of the vapor as it exists in the mixture, P_v, to the saturation pressure of the vapor at the same temperature, P_g.

$$\phi = \frac{P_v}{P_g}$$

In terms of the numbers on the T–s diagram of Fig. 12.5, the relative humidity ϕ would be

$$\phi = \frac{P_1}{P_4}$$

Since we are considering the vapor to be an ideal gas, the relative humidity can also be defined in terms of specific volume or density.

$$\phi = \frac{P_v}{P_g} = \frac{\rho_v}{\rho_g} = \frac{v_g}{v_v} \tag{12.18}$$

The humidity ratio ω of an air-water vapor mixture is defined as the ratio of the mass of water vapor m_v to the mass of dry air m_a. The term *dry air* is used to emphasize that this refers only to air and not to the water vapor. The term *specific humidity* is used synonymously with humidity ratio.

$$\omega = \frac{m_v}{m_a} \tag{12.19}$$

This definition is identical for any other gas–vapor mixture, and the subscript a refers to the gas, exclusive of the vapor. Since we consider both the vapor and the mixture to be ideal gases, a very useful expression for the humidity ratio in terms of partial pressures and molecular weights can be developed.

$$m_v = \frac{P_v V}{R_v T} = \frac{P_v V M_v}{\overline{R} T} \qquad m_a = \frac{P_a V}{R_a T} = \frac{P_a V M_a}{\overline{R} T}$$

Then

$$\omega = \frac{P_v V / R_v T}{P_a V / R_a T} = \frac{R_a P_v}{R_v P_a} = \frac{M_v P_v}{M_a P_a} \tag{12.20}$$

For an air–water vapor mixture, this reduces to

$$\omega = 0.622 \frac{P_v}{P_a} \tag{12.21}$$

The degree of saturation is defined as the ratio of the actual humidity ratio to the humidity ratio of a saturated mixture at the same temperature and total pressure.

An expression for the relation between the relative humidity ϕ and the humidity ratio ω can be found by solving Eqs. 12.18 and 12.21 for P_v and equating them. The resulting relation for an air–water vapor mixture is

$$\phi = \frac{\omega P_a}{0.622 P_g} \tag{12.22}$$

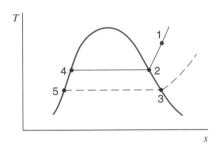

FIGURE 12.6 Temperature–entropy diagram to show the cooling of a gas–vapor mixture at a constant pressure.

A few words should also be said about the nature of the process that occurs when a gas–vapor mixture is cooled at constant pressure. Suppose that the vapor is initially superheated at state 1 in Fig. 12.6. As the mixture is cooled at constant pressure, the partial pressure of the vapor remains constant until the dew point is reached at point 2, where the vapor in the mixture is saturated. The initial condensate is at state 4, and is in equilibrium with the vapor at state 2. As the temperature is lowered further, more of the vapor condenses, which lowers the partial pressure of the vapor in the mixture. The vapor that remains in the mixture is always saturated, and the liquid or solid is in equilibrium with it. For example, when the temperature is reduced to T_3, the vapor in the mixture is at state 3, and its partial pressure is the saturation pressure corresponding to T_3. The liquid in equilibrium with it is at state 5.

EXAMPLE 12.3   Consider 100 m³ of an air–water vapor mixture at 0.1 MPa, 35°C, 70% relative humidity. Calculate the humidity ratio, dew point, mass of air, and mass of vapor.

Control mass: Mixture.

State: P, T, ϕ known; state fixed.

Analysis and Solution

From Eq. 12.18 and the steam tables,

$$\phi = 0.70 = \frac{P_v}{P_g}$$

$$P_v = 0.70(5.628) = 3.94 \text{ kPa}$$

The dew point is the saturation temperature corresponding to this pressure, which is 28.6°C.

The partial pressure of the air is

$$P_a = P - P_v = 100 - 3.94 = 96.06 \text{ kPa}$$

The humidity ratio can be calculated from Eq. 12.21.

$$\omega = 0.622 \times \frac{P_v}{P_a} = 0.622 \times \frac{3.94}{96.06} = 0.0255$$

The mass of air is

$$m_a = \frac{P_a V}{R_a T} = \frac{96.06 \times 100}{0.287 \times 308.2} = 108.6 \text{ kg}$$

The mass of the vapor can be calculated by using the humidity ratio or by using the ideal gas equation of state.

$$m_v = \omega m_a = 0.0255(108.6) = 2.77 \text{ kg}$$

$$m_v = \frac{3.94 \times 100}{0.461\,52 \times 308.2} = 2.77 \text{ kg}$$

EXAMPLE 12.3E Consider 2000 ft³ of an air–water vapor mixture at 14.7 lbf/in.², 90 F, 70% relative humidity. Calculate the humidity ratio, dew point, mass of air, and mass of vapor.

Control mass: Mixture.

State: P, T, ϕ known; state fixed.

Analysis and Solution

From Eq. 12.18 and the steam tables,

$$\phi = 0.70 = \frac{P_v}{P_g}$$

$$P_v = 0.70(0.6988) = 0.4892 \text{ lbf / in}^2$$

The dew point is the saturation temperature corresponding to this pressure, which is 78.9 F.

The partial pressure of the air is

$$P_a = P - P_v = 14.70 - 0.49 = 14.21 \text{ lbf / in}^2$$

The humidity ratio can be calculated from Eq. 12.21.

$$\omega = 0.622 \times \frac{P_v}{P_a} = 0.622 \times \frac{0.4892}{14.21} = 0.02135$$

The mass of air is

$$m_a = \frac{P_a V}{R_a T} = \frac{14.21 \times 144 \times 2000}{53.34 \times 550} = 139.6 \text{ lbm}$$

The mass of the vapor can be calculated by using the humidity ratio or by using the ideal gas equation of state.

$$m_v = \omega m_a = 0.02135(139.6) = 2.98 \text{ lbm}$$

$$m_v = \frac{0.4892 \times 144 \times 2000}{85.7 \times 550} = 2.98 \text{ lbm}$$

EXAMPLE 12.4 Calculate the amount of water vapor condensed if the mixture of Example 12.3 is cooled to 5°C in a constant-pressure process.

Control mass: Mixture.

Initial state: Known (Example 12.3).

Final state: T known.

Process: Constant pressure.

Analysis:

At the final temperature, 5°C, the mixture is saturated, since this is below the dew-point temperature. Therefore,

$$P_{v2} = P_{g2}, \qquad P_{a2} = P - P_{v2}$$

and

$$\omega_2 = 0.622 \frac{P_{v2}}{P_{a2}}$$

From the conservation of mass, it follows that the amount of water condensed is equal to the difference between the initial and final mass of water vapor, or

$$\text{Mass of vapor condensed} = m_a (\omega_1 - \omega_2)$$

Solution

$$P_{v2} = P_{g2} = 0.8721 \text{ kPa}$$

$$P_{a2} = 100 - 0.8721 = 99.128 \text{ kPa}$$

$$\omega_2 = 0.622 \times \frac{0.8721}{99.128} = 0.0055$$

$$\text{Mass of vapor condensed} = m_a (\omega_1 - \omega_2) = 108.6(0.0255 - 0.0055)$$
$$= 2.172 \text{ kg}$$

EXAMPLE 12.4E Calculate the amount of water vapor condensed if the mixture of Example 12.3E is cooled to 40 F in a constant-pressure process.

Control mass: Mixture.

Initial state: Known (Example 12.3E).

Final state: T known.

Process: Constant pressure.

Analysis:

At the final temperature, 40 F, the mixture is saturated, since this is below the dew-point temperature. Therefore,

$$P_{v2} = P_{g2}, \qquad P_{a2} = P - P_{v2}$$

and

$$\omega_2 = 0.622 \frac{P_{v2}}{P_{a2}}$$

From the conservation of mass, it follows that the amount of water condensed is equal to the difference between the initial and final mass of water vapor, or

$$\text{Mass of vapor condensed} = m_a(\omega_1 - \omega_2)$$

Solution

$$P_{v2} = P_{g2} = 0.1217 \text{ lbf / in.}^2$$

$$P_{a2} = 14.7 - 0.12 = 14.58 \text{ lbf / in.}^2$$

$$\omega_2 = 0.622 \times \frac{0.1217}{14.58} = 0.00520$$

$$\text{Mass of vapor condensed} = m_a(\omega_1 - \omega_2) = 139.6(0.02135 - 0.0052)$$

$$= 2.25 \text{ lbm}$$

12.3 THE FIRST LAW APPLIED TO GAS–VAPOR MIXTURES

In applying the first law of thermodynamics to gas–vapor mixtures, it is helpful to realize that because of our assumption that ideal gases are involved, the various components can be treated separately when calculating changes of internal energy and enthalpy. Therefore, in dealing with air–water vapor mixtures, the changes in enthalpy of the water vapor can be found from the steam tables and the ideal-gas relations can be applied to the air. This is illustrated by the examples that follow.

EXAMPLE 12.5   An air-conditioning unit is shown in Fig. 12.7, with pressure, temperature, and relative humidity data. Calculate the heat transfer per kilogram of dry air, assuming that changes in kinetic energy are negligible.

Control volume: Duct, excluding cooling coils.

Inlet state: Known (Fig. 12.7).

Exit state: Known (Fig. 12.7).

Process: SSSF, no kinetic or potential energy changes.

Model: Air—ideal gas, constant specific heat, value at 300 K. Water—steam tables. (Since the water vapor at these low pressures is being considered an ideal gas, the enthalpy of the water vapor is a function of the temperature only. Therefore the enthalpy of slightly superheated water vapor is equal to the enthalpy of saturated vapor at the same temperature.)

Analysis:

Continuity, air and water:

$$\dot{m}_{a1} = \dot{m}_{a2}$$

$$\dot{m}_{v1} = \dot{m}_{v2} + \dot{m}_{l2}$$

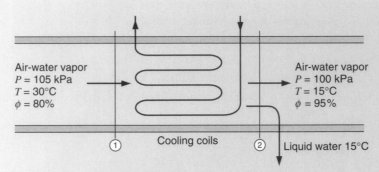

FIGURE 12.7 Sketch for Example 12.5.

First law:

$$\dot{Q}_{c.v.} + \sum \dot{m}_i h_i = \sum \dot{m}_e h_e$$

$$\dot{Q}_{c.v.} + \dot{m}_a h_{a1} + \dot{m}_{v1} h_{v1} = \dot{m}_a h_{a2} + \dot{m}_{v2} h_{v2} + \dot{m}_{l2} h_{l2}$$

If we divide this equation by $\dot{m}_a$, introduce the continuity equation for the water, and note that $\dot{m}_v = \omega \dot{m}_a$, we can write the first law in the form

$$\frac{\dot{Q}_{c.v.}}{\dot{m}_a} + h_{a1} + \omega_1 h_{v1} = h_{a2} + \omega_2 h_{v2} + (\omega_1 - \omega_2) h_{l2}$$

Solution

$$P_{v1} = \phi_1 P_{g1} = 0.80(4.246) = 3.397 \text{ kPa}$$

$$\omega_1 = \frac{R_a}{R_v} \frac{P_{v1}}{P_{a1}} = 0.622 \times \left(\frac{3.397}{105 - 3.4}\right) = 0.0208$$

$$P_{v2} = \phi_2 P_{g2} = 0.95(1.7051) = 1.620 \text{ kPa}$$

$$\omega_2 = \frac{R_a}{R_v} \times \frac{P_{v2}}{P_{a2}} = 0.622 \times \left(\frac{1.62}{100 - 1.62}\right) = 0.0102$$

Substituting:

$$\dot{Q}_{c.v.} / \dot{m}_a + h_{a1} + \omega_1 h_{v1} = h_{a2} + \omega_2 h_{v2} + (\omega_1 - \omega_2) h_{l2}$$

$$\dot{Q}_{c.v.} / \dot{m}_a = 1.0035(15 - 30) + 0.0102(2528.9)$$

$$- 0.0208(2556.3) + (0.0208 - 0.0102)(62.99)$$

$$= -41.76 \text{ kJ} / \text{kg dry air}$$

EXAMPLE 12.6   A tank has a volume of 0.5 m³ and contains nitrogen and water vapor. The temperature of the mixture is 50°C and the total pressure is 2 MPa. The partial pressure of the water vapor is 5 kPa. Calculate the heat transfer when the contents of the tank are cooled to 10°C.

Control mass: Nitrogen and water.

Initial state: P_1, T_1 known; state fixed.

Final state: T_2 known.

Process: Constant volume.

Model: Ideal gas mixture; constant specific heat for nitrogen; steam tables for water.

Analysis:

This is a constant-volume process. Since the work is zero, the first law reduces to

$$Q = U_2 - U_1 = m_{N_2} C_{v(N_2)} (T_2 - T_1) + (m_2 u_2)_v + (m_2 u_2)_l - (m_1 u_1)_v$$

This equation assumes that some of the vapor condensed. This assumption must be checked, however, as shown in the solution.

Solution

The mass of nitrogen and water vapor can be calculated using the ideal-gas equation of state.

$$m_{N_2} = \frac{P_{N_2} V}{R_{N_2} T} = \frac{1995 \times 0.5}{0.2968 \times 323.2} = 10.39 \text{ kg}$$

$$m_{v_1} = \frac{P_{v1} V}{R_v T} = \frac{5 \times 0.5}{0.461\,52 \times 323.2} = 0.016\,76 \text{ kg}$$

If condensation takes place, the final state of the vapor will be saturated vapor at 10°C. Therefore,

$$m_{v2} = \frac{P_{v2} V}{R_v T} = \frac{1.2276 \times 0.5}{0.461\,52 \times 283.2} = 0.004\,70 \text{ kg}$$

Since this amount is less than the original mass of vapor, there must have been condensation.

The mass of liquid that is condensed, m_{l2}, is

$$m_{l2} = m_{v_1} - m_{v_2} = 0.016\,76 - 0.004\,70 = 0.012\,06 \text{ kg}$$

The internal energy of the water vapor is equal to the internal energy of saturated water vapor at the same temperature. Therefore,

$$u_{v_1} = 2443.5 \text{ kJ / kg}$$
$$u_{v_2} = 2389.2 \text{ kJ / kg}$$
$$u_{l2} = 42.0 \text{ kJ / kg}$$
$$Q_{c.v.} = 10.39 \times 0.7448(10 - 50) + 0.0047(2389.2)$$
$$+ 0.012\,06(42.0) - 0.016\,76(2443.5)$$
$$= -338.8 \text{ kJ}$$

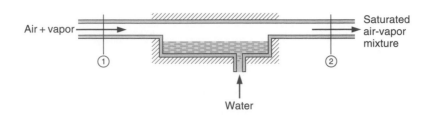

FIGURE 12.8 The adiabatic saturation process.

12.4 THE ADIABATIC SATURATION PROCESS

An important process for an air–water vapor mixture is the adiabatic saturation process. In this process, an air–vapor mixture comes in contact with a body of water in a well-insulated duct (Fig. 12.8). If the initial relative humidity is less than 100%, some of the water will evaporate and the temperature of the air–vapor mixture will decrease. If the mixture leaving the duct is saturated and if the process is adiabatic, the temperature of the mixture on leaving is known as the adiabatic saturation temperature. For this to take place as a steady-flow process, make-up water at the adiabatic saturation temperature is added at the same rate at which water is evaporated. The pressure is assumed to be constant.

Considering the adiabatic saturation process to be a steady-state, steady-flow process, and neglecting changes in kinetic and potential energy, the first law reduces to

$$h_{a1} + \omega_1 h_{v1} + (\omega_2 - \omega_1)h_{l2} = h_{a2} + \omega_2 h_{v2}$$

$$\omega_1(h_{v1} - h_{l2}) = C_{pa}(T_2 - T_1) + \omega_2(h_{v2} - h_{l2})$$

$$\omega_1(h_{v1} - h_{l2}) = C_{pa}(T_2 - T_1) + \omega_2 h_{fg2} \qquad (12.23)$$

The most significant point to be made about the adiabatic saturation process is that the adiabatic saturation temperature, the temperature of the mixture when it leaves the duct, is a function of the pressure, temperature, and relative humidity of the entering air–vapor mixture and of the exit pressure. Thus, the relative humidity and the humidity ratio of the entering air–vapor mixture can be determined from the measurements of the pressure and temperature of the air–vapor mixture entering and leaving the adiabatic saturator. Since these measurements are relatively easy to make, this is one means of determining the humidity of an air–vapor mixture.

EXAMPLE 12.7 The pressure of the mixture entering and leaving the adiabatic saturator is 0.1 MPa, the entering temperature is 30°C, and the temperature leaving is 20°C, which is the adiabatic saturation temperature. Calculate the humidity ratio and relative humidity of the air–water vapor mixture entering.

Control volume: Adiabatic saturator.

Inlet state: P_1, T_1 known.

Exit state: P_2, T_2 known; $\phi_2 = 100\%$; state fixed.

Process: SSSF, adiabatic saturation (Fig. 12.8).

Model: Ideal-gas mixture; constant specific heat for air; steam tables for water.

Analysis:

Continuity and first law, Eq. 12.23.

Solution

Since the water vapor leaving is saturated, $P_{v2} = P_{g2}$ and ω_2 can be calculated.

$$\omega_2 = 0.622 \times \left(\frac{2.339}{100 - 2.34} \right) = 0.0149$$

ω_1 can be calculated using Eq. 12.23.

$$\omega_1 = \frac{C_{pa}(T_2 - T_1) + \omega_2 h_{fg2}}{(h_{v1} - h_{l2})}$$

$$\omega_1 = \frac{1.0035(20 - 30) + 0.0149 \times 2454.1}{2556.3 - 83.96} = 0.0107$$

$$\omega_1 = 0.0107 = 0.622 \times \left(\frac{P_{v1}}{100 - P_{v1}} \right)$$

$$P_{v1} = 1.691 \text{ kPa}$$

$$\phi_1 = \frac{P_{v1}}{P_{g1}} = \frac{1.691}{4.246} = 0.398$$

EXAMPLE 12.7E   The pressure of the mixture entering and leaving the adiabatic saturator is 14.7 lbf/in.^{2}, the entering temperature is 84 F, and the temperature leaving is 70 F, which is the adiabatic saturation temperature. Calculate the humidity ratio and relative humidity of the air–water vapor mixture entering.

   Control volume: Adiabatic saturator.

   Inlet state: P_1, T_1 known.

   Exit state: P_2, T_2 known; $\phi_2 = 100\%$; state fixed.

   Process: SSSF, adiabatic saturation (Fig. 12.8).

   Model: Ideal-gas mixture; constant specific heat for air; steam tables for water.

Analysis:

Continuity and first law, Eq. 12.23.

Solution

Since the water vapor leaving is saturated, $P_{v2} = P_{g2}$ and ω_2 can be calculated.

$$\omega_2 = 0.622 \times \frac{0.3632}{14.7 - 0.36} = 0.01573$$

ω_1 can be calculated using Eq. 12.23.

$$\omega_1 = \frac{C_{pa}(T_2 - T_1) + \omega_2 h_{fg2}}{(h_{v1} - h_{l2})}$$

$$\omega_1 = \frac{0.24(70-84)+0.01573\times1054.0}{1098.1-38.1} = \frac{-3.36+16.60}{1060.0} = 0.0125$$

$$\omega_1 = 0.622\times\left(\frac{P_{v1}}{14.7-P_{v1}}\right) = 0.0125$$

$$P_{v1} = 0.289$$

$$\phi_1 = \frac{P_{v1}}{P_{g1}} = \frac{0.289}{0.584} = 0.495$$

12.5 WET-BULB AND DRY-BULB TEMPERATURES

The humidity of air–water vapor mixtures has traditionally been measured with a device called a psychrometer, which uses the flow of air past wet-bulb and dry-bulb thermometers. The bulb of the wet-bulb thermometer is covered with a cotton wick that is saturated with water. The dry-bulb thermometer is used simply to measure the temperature of the air. The airflow can be maintained by a fan, as shown in the continuous-flow psychrometer depicted in Fig. 12.9.

The processes that take place at the wet-bulb thermometer are somewhat complicated. First, if the air–water vapor mixture is not saturated, some of the water in the wick evaporates and diffuses into the surrounding air, which cools the water in the wick. As soon as the temperature of the water drops, however, heat is transferred to the water from both the air and the thermometer, with corresponding cooling. A steady state, determined by heat and mass transfer rates, will be reached, in which the wet-bulb thermometer temperature is lower than the dry-bulb temperature.

It can be argued that this evaporative cooling process is very similar, but not identical, to the adiabatic saturation process described and analyzed in Section 12.4. In fact, the adiabatic saturation temperature is often termed the *thermodynamic wet-bulb temperature*. It is clear, however, that the wet-bulb temperature as measured by a psychrometer is influenced by heat and mass transfer rates, which depend, for example, on the airflow ve-

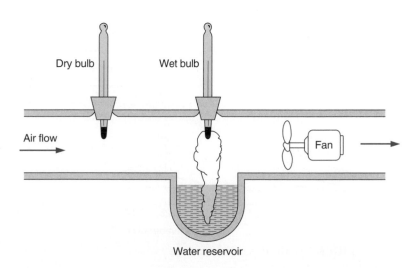

FIGURE 12.9
Steady-flow apparatus for measuring wet- and dry-bulb temperatures.

locity and not simply on thermodynamic equilibrium properties. It does happen that the two temperatures are very close for air–water vapor mixtures at atmospheric temperature and pressure, and they will be assumed to be equivalent in this text.

In recent years, humidity measurements have been made using other phenomena and other devices, primarily electronic devices for convenience and simplicity. For example, some substances tend to change in length, in shape, or in electrical capacitance, or in a number of other ways, when they absorb moisture. They are therefore sensitive to the amount of moisture in the atmosphere. An instrument making use of such a substance can be calibrated to measure the humidity of air–water vapor mixtures. The instrument output can be programmed to furnish any of the desired parameters, such as relative humidity, humidity ratio, or wet-bulb temperature.

12.6 THE PSYCHROMETRIC CHART

Properties of air–water vapor mixtures are given in graphical form on psychrometric charts. These are available in a number of different forms, and only the main features are considered here. It should be recalled that three independent properties will describe the state of this binary mixture such as pressure, temperature, and mixture composition.

A simplified version of the chart included in Appendix F, Fig. F.5, is shown in Fig. 12.10. This basic psychrometric chart is a plot of humidity ratio (ordinate) as a function of dry-bulb temperature (abscissa) with relative humidity, wet-bulb temperature and mixture enthalpy per mass of dry air as parameters. If we fix the total pressure for which the chart is to be constructed (which in our chart is 1 bar, or 100 kPa), lines of constant relative humidity and wet-bulb temperature can be drawn on the chart, because for a given dry-bulb temperature, total pressure, and humidity ratio, the relative humidity and wet-bulb temperature are fixed. The partial pressure of the water vapor is fixed by the humidity ratio and the total pressure, and therefore a second ordinate scale that indicates the

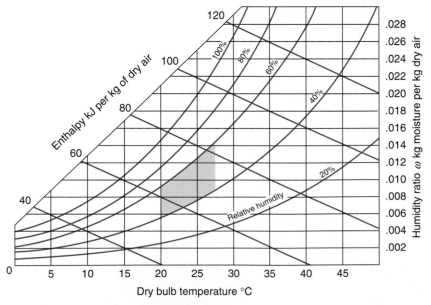

FIGURE 12.10 Psychrometric chart.

partial pressure of the water vapor could be constructed. Likewise it would also be possible to include the mixture specific volume and entropy on the chart.

Most psychrometric charts give the enthalpy of an air–vapor mixture per kilogram of dry air. The values given assume that the enthalpy of the dry air is zero at −20°C, and the enthalpy of the vapor is taken from the steam tables (which are based on the assumption that the enthalpy of saturated liquid is zero at 0°C). The value used in the psychrometric chart is then

$$\tilde{h} \equiv h_a - h_a(-20°\text{C}) + \omega h_v$$

This procedure is satisfactory because we are usually concerned only with differences in enthalpy. The fact that the lines of constant enthalpy are essentially parallel to lines of constant wet-bulb temperature is evident from the fact that the wet-bulb temperature is essentially equal to the adiabatic saturation temperature. Thus in Fig. 12.8, if we neglect the enthalpy of the liquid entering the adiabatic saturator, the enthalpy of the air–vapor mixture leaving, and a given adiabatic saturation temperature fixes the enthalpy of the mixture entering.

The chart plotted in Fig. 12.10 also indicates the human comfort zone, as the range of conditions most agreeable for human well being. An air-conditioner should then be able to maintain an environment within the comfort zone regardless of the outside atmospheric conditions to be considered adequate. Some charts are available that give corrections for variation from standard atmospheric pressures. Before using a given chart one should fully understand the assumptions made in constructing it, and that it is applicable to the particular problem at hand.

PROBLEMS

12.1 A gas mixture at 120°C, 125 kPa is 50% N_2, 30% H_2O, and 20% O_2 on a mole basis. Find the mass fractions, the mixture gas constant and the volume for 5 kg of mixture.

12.2 A 100 m^3 storage tank with fuel gases is at 20°C, 100 kPa containing a mixture of acetylene C_2H_2, propane C_3H_8, and butane C_4H_{10}. A test shows the partial pressure of the C_2H_2 is 15 kPa and that of C_3H_8 is 65 kPa. How much mass is there of each component?

12.3 A mixture of 60% N_2, 30% Ar, and 10% O_2 on a mole basis is in a cylinder at 250 kPa, 310 K, and volume 0.5 m^3. Find the mass fractions and the mass of argon.

12.4 A carbureted internal combustion engine is converted to run on methane gas (natural gas). The air–fuel ratio in the cylinder is to be 20 to 1 on a mass basis. How many moles of oxygen per mole of methane are there in the cylinder?

12.5 Weighing of masses gives a mixture at 60°C, 225 kPa with 0.5 kg O_2, 1.5 kg N_2, and 0.5 kg CH_4. Find the partial pressures of each component, the mixture specific volume (mass basis), mixture molecular weight, and the total volume.

12.6 At a certain point in a coal gasification process, a sample of the gas is taken and stored in a 1-L cylinder. An analysis of the mixture yields the following results:

Component	H_2	CO	CO_2	N_2
Percent by volume	25	40	15	20

Determine the mass fractions and total mass in the cylinder at 100 kPa, 20°C. How much heat transfer must be transferred to heat the sample at constant volume from the initial state to 100°C?

12.7 A pipe, cross-sectional area 0.1 m², carries a flow of 75% O_2 and 25% N_2 by mole with a velocity of 25 m/s at 200 kPa, 290 K. To install and operate a mass flow meter it is necessary to know the mixture density and the gas constant. What are they? What mass flow rate should the meter then show?

12.8 A pipe flows 0.05 kmol/s of a mixture with mole fractions of 40% CO_2 and 60% N_2 at 400 kPa, 300 K. Heating tape is wrapped around a section of pipe with insulation added, and 2 kW electrical power is heating the pipe flow. Find the mixture exit temperature.

12.9 A rigid insulated vessel contains 0.4 kmol of oxygen at 200 kPa, 280 K separated by a membrane from 0.6 kmol of carbon dioxide at 400 kPa, 360 K. The membrane is removed, and the mixture comes to a uniform state. Find the final temperature and pressure of the mixture.

12.10 An insulated gas turbine receives a mixture of 10% CO_2, 10% H_2O, and 80% N_2 on a mole basis at 1000 K, 500 kPa. The inlet volume flow rate is 2 m³/s, and the exhaust is at 700 K, 100 kPa. Find the power output in kW using constant specific heat from A.5 at 300 K.

12.11 Solve Problem 12.10 using the values of enthalpy from Table A.8.

12.12 Consider Problem 12.10 and find the value for the mixture heat capacity, mole basis, and the mixture ratio of specific heats, k_{mix}, both estimated at 850 K from values (differences) of h in Table A.8. With these values make an estimate for the reversible adiabatic exit temperature of the turbine at 100 kPa.

12.13 The gas mixture from Problem 12.6 is compressed in a reversible adiabatic process from the initial state in the sample cylinder to a volume of 0.2 L. Determine the final temperature of the mixture and the work done during the process.

12.14 Three SSSF flows are mixed in an adiabatic chamber at 150 kPa. Flow one is 2 kg/s of O_2 at 340 K, flow two is 4 kg/s of N_2 at 280 K, and flow three is 3 kg/s of CO_2 at 310 K. All flows are at 150 kPa the same as the total exit pressure. Find the exit temperature and the rate of entropy generation in the process.

12.15 Carbon dioxide gas at 320 K is mixed with nitrogen at 280 K in a SSSF insulated mixing chamber. Both flows are at 100 kPa, and the mole ratio of carbon dioxide to nitrogen is 2:1. Find the exit temperature and the total entropy generation per mole of the exit mixture.

12.16 A piston/cylinder contains 0.5 kg argon and 0.5 kg hydrogen at 300 K, 100 kPa. The mixture is compressed in an adiabatic process to 400 kPa by an external force on the piston. Find the final temperature, the work, and the heat transfer in the process.

12.17 Natural gas as a mixture of 75% methane and 25% ethane by volume is flowing to a compressor at 17°C, 100 kPa. The reversible adiabatic compressor brings the flow to 250 kPa. Find the exit temperature and the needed work per kg flow.

12.18 Repeat Problem 12.15 with inlet temperature of 1400 K for the carbon dioxide and 300 K for the nitrogen. First estimate the exit temperature with the specific heats from Table A.5 and use this to start iterations with values from A.8.

12.19 A mixture of 60% helium and 40% nitrogen by volume enters a turbine at 1 MPa, 800 K at a rate of 2 kg/s. The adiabatic turbine has an exit pressure of 100 kPa and an isentropic efficiency of 85%. Find the turbine work.

12.20 A mixture of 50% carbon dioxide and 50% water by mass is brought from 1500 K, 1 MPa to 500 K, 200 kPa in a polytropic process through a SSSF device. Find the necessary heat transfer and work involved using values from Table A.5.

12.21 Solve Problem 12.20 using specific heats $C_p = \Delta h/\Delta T$, from Table A.8 at 1000 K.

12.22 A 50/50 (by mole) gas mixture of methane CH_4 and ethylene C_2H_4 is contained in a cylinder/piston at the initial state 480 kPa, 330 K, 1.05 m^3. The piston is now moved, compressing the mixture in a reversible, polytropic process to the final state 260 K, 0.03 m^3. Calculate the final pressure, the polytropic exponent, the work and heat transfer, and net entropy change for the process.

12.23 A mixture of 2 kg oxygen and 2 kg of argon is in an insulated piston cylinder arrangement at 100 kPa, 300 K. The piston now compresses the mixture to half its initial volume. Find the final pressure, temperature, and the piston work.

12.24 Two insulated tanks A and B are connected by a valve. Tank A has a volume of 1 m^3 and initially contains argon at 300 kPa, 10°C. Tank B has a volume of 2 m^3 and initially contains ethane at 200 kPa, 50°C. The valve is opened and remains open until the resulting gas mixture comes to a uniform state. Determine the final pressure and temperature.

12.25 Reconsider Problem 12.24, but let the tanks have a small amount of heat transfer so the final mixture is at 400 K. Find the final pressure, the heat transfer, and the entropy change for the process.

12.26 A piston/cylinder contains helium at 110 kPa at ambient temperature 20°C, and initial volume of 20 L as shown in Fig. P12.26. The stops are mounted to give a maximum volume of 25 L, and the nitrogen line conditions are 300 kPa, 30°C. The valve is now opened, which allows nitrogen to flow in and mix with the helium. The valve is closed when the pressure inside reaches 200 kPa, at which point the temperature inside is 40°C. Is this process consistent with the second law of thermodynamics?

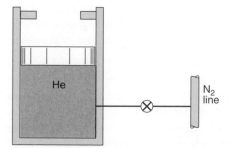

FIGURE P12.26

12.27 Repeat Problem 12.17 for an isentropic compressor efficiency of 82%.

12.28 A spherical balloon has an initial diameter of 1 m and contains argon gas at 200 kPa, 40°C. The balloon is connected by a valve to a 500-L rigid tank containing carbon dioxide at 100 kPa, 100°C. The valve is opened, and eventually the balloon and tank reach a uniform state in which the pressure is 185 kPa. The balloon pressure is directly proportional to its diameter. Take the balloon and tank as a control volume, and calculate the final temperature and the heat transfer for the process.

12.29 A large SSSF air separation plant takes in ambient air (79% N_2, 21% O_2 by volume) at 100 kPa, 20°C, at a rate of 1 kmol/s. It discharges a stream of pure O_2 gas at 200 kPa, 100°C, and a stream of pure N_2 gas at 100 kPa, 20°C. The plant operates on an electrical power input of 2000 kW. Calculate the net rate of entropy change for the process.

12.30 An insulated vertical cylinder is fitted with a frictionless constant loaded piston of cross-sectional area 0.1 m^2 and the initial cylinder height of 1.0 m. The cylinder contains methane gas at 300 K, 150 kPa, and also inside is a 5-L capsule containing neon gas at 300 K, 500 kPa. The capsule now breaks, and the two gases mix together in a constant pressure process. What is the final temperature, final cylinder height, and net entropy change for the process?

12.31 The only known sources of helium are the atmosphere (mole fraction approximately 5×10^{-6}) and natural gas. A large unit is being constructed to separate 100 m^3/s of natural gas, assumed to be 0.001 He mole fraction and 0.999 CH_4. The gas enters the unit at 150 kPa, 10°C. Pure helium exits at 100 kPa, 20°C, and pure methane exits at 150 kPa, 30°C. Any heat transfer is with the surroundings at 20°C. Is an electrical power input of 3000 kW sufficient to drive this unit?

12.32 A steady flow of 0.01 kmol/s of 50% carbon dioxide and 50% water at 1200K and 200 kPa is used in a heat exchanger where 300 kW is extracted from the flow. Find the flow exit temperature and the rate of change of entropy using Table A.8.

12.33 An insulated rigid 2 m^3 tank A contains CO_2 gas at 200°C, 1MPa. An uninsulated rigid 1 m^3 tank B contains ethane, C_2H_6, gas at 200 kPa, room temperature 20°C. The two are connected by a one-way check valve that will allow gas from A to B, but not from B to A. The valve is opened and gas flows from A to B until the pressure in B reaches 500 kPa when the valve is closed. The mixture in B is kept at room temperature due to heat transfer. Find the total number of moles and the ethane mole fraction at the final state in B. Find the final temperature and pressure in tank A and the heat transfer, to/from tank B.

12.34 A tank has two sides initially separated by a diaphragm. Side A contains 1 kg of water and side B contains 1.2 kg of air, both at 20°C, 100 kPa. The diaphragm is now broken, and the whole tank is heated to 600°C by a 700°C reservoir. Find the final total pressure, heat transfer, and total entropy generation.

12.35 A 0.2 m^3 insulated, rigid vessel is divided into two equal parts A and B by an insulated partition, as shown in Fig. P12.35. The partition will support a pressure difference of 400 kPa before breaking. Side A contains methane and side B contains carbon dioxide. Both sides are initially at 1 MPa, 30°C. A valve on side B is opened, and carbon dioxide flows out. The carbon dioxide that remains in B is assumed to undergo a reversible adiabatic expansion while there is flow out. Eventually the partition breaks, and the valve is closed. Calculate the net entropy change for the process that begins when the valve is closed.

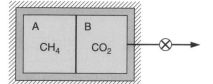

FIGURE P12.35

12.36 Consider 100 m³ of atmospheric air which is an air–water vapor mixture at 100 kPa, 15°C, and 40% relative humidity. Find the mass of water and the humidity ratio. What is the dew point of the mixture?

12.37 The products of combustion are flowing through a SSSF heat exchanger with 12% CO_2, 13% H_2O, and 75% N_2 on a volume basis at the rate 0.1 kg/s and 100 kPa. What is the dew-point temperature? If the mixture is cooled 10°C below the dew-point temperature, how long will it take to collect 10 kg of liquid water?

12.38 A new high-efficiency home heating system includes an air-to-air heat exchanger which uses energy from outgoing stale air to heat the fresh incoming air. If the outside ambient temperature is −10°C and the relative humidity is 30%, how much water will have to be added to the incoming air, if it flows in at the rate of 1 m³/s and must eventually be conditioned to 20°C and 40% relative humidity?

12.39 Consider 100 m³ of atmospheric air at 100 kPa, 25°C, and 80% relative humidity. Assume this is brought into a basement room where it cools to 15°C, 100 kPa. How much liquid water will condense out?

12.40 A flow of 2 kg/s completely dry air at T_1, 100 kPa is cooled down to 10°C by spraying liquid water at 10°C, 100 kPa into it so it becomes saturated moist air at 10°C. The process is SSSF with no external heat transfer or work. Find the exit moist air humidity ratio and the flow rate of liquid water. Find also the dry air inlet temperature T_1.

12.41 A piston/cylinder has 100 kg of saturated moist air at 100 kPa, 5°C. If it is heated to 45°C in an isobaric process, find $_1q_2$ and the final relative humidity. If it is compressed from the initial state to 200 kPa in an isothermal process, find the mass of water condensing.

12.42 A flow of moist air at 100 kPa, 40°C, 40% relative humidity is cooled to 15°C in a constant pressure SSSF device. Find the humidity ratio of the inlet and the exit flow, and the heat transfer in the device per kg dry air.

12.43 Ambient moist air enters a steady-flow air-conditioning unit at 102 kPa, 30°C, with a 60% relative humidity. The volume flow rate entering the unit is 100 L/s. The moist air leaves the unit at 95 kPa, 15°C, with a relative humidity of 100%. Liquid condensate also leaves the unit at 15°C. Determine the rate of heat transfer for this process.

12.44 A steady supply of 1.0 m³/s air at 25°C, 100 kPa, 50% relative humidity is needed to heat a building in the winter. The outdoor ambient is at 10°C, 100 kPa, 50% relative humidity. What are the required liquid water input and heat transfer rates for this purpose?

12.45 Consider a 500-L rigid tank containing an air–water vapor mixture at 100 kPa, 35°C, with a 70% relative humidity. The system is cooled until the water just begins to condense. Determine the final temperature in the tank and the heat transfer for the process.

12.46 Air in a piston/cylinder is at 35°C, 100 kPa, and a relative humidity of 80%. It is now compressed to a pressure of 500 kPa in a constant temperature process. Find the final relative and specific humidity and the volume ratio V_2/V_1.

12.47 A 300-L rigid vessel initially contains moist air at 150 kPa, 40°C, with a relative humidity of 10%. A supply line connected to this vessel by a valve carries steam at 600 kPa, 200°C. The valve is opened, and steam flows into the vessel until the relative humidity of the resultant moist air mixture is 90%. Then the valve is closed. Sufficient heat is transferred from the vessel so the temperature remains at 40°C during the process. Determine the heat transfer for the process, the mass of steam entering the vessel, and the final pressure inside the vessel.

12.48 A combination air cooler and dehumidification unit receives outside ambient air at 35°C, 100 kPa, 90% relative humidity. The moist air is first cooled to a low temperature T_2 to condense the proper amount of water; assume all the liquid leaves at T_2. The moist air is then heated and leaves the unit at 20°C, 100 kPa, relative humidity 30% with volume flow rate of 0.01 m³/s. Find the temperature T_2, the mass of liquid per kilogram of dry air, and the overall heat transfer rate.

12.49 A rigid container, 10 m³ in volume, contains moist air at 45°C, 100 kPa, $\Phi = 40\%$. The container is now cooled to 5°C. Neglect the volume of any liquid that might be present and find the final mass of water vapor, final total pressure, and the heat transfer.

12.50 A saturated air–water vapor mixture at 20°C, 100 kPa, is contained in a 5-m³ closed tank in equilibrium with 1 kg of liquid water. The tank is heated to 80°C. Is there any liquid water at the final state? Find the heat transfer for the process.

12.51 An air–water vapor mixture enters a steady flow heater humidifier unit at state 1: 10°C, 10% relative humidity, at the rate of 1 m³/s. A second air–vapor stream enters the unit at state 2: 20°C, 20% relative humidity, at the rate of 2 m³/s. Liquid water enters at state 3: 10°C, at the rate of 400 kg per hour. A single air–vapor flow exits the unit at state 4: 40°C. Calculate the relative humidity of the exit flow and the rate of heat transfer to the unit.

12.52 In a hot and dry climate, air enters an air-conditioner unit at 100 kPa, 40°C, and 5% relative humidity, at the steady rate of 1.0 m³/s. Liquid water at 20°C is sprayed into the air in the AC unit at the rate 20 kg/h, and heat is rejected from the unit at at the rate 20 kW. The exit pressure is 100 kPa. What are the exit temperature and relative humidity?

12.53 A water-filled reactor of 1 m³ is at 20 MPa, 360°C and is located inside an insulated containment room of 100 m³ that contains air at 100 kPa and 25°C. Due to a failure the reactor ruptures and the water fills the containment room. Find the final pressure.

12.54 Use the psychrometric chart to find the missing property of: Φ, ω, T_{wet}, T_{dry}
  a. $T_{dry} = 25°C$, $\Phi=80\%$ c. $T_{dry} = 20°C$, and $\omega = 0.008$
  b. $T_{dry} =15°C$, $\Phi=100\%$ d. $T_{dry} = 25°C$, $T_{wet} = 23°C$

12.55 Use the psychrometric chart to find the missing property of: Φ, ω, T_{wet}, T_{dry}
  a. $\Phi = 50\%$, $\omega = 0.012$ c. $\omega = 0.008$ and $T_{wet} =17°C$
  b. $T_{wet} =15°C$, $\Phi = 60\%$.   d. $T_{dry} = 10°C$, $\omega = 0.006$

12.56 For each of the states in Problem 12.55 find the dew point temperature.

12.57 One means of air-conditioning hot summer air is by evaporative cooling, which is a process similar to the SSSF adiabatic saturation process. Consider outdoor ambient air at 35°C, 100 kPa, 30% relative humidity. What is the maximum amount of cooling that can be achieved by such a technique? What disadvantage is there to this approach? Solve the problem using a first law analysis and repeat it using the psychrometric chart, Fig. F.5.

12.58 Use the formulas and the steam tables to find the missing property of: Φ, ω, and T_{dry}, total pressure is 100 kPa; repeat the answers using the psychrometric chart.
    a. $\Phi = 50\%$, $\omega = 0.010$ c. $T_{dry} = 25°C$, $T_{wet} = 21°C$
    b. $T_{wet} = 15°C$, $\Phi = 50\%$

12.59 Compare the weather two places where it is cloudy and breezy. At beach A it is 20°C, 103.5 kPa, relative humidity 90% and beach B has 25°C, 99 kPa, relative humidity 20%. Suppose you just took a swim and came out of the water. Where would you feel more comfortable and why?

12.60 Ambient air at 100 kPa, 30°C, 40% relative humidity goes through a constant pressure heat exchanger in a SSSF process. In one case it is heated to 45°C and in another case it is cooled until it reaches saturation. For both cases find the exit relative humidity and the amount of heat transfer per kilogram dry air.

12.61 A flow, 0.2 kg/s dry air, of moist air at 40°C, 50% relative humidity flows from the outside state 1 down into a basement where it cools to 16°C, state 2. Then it flows up to the living room where it is heated to 25°C, state 3. Find the dew point for state 1, any amount of liquid that may appear, the heat transfer that takes place in the basement, and the relative humidity in the living room at state 3.

12.62 A flow of air at 5°C, $\Phi = 90\%$, is brought into a house, where it is conditioned to 25°C, 60% relative humidity. This is done in a SSSF process with a combined heater-evaporator where any liquid water is at 10°C. Find any flow of liquid and the necessary heat transfer, both per kilogram dry air flowing. Find the dew point for the final mixture.

12.63 Atmospheric air at 35°C, relative humidity of 10%, is too warm and also too dry. An air conditioner should deliver air at 21°C and 50% relative humidity in the amount of 3600 m³/h. Sketch a setup to accomplish this, find any amount of liquid (at 20°C) that is needed or discarded and any heat transfer.

12.64 In a car's defrost/defog system atmospheric air, 21°C, relative humidity 80%, is taken in and cooled such that liquid water drips out. The now dryer air is heated to 41°C and then blown onto the windshield, where it should have a maximum of 10% relative humidity to remove water from the windshield. Find the dew point of the atmospheric air, specific humidity of air onto the windshield, the lowest temperature, and the specific heat transfer in the cooler.

12.65 Two moist air streams with 85% relative humidity, both flowing at a rate of 0.1 kg/s of dry air, are mixed in a SSSF setup. One inlet flowstream is at 32.5°C and the other at 16°C. Find the exit relative humidity.

12.66 A flow of moist air at 21°C, 60% relative humidity should be produced from mixing two different moist air flows. Flow 1 is at 10°C, relative humidity 80% and flow 2 is at 32°C and has $T_{wet} = 27$°C. The mixing chamber can be followed by a heater or a cooler. No liquid water is added, and $P = 100$ kPa. Find the two controls—one is the ratio of the two mass flow rates m_{a1}/m_{a2} and the other is the heat transfer in the heater/cooler per kg dry air.

12.67 Consider two states of atmospheric air. (1) 35°C, $T_{wet} = 18$°C and (2) 26.5°C, $\Phi = 60\%$. Suggest a system of devices that will allow air in a SSSF process to change from (1) to (2) and from (2) to (1). Heaters, coolers, (de)humidifiers, liquid traps, etc. are available, and any liquid/solid flowing is assumed to be at the lowest temperature seen in the process. Find the specific and relative humidity for state 1, dew point for state 2 and the heat transfer per kilogram dry air in each component in the systems.

12.68 An insulated tank has an air inlet, $\omega_1 = 0.0084$, and an outlet, $T_2 = 22$°C, $\Phi_2 = 90\%$ both at 100 kPa. A third line sprays 0.25 kg/s of water at 80°C, 100 kPa. For a SSSF operation find the outlet specific humidity, the mass flow rate of air needed and the required air inlet temperature, T_1.

12.69 You have just washed your hair and now blow dry it in a room with 23°C, $\Phi = 60\%$, (1). The dryer, 500 W, heats the air to 49°C, (2), blows it through your hair where the air becomes saturated (3), and then flows on to hit a window where it cools to 15°C (4). Find the relative humidity at state 2, the heat transfer per kilogram of dry air in the dryer, the air flow rate, and the amount of water condensed on the window, if any.

12.70 A water-cooling tower for a power plant cools 45°C liquid water by evaporation. The tower receives air at 19.5°C, $\Phi = 30\%$, 100 kPa that is blown through/over the water such that it leaves the tower at 25°C, $\Phi = 70\%$. The remaining liquid water flows back to the condenser at 30°C having given off 1 MW. Find the mass flow rate of air, and the amount of water that evaporates.

12.71 An indoor pool evaporates 1.512 kg/h of water, which is removed by a dehumidifier to maintain 21°C, $\Phi = 70\%$ in the room. The dehumidifier, shown in Fig. P12.71, is a refrigeration cycle in which air flowing over the evaporator cools such that liquid water drops out, and the air continues flowing over the condenser. For an air flow rate of 0.1 kg/s the unit requires 1.4 kW input to a motor driving a fan and the compressor and it has a coefficient of performance, $\beta = \dot{Q}_L/\dot{W}_c = 2.0$. Find the state of the air as it returns to the room and the compressor work input.

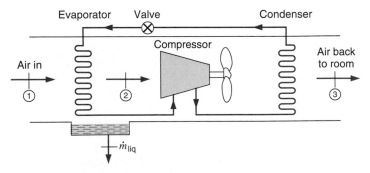

FIGURE P12.71

12.72 To refresh air in a room, a counterflow heat exchanger, see Fig. P12.72, is mounted in the wall, drawing in outside air at 0.5°C, 80% relative humidity and pushing out room air, 40°C, 50% relative humidity. Assume an exchange of 3 kg/min dry air in a SSSF device, and also that the room air exits the heat exchanger to the atmosphere at 23°C. Find the net amount of water removed from the room, any liquid flow in the heat exchanger, and (*T*, *Φ*) for the fresh air entering the room.

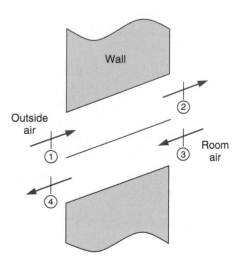

FIGURE P12.72

12.73 Steam power plants often utilize large cooling towers to cool the condenser cooling water so it can be recirculated; see Fig. P12.73. The process is essentially evaporative adiabatic cooling, in which part of the water is lost and must therefore be replenished. Consider the setup shown in Fig. P12.73, in which 1000 kg/s of warm water at 32°C from the condenser enters the top of the cooling tower and the cooled water leaves the bottom at 20°C. The moist ambient air enters the bottom at 100 kPa, dry bulb temperature of 18°C and a wet bulb temperature of 10°C. The moist air leaves the tower at 95 kPa, 30°C, and relative humidity of 85%. Determine the required mass flow rate of dry air, and the fraction of the incoming water that evaporates and is lost.

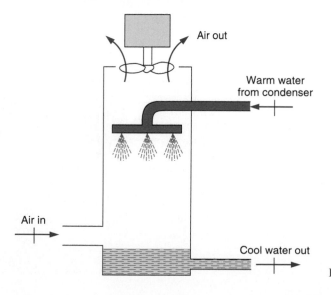

FIGURE P12.73

ADVANCED PROBLEMS

12.74 A semipermeable membrane is used for the partial removal of oxygen from air that is blown through a grain elevator storage facility. Ambient air (79% nitrogen, 21% oxygen on a mole basis) is compressed to an appropriate pressure, cooled to ambient temperature 25°C, and then fed through a bundle of hollow polymer fibers that selectively absorb oxygen, so the mixture leaving at 120 kPa, 25°C, contains only 5% oxygen. The absorbed oxygen is bled off through the fiber walls at 40 kPa, 25°C, to a vacuum pump. Assume the process to be reversible and adiabatic and determine the minimum inlet air pressure to the fiber bundle.

12.75 A 100-L insulated tank contains N_2 gas at 200 kPa and ambient temperature 25°C. The tank is connected by a valve to a supply line flowing CO_2 at 1.2 MPa, 90°C. A mixture of 50% N_2, 50% CO_2 by mole should be obtained by opening the valve and allowing CO_2 to flow in until an appropriate pressure is reached, when the valve is closed. What is the pressure? The tank eventually cools to ambient temperature. Find the net entropy change for the overall process.

12.76 A cylinder/piston loaded with a linear spring contains saturated moist air at 120 kPa, 0.1 m^3 volume and also 0.01 kg of liquid water, all at ambient temperature 20°C. The piston area is 0.2 m^2, and the spring constant is 20 kN/m. This cylinder is attached by a valve to a line flowing dry air at 800 kPa, 80°C. The valve is opened, and air flows into the cylinder until the pressure reaches 200 kPa, at which point the temperature is 40°C. Determine the relative humidity at the final state, the mass of air entering the cylinder, and the work done during the process.

12.77 Consider the previous problem and additionally determine the heat transfer. Show that the process does not violate the second law.

12.78 The air-conditioning by evaporative cooling in Problem 12.57 is modified by adding a dehumidification process before the water spray cooling process. This dehumidification is achieved as shown in Fig. P12.78 by using a desiccant material, which absorbs water on one side of a rotating drum heat exchanger. The desiccant is regenerated by heating on the other side of the drum to drive the water out. The pressure is 100 kPa everywhere, and other properties are on the diagram. Calculate the relative humidity of the cool air supplied to the room at state 4, and the heat transfer per unit mass of air that needs to be supplied to the heater unit.

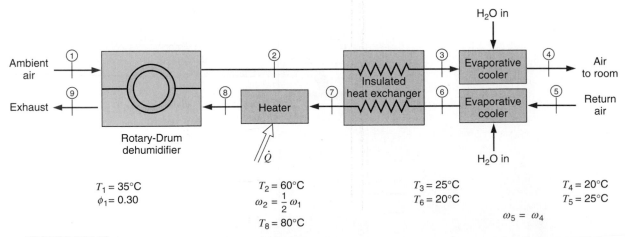

FIGURE P12.78

12.79 A vertical cylinder is fitted with a piston held in place by a pin, as shown in Fig. P12.79. The initial volume is 200 L, and the cylinder contains moist air at 100 kPa, 25°C, with wet-bulb temperature of 15°C. The pin is removed, and at the same time a valve on the bottom of the cylinder is opened, allowing the mixture to flow out. A cylinder pressure of 150 kPa is required to balance the piston. The valve is closed when the cylinder volume reaches 100 L, at which point the temperature is that of the surroundings, 15°C.

   a. Is there any liquid water in the cylinder at the final state?

   b. Calculate the heat transfer to the cylinder during the process.

   c. Take a control volume around the cylinder, and calculate the entropy change of the control volume and that of the surroundings.

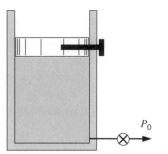

P_0

FIGURE P12.79

12.80 Ambient air is at a condition of 100 kPa, 35°C, 50% relative humidity. A steady stream of air at 100 kPa, 23°C, 70% relative humidity is to be produced by first cooling one stream to an appropriate temperature to condense out the proper amount of water and then mix this stream adiabatically with the second one at ambient conditions. What is the ratio of the two flow rates? To what temperature must the first stream be cooled?

ENGLISH UNIT PROBLEMS

12.81E A gas mixture at 250 F, 18 lbf/in.^{2} is 50% N_2, 30% H_2O, and 20% O_2 on a mole basis. Find the mass fractions, the mixture gas constant, and the volume for 10 lbm of mixture.

12.82E Weighing of masses gives a mixture at 60°C, 225 kPa with 1 lbm O_2, 3 lbm N_2, and 1 lbm CH_4. Find the partial pressures of each component, the mixture specific volume (mass basis), mixture molecular weight, and the total volume.

12.83E A pipe flows 0.15 lb mol/s of a mixture with mole fractions of 40% CO_2 and 60% N_2 at 60 lbf/in.^{2}, 540 R. Heating tape is wrapped around a section of pipe with insulation added, and 2 Btu/s electrical power is heating the pipe flow. Find the mixture exit temperature.

12.84E An insulated gas turbine receives a mixture of 10% CO_2, 10% H_2O, and 80% N_2 on a mole basis at 1800 R, 75 lbf/in.^{2}. The inlet volume flow rate is 70 ft^3/s, and the exhaust is at 1300 R, 15 lbf/in.^{2}. Find the power output in Btu/s using constant specific heat from C.4 at 540 R.

12.85E Solve Problem 12.84 using the values of enthalpy from Table C.7.

12.86E Carbon dioxide gas at 580 R is mixed with nitrogen at 500 R in a SSSF insulated mixing chamber. Both flows are at 14.7 lbf/in.^{2}, and the mole ratio of carbon dioxide to nitrogen is $2:1$. Find the exit temperature and the total entropy generation per mole of the exit mixture.

12.87E A mixture of 60% helium and 40% nitrogen by volume enters a turbine at 150 lbf/in.^{2}, 1500 R at a rate of 4 lbm/s. The adiabatic turbine has an exit pressure of 15 lbf/in.^{2} and an isentropic efficiency of 85%. Find the turbine work.

12.88E A mixture of 50% carbon dioxide and 50% water by mass is brought from 2800 R, 150 lbf/in.^{2} to 900 R, 30 lbf/in.^{2} in a polytropic process through a SSSF device. Find the necessary heat transfer and work involved using values from C.4.

12.89E A mixture of 4 lbm oxygen and 4 lbm of argon is in an insulated piston cylinder arrangement at 14.7 lbf/in.^{2}, 540 R. The piston now compresses the mixture to half its initial volume. Find the final pressure, temperature, and the piston work.

12.90E Two insulated tanks A and B are connected by a valve. Tank A has a volume of 30 ft^3 and initially contains argon at 50 lbf/in.^{2}, 50 F. Tank B has a volume of 60 ft^3 and initially contains ethane at 30 lbf/in.^{2}, 120 F. The valve is opened and remains open until the resulting gas mixture comes to a uniform state. Find the final pressure and temperature and the entropy change for the process.

12.91E A large SSSF air separation plant takes in ambient air (79% N_2, 21% O_2 by volume) at 14.7 lbf/in.^{2}, 70 F, at a rate of 2 lb mol/s. It discharges a stream of pure O_2 gas at 30 lbf/in.^{2}, 200 F, and a stream of pure N_2 gas at 14.7 lbf/in.^{2}, 70 F. The plant operates on an electrical power input of 2000 kW. Calculate the net rate of entropy change for the process.

12.92E A tank has two sides initially separated by a diaphragm. Side A contains 2 lbm of water, and side B contains 2.4 lbm of air—both at 68 F, 14.7 lbf/in.^{2}. The diaphragm is now broken, and the whole tank is heated to 1100 F by a 1300 F reservoir. Find the final total pressure, heat transfer, and total entropy generation.

12.93E Consider a volume of 2000 ft^3 that contains an air–water vapor mixture at 14.7 lbf/in.^{2}, 60 F, and 40% relative humidity. Find the mass of water and the humidity ratio. What is the dew point of the mixture?

12.94E Consider a 10-ft^3 rigid tank containing an air–water vapor mixture at 14.7 lbf/in.^{2}, 90 F, with a 70% relative humidity. The system is cooled until the water just begins to condense. Determine the final temperature in the tank and the heat transfer for the process.

12.95E Air in a piston/cylinder is at 95 F, 15 lbf/in.^{2} and a relative humidity of 80%. It is now compressed to a pressure of 75 lbf/in.^{2} in a constant temperature process. Find the final relative and specific humidity and the volume ratio V_2/V_1.

12.96E A 10-ft^3 rigid vessel initially contains moist air at 20 lbf/in.^{2}, 100 F, with a relative humidity of 10%. A supply line connected to this vessel by a valve carries steam at 100 lbf/in.^{2}, 400 F. The valve is opened, and steam flows into the vessel until the relative humidity of the resultant moist air mixture is 90%. Then the valve is closed. Sufficient heat is transferred from the vessel so the temperature remains at 100 F during the process. Determine the heat transfer for the process, the mass of steam entering the vessel, and the final pressure inside the vessel.

12.97E A water-filled reactor of 50 ft^3 is at 2000 lbf/in.^{2}, 550 F and located inside an insulated containment room of 5000 ft^3 that has air at 1 atm. and 77 F. Due to a failure, the reactor ruptures and the water fills the containment room. Find the final pressure.

12.98E Atmospheric air at 95 F, relative humidity of 10%, is too warm and also too dry. An air conditioner should deliver air at 70 F and 50% relative humidity in the amount of 3600 ft^3 per hour. Sketch a setup to accomplish this, find any amount of liquid (at 68 F) that is needed or discarded and any heat transfer.

12.99E Two moist air streams with 85% relative humidity, both flowing at a rate of 0.2 lbm/s of dry air are mixed in a SSSF setup. One inlet flowstream is at 90 F, and the other at 61 F. Find the exit relative humidity.

12.100E An indoor pool evaporates 3 lbm/h of water, which is removed by a dehumidifier to maintain 70 F, $\Phi = 70\%$ in the room. The dehumidifier is a refrigeration cycle in which air flowing over the evaporator cools such that liquid water drops out, and the air continues flowing over the condenser, as shown in Fig. P12.71. For an air flow rate of 0.2 lbm/s the unit requires 1.2 Btu/s input to a motor driving a fan and the compressor and it has a coefficient of performance, $\beta = \dot{Q}_L / \dot{W}_c = 2.0$. Find the state of the air after the evaporator, T_2, ω_2, Φ_2, and the heat, rejected. Find the state of the air as it returns to the room and the compressor work input.

12.101E To refresh air in a room, a counterflow heat exchanger is mounted in the wall, as shown in Fig. P12.72. It draws in outside air at 33 F, 80% relative humidity and draws room air, 104 F, 50% relative humidity, out. Assume an exchange of 6 lbm/min dry air in a SSSF device, and also that the room air exits the heat exchanger to the atmosphere at 72 F. Find the net amount of water removed from the room, any liquid flow in the heat exchanger, and (T, Φ) for the fresh air entering the room.

12.102E A 4-ft^3 insulated tank contains nitrogen gas at 30 lbf/in.^{2} and ambient temperature 77 F. The tank is connected by a valve to a supply line flowing carbon dioxide at 180 lbf/in.^{2}, 190 F. A mixture of 50 mole percent nitrogen and 50 mole percent carbon dioxide is to be obtained by opening the valve and allowing flow into the tank until an appropriate pressure is reached, when the valve is closed. What is the pressure? The tank eventually cools to ambient temperature. Calculate the net entropy change for the overall process.

12.103E Ambient air is at a condition of 14.7 lbf/in.^{2}, 95 F, 50% relative humidity. A steady stream of air at 14.7 lbf/in.^{2}, 73 F, 70% relative humidity is to be produced by first cooling one stream to an appropriate temperature to condense out the proper amount of water and then mix this stream adiabatically with the second one at ambient conditions. What is the ratio of the two flow rates? To what temperature must the first stream be cooled?

COMPUTER, DESIGN, AND OPEN-ENDED PROBLEMS

12.104 Write a program to solve the general case of Problem 12.24/25 in which the two volumes and the initial state properties of the argon and the ethane are input variables. Use constant specific heat from Table A.5.

12.105 Mixing of CO_2 and N_2 in a SSSF setup was given in Problem 12.15. If the temperatures are very different an assumption of constant specific heat is inappropriate. Study the problem assuming the CO_2 enters at 300 K, 100 kPa as a function of the N_2 inlet temperature using specific heat from Table A.7 or the menu-driven software. Give the nitrogen inlet temperature for which the constant specific heat assumption starts to be more than 1%, 5%, and 10% wrong for the exit mixture temperature.

12.106 The setup in Problem 12.68 is similar to a process which can be used to produce dry powder from a slurry of water and dry material as coffee or milk. The water flow at state 3 is a mixture of 80% liquid water and 20% dry material on a mass basis with $C_{dry} = 0.4$ kJ/kg K. After the water is evaporated the dry material falls to the bottom and is removed in an additional line, $\dot{m}_{dry}$ exit at state 4. Assume a reasonable T_4 and that state 1 is heated atmospheric air. Investigate the inlet flow temperature as a function of state 1 humidity ratio.

12.107 A dehumidifier for household applications is similar to the system shown in Fig. P12.71. Study the requirements to the refrigeration cycle as a function of the atmospheric conditions and include a worst case estimation.

12.108 A clothes dryer has a 60°C, $\Phi = 90\%$ air flow out at a rate of 3 kg/min. The atmospheric conditions are 20°C, relative humidity of 50%. How much water is carried away and how much power is needed? To increase the efficiency, a counterflow heat exchanger is installed to preheat the incoming atmospheric air up with the hot exit flow. Estimate suitable exit temperatures from the heat exchanger and investigate the design changes to the clothes dryer. (What happens to the condensed water?) How much energy can be saved this way?

12.109 Addition of steam to combustors in gas turbines and to internal combustion engines reduces the peak temperatures and lowers emission of NO_x. Consider a modification to a gas turbine, as shown in Fig. P12.109, where the modified cycle is called the Cheng cycle. In this example it is used for a cogenerating power

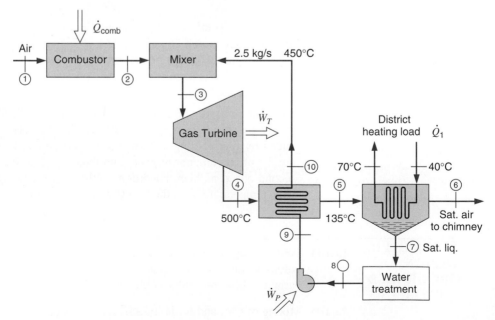

FIGURE P12.109

plant. Assume 12 kg/s air with state 2 at 1.25 Mpa, unknown temperature, is mixed with 2.5 kg/s water at 450°C at constant pressure before the inlet to the turbine. The turbine exit temperature is $T_4 = 500°C$, and the pressure is 125 kPa. For a reasonable turbine efficiency, estimate the required air temperature at state 2. Compare the result to the case where no steam is added to the mixing chamber and only air runs through the turbine.

12.110 Consider the district water heater acting as the condenser for part of the water between states 5 and 6 in Fig. P12.109. If the temperature of the mixture (12 kg/s air, 2.5 kg/s steam) at state 5 is 135°C, make a study of the district heating load, $\dot{Q}_1$, as a function of the exit temperature T_6. Study also the sensitivity of the results with respect to the assumption that state 6 is saturated moist air.

12.111 The cogeneration gas turbine cycle can be augmented with a heat pump to extract more energy from the turbine exhaust gas, as shown in Fig. P12.111. The heat pump upgrades the energy to be delivered at the 70°C line for district heating. In the modified application the first heat exchanger has exit temperature $T_{6a} = T_{7a} = 45°C$, and the second one has $T_{6b} = T_{7b} = 36°C$. Assume the district heating line has the same exit temperature as before so this arrangement allows for a higher flow rate. Estimate the increase in the district heating load that can be obtained and the necessary work input to the heat pump.

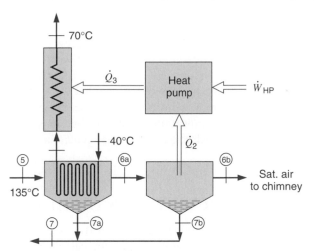

FIGURE P12.111

12.112 Several applications of dehumidification do not rely on water condensation by cooling. A desiccant with a greater affinity to water can absorb water directly from the air accompanied by a heat release. The desiccant is then regenerated by heating, driving the water out. Make a list of several such materials as liquids, gels, and solids and show examples of their use.

13 THERMODYNAMIC RELATIONS

We have already defined and used several thermodynamic properties. Among these are pressure, specific volume, density, temperature, mass, internal energy, enthalpy, entropy, constant-pressure and constant-volume specific heats, and the Joule–Thomson coefficient. Two other properties, the Helmholtz function and the Gibbs function, will also be introduced and will be used more extensively in the following chapters. We have also had occasion to use tables of thermodynamic properties for a number of different substances.

One important question is now raised: Which of the thermodynamic properties can be experimentally measured? We can answer this question by considering the measurements we can make in the laboratory. Some of the properties such as internal energy and entropy cannot be measured directly and must be calculated from other experimental data. If we carefully consider all these thermodynamic properties, we conclude that there are only four that can be directly measured: pressure, temperature, volume, and mass.

This leads to a second question: How can values of the thermodynamic properties that cannot be measured be determined from experimental data on those properties that can be measured? In answering this question we will develop certain general thermodynamic relations. In view of the fact that there are millions of such equations that can be written, our study will be limited to certain basic considerations, with particular reference to the determination of thermodynamic properties from experimental data. We will also consider such related matters as generalized charts and equations of state.

13.1 THE CLAPEYRON EQUATION

In calculating thermodynamic properties such as enthalpy or entropy in terms of other properties that can be measured, the calculations fall into two broad categories, differences in properties between two different phases, and changes within a single, homogeneous phase. In this section, we focus our attention on the first of these categories, that of different phases. Let us assume that the two phases are liquid and vapor, but we will see that the results apply to other differences as well.

Consider a Carnot cycle heat engine operating across a small temperature difference between reservoirs at T and $T - \Delta T$. The corresponding saturation pressures are P and $P - \Delta P$. The Carnot cycle operates with four steady-state, steady-flow devices. In the high temperature heat transfer process, the working fluid changes from saturated liquid at 1 to saturated vapor at 2, as shown in the two diagrams of Fig. 13.1.

From Fig. 13.1(a), for reversible heat transfers,

$$q_H = Ts_{fg}; \qquad q_L = (T - \Delta T)s_{fg}$$

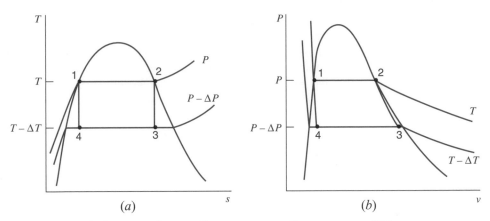

FIGURE 13.1 A Carnot cycle operating across a small temperature difference.

so that

$$w_{NET} = q_H - q_L = \Delta T s_{fg} \tag{13.1}$$

From Fig. 13.1(*b*), each process is SSSF and reversible, such that the work in each process is given by Eq. 9.19,

$$w = -\int v\,dP$$

Overall, for the four processes in the cycle,

$$w_{NET} = 0 - \int_2^3 v\,dP + 0 - \int_4^1 v\,dP$$

$$\approx -\left(\frac{v_2 + v_3}{2}\right)(P - \Delta P - P) - \left(\frac{v_1 + v_4}{2}\right)(P - P + \Delta P)$$

$$\approx \Delta P\left[\left(\frac{v_2 + v_3}{2}\right) - \left(\frac{v_1 + v_4}{2}\right)\right] \tag{13.2}$$

(The smaller the ΔP, the better the approximation.)

Now, comparing Eqs. 13.1 and 13.2 and rearranging,

$$\frac{\Delta P}{\Delta T} \approx \frac{s_{fg}}{\left(\dfrac{v_2 + v_3}{2}\right) - \left(\dfrac{v_1 + v_4}{2}\right)}$$

In the limit as $\Delta T \to 0$: $v_3 \to v_2 = v_G$, $v_4 \to v_1 = v_F$ which results in

$$\lim_{\Delta T \to 0}\frac{\Delta P}{\Delta T} = \frac{dP_{\text{sat}}}{dT} = \frac{s_{fg}}{v_{fg}} \tag{13.3}$$

Since the heat addition process $1 - 2$ is at constant pressure as well as constant temperature,

$$q_H = h_{fg} = T s_{fg}$$

and the general result of Eq. 13.3 is the expression

$$\frac{dP_{\text{sat}}}{dT} = \frac{s_{fg}}{v_{fg}} = \frac{h_{fg}}{T v_{fg}} \tag{13.4}$$

which is called the Clapeyron equation. This is a very simple relation, and yet an extremely powerful one. We can experimentally determine the left-hand side of Eq. 13.4, which is the slope of the vapor pressure as a function of temperature. We can also measure the specific volumes of saturated vapor and saturated liquid at the given temperature, which means that the enthalpy change and entropy change of vaporization can both be calculated from Eq. 13.4. This establishes the means to cross from one phase to another in first- or second-law calculations, which was the goal of this development.

We could proceed along the same lines for the change of phase solid to liquid or for solid to vapor. In each case, the result is the Clapeyron equation, in which the appropriate saturation pressure, specific volumes, entropy change, and enthalpy change are involved. For solid i to liquid f, the process occurs along the fusion line, and the result is

$$\frac{dP_{\text{fus}}}{dT} = \frac{s_{if}}{v_{if}} = \frac{h_{if}}{Tv_{if}} \tag{13.5}$$

We note that $v_{if} = v_f - v_i$ is typically a very small number, such that the slope of the fusion line is very steep (in the case of water, v_{if} is a negative number, which is highly unusual, and the slope of the fusion line is not only steep, it is also negative.)

For sublimation, the change from solid i directly to vapor g, the Clapeyron equation has the values

$$\frac{dP_{\text{sub}}}{dT} = \frac{s_{ig}}{v_{ig}} = \frac{h_{ig}}{Tv_{ig}} \tag{13.6}$$

A special case of the Clapeyron equation involving the vapor phase occurs at low temperatures when the saturation pressure becomes very small. The specific volume v_g is then not only much larger than that of the condensed phase, liquid in Eq. 13.4 or solid in Eq. 13.6, but is also closely represented by the ideal gas equation of state. The Clapeyron equation then reduces to the form

$$\frac{dP_{\text{sat}}}{dT} = \frac{h_{fg}}{Tv_{fg}} = \frac{h_{fg}P_{\text{sat}}}{RT^2} \tag{13.7}$$

At low temperatures (not near the critical temperature), h_{fg} does not change very much with temperature. If it is assumed to be constant, then Eq. 13.7 can be rearranged and integrated over a range of temperatures to calculate a saturation pressure at a temperature at which it is not known. This point is illustrated by the following example.

EXAMPLE 13.1 Determine the sublimation pressure of water vapor at −60°C using data available in the steam tables.

Control mass: Water.

Solution

Table 6 of the steam tables (Appendix Table B.1.5) does not give saturation pressures for temperatures less than −40°C. However, we do notice that h_{ig} is relatively constant in this range and, therefore, we proceed to Eq. 13.7 and integrate between the limits

−40°C and −60°C.

$$\int_1^2 \frac{dP}{P} = \int_1^2 \frac{h_{ig}}{R} \frac{dT}{T^2} = \frac{h_{ig}}{R} \int_1^2 \frac{dT}{T^2}$$

$$\ln \frac{P_2}{P_1} = \frac{h_{ig}}{R} \left(\frac{T_2 - T_1}{T_1 T_2} \right)$$

Let

$$P_2 = 0.0129 \text{ kPa} \qquad T_2 = 233.2 \text{ K} \qquad T_1 = 213.2 \text{ K}$$

Then

$$\ln \frac{P_2}{P_1} = \frac{2838.9}{0.461\ 52} \left(\frac{233.2 - 213.2}{233.2 \times 213.2} \right) = 2.4744$$

$$P_1 = 0.001\ 09 \text{ kPa}$$

EXAMPLE 13.1E Determine the sublimation pressure of water vapor at −70 F using data available in the steam tables.

Control mass: Water.

Solution

Table 6 of the steam tables (Appendix Table C.5) does not give saturation pressures for temperatures less than −40 F. However, we do notice that h_{ig} is relatively constant in this range and, therefore, we proceed to use Eq. 13.7 and integrate between the limits −40 F and −70 F.

$$\int_1^2 \frac{dP}{P} = \int_1^2 \frac{h_{ig}}{R} \frac{dT}{T^2} = \frac{h_{ig}}{R} \int_1^2 \frac{dT}{T^2}$$

$$\ln \frac{P_2}{P_1} = \frac{h_{ig}}{R} \left(\frac{T_2 - T_1}{T_1 T_2} \right)$$

Let

$$P_2 = 0.0019 \text{ lbf / in.}^2 \qquad T_2 = 419.7 \text{ R} \qquad T_1 = 389.7 \text{ R}$$

Then,

$$\ln \frac{P_2}{P_1} = \frac{1218.7 \times 778}{85.76} \left(\frac{419.7 - 389.7}{419.7 \times 389.7} \right) = 2.0279$$

$$P_1 = 0.000\ 25 \text{ lbf / in.}^2$$

13.2 MATHEMATICAL RELATIONS FOR A HOMOGENEOUS PHASE

In the preceding section, we established the means to calculate differences in enthalpy (and therefore internal energy) and entropy between different phases, in terms of properties that are readily measured. In the following sections, we wil develop expressions for calculating differences in these properties within a single, homogeneous phase (gas, liquid, or solid), assuming a simple compressible substance. In order to develop such expressions, it is first necessary to present two mathematical relations that will prove useful in this procedure.

Consider a variable (thermodynamic property) that is a continuous function of x and y.

$$z = f(x, y)$$

$$dz = \left(\frac{\partial z}{\partial x}\right)_y dx + \left(\frac{\partial z}{\partial y}\right)_x dy$$

It is convenient to write this function in the form

$$dz = M\, dx + N\, dy \tag{13.8}$$

$$M = \left(\frac{\partial z}{\partial x}\right)_y$$

= partial derivative of z with respect to x (the variable y being held constant)

$$N = \left(\frac{\partial z}{\partial y}\right)_x$$

= partial derivative of z with respect to y (the variable x being held constant)

The physical significance of partial derivatives as they relate to the properties of a pure substance can be explained by referring to Fig. 13.2, which shows a P–v–T surface of the superheated vapor region of a pure substance. It shows constant-temperature, constant-pressure, and constant-specific volume planes that intersect at point b on the surface. Thus, the partial derivative $(\partial P/\partial v)_T$ is the slope of curve abc at point b. Line de represents the tangent to curve abc at point b. A similar interpretation can be made of the partial derivatives $(\partial P/\partial T)_v$ and $(\partial v/\partial T)_p$.

If we wish to evaluate the partial derivative along a constant-temperature line the rules for ordinary derivatives can be applied. Thus, we can write for a constant-temperature process:

$$\left(\frac{\partial P}{\partial v}\right)_T = \frac{dP_T}{dv_T}$$

and the integration can be performed as usual. This point will be demonstrated later in a number of examples.

Let us return to the consideration of the relation

$$dz = M\, dx + N\, dy$$

If x, y, and z are all point functions (that is, quantities that depend only on the state and are independent of the path), the differentials are exact differentials. If this is the case, the following important relation holds:

$$\left(\frac{\partial M}{\partial y}\right)_x = \left(\frac{\partial N}{\partial x}\right)_y \tag{13.9}$$

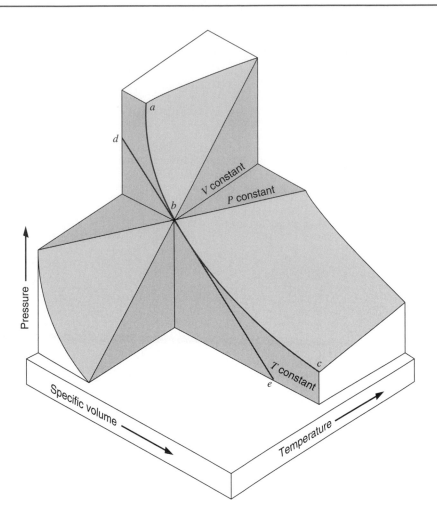

FIGURE 13.2
Schematic representation
of partial derivatives.

The proof of this is

$$\left(\frac{\partial M}{\partial y}\right)_x = \frac{\partial^2 z}{\partial x \partial y}$$

$$\left(\frac{\partial N}{\partial x}\right)_y = \frac{\partial^2 z}{\partial y \partial x}$$

Since the order of differentiation makes no difference when point functions are involved, it follows that

$$\frac{\partial^2 z}{\partial x \partial y} = \frac{\partial^2 z}{\partial y \partial x}$$

$$\left(\frac{\partial M}{\partial y}\right)_x = \left(\frac{\partial N}{\partial x}\right)_y$$

The second important mathematical relation is

$$\left(\frac{\partial x}{\partial y}\right)_z \left(\frac{\partial y}{\partial z}\right)_x \left(\frac{\partial z}{\partial x}\right)_y = -1 \qquad (13.10)$$

The proof of this relation is as follows. Consider three variables x, y, and z. Suppose there exists a relation between the variables of the form

$$x = f(y, z)$$

Then

$$dx = \left(\frac{\partial x}{\partial y}\right)_z dy + \left(\frac{\partial x}{\partial z}\right)_y dz \tag{13.11}$$

If this relationship between the three variables is written in the form

$$y = f(x, z)$$

it follows that

$$dy = \left(\frac{\partial y}{\partial x}\right)_z dx + \left(\frac{\partial y}{\partial z}\right)_x dz \tag{13.12}$$

Substituting Eq. 13.12 into Eq. 13.11, we have

$$dx = \left(\frac{\partial x}{\partial y}\right)_z \left[\left(\frac{\partial y}{\partial x}\right)_z dx + \left(\frac{\partial y}{\partial z}\right)_x dz\right] + \left(\frac{\partial x}{\partial z}\right)_y dz$$

$$= \left(\frac{\partial x}{\partial y}\right)_z \left(\frac{\partial y}{\partial x}\right)_z dx + \left[\left(\frac{\partial x}{\partial y}\right)_z \left(\frac{\partial y}{\partial z}\right)_x + \left(\frac{\partial x}{\partial z}\right)_y\right] dz$$

There are two independent variables, and we select x and z as these variables. Suppose that $dz = 0$ and $dx \neq 0$. It then follows that

$$\left(\frac{\partial x}{\partial y}\right)_z \left(\frac{\partial y}{\partial x}\right)_z = 1 \tag{13.13}$$

Similarly, suppose that $dx = 0$ and $dz \neq 0$. It then follows that

$$\left(\frac{\partial x}{\partial y}\right)_z \left(\frac{\partial y}{\partial z}\right)_x + \left(\frac{\partial x}{\partial z}\right)_y = 0$$

$$\left(\frac{\partial x}{\partial y}\right)_z \left(\frac{\partial y}{\partial z}\right)_x = -\left(\frac{\partial x}{\partial z}\right)_y$$

$$\left(\frac{\partial x}{\partial y}\right)_z \left(\frac{\partial y}{\partial z}\right)_x \left(\frac{\partial z}{\partial x}\right)_y = -1$$

This is Eq. 13.10, which we set out to derive.

13.3 THE MAXWELL RELATIONS

Consider a simple compressible control mass of fixed chemical composition. The Maxwell relations, which can be written for such a system, are four equations relating the properties P, v, T, and s. These wil be found to be useful in the calculation of entropy in terms of the other, measurable properties.

The Maxwell relations are most easily derived by considering the different forms of the thermodynamic property relation, which was the subject of Section 8.5. The two forms of this expression are rewritten here as

$$du = T\, ds - P\, dv \tag{13.14}$$

and

$$dh = T \, ds - v \, dP \tag{13.15}$$

Note that in the mathematical representation of Eq. 13.8, these expressions are of the form

$$u = u(s, v), \qquad h = h(s, P)$$

in both of which entropy is used as one of the two independent properties. This is an undesirable situation, in that entropy is one of the properties that cannot be measured. We can, however, eliminate entropy as an independent property by introducing two new properties, and thereby two new forms of the thermodynamic property relation. The first of these is the Helmholtz function A,

$$A = U - TS, \qquad a = u - Ts \tag{13.16}$$

Differentiating and substituting Eq. 13.14 results in

$$da = du - T \, ds - s \, dT$$
$$= -s \, dT - P \, dv \tag{13.17}$$

which we note is a form of the property relation utilizing T and v as the independent properties. The second new property is the Gibbs function G,

$$G = H - TS, \qquad g = h - Ts \tag{13.18}$$

Differentiating and substituting Eq. 13.15,

$$dg = dh - T \, ds - s \, dt$$
$$= -s \, dT + v \, dP \tag{13.19}$$

a fourth form of the property relation, this form using T and P as the independent properties.

Since Eqs. 13.14, 13.15, 13.17, and 13.19 are all relations involving only properties, we conclude that these are exact differentials and, therefore, are of the general form of Eq. 13.8,

$$dz = M \, dx + N \, dy$$

in which Eq. 13.9 relates the coefficients M and N,

$$\left(\frac{\partial M}{\partial y}\right)_x = \left(\frac{\partial N}{\partial x}\right)_y$$

It follows from Eq. 13.14 that

$$\left(\frac{\partial T}{\partial v}\right)_s = -\left(\frac{\partial P}{\partial s}\right)_v \tag{13.20}$$

Similarly, from Eqs. 13.15, 13.17, and 13.19 we can write

$$\left(\frac{\partial T}{\partial P}\right)_s = \left(\frac{\partial v}{\partial s}\right)_P \tag{13.21}$$

$$\left(\frac{\partial P}{\partial T}\right)_v = \left(\frac{\partial s}{\partial v}\right)_T \tag{13.22}$$

$$\left(\frac{\partial v}{\partial T}\right)_P = -\left(\frac{\partial s}{\partial P}\right)_T \tag{13.23}$$

These four equations are known as the Maxwell relations for a simple compressible mass, and the great utility of these equations will be demonstrated in later sections of this chapter. As was noted earlier, these relations will enable us to calculate entropy changes in terms of the measurable properties pressure, temperature, and specific volume.

There are a number of other useful relations that can be derived from Eqs. 13.14, 13.15, 13.17, and 13.19. For example, from Eq. 13.14, we can write the relations

$$\left(\frac{\partial u}{\partial s}\right)_v = T, \qquad \left(\frac{\partial u}{\partial v}\right)_s = -P \tag{13.24}$$

Similarly, from the other three equations, we have the following

$$\left(\frac{\partial h}{\partial s}\right)_P = T, \qquad \left(\frac{\partial h}{\partial P}\right)_s = v$$
$$\left(\frac{\partial a}{\partial v}\right)_T = -P, \qquad \left(\frac{\partial a}{\partial T}\right)_v = -s$$
$$\left(\frac{\partial g}{\partial P}\right)_T = v, \qquad \left(\frac{\partial g}{\partial T}\right)_P = -s \tag{13.25}$$

As already noted, the Maxwell relations just presented are written for a simple compressible substance. It is readily evident, however, that similar Maxwell relations can be written for substances involving other effects, such as surface or electrical effects. For example, Eq. 8.9 can be written in the form

$$dU = T\,dS - P\,dV + \mathscr{T}\,dL + \mathscr{S}\,d\mathscr{A} + \mathscr{E}\,dZ + \cdots \tag{13.26}$$

Thus, for a substance involving only surface effects, we can write

$$dU = T\,dS + \mathscr{S}\,d\mathscr{A}$$

and it follows that for such a substance

$$\left(\frac{\partial T}{\partial \mathscr{A}}\right)_S = \left(\frac{\partial \mathscr{S}}{\partial S}\right)_{\mathscr{A}}$$

Other Maxwell relations could also be written for such a substance, by writing the property relation in terms of different variables, and this approach could also be extended to systems having multiple effects. This matter also becomes more complex when we consider applying the property relation to a system of variable composition, a topic that will be taken up in Section 13.11.

EXAMPLE 13.2

From an examination of the properties of compressed liquid water, as given in Table B.1.4 of the Appendix, we find that the entropy of compressed liquid is greater than the entropy of saturated liquid for a temperature of 0°C, and is less than that of saturated liquid for all the other temperatures listed. Explain why this follows from other thermodynamic data.

Control mass: Water.

Solution

Suppose we increase the pressure of liquid water that is initially saturated, while keeping the temperature constant. The change of entropy for the water during this process

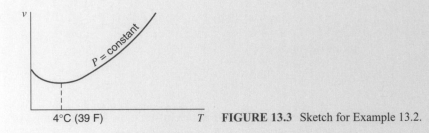

FIGURE 13.3 Sketch for Example 13.2.

can be found by integrating the following Maxwell relation, Eq. 13.23:

$$\left(\frac{\partial s}{\partial P}\right)_T = -\left(\frac{\partial v}{\partial T}\right)_P$$

Therefore, the sign of the entropy change depends on the sign of the term $(\partial v/\partial T)_P$. The physical significance of this is that it involves the change in specific volume of water as the temperature changes while the pressure remains constant. As water at moderate pressures and 0°C is heated in a constant-pressure process, the specific volume decreases until the point of maximum density is reached at approximately 4°C, after which it increases. This is shown on a v–T diagram in Fig. 13.3. Thus, the quantity $(\partial v/\partial T)_P$ is the slope of the curve in Fig. 13.3. Since this slope is negative at 0°C, the quantity $(\partial s/\partial P)_T$ is positive at 0°C. At the point of maximum density the slope is zero and, therefore, the constant-pressure line shown in Fig. 8.7 crosses the saturated-liquid line at the point of maximum density.

13.4 SOME THERMODYNAMIC RELATIONS INVOLVING ENTHALPY, INTERNAL ENERGY, AND ENTROPY

Let us first derive two equations, one involving C_p and the other involving C_v.

We have defined C_p as

$$C_p \equiv \left(\frac{\partial h}{\partial T}\right)_P$$

We have also noted that for a pure substance

$$T\,ds = dh - v\,dP$$

Therefore,

$$C_p = \left(\frac{\partial h}{\partial T}\right)_P = T\left(\frac{\partial s}{\partial T}\right)_P \tag{13.27}$$

Similarly, from the definition of C_v,

$$C_v \equiv \left(\frac{\partial u}{\partial T}\right)_v$$

and the relation

$$T\,ds = du + P\,dv$$

it follows that

$$C_v = \left(\frac{\partial u}{\partial T}\right)_v = T\left(\frac{\partial s}{\partial T}\right)_v \tag{13.28}$$

We will now derive a general relation for the change of enthalpy of a pure substance. We first note that for a pure substance

$$h = h(T, P)$$

Therefore,

$$dh = \left(\frac{\partial h}{\partial T}\right)_P dT + \left(\frac{\partial h}{\partial P}\right)_T dP$$

From the relation

$$T\,ds = dh - v\,dP$$

it follows that

$$\left(\frac{\partial h}{\partial P}\right)_T = v + T\left(\frac{\partial s}{\partial P}\right)_T$$

Substituting the Maxwell relation, Eq. 13.23, we have

$$\left(\frac{\partial h}{\partial P}\right)_T = v - T\left(\frac{\partial v}{\partial T}\right)_P \tag{13.29}$$

On substituting this equation and Eq. 13.27, we have

$$dh = C_p\,dT + \left[v - T\left(\frac{\partial v}{\partial T}\right)_P\right]dP \tag{13.30}$$

Along an isobar we have

$$dh_p = C_p\,dT_p$$

and along an isotherm,

$$dh_T = \left[v - T\left(\frac{\partial v}{\partial T}\right)_P\right]dP_T \tag{13.31}$$

The significance of Eq. 13.30 is that this equation can be integrated to give the change in enthalpy associated with a change of state

$$h_2 - h_1 = \int_1^2 C_p\,dT + \int_1^2 \left[v - T\left(\frac{\partial v}{\partial T}\right)_P\right]dP \tag{13.32}$$

The information needed to integrate the first term is a constant-pressure specific heat along one (and only one) isobar. The integration of the second integral requires that an equation of state giving the relation between P, v, and T be known. Furthermore, it is advantageous to have this equation of state explicit in v, for then the derivative $(\partial v/\partial T)_P$ is readily evaluated.

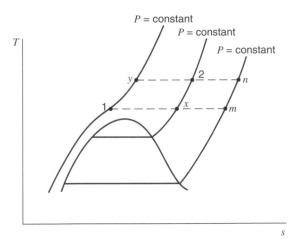

FIGURE 13.4 Sketch showing various paths by which a given change of state can take place.

This matter can be further illustrated by reference to Fig. 13.4. Suppose we wish to know the change of enthalpy between states 1 and 2. We might determine this change along path 1–x–2, which consists of one isotherm, 1–x, and one isobar, x–2. Thus, we could integrate Eq. 13.32:

$$h_2 - h_1 = \int_{T_1}^{T_2} C_p \, dT + \int_{P_1}^{P_2} \left[v - T \left(\frac{\partial v}{\partial T} \right)_P \right] dP$$

Since $T_1 = T_x$ and $P_2 = P_x$, this can be written

$$h_2 - h_1 = \int_{T_x}^{T_2} C_p \, dT + \int_{P_1}^{P_x} \left[v - T \left(\frac{\partial v}{\partial T} \right)_P \right] dP$$

The second term in this equation gives the change in enthalpy along the isotherm 1–x and the first term the change in enthalpy along the isobar x–2. When these are added together, the result is the net change in enthalpy between 1 and 2. Therefore, the constant-pressure specific heat must be known along the isobar passing through 2 and x. The change in enthalpy could also be found by following path 1–y–2, in which case the constant-pressure specific heat must be known along the 1–y isobar. If the constant-pressure specific heat is known at another pressure, say, the isobar passing through m–n, the change in enthalpy can be found by following path 1–m–n–2. This involves calculating the change of enthalpy along two isotherms—1–m and n–2.

Let us now derive a similar relation for the change of internal energy. All the steps in this derivation are given, but without detailed comment. Note that the starting point is to write $u = u(T, v)$, whereas in the case of enthalpy the starting point was $h = h(T, P)$.

$$u = f(T, v)$$

$$du = \left(\frac{\partial u}{\partial T} \right)_v dT + \left(\frac{\partial u}{\partial v} \right)_T dv$$

$$T \, ds = du + P \, dv$$

Therefore,

$$\left(\frac{\partial u}{\partial v} \right)_T = T \left(\frac{\partial s}{\partial v} \right)_T - P \tag{13.33}$$

Substituting the Maxwell relation, Eq. 13.22, we have

$$\left(\frac{\partial u}{\partial v}\right)_T = T\left(\frac{\partial P}{\partial T}\right)_v - P$$

Therefore,

$$du = C_v\,dT + \left[T\left(\frac{\partial P}{\partial T}\right)_v - P\right]dv \tag{13.34}$$

Along an isometric this reduces to

$$du_v = C_v\,dT_v$$

and along an isotherm we have

$$du_T = \left[T\left(\frac{\partial P}{\partial T}\right)_v - P\right]dv_T \tag{13.35}$$

In a manner similar to that outlined above for changes in enthalpy, the change of internal energy for a given change of state for a pure substance can be determined from Eq. 13.34 if the constant-volume specific heat is known along one isometric and an equation of state explicit in P [to obtain the derivative $(\partial P/\partial T)_v$] is available in the region involved. A diagram similar to Fig. 13.4 could be drawn, with the isobars replaced with isometrics, and the same general conclusions would be reached.

To summarize, we have derived Eqs. 13.30 and 13.34:

$$dh = C_p\,dT + \left[v - T\left(\frac{\partial v}{\partial T}\right)_P\right]dP$$

$$du = C_v\,dT + \left[T\left(\frac{\partial P}{\partial T}\right)_v - P\right]dv$$

The first of these equations concerns the change of enthalpy, the constant-pressure specific heat, and is particularly suited to an equation of state explicit in v. The second equation concerns the change of internal energy and the constant-volume specific heat, and is particularly suited to an equation of state explicit in P. If the first of these equations is used to determine the change of enthalpy, the internal energy is readily found by noting that

$$u_2 - u_1 = h_2 - h_1 - (P_2 v_2 - P_1 v_1)$$

If the second equation is used to find changes of internal energy, the change of enthalpy is readily found from this same relation. Which of these two equations is used to determine changes in internal energy and enthalpy will depend on the information available for specific heat and an equation of state (or other P–v–T data).

Two parallel expressions can be found for the change of entropy.

$$s = s(T, P)$$

$$ds = \left(\frac{\partial s}{\partial T}\right)_P dT + \left(\frac{\partial s}{\partial P}\right)_T dP$$

Substituting Eqs. 13.27 and 13.23, we have

$$ds = C_p \frac{dT}{T} - \left(\frac{\partial v}{\partial T} \right)_P dP \tag{13.36}$$

$$s_2 - s_1 = \int_1^2 C_p \frac{dT}{T} - \int_1^2 \left(\frac{\partial v}{\partial T} \right)_P dP \tag{13.37}$$

Along an isobar we have

$$(s_2 - s_1)_P = \int_1^2 C_p \frac{dT_P}{T}$$

and along an isotherm

$$(s_2 - s_1)_T = -\int_1^2 \left(\frac{\partial v}{\partial T} \right)_P dP$$

Note from Eq. 13.37 that if a constant-pressure specific heat is known along one isobar and an equation of state explicit in v is available, the change of entropy can be evaluated. This is analogous to the expression for the change of enthalpy given in Eq. 13.30.

$$s = s(T, v)$$

$$ds = \left(\frac{\partial s}{\partial T} \right)_v dT + \left(\frac{\partial s}{\partial v} \right)_T dv$$

Substituting Eqs. 13.28 and 13.22 gives

$$ds = C_v \frac{dT}{T} + \left(\frac{\partial P}{\partial T} \right)_v dv \tag{13.38}$$

$$s_2 - s_1 = \int_1^2 C_v \frac{dT}{T} + \int_1^2 \left(\frac{\partial P}{\partial T} \right)_v dv \tag{13.39}$$

This expression for change of entropy concerns the change of entropy along an isometric where the constant-volume specific heat is known and along an isotherm where an equation of state explicit in P is known. Thus, it is analogous to the expression for change of internal energy given in Eq. 13.34.

EXAMPLE 13.3   Over a certain small range of pressures and temperatures, the equation of state of a certain substance is given with reasonable accuracy by the relation

$$\frac{Pv}{RT} = 1 - C' \frac{P}{T^4}$$

or

$$v = \frac{RT}{P} - \frac{C}{T^3}$$

where C and C' are constants.

Derive an expression for the change of enthalpy and entropy of this substance in an isothermal process.

Control mass: Gas.

Solution

Since the equation of state is explicit in v, Eq. 13.31 is particularly relevant to the change in enthalpy. On integrating this equation, we have

$$(h_2 - h_1)_T = \int_1^2 \left[v - T \left(\frac{\partial v}{\partial T} \right)_P \right] dP_T$$

From the equation of state,

$$\left(\frac{\partial v}{\partial T} \right)_P = \frac{R}{P} + \frac{3C}{T^4}$$

Therefore,

$$(h_2 - h_1)_T = \int_1^2 \left[v - T \left(\frac{R}{P} + \frac{3C}{T^4} \right) \right] dP_T$$

$$= \int_1^2 \left[\frac{RT}{P} - \frac{C}{T^3} - \frac{RT}{P} - \frac{3C}{T^3} \right] dP_T$$

$$(h_2 - h_1)_T = \int_1^2 -\frac{4C}{T^3} \, dP_T = -\frac{4C}{T^3} (P_2 - P_1)_T$$

For the change in entropy we use Eq. 13.37, which is particularly relevant for an equation of state explicit in v.

$$(s_2 - s_1)_T = -\int_1^2 \left(\frac{\partial v}{\partial T} \right)_P dP_T = -\int_1^2 \left(\frac{R}{P} + \frac{3C}{T^4} \right) dP_T$$

$$(s_2 - s_1)_T = -R \ln \left(\frac{P_2}{P_1} \right)_T - \frac{3C}{T^4} (P_2 - P_1)_T$$

13.5 VOLUME EXPANSIVITY AND ISOTHERMAL AND ADIABATIC COMPRESSIBILITY

The student has most likely encountered the coefficient of linear expansion in his or her studies of strength of materials. This coefficient indicates how the length of a solid body is influenced by a change in temperature while the pressure remains constant. In terms of the notation of partial derivatives, the *coefficient of linear expansion, δ_T,* is defined as

$$\delta_T = \frac{1}{L} \left(\frac{\partial L}{\partial T} \right)_P \tag{13.40}$$

A similar coefficient can be defined for changes in volume. Such a coefficient is applicable to liquids and gases as well as to solids. This coefficient of volume expansion, α_P, also called the volume expansivity, is an indication of the change in volume as temperature changes while the pressure remains constant. The definition of *volume expansivity* is

$$\alpha_P \equiv \frac{1}{V}\left(\frac{\partial V}{\partial T}\right)_P = \frac{1}{v}\left(\frac{\partial v}{\partial T}\right)_P \tag{13.41}$$

The isothermal compressibility, β_T, is an indication of the change in volume as pressure changes while the temperature remains constant. The definition of the *isothermal compressibility* is

$$\beta_T \equiv -\frac{1}{V}\left(\frac{\partial V}{\partial P}\right)_T = -\frac{1}{v}\left(\frac{\partial v}{\partial P}\right)_T \tag{13.42}$$

The reciprocal of the isothermal compressibility is called the *isothermal bulk modulus, B_T*.

$$B_T \equiv -v\left(\frac{\partial P}{\partial v}\right)_T \tag{13.43}$$

The *adiabatic compressibility, β_s*, is an indication of the change in volume as pressure changes while the entropy remains constant; it is defined as

$$\beta_s \equiv -\frac{1}{v}\left(\frac{\partial v}{\partial P}\right)_s \tag{13.44}$$

The *adiabatic bulk modulus, B_s*, is the reciprocal of the adiabatic compressibility.

$$B_s \equiv -v\left(\frac{\partial P}{\partial v}\right)_s \tag{13.45}$$

Both the volume expansivity and isothermal compressibility are thermodynamic properties of a substance, and for a simple compressible substance are functions of two independent properties. Values of these properties are found in the standard handbooks of physical properties. The following examples give an indication of the use and significance of the volume expansivity and isothermal compressibility.

EXAMPLE 13.4   The pressure on a block of copper having a mass of 1 kg is increased in a reversible process from 0.1 to 100 MPa while the temperature is held constant at 15°C. Determine the work done on the copper during this process, the change in entropy per kilogram of copper, the heat transfer, and the change of internal energy per kilogram.

Over the range of pressure and temperature in this problem, the following data can be used:

> Volume expansivity = $\alpha_p = 5.0 \times 10^{-5} \mathrm{K}^{-1}$
> Isothermal compressibility = $\beta_T = 8.6 \times 10^{-12} \mathrm{m^2/N}$
> Specific volume = 0.000 114 m³/kg

Analysis:

Control mass: Copper block.

States: Initial and final states known.

Process: Constant temperature, reversible.

The work done during the isothermal compression is

$$w = \int P\, dv_T$$

The isothermal compressibility has been defined as

$$\beta_T = -\frac{1}{v}\left(\frac{\partial v}{\partial P}\right)_T$$

$$v\beta_T\, dP_T = -dv_T$$

Therefore, for this isothermal process,

$$w = -\int_1^2 v\beta_T P\, dP_T$$

Since v and β_T remain essentially constant, this is readily integrated:

$$w = -\frac{v\beta_T}{2}(P_2^2 - P_1^2)$$

The change of entropy can be found by considering the Maxwell relation, Eq. 13.23, and the definition of volume expansivity.

$$\left(\frac{\partial s}{\partial P}\right)_T = -\left(\frac{\partial v}{\partial T}\right)_P = -\frac{v}{v}\left(\frac{\partial v}{\partial T}\right)_P = -v\alpha_P$$

$$ds_T = -v\alpha_P\, dP_T$$

This equation can be readily integrated, if we assume that v and α_P remain constant:

$$(s_2 - s_1)_T = -v\alpha_p(P_2 - P_1)_T$$

The heat transfer for this reversible isothermal process is

$$q = T(s_2 - s_1)$$

The change in internal energy follows directly from the first law.

$$(u_2 - u_1) = q - w$$

Solution

$$w = -\frac{v\beta_T}{2}(P_2^2 - P_1^2)$$

$$= -\frac{0.000\,114 \times 8.6\times10^{-12}}{2}(100^2 - 0.1^2)\times10^{12}$$

$$= -4.9 \text{ J/kg}$$

$$(s_2 - s_1)_T = -v\alpha_p(P_2 - P_1)_T$$

$$= -0.000\,114 \times 5.0\times10^{-5}(100 - 0.1)\times10^6$$

$$= -0.5694 \text{ J/kg K}$$

$$q = T(s_1 - s_2) = -288.2 \times 0.5694 = -164.1 \text{ J/kg}$$

$$(u_2 - u_1) = q - w = -164.1 - (-4.9) = -159.2 \text{ J/kg}$$

13.6 DEVELOPING TABLES OF THERMODYNAMIC PROPERTIES FROM EXPERIMENTAL DATA

There are many ways in which tables of thermodynamic properties can be developed from experimental data. The purpose of this section is to convey some general principles and concepts by considering only the liquid and vapor phases.

Let us assume that the following data for a pure substance have been obtained in the laboratory.

1. Vapor-pressure data. That is, saturation pressures and temperatures have been measured over a wide range.
2. Pressure, specific volume, and temperature data in the vapor region. These data are usually obtained by determining the mass of the substance in a closed vessel (which means a fixed specific volume) and then measuring the pressure as the temperature is varied. This is done for a large number of specific volumes.
3. Density of the saturated liquid and the critical pressure and temperature.
4. Zero-pressure specific heat for the vapor. This might be obtained either calorimetrically or from spectroscopic data and statistical thermodynamics.

From these data a complete set of thermodynamic tables for the saturated liquid, saturated vapor, and superheated vapor can be calculated. The first step is to determine an equation for the vapor-pressure curve that accurately fits the data. It may be necessary to use one equation for one portion of the vapor-pressure curve and a different equation for another portion of the curve.

One form of equation that has been used is

$$\ln P_{\text{sat}} = A + \frac{B}{T} + C \ln T + DT$$

Once an equation has been found that accurately represents the data, the saturation pressure for any given temperature can be found by solving this equation. Thus, the saturation pressures in Table 1 of the Steam Tables would be determined for the given temperatures. The second step is to determine an equation of state for the vapor region that accurately represents the P–v–T data. There are many possible forms of the equation of state that may be selected. The important considerations are that the equation of state accurately represents the data, and that it be of such a form that the differentiations required can be performed (that is, in some cases it may be desirable to have an equation of state explicit in v, whereas on other occasions an equation of state that is explicit in P may be more desirable).

Once an equation of state has been determined, the specific volume of superheated vapor at given pressures and temperatures can be determined by solving the equation and tabulating the results as in the superheat tables for steam, ammonia, and the other substances listed in the appendix. The specific volume of saturated vapor at a given temperature may be found by finding the saturation pressure from the vapor-pressure curve and substituting this saturation pressure and temperature into the equation of state.

The procedure followed in determining enthalpy and entropy is best explained with the aid of Fig. 13.5. Let us assume that the enthalpy and entropy of saturated liquid in state 1 are zero. The enthalpy of saturated vapor in state 2 can be found from the Clapeyron equation.

$$\left(\frac{dP}{dT}\right)_{\text{sat}} = \frac{h_{fg}}{T(v_g - v_f)}$$

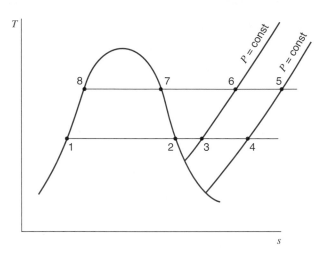

FIGURE 13.5 Sketch showing procedure for developing a table of thermodynamic properties from experimental data.

The left side of this equation is found by differentiating the vapor-pressure curve. The specific volume of the saturated vapor is found by the procedure outlined in the last paragraph, and it is assumed the specific volume of the saturated liquid has been measured. Thus the enthalpy of evaporation, h_{fg}, can be found for this particular temperature, and the enthalpy at state 2 is equal to the enthalpy of evaporation (since the enthalpy in state 1 is assumed to be zero). The entropy at state 2 is readily found, since

$$s_{fg} = \frac{h_{fg}}{T}$$

From state 2 we proceed along this isotherm into the superheated vapor region. The specific volume at 3 is found from the equation of state at this pressure, while the enthalpy and entropy are determined by integrating Eqs. 13.31 and 13.37:

$$h_3 - h_2 = \int_2^3 \left[v - T \left(\frac{\partial v}{\partial T} \right)_P \right] dP_T$$

$$s_3 - s_2 = \int_2^3 -\left(\frac{\partial v}{\partial T} \right)_P dP_T$$

The properties at point 4 are found in exactly the same manner. Pressure P_4 is sufficiently low that the real superheated vapor behaves essentially as an ideal gas (perhaps 10 kPa). Thus we use this constant-pressure line to make all temperature changes for our calculations, as for example to point 5. Since the specific heat C_{po} is known as a function of temperature, the enthalpy and entropy at 5 are found by integrating the ideal-gas relations

$$(h_5 - h_4)_P = \int_4^5 C_{po} dT_P$$

$$(s_5 - s_4)_P = \int_4^5 C_{po} \frac{dT_P}{T}$$

The properties at points 6 and 7 are found from those at 5 in the same manner as those at points 3 and 4 were found from 2 (the saturation pressure P_7 is calculated from the vapor-pressure equation). Finally, the enthalpy and entropy for saturated liquid at point 8 are found from the properties at point 7 by applying the Clapeyron equation.

Thus values for the pressure, temperature, specific volume, enthalpy, entropy, and internal energy of saturated liquid, saturated vapor, and superheated vapor can be tabulated for the entire region for which experimental data were obtained. The accuracy of such a table depends both on the accuracy of the experimental data and the degree to which the equation for the vapor pressure and the equation of state represent the experimental data.

13.7 REAL GAS BEHAVIOR AND EQUATIONS OF STATE

In Section 3.4, we examined the $P-v-T$ behavior of gases, and defined the compressibility factor in Eq. 3.6,

$$Z = \frac{Pv}{RT}$$

We then proceeded to develop the generalized compressibility chart, presented in Appendix Fig. D.1 in terms of the reduced pressure and temperature. The generalized chart does not apply specifically to any one substance, but is instead an approximate relation that is reasonably accurate for many substances, especially those that are fairly simple in molecular structure. In this sense, the generalized compressibility chart can be viewed as one aspect of generalized behavior of substances, and also as a graphical form of equation of state representing real behavior of gases and liquids over a broad range of variables.

To gain additional insight into the behavior of gases at low density, let us examine the low-pressure portion of the generalized compressibility chart in greater detail. This behavior is as shown in Fig. 13.6. The isotherms are essentially straight lines in this region, and their slope is of particular importance. Note that the slope increases as T_r increases until a maximum value is reached at a T_r of about 5, and then the slope decreases toward the $Z = 1$ line for higher temperatures. That single temperature, about 2.5 times the critical temperature, for which

$$\lim_{P \to 0}\left(\frac{\partial Z}{\partial P}\right)_T = 0 \tag{13.46}$$

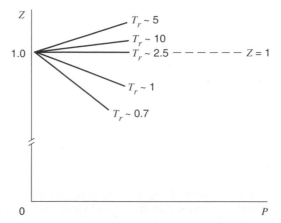

FIGURE 13.6 Low-pressure region of compressibility chart.

is defined as the Boyle temperature of the substance. This is the only temperature at which a gas really behaves exactly as an ideal gas at low, but finite pressures, since all other isotherms go to zero pressure on Fig. 13.6 with a nonzero slope. To amplify this point, let us consider the residual volume α,

$$\alpha = \frac{\overline{R}T}{P} - \overline{v} \tag{13.47}$$

Multiplying this equation by P we have

$$\alpha P = \overline{R}T - P\overline{v}$$

Thus the quantity αP is the difference between $\overline{R}T$ and $P\overline{v}$. Now as $P \to 0$, $P\overline{v} \to \overline{R}T$. However, it does not necessarily follow that $\alpha \to 0$ as $P \to 0$. Instead, it is only required that α remain finite. The derivative in Eq. 13.46 can be written as

$$\lim_{P \to 0} \left(\frac{\partial Z}{\partial P} \right)_T = \lim_{P \to 0} \left(\frac{Z-1}{P-0} \right)$$

$$= \lim_{P \to 0} \frac{1}{\overline{R}T} \left(\overline{v} - \frac{\overline{R}T}{P} \right)$$

$$= -\frac{1}{\overline{R}T} \lim_{P \to 0}(\alpha) \tag{13.48}$$

from which we find that α tends to zero as $P \to 0$ only at the Boyle temperature, since that is the only temperature for which the isothermal slope is zero on Fig. 13.6. It is perhaps a somewhat surprising result that in the limit as $P \to 0$, $P\overline{v} \to \overline{R}T$ but in general the quantity $(\overline{R}T/P - \overline{v})$ does not go to zero, but is instead a small difference between two large values. This does have an effect on certain other properties of the gas.

The compressibility behavior of low-density gases as noted in Fig. 13.6 is the result of intermolecular interactions, and can be expressed in the form of equation of state called the virial equation, which is derived from statistical thermodynamics. The result is

$$Z = \frac{P\overline{v}}{\overline{R}T} = 1 + \frac{B(T)}{\overline{v}} + \frac{C(T)}{\overline{v}^2} + \frac{D(T)}{\overline{v}^3} + \cdots \tag{13.49}$$

where $B(T)$, $C(T)$, $D(T)$ are temperature dependent and are called virial coefficients. $B(T)$ is termed the second virial coefficient and is due to binary interactions on the molecular level. The general temperature dependence of the second virial coefficient is as shown for nitrogen in Fig. 13.7. If we multiply Eq. 13.49 by $\overline{R}T/P$, the result can be rearranged to the form

$$\frac{\overline{R}T}{P} - \overline{v} = \alpha = -B(T)\frac{\overline{R}T}{P\overline{v}} - C(T)\frac{\overline{R}T}{P\overline{v}^2} \cdots \tag{13.50}$$

In the limit, as $P \to 0$,

$$\lim_{P \to 0} \alpha = -B(T) \tag{13.51}$$

and we conclude from Eqs. 13.46 and 13.48 that the single temperature at which $B(T) = 0$, Fig. 13.7, is the Boyle temperature. The second virial coefficient can be viewed as the first-order correction for nonideality of the gas, and consequently becomes of considerable importance and interest. In fact, the low-density behavior of the isotherms shown in Fig. 13.6 is directly attributable to the second virial coefficient.

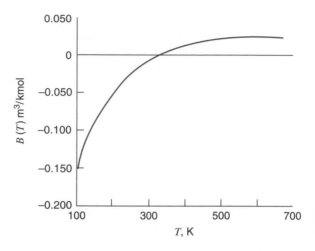

FIGURE 13.7 The second virial coefficient for nitrogen.

Another aspect of generalized behavior of gases is the behavior of isotherms in the vicinity of the critical point. If we plot experimental data on P–v coordinates, it is found that the critical isotherm is unique in that it goes through a horizontal inflection point at the critical point as shown in Fig. 13.8. Mathematically, this means that the first two derivatives are zero at the critical point,

$$\left(\frac{\partial P}{\partial v}\right)_{T_c} = 0 \qquad \text{at C.P.} \tag{13.52}$$

$$\left(\frac{\partial^2 P}{\partial v^2}\right)_{T_c} = 0 \qquad \text{at C.P.} \tag{13.53}$$

a feature that is used to constrain many equations of state.

The Joule–Thomson coefficient, defined in connection with the throttling process, can be expressed in terms of P–v–T behavior as

$$\mu_J = \left(\frac{\partial T}{\partial P}\right)_h = \frac{T\left(\dfrac{\partial v}{\partial T}\right)_P - v}{C_p} = \frac{RT^2}{PC_p}\left(\frac{\partial Z}{\partial T}\right)_P \tag{13.54}$$

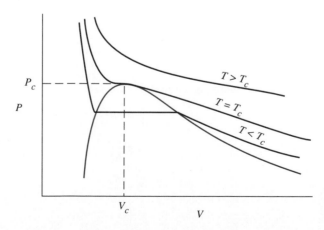

FIGURE 13.8 Plot of isotherms in the region of the critical point on pressure–volume coordinates for a typical pure substance.

From Eq. 13.54 and Fig. 13.6 it is evident that at low pressure μ_J is zero only at one temperature, T_r about 5, and is positive at lower temperatures and negative at higher temperatures. By examining higher density P–v–T data, we can determine the locus of points at which μ_J is zero, which is called the Joule–Thomson inversion curve. The result is as shown in Fig. 13.9. Inside the dome μ_J is positive, which means that a substance cools upon being throttled to a lower pressure. Outside the dome μ_J is negative, and the temperature increases during a throttling process.

To this point, we have discussed the generalized compressibility chart, a graphical form of equation of state, and the virial equation, a theoretically founded equation of state. We now proceed to discuss other analytical equations of state, which may be either generalized behavior in form, or empirical equations, relying on specific P−v−T data for their constants. The oldest generalized equation, the van de Waals equation, was presented in 1873 as a semitheoretical improvement over the ideal gas model. The van der Waals equation of state has two constants and is written as

$$P = \frac{RT}{v-b} - \frac{a}{v^2} \tag{13.55}$$

The constant b is intended to correct for the volume occupied by the molecules, and the term a/v^2 is a correction that accounts for the intermolecular forces of attraction. As might be expected in the case of a generalized equation, the constants a and b are evaluated from the general behavior of gases. In particular these constants are evaluated by noting that the critical isotherm passes through a point of inflection at the critical point, and that the slope is zero at this point. Thus, for the van der Waals equation of state we have

$$\left(\frac{\partial P}{\partial v}\right)_T = -\frac{RT}{(v-b)^2} + \frac{2a}{v^3} \tag{13.56}$$

$$\left(\frac{\partial^2 P}{\partial v^2}\right)_T = \frac{2RT}{(v-b)^3} - \frac{6a}{v^4} \tag{13.57}$$

Since both of these derivatives are equal to zero at the critical point we can write

$$-\frac{RT_c}{(v_c-b)^2} + \frac{2a}{v_c^3} = 0$$

$$\frac{2RT_c}{(v_c-b)^3} - \frac{6a}{v_c^4} = 0 \tag{13.58}$$

$$P_c = \frac{RT_c}{(v_c-b)} - \frac{a}{v_c^2}$$

Solving these three equations we find

$$v_c = 3b$$

$$a = \frac{27}{64} \frac{R^2 T_c^2}{P_c} \tag{13.59}$$

$$b = \frac{RT_c}{8P_c}$$

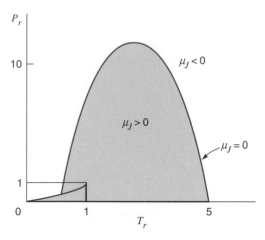

FIGURE 13.9 Joule–Thomson inversion curve.

The compressibility factor at the critical point for the van der Waals equation is

$$Z_c = \frac{P_c v_c}{RT_c} = \frac{3}{8}$$

which is considerably higher than the actual value for any substance.

A simple equation of state that is considerably more accurate than the van der Waals equation is that proposed by Redlich and Kwong in 1949.

$$P = \frac{\overline{R}T}{\overline{v}-b} - \frac{a}{\overline{v}(\overline{v}+b)T^{1/2}} \tag{13.60}$$

with

$$a = 0.427\ 48\frac{\overline{R}^2 T_c^{5/2}}{P_c} \tag{13.61}$$

$$b = 0.086\ 64\frac{\overline{R}T_c}{P_c} \tag{13.62}$$

The numerical values in the constants have been determined by a procedure similar to that followed in the van der Waals equation. Because of its simplicity this equation could not be expected to be sufficiently accurate to find use in the calculation of precision tables of thermodynamic properties. It has, however, been used frequently for mixture calculations and phase equilibrium correlations with reasonably good success. A number of modified versions of this equation have also been utilized in recent years.

One of the best known empirical equations of state is the Benedict-Webb-Rubin equation, often termed the BWR equation. The original equation, proposed in 1940, containing eight empirical constants, was given in Chapter 3 as Eq. 3.7. The constants for a number of substances are given in Appendix Table D.2. This equation, and particularly a number of modifications to it, have been widely used over the years.

One particularly interesting modification of the BWR equation of state is the Lee–Kesler equation, which was proposed in 1975. This equation has 12 constants and is written in terms of generalized properties as

$$Z = \frac{P_r v_r'}{T_r} = 1 + \frac{B}{v_r'} + \frac{C}{v_r'^2} + \frac{D}{v_r'^5} + \frac{c_4}{T_r^3 v_r'^2}\left(\beta + \frac{\gamma}{v_r'^2}\right)\exp\left(-\frac{\gamma}{v_r'^2}\right)$$

$$B = b_1 - \frac{b_2}{T_r} - \frac{b_3}{T_r^2} - \frac{b_4}{T_r^3}$$

$$C = c_1 - \frac{c_2}{T_r} + \frac{c_3}{T_r^3} \tag{13.63}$$

$$D = d_1 + \frac{d_2}{T_r}$$

in which the variable v_r' is not the true reduced specific volume, but is instead defined as

$$v_r' = \frac{v}{RT_c/P_c} \tag{13.64}$$

Empirical constants for simple fluids for this equation are also given the Appendix Table D.3.

13.8 THE GENERALIZED CHART FOR CHANGES OF ENTHALPY AT CONSTANT TEMPERATURE

In Section 13.4, Eq. 13.31 was derived for the change of enthalpy at constant temperature.

$$(h_2 - h_1)_T = \int_1^2 \left[v - T\left(\frac{\partial v}{\partial T}\right)_P\right]dP_T$$

This equation is appropriately used when a volume-explicit equation of state is known. Otherwise, it is more convenient to calculate the isothermal change in internal energy from Eq. 13.35

$$(u_2 - u_1)_T = \int_1^2 \left[T\left(\frac{\partial P}{\partial T}\right)_v - P\right]dv_T$$

and then calculate the change in enthalpy from its definition as

$$(h_2 - h_1) = (u_2 - u_1) + (P_2 v_2 - P_1 v_1)$$
$$= (u_2 - u_1) + RT(Z_2 - Z_1)$$

To determine the change in enthalpy behavior consistent with the generalized chart, Fig. D.1, we follow the second of these approaches, since the Lee–Kesler generalized equation of state, Eq. 13.63, is a pressure-explicit form in terms of specific volume and temperature. Equation 13.63 is expressed in terms of the compressibility factor Z, so we write

$$P = \frac{ZRT}{v}, \qquad \left(\frac{\partial P}{\partial T}\right)_v = \frac{ZR}{v} + \frac{RT}{v}\left(\frac{\partial Z}{\partial T}\right)_v$$

Therefore, substituting into Eq. 13.35, we have

$$du = \frac{RT^2}{v}\left(\frac{\partial Z}{\partial T}\right)_v dv$$

But

$$\frac{dv}{v} = \frac{dv_r'}{v_r'} \qquad \frac{dT}{T} = \frac{dT_r}{T_r}$$

so that, in terms of reduced variables,

$$\frac{1}{RT_c}du = \frac{T_r^2}{v_r'}\left(\frac{\partial Z}{\partial T_r}\right)_{v_r'} dv_r'$$

This expression is now integrated at constant temperature from any given state (P_r, v_r') to the ideal-gas limit $(P_r^* \to 0, v_r'^* \to \infty)$ (the superscript * will always denote an ideal-gas state or property), causing an internal energy change or departure from the ideal-gas value at the given state,

$$\frac{u^*-u}{RT_c} = \int_{v_r'}^{\infty} \frac{T_r^2}{v_r'}\left(\frac{\partial Z}{\partial T_r}\right)_{v_r'} dv_r' \tag{13.65}$$

The integral on the right-hand side of Eq. 13.65 can be evaluated from the Lee–Kesler equation, Eq. 13.63. The corresponding enthalpy departure at the given state (P_r, v_r') is then found from integrating Eq. 13.65 to be

$$\frac{h^*-h}{RT_c} = \frac{u^*-u}{RT_c} + T_r(1-Z) \tag{13.66}$$

Following the same procedure as for the compressibility factor, we can evaluate Eq. 13.66 with the set of Lee–Kesler simple-fluid constants to give a simple-fluid enthalpy departure. The values for the enthalpy departure are shown graphically in Fig. D.2. Use of the enthalpy departure function is illustrated in the following example.

EXAMPLE 13.5   Nitrogen is throttled from 20 MPa, −70°C, to 2 MPa in an adiabatic, steady-state, steady-flow process. Determine the final temperature of the nitrogen.

Control volume: Throttling valve.

Inlet state: P_1, T_1 known; state fixed.

Exit state: P_2 known.

Process: SSSF, throttling process.

Diagram: Figure 13.10.

Model: Generalized charts, Fig. D.2

Analysis:

First law:

$$h_1 = h_2$$

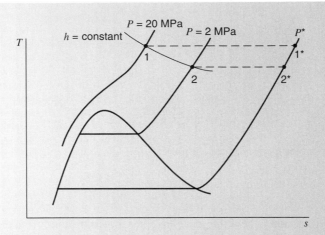

FIGURE 13.10 Sketch for Example 13.9.

Solution

Using values from Table A.2, we have

$$P_1 = 20 \text{ MPa} \qquad P_{r1} = \frac{20}{3.39} = 5.9$$

$$T_1 = 203.2 \text{ K} \qquad T_{r1} = \frac{203.2}{126.2} = 1.61$$

$$P_2 = 2 \text{ MPa} \qquad P_{r2} = \frac{2}{3.39} = 0.59$$

From the generalized charts, Fig. D.2, for the change in enthalpy at constant temperature, we have

$$\frac{h_1^* - h_1}{RT_c} = 2.1$$

$$h_1^* - h_1 = 2.1 \times 0.2968 \times 126.2 = 78.7 \text{ kJ / kg}$$

It is now necessary to assume a final temperature and to check whether the net change in enthalpy for the process is zero. Let us assume that $T_2 = 146$ K. Then the change in enthalpy between 1* and 2* can be found from the zero-pressure, specific-heat data.

$$h_1^* - h_2^* = C_{p0}(T_1^* - T_2^*) = 1.0416(203.2 - 146) = +59.6 \text{ kJ / kg}$$

(The variation in C_{p0} with temperature can be taken into account when necessary.)

We now find the enthalpy change between 2* and 2.

$$T_{r2} = \frac{146}{126.2} = 1.157 \qquad P_{r2} = 0.59$$

Therefore, from the enthalpy departure chart, Fig. D.2, at this state

$$\frac{h_2^* - h_2}{RT_c} = 0.5$$

$$h_2^* - h_2 = 0.5 \times 0.2968 \times 126.2 = 19.5 \text{ kJ / kg}$$

We now check to see whether the net change in enthalpy for the process is zero.

$$h_1 - h_2 = 0 = -(h_1^* - h_1) + (h_1^* - h_2^*) + (h_2^* - h_2)$$

$$= -78.7 + 59.6 + 19.5 \approx 0$$

It essentially checks. We conclude that the final temperature is approximately 146 K. It is interesting that the thermodynamic tables for nitrogen, Table B.6, give essentially this same value for the final temperature.

13.9 THE GENERALIZED CHART FOR CHANGES OF ENTROPY AT CONSTANT TEMPERATURE

In this section we wish to develop a generalized chart giving entropy departures from ideal gas values at a given temperature and pressure, in a manner similar to that followed for enthalpy in the previous section. Once again, we have two alternatives. From Eq. 13.36, at constant temperature,

$$ds_T = -\left(\frac{\partial v}{\partial T}\right)_P dP_T$$

which is convenient for use with a volume-explicit equation of state. The Lee–Kesler expression, Eq. 13.63, is, however, a pressure-explicit equation. It is therefore more appropriate to use Eq. 13.38, which is, along an isotherm,

$$ds_T = \left(\frac{\partial P}{\partial T}\right)_v dv_T$$

In the Lee–Kesler form, in terms of reduced properties, this equation becomes

$$\frac{ds}{R} = \left(\frac{\partial P_r}{\partial T_r}\right)_{v_r'} dv_r'$$

When this expression is integrated from a given state (P_r, v_r') to the ideal-gas limit $(P_r^* \to 0, v_r'^* \to \infty)$, there is a problem because ideal-gas entropy is a function of pressure and approaches infinity as the pressure approaches zero. We can eliminate this problem with a two-step procedure. First, the integral is taken only to a certain finite $P_r^*, v_r'^*$, which gives the entropy change

$$\frac{s_{p*}^* - s_p}{R} = \int_{v_r'}^{v_r'^*} \left(\frac{\partial P_r}{\partial T_r}\right)_{v_r'} dv_r' \qquad (13.67)$$

This integration by itself is not entirely acceptable, because it contains the entropy at some arbitrary, low reference pressure. A value for the reference pressure would have to be specified. Let us now repeat the integration over the same change of state, except this time for a hypothetical ideal gas. The entropy change for this integration is

$$\frac{s_{p*}^* - s_p^*}{R} = +\ln\frac{P}{P^*} \qquad (13.68)$$

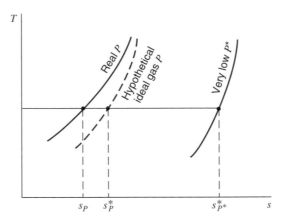

FIGURE 13.11 Real and ideal gas states and entropies.

If we now subtract Eq. 13.68 from Eq. 13.67, the result is the difference in entropy of a hypothetical ideal gas at a given state (T_r, P_r) and that of the real substance at the same state, or

$$\frac{s_p^* - s_p}{R} = -\ln\frac{P}{P^*} + \int_{v_r'}^{v_r^* \to \infty}\left(\frac{\partial P_r}{\partial T_r}\right)_{v_r'} dv_r' \tag{13.69}$$

Here the values associated with the arbitrary reference state P_r^*, $v_r'^*$, cancel out of the right-hand side of the equation. (The first term of the integral includes the term $+\ln(P/P^*)$, which cancels the other term. The three different states associated with the development of Eq. 13.69 are shown in Fig. 13.11.

The same procedure that was given in Section 13.8 for enthalpy departure values is followed for generalized entropy departure values. The Lee–Kesler simple-fluid constants are used in evaluating the integral of Eq. 13.69 and yield a simple-fluid entropy departure. The values for the entropy departure are shown graphically in Fig. D.3.

EXAMPLE 13.6   Nitrogen at 8 MPa, 150 K, is throttled to 0.5 MPa. After the gas passes through a short length of pipe, its temperature is measured and found to be 125 K. Determine the heat transfer and the change of entropy using the generalized charts. Compare these results with those obtained by using the nitrogen tables.

> *Control volume:*   Throttle and pipe.
>
> *Inlet state:* P_1, T_1 known; state fixed.
>
> *Exit state:* P_2, T_2 known; state fixed.
>
> *Process:*   SSSF.
>
> *Diagram:*   Figure 13.12.
>
> *Model:*   Generalized charts, results to be compared with those obtained with nitrogen tables.

Analysis:

There is no work done, and we neglect changes in kinetic and potential energies.

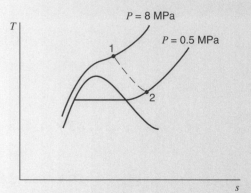

FIGURE 13.12 Sketch for Example 13.6.

Therefore, per kilogram,
First law:

$$q + h_1 = h_2$$

$$q = h_2 - h_1 = -(h_2^* - h_2) + (h_2^* - h_1^*) + (h_1^* - h_1)$$

Solution

Using values from Table A.2, we have

$$P_{r1} = \frac{8}{3.39} = 2.36 \qquad T_{r1} = \frac{150}{126.2} = 1.189$$

$$P_{r2} = \frac{0.5}{3.39} = 0.147 \qquad T_{r2} = \frac{125}{126.2} = 0.99$$

From Fig. D.2,

$$\frac{h_1^* - h_1}{RT_c} = 2.5$$

$$h_1^* - h_1 = 2.5 \times 0.2968 \times 126.2 = 93.6 \text{ kJ / kg}$$

$$\frac{h_2^* - h_2}{RT_c} = 0.15$$

$$h_2^* - h_2 = 0.15 \times 0.2968 \times 126.2 = 5.6 \text{ kJ / kg}$$

Assuming a constant specific heat for the ideal gas, we have

$$h_2^* - h_1^* = C_{p0}(T_2 - T_1) = 1.0416(125 - 150) = -26.0 \text{ kJ / kg}$$

$$q = -5.6 - 26.0 + 93.6 = 62.0 \text{ kJ / kg}$$

From the nitrogen tables, Table B.6, we can find the change of enthalpy directly.

$$q = h_2 - h_1 = 123.77 - 61.92 = 61.85 \text{ kJ / kg}$$

To calculate the change of entropy using the generalized charts, we proceed as follows:

$$s_2 - s_1 = -(s^*_{P_2,T_2} - s_2) + (s^*_{P_2,T_2} - s^*_{P_1,T_1}) + (s^*_{P_1,T_1} - s_1)$$

From Fig. D.3

$$\frac{s^*_{P_1,T_1} - s_{P_1,T_1}}{R} = 1.6$$

$$s^*_{P_1,T_1} - s_{P_1,T_1} = 1.6 \times 0.2968 = 0.475 \text{ kJ / kg K}$$

$$\frac{s^*_{P_2,T_2} - s_{P_2,T_2}}{R} = 0.1$$

$$s^*_{P_2,T_2} - s_{P_2,T_2} = 0.1 \times 0.2968 = 0.0297 \text{ kJ / kg K}$$

Assuming a constant specific heat for the ideal gas, we have

$$s^*_{P_2,T_2} - s^*_{P_1,T_1} = C_{p0} \ln \frac{T_2}{T_1} - R \ln \frac{P_2}{P_1}$$

$$= 1.0416 \ln \frac{125}{150} - 0.2968 \ln \frac{0.5}{8}$$

$$= 0.6330 \text{ kJ / kg K}$$

$$s_2 - s_1 = -0.0297 + 0.6330 + 0.475$$

$$= 1.078 \text{ kJ / kg K}$$

From the nitrogen tables, Table B.6,

$$s_2 - s_1 = -5.4282 - 4.3522 = 1.0760 \text{ kJ / kg K}$$

13.10 THE PROPERTY RELATION FOR MIXTURES

In Chapter 12 the partial molal property was briefly introduced, but, the consideration of mixtures was limited to ideal gases. There was no need at that point for further expansion of the subject. We now continue this subject with a view toward developing the property relations for mixtures. This subject will be particularly relevant to our consideration of chemical equilibrium in Chapter 15.

For a mixture, any extensive property X is a function of the temperature and pressure of the mixture and the number of moles of each component. Thus, for a mixture of two components,

$$X = f(T, P, n_A, n_B)$$

Therefore,

$$dX_{T,P} = \left(\frac{\partial X}{\partial n_A}\right)_{T,P,n_B} dn_A + \left(\frac{\partial X}{\partial n_B}\right)_{T,P,n_A} dn_B \qquad (13.70)$$

Since at constant temperature and pressure an extensive property is directly proportional to the mass, Eq. 13.70 can be integrated to give

$$X_{T,P} = \overline{X}_A n_A + \overline{X}_B n_B \qquad (13.71)$$

where

$$\overline{X}_A = \left(\frac{\partial X}{\partial n_A}\right)_{T,P,n_B}, \qquad \overline{X}_B = \left(\frac{\partial X}{\partial n_B}\right)_{T,P,n_A}$$

Here $\overline{X}$ is defined as the partial molal property for a component in a mixture. It is particularly important to note that the partial molal property is defined under conditions of constant temperature and pressure. Note that Eq. 13.70 has the same form as Eq. 12.6.

The partial molal property is particularly significant when a mixture undergoes a chemical reaction. Suppose a mixture consists of components A and B, and a chemical reaction takes place so that the number of moles of A is changed by dn_A and the number of moles of B by dn_B. The temperature and the pressure remain constant. What is the change in internal energy of the mixture during this process? From Eq. 13.71 we conclude that

$$dU_{T,P} = \overline{U}_A dn_A + \overline{U}_B dn_B \qquad (13.72)$$

where $\overline{U}_A$ and $\overline{U}_B$ are the partial molal internal energy of A and B, respectively. Equation 13.72 suggests that the partial molal internal energy of each component can also be defined as the internal energy of the component as it exists in the mixture.

In Section 13.3 we considered a number of property relations for systems of fixed mass such as

$$dU = T\,dS - P\,dV$$

We note that in this equation, temperature is the intensive property or potential function associated with entropy, and pressure is the intensive property associated with volume. Suppose we have a chemical reaction such as described in the last paragraph. How would we modify this property relation for this situation? Intuitively we might write the equation

$$dU = T\,dS - P\,dV + \mu_A\,dn_A + \mu_B\,dn_B \qquad (13.73)$$

where μ_A is the intensive property or potential function associated with n_A, and similarly μ_B for n_B. This potential function is called the chemical potential.

To derive an expression for this chemical potential, we examine Eq. 13.73 and conclude that it might be reasonable to write an expression for U in the form

$$U = f(S, V, n_A, n_B)$$

Therefore,

$$dU = \left(\frac{\partial U}{\partial S}\right)_{V,n_A,n_B} dS + \left(\frac{\partial U}{\partial V}\right)_{S,n_A,n_B} dV + \left(\frac{\partial U}{\partial n_A}\right)_{S,V,n_B} dn_A + \left(\frac{\partial U}{\partial n_B}\right)_{S,V,n_A} dn_B$$

Since the expressions

$$\left(\frac{\partial U}{\partial S}\right)_{V,n_A,n_B} \quad \text{and} \quad \left(\frac{\partial U}{\partial V}\right)_{S,n_A,n_B}$$

imply constant composition, it follows from Eq. 13.24 that

$$\left(\frac{\partial U}{\partial S}\right)_{V,n_A,n_B} = T \quad \text{and} \quad \left(\frac{\partial U}{\partial V}\right)_{S,n_A,n_B} = -P$$

Thus

$$dU = T\,dS - P\,dV + \left(\frac{\partial U}{\partial n_A}\right)_{S,V,n_B} dn_A + \left(\frac{\partial U}{\partial n_B}\right)_{S,V,n_A} dn_B \qquad (13.74)$$

On comparing this equation with Eq. 13.73, we find that the chemical potential can be defined by the relation

$$\mu_A = \left(\frac{\partial U}{\partial n_A}\right)_{S,V,n_B}, \qquad \mu_B = \left(\frac{\partial U}{\partial n_B}\right)_{S,V,n_A} \qquad (13.75)$$

We can also relate the chemical potential to the partial molal Gibbs function. We proceed as follows.

$$G = U + PV - TS$$
$$dG = dU + P\,dV + V\,dP - T\,dS - S\,dT$$

Substituting Eq. 13.73 into this relation, we have

$$dG = -S\,dT + V\,dP + \mu_A\,dn_A + \mu_B\,dn_B \qquad (13.76)$$

This equation suggests that we write an expression for G in the following form.

$$G = f(T, P, n_A, n_B)$$

Proceeding as we did for a similar expression for internal energy, Eq. 13.78, we have

$$dG = \left(\frac{\partial G}{\partial T}\right)_{P,n_A,n_B} dT + \left(\frac{\partial G}{\partial P}\right)_{T,n_A,n_B} dP + \left(\frac{\partial G}{\partial n_A}\right)_{T,P,n_B} dn_A + \left(\frac{\partial G}{\partial n_B}\right)_{T,P,n_A} dn_B$$

$$= -S\,dT + V\,dP + \left(\frac{\partial G}{\partial n_A}\right)_{T,P,n_B} dn_A + \left(\frac{\partial G}{\partial n_B}\right)_{T,P,n_A} dn_B$$

When this equation is compared with Eq. 13.76, it follows that

$$\mu_A = \left(\frac{\partial G}{\partial n_A}\right)_{T,P,n_B}, \qquad \mu_B = \left(\frac{\partial G}{\partial n_B}\right)_{T,P,n_A}$$

Because partial molal properties are defined at constant temperature and pressure, the quantities $(G/n_A)_{T,P,n_B}$ and $(G/n_B)_{T,P,n_A}$ are the partial molal Gibbs functions for the two components. That is, the chemical potential is equal to the partial molal Gibbs function.

$$\mu_A = \overline{G}_A = \left(\frac{\partial G}{\partial n_A}\right)_{T,P,n_B}, \qquad \mu_B = \overline{G}_B = \left(\frac{\partial G}{\partial n_B}\right)_{T,P,n_A} \qquad (13.77)$$

Although μ can also be defined in terms of other properties, such as in Eq. 13.75, this expression is not the partial molal internal energy, since the pressure and temperature are not constant in this partial derivative. The partial molal Gibbs function is an extremely important property in the thermodynamic analysis of chemical reactions, for at constant temperature and pressure (the conditions under which many chemical reactions occur) it is a measure of the chemical potential or the driving force that tends to make a chemical reaction take place.

13.11 PSEUDOPURE SUBSTANCE MODELS FOR REAL GAS MIXTURES

A basic prerequisite to the treatment of real gas mixtures in terms of pseudopure substance models is the concept and use of appropriate reference states. As an introduction to this topic, let us consider several preliminary reference state questions for a pure substance undergoing a change of state, for which it is desired to calculate the entropy change. We can express the entropy at the initial state 1 and also at the final state 2 in terms of a reference state 0, in a manner similar to that followed when dealing with the generalized-charts corrections. It follows that

$$s_1 = s_0 + (s^*_{P_0 T_0} - s_0) + (s^*_{P_1 T_1} - s^*_{P_0 T_0}) + (s_1 - s^*_{P_1 T_1}) \qquad (13.78)$$

$$s_2 = s_0 + (s^*_{P_0 T_0} - s_0) + (s^*_{P_2 T_2} - s^*_{P_0 T_0}) + (s_2 - s^*_{P_2 T_2}) \qquad (13.79)$$

These are entirely general expressions for the entropy at each state in terms of an arbitrary reference state value and a set of consistent calculations from that state to the actual desired state. One simplification of these equations would result from choosing the reference state to be a hypothetical ideal gas state at P_0 and T_0, thereby making the term

$$(s^*_{P_0 T_0} - s_0) = 0 \qquad (13.80)$$

in each equation, which results in

$$s_0 = s^*_0 \qquad (13.81)$$

It should be apparent that this choice is a reasonable one, since whatever value is chosen for the correction term, Eq. 13.80, it will cancel out of the two equations when the change $s_2 - s_1$ is calculated, and the simplest value to choose is zero. In a similar manner, the simplest value to choose for the ideal gas reference value, Eq. 13.81, is zero, and we would commonly do that if there are no restrictions on choice, such as occur in the case of a chemical reaction.

Another point to be noted concerning reference states is related to the choice of P_0 and T_0, and for this purpose, let us substitute Eqs. 13.80 and 13.81 into Eqs. 13.78 and 13.79, and also assume constant specific heat, such that those equations can be written in the form

$$s_1 = s^*_0 + C_{po} \ln\left(\frac{T_1}{T_0}\right) - R \ln\left(\frac{P_1}{P_0}\right) + (s_1 - s^*_{P_1 T_1}) \qquad (13.82)$$

$$s_2 = s^*_0 + C_{po} \ln\left(\frac{T_2}{T_0}\right) - R \ln\left(\frac{P_2}{P_0}\right) + (s_2 - s^*_{P_2 T_2}) \qquad (13.83)$$

Since the choice for P_0 and T_0 is arbitrary if there are no restrictions, such as would be the case with chemical reactions, it should be apparent from examining Eqs. 13.82 and 13.83 that the simplest choice would be for

$$P_0 = P_1 \text{ or } P_2 \qquad T_0 = T_1 \text{ or } T_2$$

It should be emphasized that inasmuch as the reference state was chosen as a hypothetical ideal gas at P_0, T_0, Eq. 13.80, it is immaterial how the real substance behaves at that pressure and temperature. As a result, there is no need to select a low value for the reference state pressure P_0.

Let us now extend these reference state developments to include real gas mixtures. Consider the mixing process shown in Fig. 13.13, with the states and amounts of each substance as given on the diagram. Proceeding with entropy expressions as was done

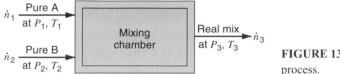

FIGURE 13.13 Example of mixing process.

above, we have

$$\bar{s}_1 = \bar{s}_{A_0}^* + \overline{C}_{po_A} \ln\left(\frac{T_1}{T_0}\right) - \overline{R} \ln\left(\frac{P_1}{P_0}\right) + (\bar{s}_1 - \bar{s}_{P_1 T_1}^*)_A \qquad (13.84)$$

$$\bar{s}_2 = \bar{s}_{B_0}^* + \overline{C}_{po_B} \ln\left(\frac{T_2}{T_0}\right) - \overline{R} \ln\left(\frac{P_2}{P_0}\right) + (\bar{s}_2 - \bar{s}_{P_2 T_2}^*)_B \qquad (13.85)$$

$$\bar{s}_3 = \bar{s}_{\text{mix}_0}^* + \overline{C}_{po_{\text{mix}}} \ln\left(\frac{T_3}{T_0}\right) - \overline{R} \ln\left(\frac{P_3}{P_0}\right) + (\bar{s}_3 - \bar{s}_{P_3 T_3}^*)_{\text{mix}} \qquad (13.86)$$

in which

$$\bar{s}_{\text{mix}_0}^* = y_A \bar{s}_{A_0}^* + y_B \bar{s}_{B_0}^* - \overline{R}(y_A \ln y_A + y_B \ln y_B) \qquad (13.87)$$

$$\overline{C}_{po_{\text{mix}}} = y_A \overline{C}_{po_A} + y_B \overline{C}_{po_B} \qquad (13.88)$$

We note that when Eqs. 13.84–13.86 are substituted into the equation for the entropy change,

$$n_3 \bar{s}_3 - n_1 \bar{s}_1 - n_2 \bar{s}_2$$

the arbitrary reference values s_{A0}^*, s_{B0}^*, P_0, and T_0 all cancel out of the result, which is, of course, necessary in view of their arbitrary nature. An ideal gas entropy of mixing expression, the final term in Eq. 13.87, remains in the result, establishing, in effect, the mixture reference value relative to its components. The remarks made earlier concerning the choices for reference state and the reference state entropies apply in this situation as well.

To summarize the development to this point, we find that a calculation of real mixture properties, as, for example, using Eq. 13.86, requires the establishment of a hypothetical ideal gas reference state, a consistent ideal gas calculation to the conditions of the real mixture, and finally a correction that accounts for the real behavior of the mixture at that state. We note that this last term is the only place that the real behavior is introduced, and this is therefore the term that must be calculated by the pseudopure substance model to be used.

In treating a real gas mixture as a pseudopure substance, there are two approaches that we will follow to represent the $P-v-T$ behavior: use of the generalized charts, and use of an analytical equation of state. With the generalized charts, we need to have a model that provides a set of pseudocritical pressure and temperature in terms of the mixture component values. Many such models have been proposed and utilized over the years, but the simplest is that suggested by W. B. Kay in 1936, in which

$$(P_c)_{\text{mix}} = \sum_i y_i P_{ci}, \qquad (T_c)_{\text{mix}} = \sum_i y_i T_{ci} \qquad (13.89)$$

This is the only pseudocritical model that we will consider in this chapter. Other models are somewhat more complicated to evaluate and use, but considerably more accurate.

The other approach to be considered is that of using an analytical equation of state, in which the equation for the mixture must be developed from that for the components. In other words, for an equation in which the constants are known for each of the components, we must develop a set of empirical combining rules that will then give a set of constants for the mixture as though it were a pseudopure substance. This problem has been studied for many equations of state, using experimental data for the real gas mixtures, and various empirical rules have been proposed. For example, for both the van der Waals equation, Eq. 13.55, and the Redlich–Kwong equation, Eq. 13.60, the two pure substance constants a and b are commonly combined according to the relations

$$a_m = \left(\sum_i y_i a_i^{1/2} \right)^2 \qquad b_m = \sum_i y_i b_i \qquad (13.90)$$

The following example illustrates the use of these two approaches to treating real gas mixtures as pseudopure substances.

EXAMPLE 13.7 A mixture of 59.39% CO_2 and 40.61% CH_4 (mole basis) is maintained at 310.94 K, 86.19 bar, at which condition the specific volume has been measured as 0.2205 m^3/kmol. Calculate the percent deviation if the specific volume had been calculated by (a) Kay's rule and (b) van der Waals' equation of state.

  Control mass: Gas mixture.

  State: P, v, T known.

  Model: (a) Kay's rule. (b) van der Waals' equation.

Solution

(a) For convenience, let

$$CO_2 = A \qquad CH_4 = B$$

Then

$$T_{c_A} = 304.1 \text{ K} \qquad P_{cA} = 7.38 \text{ MPa}$$
$$T_{c_B} = 190.4 \text{ K} \qquad P_{cB} = 4.60 \text{ MPa}$$

For Kay's rule, Eq. 13.89,

$$T_{c_m} = \sum_i y_i T_{c_i} = y_A T_{c_A} + y_B T_{c_B}$$
$$= 0.5939(304.1) + 0.4061(190.4)$$
$$= 257.9 \text{ K}$$
$$P_{c_m} = \sum_i y_i P_{c_i} = y_A P_{c_A} + y_B P_{c_B}$$
$$= 0.5939(7.38) + 0.4061(4.60)$$
$$= 6.257 \text{ MPa}$$

Therefore, the pseudoreduced properties of the mixture are

$$T_{r_m} = \frac{T}{T_{c_m}} = \frac{310.94}{257.9} = 1.206$$

$$P_{r_m} = \frac{P}{P_{c_m}} = \frac{8.619}{6.251} = 1.379$$

From the generalized chart, Fig. D.1

$$Z_m = 0.7$$

and

$$\bar{v} = \frac{Z_m \bar{R} T}{P} = \frac{0.7 \times 8.3145 \times 310.94}{8619}$$

$$= 0.21 \; \text{m}^3 / \text{kmol}$$

The percent deviation from the experimental value is

$$\text{Percent deviation} = \left(\frac{0.2205 - 0.21}{0.2205} \right) \times 100 = 4.8\%$$

The major factor contributing to this 5% error is the use of the linear Kay's rule pseudocritical model, Eq. 13.89. Use of an accurate pseudocritical model and the generalized chart would reduce the error to approximately 1%.

(b) For van der Waals' equation, the pure substance constants are

$$a_A = \frac{27 \bar{R}^2 T_{cA}^2}{64 P_{cA}} = 365.454 \frac{\text{kPa m}^6}{\text{kmol}^2}$$

$$b_A = \frac{\bar{R} T_{cA}}{8 P_{cA}} = 0.042 \, 83 \; \text{m}^3 / \text{kmol}$$

and

$$a_B = \frac{27 \bar{R}^2 T_{cB}^2}{64 P_{cB}} = 229.843 \frac{\text{kPa m}^6}{\text{kmol}^2}$$

$$b_B = \frac{\bar{R} T_{cB}}{8 P_{cB}} = 0.043 \, 02 \; \text{m}^3 / \text{kmol}$$

Therefore, for the mixture, from Eq. 13.90,

$$a_m = \left(y_A \sqrt{a_A} + y_B \sqrt{a_B} \right)^2$$

$$= \left(0.5939 \sqrt{365.454} + 0.4061 \sqrt{229.843} \right)^2$$

$$= 306.607 \frac{\text{kPa m}^6}{\text{kmol}^2}$$

$$b_m = y_A b_A + y_B b_B$$

$$= (0.5939 \times 0.042 \, 83 + 0.4061 \times 0.04302)$$

$$= 0.042 \, 91 \; \text{m}^3 / \text{kmol}$$

The equation of state for the mixture of this composition is

$$P = \frac{\bar{R} T}{\bar{v} - b_m} - \frac{a_m}{\bar{v}^2}$$

$$8619 = \frac{8.3145 \times 310.94}{\bar{v} - 0.042 \, 91} - \frac{306.607}{\bar{v}^2}$$

Solving for $\bar{v}$ by trial and error,

$$\bar{v} = 0.2063 \text{ m}^3 / \text{kmol}$$

$$\text{Percent deviation} = \left(\frac{0.2205 - 0.2063}{0.2205} \right) \times 100 = 6.4\%$$

As a point of interest from the ideal-gas law, $\bar{v} = 0.300$ m³/kmol, which is a deviation of 36% from the measured value. Also, if we use the Redlich–Kwong equation of state, and follow the same procedure as for the van der Waals equation, the calculated specific volume of the mixture is 0.2127 m³/kmol, which is in error by 3.5%.

We must be careful not to draw too general a conclusion from the results of this example. We have calculated percent deviation in v at only a single point for only one mixture. We do note, however, that the various methods used give quite different results. From a more general study of these models for a number of mixtures, we find that the results found here are fairly typical, at least qualitatively. Kay's rule is very useful because it is fairly accurate and yet relatively simple. The van der Waals equation is too simplified an expression to accurately represent P–v–T behavior, but it is useful to demonstrate the procedures followed in utilizing more complex analytical equations of state. The Redlich–Kwong equation is considerably better, and is still relatively simple to use.

As noted in the example, the more sophisticated generalized behavior models and empirical equations of state will represent mixture P–v–T behavior to within about 1 percent over a wide range of density, but they are of course more difficult to use than the methods considered in Example 13.7. The generalized models have the advantage of being easier to use, and they are suitable for hand computations. Calculations with the complex empirical equations of state become very involved, but have the advantage of expressing the P–v–T composition relations in analytical form, which is of great value when using a digital computer for such calculations.

PROBLEMS

13.1 A special application requires R-12 at −140°C. It is known that the triple-point temperature is −157°C. Find the pressure and specific volume of the saturated vapor at the required condition.

13.2 In a Carnot heat engine, the heat addition changes the working fluid from saturated liquid to saturated vapor at T, P. The heat rejection process occurs at lower temperature and pressure $(T − \Delta T)$, $(P − \Delta P)$. The cycle takes place in a piston cylinder arrangement where the work is boundary work. Apply both the first and second law with simple approximations for the integral equal to work. Then show that the relation between ΔP and ΔT results in the Clapeyron equation in the limit $\Delta T \to dT$.

13.3 Ice (solid water) at −3°C, 100 kPa is compressed isothermally until it becomes liquid. Find the required pressure.

13.4 Calculate the values h_{fg} and s_{fg} for nitrogen at 70 K and at 110 K from the Clapeyron equation, using the necessary pressure and specific volume values from Table B.6.1.

13.5 Using thermodynamic data for water from Tables B.1.1 and B.1.5, estimate the freezing temperature of liquid water at a pressure of 30 MPa.

13.6 Helium boils at 4.22 K at atmospheric pressure, 101.3 kPa, with h_{fg} = 83.3 kJ/kmol. By pumping a vacuum over liquid helium, the pressure can be lowered and it may then boil at a lower temperature. Estimate the necessary pressure to produce a boiling temperature of 1 K and one of 0.5 K.

13.7 A certain refrigerant vapor enters a SSSF constant pressure condenser at 150 kPa, 70°C, at a rate of 1.5 kg/s, and it exits as saturated liquid. Calculate the rate of heat transfer from the condenser. It may be assumed that the vapor is an ideal gas, and also that at saturation, $v_f \ll v_g$. The following is known

$$\ln P_g = 8.15 - 1000/T \qquad C_{po} = 0.7 \text{ kJ/kg K}$$

with pressure in kPa and temperature in K. The molecular weight is 100.

13.8 A container has a double wall where the wall cavity is filled with carbon dioxide at room temperature and pressure. When the container is filled with a cryogenic liquid at 100 K the carbon dioxide will freeze so the wall cavity has a mixture of solid and vapor carbon dioxide at the sublimation pressure. Assume that we do not have data for CO_2 at 100 K, but it is known that at −90°C: P_{sub} = 38.1 kPa, h_{ig} = 574.5 kJ/kg. Estimate the pressure in the wall cavity at 100 K.

13.9 Small solid particles formed in combustion should be investigated. We would like to know the sublimation pressure as a function of temperature. The only information available is T, h_{fg} for boiling at 101.3 kPa and T, h_{if} for melting at 101.3 kPa. Develop a procedure that will allow a determination of the sublimation pressure, $P_{sub}(T)$.

13.10 Derive expressions for $(\partial T/\partial v)_u$ and for $(\partial h/\partial s)_v$ that do not contain the properties h, u, or s.

13.11 Derive expressions for $(\partial h/\partial v)_T$ and for $(\partial h/\partial T)_v$ that do not contain the properties h, u, or s.

13.12 Develop an expression for the variation in temperature with pressure in a constant entropy process, $(\partial T/\partial P)_s$, that only includes the properties P–v–T and the specific heat, C_p.

13.13 Determine the volume expansivity, α_P, and the isothermal compressibility, β_T, for water at 20°C, 5 MPa and at 300°C, 15 MPa using the steam tables.

13.14 Sound waves propagate through media as pressure waves that cause the media to go through isentropic compression and expansion processes. The speed of sound c is defined by $c^2 = (\partial P/\partial \rho)_s$ and it can be related to the adiabatic compressibility, which for liquid ethanol at 20°C is 940 μm^2/N. Find the speed of sound at this temperature.

13.15 Consider the speed of sound as defined in Problem 13.14. Calculate the speed of sound for liquid water at 20°C, 2.5 MPa and for water vapor at 200°C, 300 kPa using the steam tables.

13.16 Find the speed of sound for air at 20°C, 100 kPa using the definition in Problem 13.14 and relations for polytropic processes in ideal gases.

13.17 A cylinder fitted with a piston contains liquid methanol at 20°C, 100 kPa and volume 10 L. The piston is moved, compressing the methanol to 20 MPa at constant temperature. Calculate the work required for this process. The isothermal compressibility of liquid methanol at 20°C is 1220 $\mu m^2/N$.

13.18 A piston/cylinder contains 5 kg of butane gas at 500 K, 5 MPa. The butane expands in a reversible polytropic process with polytropic exponent, $n = 1.05$, until the final pressure is 3 MPa. Determine the final temperature and the work done during the process.

13.19 Show that the two expressions for the Joule–Thomson coefficient μ_J given by Eq. 13.54 are valid.

13.20 A 200-L rigid tank contains propane at 9 MPa, 280°C. The propane is then allowed to cool to 50°C as heat is transferred with the surroundings. Determine the quality at the final state and the mass of liquid in the tank, using the generalized compressibility chart, Fig. D.1.

13.21 A rigid tank contains 5 kg of ethylene at 3 MPa, 30°C. It is cooled until the ethylene reaches the saturated vapor curve. What is the final temperature?

13.22 Two uninsulated tanks of equal volume are connected by a valve. One tank contains a gas at a moderate pressure P_1, and the other tank is evacuated. The valve is opened and remains open for a long time. Is the final pressure P_2 greater than, equal to, or less than $P_1/2$?

13.23 Show that van der Waals equation can be written as a cubic equation in the compressibility factor involving the reduced pressure and reduced temperature as

$$Z^3 - \left(\frac{P_r}{8T_r} + 1\right)Z^2 + \left(\frac{27P_r}{64T_r^2}\right)Z - \frac{27P_r^2}{512T_r^3} = 0$$

13.24 Develop expressions for isothermal changes in enthalpy and in entropy for both van der Waals equation and Redlich–Kwong equation of state.

13.25 Determine the reduced Boyle temperature as predicted by an equation of state (the experimentally observed value is about 2.5), using the van der Waals equation and the Redlich–Kwong equation. *Note:* It is helpful to use Eqs. 13.47 and 13.48 in addition to Eq. 13.46.

13.26 Consider a straight line connecting the point $P = 0$, $Z = 1$ to the critical point $P = P_c$, $Z = Z_c$ on a Z versus P compressibility diagram. This straight line will be tangent to one particular isotherm at low pressure. (The experimentally determined value is about 0.8 T_c). Determine what value of reduced temperature is predicted by an equation of state, using the van der Waals equation and the Redlich–Kwong equation. See also note for Problem 13.25.

13.27 Determine the second virial coefficient $B(T)$ using the van der Waals equation and the Redlich–Kwong equation of state. Find also its value at the critical temperature (the experimentally observed value is about $-0.34\ RT_c/P_c$).

13.28 One early attempt to improve on the van der Waals equation of state was an expression of the form

$$P = \frac{RT}{v - b} - \frac{a}{v^2 T}$$

Solve for the constants a, b, and v_c using the same procedure as for the van der Waals equation.

13.29 Use the equation of state from the previous problem and determine the Boyle temperature.

13.30 Calculate the difference in internal energy of the ideal-gas value and the real-gas value for carbon dioxide at the state 20°C, 1 MPa, as determined using the virial equation of state, including second virial coefficient terms. For carbon dioxide we have: $B = -0.128$ m^3/kmol, $T(dB/dT) = 0.266$ m^3/kmol, both at 20°C.

13.31 Refrigerant-123, dichlorotrifluoroethane, which is currently under development as a potential replacement for environmentally hazardous refrigerants, undergoes an isothermal SSSF process in which the R-123 enters a heat exchanger as saturated liquid at 40°C and exits at 100 kPa. Calculate the heat transfer per kilogram of R-123, using the generalized charts, Fig. D.2.

13.32 Calculate the heat transfer during the process described in Problem 13.18.

13.33 Saturated vapor R-22 at 30°C is throttled to 200 kPa in an SSSF process. Calculate the exit temperature assuming no changes in the kinetic energy, using the generalized charts, Fig. D.2 and the R-22 tables, Table B.4

13.34 A 250-L tank contains propane at 30°C, 90% quality. The tank is heated to 300°C. Calculate the heat transfer during the process.

13.35 The new refrigerant fluid R-123 (see Table A.2) is used in a refrigeration system that operates in the ideal refrigeration cycle, except the compressor is neither reversible nor adiabatic. Saturated vapor at −26.5°C enters the compressor, and superheated vapor exits at 65°C. Heat is rejected from the compressor as 1 kW and the R-123 flow rate is 0.1 kg/s. Saturated liquid exits the condenser at 37.5°C. Specific heat for R-123 is $C_{po} = 0.6$ kJ/kg K. Find the coefficient of performance.

13.36 A cylinder contains ethylene, C_2H_4, at 1.536 MPa, −13°C. It is now compressed in a reversible isobaric (constant P) process to saturated liquid. Find the specific work and heat transfer.

13.37 A piston/cylinder initially contains propane at $T_1 = -7$°C, quality 50%, and volume 10L. A valve connecting the cylinder to a line flowing nitrogen gas at $T_i = 20$°C, $P_i = 1$ MPa is opened and nitrogen flows in. When the valve is closed the cylinder contains a gas mixture of 50% nitrogen, 50% propane on a mole basis at $T_2 = 20$°C, $P_2 = 500$ kPa. What is the cylinder volume at the final state and how much heat transfer took place?

13.38 An ordinary lighter is nearly full of liquid propane with a small amount of vapor, the volume is 5 cm^3 and temperature is 23°C. The propane is now discharged slowly such that heat transfer keeps the propane and valve flow at 23°C. Find the initial pressure and mass of propane and the total heat transfer to empty the lighter.

13.39 An uninsulated piston/cylinder contains propene, C_3H_6, at ambient temperature, 19°C, with a quality of 50% and a volume of 10-L. The propene now expands slowly until the pressure drops to 460 kPa. Calculate the mass of propene, the work and heat transfer for this process.

13.40 A 200-L rigid tank contains propane at 400 K, 3.5 MPa. A valve is opened, and propane flows out until half the initial mass has escaped, at which point the valve is closed. During this process the mass remaining inside the tank expands according to the relation $Pv^{1.4}$ = constant. Calculate the heat transfer to the tank during the process.

13.41 A newly developed compound is being considered for use as the working fluid in a small Rankine-cycle power plant driven by a supply of waste heat. Assume the cycle is ideal, with saturated vapor at 200°C entering the turbine and saturated liquid at 20°C exiting the condenser. The only properties known for this compound are molecular weight of 80 kg/kmol, ideal gas heat capacity $C_{po} = 0.80$ kJ/kg K and $T_c = 500$ K, $P_c = 5$ MPa. Calculate the work input, per kilogram, to the pump and the cycle thermal efficiency.

13.42 A geothermal power plant on the Raft river uses isobutane as the working fluid. The fluid enters the reversible adiabatic turbine, as shown in Fig. P13.42, at 160°C, 5.475 MPa and the condenser exit condition is saturated liquid at 33°C. Isobutane has the properties $T_c = 408.14$ K, $P_c = 3.65$ MPa, $C_{po} = 1.664$ kJ/kg K and ratio of specific heats $k = 1.094$ with a molecular weight as 58.124. Find the specific turbine work and the specific pump work.

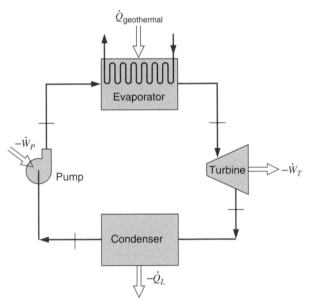

FIGURE P13.42

13.43 Carbon dioxide collected from a fermentation process at 5°C, 100 kPa should be brought to 243 K, 4 MPa in a SSSF process. Find the minimum amount of work required and the heat transfer. What devices are needed to accomplish this change of state?

13.44 An insulated piston/cylinder contains saturated vapor carbon dioxide at 0°C and a volume of 20 L. The external force on the piston is slowly decreased allowing the carbon dioxide to expand until the temperature reaches −30°C. Calculate the work done by the CO_2 during this process.

13.45 An evacuated 100-L rigid tank is connected to a line flowing R-142b gas, chloro-difluoroethane, at 2 MPa, 100°C. The valve is opened, allowing the gas to flow into the tank for a period of time and then it is closed. Eventually, the tank cools to ambient temperature, 20°C, at which point it contains 50% liquid, 50% vapor, by volume. Calculate the quality at the final state and the heat transfer for the process. The ideal-gas specific heat of R-142b is $C_p = 0.787$ kJ/kg K.

13.46 A piston/cylinder contains propane initially at 67°C and 50% quality with a volume of 2 L. The piston cross-sectional area is 0.2 m². The external force on the piston is gradually reduced to a final value of 85 kN during which process the propane expands to ambient temperature, 4°C. Any heat transfer to the propane comes from a constant temperature reservoir at 67°C while any heat transfer from the propane goes to the ambient. It is claimed that the propane does 30 kJ of work during the process. Does this violate the second law?

13.47 Consider the following equation of state, expressed in terms of reduced pressure and temperature:

$$Z = 1 + (P_r/14T_r) [1-6T_r^{-2}]$$

What does this equation predict for enthalpy departure from the ideal gas value at the state $P_r = 0.4$, $T_r = 0.9$? What does it predict for the reduced Boyle temperature?

13.48 Saturated liquid ethane at 2.44 MPa enters (SSSF) a heat exchanger and is brought to 611 K at constant pressure, after which it enters a reversible adiabatic turbine where it expands to 100 kPa. Find the heat transfer in the heat exchanger, the turbine exit temperature, and turbine work.

13.49 A flow of oxygen at 230 K, 5 MPa is throttled to 100 kPa in a SSSF process. Find the exit temperature and the entropy generation.

13.50 A cylinder contains ethylene, C_2H_4, at 1.536 MPa, −13°C. It is now compressed isothermally in a reversible process to 5.12 MPa. Find the specific work and heat transfer.

13.51 A control mass of 10 kg butane gas initially at 80°C, 500 kPa, is compressed in a reversible isothermal process to one-fifth of its initial volume. What is the heat transfer in the process?

13.52 An uninsulated compressor delivers ethylene, C_2H_4, to a pipe, $D = 10$ cm, at 10.24 MPa, 94°C, and velocity 30 m/s. The ethylene enters the compressor at 6.4 MPa, 20.5°C, and the work input required is 300 kJ/kg. Find the mass flow rate, the total heat transfer and entropy generation, assuming the surroundings are at 25°C.

13.53 A distributor of bottled propane, C_3H_8, needs to bring propane from 350 K, 100 kPa to saturated liquid at 290 K in a SSSF process. If this should be accomplished in a reversible setup given the surroundings at 300 K, find the ratio of the volume flow rates V_{in}/V_{out}, the heat transfer and the work involved in the process.

13.54 Saturated liquid ethane at $T_1 = 14$°C is throttled into a SSSF mixing chamber at the rate of 0.25 kmol/s. Argon gas at $T_2 = 25$°C, 800 kPa enters the chamber at the rate 0.75 kmol/s. Heat is transferred to the chamber from a constant temperature source at 150°C at a rate such that a gas mixture exits the chamber at $T_3 = 120$°C, 800 kPa. Find the rate of heat transfer and the rate of entropy generation.

13.55 One kilogram per second water enters a solar collector at 40°C and exits at 190°C, as shown in Fig. P13.55. The hot water is sprayed into a direct-contact heat exchanger (no mixing of the two fluids) used to boil the liquid butane. Pure saturated-vapor butane exits at the top at 80°C and is fed to the turbine. If the butane condenser temperature is 30°C and the turbine and pump isentropic efficiencies are each 80%, determine the net power output of the cycle.

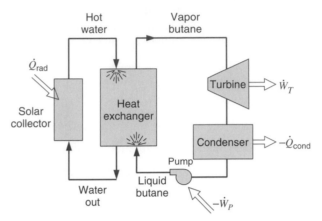

FIGURE P13.55

13.56 A line with a steady supply of octane, C_8H_{18}, is at 400°C, 3 MPa. What is your best estimate for the availability in a SSSF setup where changes in potential and kinetic energies may be neglected?

13.57 A piston/cylinder contains ethane gas initially at 500 kPa, 100 L and at ambient temperature 0°C. The piston is moved, compressing the ethane until it is at 20°C with a quality of 50%. The work required is 25% more than would have been needed for a reversible polytropic process between the same initial and final states. Calculate the heat transfer and the net entropy change for the process.

13.58 The environmentally safe refrigerant R-152a (see Problem 13.65) is to be evaluated as the working fluid for a heat pump system that will heat winter households in two different climates. In the colder climate the cycle evaporator temperature is −20°C, and the more moderate climate the evaporator temperature is 0°C. In both climates the cycle condenser temperature is 30°C. For this study assume all processes are ideal. Determine the cycle coefficient of performance for the two climates.

13.59 Repeat the calculation for the coefficient of performance of the heat pump in the two climates as described in Problem 13.58 using R-12 as the working fluid. Compare the two results.

13.60 One kmol/s of saturated liquid methane, CH_4, at 1 MPa and 2 kmol/s of ethane, C_2H_6, at 250°C, 1 MPa are fed to a mixing chamber with the resultant mixture exiting at 50°C, 1 MPa. Assume that Kay's rule applies to the mixture and determine the heat transfer in the process.

13.61 Consider the following reference state conditions: the entropy of real saturated liquid methane at −100°C is to be taken as 100 kJ/kmol K, and the entropy of hypothetical ideal gas ethane at −100°C is to be taken as 200 kJ/kmol K. Calculate the entropy per kmol of a real gas mixture of 50% methane, 50% ethane (mole basis) at 20°C, 4 MPa, in terms of the specified reference state values, and assuming Kay's rule for the real mixture behavior.

ADVANCED PROBLEMS

13.62 An experiment is conducted at −100°C inside a rigid sealed tank containing liquid R-22 with a small amount of vapor at the top. When the experiment is done, the container and the R-22 warm up to room temperature of 20°C. What is the pressure inside the tank during the experiment? If the pressure at room temperature should not exceed 1 MPa, what is the maximum percent of liquid by volume that can be used during the experiment?

13.63 Determine the low-pressure Joule–Thomson inversion temperature from the condition in Eq.13.54 ($\lim \mu_J = 0$ for $P \rightarrow 0$), as predicted by an equation of state, using the van der Waals equation and the Redlich–Kwong equation.

13.64 Suppose the following information is available for a given pure substance:

a. The liquid-vapor saturation pressure, $P_{sat}(T)$

b. An equation of state for the vapor, $P = \text{fn}(T, v)$

c. The saturated liquid specific volume, $v_f(T)$

d. Critical pressure and temperature, P_c, T_c

e. The constant volume specific heat for vapor, C_v, at v_x

Outline the procedure that should be followed to develop a table of thermodynamic properties comparable to Tables 1, 2, and 3 of the steam tables.

13.65 The refrigerant R-152a, difluoroethane, is tested by the following procedure. A 10-L evacuated tank is connected to a line flowing saturated-vapor R-152a at 40°C. The valve is opened, and the fluid flows in rapidly, so the process is essentially adiabatic. The valve is to be closed when the pressure reaches a certain value P_2, and the tank will then be disconnected from the line. After a period of time, the temperature inside the tank will return to ambient temperature, 25°C, through heat transfer with the surroundings. At this time, the pressure inside the tank must be 500 kPa. What is the pressure P_2 at which the valve should be closed during the filling process? The ideal gas specific heat of R-152a is $C_p = 0.996$ kJ/kg K.

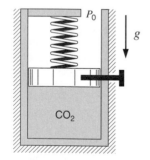

FIGURE P13.66

13.66 An insulated cylinder has a piston loaded with a linear spring (spring constant of 600 kN/m) and held by a pin, as shown in Fig. P13.66. The cylinder cross-sectional area is 0.2 m², the initial volume is 0.1 m³, and it contains carbon dioxide at 2.5 MPa, 0°C. The piston mass and outside atmosphere add a force per unit area of 250 kPa, and the spring force would be zero at a cylinder volume of 0.05 m³. Now the pin is pulled out. What is the final pressure inside the cylinder?

13.67 A 10-m³ storage tank contains methane at low temperature. The pressure inside is 700 kPa, and the tank contains 25% liquid and 75% vapor, on a volume basis. The tank warms very slowly because heat is transferred from the ambient.

a. What is the temperature of the methane when the pressure reaches 10 MPa?

b. Calculate the heat transferred in the process, using the generalized charts.

c. Repeat parts (a) and (b), using the methane tables, Table B.7. Discuss the differences in the results.

13.68 Calculate the difference in entropy of the ideal-gas value and the real-gas value for carbon dioxide at the state 20°C, 1 MPa, as determined using the virial equation of state. Use numerical values given in Problem 13.30.

13.69 Carbon dioxide gas enters a turbine at 5 MPa, 100°C, and exits at 1 MPa. If the isentropic efficiency of the turbine is 75%, determine the exit temperature and the second-law efficiency.

13.70 A 4-m³ uninsulated storage tank, initially evacuated, is connected to a line flowing ethane gas at 10 MPa, 100°C. The valve is opened, and ethane flows into the tank for a period of time, after which the valve is closed. Eventually, the whole system cools to ambient temperature, 0°C, at which time it contains one-fourth liquid and three-fourths vapor, by volume. For the overall process, calculate the heat transfer from the tank and the net change of entropy.

13.71 The environmentally safe refrigerant R-142b (see Problem 13.45) is to be evaluated as the working fluid in a portable, closed-cycle power plant, as shown in Fig. P13.71. The air-cooled condenser temperature is fixed at 50°C, and the maximum cycle temperature is fixed at 180°C, because of concerns about thermal stability. The isentropic efficiency of the expansion engine is estimated to be 80%, and the minimum allowable quality of the fluid exiting the expansion engine is 90%. Calculate the heat transfer from the condenser, assuming saturated liquid at the exit. Determine the maximum cycle pressure, based on the specifications listed.

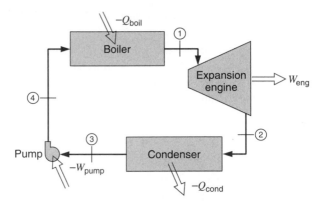

FIGURE P13.71

13.72 The refrigerant fluid R-21 (see Table A.2) is to be used as the working fluid in a solar energy powered Rankine cycle type power plant. Saturated liquid R-21 enters the pump, state 1, at 25°C and saturated vapor enters the turbine, state 3, at 88°C. For R-21 use $C_{po} = 0.582$ kJ/kg K and find the boiler heat transfer q_{23} and the thermal efficiency of the cycle.

13.73 A cylinder/piston contains a gas mixture, 50% CO_2 and 50% C_2H_6 (mole basis) at 700 kPa, 35°C, at which point the cylinder volume is 5 L. The mixture is now compressed to 5.5 MPa in a reversible isothermal process. Calculate the heat transfer and work for the process, using the following model for the gas mixture:

a. Ideal gas mixture.

b. Kay's rule and the generalized charts.

c. The van der Waals equation of state.

13.74 A gas mixture of a known composition is required for the calibration of gas analyzers. It is desired to prepare a gas mixture of 80% ethylene and 20% carbon dioxide (mole basis) at 10 MPa, 25°C in an uninsulated, rigid 50-L tank. The tank is initially to contain CO_2 at 25°C and some pressure P_1. The valve to a line flowing C_2H_4 at 25°C, 10 MPa, is now opened slightly, and remains open until the tank reaches 10 MPa, at which point the temperature can be assumed to be 25°C. Assume that the gas mixture so prepared can be represented by Kay's rule and the generalized charts. Given the desired final state, what is the initial pressure of the carbon dioxide, P_1? Determine the heat transfer and the net entropy change for the process of charging ethylene into the tank.

ENGLISH UNIT PROBLEMS

13.75E A special application requires R-22 at −150 F. It is known that the triple-point temperature is less than −150 F. Find the pressure and specific volume of the saturated vapor at the required condition.

13.76E Ice (solid water) at 27 F, 1 atm is compressed isothermally until it becomes liquid. Find the required pressure.

13.77E Using thermodynamic data for water from Tables C.8.1 and C.8.5, estimate the freezing temperature of liquid water at a pressure of 5000 lbf/in.^{2}.

13.78E Determine the volume expansivity, α_P, and the isothermal compressibility, β_T, for water at 50 F, 500 lbf/in.^{2} and at 500 F, 1500 lbf/in.^{2} using the steam tables.

13.79E Sound waves propagate through media as pressure waves that cause the media to go through isentropic compression and expansion processes. The speed of sound c is defined by $c^2 = (\partial P/\partial \rho)_s$, and it can be related to the adiabatic compressibility, which for liquid ethanol at 70 F is 6.4 in.^{2}/lbf. Find the speed of sound at this temperature.

13.80E Consider the speed of sound as defined in Problem 13.79. Calculate the speed of sound for liquid water at 50 F, 250 lbf/in.^{2} and for water vapor at 400 F, 80 lbf/in.^{2} using the steam tables.

13.81E A cylinder fitted with a piston contains liquid methanol at 70 F, 15 lbf/in.^{2} and volume 1 ft^3. The piston is moved, compressing the methanol to 3000 lbf/in.^{2} at constant temperature. Calculate the work required for this process. The isothermal compressibility of liquid methanol at 70 F is 8.3×10^{-3} in.^{2}/lbf.

13.82E A piston/cylinder contains 10 lbm of butane gas at 900 R, 750 lbf/in.^{2}. The butane expands in a reversible polytropic process with polytropic exponent, $n = 1.05$, until the final pressure is 450 lbf/in.^{2}. Determine the final temperature and the work done during the process.

13.83E A 7-ft^3 rigid tank contains propane at 1300 lbf/in.^{2}, 540 F. The propane is then allowed to cool to 120 F as heat is transferred with the surroundings. Determine the quality at the final state and the mass of liquid in the tank, using the generalized compressibility chart.

13.84E A rigid tank contains 5 lbm of ethylene at 450 lbf/in.^{2}, 90 F. It is cooled until the ethylene reaches the saturated vapor curve. What is the final temperature?

13.85E Calculate the difference in internal energy of the ideal-gas value and the real-gas value for carbon dioxide at the state 70 F, 150 lbf/in.^{2}, as determined using the virial equation of state. At this state $B = -2.036$ ft^3/lb mol, $T(dB/dT) = 4.236$ ft^3/lb mol.

13.86E Calculate the heat transfer during the process described in Problem 13.82.

13.87E Saturated vapor R-22 at 90 F is throttled to 30 lbf/in.^{2} in a SSSF process. Calculate the exit temperature assuming no changes in the kinetic energy, using the generalized charts, Fig. D.2 and repeat using the R-22 tables, Table C.10.

13.88E A 10-ft^3 tank contains propane at 90 F, 90% quality. The tank is heated to 600 F. Calculate the heat transfer during the process.

13.89E A newly developed compound is being considered for use as the working fluid in a small Rankine-cycle power plant driven by a supply of waste heat. Assume the cycle is ideal, with saturated vapor at 400 F entering the turbine and saturated liquid at 70 F exiting the condenser. The only properties known for this compound are molecular weight of 80 lbm/lbmol, ideal gas heat capacity $C_{po} = 0.20$ Btu/lbm R and $T_c = 900$ R, $P_c = 750$ lbf/in.^{2}. Calculate the work input, per lbm, to the pump and the cycle thermal efficiency.

13.90E A 7-ft^3 rigid tank contains propane at 730 R, 500 lbf/in.^{2}. A valve is opened, and propane flows out until half the initial mass has escaped, at which point the valve is closed. During this process the mass remaining inside the tank expands according to the relation $Pv^{1.4}$ = constant. Calculate the heat transfer to the tank during the process.

13.91E A geothermal power plant on the Raft river uses isobutane as the working fluid as shown in Fig. P13.42. The fluid enters the reversible adiabatic turbine at 320 F, 805 lbf/in.^{2}, and the condenser exit condition is saturated liquid at 91 F. Isobutane has the properties $T_c = 734.65$ R, $P_c = 537$ lbf/in.^{2}, $C_{po} = 0.3974$ Btu/lbm R and ratio of specific heats $k = 1.094$ with a molecular weight as 58.124. Find the specific turbine work and the specific pump work.

13.92E Carbon dioxide collected from a fermentation process at 40 F, 15 lbf/in.^{2} should be brought to 438 R, 590 lbf/in.^{2} in a SSSF process. Find the minimum amount of work required and the heat transfer. What devices are needed to accomplish this change of state?

13.93E A control mass of 10 lbm butane gas initially at 180 F, 75 lbf/in.^{2}, is compressed in a reversible isothermal process to one-fifth of its initial volume. What is the heat transfer in the process?

13.94E A cylinder contains ethylene, C_2H_4, at 222.6 lbf/in.^{2}, 8 F. It is now compressed isothermally in a reversible process to 742 lbf/in.^{2}. Find the specific work and heat transfer.

13.95E A cylinder contains ethylene, C_2H_4, at 222.6 lbf/in.^{2}, 8 F. It is now compressed in a reversible isobaric (constant P) process to saturated liquid. Find the specific work and heat transfer.

13.96E A distributor of bottled propane, C_3H_8, needs to bring propane from 630 R, 14.7 lbf/in.^{2} to saturated liquid at 520 R in a SSSF process. If this should be accomplished in a reversible setup given the surroundings at 540 R, find the ratio of the volume flow rates $\dot{V}_{in}/\dot{V}_{out}$, the heat transfer, and the work involved in the process.

13.97E A line with a steady supply of octane, C_8H_{18}, is at 750 F, 440 lbf/in.^{2}. What is your best estimate for the availability in a SSSF setup where changes in potential and kinetic energies may be neglected?

COMPUTER, DESIGN, AND OPEN-ENDED PROBLEMS

13.98 Write a program to obtain a plot of pressure versus specific volume at various temperatures (all on a generalized reduced basis) as predicted by the van der Waals equation of state. Temperatures less than the critical temperature should be included in the results.

13.99 We wish to determine the isothermal compressibility, β_T, for a range of states of liquid water. Use the menu-driven software or write a program to determine this at a pressure of 1 MPa and at 25 MPa for temperatures of 0°C, 100°C, and 300°C.

13.100 Consider the small Rankine-cycle power plant in Problem 13.44. What single change would you suggest to make the power plant more realistic?

13.101 Supercritical fluid chromatography is an experimental technique for analyzing compositions of mixtures. It utilizes a carrier fluid, often CO_2, in the dense fluid region just above the critical temperature. Write a program to express the fluid density as a function of reduced temperature and pressure in the region of $1.0 \leq T_r \leq 1.2$ in reduced temperature and $2 \leq P_r \leq 8$ in reduced pressure. The relation should be an expression curve-fitted to values consistent with the generalized compressibility charts.

13.102 It is desired to design a portable breathing system for an average-sized adult. The breather will store liquid oxygen sufficient for a 24-hour supply, and include a heater for delivering oxygen gas at ambient temperature. Determine the size of the system container and the heat exchanger.

13.103 Liquid nitrogen is used in cryogenic experiments and applications where a non-oxidizing gas is desired. Size a tank to hold 500 kg to be placed next to a building and estimate the size of an environmental (to atmospheric air) heat exchanger that can deliver nitrogen gas at a rate of 10 kg/h at roughly ambient tempertature.

13.104 List a number of requirements for a substance that should be used as the working fluid in a refrigerator. Discuss the choices and explain the requirements.

13.105 The speed of sound is used in many applications. Make a list of the speed of sound at P_o, T_o for gases, liquids, and solids. Find at least three different substances for each phase. List a number of applications where the knowledge about the speed of sound can be used to estimate other quantities of interest.

13.106 Propane is used as a fuel distributed to the end consumer in a steel bottle. Make a list of design specifications for these bottles and give characteristic sizes and the amount of propane they can hold.

13.107 Carbon dioxide is used in soft drinks and comes in a separate bottle for large volume users such as restaurants. Find typical sizes of these, the pressure they should withstand, and the amount of carbon dioxide they can hold.

CHEMICAL REACTIONS 14

Many thermodynamic problems involve chemical reactions. Among the most familiar of these is the combustion of hydrocarbon fuels, for this process is utilized in most of our power-generating devices. However, we can all think of a host of other processes involving chemical reactions, including those that occur in the human body.

It is our purpose in this chapter to consider a first and second law analysis of systems undergoing a chemical reaction. In many respects, this chapter is simply an extension of our previous consideration of the first and second laws. However, a number of new terms are introduced, and it will also be necessary to introduce the third law of thermodynamics.

In this chapter the combustion process is considered in detail. There are two reasons for this emphasis. The first reason is that the combustion process is of great significance in many problems and devices with which the engineer is concerned. The second reason is that the combustion process provides an excellent vehicle for teaching the basic principles of the thermodynamics of chemical reactions. The student should keep both of these objectives in mind as the study of this chapter progresses.

Chemical equilibrium will be considered in Chapter 15 and, therefore, the matter of dissociation will be deferred until then.

14.1 FUELS

A thermodynamics textbook is not the place for a detailed treatment of fuels. However, some knowledge of them is a prerequisite to a consideration of combustion, and this section is therefore devoted to a brief discussion of some of the hydrocarbon fuels. Most fuels fall into one of three categories—coal, liquid hydrocarbons, or gaseous hydrocarbons.

Coal consists of the remains of vegetation deposits of past geologic ages, after subjection of biochemical actions, high pressure, temperature, and submersion. The characteristics of coal vary considerably with location, and even within a given mine there is some variation in composition.

The analysis of a sample of coal is given on one of two bases: the proximate analysis specifies, on a mass basis, the relative amounts of moisture, volatile matter, fixed carbon, and ash; the ultimate analysis specifies, on a mass basis, the relative amounts of carbon, sulfur, hydrogen, nitrogen, oxygen, and ash. The ultimate analysis may be given on an "as received" basis or on a dry basis. In the latter case the ultimate analysis does not include the moisture as determined by the proximate analysis.

There are also a number of other properties of coal that are important in evaluating a coal for a given use. Some of these are the fusibility of the ash, the grindability or ease of pulverization, the weathering characteristics, and size.

Table 14.1
Characteristics of Some of the Hydrocarbon Families

Family	Formula	Structure	Saturated
Paraffin	C_nH_{2n+2}	Chain	Yes
Olefin	C_nH_{2n}	Chain	No
Diolefin	C_nH_{2n-2}	Chain	No
Naphthene	C_nH_{2n}	Ring	Yes
Aromatic			
Benzene	C_nH_{2n-6}	Ring	No
Naphthene	C_nH_{2n-12}	Ring	No

Most liquid and gaseous hydrocarbon fuels are a mixture of many different hydrocarbons. For example, gasoline consists primarily of a mixture of about 40 hydrocarbons, with many others present in very small quantities. In discussing hydrocarbon fuels, therefore, brief consideration should be given to the most important families of hydrocarbons, which are summarized in Table 14.1.

Three concepts should be defined. The first pertains to the structure of the molecule. The important types are the ring and chain structures; the difference between the two is illustrated in Fig. 14.1. The same figure illustrates the definition of saturated and unsaturated hydrocarbons. An unsaturated hydrocarbon has two or more adjacent carbon atoms joined by a double or triple bond, whereas in a saturated hydrocarbon all the carbon atoms are joined by a single bond. The third term to be defined is an isomer. Two hydrocarbons with the same number of carbon and hydrogen atoms and different structures are called isomers. Thus there are several different octanes (C_8H_{18}), each having 8 carbon atoms and 18 hydrogen atoms, but each with a different structure.

The various hydrocarbon families are identified by a common suffix. The compounds comprising the paraffin family all end in "-ane" (as propane and octane). Similarly, the compounds comprising the olefin family end in "-ylene" or "-ene" (as propene and octene), and the diolefin family ends in "-diene" (as butadiene). The naphthene family has the same chemical formula as the olefin family, but has a ring rather than chain structure. The hydrocarbons in the naphthene family are named by adding the prefix "cyclo-" (as cyclopentane).

The aromatic family includes the benzene series (C_nH_{2n-6}) and the naphthalene series (C_nH_{2n-12}). The benzene series has a ring structure and is unsaturated.

Alcohols are sometimes used as a fuel in internal combustion engines. The characteristic feature of the alcohol family is that one of the hydrogen atoms is replaced by an OH radical. Thus methyl alcohol, also called methanol, is CH_3OH.

Most liquid hydrocarbon fuels are mixtures of hydrocarbons that are derived from crude oil through distillation and cracking processes. Thus, from a given crude oil, a vari-

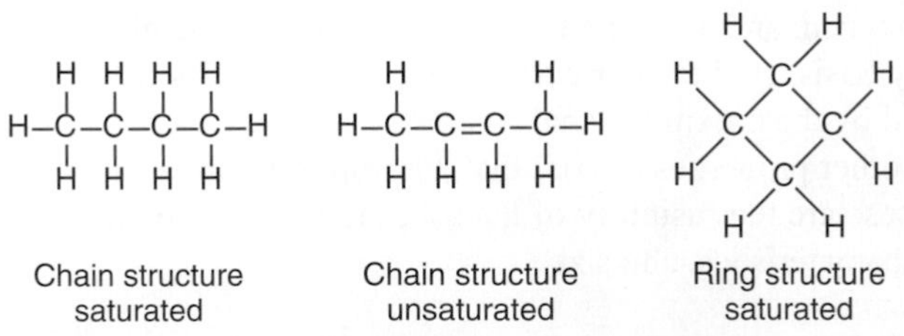

Chain structure
saturated

Chain structure
unsaturated

Ring structure
saturated

FIGURE 14.1 Molecular structure of some hydrocarbon fuels.

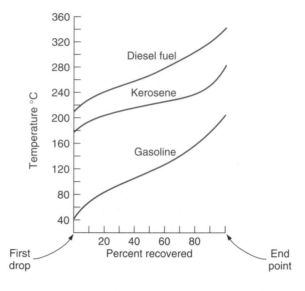

FIGURE 14.2 Typical distillation curves of some hydrocarbon fuels.

ety of different fuels can be produced, some of the common ones being gasoline, kerosene, diesel fuel, and fuel oil. Within each of these classifications there is a wide variety of grades, and each is made up of a large number of different hydrocarbons. The important distinction between these fuels is the distillation curve, Fig. 14.2. The distillation curve is obtained by slowly heating a sample of fuel so that it vaporizes. The vapor is then condensed and the amount measured. The more volatile hydrocarbons are vaporized first, and thus the temperature of the nonvaporized fraction increases during the process. The distillation curve, which is a plot of the temperature of the nonvaporized fraction versus the amount of vapor condensed, is an indication of the volatility of the fuel.

TABLE 14.2
Volumetric Analyses of Some Typical Gaseous Fuels

Constituent	Various Natural Gases				Producer Gas from Bituminous Coal	Carbureted Water Gas	Coke-Oven Gas
	A	B	C	D			
Methane	93.9	60.1	67.4	54.3	3.0	10.2	32.1
Ethane	3.6	14.8	16.8	16.3			
Propane	1.2	13.4	15.8	16.2			
Butanes plus[a]	1.3	4.2		7.4			
Ethene						6.1	3.5
Benzene						2.8	0.5
Hydrogen					14.0	40.5	46.5
Nitrogen		7.5		5.8	50.9	2.9	8.1
Oxygen					0.6	0.5	0.8
Carbon monoxide					27.0	34.0	6.3
Carbon dioxide					4.5	3.0	2.2

[a]This includes butane and all heavier hydrocarbons

For the combustion of liquid fuels, it is convenient to express the composition in terms of a single hydrocarbon, even though it is a mixture of many hydrocarbons. Thus gasoline is usually considered to be octane, C_8H_{18}, and diesel fuel is considered to be dodecane, $C_{12}H_{26}$. The composition of a hydrocarbon fuel may also be given in terms of percentage of carbon and hydrogen.

The two primary sources of gaseous hydrocarbon fuels are natural gas wells and certain chemical manufacturing processes. Table 14.2 gives the composition of a number of gaseous fuels. The major constituent of natural gas is methane, which distinguishes it from manufactured gas.

At the present time, there is a considerable effort being devoted to develop more economical processes for producing gaseous and also liquid hydrocarbon fuels from coal, and also from oil shale and tar sands deposits. Several alternative techniques have been demonstrated to be feasible, and these resources promise to provide an increasing proportion of our fuel supply in future years.

14.2 THE COMBUSTION PROCESS

The combustion process consists of the oxidation of constituents in the fuel that are capable of being oxidized, and can therefore be represented by a chemical equation. During a combustion process the mass of each element remains the same. Thus, writing chemical equations and solving problems concerning quantities of the various constituents basically involves the conservation of mass of each element. A brief review of this subject, particularly as it applies to the combustion process, is presented in this chapter.

Consider first the reaction of carbon with oxygen.

Reactants   Products

$$C + O_2 \rightarrow CO_2$$

This equation states that 1 kmol of carbon reacts with 1 kmol of oxygen to form 1 kmol of carbon dioxide. This also means that 12 kg of carbon react with 32 kg of oxygen to form 44 kg of carbon dioxide. All the initial substances that undergo the combustion process are called the reactants, and the substances that result from the combustion process are called the products.

When a hydrocarbon fuel is burned, both the carbon and the hydrogen are oxidized. Consider the combustion of methane as an example.

$$CH_4 + 2O_2 \rightarrow CO_2 + 2H_2O \tag{14.1}$$

Here the products of combustion include both carbon dioxide and water. The water may be in the vapor, liquid, or solid phases, depending on the temperature and pressure of the products of combustion.

It should be pointed out that in the combustion process many intermediate products are formed during the chemical reaction. In this book we are concerned with the initial and final products and not with the intermediate products, but this aspect is very important in a detailed consideration of combustion.

In most combustion processes the oxygen is supplied as air rather than as pure oxygen. The composition of air on a molal basis is approximately 21% oxygen, 78% nitrogen, and 1% argon. We assume that the nitrogen and the argon do not undergo chemical reaction (except for dissociation, which will be considered in Chapter 15). They do leave at the same temperature as the other products, however, and therefore undergo a change of state if the products are at a temperature other than the original air temperature. It should be pointed out that at the high temperatures achieved in internal-combustion engines, there is actually some reaction between the nitrogen and oxygen, and this gives rise to the air pollution problem associated with the oxides of nitrogen in the engine exhaust.

In combustion calculations concerning air, the argon is usually neglected, and the air is considered to be composed of 21% oxygen and 79% nitrogen by volume. When this assumption is made, the nitrogen is sometimes referred to as atmospheric nitrogen. Atmospheric nitrogen has a molecular weight of 28.16 (which takes the argon into account) as compared to 28.013 for pure nitrogen. This distinction will not be made in this text, and we will consider the 79% nitrogen to be pure nitrogen.

The assumption that air is 21.0% oxygen and 79.0% nitrogen by volume leads to the conclusion that for each mole of oxygen, $79.0/21.0 = 3.76$ moles of nitrogen are involved. Therefore, when the oxygen for the combustion of methane is supplied as air, the reaction can be written

$$CH_4 + 2O_2 + 2(3.76)N_2 \rightarrow CO_2 + 2H_2O + 7.52N_2 \qquad (14.2)$$

The minimum amount of air that supplies sufficient oxygen for the complete combustion of all the carbon, hydrogen, and any other elements in the fuel that may oxidize is called the theoretical air. When complete combustion is achieved with theoretical air, the products contain no oxygen. A general combustion reaction with a hydrocarbon fuel and air is thus written

$$C_xH_y + v_{O_2}(O_2 + 3.76N_2) \rightarrow v_{CO_2}CO_2 + v_{H_2O}H_2O + v_{N_2}N_2 \qquad (14.3)$$

with the coefficients to the substances called stoichiometric coefficients. The balance of atoms yields the theoretical amount of air as

$$
\begin{aligned}
&C: &v_{CO_2} &= x \\
&H: &2v_{H_2O} &= y \\
&N_2: &v_{N_2} &= 3.76 \times v_{O_2} \\
&O_2: &v_{O_2} &= v_{CO_2} + v_{H_2O}/2 = x + y/4
\end{aligned}
$$

and the total number of moles of air for 1 mole of fuel becomes

$$n_{air} = v_{O_2} \times 4.76 = 4.76(x + y/4)$$

This amount of air is equal to 100% theoretical air. In practice, complete combustion is not likely to be achieved unless the amount of air supplied is somewhat greater than the theoretical amount. Two important parameters often used to express the ratio of fuel and air are the air–fuel ratio (designated AF) and its reciprocal, the fuel–air ratio (designated FA). These ratios are usually expressed on a mass basis, but a mole basis is used at times.

$$AF_{mass} = \frac{m_{air}}{m_{fuel}} \qquad (14.4)$$

$$AF_{mole} = \frac{n_{air}}{n_{fuel}} \qquad (14.5)$$

They are related through the molecular weights as

$$AF_{mass} = \frac{m_{air}}{m_{fuel}} = \frac{n_{air}M_{air}}{n_{fuel}M_{fuel}} = AF_{mole}\frac{M_{air}}{M_{fuel}}$$

and a subscript *s* is used to indicate the ratio for 100% theoretical air, also called a stoichiometric mixture. In an actual combustion process, an amount of air is expressed as a fraction of the theoretical amount, called percent theoretical air. A similar ratio named the equivalence ratio equals the actual fuel–air ratio divided by the theoretical fuel–air ratio as

$$\Phi = FA \, / \, FA_s = AF_s \, / \, AF \qquad (14.6)$$

the reciprocal of percent theoretical air. Since the percent theoretical air and the equivalence ratio both are ratios of the stoichiometric air–fuel ratio and the actual air–fuel ratio the molecular weights cancel out and they are the same whether a mass basis or a mole basis is used.

Thus, 150% theoretical air means that the air actually supplied is 1.5 times the theoretical air and the equivalence ratio is $^2/_3$. The complete combustion of methane with 150% theoretical air is written

$$CH_4 + 1.5 \times 2(O_2 + 3.76N_2) \rightarrow CO_2 + 2H_2O + O_2 + 11.28N_2 \qquad (14.7)$$

having balanced all the stoichiometric coefficients from conservation of all the atoms.

The amount of air actually supplied may also be expressed in terms of percent excess air. The excess air is the amount of air supplied over and above the theoretical air. Thus, 150% theoretical air is equivalent to 50% excess air. The terms "theoretical air," "excess air" and "equivalence ratio" are all in current usage and give an equivalent information about the reactant mixture of fuel and air.

When the amount of air supplied is less than the theoretical air required, the combustion is incomplete. If there is only a slight deficiency of air, the usual result is that some of the carbon unites with the oxygen to form carbon monoxide (CO) instead of carbon dioxide (CO_2). If the air supplied is considerably less than the theoretical air, there may also be some hydrocarbons in the products of combustion.

Even when some excess air is supplied there may be small amounts of carbon monoxide present, the exact amount depending on a number of factors including the mixing and turbulence during combustion. Thus the combustion of methane with 110% theoretical air might be as follows:

$$CH_4 + 2(1.1)O_2 + 2(1.1)3.76N_2 \rightarrow$$

$$+ 0.95CO_2 + 0.05CO + 2H_2O + 0.225O_2 + 8.27N_2 \qquad (14.8)$$

The material covered so far in this section is illustrated by the following examples.

EXAMPLE 14.1 Calculate the theoretical air–fuel ratio for the combustion of octane, C_8H_{18}.

Solution

The combustion equation is

$$C_8H_{18} + 12.5O_2 + 12.5(3.76)N_2 \rightarrow 8CO_2 + 9H_2O + 47.0N_2$$

The air–fuel ratio on a mole basis is

$$AF = \frac{12.5 + 47.0}{1} = 59.5 \text{ kmol air} \, / \, \text{kmol fuel}$$

The theoretical air–fuel ratio on a mass basis is found by introducing the molecular weight of the air and fuel.

$$AF = \frac{59.5(28.97)}{114.2} = 15.0 \text{ kg air / kg fuel}$$

EXAMPLE 14.2 Determine the molal analysis of the products of combustion when octane, C_8H_{18}, is burned with 200% theoretical air, and determine the dew point of the products if the pressure is 0.1 MPa.

Solution

The equation for the combustion of octane with 200% theoretical air is

$$C_8H_{18} + 12.5(2)O_2 + 12.5(2)(3.76)N_2 \rightarrow 8CO_2 + 9H_2O + 12.5O_2 + 94.0N_2$$

Total kmols of product = 8 + 9 + 12.5 + 94.0 = 123.5
Molal analysis of products:

$$
\begin{aligned}
CO_2 &= 8/123.5 &=& \quad 6.47\% \\
H_2O &= 9/123.5 &=& \quad 7.29 \\
O_2 &= 12.5/123.5 &=& \quad 10.12 \\
N_2 &= 94/123.5 &=& \quad 76.12 \\
\hline
& & & 100.00\%
\end{aligned}
$$

The partial pressure of the water is 100(0.0729) = 7.29 kPa.

The saturation temperature corresponding to this pressure is 39.7°C, which is also the dew-point temperature.

The water condensed from the products of combustion usually contains some dissolved gases and therefore may be quite corrosive. For this reason the products of combustion are often kept above the dew point until discharged to the atmosphere.

EXAMPLE 14.2E Determine the molal analysis of the products of combustion when octane, C_8H_{18}, is burned with 200% theoretical air, and determine the dew point of the products if the pressure is 14.7 lbf/in.^{2}.

Solution

The equation for the combustion of octane with 200% theoretical air is

$$C_8H_{18} + 12.5(2)O_2 + 12.5(2)(3.76)N_2 \rightarrow 8CO_2 + 9H_2O + 12.5O_2 + 94.0N_2$$

Total moles of product = 8 + 9 + 12.5 + 94.0 = 123.5

Molal analysis of products:

$$CO_2 = 8/123.5 \quad = \quad 6.47\%$$
$$H_2O = 9/123.5 \quad = \quad 7.29$$
$$O_2 = 12.5/123.5 = \quad 10.12$$
$$N_2 = 94/123.5 \quad = \quad 76.12$$
$$\overline{\qquad\qquad}$$
$$100.00\%$$

The partial pressure of the H_2O is $14.7(0.0729) = 1.072$ lbf/in.^{2}

The saturation temperature corresponding to this pressure is 104 F, which is also the dew-point temperature.

The water condensed from the products of combustion usually contains some dissolved gases and therefore may be quite corrosive. For this reason the products of combustion are often kept above the dew point until discharged to the atmosphere.

EXAMPLE 14.3   Producer gas from bituminous coal (see Table 14.2) is burned with 20% excess air. Calculate the air–fuel ratio on a volumetric basis and on a mass basis.

Solution

To calculate the theoretical air requirement, let us write the combustion equation for the combustible substances in 1 kmol of fuel.

$$0.14H_2 + 0.070O_2 \rightarrow 0.14H_2O$$
$$0.27CO + 0.135O_2 \rightarrow 0.27CO_2$$
$$0.03CH_4 + 0.06O_2 \rightarrow 0.03CO_2 + 0.06H_2O$$
$$\overline{\qquad\qquad}$$
$$0.265 = \text{kmol oxygen required / kmol fuel}$$
$$-0.006 = \text{oxygen in fuel / kmol fuel}$$
$$\overline{\qquad\qquad}$$
$$0.259 = \text{kmol oxygen required from air / kmol fuel}$$

Therefore, the complete combustion equation for 1 kmol of fuel is

$$\overbrace{0.14H_2 + 0.27CO + 0.03CH_4 + 0.006O_2 + 0.509N_2 + 0.045CO_2}^{\text{fuel}}$$

$$\overbrace{+0.259O_2 + 0.259(3.76)N_2}^{\text{air}} \rightarrow 0.20H_2O + 0.345CO_2 + 1.482N_2$$

$$\left(\frac{\text{kmol air}}{\text{kmol fuel}}\right)_{\text{theo}} = 0.259 \times \frac{1}{0.21} = 1.233$$

If the air and fuel are at the same pressure and temperature, this also represents the ratio of the volume of air to the volume of fuel.

For 20% excess air, $\dfrac{\text{kmol air}}{\text{kmol fuel}} = 1.233 \times 1.200 = 1.48$

The air–fuel ratio on a mass basis is

$$AF = \frac{1.48(28.97)}{0.14(2)+0.27(28)+0.03(16)+0.006(32)+0.509(28)+0.045(44)}$$

$$= \frac{1.48(28.97)}{24.74} = 1.73 \text{ kg air / kg fuel}$$

An analysis of the products of combustion affords a very simple method for calculating the actual amount of air supplied in a combustion process. There are various experimental methods by which such an analysis can be made. Some yield results on a "dry" basis; that is, the fractional analysis of all the components, except for water vapor. Other experimental procedures give results that include the water vapor. In this presentation we are not concerned with the experimental devices and procedures, but rather with the use of such information in a thermodynamic analysis of the chemical reaction. The following examples illustrate how an analysis of the products can be used to determine the chemical reaction and the composition of the fuel.

The basic principle in using the analysis of the products of combustion to obtain the actual fuel–air ratio is conservation of the mass of each of the elements. Thus, in changing from reactants to products, we can make a carbon balance, hydrogen balance, oxygen balance, and nitrogen balance (plus any other elements that may be involved). Furthermore, we recognize that there is a definite ratio between the amounts of some of these elements. Thus, the ratio between the nitrogen and oxygen supplied in the air is fixed, as well as the ratio between carbon and hydrogen if the composition of a hydrocarbon fuel is known.

EXAMPLE 14.4 Methane (CH_4) is burned with atmospheric air. The analysis of the products on a dry basis is as follows:

CO_2	10.00%
O_2	2.37
CO	0.53
N_2	87.10
	100.00%

Calculate the air–fuel ratio and the percent theoretical air, and determine the combustion equation.

Solution

The solution consists of writing the combustion equation for 100 kmol of dry products, introducing letter coefficients for the unknown quantities, and then solving for them.

From the analysis of the products, the following equation can be written, keeping in mind that this analysis is on a dry basis.

$$a CH_4 + b O_2 + c N_2 \rightarrow 10.0 CO_2 + 0.53 CO + 2.37 O_2 + d H_2O + 87.1 N_2$$

A balance for each of the elements will enable us to solve for all the unknown coefficients:

$$\text{Nitrogen balance: } c = 87.1$$

Since all the nitrogen comes from the air,

$$\frac{c}{b} = 3.76 \qquad b = \frac{87.1}{3.76} = 23.16$$

Carbon balance: $a = 10.00 + 0.53 = 10.53$
Hydrogen balance: $d = 2a = 21.06$

Oxygen balance: All the unknown coefficients have been solved for, and therefore the oxygen balance provides a check on the accuracy. Thus, b can also be determined by an oxygen balance

$$b = 10.00 + \frac{0.53}{2} + 2.37 + \frac{21.06}{2} = 23.16$$

Substituting these values for a, b, c, and d we have

$$10.53CH_4 + 23.16O_2 + 87.1N_2$$
$$\rightarrow 10.0CO_2 + 0.53CO + 2.37O_2 + 21.06H_2O + 87.1N_2$$

Dividing through by 10.53 yields the combustion equation per kmol of fuel.

$$CH_4 + 2.2O_2 + 8.27N_2 \rightarrow 0.95CO_2 + 0.05CO + 2H_2O + 0.225O_2 + 8.27N_2$$

The air–fuel ratio on a mole basis is

$$2.2 + 8.27 = 10.47 \text{ kmol air / kmol fuel}$$

The air–fuel ratio on a mass basis is found by introducing the molecular weights.

$$AF = \frac{10.47 \times 28.97}{16.0} = 18.97 \text{ kg air / kg fuel}$$

The theoretical air–fuel ratio is found by writing the combustion equation for theoretical air.

$$CH_4 + 2O_2 + 2(3.76)N_2 \rightarrow CO_2 + 2H_2O + 7.52N_2$$

$$AF_{theo} = \frac{(2 + 7.52)28.97}{16.0} = 17.23 \text{ kg air / kg fuel}$$

The percent theoretical air is $\dfrac{18.97}{17.23} = 110\%$

EXAMPLE 14.5 Coal from Jenkin, Kentucky, has the following ultimate analysis on a dry basis, percent by mass:

Component	Percent by Mass
Sulfur	0.6
Hydrogen	5.7
Carbon	79.2
Oxygen	10.0
Nitrogen	1.5
Ash	3.0

This coal is to be burned with 30% excess air. Calculate the air–fuel ratio on a mass basis.

Solution

One approach to this problem is to write the combustion equation for each of the combustible elements per 100 kg of fuel. The molal composition per 100 kg of fuel is found first.

$$\text{kmol S}/100 \text{ kg fuel} = \frac{0.6}{32} = 0.02$$

$$\text{kmol H}_2/100 \text{ kg fuel} = \frac{5.7}{2} = 2.85$$

$$\text{kmol C}/100 \text{ kg fuel} = \frac{79.2}{12} = 6.60$$

$$\text{kmol O}_2/100 \text{ kg fuel} = \frac{10}{32} = 0.31$$

$$\text{kmol N}_2/100 \text{ kg fuel} = \frac{1.5}{28} = 0.05$$

The combustion equations for the combustible elements are now written, which enables us to find the theoretical oxygen required.

$$0.02S + 0.02O_2 \rightarrow 0.02SO_2$$
$$2.85H_2 + 1.42O_2 \rightarrow 2.85H_2O$$
$$6.60C + 6.60O_2 \rightarrow 6.60CO_2$$

$$8.04 \text{ kmol O}_2 \text{ required}/100 \text{ kg fuel}$$
$$-0.31 \text{ kmol O}_2 \text{ in fuel}/100 \text{ kg fuel}$$
$$7.73 \text{ kmol O}_2 \text{ from air}/100 \text{ kg fuel}$$

$$AF_{\text{theo}} = \frac{[7.73 + 7.73(3.76)]28.97}{100} = 10.63 \text{ kg air}/\text{kg fuel}$$

For 30% excess air the air–fuel ratio is

$$AF = 1.3 \times 10.63 = 13.82 \text{ kg air}/\text{kg fuel}$$

14.3 ENTHALPY OF FORMATION

In the first thirteen chapters of this book the problems always concerned a fixed chemical composition and never a change of composition through a chemical reaction. Therefore, in dealing with a thermodynamic property, we used tables of thermodynamic properties for the given substance, and in each of these tables the thermodynamic properties were given relative to some arbitrary base. In the steam tables, for example, the internal energy of saturated liquid at 0.01°C is assumed to be zero. This procedure is quite adequate when there is no change in composition, because we are concerned with the changes in the properties of a given substance. The properties at the condition of the reference state cancel out in the calculation. When dealing with the matter of reference states in Section 13.11, we noted that for a given substance (perhaps a component of a mixture), we are free to choose a reference state condition, for example, a hypothetical ideal gas, as long as we then carry out a consistent calculation from that state and condition to the real desired state. We also noted that we are free to choose a reference state value, as long as there is no subsequent inconsistency in the calculation of the change in a property because of a chemical reaction with a resulting change in the amount of a given substance. Now that we are to include the possibility of a chemical reaction, it will become necessary to choose these reference state values on a common and consistent basis. We will use as our reference state a temperature of 25°C, a pressure of 0.1 MPa, and a hypothetical ideal gas condition for those substances that are gases.

Consider the simple SSSF combustion process shown in Fig. 14.3. This idealized reaction involves the combustion of solid carbon with gaseous (ideal gas) oxygen, each of which enters the control volume at the reference state, 25°C and 0.1 MPa. The carbon dioxide (ideal gas) formed by the reaction leaves the chamber at the reference state, 25°C and 0.1 MPa. If the heat transfer could be accurately measured, it would be found to be −393 522 kJ/kmol of carbon dioxide formed. The chemical reaction can be written

$$C + O_2 \rightarrow CO_2$$

Applying the first law to this process we have

$$Q_{c.v.} + H_R = H_P \tag{14.9}$$

where the subscripts R and P refer to the reactants and products, respectively. We will find it convenient to also write the first law for such a process in the form

$$Q_{c.v.} + \sum_R n_i \bar{h}_i = \sum_P n_e \bar{h}_e \tag{14.10}$$

where the summations refer, respectively, to all the reactants or all the products.

Thus, a measurement of the heat transfer would give us the difference between the enthalpy of the products and the reactants, where each is at the reference state condition.

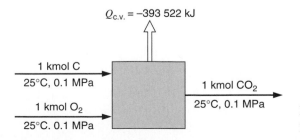

FIGURE 14.3 Example of combustion process.

Suppose, however, that we assign the value of zero to the enthalpy of all the elements at the reference state. In this case, the enthalpy of the reactants is zero, and

$$Q_{c.v.} = H_P = -393\ 522\ kJ\ /\ kmol$$

The enthalpy of (hypothetical) ideal gas carbon dioxide at 25°C, 0.1 MPa pressure (with reference to this arbitrary base in which the enthalpy of the elements is chosen to be zero), is called the enthalpy of formation. We designate this with the symbol $\bar{h}_f$. Thus, for carbon dioxide

$$\bar{h}_f^\circ = -393\ 522\ kJ\ /\ kmol$$

The enthalpy of carbon dioxide in any other state, relative to this base in which the enthalpy of the elements is zero, would be found by adding the change of enthalpy between ideal gas at 25°C, 0.1 MPa, and the given state to the enthalpy of formation. That is, the enthalpy at any temperature and pressure, $h_{T,P}$, is

$$\bar{h}_{T,P} = (\bar{h}_f^\circ)_{298,0.1\text{MPa}} + (\Delta\bar{h})_{298,0.1\text{MPa}\rightarrow T,P} \tag{14.11}$$

where the term $(\Delta\bar{h})_{298,\ 0.1\text{MPa}\rightarrow T,P}$ represents the difference in enthalpy between any given state and the enthalpy of ideal gas at 298.15 K, 0.1 MPa. For convenience we usually drop the subscripts in the examples that follow.

The procedure that we have demonstrated for carbon dioxide can be applied to any compound.

Table A.9 gives values of the enthalpy of formation for a number of substances in the units kJ/kmol (or Btu/lb mol in C.12).

Three further observations should be made in regard to enthalpy of formation.

1. We have demonstrated the concept of enthalpy of formation in terms of the measurement of the heat transfer in an idealized chemical reaction in which a compound is formed from the elements. Actually, the enthalpy of formation is usually found by the application of statistical thermodynamics, using observed spectroscopic data.

2. The justification of this procedure of arbitrarily assigning the value of zero to the enthalpy of the elements at 25°C, 0.1 MPa, rests on the fact that in the absence of nuclear reactions the mass of each element is conserved in a chemical reaction. No conflicts or ambiguities arise with this choice of reference state, and it proves to be very convenient in studying chemical reactions from a thermodynamic point of view.

3. In certain cases an element or compound can exist in more than one state at 25°C, 0.1 MPa. Carbon, for example, can be in the form of graphite or diamond. It is essential that the state to which a given value is related be clearly identified. Thus, in Table A.9, the enthalpy of formation of graphite is given the value of zero, and the enthalpy of each substance that contains carbon is given relative to this base. Another example is that oxygen may exist in the monatomic or diatomic form, and also as ozone, O_3. The value chosen as zero is for the form that is chemically stable at the reference state, which in the case of oxygen is the diatomic form. Then each of the other forms must have an enthalpy of formation consistent with the chemical reaction and heat transfer for the reaction that produces that form of oxygen.

It will be noted from Table A.9 that two values are given for the enthalpy of formation for water; one is for liquid water and the other for gaseous (hypothetical ideal gas) water, both at the reference state of 25°C, 0.1 MPa. It is convenient to use the hypothetical ideal-gas reference in connection with the ideal-gas table property changes given in Table A.8, and to use the real liquid reference in connection with real water property changes as given in the steam tables, Table B.1. The real-liquid reference state properties are obtained from those at the hypothetical ideal-gas reference by following the procedure of calculation described in Section 13.11. The same procedure can be followed for other substances that have a saturation pressure less than 0.1 MPa at the reference temperature of 25°C.

Frequently students are bothered by the minus sign when the enthalpy of formation is negative. For example, the enthalpy of formation of carbon dioxide is negative. This is quite evident because the heat transfer is negative during the steady-flow chemical reaction, and the enthalpy of the carbon dioxide must be less than the sum of enthalpy of the carbon and oxygen initially, both of which are assigned the value of zero. This is quite analogous to the situation we would have in the steam tables if we let the enthalpy of saturated vapor be zero at 0.1 MPa pressure, for in this case the enthalpy of the liquid would be negative, and we would simply use the negative value for the enthalpy of the liquid when solving problems.

14.4 FIRST LAW ANALYSIS OF REACTING SYSTEMS

The significance of the enthalpy of formation is that it is most convenient in performing a first law analysis of a reacting system, for the enthalpies of different substances can be added or subtracted, since they are all given relative to the same base.

In such problems we will write the first law for a steady-state, steady-flow process in the form

$$Q_{c.v.} + H_R = W_{c.v.} + H_P$$

or

$$Q_{c.v.} + \sum_R n_i \bar{h}_i = W_{c.v.} + \sum_P n_e \bar{h}_e$$

where R and P refer to the reactants and products, respectively. In each problem it is necessary to choose one parameter as the basis of the solution. Usually this is taken as 1 kmol of fuel.

Example 14.6 Consider the following reaction, which occurs in a steady-state, steady-flow process.

$$CH_4 + 2O_2 \rightarrow CO_2 + 2H_2O(l)$$

The reactants and products are each at a total pressure of 0.1 MPa and 25°C. Determine the heat transfer per kilomole of fuel entering the combustion chamber.

Control volume: Combustion chamber.

Inlet state: P and T known; state fixed.

Exit state: P and T known; state fixed.

Process: SSSF.

Model: Three gases ideal gases; real liquid water.

Analysis:

First law:

$$Q_{c.v.} + \sum_R n_i \bar{h}_i = \sum_P n_e \bar{h}_e$$

Solution

Using values from Table A.9, we have

$$\sum_R n_i \bar{h}_i = (\bar{h}_f^\circ)_{CH_4} = -74\,873 \text{ kJ}$$

$$\sum_P n_e \bar{h}_e = (\bar{h}_f^\circ)_{CO_2} + 2(\bar{h}_f^\circ)_{H_2O(l)}$$

$$= -393\,522 + 2(-285\,830) = -965\,182 \text{ kJ}$$

$$Q_{c.v.} = -965\,182 - (-74\,873) = -890\,309 \text{ kJ}$$

In most instances, however, the substances that comprise the reactants and products in a chemical reaction are not at a temperature of 25°C and a pressure of 0.1 MPa (the state at which the enthalpy of formation is given). Therefore, the change of enthalpy between 25°C and 0.1 MPa and the given state must be known. For a solid or liquid, this change of enthalpy can usually be found from a table of thermodynamic properties or from specific heat data. For gases, this change of enthalpy can usually be found by one of the following procedures.

1. Assume ideal-gas behavior between 25°C, 0.1 MPa, and the given state. In this case, the enthalpy is a function of the temperature only, and can be found by an equation of $\bar{C}_{po}$ or from tabulated values of enthalpy as a function of temperature (which assumes ideal gas behavior). Table A.6 gives an equation for $\bar{C}_{po}$ for a number of substances and Table A.8 gives values of $(\bar{h}^\circ - \bar{h}^\circ_{298})$ (that is, the $\Delta\bar{h}$ of Eq. 14.11) in kJ/kmol, ($\bar{h}^\circ_{298}$ refers to 25°C or 298.15 K. For simplicity this is designated $\bar{h}^\circ_{298}$.) The superscript $^\circ$ is used to designate that this is the enthalpy at 0.1 MPa pressure, based on ideal-gas behavior, that is, the standard-state enthalpy.

2. If a table of thermodynamic properties is available, $\Delta\bar{h}$ can be found directly from these tables if a real substance behavior reference state is being used, such as that described above for liquid water. If a hypothetical ideal gas reference state is being used, then it is necessary to account for the real substance correction to properties at that state to gain entry to the tables.

3. If the deviation from ideal-gas behavior is significant, but no tables of thermodynamic properties are available, the value for $\Delta\bar{h}$ can be found from the generalized tables or charts and the values for $\bar{C}_{po}$ or $\Delta\bar{h}$ at 0.1 MPa pressure as indicated above.

Thus, in general, for applying the first law to a steady-state, steady-flow process involving a chemical reaction and negligible changes in kinetic and potential energy, we can write

$$Q_{c.v.} + \sum_R n_i (\bar{h}_f^\circ + \Delta\bar{h})_i = W_{c.v.} + \sum_P n_e (\bar{h}_f^\circ + \Delta\bar{h})_e \qquad (14.12)$$

EXAMPLE 14.7 Calculate the enthalpy of water (on a kilomole basis) at 3.5 MPa, 300°C, relative to the 25°C and 0.1 MPa base, using the following procedures.

1. Assume the steam to be an ideal gas with the value of $\bar{C}_{po}$ given in the Appendix, Table A.6.
2. Assume the steam to be an ideal gas with the value for $\Delta\bar{h}$ as given in the Appendix, Table A.8.
3. The steam tables.
4. The specific heat behavior given in 2 above and the generalized charts.

Solution

For each of these procedures, we can write

$$\bar{h}_{T,P} = \left(\bar{h}_f^\circ + \Delta\bar{h}\right)$$

The only difference is in the procedure by which we calculate $\Delta\bar{h}$. From Table A.9 we note that

$$(\bar{h}_f^\circ)_{H_2O(g)} = -241\,826 \text{ kJ / kmol}$$

1. Using the specific heat equation for $H_2O(g)$ from Table A.6,

$$\bar{C}_{po} = 143.05 - 183.54\,\theta^{0.25} + 82.751\,\theta^{0.5} - 3.6989\,\theta \text{ kJ / kmol K}$$

where

$$\theta = T / 100$$

Therefore,

$$\Delta\bar{h} = \int_{298.15}^{573.15} \bar{C}_{po}(T)dT$$
$$= \int_{2.9815}^{5.7315} \bar{C}_{po}(\theta)100\,d\theta$$
$$= 9517 \text{ kJ / kmol}$$
$$\bar{h}_{T,P} = -241\,826 + 9517 = -232\,309 \text{ kJ / kmol}$$

2. Using Table A.8 for $H_2O(g)$,

$$\Delta\bar{h} = 9494 \text{ kJ / kmol}$$
$$\bar{h}_{T,P} = -241\,826 + 9494 = -232\,332 \text{ kJ / kmol}$$

3. Using the steam tables, either the liquid reference or the gaseous reference state may be used.

   For the liquid,

$$\Delta \overline{h} = 18.015(2977.5 - 104.9) = 51\,750 \text{ kJ / kmol}$$

$$\overline{h}_{T,P} = -285\,830 + 51\,750 = -234\,080 \text{ kJ / kmol}$$

For the gas,

$$\Delta \overline{h} = 18.015(2977.5 - 2547.2) = 7752 \text{ kJ / kmol}$$

$$\overline{h}_{T,P} = -241\,826 + 7752 = -234\,074 \text{ kJ / kmol}$$

The very small difference results from using the enthalpy of saturated vapor at 25°C (which is almost but not exactly an ideal gas) in calculating the $\Delta \overline{h}$.

4. When using the generalized charts we use the notation introduced in Chapter 13.

$$\overline{h}_{T,P} = \overline{h}_f^{\circ} - (\overline{h}_2^* - \overline{h}_2) + (\overline{h}_2^* - \overline{h}_1^*) + (\overline{h}_1^* - \overline{h}_1)$$

where subscript 2 refers to the state at 3.5 MPa, 300°C, and state 1 refers to the state at 0.1 MPa, 25°C.
From part 2, $\overline{h}_2 - \overline{h}_1^* = 9494$ kJ/kmol.

$$\overline{h}_1^* - \overline{h}_1 = 0 \qquad \text{(ideal gas reference)}$$

$$P_{r2} = \frac{3.5}{22.09} = 0.158 \qquad T_{r2} = \frac{573.2}{647.3} = 0.886$$

From the generalized enthalpy chart, Fig. D.2.

$$\frac{\overline{h}_2^* - \overline{h}_2}{\overline{R} T_c} = 0.21, \quad \overline{h}_2^* - \overline{h}_2 = 0.21 \times 8.3145 \times 647.3 = 1130 \text{ kJ / kmol}$$

$$\overline{h}_{T,P} = -241\,826 - 1130 + 9494 = -233\,462 \text{ kJ / mol}$$

The particular approach that is used in a given problem will depend on the data available for the given substance.

EXAMPLE 14.8   A small gas turbine uses $C_8H_{18}(l)$ for fuel, and 400% theoretical air. The air and fuel enter at 25°C and the products of combustion leave at 900 K. The output of the engine and the fuel consumption are measured, and it is found that the specific fuel consumption is 0.25 kg/s of fuel per megawatt output. Determine the heat transfer from the engine per kilomole of fuel. Assume complete combustion.

Control volume: Gas turbine engine.

Inlet states: *T* known for fuel and air.

Exit state: *T* known for combustion products.

Process: SSSF.

Model: All gases ideal gases, Table A.8; liquid octane, Table A.9.

Analysis:

The combustion equation is

$$C_8H_{18}(l) + 4(12.5)O_2 + 4(12.5)(3.76)N_2 \rightarrow 8CO_2 + 9H_2O + 37.5O_2 + 188.0N_2$$

First law:

$$Q_{c.v.} + \sum_R n_i(\overline{h}_f^\circ + \Delta\overline{h})_i = W_{c.v.} + \sum_P n_e(\overline{h}_f^\circ + \Delta\overline{h})_e$$

Solution

Since the air is composed of elements and enters at 25°C, the enthalpy of the reactants is equal to that of the fuel,

$$\sum_R n_i(\overline{h}_f^\circ + \Delta\overline{h})_i = (\overline{h}_f^\circ)_{C_8H_{18}(l)} = -250\,105 \text{ kJ / kmol fuel}$$

Considering the products, we have

$$\sum_P n_e(\overline{h}_f^\circ + \Delta\overline{h})_e = n_{CO_2}(\overline{h}_f^\circ + \Delta\overline{h})_{CO_2} + n_{H_2O}(\overline{h}_f^\circ + \Delta\overline{h})_{H_2O}$$

$$+ n_{O_2}(\Delta\overline{h})_{O_2} + n_{N_2}(\Delta\overline{h})_{N_2}$$

$$= 8(-393\,522 + 28\,030) + 9(-241\,826 + 21\,892)$$

$$+ 37.5(19\,249) + 188(18\,222)$$

$$= -755\,769 \text{ kJ / kmol fuel}$$

$$W_{c.v.} = \frac{1000 \text{ kJ / s}}{0.25 \text{ kg / s}} \times \frac{114.23 \text{ kg}}{\text{kmol}} = 456\,920 \text{ kJ / kmol fuel}$$

Therefore, from the first law,

$$Q_{c.v.} = -755\,769 + 456\,920 - (-250\,105)$$

$$= -48\,744 \text{ kJ / kmol fuel}$$

EXAMPLE 14.8E   A small gas turbine uses $C_8H_{18}(l)$ for fuel, and 400% theoretical air. The air and fuel enter at 77 F, and the products of combustion leave at 1100 F. The output of the engine and the fuel consumption are measured and it is found that the specific fuel consumption is one pound of fuel per horsepower-hour. Determine the heat transfer from the engine per pound mole of fuel. Assume complete combustion.

Control volume: Gas turbine engine.

Inlet states: T known for fuel and air.

Exit state: T known for combustion products.

Process: SSSF.

Model: All gases ideal gases, Table C.7; liquid octane, Table C.12

Analysis:

The combustion equation is

$$C_8H_{18}(l) + 4(12.5)O_2 + 4(12.5)(3.76)N_2 \rightarrow 8CO_2 + 9H_2O + 37.5O_2 + 188.0N_2$$

First law:

$$Q_{c.v.} + \sum_R n_i (\bar{h}_f^\circ + \Delta\bar{h})_i = W_{c.v.} + \sum_P n_e (\bar{h}_f^\circ + \Delta\bar{h})_e$$

Solution

Since the air is composed of elements and enters at 77 F, the enthalpy of the reactants is equal to that of the fuel.

$$\sum_R n_i \left[\bar{h}_f^\circ + \Delta\bar{h}\right]_i = (\bar{h}_f^\circ)_{C_8H_{18}(l)} = -107\,526 \text{ Btu / lb mol.}$$

Considering the products

$$\sum_P n_e (\bar{h}_f^\circ + \Delta\bar{h})_e = n_{CO_2}(\bar{h}_f^\circ + \Delta\bar{h})_{CO_2} + n_{H_2O}(\bar{h}_f^\circ + \Delta\bar{h})_{H_2O} + n_{O_2}(\Delta\bar{h})_{O_2} + n_{N_2}(\Delta\bar{h})_{N_2}$$

$$= 8(-169\,184 + 11\,291) + 9(-103\,966 + 8858)$$

$$+ 37.5(7778) + 188(7374)$$

$$= -441\,129 \text{ Btu / lb mol fuel.}$$

$$W_{c.v.} = 2544 \times 114.23 = 290\,601 \text{ Btu / lb mol fuel.}$$

Therefore, from the first law,

$$Q_{c.v.} = -441\,129 + 290\,601 - (-107\,526)$$

$$= -43\,002 \text{ Btu / lb mol fuel.}$$

EXAMPLE 14.9 A mixture of 1 kmol of gaseous ethene and 3 kmol of oxygen at 25°C reacts in a constant-volume bomb. Heat is transferred until the products are cooled to 600 K. Determine the amount of heat transfer from the system.

Control mass: Constant-volume bomb.

Initial state: T known.

Final state: T known.

Process: Constant volume.

Model: Ideal gas mixtures, Tables A.8, A.9,

Analysis:

The chemical reaction is

$$C_2H_4 + 3O_2 \rightarrow 2CO_2 + 2H_2O(g)$$

First law:

$$Q + U_R = U_P$$

$$Q + \sum_R n(\bar{h}_f^\circ + \Delta\bar{h} - \bar{R}T) = \sum_P n(\bar{h}_f^\circ + \Delta\bar{h} - \bar{R}T)$$

Solution

Using values from Tables A.8 and A.9, gives

$$\sum_R n(\bar{h}_f^\circ + \Delta\bar{h} - \bar{R}T) = (\bar{h}_f^\circ - \bar{R}T)_{C_2H_4} - n_{O_2}(\bar{R}T)_{O_2} = (\bar{h}_f^\circ)_{C_2H_4} - 4\bar{R}T$$

$$= 52\,467 - 4\times8.3145\times298.2 = 42\,550 \text{ kJ}$$

$$\sum_P n(\bar{h}_f^\circ + \Delta\bar{h} - \bar{R}T) = 2\left[(\bar{h}_f^\circ)_{CO_2} + \Delta\bar{h}_{CO_2}\right] + 2\left[(\bar{h}_f^\circ)_{H_2O(g)} + \Delta\bar{h}_{H_2O(g)}\right] - 4\bar{R}T$$

$$= 2(-393\,522 + 12\,899) + 2(-241\,826 + 10\,463)$$

$$-4\times8.3145\times600$$

$$= -1\,243\,927 \text{ kJ}$$

Therefore,

$$Q = -1\,243\,927 - 42\,550 = -1\,286\,477 \text{ kJ}$$

For a real gas mixture, a pseudocritical method such as Kay's rule, Eq. 13.89, could be used to evaluate the nonideal gas contribution to enthalpy at the temperature and pressure of the mixture and this value added to the ideal gas mixture enthalpy at that temperature, as in the procedure developed in Section 13.11.

14.5 ADIABATIC FLAME TEMPERATURE

Consider a given combustion process that takes place adiabatically and with no work or changes in kinetic or potential energy involved. For such a process the temperature of the products is referred to as the adiabatic flame temperature. With the assumptions of no work and no changes in kinetic or potential energy, this is the maximum temperature that can be achieved for the given reactants because any heat transfer from the reacting substances and any incomplete combustion would tend to lower the temperature of the products.

For a given fuel and given pressure and temperature of the reactants, the maximum adiabatic flame temperature that can be achieved is with a stoichiometric mixture. The adiabatic flame temperature can be controlled by the amount of excess air that is used. This is important, for example, in gas turbines, where the maximum permissible temperature is determined by metallurgical considerations in the turbine, and close control of the temperature of the products is essential.

Example 14.10 shows how the adiabatic flame temperature may be found. The dissociation that takes place in the combustion products, which has a significant effect on the adiabatic flame temperature, will be considered in the next chapter.

EXAMPLE 14.10 Liquid octane at 25°C is burned with 400% theoretical air at 25°C in a steady-flow process. Determine the adiabatic flame temperature.

Control volume: Combustion chamber.

Inlet states: *T* known for fuel and air.

Process: SSSF.

Model: Gases ideal gases, Table A.8; liquid octane, Table A.9.

Analysis:

The reaction is

$$C_8H_{18}(l) + 4(12.5)O_2 + 4(12.5)(3.76)N_2 \rightarrow$$

$$8CO_2 + 9H_2O(g) + 37.5O_2 + 188.0N_2$$

First law: Since the process is adiabatic,

$$H_R = H_P$$

$$\sum_R n_i(\bar{h}_f^{\circ} + \Delta\bar{h})_i = \sum_P n_e(\bar{h}_f^{\circ} + \Delta\bar{h})_e$$

where $\Delta\bar{h}_e$ refers to each constituent in the products at the adiabatic flame temperature.

Solution

From Tables A.8 and A.9,

$$H_R = \sum_R n_i(\bar{h}_f^{\circ} + \Delta\bar{h})_i = (\bar{h}_f^{\circ})_{C_8H_{18}(l)} = -250\,105 \text{ kJ / kmol fuel}$$

$$H_P = \sum_P n_e(\bar{h}_f^{\circ} + \Delta\bar{h})_e$$

$$= 8(-393\,522 + \Delta\bar{h}_{CO_2}) + 9(-241\,826 + \Delta\bar{h}_{H_2O}) + 37.5\Delta\bar{h}_{O_2} + 188.0\Delta\bar{h}_{N_2}$$

By trial-and-error solution, a temperature of the products is found that satisfies this equation. Assume that

$$T_P = 900 \text{ K}$$

$$H_P = \sum_P n_e(\bar{h}_f^{\circ} + \Delta\bar{h})_e$$

$$= 8(-393\,522 + 28\,030) + 9(-241\,826 + 21\,892)$$

$$+ 37.5(19\,249) + 188(18\,222)$$

$$= -755\,769 \text{ kJ / kmol fuel}$$

Assume that

$$T_P = 1000 \text{ K}$$

$$H_P = \sum_P n_e(\bar{h}_f^{\circ} + \Delta\bar{h})_e$$

$$= 8(-393\,522 + 33\,400) + 9(-241\,826 + 25\,956)$$

$$+ 37.5(22\,710) + 188(21\,461)$$

$$= 62\,487 \text{ kJ / kmol fuel}$$

Since $H_P = H_R = -250\,105$ kJ, we find by linear interpolation that the adiabatic flame temperature is 961.8 K. Because the ideal-gas enthalpy is not really a linear function of temperature, the true answer will be slightly different from this value.

14.6 ENTHALPY AND INTERNAL ENERGY OF COMBUSTION; HEAT OF REACTION

The enthalpy of combustion, h_{RP}, is defined as the difference between the enthalpy of the products and the enthalpy of the reactants when complete combustion occurs at a given temperature and pressure. That is,

$$\bar{h}_{RP} = H_P - H_R$$
$$\bar{h}_{RP} = \sum_P n_e(\bar{h}_f^\circ + \Delta\bar{h})_e - \sum_R n_i(\bar{h}_f^\circ + \Delta\bar{h})_i \tag{14.13}$$

The usual parameter for expressing the enthalpy of combustion is a unit mass of fuel, such as a kilogram (h_{RP}) or a kilomole ($\bar{h}_{RP}$) of fuel.

The tabulated values of the enthalpy of combustion of fuels are usually given for a temperature of 25°C and a pressure of 0.1 MPa. The enthalpy of combustion for a number of hydrocarbon fuels at this temperature and pressure, which we designate h_{RP0}, is given in Table 14.3.

The internal energy of combustion is defined in a similar manner.

$$\bar{u}_{RP} = U_P - U_R$$
$$= \sum_P n_e(\bar{h}_f^\circ + \Delta\bar{h} - P\bar{v})_e - \sum_R n_i(\bar{h}_f^\circ + \Delta\bar{h} - P\bar{v})_i \tag{14.14}$$

When all the gaseous constituents can be considered as ideal gases, and the volume of the liquid and solid constituents is negligible compared to the value of the gaseous constituents, this relation for $\bar{u}_{RP}$ reduces to

$$\bar{u}_{RP} = \bar{h}_{RP} - \bar{R}T(n_{\text{gaseous products}} - n_{\text{gaseous reactants}}) \tag{14.15}$$

Frequently the term "heating value" or "heat of reaction" is used. This represents the heat transferred from the chamber during combustion or reaction at constant temperature. In the case of a constant pressure or steady-flow process, we conclude from the first law of thermodynamics that it is equal to the negative of the enthalpy of combustion. For this reason this heat transfer is sometimes designated the constant-pressure heating value for combustion processes.

In the case of a constant-volume process, the heat transfer is equal to the negative of the internal energy of combustion. This is sometimes designated the constant-volume heating value in the case of combustion.

When the term heating value is used, the terms "higher" and "lower" heating value are used. The higher heating value is the heat transfer with liquid water in the products, and the lower heating value is the heat transfer with vapor water in the products.

EXAMPLE 14.11 Calculate the enthalpy of combustion of propane at 25°C on both a kilomole and kilogram basis under the following conditions.

1. Liquid propane with liquid water in the products.
2. Liquid propane with gaseous water in the products.
3. Gaseous propane with liquid water in the products.
4. Gaseous propane with gaseous water in the products.

TABLE 14.3

Enthalpy of Combustion of Some Hydrocarbons at 25°C

Hydrocarbon	UNITS: $\frac{kJ}{kg}$ Formula	LIQUID H_2O IN PRODUCTS		GAS H_2O IN PRODUCTS	
		Liq. HC	Gas HC	Liq. HC	Gas HC
Paraffins	C_nH_{2n+2}				
Methane	CH_4		−55 496		−50 010
Ethane	C_2H_6		−51 875		−47 484
Propane	C_3H_8	−49 973	−50 343	−45 982	−46 352
n-Butane	C_4H_{10}	−49 130	−49 500	−45 344	−45 714
n-Pentane	C_5H_{12}	−48 643	−49 011	−44 983	−45 351
n-Hexane	C_6H_{14}	−48 308	−48 676	−44 733	−45 101
n-Heptane	C_7H_{16}	−48 071	−48 436	−44 557	−44 922
n-Octane	C_8H_{18}	−47 893	−48 256	−44 425	−44 788
n-Decane	$C_{10}H_{22}$	−47 641	−48 000	−44 239	−44 598
n-Dodecane	$C_{12}H_{26}$	−47 470	−47 828	−44 109	−44 467
n-Cetane	$C_{16}H_{34}$	−47 300	−47 658	−44 000	−44 358
Olefins	C_nH_{2n}				
Ethene	C_2H_4		−50 296		−47 158
Propene	C_3H_6		−48 917		−45 780
Butene	C_4H_8		−48 453		−45 316
Pentene	C_5H_{10}		−48 134		−44 996
Hexene	C_6H_{12}		−47 937		−44 800
Heptene	C_7H_{14}		−47 800		−44 662
Octene	C_8H_{16}		−47 693		−44 556
Nonene	C_9H_{18}		−47 612		−44 475
Decene	$C_{10}H_{20}$		−47 547		−44 410
Alkylbenzenes	$C_{6+n}H_{6+2n}$				
Benzene	C_6H_6	−41 831	−42 266	−40 141	−40 576
Methylbenzene	C_7H_8	−42 437	−42 847	−40 527	−40 937
Ethylbenzene	C_8H_{10}	−42 997	−43 395	−40 924	−41 322
Propylbenzene	C_9H_{12}	−43 416	−43 800	−41 219	−41 603
Butylbenzene	$C_{10}H_{14}$	−43 748	−44 123	−41 453	−41 828
Other fuels					
Gasoline	C_7H_{17}	−48 201	−48 582	−44 506	−44 886
Diesel	$C_{14.4}H_{24.9}$	−45 700	−46 074	−42 934	−43 308
Methanol	CH_3OH	−22 657	−23 840	−19 910	−21 093
Ethanol	C_2H_5OH	−29 676	−30 596	−26 811	−27 731
Nitromethane	CH_3NO_2	−11 618	−12 247	−10 537	−11 165
Phenol	C_6H_5OH	−32 520	−33 176	−31 117	−31 774
Hydrogen	H_2		−60955		−51 750

This example is designed to show how the enthalpy of combustion can be determined from enthalpies of formation. The enthalpy of evaporation of propane is 370 kJ/kg.

Analysis and Solution

The basic combustion equation is

$$C_3H_8 + 5O_2 \rightarrow 3CO_2 + 4H_2O$$

From Table A.9, $(\overline{h}_f^\circ)_{C_3H_8(g)} = -103\,900$ kJ/kmol. Therefore,

$$(\overline{h}_f^\circ)_{C_3H_8(l)} = -103\,900 - 44.097(370) = -120\,216 \text{ kJ / kmol}$$

1. Liquid propane–liquid water:

$$\overline{h}_{RP_0} = 3(\overline{h}_f^\circ)_{CO_2} + 4(\overline{h}_f^\circ)_{H_2O(l)} - (\overline{h}_f^\circ)_{C_3H_8(l)}$$
$$= 3(-393\,522) + 4(-285\,830) - (-120\,216)$$
$$= -2\,203\,670 \text{ kJ / kmol} = -\frac{2\,203\,670}{44.097} = -49\,973 \text{ kJ / kg}$$

The higher heating value of liquid propane is 49 973 kJ/kg.

2. Liquid propane–gaseous water:

$$\overline{h}_{RP_0} = 3(\overline{h}_f^\circ)_{CO_2} + 4(\overline{h}_f^\circ)_{H_2O(g)} - (\overline{h}_f^\circ)_{C_3H_8(l)}$$
$$= 3(-393\,522) + 4(-241\,826) - (-120\,216)$$
$$= -2\,027\,654 \text{ kJ / kmol} = -\frac{2\,027\,654}{44.097} = -45\,982 \text{ kJ / kg}$$

The lower heating value of liquid propane is 45 982 kJ/kg.

3. Gaseous propane–liquid water:

$$\overline{h}_{RP_0} = 3(\overline{h}_f^\circ)_{CO_2} + 4(\overline{h}_f^\circ)_{H_2O(l)} - (\overline{h}_f^\circ)_{C_3H_8(g)}$$
$$= 3(-393\,522) + 4(-285\,830) - (-103\,900)$$
$$= -2\,219\,986 \text{ kJ / kmol} = -\frac{2\,219\,986}{44.097} = -50\,343 \text{ kJ / kg}$$

The higher heating value of gaseous propane is 50 343 kJ/kg.

4. Gaseous propane–gaseous water:

$$\overline{h}_{RP_0} = 3(\overline{h}_f^\circ)_{CO_2} + 4(\overline{h}_f^\circ)_{H_2O(g)} - (\overline{h}_f^\circ)_{C_3H_8(g)}$$
$$= 3(-393\,522) + 4(-241\,826) - (-103\,900)$$
$$= -2\,043\,970 \text{ kJ / kmol} = -\frac{2\,043\,970}{44.097} = -46\,352 \text{ kJ / kg}$$

The lower heating value of gaseous propane is 46 352 kJ/kg.

Each of the four values calculated in this example corresponds to the appropriate value given in Table 14.3.

EXAMPLE 14.12 Calculate the enthalpy of combustion of gaseous propane at 500 K. (At this temperature all the water formed during combustion will be vapor.) This example will demonstrate how the enthalpy of combustion of propane varies with temperature. The average constant-pressure specific heat of propane between 25°C and 500 K is 2.1 kJ/kg K.

Analysis:

The combustion equation is

$$C_3H_8(g) + 5O_2 \rightarrow 3CO_2 + 4H_2O(g)$$

The enthalpy of combustion is, from Eq. 14.13,

$$(\bar{h}_{RP})_T = \sum_P n_e (\bar{h}_f^\circ + \Delta \bar{h})_e - \sum_R n_i (\bar{h}_f^\circ + \Delta \bar{h})_i$$

Solution

$$\bar{h}_{R_{500}} = \left[\bar{h}_f^\circ + \bar{C}_{p\,av}(\Delta T) \right]_{C_3H_8(g)} + n_{O_2}(\Delta \bar{h})_{O_2}$$

$$= -103\,900 + 2.1 \times 44.097(500 - 298.2) + 5(6095)$$

$$= -54\,738 \text{ kJ}$$

$$\bar{h}_{P_{500}} = n_{CO_2}(\bar{h}_f^\circ + \Delta \bar{h})_{CO_2} + n_{H_2O}(\bar{h}_f^\circ + \Delta \bar{h})_{H_2O}$$

$$= 3(-393\,522 + 8297) + 4(-241\,826 + 6896)$$

$$= -2\,095\,395 \text{ kJ}$$

$$\bar{h}_{RP_{500}} = -2\,095\,395 - (-54\,738) = -2\,040\,657 \text{ kJ / kmol}$$

$$h_{RP_{500}} = \frac{-2\,040\,657}{44.097} = -46\,277 \text{ kJ / kg}$$

This compares with a value of −46 352 at 25°C.

This problem could also have been solved using the given value of the enthalpy of combustion at 25°C by noting that

$$\bar{h}_{RP_{500}} = (H_P)_{500} - (H_R)_{500}$$

$$= n_{CO_2}(\bar{h}_f^\circ + \Delta \bar{h})_{CO_2} + n_{H_2O}(\bar{h}_f^\circ + \Delta \bar{h})_{H_2O}$$

$$- \left[\bar{h}_f^\circ + \bar{C}_{p\,av}(\Delta T) \right]_{C_3H_8(g)} - n_{O_2}(\Delta \bar{h})_{O_2}$$

$$= \bar{h}_{RP_0} + n_{CO_2}(\Delta \bar{h})_{CO_2} + n_{H_2O}(\Delta \bar{h})_{H_2O}$$

$$- \bar{C}_{p\,av}(\Delta T)_{C_3H_8(g)} - n_{O_2}(\Delta \bar{h})_{O_2}$$

$$\bar{h}_{RP_{500}} = -46\,352 \times 44.097 + 3(8297) + 4(6896)$$

$$- 2.1 \times 44.097(500 - 298.2) - 5(6095)$$

$$= -2\,040\,657 \text{ kJ / kmol}$$

$$h_{RP_{500}} = \frac{-2\,040\,657}{44.097} = -46\,277 \text{ kJ / kg}$$

14.7 The Third Law of Thermodynamics and Absolute Entropy

As we consider a second-law analysis of chemical reactions, we face the same problem we had with the first law: What base should be used for the entropy of the various substances? This problem leads directly to a consideration of the third law of thermodynamics.

The third law of thermodynamics was formulated during the early part of the twentieth century. The initial work was done primarily by W. H. Nernst (1864–1941) and Max Planck (1858–1947). The third law deals with the entropy of substances at the absolute zero of temperature, and in essence states that the entropy of a perfect crystal is zero at absolute zero. From a statistical point of view, this means that the crystal structure has the maximum degree of order. Further, because the temperature is absolute zero, the thermal energy is minimum. It also follows that a substance that does not have a perfect crystalline structure at absolute zero, but instead has a degree of randomness, such as a solid solution or a glassy solid, has a finite value of entropy at absolute zero. The experimental evidence on which the third law rests is primarily data on chemical reactions at low temperatures and measurements of heat capacity at temperatures approaching absolute zero. In contrast to the first and second laws, which lead, respectively, to the properties of internal energy and entropy, the third law deals only with the question of entropy at absolute zero. However, the implications of the third law are quite profound, particularly in respect to chemical equilibrium.

The particular relevance of the third law is that it provides an absolute base from which to measure the entropy of each substance. The entropy relative to this base is termed the absolute entropy. The increase in entropy between absolute zero and any given state can be found either from calorimetric data or by procedures based on statistical thermodynamics. The calorimetric method gives precise measurements of specific-heat data over the temperature range, as well as of the energy associated with phase transformations. These measurements are in agreement with the calculations based on statistical thermodynamics and observed molecular data.

Table A.9 gives the absolute entropy at 25°C and 0.1-MPa pressure for a number of substances. Table A.8 gives the absolute entropy for a number of gases at 0.1-MPa pressure and various temperatures. For gases the numbers in all these tables are the hypothetical ideal-gas values. The pressure $P°$ of 0.1 MPa is termed the standard-state pressure, and the absolute entropy as given in these tables is designated $\bar{s}°$. The temperature is designated in kelvins with a subscript such as $\bar{s}°_{1000}$.

If the value of the absolute entropy is known at the standard-state pressure of 0.1 MPa and a given temperature, it is a straightforward procedure to calculate the entropy change from this state (whether hypothetical ideal gas or real substance) to another desired state following the procedure described in Section 13.11. If the substance is listed in Table A.8, then

$$\bar{s}_{T,P} = \bar{s}°_T - \bar{R} \ln \frac{P}{P°} + (\bar{s}_{T,P} - \bar{s}^*_{T,P}) \tag{14.16}$$

In this expression, the first term on the right side is the value from Table A.8, the second is the ideal-gas term to account for a change in pressure from $P°$ to P, and the third is the term that corrects for real-substance behavior, as given in the generalized entropy chart in the Appendix. If the real-substance behavior is to be evaluated from an equation of state or thermodynamic table of properties, the term for the change in pressure should be made to a low pressure $P*$, at which ideal-gas behavior is a reasonable assumption, but it is also listed in the tables. Then

$$\bar{s}_{T,P} = \bar{s}_T^\circ - \bar{R} \ln \frac{P*}{P^\circ} + (\bar{s}_{T,P} - \bar{s}_{T,P*}^*) \tag{14.17}$$

If the substance is not one of those listed in Table A.8, and the absolute entropy is known only at one temperature T_0, as given in Table A.9 for example, then it will be necessary to calculate $\bar{s}_T^\circ$ from

$$\bar{s}_T^\circ = \bar{s}_{T_0}^\circ + \int_{T_0}^T \frac{\overline{C}_{p0}}{T} dT \tag{14.18}$$

and then proceed with the calculation of Eq. 14.16 or 14.17.

It should be noted that if Eq. 14.16 is being used to calculate the absolute entropy of a substance in a region in which the ideal-gas model is a valid representation of the behavior of that substance, then the last term on the right-side of Eq. 14.16 simply drops out of the calculation.

For calculation of the absolute entropy of a mixture of ideal gases at T, P, the mixture entropy is given in terms of the component partial entropies as

$$\bar{s}_{\text{mix}}^* = \sum_i y_i \bar{S}_i^* \tag{14.19}$$

where

$$\bar{S}_i^* = \bar{s}_{Ti}^\circ - \bar{R} \ln \frac{P}{P^\circ} - \bar{R} \ln y_i = \bar{s}_{Ti}^\circ - \bar{R} \ln \frac{y_i P}{P^\circ} \tag{14.20}$$

For a real-gas mixture, a correction can be added to the ideal-gas entropy calculated from Eqs. 14.19 and 14.20 by using a pseudocritical method such as was discussed in Section 13.11. The corrected expression is

$$\bar{s}_{\text{mix}} = \bar{s}_{\text{mix}}^* + (\bar{s} - \bar{s}*)_{T,P} \tag{14.21}$$

in which the second term on the right side is the correction term from the generalized entropy chart.

14.8 SECOND-LAW ANALYSIS OF REACTING SYSTEMS

The concepts of reversible work, irreversibility, and availability were introduced in Chapter 10. These concepts included both the first and second laws of thermodynamics. We proceed now to develop this matter further, and we will be particularly concerned with determining the maximum work (availability) that can be done through a combustion process and with examining the irreversibilities associated with such processes.

The reversible work for a steady-state, steady-flow process in which there is no heat transfer with reservoirs other than the surroundings, and also in the absence of changes in kinetic and potential energy is, from Eq. 10.12 on a total mass basis,

$$W^{\text{rev}} = \sum m_i (h_i - T_0 s_i) - \sum m_e (h_e - T_0 s_e)$$

Applying this equation to an SSSF process that involves a chemical reaction, and introducing the symbols from this chapter, we have

$$W^{\text{rev}} = \sum_R n_i(\bar{h}_f^\circ + \Delta\bar{h} - T_0\bar{s})_i - \sum_P n_e(\bar{h}_f^\circ + \Delta\bar{h} - T_0\bar{s})_e \tag{14.22}$$

Similarly, the irreversibility for such a process can be written as

$$I = W^{\text{rev}} - W = \sum_P n_e T_0\bar{s}_e - \sum_R n_i T_0\bar{s}_i - Q_{\text{c.v.}} \tag{14.23}$$

The availability, ψ, for an SSSF process, in the absence of kinetic and potential energy changes, is given by Eq. 10.19 as

$$\psi = (h - T_0 s) - (h_0 - T_0 s_0)$$

We further note that if a SSSF chemical reaction takes place in such a manner that both the reactants and products are in temperature equilibrium with the surroundings, the Gibbs function ($g = h - Ts$), defined in Eq. 13.18, becomes a significant variable. For such a process, in the absence of changes in kinetic and potential energy, the reversible work is given by the relation

$$W^{\text{rev}} = \sum_R n_i\bar{g}_i - \sum_P n_e\bar{g}_e = -\Delta G \tag{14.24}$$

in which

$$\Delta G = \Delta H - T\Delta S \tag{14.25}$$

We should keep in mind that Eq. 14.24 is a special case, and that the reversible work is given by Eq. 14.22 if the reactants and products are not in temperature equilibrium with the surroundings.

Let us now consider the question of the maximum work that can be done during a chemical reaction. For example, consider 1 kmol of hydrocarbon fuel and the necessary air for complete combustion, each at 0.1-MPa pressure and 25°C, the pressure and temperature of the surroundings. What is the maximum work that can be done as this fuel reacts with the air? From the considerations covered in Chapter 10, we conclude that the maximum work would be done if this chemical reaction took place reversibly and the products were finally in pressure and temperature equilibrium with the surroundings. We conclude that this reversible work could be calculated from the relation in Eq. 14.24,

$$W^{\text{rev}} = \sum_R n_i\bar{g}_i - \sum_P n_e\bar{g}_e = -\Delta G$$

However, since the final state is in equilibrium with the surroundings, we could consider this amount of work to be the availability of the fuel and air.

EXAMPLE 14.13 Ethene (g) at 25°C and 0.1-MPa pressure is burned with 400% theoretical air at 25°C and 0.1-MPa pressure. Assume that this reaction takes place reversibly at 25°C and that the products leave at 25°C and 0.1-MPa pressure. To simplify this problem further, assume that the oxygen and nitrogen are separated before the reaction takes place (each at 0.1 MPa, 25°C), that the constituents in the products are separated, and that each is at 25°C and 0.1 MPa. Thus, the reaction takes place as shown in Fig. 14.4. For purposes of comparison between this and the two subsequent examples, we consider all the water in the products to be a gas (a hypothetical situation in this example and Example 14.15).

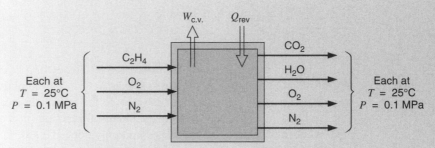

FIGURE 14.4 Sketch for Example 14.13.

Determine the reversible work for this process (that is, the work that would be done if this chemical action took place reversibly and isothermally).

Control volume: Combustion chamber.

Inlet states: P, T known for each gas.

Exit states: P, T known for each gas.

Model: All ideal gases, Tables A.8 and A.9.

Sketch: Figure 14.4.

Analysis:

The equation for this chemical reaction is

$$C_2H_4(g) + 3(4)O_2 + 3(4)(3.76)N_2 \rightarrow 2CO_2 + 2H_2O(g) + 9O_2 + 45.1N_2$$

The reversible work for this process is equal to the decrease in Gibbs function during this reaction, Eq. 14.24. Since each component is at the standard-state pressure P°, we write Eqs. 14.24 and 14.25 as

$$W^{\text{rev}} = -\Delta G^\circ, \quad \Delta G^\circ = \Delta H^\circ - T\,\Delta S^\circ$$

We also note that the $45.1N_2$ cancels out of both sides in these expressions, as does 9 of the $12O_2$.

Solution

Using values from Tables A.8 and A.9 at 25°C,

$$\Delta H^\circ = 2\,\overline{h}^{\,\circ}_{f\,CO_2} + 2\,\overline{h}^{\,\circ}_{f\,H2O(g)} - \overline{h}^{\,\circ}_{f\,C2H4} - 3\,\overline{h}^{\,\circ}_{f\,O_2}$$

$$= 2(-393\ 522) + 2(-241\ 826) - (+52\ 467) - 3(0)$$

$$= -1\ 323\ 163\ \text{kJ}$$

$$\Delta S = 2\overline{s}^{\,\circ}_{CO_2} + 2\overline{s}^{\,\circ}_{H_2O(g)} - \overline{s}^{\,\circ}_{C_2H_4} - 3\overline{s}^{\,\circ}_{O_2}$$

$$= 2(213.795) + 2(188.834) - (219.330) - 3(205.148)$$

$$= -29.516\ \text{kJ/K}$$

$$\Delta G^\circ = -1\ 323\ 163 - 298.15(-29.516)$$

$$= -1\ 314\ 363\ \text{kJ/kmol}\ C_2H_4$$

$$W^{\text{rev}} = -\Delta G^\circ = 1\ 314\ 363\ \text{kJ/kmol}\ C_2H_4$$

$$= \frac{1\ 314\ 363}{28.054} = 46\ 851\ \text{kJ/kg}$$

Therefore, we might say that when 1 kg of ethene is at 25°C, and the standard-state pressure, 0.1 MPa, it has an availability of 46 851 kJ.

Thus, it would seem logical to rate the efficiency of a device designed to do work by utilizing a combustion process, such as an internal-combustion engine or a steam power plant, as the ratio of the actual work to the reversible work, or in Example 14.13, the decrease in Gibbs function for the chemical reaction, instead of comparing the actual work to the heating value, as is commonly done. This is, in fact, the basic principle of the second-law efficiency, which was introduced in connection with availability analysis in Chapter 10. As noted from Example 14.13, the difference between the decrease in Gibbs function and the heating value is small, which is typical for hydrocarbon fuels. The difference in the two types of efficiencies will, therefore, not usually be large. We must always be careful, however, when discussing efficiencies, to note the definition of the efficiency under consideration.

It is of particular interest to study the irreversibility that takes place during a combustion process. The following examples illustrate this matter. We consider the same hydrocarbon fuel that was used in Example 14.13—ethene (g) at 25°C and 0.1 MPa. We determined its availability and found it to be 46 851 kJ/kg. Now let us burn this fuel with 400% theoretical air in a steady-state, steady-flow adiabatic process. We can determine the irreversibility of this process in two ways. The first way is to calculate the increase in entropy for the process. Since the process is adiabatic, the increase in entropy is due entirely to the irreversibilities for the process, and we can find the irreversibility from Eq. 14.23. We can also calculate the availabilities of the products of combustion at the adiabatic flame temperature, and note that they are less than the availability of the fuel and air before the combustion process. The difference is the irreversibility that occurs during the combustion process.

EXAMPLE 14.14

Consider the same combustion process as in Example 14.13, but let it take place adiabatically. Assume that each constituent in the products is at 0.1-MPa pressure and at the adiabatic flame temperature. This combustion process is shown schematically in Fig. 14.5. The temperature of the surroundings is 25°C.

For this combustion process, determine (1) the increase in entropy during combustion and (2) the availability of the products of combustion.

Control volume: Combustion chamber.

Inlet states: P, T known for each gas.

Exit states: P known for each gas.

Sketch: Figure 14.5.

Model: All ideal gases, Table A.8; Table A.9 for ethene.

Analysis:

The combustion equation is

$$C_2H_4(g) + 12O_2 + 12(3.76)N_2 \rightarrow 2CO_2 + 2H_2O(g) + 9O_2 + 45.1N_2$$

The adiabatic flame temperature is determined first.

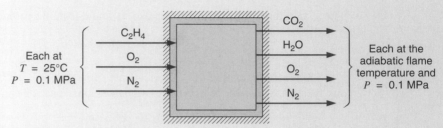

FIGURE 14.5 Sketch for Example 14.14

First law:

$$H_R = H_P$$

$$\sum_R n_i (\bar{h}_f^{\circ})_i = \sum_P n_e (\bar{h}_f^{\circ} + \Delta \bar{h})_e$$

Solution

$$52\,467 = 2(-393\,522 + \Delta \bar{h}_{CO_2}) + 2(-241\,826 + \Delta \bar{h}_{H_2O(g)}) + 9\Delta \bar{h}_{O_2} + 45.1\Delta \bar{h}_{N_2}$$

By a trial-and-error solution we find the adiabatic flame temperature to be 1016 K.

We now proceed to find the change in entropy during this adiabatic combustion process.

$$S_R = \sum_R (n_i \bar{s}_i^{\circ})_{298} = (\bar{s}_{C_2H_4}^{\circ} + 12\bar{s}_{O_2}^{\circ} + 45.1\bar{s}_{N_2}^{\circ})_{298}$$

$$= 219.330 + 12(205.147) + 45.1(191.610)$$

$$= 11\,322.705 \text{ kJ / kmol (fuel) K}$$

$$S_P = \sum_P (n_e \bar{s}_e^{\circ})_{1016} = (2\bar{s}_{CO_2}^{\circ} + 2\bar{s}_{H_2O(g)}^{\circ} + 9\bar{s}_{O_2}^{\circ} + 45.1\bar{s}_{N_2}^{\circ})_{1016}$$

$$= 2(270.194) + 2(233.355) + 9(244.135) + 45.1(228.691)$$

$$= 13\,518.277 \text{ kJ / kmol (fuel) K}$$

$$S_P - S_R = 2195.572 \text{ kJ / kmol (fuel) K}$$

Since this is an adiabatic process, the increase in entropy indicates the irreversibility of the adiabatic combustion process. This irreversibility can be found from Eq. 14.23.

$$I = T_0 \left(\sum_P n_e \bar{s}_e - \sum_R n_i \bar{s}_i \right)$$

$$= 298.15 \times 2195.572 = 654\,610 \text{ kJ / kmol}$$

$$= \frac{654\,610}{28.054} = 23\,334 \text{ kJ / kg fuel}$$

Therefore, the availability after the combustion process is

$$\psi_P = 46\,851 - 23\,334 = 23\,517 \text{ kJ / kg}$$

The availability of the products can also be found from the relation

$$\psi_P = \sum_P \left[(\bar{h}_e - T_0 \bar{s}_e) - (\bar{h}_0 - T_0 \bar{s}_0) \right]$$

Since in this problem the products are separated, and each is at 0.1-MPa pressure and the adiabatic flame temperature of 1016 K, this equation can be evaluated, yielding

$$\psi_P = \sum_P n_e \left[(\bar{h}_e^\circ - \bar{h}_0^\circ) - T_0 (\bar{s}_e^\circ - \bar{s}_0^\circ) \right]$$

$$= 2(34\,271) + 2(26\,618) + 9(23\,268) + 45.1(21\,985)$$

$$- 298.15 \left[2(270.194 - 213.795) + 2(233.355 - 188.834) \right.$$

$$+ 9(244.135 - 205.147) + 45.1(228.691 - 191.610) \big]$$

$$= 659\,746 \text{ kJ / kmol} = 23\,517 \text{ kJ / kg}$$

In other words, if every process after the adiabatic combustion process is reversible, the maximum amount of work that could be done is 23 517 kJ/kg fuel. This compares to a value of 46 851 kJ/kg for the reversible isothermal reaction. This means that if we had an engine with the indicated adiabatic combustion process, and if all other processes were completely reversible, the efficiency would be about 50%.

In the two prior examples we assumed, for purposes of simplifying the calculation, that the constituents in the reactants and products were separated, and each was at 0.1-MPa pressure. This of course is not a realistic problem. In the following example, Example 14.13 is repeated with the assumption that the reactants and products each consist of a mixture at 0.1-MPa pressure.

EXAMPLE 14.15   Consider the same combustion process as in Example 14.13, but assume that the reactants consist of a mixture at 0.1-MPa pressure and 25°C and that the products also consist of a mixture at 0.1 MPa and 25°C. Thus, the combustion process is as shown in Fig. 14.6.

Determine the work that would be done if this combustion process took place reversibly and in pressure and temperature equilibrium with the surroundings.

Control volume: Combustion chamber.

Inlet state: P, T known.

Exit state: P, T known.

Sketch: Figure 14.6.

Model: Reactants—ideal-gas mixture, Table A.8. Products—ideal-gas mixture, Table A.8.

Analysis:

The combustion equation, as noted previously, is

$$C_2H_4(g) + 3(4)O_2 + 3(4)(3.76)N_2 \rightarrow 2CO_2 + 2H_2O(g) + 9O_2 + 45.1N_2$$

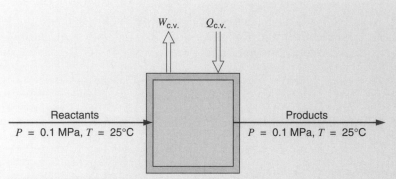

FIGURE 14.6 Sketch for Example 14.15.

In this case we must find the entropy of each substance as it exists in the mixture; that is, at its partial pressure and the given temperature of 25°C. Because the absolute entropies given in Tables A.8 and A.9 are at 0.1-MPa pressure and 25°C, the entropy of each constituent in the mixture can be found, using Eq. 14.20 from the relation

$$\bar{S}^* = \bar{s}^{\circ} - \bar{R} \ln y \frac{P}{P^{\circ}}$$

where S^* = partial entropy of the constituent in the mixture
$\bar{s}^{\circ}$ = absolute entropy at the same temperature and 0.1-MPa pressure
P = pressure of the mixture
P° = 0.1-MPa pressure
y = mole fraction of the constituent

Since P° and the pressure of the mixture are both 0.1 MPa, the partial entropy of each constituent can be found by the relation

$$\bar{S}^* = \bar{s}^{\circ} - \bar{R} \ln y = \bar{s}^{\circ} + \bar{R} \ln \frac{1}{y}$$

Solution

For the reactants:

	n	$1/y$	$\bar{R} \ln 1/y$	$\bar{s}^{\circ}$	$\bar{S}^*$
C_2H_4	1	58.1	33.774	219.330	253.104
O_2	12	4.842	13.114	205.147	218.261
N_2	45.1	1.288	2.104	191.610	193.714
	58.1				

For the products:

	n	$1/y$	$\bar{R} \ln 1/y$	$\bar{s}^{\circ}$	$\bar{S}^*$
CO_2	2	29.05	28.011	213.795	241.806
H_2O	2	29.05	28.011	188.834	216.845
O_2	9	6.456	15.506	205.147	220.653
N_2	45.1	1.288	2.104	191.610	193.714
	58.1				

With the assumption of ideal-gas behavior, the enthalpy of each constituent is equal to the enthalpy of formation at 25°C. The values of entropy are as just calculated. Therefore, from Eq. 14.22,

$$W^{\text{rev}} = \sum_R n_i (\bar{h}_f^\circ)_i - \sum_P n_e (\bar{h}_f^\circ)_e - T_0 \left(\sum_R n_i \bar{s}_i - \sum_P n_e \bar{s}_e \right)$$

$$= (\bar{h}_f^\circ)_{C_2H_4} - 2(\bar{h}_f^\circ)_{CO_2} - 2(\bar{h}_f^\circ)_{H_2O(g)}$$

$$\quad -298.15\left(\bar{S}_{C_2H_4}^* + 12\bar{S}_{O_2}^* + 45.1\bar{S}_{N_2}^* - 2\bar{S}_{CO_2}^* - 2\bar{S}_{H_2O(g)}^* - 9\bar{S}_{O_2}^* - 45.1\bar{S}_{N_2}^* \right)$$

$$= 52\,467 - 2(-393\,522) - 2(-241\,826)$$

$$\quad -298.15[253.104 + 12(218.264) + 45.1(193.714)$$

$$\quad -2(241.806) - 2(216.845) - 9(220.653) - 45.1(193.714)]$$

$$= 1\,332\,378 \text{ kJ / kmol}$$

$$= \frac{1\,332\,378}{28.054} = 47\,493 \text{ kJ / kg}$$

Note that this value is essentially the same as the value that was obtained in Example 14.13, when the reactants and products were each separated and at 0.1-MPa pressure.

These examples raise the question of the possibility of a reversible chemical reaction. Some reactions can be made to approach reversibility by having them take place in an electrolytic cell, as described in Chapter 1. When a potential exactly equal to the electromotive force of the cell is applied, no reaction takes place. When the applied potential is increased slightly, the reaction proceeds in one direction, and if the applied potential is decreased slightly, the reaction proceeds in the opposite direction. The work done is the electrical energy supplied or delivered.

Consider a reversible reaction occuring at constant temperature equal to that of its environment. The work output of the fuel cell is

$$W = -\left(\sum n_e \bar{g}_e - \sum n_i \bar{g}_i \right) = -\Delta G$$

where ΔG is the change in Gibbs function for the overall chemical reaction. We also realize that the work is given in terms of the charged electrons flowing through an electrical potential $\mathscr{E}$ as

$$W = \mathscr{E} n_e N_0 e$$

in which n_e is the number of kilomoles of electrons flowing through the external circuit, and

$$N_0 e = 6.022\,136 \times 10^{26} \text{ elec / kmol} \times 1.602\,177 \times 10^{-22} \text{ kJ / elec V}$$

$$= 96\,485 \text{ kJ / kmol V}$$

Thus, for a given reaction, the maximum (reversible reaction) electrical potential $\mathscr{E}^\circ$ of a fuel cell at a given temperature is

$$\mathscr{E}^\circ = \frac{-\Delta G}{96\,485 n_e} \tag{14.26}$$

EXAMPLE 14.16 Calculate the reversible electromotive force (EMF) at 25°C for the hydrogen–oxygen fuel cell described in Section 1.2.

Solution

The anode side reaction was stated to be

$$2H_2 \rightarrow 4H^+ + 4e^-$$

and the cathode side reaction is

$$4H^+ + 4e^- + O_2 \rightarrow 2H_2O$$

Therefore, the overall reaction is, in kilomoles,

$$2H_2 + O_2 \rightarrow 2H_2O$$

for which 4 kmol of electrons flow through the external circuit. Let us assume that each component is at its standard-state pressure of 0.1 MPa and that the water formed is liquid. Then

$$\Delta H° = 2\bar{h}_{f_{H_2O(l)}}^° - 2\bar{h}_{f_{H_2}}^° - \bar{h}_{f_{O_2}}^°$$
$$= 2(-285\,830) - 2(0) - 1(0) = -571\,660 \text{ kJ}$$
$$\Delta S° = 2\bar{s}_{H_2O(l)}^° - 2\bar{s}_{H_2}^° - \bar{s}_{O_2}^°$$
$$= 2(69.950) - 2(130.678) - 1(205.148) = -326.604 \text{ kJ/K}$$
$$\Delta G° = -571\,660 - 298.15(-326.604) = -474\,283 \text{ kJ}$$

Therefore, from Eq. 14.26,

$$\mathscr{E}° = \frac{-(-474\,283)}{96\,485 \times 4} = 1.229 \text{ V}$$

Much effort is being directed toward the development of fuel cells in which carbon, hydrogen, or hydrocarbons will react with oxygen and produce electricity directly. If a fuel cell can be developed that utilizes a hydrocarbon fuel and has a sufficiently high efficiency and capacity (for a given volume or weight), our techniques for generating electricity on a commercial scale will undergo drastic changes. At the present time, however, fuel cells cannot compete with conventional power plants for the large-scale production of electricity.

14.9 EVALUATION OF ACTUAL COMBUSTION PROCESSES

A number of different parameters can be defined for evaluating the performance of an actual combustion process, depending on the nature of the process and the system considered. In the combustion chamber of a gas turbine, for example, the objective is to raise the temperature of the products to a given temperature (usually the maximum temperature the metals in the turbine can withstand). If we had a combustion process that achieved complete combustion and that was adiabatic, the temperature of the products would be the adiabatic flame temperature. Let us designate the fuel–air ratio needed to reach a given temperature under these conditions as the ideal fuel–air ratio. In the actual combustion chamber the combustion will be incomplete to some extent, and there will be some heat transfer to the surroundings. Therefore, more fuel will be required to reach the given temperature, and this we designate as the actual fuel–air ratio. The combustion efficiency, η_{comb}, is defined here as

$$\eta_{comb} = \frac{FA_{ideal}}{FA_{actual}} \tag{14.27}$$

On the other hand, in the furnace of a steam generator (boiler), the purpose is to transfer the maximum possible amount of heat to the steam (water). In practice, the efficiency of a steam generator is defined as the ratio of the heat transferred to the steam to the higher heating value of the fuel. For a coal this is the heating value as measured in a bomb calorimeter, which is the constant-volume heating value, and it corresponds to the internal energy of combustion. We observe a minor inconsistency, since the boiler involves a flow process, and the change in enthalpy is the significant factor. In most cases, however, the error thus introduced is less than the experimental error involved in measuring the heating value, and the efficiency of a steam generator is defined by the relation

$$\eta_{steam\ generator} = \frac{\text{heat transferred to steam}\,/\,\text{kg fuel}}{\text{higher heating value of the fuel}} \tag{14.28}$$

In an internal-combustion engine the purpose is to do work. The logical way to evaluate the performance of an internal-combustion engine would be to compare the actual work done to the maximum work that would be done by a reversible change of state from the reactants to the products. This, as we noted previously, is called the second-law efficiency.

However, in practice the efficiency of an internal-combustion engine is defined as the ratio of the actual work to the negative of the enthalpy of combustion of the fuel (that is, the constant-pressure heating value). This ratio is usually called the thermal efficiency, η_{th}:

$$\eta_{th} = \frac{w}{-h_{RP_0}} = \frac{w}{\text{heating value}} \tag{14.29}$$

The overall efficiency of a gas turbine or steam power plant is defined in the same way. It should be pointed out that in an internal-combustion engine or fuel-burning steam power plant, the fact that the combustion is itself irreversible is a significant factor in the relatively low thermal efficiency of these devices.

One other factor should be pointed out regarding efficiency. We have noted that the enthalpy of combustion of a hydrocarbon fuel varies considerably with the phase of the water in the products, which leads to the concept of higher and lower heating values. Therefore, when we consider the thermal efficiency of an engine, the heating value used to determine this efficiency must be borne in mind. Two engines made by different manufacturers may have identical performance, but if one manufacturer bases his or her efficiency on the higher heating value and the other on the lower heating value, the latter will be able to claim a higher thermal efficiency. This claim is not significant, of course, as the performance is the same, and a consideration of how the efficiency was defined would reveal this.

The whole matter of the efficiencies of devices that undergo combustion processes is treated in detail in textbooks dealing with particular applications, and our discussion is intended only as an introduction to the subject. Two examples are given, however, to illustrate these remarks.

EXAMPLE 14.17 The combustion chamber of a gas turbine uses a liquid hydrocarbon fuel that has an approximate composition of C_8H_{18}. During testing the following data are obtained.

$$T_{air} = 400 \text{ K} \qquad T_{products} = 1100 \text{K}$$

$$\mathbf{V}_{air} = 100 \text{ m/s} \qquad \mathbf{V}_{products} = 150 \text{ m/s}$$

$$T_{fuel} = 50°C \qquad FA_{actual} = 0.0211 \text{ kg fuel/kg air}$$

Calculate the combustion efficiency for this process.

Control volume: Combustion chamber.

Inlet states: T known for air and fuel.

Exit state: T known.

Model: Air and products—ideal gas, Table A.8. Fuel—Table A.9.

Analysis:

For the ideal chemical reaction the heat transfer is zero. Therefore, writing the first law for a control volume that includes the combustion chamber, we have

$$H_R + KE_R = H_P + KE_P$$

$$H_R + KE_R = \sum_R n_i \left(\overline{h}_f^\circ + \Delta\overline{h} + \frac{M\mathbf{V}^2}{2} \right)_i$$

$$= \left[\overline{h}_f^\circ + \overline{C}_p(50-25) \right]_{C_8H_{18}(l)} + n_{O_2} \left(\Delta\overline{h} + \frac{M\mathbf{V}^2}{2} \right)_{O_2}$$

$$+ 3.76 n_{O_2} \left(\Delta\overline{h} + \frac{M\mathbf{V}^2}{2} \right)_{N_2}$$

$$H_P + KE_P = \sum_P n_e \left(\overline{h}_f^\circ + \Delta\overline{h} + \frac{M\mathbf{V}^2}{2} \right)_e$$

$$= 8 \left(\overline{h}_f^\circ + \Delta\overline{h} + \frac{M\mathbf{V}^2}{2} \right)_{CO_2} + 9 \left(\overline{h}_f^\circ + \Delta\overline{h} + \frac{M\mathbf{V}^2}{2} \right)_{H_2O}$$

$$+ (n_{O_2} - 12.5) \left(\Delta\overline{h} + \frac{M\mathbf{V}^2}{2} \right)_{O_2} + 3.76 n_{O_2} \left(\Delta\overline{h} + \frac{M\mathbf{V}^2}{2} \right)_{N_2}$$

Solution

$$H_R + KE_R = -250\,105 + 1.7113 \times 114.23(50-25)$$

$$+ n_{O_2}\left[3034 + \frac{32 \times (100)^2}{2 \times 1000}\right]$$

$$+ 3.76 n_{O_2}\left[2971 + \frac{28.02 \times (100)^2}{2 \times 1000}\right]$$

$$= -245\,218 + 14\,892 n_{O_2}$$

$$H_P + KE_P = 8\left[-393\,522 + 38\,891 + \frac{44.01 \times (150)^2}{2 \times 1000}\right]$$

$$+ 9\left[-241\,826 + 30\,147 + \frac{18.02 \times (150)^2}{2 \times 1000}\right]$$

$$+ (n_{O_2} - 12.5)\left[26\,218 + \frac{32 \times (150)^2}{2 \times 1000}\right]$$

$$+ 3.76 n_{O_2}\left[24\,758 + \frac{28.02 \times (150)^2}{2 \times 1000}\right]$$

$$= -5\,068\,599 + 120\,853 n_{O_2}$$

Therefore,

$$-245\,218 + 14\,892 n_{O_2} = -5\,068\,599 + 120\,853 n_{O_2}$$

$$n_{O_2} = 45.52 \text{ kmol O}_2 \text{ / kmol fuel}$$

$$\text{kmol air / kmol fuel} = 4.76(45.52) = 216.675$$

$$FA_{\text{ideal}} = \frac{114.23}{216.675 \times 28.97} = 0.0182 \text{ kg fuel / kg air}$$

$$\eta_{\text{comb}} = \frac{0.0182}{0.0211} \times 100 = 86.2 \text{ percent}$$

EXAMPLE 14.18   In a certain steam power plant 325 000 kg of water per hour enters the boiler at a pressure of 12.5 MPa and a temperature of 200°C. Steam leaves the boiler at 9 MPa, 500°C. The power output of the turbine is 81 000 kW. Coal is used at the rate of 26 700 kg/h and has a higher heating value of 33 250 kJ/kg. Determine the efficiency of the steam generator and the overall thermal efficiency of the plant.

In power plants the efficiency of both the boiler and the overall efficiency of the plant are based on the higher heating value of the fuel.

Solution

The efficiency of the boiler is defined by Eq. 14.28 as

$$\eta_{\text{steam generator}} = \frac{\text{heat transferred to } H_2O / \text{kg fuel}}{\text{higher heating value}}$$

Therefore

$$\eta_{\text{steam generator}} = \frac{325\,000(3386.1 - 857.1)}{26\,700 \times 33\,250} \times 100 = 92.6\%$$

The thermal efficiency is defined by Eq. 14.29,

$$\eta_{\text{th}} = \frac{w}{\text{heating value}} = \frac{81\,000 \times 3600}{26\,700 \times 33\,250} \times 100 = 32.8\%$$

PROBLEMS

14.1 Calculate the theoretical air–fuel ratio on a mass and mole basis for the combustion of ethanol, C_2H_5OH.

14.2 In a combustion process with decane, $C_{10}H_{22}$, and air, the dry product mole fractions are 86.9% N_2, 1.163% O_2, 10.975% CO_2, and 0.954% CO. Find the equivalence ratio and the percent theoretical air of the reactants.

14.3 A certain fuel oil has the composition $C_{10}H_{22}$. If this fuel is burned with 150% theoretical air, what is the composition of the products of combustion?

14.4 Natural gas B from Table 14.2 is burned with 20% excess air. Determine the composition of the products.

14.5 A Pennsylvania coal contains 74.2% C, 5.1% H, 6.7% O, (dry basis, mass percent) plus ash and small percentages of N and S. This coal is fed into a gasifier along with oxygen and steam, as shown in Fig. P14.5. The exiting product gas composition is measured on a mole basis to: 39.9% CO, 30.8% H_2, 11.4% CO_2, 16.4% H_2O plus small percentages of CH_4, N_2, and H_2S. How many kilograms of coal are required to produce 100 kmol of product gas? How much oxygen and steam are required?

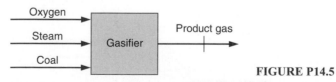

FIGURE P14.5

14.6 Repeat Problem 14.5 for a certain Utah coal that contains, according to the coal analysis, 68.2% C, 4.8% H, and 15.7% O on a mass basis. The exiting product gas contains 30.9% CO, 26.7% H_2, 15.9% CO_2, and 25.7% H_2O on a mole basis.

14.7 A sample of pine bark has the following ultimate analysis on a dry basis, percent by mass: 5.6% H, 53.4% C, 0.1% S, 0.1% N, 37.9% O, and 2.9% ash. This bark will be used as a fuel by burning it with 100% theoretical air in a furnace. Determine the air–fuel ratio on a mass basis.

14.8 Liquid propane is burned with dry air. A volumetric analysis of the products of combustion yields the following volume percent composition on a dry basis: 8.6% CO_2, 0.6% CO, 7.2% O_2, and 83.6% N_2. Determine the percent of theoretical air used in this combustion process.

14.9 A fuel, C_xH_y, is burned with dry air, and the product composition is measured on a dry basis to be: 9.6% CO_2, 7.3% O_2, and 83.1% N_2. Find the fuel composition (x/y) and the percent theoretical air used.

14.10 Many coals from the western United States have a high moisture content. Consider the following sample of Wyoming coal, for which the ultimate analysis on an as-received basis is, by mass:

Component	Moisture	H	C	S	N	O	Ash
% mass	28.9	3.5	48.6	0.5	0.7	12.0	5.8

This coal is burned in the steam generator of a large power plant with 150% theoretical air. Determine the air–fuel ratio on a mass basis.

14.11 Pentane is burned with 120% theoretical air in a constant pressure process at 100 kPa. The products are cooled to ambient temperature, 20°C. How much mass of water is condensed per kilogram of fuel? Repeat the answer, assuming that the air used in the combustion has a relative humidity of 90%.

14.12 The coal gasifier in an integrated gasification combined cycle (IGCC) power plant produces a gas mixture with the following volumetric percent composition:

Product	CH_4	H_2	CO	CO_2	N_2	H_2O	H_2S	NH_3
% vol.	0.3	29.6	41.0	10.0	0.8	17.0	1.1	0.2

This gas is cooled to 40°C, 3 MPa, and the H_2S and NH_3 are removed in water scrubbers. Assuming that the resulting mixture, which is sent to the combustors, is saturated with water, determine the mixture composition and the theoretical air–fuel ratio in the combustors.

14.13 The hot exhaust gas from an internal combustion engine is analyzed and found to have the following percent composition on a volumetric basis at the engine exhaust manifold. 10% CO_2, 2% CO, 13% H_2O, 3% O_2, and 72% N_2. This gas is fed to an exhaust gas reactor and mixed with a certain amount of air to eliminate the carbon monoxide, as shown in Fig. P14.13. It has been determined that a mole fraction of 10% oxygen in the mixture at state 3 will ensure that no CO remains. What must be the ratio of flows entering the reactor?

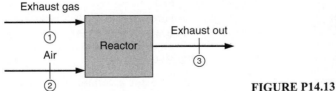

FIGURE P14.13

14.14 Methanol, CH_3OH, is burned with 200% theoretical air in an engine, and the products are brought to 100 kPa, 30°C. How much water is condensed per kilogram of fuel?

14.15 The output gas mixture of a certain air-blown coal gasifier has the composition of producer gas as listed in Table 14.2. Consider the combustion of this gas with

120% theoretical air at 100 kPa pressure. Determine the dew point of the products and find how many kilograms of water will be condensed per kilogram of fuel if the products are cooled 10°C below the dew-point temperature.

14.16 Pentene, C_5H_{10}, is burned with pure oxygen in a SSSF process. The products at one point are brought to 700 K and used in a heat exchanger, where they are cooled to 25°C. Find the specific heat transfer in the heat exchanger.

14.17 Butane gas and 200% theoretical air, both at 25°C, enter a SSSF combustor. The products of combustion exit at 1000 K. Calculate the heat transfer from the combustor per kmol of butane burned.

14.18 Liquid pentane is burned with dry air, and the products are measured on a dry basis as: 10.1% CO_2, 0.2% CO, 5.9% O_2 remainder N_2. Find the enthalpy of formation for the fuel and the actual equivalence ratio.

14.19 A rigid vessel initially contains 2 kmol of carbon and 2 kmol of oxygen at 25°C, 200 kPa. Combustion occurs, and the resulting products consist of 1 kmol of carbon dioxide, 1 kmol of carbon monoxide, and excess oxygen at a temperature of 1000 K. Determine the final pressure in the vessel and the heat transfer from the vessel during the process.

14.20 In a test of rocket propellant performance, liquid hydrazine (N_2H_4) at 100 kPa, 25°C, and oxygen gas at 100 kPa, 25°C, are fed to a combustion chamber in the ratio of 0.5 kg O_2/kg N_2H_4. The heat transfer from the chamber to the surroundings is estimated to be 100 kJ/kg N_2H_4. Determine the temperature of the products exiting the chamber. Assume that only H_2O, H_2, and N_2 are present. The enthalpy of formation of liquid hydrazine is +50417 kJ/kmol.

14.21 Repeat the previous problem, but assume that saturated-liquid oxygen at 90 K is used instead of 25°C oxygen gas in the combustion process. Use the generalized charts to determine the properties of liquid oxygen.

14.22 The combustion of heptane C_7H_{16} takes place in a SSSF burner where fuel and air are added as gases at P_o, T_o. The mixture has 125% theoretical air, and the products pass through a heat exchanger where they are cooled to 600 K. Find the heat transfer from the heat exchanger per kmol of heptane burned.

14.23 Ethene, C_2H_4, and propane, C_3H_8, in a 1:1 mole ratio as gases are burned with 120% theoretical air in a gas turbine. Fuel is added at 25°C, 1 MPa, and the air comes from the atmosphere, 25°C, 100 kPa through a compressor to 1 MPa and mixed with the fuel. The turbine work is such that the exit temperature is 800 K with an exit pressure of 100 kPa. Find the mixture temperature before combustion, and also the work, assuming an adiabatic turbine.

14.24 One alternative to using petroleum or natural gas as fuels is ethanol (C_2H_5OH), which is commonly produced from grain by fermentation. Consider a combustion process in which liquid ethanol is burned with 120% theoretical air in a SSSF process. The reactants enter the combustion chamber at 25°C, and the products exit at 60°C, 100 kPa. Calculate the heat transfer per kilomole of ethanol, using the enthalpy of formation of ethanol gas plus the generalized charts.

14.25 Another alternative to using petroleum or natural gas as fuels is methanol, (CH_3OH), which can be produced from coal. Both methanol and ethanol have been used in automotive engines. Repeat the previous problem using liquid methanol as the fuel instead of ethanol.

14.26 Another alternative fuel to be seriously considered is hydrogen. It can be produced from water by various techniques that are under extensive study. Its biggest problem at the present time is cost, storage, and safety. Repeat Problem 14.24 using hydrogen gas as the fuel instead of ethanol.

14.27 Hydrogen peroxide, H_2O_2, enters a gas generator at 25°C, 500 kPa at the rate of 0.1 kg/s and is decomposed to steam and oxygen exiting at 800 K, 500 kPa. The resulting mixture is expanded through a turbine to atmospheric pressure, 100 kPa, as shown in Fig. P14.27. Determine the power output of the turbine, and the heat transfer rate in the gas generator. The enthalpy of formation of liquid H_2O_2 is −187 583 kJ/kmol.

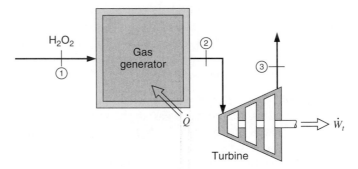

FIGURE P14.27

14.28 In a new high-efficiency furnace, natural gas, assumed to be 90% methane and 10% ethane (by volume) and 110% theoretical air each enter at 25°C, 100 kPa, and the products (assumed to be 100% gaseous) exit the furnace at 40°C, 100 kPa. What is the heat transfer for this process? Compare this to an older furnace where the products exit at 250°C, 100 kPa.

14.29 Repeat the previous problem, but take into account the actual phase behavior of the products exiting the furnace.

14.30 Methane, CH_4, is burned in a SSSF process with two different oxidizers: Case **A**: Pure oxygen, O_2 and case **B**: A mixture of $O_2 + x$ Ar. The reactants are supplied at T_0, P_0 and the products should for both cases be at 1800 K. Find the required equivalence ratio in case (A) and the amount of Argon, x, for a stoichiometric ratio in case (B).

14.31 Butane gas at 25°C is mixed with 150% theoretical air at 600 K and is burned in an adiabatic SSSF combustor. What is the temperature of the products exiting the combustor?

14.32 In a rocket, hydrogen is burned with air, both reactants supplied as gases at P_o, T_o. The combustion is adiabatic, and the mixture is stoichiometric (100% theoretical air). Find the products' dew point and the adiabatic flame temperature (~2500 K).

14.33 Liquid butane at 25°C is mixed with 150% theoretical air at 600 K and is burned in an adiabatic SSSF combustor. Use the generalized charts for the liquid fuel and find the temperature of the products exiting the combustor.

14.34 A stoichiometric mixture of benzene, C_6H_6, and air is mixed from the reactants flowing at 25°C, 100 kPa. Find the adiabatic flame temperature. What is the error if constant specific heat at T_0 for the products from Table A.5 are used?

14.35 Liquid n-butane at T_0, is sprayed into a gas turbine with primary air flowing at 1.0 MPa, 400 K in a stoichiometric ratio. After complete combustion, the products are at the adiabatic flame temperature, which is too high, so secondary air at 1.0 MPa, 400 K is added, with the resulting mixture being at 1400 K. Show that $T_{ad} > 1400$ K and find the ratio of secondary to primary air flow.

14.36 Consider the gas mixture fed to the combustors in the integrated gasification combined cycle power plant, as described in Problem 14.12. If the adiabatic flame temperature should be limited to 1500 K, what percent theoretical air should be used in the combustors?

14.37 Acetylene gas at 25°C, 100 kPa is fed to the head of a cutting torch. Calculate the adiabatic flame temperature if the acetylene is burned with

a. 100% theoretical air at 25°C.   b. 100% theoretical oxygen at 25°C.

14.38 Ethene, C_2H_4, burns with 150% theoretical air in a SSSF constant-pressure process with reactants entering at P_0, T_0. Find the adiabatic flame temperature.

14.39 Solid carbon is burned with stoichiometric air in a SSSF process. The reactants at T_0, P_0 are heated in a preheater to $T_2 = 500$ K, as shown in Fig. P14.39, with the energy given by the product gases before flowing to a second heat exchanger, which they leave at T_0. Find the temperature of the products T_4, and the heat transfer per kmol of fuel (4 to 5) in the second heat exchanger.

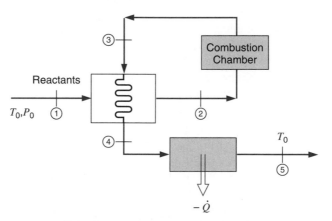

FIGURE P14.39

14.40 A study is to be made using liquid ammonia as the fuel in a gas-turbine engine. Consider the compression and combustion processes of this engine.

a. Air enters the compressor at 100 kPa, 25°C and is compressed to 1600 kPa, where the isentropic compressor efficiency is 87%. Determine the exit temperature and the work input per kilomole.

b. Two kilomoles of liquid ammonia at 25°C and x times theoretical air from the compressor enter the combustion chamber. What is x if the adiabatic flame temperature is to be fixed at 1600 K?

14.41 A closed, insulated container is charged with a stoichiometric ratio of oxygen and hydrogen at 25°C and 150 kPa. After combustion, liquid water at 25°C is sprayed in such a way that the final temperature is 1200 K. What is the final pressure?

14.42 Wet biomass waste from a food-processing plant is fed to a catalytic reactor, where in a SSSF process it is converted into a low-energy fuel gas suitable for firing the processing plant boilers. The fuel gas has a composition of 50% methane, 45% carbon dioxide, and 5% hydrogen on a volumetric basis. Determine the lower heating value of this fuel gas mixture per unit volume.

14.43 Determine the lower heating value of the gas generated from coal, as described in Problem 14.12. Do not include the components removed by the water scrubbers.

14.44 Propylbenzene, C_9H_{12}, is listed in Table 14.2, but not in Table A.9. No molecular weight is listed in the book. Find the molecular weight, the enthalpy of formation for the liquid fuel, and the enthalpy of evaporation.

14.45 Determine the higher heating value of the sample Wyoming coal as specified in Problem 14.10.

14.46 Consider natural gas A and natural gas D, both of which are listed in Table 14.2. Calculate the enthalpy of combustion of each gas at 25°C, assuming that the products include vapor water. Repeat the answer for liquid water in the products.

14.47 Blast furnace gas in a steel mill is available at 250°C to be burned for the generation of steam. The composition of this gas is, on a volumetric basis,

Component	CH_4	H_2	CO	CO_2	N_2	H_2O
Percent by volume	0.1	2.4	23.3	14.4	56.4	3.4

Find the lower heating value (kJ/m^3) of this gas at 250°C and ambient pressure.

14.48 The enthalpy of formation of magnesium oxide, MgO(s), is −601 827 kJ/kmol at 25°C. The melting point of magnesium oxide is approximately 3000 K, and the increase in enthalpy between 298 and 3000 K is 128 449 kJ/kmol. The enthalpy of sublimation at 3000 K is estimated at 418 000 kJ/kmol, and the specific heat of magnesium oxide vapor above 3000 K is estimated at 37.24 kJ/kmol K.

   a. Determine the enthalpy of combustion per kilogram of magnesium.

   b. Estimate the adiabatic flame temperature when magnesium is burned with theoretical oxygen.

14.49 A rigid container is charged with butene, C_4H_8, and air in a stoichiometric ratio at P_0, T_0. The charge burns in a short time with no heat transfer to state 2. The products then cool with time to 1200 K, state 3. Find the final pressure, P_3, the total heat transfer, $_1Q_3$, and the temperature immediately after combustion, T_2.

14.50 In an engine a mixture of liquid octane and ethanol, mole ratio 9 : 1, and stoichiometric air are taken in at T_0, P_0. In the engine the enthalpy of combustion is used so that 30% goes out as work, 30% goes out as heat loss, and the rest goes out the exhaust. Find the work and heat transfer per kilogram of fuel mixture and also the exhaust temperature.

14.51 Consider the same situation as in the previous problem. Find the dew point temperature of the products. If the products in the exhaust are cooled to 10°C, find the mass of water condensed per kilogram of fuel mixture.

14.52 Calculate the irreversibility for the process described in Problem 14.19.

14.53 Pentane gas at 25°C, 150 kPa enters an insulated SSSF combustion chamber. Sufficient excess air to hold the combustion products temperature to 1800 K enters separately at 500 K, 150 kPa. Calculate the percent theoretical air required and the irreversibility of the process per kmol of pentane burned.

14.54 Consider the combustion of methanol, CH_3OH, with 25% excess air. The combustion products are passed through a heat exchanger and exit at 200 kPa, 40°C. Calculate the absolute entropy of the products exiting the heat exchanger per kilomole of methanol burned, using appropriate reference states as needed.

14.55 The turbine in Problem 14.23 is adiabatic. Is it reversible, irreversible, or impossible?

14.56 Saturated liquid butane enters an insulated constant-pressure combustion chamber at 25°C, and x times theoretical oxygen gas enters at the same P and T. The combustion products exit at 3400 K. With complete combustion find x. What is the pressure at the chamber exit? What is the irreversibility of the process?

14.57 An inventor claims to have built a device that will take 0.001 kg/s of water from the faucet at 10°C, 100 kPa, and produce separate streams of hydrogen and oxygen gas, each at 400 K, 175 kPa. It is stated that this device operates in a 25°C room on 10-kW electrical power input. How do you evaluate this claim?

14.58 Two kilomoles of ammonia are burned in an SSSF process with x kmol of oxygen. The products, consisting of H_2O, N_2, and the excess O_2, exit at 200°C, 7 MPa.

a. Calculate x if half the water in the products is condensed.

b. Calculate the absolute entropy of the products at the exit conditions.

14.59 Consider the SSSF combustion of propane at 25°C with air at 400 K. The products exit the combustion chamber at 1200 K. It may be assumed that the combustion efficiency is 90%, and that 95% of the carbon in the propane burns to form carbon dioxide; the remaining 5% forms carbon monoxide. Determine the ideal fuel–air ratio and the heat transfer from the combustion chamber.

14.60 Graphite, C, at P_0, T_0 is burned with air coming in at P_0, 500 K in a ratio so the products exit at P_0, 1200 K. Find the equivalence ratio, the percent theoretical air, and the total irreversibility.

14.61 A gasoline engine is converted to run on propane as shown in Fig. P14.61. Assume the propane enters the engine at 25°C, at the rate 40 kg/h. Only 90% theoretical air enters at 25°C such that 90% of the C burns to form CO_2, and 10% of the C burns to form CO. The combustion products, also including H_2O, H_2, and N_2, exit the exhaust pipe at 1000 K. Heat loss from the engine (primarily to the cooling water) is 120 kW. What is the power output of the engine? What is the thermal efficiency?

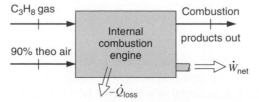

FIGURE P14.61

14.62 A small air-cooled gasoline engine is tested, and the output is found to be 1.0 kW. The temperature of the products is measured as 600 K. The products are analyzed on a dry volumetric basis, with the result: 11.4% CO_2, 2.9% CO, 1.6% O_2, and 84.1% N_2. The fuel may be considered to be liquid octane. The fuel and air enter the engine at 25°C, and the flow rate of fuel to the engine is 1.5×10^{-4} kg/s. Determine the rate of heat transfer from the engine and its thermal efficiency.

14.63 A gasoline engine uses liquid octane and air, both supplied at P_0, T_0, in a stoichiometric ratio. The products (complete combustion) flow out of the exhaust valve at 1100 K. Assume that the heat loss carried away by the cooling water, at 100°C, is equal to the work output. Find the efficiency of the engine expressed as (work/lower heating value) and the second law efficiency.

14.64 In Example 14.16, a basic hydrogen–oxygen fuel cell reaction was analyzed at 25°C, 100 kPa. Repeat this calculation, assuming that the fuel cell operates on air at 25°C, 100 kPa, instead of on pure oxygen at this state.

14.65 Consider a methane-oxygen fuel cell in which the reaction at the anode is

$$CH_4 + 2H_2O \rightarrow CO_2 + 8e^- + 8H^+$$

The electrons produced by the reaction flow through the external load, and the positive ions migrate through the electrolyte to the cathode, where the reaction is

$$8\,e^- + 8\,H^+ + 2\,O_2 \rightarrow 4\,H_2O$$

a. Calculate the reversible work and the reversible EMS for the fuel cell operating at 25°C, 100 kPa.

b. Repeat part (a), but assume that the fuel cell operates at 600 K instead of at room temperature.

ADVANCED PROBLEMS

14.66 A gas mixture of 50% ethane and 50% propane by volume enters a combustion chamber at 350 K, 10 MPa. Determine the enthalpy per kilomole of this mixture relative to the thermochemical base of enthalpy using Kay's rule.

14.67 A mixture of 80% ethane and 20% methane on a mole basis is throttled from 10 MPa, 65°C, to 100 kPa and is fed to a combustion chamber where it undergoes complete combustion with air, which enters at 100 kPa, 600 K. The amount of air is such that the products of combustion exit at 100 kPa, 1200 K. Assume that the combustion process is adiabatic and that all components behave as ideal gases except the fuel mixture, which behaves according to the generalized charts, with Kay's rule for the pseudocritical constants. Determine the percentage of theoretical air used in the process and the dew-point temperature of the products.

14.68 Gaseous propane mixes with air, both supplied at 500 K, 0.1 MPa. The mixture goes into a combustion chamber, and products of combustion exit at 1300 K, 0.1 MPa. The products analyzed on a dry basis are 11.42% CO_2, 0.79% CO, 2.68% O_2, and 85.11% N_2 on a volume basis. Find the equivalence ratio and the heat transfer per kmol of fuel.

14.69 A closed, rigid container is charged with propene, C_3H_6, and 150% theoretical air at 100 kPa, 298 K. The mixture is ignited and burns with complete combustion. Heat is transferred to a reservoir at 500 K so the final temperature of the products is 700 K. Find the final pressure, the heat transfer per kmol fuel and the total entropy generated per kmol fuel in the process.

14.70 Consider one cylinder of a spark-ignition, internal-combustion engine. Before the compression stroke, the cylinder is filled with a mixture of air and methane. Assume that 110% theoretical air has been used, that the state before compression is 100 kPa, 25°C. The compression ratio of the engine is 9 to 1.

    a. Determine the pressure and temperature after compression, assuming a reversible adiabatic process.

    b. Assume that complete combustion takes place while the piston is at top dead center (at minimum volume) in an adiabatic process. Determine the temperature and pressure after combustion, and the increase in entropy during the combustion process.

    c. What is the irreversibility for this process?

14.71 Consider the combustion process described in Problem 14.67.

    a. Calculate the absolute entropy of the fuel mixture before it is throttled into the combustion chamber.

    b. Calculate the irreversibility for the overall process.

14.72 Liquid acetylene, C_2H_2, is stored in a high-pressure storage tank at ambient temperature, 25°C. The liquid is fed to an insulated combustor/steam boiler at the steady rate of 1 kg/s, along with 140% theoretical oxygen, O_2, which enters at 500 K, as shown in Fig. P14.72. The combustion products exit the unit at 500 kPa, 350 K. Liquid water enters the boiler at 10°C, at the rate of 15 kg/s, and superheated steam exits at 200 kPa.

    a. Calculate the absolute entropy, per kmol, of liquid acetylene at the storage tank state.

    b. Determine the phase(s) of the combustion products exiting the combustor boiler unit, and the amount of each, if more than one.

    c. Determine the temperature of the steam at the boiler exit.

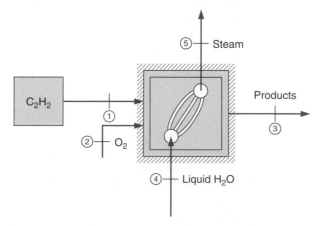

FIGURE P14.72

14.73 Natural gas (approximate it as methane) at a ratio of 0.3 kg/s is burned with 250% theoretical air in a combustor at 1 MPa where the reactants are supplied at T_0. Steam at 1 MPa, 450°C at a rate of 2.5 kg/s is added to the products before they enter an adiabatic turbine with an exhaust pressure of 150 kPa. Determine the turbine inlet temperature and the turbine work assuming the turbine is reversible.

14.74 Liquid hexane enters a combustion chamber at 31°C, 200 kPa, at the rate 1 kmol/s. 200% theoretical air enters separately at 500 K, 200 kPa, and the combustion products exit at 1000 K, 200 kPa. The specific heat of ideal gas hexane is $C_{po} = 143$ kJ/kmol K. Calculate the rate of irreversibility of the process.

14.75E Pentane is burned with 120% theoretical air in a constant pressure process at 14.7 lbf/in.^{2}. The products are cooled to ambient temperature, 70 F. How much mass of water is condensed per pound-mass of fuel? Repeat the problem, assuming that the air used in the combustion has a relative humidity of 90%.

14.76E The output gas mixture of a certain air-blown coal gasifier has the composition of producer gas as listed in Table 14.2. Consider the combustion of this gas with 120% theoretical air at 14.7 lbf/in.^{2} pressure. Find the dew point of the products and the mass of water condensed per pound-mass of fuel if the products are cooled 20 F below the dew point temperature?

14.77E Pentene, C_5H_{10} is burned with pure oxygen in a SSSF process. The products at one point are brought to 1300 R and used in a heat exchanger, where they are cooled to 77 F. Find the specific heat transfer in the heat exchanger.

14.78E A rigid vessel initially contains 2 pound moles of carbon and 2 pound moles of oxygen at 77 F, 30 lbf/in.^{2}. Combustion occurs, and the resulting products consist of 1 pound mole of carbon dioxide, 1 pound mole of carbon monoxide, and excess oxygen at a temperature of 1800 R. Determine the final pressure in the vessel and the heat transfer from the vessel during the process.

14.79E In a test of rocket propellant performance, liquid hydrazine (N_2H_4) at 14.7 lbf/in.^{2}, 77 F, and oxygen gas at 14.7 lbf/in.^{2}, 77 F, are fed to a combustion chamber in the ratio of 0.5 lbm O_2/lbm N_2H_4. The heat transfer from the chamber to the surroundings is estimated to be 45 Btu/lbm N_2H_4. Determine the temperature of the products exiting the chamber. Assume that only H_2O, H_2, and N_2 are present. The enthalpy of formation of liquid hydrazine is +21647 Btu/lb mole.

14.80E Repeat the previous problem, but assume that saturated-liquid oxygen at 170 R is used instead of 77 F oxygen gas in the combustion process. Use the generalized charts to determine the properties of liquid oxygen.

14.81E Ethene, C_2H_4, and propane, C_3H_8, in a 1:1 mole ratio as gases are burned with 120% theoretical air in a gas turbine. Fuel is added at 77 F, 150 lbf/in.^{2}, and the air comes from the atmosphere, 77 F, 15 lbf/in.^{2} through a compressor to 150 lbf/in.^{2} and mixed with the fuel. The turbine work is such that the exit temperature is 1500 R with an exit pressure of 14.7 lbf/in.^{2}. Find the mixture temperature before combustion, and also the work, assuming an adiabatic turbine.

14.82E One alternative to using petroleum or natural gas as fuels is ethanol (C_2H_5OH), which is commonly produced from grain by fermentation. Consider a combustion process in which liquid ethanol is burned with 120% theoretical air in a SSSF process. The reactants enter the combustion chamber at 77 F, and the products exit at 140 F, 14.7 lbf/in.^{2}. Calculate the heat transfer per pound mole of ethanol, using the enthalpy of formation of ethanol gas plus the generalized tables or charts.

14.83E Hydrogen peroxide, H_2O_2, enters a gas generator at 77 F, 75 lbf/in.^{2} at the rate of 0.2 lbm/s and is decomposed to steam and oxygen exiting at 1500 R, 75 lbf/in.^{2}. The resulting mixture is expanded through a turbine to atmospheric pressure, 14.7 lbf/in.^{2}, as shown in Fig. P14.27. Determine the power output of the turbine, and the heat transfer rate in the gas generator. The enthalpy of formation of liquid H_2O_2 is −80 541 Btu/lb mol.

14.84E In a new high-efficiency furnace, natural gas, assumed to be 90% methane and 10% ethane (by volume) and 110% theoretical air each enter at 77 F, 14.7 lbf/in.^{2}, and the products (assumed to be 100% gaseous) exit the furnace at 100 F, 14.7 lbf/in.^{2}. What is the heat transfer for this process? Compare this to an older furnace where the products exit at 450 F, 14.7 lbf/in.^{2}.

14.85E Repeat the previous problem, but take into account the actual phase behavior of the products exiting the furnace.

14.86E Methane, CH_4, is burned in a SSSF process with two different oxidizers: **A** Pure oxygen, O_2 and **B** a mixture of $O_2 + x$ Ar. The reactants are supplied at T_0, P_0, and the products in are at 3200 R both cases. Find the required equivalence ratio in case **A** and the amount of Argon, x, for a stoichiometric ratio in case **B**.

14.87E Butane gas at 77 F is mixed with 150% theoretical air at 1000 R and is burned in an adiabatic SSSF combustor. What is the temperature of the products exiting the combustor?

14.88E Liquid n-butane at T_0, is sprayed into a gas turbine with primary air flowing at 150 lbf/in.^{2}, 700 R in a stoichiometric ratio. After complete combustion, the products are at the adiabatic flame temperature, which is too high so secondary air at 150 lbf/in.^{2}, 700 R is added, with the resulting mixture being at 2500 R. Show that $T_{ad} > 2500$ R and find the ratio of secondary to primary air flow.

14.89E Acetylene gas at 77 F, 14.7 lbf/in.^{2} is fed to the head of a cutting torch. Calculate the adiabatic flame temperature if the acetylene is burned with 100% theoretical air at 77 F. Repeat the answer for 100% theoretical oxygen at 77 F.

14.90E Ethene, C_2H_4, burns with 150% theoretical air in a SSSF constant-pressure process with reactants entering at P_0, T_0. Find the adiabatic flame temperature.

14.91E Solid carbon is burned with stoichiometric air in a SSSF process, as shown in Fig. P14.39. The reactants at T_0, P_0 are heated in a preheater to $T_2 = 900$ R with the energy given by the products before flowing to a second heat exchanger, which they leave at T_0. Find the temperature of the products T_4, and the heat transfer per lb mol of fuel (4 to 5) in the second heat exchanger.

14.92E A closed, insulated container is charged with a stoichiometric ratio of oxygen and hydrogen at 77 F and 20 lbf/in.^{2}. After combustion, liquid water at 77 F is sprayed in such a way that the final temperature is 2100 R. What is the final pressure?

14.93E Blast furnace gas in a steel mill is available at 500 F to be burned for the generation of steam. The composition of this gas is, on a volumetric basis,

Component	CH_4	H_2	CO	CO_2	N_2	H_2O
Percent by volume	0.1	2.4	23.3	14.4	56.4	3.4

Find the lower heating value (Btu/ft^3) of this gas at 500 F and P_0.

14.94E Two pound moles of ammonia are burned in a SSSF process with x lb mol of oxygen. The products, consisting of H_2O, N_2, and the excess O_2, exit at 400 F, 1000 lbf/in.^{2}.

a. Calculate x if half the water in the products is condensed.

b. Calculate the absolute entropy of the products at the exit conditions.

14.95E Graphite, C, at P_0, T_0 is burned with air coming in at P_0, 900 R in a ratio so the products exit at P_0, 2200 R. Find the equivalence ratio, the percent theoretical air, and the total irreversibility.

14.96E A small, air-cooled gasoline engine is tested, and the output is found to be 2.0 hp. The temperature of the products is measured and found to be 730 F. The products are analyzed on a dry volumetric basis, with the following result: 11.4% CO_2, 2.9% CO, 1.6% O_2, and 84.1% N_2. The fuel may be considered to be liquid octane. The fuel and air enter the engine at 77 F, and the flow rate of fuel to the engine is 1.8 lbm/h. Determine the rate of heat transfer from the engine and its thermal efficiency.

14.97E A gasoline engine uses liquid octane and air, both supplied at P_0, T_0, in a stoichiometric ratio. The products (complete combustion) flow out of the exhaust valve at 2000 R. Assume that the heat loss carried away by the cooling water, at 200 F, is equal to the work output. Find the efficiency of the engine expressed as (work/lower heating value) and the second-law efficiency.

14.98E In Example 14.16, a basic hydrogen–oxygen fuel cell reaction was analyzed at 25°C, 100 kPa. Repeat this calculation, assuming that the fuel cell operates on air at 77 F, 14.7 lbf/in.^{2}, instead of on pure oxygen at this state.

Computer, Design, and Open-Ended Problems

14.99 Write a program to solve the general case of Problem 14.4 for any hydrocarbon fuel C_xH_y, where x and y are input parameters. We wish to calculate the percentage of theoretical air for any given percentages of combustion products.

14.100 Write a program to solve the general case of Problem 14.10 for different percentages of the components given in the ultimate analysis of the coal.

14.101 Use the menu-driven program for the ideal gas properties to do Problem 14.39.

14.102 Write a program to study the effect of the percentage of theoretical air on the adiabatic flame temperature for a (variable) hydrocarbon fuel. Assume reactants enter the combustion chamber at 25°C, and complete combustion. Use constant specific heat of the various products of combustion and let the fuel composition and its enthalpy of formation be program inputs.

14.103 Power plants may use off-peak power to compress air into a large storage facility (see Problem 9.31). The compressed air is then used as the air supply to a gas-turbine system where it is burned with some fuel, usually natural gas. The system is then used to produce power at peak load times. Investigate such a setup and estimate the power generated with conditions given in Problem 9.31 and combustion with 200–300% theoretical air and exhaust to the atmosphere.

14.104 A car that runs on natural gas has it stored in a heavy tank with a maximum pressure of 3600 psi (25 MPa). Size the tank for a range of 300 miles (500 km) assuming a car engine that has a 30% efficiency requiring about 25 hp (20 kW) to drive the car at 55 mi/h (90 km/h).

14.105 The Cheng cycle, shown in Fig. P12.109, is powered by the combustion of natural gas (essentially methane) being burned with 250–300% theoretical air. In the case with a single water-condensing heat exchanger, where $T_6 = 40°C$ and $\Phi_6 = 100\%$, is any make-up water needed at state 8 or is there a surplus? Does the humidity in the compressed atmospheric air at state 1 make any difference? Study the problem over a range of air–fuel ratios.

14.106 The cogenerating powerplant shown in Problem 11.44 burns 170 kg/s air with natural gas, CH_4. The setup is shown in Fig. P14.106 where a fraction of the air flow out of the compressor with compression ratio 15.8:1 is used to preheat the feedwater in the steam cycle. The fuel flow rate is 3.2 kg/s. Make an analysis of the system determining the total heat transfer to the steam cycle from the turbine exhaust gases, the heat transfer in the preheater, and the gas turbine inlet temperature.

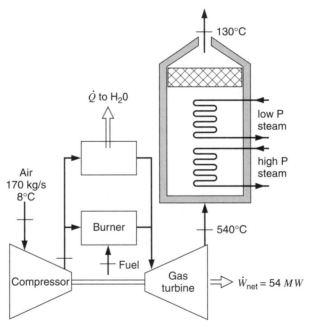

FIGURE P14.106

14.107 Consider the combustor in the Cheng cycle (see Problems 12.109 and 14.73). Atmospheric air is compressed to 1.25 MPa, state 1. It is burned with natural gas, CH_4, with the products leaving at state 2. The fuel should add a total of about 15 MW to the cycle, with an air flow of 12 kg/s. For a compressor with an intercooler estimate the temperatures T_1, T_2, and the fuel flow rate.

14.108 Study the coal gasification process that will produce methane, CH_4, or methanol CH_3OH. What is involved in such a process? Compare the heating values of the gas products with that of the original coal. Discuss the merits of this conversion.

14.109 Ethanol, C_2H_5OH, can be produced from corn or biomass. Investigate the process and the chemical reactions that occur. For different raw materials estimate the amount of ethanol that can be obtained per mass of the raw material.

14.110 A Diesel engine is used as a stationary power plant in remote locations such as a ship, oil drilling rig, farm, etc. Assume diesel fuel is used with 300% theoretical air in a 1000-hp diesel engine. Estimate the amount of fuel used, the efficiency, and the potential use of the exhaust gases for heating of rooms or water. Investigate if other fuels can be used.

14.111 When a power plant burns coal or some blends of oil, the combustion process can generate pollutants as SO_x and NO_x. Investigate the use of scrubbers to remove these. Explain the processes that take place and the effect on the power plant operation (energy, exhaust pressures, etc.).

14.112 For a number of fuels listed in Table 14.3 estimate their adiabatic flame temperature when they are burned with 200% theoretical air. Assume a power generating device like a gasoline or diesel engine or a gas-turbine with reasonable choices for their operating conditions. Find the power that can be generated as a fraction of the enthalpy of combustion. Does a ranking of the fuels follow the magnitude of the enthalpy of combustion?

Introduction to Phase and Chemical Equilibrium 15

Up to this point we have assumed that we are dealing either with systems that are in equilibrium or with those in which the deviation from equilibrium is infinitesimal, as in a quasiequilibrium or reversible process. For irreversible processes we made no attempt to describe the state of the system during the process but dealt only with the initial and final states of the system, in the case of a control mass, or the inlet and exit states as well in the case of a control volume. For any case, we either considered the system to be in equilibrium throughout, or at least made the assumption of local equilibrium.

In this chapter we examine the criteria for equilibrium and from them derive certain relations that will enable us, under certain conditions, to determine the properties of a system when it is in equilibrium. The specific case we will consider is that involving chemical equilibrium in a single phase (homogeneous equilibrium) as well as certain related topics.

15.1 Requirements for Equilibrium

As a general requirement for equilibrium we postulate that a system is in equilibrium when there is no possibility that it can do any work when it is isolated from its surroundings. In applying this criterion it is helpful to divide the system into two or more subsystems, and consider the possibility of doing work by any conceivable interaction between these two subsystems. For example, in Fig. 15.1 a system has been divided into two systems and an engine, of any conceivable variety, placed between these subsystems. A system may be so defined as to include the immediate surroundings. In this case we can let the immediate surroundings be a subsystem and thus consider the general case of the equilibrium between a system and its surroundings.

The first requirement for equilibrium is that the two subsystems have the same temperature, for otherwise we could operate a heat engine between the two systems and do work. Thus we conclude that one requirement for equilibrium is that a system must be at a uniform temperature to be in equilibrium. It is also evident that there must be no unbalanced mechanical forces between the two systems, or else one could operate a turbine or piston engine between the two systems and do work.

However, we would like to establish general criteria for equilibrium that would apply to all simple compressible substances, including those that undergo chemical reactions. We will find that the Gibbs function is a particularly significant property in defining the criteria for equilibrium.

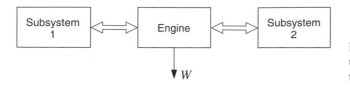

FIGURE 15.1 Two subsystems that communicate through an engine.

Let us first consider a qualitative example to illustrate this point. Consider a natural gas well that is 1 km deep, and let us assume that the temperature of the gas is constant throughout the gas well. Suppose we have analyzed the composition of the gas at the top of the well, and we would like to know the composition of the gas at the bottom of the well. Furthermore, let us assume that equilibrium conditions prevail in the well. If this is true we would expect that an engine such as that shown in Fig. 15.2 (which operates on the basis of the pressure and composition change with elevation and does not involve combustion) would not be capable of doing any work.

If we consider a steady-state, steady-flow process for a control volume around this engine, the reversible work for the change of state from i to e is given by Eq. 10.12 on a total mass basis

$$\dot{W}^{\text{rev}} = \dot{m}_i \left(h_i + \frac{\mathbf{V}_i^2}{2} + gZ_i - T_0 s_i \right) - \dot{m}_e \left(h_e + \frac{\mathbf{V}_e^2}{2} + gZ_e - T_0 s_e \right)$$

Further, since $T_i = T_e = T_0 = $ constant, this reduces to the form of the Gibbs function $g = h - Ts$, Eq. 13.18, and the reversible work is

$$\dot{W}^{\text{rev}} = \dot{m}_i \left(g_i + \frac{\mathbf{V}_i^2}{2} + gZ_i \right) - \dot{m}_e \left(g_e + \frac{\mathbf{V}_e^2}{2} + gZ_e \right)$$

However,

$$\dot{W}^{\text{rev}} = 0, \qquad \dot{m}_i = \dot{m}_e \qquad \text{and} \qquad \frac{\mathbf{V}_i^2}{2} = \frac{\mathbf{V}_e^2}{2}$$

Then we can write

$$g_i + gZ_i = g_e + gZ_e$$

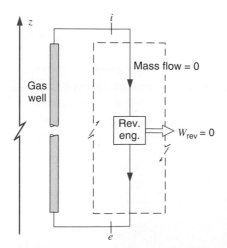

FIGURE 15.2 Illustration showing the relation between reversible work and the criteria for equilibrium.

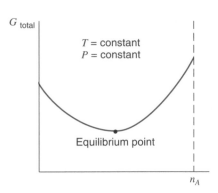

FIGURE 15.3 Illustration of the requirement for chemical equilibrium.

and the requirement for equilibrium in the well between two levels that are a distance dZ apart would be

$$dg_T + g\,dZ_T = 0$$

In contrast to a deep gas well, most of the systems that we consider are of such size that ΔZ is negligibly small, and therefore we consider the pressure to be uniform throughout.

This leads to the general statement of equilibrium that applies to simple compressible substances that may undergo a change in chemical composition, namely, that at equilibrium

$$dG_{T,P} = 0 \tag{15.1}$$

In the case of a chemical reaction, it is helpful to think of the equilibrium state as the state in which the Gibbs function is a minimum. For example, consider a control mass consisting initially of n_A moles of substance A and n_B moles of substance B, which react in accordance with the relation

$$v_A A + v_B B \rightleftharpoons v_C C + v_D D$$

Let the reaction take place at constant pressure and temperature. If we plot G for this control mass as a function of n_A, the number of moles of A present, we would have a curve as shown in Fig. 15.3. At the minimum point on the curve, $dG_{T,P} = 0$, and this will be the equilibrium composition for this system at the given temperature and pressure. The subject of chemical equilibrium will be developed further in Section 15.4.

15.2 EQUILIBRIUM BETWEEN TWO PHASES OF A PURE SUBSTANCE

As another example of this requirement for equilibrium, let us consider the equilibrium between two phases of a pure substance. Consider a control mass consisting of two phases of a pure substance at equilibrium. We know that under these conditions the two phases are at the same pressure and temperature. Consider the change of state associated with a transfer of dn moles from phase 1 to phase 2 while the temperature and pressure remain constant. That is,

$$dn^1 = -dn^2$$

The Gibbs function of this control mass is given by

$$G = f(T, P, n^1, n^2)$$

where n^1 and n^2 designate the number of moles in each phase. Therefore,

$$dG = \left(\frac{\partial G}{\partial T}\right)_{P,n^1,n^2} dT + \left(\frac{\partial G}{\partial P}\right)_{T,n^1,n^2} dP + \left(\frac{\partial G}{\partial n^1}\right)_{T,P,n^2} dn^1 + \left(\frac{\partial G}{\partial n^2}\right)_{T,P,n^1} dn^2$$

By definition,

$$\left(\frac{\partial G}{\partial n^1}\right)_{T,P,n^2} = \bar{g}^1 \qquad \left(\frac{\partial G}{\partial n^2}\right)_{T,P,n^1} = \bar{g}^2$$

Therefore, at constant temperature and pressure,

$$dG = \bar{g}^1 dn^1 + \bar{g}^2 dn^2 = dn^1(\bar{g}^1 - \bar{g}^2)$$

Now at equilibrium (Eq. 15.1)

$$dG_{T,P} = 0$$

Therefore, at equilibrium, we have

$$\bar{g}^1 = \bar{g}^2 \tag{15.2}$$

That is, under equilibrium conditions, the Gibbs function of each phase of a pure substance is equal. Let us check this by determining the Gibbs function of saturated liquid (water) and saturated vapor (steam) at 300 kPa. From the steam tables:

For the liquid:

$$g_f = h_f - T s_f = 561.47 - 406.7 \times 1.6718 = -118.4 \text{ kJ}/\text{kg}$$

For the vapor:

$$g_g = h_g - T s_g = 2725.3 - 406.7 \times 6.9919 = -118.4 \text{ kJ}/\text{kg}$$

Equation 15.2 can also be derived by applying the relation

$$T\,ds = dh - v\,dP$$

to the change of phase that takes place at constant pressure and temperature. For this process this relation can be integrated as follows:

$$\int_f^g T\,ds = \int_f^g dh$$

$$T(s_g - s_f) = (h_g - h_f)$$

$$h_f - T s_f = h_g - T s_g$$

$$g_f = g_g$$

The Clapeyron equation, which was derived in Section 13.1, can be derived by an alternate method by considering the fact that the Gibbs functions of two phases in equilibrium are equal. In Chapter 13 we considered the relation (Eq. 13.19) for a simple compressible substance:

$$dg = v\,dP - s\,dT$$

Consider a control mass that consists of a saturated liquid and a saturated vapor in equilibrium, and let this system undergo a change of pressure dP. The corresponding change in temperature, as determined from the vapor-pressure curve, is dT. Both phases will undergo the change in Gibbs function, dg, but since the phases always have the same value of the Gibbs function when they are in equilibrium, it follows that

$$dg_f = dg_g$$

But, from Eq. 13.19,

$$dg = v \, dP - s \, dT$$

it follows that

$$dg_f = v_f dP - s_f dT$$

$$dg_g = v_g dP - s_g dT$$

Since

$$dg_f = dg_g$$

it follows that

$$v_f dP - s_f dT = v_g dP - s_g dT$$

$$dP(v_g - v_f) = dT(s_g - s_f) \tag{15.3}$$

$$\frac{dP}{dT} = \frac{s_{fg}}{v_{fg}} = \frac{h_{fg}}{Tv_{fg}}$$

In summary, when different phases of a pure substance are in equilibrium, each phase has the same value of the Gibbs function per unit mass. This fact is relevant to different solid phases of a pure substance and is important in metallurgical applications of thermodynamics. Example 15.1 illustrates this principle.

EXAMPLE 15.1   What pressure is required to make diamonds from graphite at a temperature of 25°C? The following data are given for a temperature of 25°C and a pressure of 0.1 MPa.

	Graphite	Diamond
g	0	2867.8 kJ/mol
v	0.000 444 m³/kg	0.000 284 m³/kg
β_T	0.304×10^{-6} 1/MPa	0.016×10^{-6} 1/MPa

Analysis and Solution

The basic principle in the solution is that graphite and diamond can exist in equilibrium when they have the same value of the Gibbs function. At 0.1 MPa pressure the Gibbs function of the diamond is greater than that of the graphite. However, the rate of increase in Gibbs function with pressure is greater for the graphite than for the diamond and, therefore, at some pressure they can exist in equilibrium, and our problem is to find this pressure.

We have already considered the relation

$$dg = v \, dP - s \, dT$$

Since we are considering a process that takes place at constant temperature, this reduces to

$$dg_T = v \, dP_T \qquad \text{(a)}$$

Now at any pressure P and the given temperature, the specific volume can be found from the following relation, which utilizes the isothermal compressibility factor.

$$v = v^\circ + \int_{P=0.1}^{P} \left(\frac{\partial v}{\partial P} \right)_T dP = v^\circ + \int_{P=0.1}^{P} \frac{v}{v} \left(\frac{\partial v}{\partial P} \right)_T dP$$

$$= v^\circ - \int_{P=0.1}^{P} v\beta_T dP \qquad \text{(b)}$$

The superscript $^\circ$ will be used in this example to indicate the properties at a pressure of 0.1 MPa and a temperature of 25°C.

The specific volume changes only slightly with pressure, so that $v \approx v^\circ$. Also, we assume that β_T is constant and that we are considering a very high pressure. With these assumptions this equation can be integrated to give

$$v = v^\circ - v^\circ \beta_T P = v^\circ (1 - \beta_T P) \qquad \text{(c)}$$

We can now substitute this into Eq. a to give the relation

$$dg_T = [v^\circ (1 - \beta_T P)] dP_T$$

$$g - g^\circ = v^\circ (P - P^\circ) - v^\circ \beta_T \frac{(P^2 - P^{\circ 2})}{2} \qquad \text{(d)}$$

If we assume that $P^\circ \ll P$, this reduces to

$$g - g^\circ = v^\circ \left(P - \frac{\beta_T P^2}{2} \right) \qquad \text{(e)}$$

For the graphite, $g^\circ = 0$ and we can write

$$g_G = v_G^\circ \left[P - (\beta_T)_G \frac{P^2}{2} \right]$$

For the diamond, g° has a definite value and we have

$$g_D = g_D^\circ + v_D^\circ \left[P - (\beta_T)_D \frac{P^2}{2} \right]$$

But, at equilibrium the Gibbs function of the graphite and diamond are equal

$$g_G = g_D$$

Therefore,

$$v_G^\circ\left[P-(\beta_T)_G\,\frac{P^2}{2}\right]=g_D^\circ+v_D^\circ\left[P-(\beta_T)_D\,\frac{P^2}{2}\right]$$

$$(v_G^\circ-v_D^\circ)P-[v_G^\circ(\beta_T)_G-v_D^\circ(\beta_T)_D]\frac{P^2}{2}=g_D^\circ$$

$$(0.000\,444-0.000\,284\,)P$$

$$-\,(0.000\,444\times0.304\times10^{-6}-0.000\,284\times0.016\times10^{-6}\,)P^2\,/\,2=\frac{2867.8}{12.011\times1000}$$

Solving this for P we find

$$P=1493\ \text{MPa}$$

That is, at 1493 MPa, 25°C, graphite and diamond can exist in equilibrium, and the possibility exists for conversion from graphite to diamonds.

15.3 METASTABLE EQUILIBRIUM

Although the limited scope of this book precludes an extensive treatment of metastable equilibrium, a brief introduction to the subject is presented in this section. Let us first consider an example of metastable equilibrium.

Consider a slightly superheated vapor, such as steam, expanding in a convergent-divergent nozzle, as shown in Fig. 15.4. Assuming the process is reversible and adiabatic,

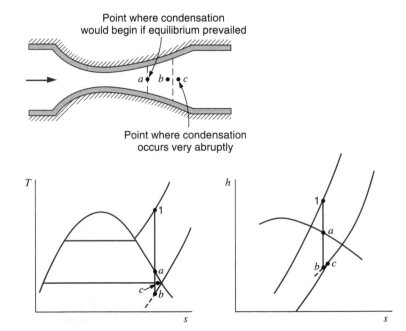

FIGURE 15.4
Illustration of the phenomenon of supersaturation in a nozzle.

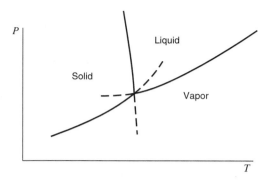

FIGURE 15.5 Metastable states for solid–liquid–vapor equilibrium.

the steam will follow path 1-*a* on the *T*–*s* diagram, and at point *a* we would expect condensation to occur. However, if point *a* is reached in the divergent section of the nozzle, it is observed that no condensation occurs until point *b* is reached, and at this point the condensation occurs very abruptly in what is referred to as a condensation shock. Between points *a* and *b* the steam exists as a vapor, but the temperature is below the saturation temperature for the given pressure. This is known as a metastable state. The possibility of a metastable state exists with any phase transformation. The dotted lines on the equilibrium diagram shown in Fig. 15.5 represent possible metastable states for solid–liquid–vapor equilibrium.

The nature of a metastable state is often pictured schematically by the kind of diagram shown in Fig. 15.6. The ball is in a stable position (the "metastable state") for small displacements, but with a large displacement it moves to a new equilibrium position. The steam expanding in the nozzle is in a metastable state between *a* and *b*. This means that droplets smaller than a certain critical size will reevaporate, and only when droplets of larger than this critical size have formed (this corresponds to moving the ball out of the depression) will the new equilibrium state appear.

15.4 CHEMICAL EQUILIBRIUM

We now turn our attention to chemical equilibrium and consider first a chemical reaction involving only one phase. This is referred to as a homogeneous chemical reaction. It may be helpful to visualize this as a gaseous phase, but the basic considerations apply to any phase.

Consider a vessel, Fig. 15.7, that contains four compounds, *A*, *B*, *C*, and *D*, which are in chemical equilibrium at a given pressure and temperature. For example, these might consist of CO_2, H_2, CO, and H_2O in equilibrium. Let the number of moles of each component be designated n_A, n_B, n_C, and n_D. Further, let the chemical reaction that takes

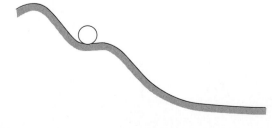

FIGURE 15.6 Schematic diagram illustrating a metastable state.

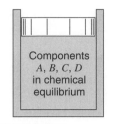

FIGURE 15.7
Schematic diagram for consideration of chemical equilibrium.

place between these four constituents be

$$v_A A + v_B B \rightleftharpoons v_C C + v_D D \qquad (15.4)$$

where the v's are the stoichiometric coefficients. It should be emphasized that there is a very definite relation between the v's (the stoichiometric coefficients), whereas the n's (the number of moles present) for any constituent can be varied simply by varying the amount of that component in the reaction vessel.

Let us now consider how the requirement for equilibrium, namely, that $dG_{T,P} = 0$ at equilibrium, applies to a homogeneous chemical reaction. Let us assume that the four components are in chemical equilibrium and then assume that from this equilibrium state, while the temperature and pressure remain constant, the reaction proceeds an infinitesimal amount toward the right as Eq. 15.4 is written. This results in a decrease in the moles of A and B and an increase in the moles of C and D. Let us designate the degree of reaction by ε, and define the degree of reaction by the relations

$$dn_A = -v_A d\varepsilon$$
$$dn_B = -v_B d\varepsilon$$
$$dn_C = +v_C d\varepsilon$$
$$dn_D = +v_D d\varepsilon \qquad (15.5)$$

That is, the change in the number of moles of any component during a chemical reaction is given by the product of the stoichiometric coefficients (the v's) and the degree of reaction.

Let us evaluate the change in the Gibbs function associated with this chemical reaction that proceeds to the right in the amount $d\varepsilon$. In doing so we use, as would be expected, the Gibbs function of each component in the mixture—the partial molal Gibbs function (or its equivalent, the chemical potential):

$$dG_{T,P} = \overline{G}_C dn_C + \overline{G}_D dn_D + \overline{G}_A dn_A + \overline{G}_B dn_B$$

Substituting Eq. 15.5, we have

$$dG_{T,P} = (v_C \overline{G}_C + v_D \overline{G}_D - v_A \overline{G}_A - v_B \overline{G}_B)\, d\varepsilon \qquad (15.6)$$

We now need to develop expressions for the partial molal Gibbs functions in terms of properties that we are able to calculate. From the definition of Gibbs function, Eq. 13.18,

$$G = H - TS$$

For a mixture of two components A and B, we differentiate this equation with respect to n_A at constant T, P, and n_B, which results in

$$\left(\frac{\partial G}{\partial n_A}\right)_{T,P,n_B} = \left(\frac{\partial H}{\partial n_A}\right)_{T,P,n_B} - T\left(\frac{\partial S}{\partial n_A}\right)_{T,P,n_B}$$

All three of these quantities satisfy the definition of partial molal properties according to Eq. 13.71, such that

$$\overline{G}_A = \overline{H}_A - T\overline{S}_A \qquad (15.7)$$

For an ideal gas mixture, enthalpy is not a function of pressure, and

$$\overline{H}_A = \overline{h}_{A\,TP} = \overline{h}^0_{A\,TP^0} \qquad (15.8)$$

Entropy is, however, a function of pressure, so that the partial entropy of A can be expressed by Eq. 14.20,

$$\overline{S}_A = \overline{s}_{A\,TP_A = y_A P}$$

$$= \overline{s}^0_{A\,TP^0} - \overline{R}\,\ln\!\left(\frac{y_A P}{P^0}\right) \tag{15.9}$$

Now, substituting Eqs. 15.8 and 15.9 into Eq. 15.7,

$$\overline{G}_A = \overline{h}^0_{A\,TP^0} - T\overline{s}^0_{A\,TP^0} + \overline{R}T\,\ln\!\left(\frac{y_A P}{P^0}\right)$$

$$= \overline{g}^0_{A\,TP^0} + \overline{R}T\,\ln\!\left(\frac{y_A P}{P^0}\right) \tag{15.10}$$

Equation 15.10 is an expression for the partial Gibbs function of a component in a mixture in terms of a specific reference value, the pure-substance standard-state Gibbs function at the same temperature, and a function of the temperature, pressure, and composition of the mixture. This expression can be applied to each of the components in Eq. 15.6, resulting in

$$dG_{TP} = \left\{ v_C\!\left[\overline{g}^0_C + \overline{R}\,T\,\ln\!\left(\frac{y_C P}{P^0}\right)\right] + v_D\!\left[\overline{g}^0_D + \overline{R}\,T\,\ln\!\left(\frac{y_D P}{P^0}\right)\right]\right.$$

$$\left. - v_A\!\left[\overline{g}^0_A + \overline{R}\,T\,\ln\!\left(\frac{y_A P}{P_0}\right)\right] - v_B\!\left[\overline{g}^0_B + \overline{R}\,T\,\ln\!\left(\frac{y_B P}{P^0}\right)\right]\right\} d\varepsilon \tag{15.11}$$

Let us define ΔG° as follows:

$$\Delta G^\circ = v_C \overline{g}^\circ_C + v_D \overline{g}^\circ_D - v_A \overline{g}^\circ_A - v_B \overline{g}^\circ_B \tag{15.12}$$

That is, ΔG° is the change in the Gibbs function that would occur if the chemical reaction given by Eq. 15.4 (which involves the stoichiometric amounts of each component) proceeded completely from left to right, with the reactants A and B initially separated and at temperature T and the standard state pressure and the products C and D finally separated and at temperature T and the standard state pressure. Note also that ΔG° for a given reaction is a function of only the temperature. This will be most important to bear in mind as we proceed with our developments of homogeneous chemical equilibrium. Let us now digress from our development to consider an example involving the calculation of ΔG°.

EXAMPLE 15.2 Determine the value of ΔG° for the reaction $2H_2O \rightleftharpoons 2H_2 + O_2$ at 25°C and at 2000 K, with the water in the gaseous phase.

Solution

At any given temperature, the standard-state Gibbs function change of Eq. 15.12 can be calculated from the relation

$$\Delta G^\circ = \Delta H^\circ - T\Delta S^\circ$$

At 25°C,

$$\Delta H^\circ = 2\bar{h}^\circ_{fH_2} + \bar{h}^\circ_{fO_2} - 2\bar{h}^\circ_{fH_2O(g)}$$
$$= 2(0) + 1(0) - 2(-241\,826) = 483\,652 \text{ kJ}$$

$$\Delta S^\circ = 2\bar{s}^\circ_{H_2} + \bar{s}^\circ_{O_2} - 2\bar{s}^\circ_{H_2O(g)}$$
$$= 2(130.678) + 1(205.148) - 2(188.834) = 88.836 \text{ kJ/K}$$

Therefore, at 25°C,

$$\Delta G^\circ = 483\,652 - 298.15(88.836) = 457\,166 \text{ kJ}$$

At 2000 K,

$$\Delta H^\circ = 2(\bar{h}^\circ_{2000} - \bar{h}^\circ_{298})_{H_2} + (\bar{h}^\circ_{2000} - \bar{h}^\circ_{298})_{O_2} - 2(\bar{h}^\circ_f + \bar{h}^\circ_{2000} - \bar{h}^\circ_{298})_{H_2O}$$

$$= 2(52\,942) + (59\,176) - 2(-241\,826 + 72\,788)$$

$$= 503\,136 \text{ kJ}$$

$$\Delta S^\circ = 2(\bar{s}^\circ_{2000})_{H_2} + (\bar{s}^\circ_{2000})_{O_2} - 2(\bar{s}^\circ_{2000})_{H_2O}$$

$$= 2(188.419) + 268.748 - 2(264.769)$$

$$= 116.048 \text{ kJ/K}$$

Therefore,

$$\Delta G^\circ = 503\,136 - 2000 \times 116.048 = 271\,040 \text{ kJ}$$

Returning now to our development, substituting Eq. 15.12 into Eq. 15.11 and rearranging we can write

$$dG_{T,P} = \left\{ \Delta G^\circ + \bar{R}T \ln\left[\frac{y_C^{v_C} y_D^{v_D}}{y_A^{v_A} y_B^{v_B}} \right] \right\} d\varepsilon \tag{15.13}$$

At equilibrium $dG_{T,P} = 0$. Therefore, since $d\varepsilon$ is arbitrary,

$$\ln\left[\frac{y_C^{v_C} y_D^{v_D}}{y_A^{v_A} y_B^{v_B}} \left(\frac{P}{P^0} \right)^{v_C + v_D - v_A - v_B} \right] = -\frac{\Delta G^\circ}{\bar{R}T} \tag{15.14}$$

For convenience, we define the equilibrium constant K as

$$\ln K = -\frac{\Delta G^\circ}{\bar{R}T} \tag{15.15}$$

which we note must be a function of temperature only for a given reaction, since ΔG° is given by Eq. 15.12 in terms of the properties of the pure substances at a given temperature and the standard-state pressure.

Combining Eqs. 15.14 and 15.15, we have

$$K = \frac{y_C^{v_C} y_D^{v_D}}{y_A^{v_A} y_B^{v_B}} \left(\frac{P}{P^0} \right)^{v_C + v_D - v_A - v_B} \tag{15.16}$$

which is the chemical equilibrium equation corresponding to the reaction equation, Eq. 15.4.

EXAMPLE 15.3   Determine the equilibrium constant K, expressed as in K, for the reaction $2H_2O \rightleftharpoons 2H_2 + O_2$ at 25°C and at 2000 K.

Solution

We have already found, in Example 15.2, $\Delta G°$ for this reaction at these two temperatures. Therefore, at 25°C,

$$(\ln K)_{298} = -\frac{\Delta G°_{298}}{\overline{R}T} = \frac{-457\,166}{8.3145 \times 298.15} = -184.42$$

At 2000 K, we have

$$(\ln K)_{2000} = -\frac{\Delta G°_{2000}}{\overline{R}T} = \frac{-271\,040}{8.3145 \times 2000} = -16.299$$

Table A.10 gives the values of the equilibrium constant for a number of reactions. Note again that for each reaction the value of the equilibrium constant is determined from the properties of each of the pure constituents at the standard-state pressure and is a function of temperature only.

For other reaction equations, the chemical equilibrium constant can be calculated as in Example 15.3.

We now consider a number of examples that illustrate the procedure for determining the equilibrium composition for a homogeneous reaction and the influence of certain variables on the equilibrium composition.

EXAMPLE 15.4   One kilomole of carbon at 25°C and 0.1 MPa pressure reacts with 1 kmol of oxygen at 25°C and 0.1 MPa pressure to form an equilibrium mixture of CO_2, CO, and O_2 at 3000 K, 0.1 MPa pressure, in a steady-flow process. Determine the equilibrium composition and the heat transfer for this process.

> *Control volume:*   Combustion chamber.
>
> *Inlet states:* P, T known for carbon and for oxygen.
>
> *Exit state:* P, T known.
>
> *Process:*   SSSF.
>
> *Sketch:*   Fig. 15.8.
>
> *Model:*   Table A.9 for carbon; ideal gases, Tables A.8 and A.9.

Analysis and Solution

It is convenient to view the overall process as though it occurs in two separate steps, a combustion process followed by a heating and dissociation of the combustion product carbon dioxide, as indicated in Fig. 15.8. This two-step process is represented as

$$\text{Combustion:} \qquad C + O_2 \rightarrow CO_2$$
$$\text{Dissociation reaction:} \qquad 2CO_2 \rightleftharpoons 2CO + O_2$$

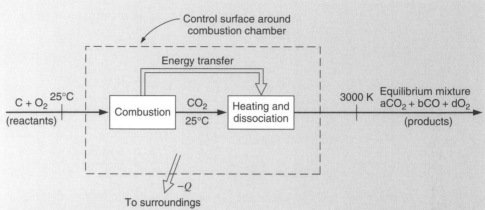

FIGURE 15.8 Sketch for Example 15.4.

That is, the energy released by the combustion of C and O_2 heats the CO_2 formed to high temperature, which causes dissociation of part of the CO_2 to CO and O_2. Thus, the overall reaction can be written

$$C + O_2 \rightarrow aCO_2 + bCO + dO_2$$

where the unknown coefficients a, b, and d must be found by solution of the equilibrium equation associated with the dissociation reaction. Once this is accomplished, we can write the first law for a control volume around the combustion chamber to calculate the heat transfer.

From the combustion equation we find that the initial composition for the dissociation reaction is 1 kmol CO_2. Therefore, letting $2z$ be the number of kilomoles of CO_2 dissociated, we find

	$2CO_2 \rightleftharpoons 2CO + O_2$		
Initial:	1	0	0
Change:	$-2z$	$+2z$	$+z$
At equilibrium:	$(1-2z)$	$2z$	z

Therefore, the overall reaction is

$$C + O_2 \rightarrow (1-2z)CO_2 + 2z\,CO + z\,O_2$$

and the total number of kilomoles at equilibrium is

$$n = (1-2z) + 2z + z = 1 + z$$

The equilibrium mole fractions are

$$y_{CO_2} = \frac{1-2z}{1+z} \qquad y_{CO} = \frac{2z}{1+z} \qquad y_{O_2} = \frac{z}{1+z}$$

From Table A.10 we find that the value of the equilibrium constant at 3000 K for the dissociation reaction considered here is

$$\ln K = -2.217 \qquad K = 0.1089$$

Substituting these quantities along with $P = 0.1$ MPa into Eq. 15.16, we have the equilibrium equation,

$$K = 0.1089 = \frac{y^2_{CO} y_{O_2}}{y^2_{CO_2}} \left(\frac{P}{P^\circ} \right)^{2+1-2} = \frac{\left(\frac{2z}{1+z} \right)^2 \left(\frac{z}{1+z} \right)}{\left(\frac{1-2z}{1+z} \right)^2} \quad (1)$$

or, in more convenient form,

$$\frac{K}{P / P^\circ} = \frac{0.1089}{1} = \left(\frac{2z}{1-2z} \right)^2 \left(\frac{z}{1+z} \right)$$

To obtain the physically meaningful root of this mathematical relation, we note that the number of moles of each component must be greater than zero. Thus, the root of interest to us must lie in the range

$$0 \leq z \leq 0.5$$

Solving the equilibrium equation by trial and error, we find

$$z = 0.2189$$

Therefore, the overall process is

$$C + O_2 \rightarrow 0.5622 CO_2 + 0.4378 CO + 0.2189 O_2$$

where the equilibrium mole fractions are

$$y_{CO_2} = \frac{0.5622}{1.2189} = 0.4612$$

$$y_{CO} = \frac{0.4378}{1.2189} = 0.3592$$

$$y_{O_2} = \frac{0.2189}{1.2189} = 0.1796$$

The heat transfer from the combustion chamber to the surroundings can be calculated using the enthalpies of formation and Table A.8. For this process

$$H_R = (\bar{h}^\circ_f)_C + (\bar{h}^\circ_f)_{O_2} = 0 + 0 = 0$$

The equilibrium products leave the chamber at 3000 K. Therefore,

$$H_P = n_{CO_2}(\bar{h}^\circ_f + \bar{h}^\circ_{3000} - \bar{h}^\circ_{298})_{CO_2}$$

$$+ n_{CO}(\bar{h}^\circ_f + \bar{h}^\circ_{3000} - \bar{h}^\circ_{298})_{CO}$$

$$+ n_{O_2}(\bar{h}^\circ_f + \bar{h}^\circ_{3000} - \bar{h}^\circ_{298})_{O_2}$$

$$= 0.5622(-393\,522 + 152\,853)$$

$$+ 0.4378(-110\,527 + 93\,504)$$

$$+ 0.2189(98\,013)$$

$$= -121\,302 \text{ kJ}$$

Substituting into the first law gives

$$Q_{c.v.} = H_P - H_g$$
$$= -121\,302 \text{ kJ / kmol C burned}$$

EXAMPLE 15.5 One kilomole of carbon at 25°C reacts with 2 kmol of oxygen at 25°C to form an equilibrium mixture of CO_2, CO, and O_2 at 3000 K, 0.1 MPa pressure. Determine the equilibrium composition.

   Control volume: Combustion chamber.

   Inlet states: T known for carbon and for oxygen.

   Exit state: P, T known.

   Process: SSSF.

   Model: Ideal gas mixture at equilibrium.

Analysis and Solution

The overall process can be imagined to occur in two steps as in the previous example. The combustion process is

$$C + 2O_2 \rightarrow CO_2 + O_2$$

and the subsequent dissociation reaction is

	$2CO_2$	$\rightleftharpoons$	$2CO$	$+ O_2$
Initial:	1		0	1
Change:	$-2z$		$+2z$	$+z$
At equilibrium:	$(1-2z)$		$2z$	$(1+z)$

We find that in this case the overall process is

$$C + 2O_2 \rightarrow (1 - 2z)CO_2 + 2z\,CO + (1+z)O_2$$

and the total number of kilomoles at equilibrium is

$$n = (1 - 2z) + 2z + (1 + z) = 2 + z$$

The mole fractions are

$$y_{CO_2} = \frac{1 - 2z}{2 + z} \qquad y_{CO} = \frac{2z}{2 + z} \qquad y_{O_2} = \frac{1 + z}{2 + z}$$

The equilibrium constant for the reaction $2CO_2 \rightleftharpoons 2CO + O_2$ at 3000 K was found in Example 15.4 to be 0.1089. Therefore, with these expressions, quantities, and $P = 0.1$

MPa substituted, the equilibrium equation is

$$K = 0.1089 = \frac{y_{CO}^2 y_{O_2}}{y_{CO_2}^2}\left(\frac{P}{P^\circ}\right)^{2+1-2}$$

$$= \frac{\left(\dfrac{2z}{2+z}\right)^2\left(\dfrac{1+z}{2+z}\right)}{\left(\dfrac{1-2z}{2+z}\right)^2}(1)$$

or

$$\frac{K}{P/P^\circ} = \frac{0.1089}{1} = \left(\frac{2z}{1-2z}\right)^2\left(\frac{1+z}{2+z}\right)$$

We note that in order for the number of kilomoles of each component to be greater than zero,

$$0 \le z \le 0.5$$

Solving the equilibrium equation for z, we find

$$z = 0.1553$$

so that the overall process is

$$C + 2O_2 \rightarrow 0.6894CO_2 + 0.3106CO + 1.1553O_2$$

The mole fractions of the components in the equilibrium mixture are

$$y_{CO_2} = \frac{0.6894}{2.1553} = 0.320$$

$$y_{CO} = \frac{0.3106}{2.1553} = 0.144$$

$$y_{O_2} = \frac{1.1553}{2.1553} = 0.536$$

The heat transferred from the chamber in this process could be found by the same procedure followed in Example 15.4, considering the overall process.

15.5 SIMULTANEOUS REACTIONS

In developing the equilibrium equation and equilibrium constant expressions of Section 15.4, it was assumed that there was only a single chemical reaction equation relating the substances present in the system. To demonstrate the more general situation in which there is more than one chemical reaction, we will now analyze a case involving two simultaneous reactions by a procedure analogous to that followed in Section 15.4. These results are then readily extended to systems involving several simultaneous reactions.

Consider a mixture of substances A, B, C, D, L, M, and N as indicated in Fig. 15.9. These substances are assumed to exist at a condition of chemical equilibrium at tempera-

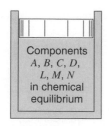

FIGURE 15.9 Sketch demonstrating simultaneous reactions.

ture T and pressure P, and are related by the two independent reactions

$$(1) \quad v_{A_1}A + v_B B \rightleftharpoons v_C C + v_D D \tag{15.17}$$

$$(2) \quad v_{A_2}A + v_L L \rightleftharpoons v_M M + v_N N \tag{15.18}$$

We have considered the situation where one of the components (substance A) is involved in each of the reactions in order to demonstrate the effect of this condition on the resulting equations. As in the previous section, the changes in amounts of the components are related by the various stoichiometric coefficients (which are not the same as the number of moles of each substance present in the vessel). We also realize that the coefficients v_{A1} and v_{A2} are not necessarily the same. That is, substance A does not in general take part in each of the reactions to the same extent.

Development of the requirement for equilibrium is completely analogous to that of Section 15.4. We consider that each reaction proceeds an infinitesimal amount toward the right side. This results in a decrease in the number of moles of A, B, and L, and an increase in the moles of C, D, M, and N. Letting the degrees of reaction be ε_1 and ε_2 for reactions 1 and 2, respectively, the changes in the number of moles are, for infinitesimal shifts from the equilibrium composition,

$$dn_A = -v_{A_1}d\varepsilon_1 - v_{A_2}d\varepsilon_2$$

$$dn_B = -v_B d\varepsilon_1$$

$$dn_L = -v_L d\varepsilon_2$$

$$dn_C = +v_C d\varepsilon_1$$

$$dn_D = +v_D d\varepsilon_1$$

$$dn_M = +v_M d\varepsilon_2$$

$$dn_N = +v_N d\varepsilon_2 \tag{15.19}$$

The change in Gibbs function for the mixture in the vessel at constant temperature and pressure is

$$dG_{T,P} = \overline{G}_A dn_A + \overline{G}_B dn_B + \overline{G}_C dn_C + \overline{G}_D dn_D + \overline{G}_L dn_L + \overline{G}_M dn_M + \overline{G}_N dn_N$$

Substituting the expressions of Eq. 15.19 and collecting terms,

$$dG_{T,P} = (v_C \overline{G}_C + v_D \overline{G}_D - v_{A_1}\overline{G}_A - v_B \overline{G}_B)d\varepsilon_1$$

$$+ (v_M \overline{G}_M + v_N \overline{G}_N - v_{A_2}\overline{G}_A - v_L \overline{G}_L)d\varepsilon_2 \tag{15.20}$$

It is convenient to again express each of the partial molal Gibbs functions in terms of

$$\overline{G}_i = \overline{g}_i^\circ + \overline{R}T \ln\left(\frac{y_i P}{P^\circ}\right)$$

Equation 15.20 written in this form becomes

$$dG_{T,P} = \left\{ \Delta G_1^\circ + \overline{R}T \ln\left[\frac{y_C^{v_C} y_D^{v_D}}{y_A^{v_{A1}} y_B^{v_B}} \left(\frac{P}{P^\circ}\right)^{v_C + v_D - v_{A1} - v_B} \right] \right\} d\varepsilon_1$$

$$+ \left\{ \Delta G_2^\circ + \overline{R}T \ln\left[\frac{y_M^{v_M} y_N^{v_N}}{y_A^{v_{A2}} y_L^{v_L}} \left(\frac{P}{P^\circ}\right)^{v_M + v_N - v_{A2} - v_L} \right] \right\} d\varepsilon_2 \tag{15.21}$$

In this equation the standard-state change in Gibbs function for each reaction is defined as

$$\Delta G_1^\circ = v_C \bar{g}_C^\circ + v_D \bar{g}_D^\circ - v_{A1} \bar{g}_A^\circ - v_B \bar{g}_B^\circ \tag{15.22}$$

$$\Delta G_2^\circ = v_M \bar{g}_M^\circ + v_N \bar{g}_N^\circ - v_{A2} \bar{g}_A^\circ - v_L \bar{g}_L^\circ \tag{15.23}$$

Equation 15.21 expresses the change in Gibbs function of the system at constant T, P, for infinitesimal degrees of reaction of both reactions 1 and 2, Eqs. 15.17 and 15.18. The requirement for equilibrium is that $dG_{T,P} = 0$. Therefore, since reactions 1 and 2 are independent, $d\varepsilon_1$ and $d\varepsilon_2$ can be independently varied. It follows that at equilibrium each of the bracketed terms of Eq. 15.21 must be zero. Defining equilibrium constants for the two reactions by

$$\ln K_1 = -\frac{\Delta G_1^\circ}{\bar{R}T} \tag{15.24}$$

and

$$\ln K_2 = -\frac{\Delta G_2^\circ}{\bar{R}T} \tag{15.25}$$

we find that, at equilibrium

$$K_1 = \frac{y_C^{v_C} y_D^{v_D}}{y_A^{v_{A1}} y_B^{v_B}} \left(\frac{P}{P^\circ} \right)^{v_C + v_D - v_{A1} - v_B} \tag{15.26}$$

and

$$K_2 = \frac{y_M^{v_M} y_N^{v_N}}{y_A^{v_{A2}} y_L^{v_L}} \left(\frac{P}{P^\circ} \right)^{v_M + v_N - v_{A1} 2 - v_L} \tag{15.27}$$

These expressions for the equilibrium composition of the mixture must be solved simultaneously. The following example demonstrates and clarifies this procedure.

EXAMPLE 15.6   One kilomole of water vapor is heated to 3000 K, 0.1 MPa pressure. Determine the equilibrium composition, assuming that H_2O, H_2, O_2, and OH are present.

Control volume: Heat exchanger.

Exit state: P, T known.

Model: Ideal gas mixture at equilibrium.

Analysis and Solution

There are two independent reactions relating the four components of the mixture at equilibrium. These can be written as

(1) $2H_2O \rightleftharpoons 2H_2 + O_2$

(2) $2H_2O \rightleftharpoons H_2 + 2OH$

Let $2a$ be the number of kilomoles of water dissociating according to reaction 1 during the heating, and let $2b$ be the number of kilomoles of water dissociating according to reaction 2. Since the initial composition is 1 kmol water, the changes according to the two reactions are

Change:

$$(1) \quad 2H_2O \rightleftharpoons 2H_2 + O_2$$
$$ -2a \quad +2a \quad +a$$

Change:

$$(2) \quad 2H_2O \rightleftharpoons H_2 + 2OH$$
$$ -2b \quad +b \quad +2b$$

Therefore, the number of kilomoles of each component at equilibrium is its initial number plus the change, so that at equilibrium

$$n_{H_2O} = 1 - 2a - 2b$$
$$n_{H_2} = 2a + b$$
$$n_{O_2} = a$$
$$\underline{n_{OH} = 2b}$$
$$n = 1 + a + b$$

The overall chemical reaction that occurs during the heating process can be written

$$H_2O \rightarrow (1-2a-2b)H_2O + (2a+b)H_2 + aO_2 + 2bOH$$

The right-hand side of this expression is the equilibrium composition of the system. Since the number of kilomoles of each substance must necessarily be greater than zero, we find that the possible values of a and b are restricted to

$$a \geq 0$$
$$b \geq 0$$
$$(a+b) \leq 0.5$$

The two equilibrium equations are, assuming that the mixture behaves as an ideal gas,

$$K_1 = \frac{y_{H_2}^2 y_{O_2}}{y_{H_2O}^2} \left(\frac{P}{P^\circ} \right)^{2+1-2}$$

$$K_2 = \frac{y_{H_2} y_{OH}^2}{y_{H_2O}^2} \left(\frac{P}{P^\circ} \right)^{1+2-2}$$

Since the mole fraction of each component is the ratio of the number of kilomoles of the component to the total number of kilomoles of the mixture, these equations can be written in the form

$$K_1 = \frac{\left(\dfrac{2a+b}{1+a+b} \right)^2 \left(\dfrac{a}{1+a+b} \right)}{\left(\dfrac{1-2a-2b}{1+a+b} \right)^2} \left(\frac{P}{P^\circ} \right)$$

$$= \left(\frac{2a+b}{1-2a-2b} \right)^2 \left(\frac{a}{1+a+b} \right) \left(\frac{P}{P^\circ} \right)$$

and

$$K_2 = \frac{\left(\dfrac{2a+b}{1+a+b}\right)\left(\dfrac{2b}{1+a+b}\right)^2}{\left(\dfrac{1-2a-2b}{1+a+b}\right)^2}\left(\frac{P}{P^\circ}\right)$$

$$= \left(\frac{2a+b}{1+a+b}\right)\left(\frac{2b}{1-2a-2b}\right)^2\left(\frac{P}{P^\circ}\right)$$

giving two equations in the two unknowns a and b, since $P = 0.1$ MPa and the values of K_1, K_2 are known. From Table A.10 at 3000 K, we find

$$K_1 = 0.002\,062 \qquad K_2 = 0.002\,893$$

Therefore, the equations can be solved simultaneously for a and b. The values satisfying the equations are

$$a = 0.0534 \qquad b = 0.0551$$

Substituting these values into the expressions for the number of kilomoles of each component and of the mixture, we find the equilibrium mole fractions to be

$$y_{H_2O} = 0.7063$$

$$y_{H_2} = 0.1461$$

$$y_{O_2} = 0.0482$$

$$y_{OH} = 0.0994$$

The methods used in this section can readily be extended to equilibrium systems having more than two independent reactions. In each case, the number of simultaneous equilibrium equations is equal to the number of independent reactions. The solution of a large set of nonlinear simultaneous equations naturally becomes quite difficult, however, and is not easily accomplished by hand calculations. These problems are normally solved using iterative procedures on a digital computer.

15.6 Ionization

In the final section of this chapter, we consider the equilibrium of systems that are made up of ionized gases, or plasmas, a field that has been studied and applied increasingly in recent years. In previous sections we discussed chemical equilibrium, with a particular emphasis on molecular dissociation, as for example the reaction

$$N_2 \rightleftharpoons 2N$$

which occurs to an appreciable extent for most molecules only at high temperature, of the order of magnitude 3000 to 10 000 K. At still higher temperatures, such as those found in electric arcs, the gas becomes ionized. That is, some of the atoms lose an electron, according to the reaction

$$N \rightleftharpoons N^+ + e^-$$

where N^+ denotes a singly ionized nitrogen atom, one that has lost one electron and consequently has a positive charge, and e^- represents the free electron. As the temperature rises still higher, many of the ionized atoms lose another electron, according to the reaction

$$N^+ \rightleftharpoons N^{++} + e^-$$

and thus becomes doubly ionized. As the temperature continues to rise, the process continues until a temperature is reached at which all the electrons have been stripped from the nucleus.

Ionization generally is appreciable only at high temperature. However, dissociation and ionization both tend to occur to greater extents at low pressure, and consequently dissociation and ionization may be appreciable in such environments as the upper atmosphere, even at moderate temperature. Other effects such as radiation will also cause ionization, but these effects are not considered here.

The problems of analyzing the composition in a plasma become much more difficult than for an ordinary chemical reaction, for in an electric field the free electrons in the mixture do not exchange energy with the positive ions and neutral atoms at the same rate that they do with the field. Consequently, in a plasma in an electric field, the electron gas is not at exactly the same temperature as the heavy particles. However, for moderate fields, assuming a condition of thermal equilibrium in the plasma is a reasonable approximation, at least for preliminary calculations. Under this condition we can treat the ionization equilibrium in exactly the same manner as an ordinary chemical equilibrium analysis.

At these extremely high temperatures, we may assume that the plasma behaves as an ideal-gas mixture of neutral atoms, positive ions, and electron gas. Thus, for the ionization of some atomic species A,

$$A \rightleftharpoons A^+ + e^- \tag{15.28}$$

we may write the ionization equilibrium equation in the form

$$K = \frac{y_{A^+} y_{e^-}}{y_A} \left(\frac{P}{P^\circ} \right)^{1+1-1} \tag{15.29}$$

The ionization-equilibrium constant K is defined in the ordinary manner

$$\ln K = -\frac{\Delta G^\circ}{\overline{R} T} \tag{15.30}$$

and is a function of temperature only. The standard-state Gibbs function change for reaction 15.28 is found from

$$\Delta G^\circ = \overline{g}_{A^+}^{\,\circ} + \overline{g}_{e^-}^{\,\circ} - \overline{g}_A^{\,\circ} \tag{15.31}$$

The standard-state Gibbs function for each component at the given plasma temperature can be calculated using the procedures of statistical thermodynamics, so that ionization–equilibrium constants can be tabulated as functions of temperature.

Solution of the ionization–equilibrium equation, Eq. 15.29, is then accomplished in the same manner as for an ordinary chemical-reaction equilibrium.

EXAMPLE 15.7 Calculate the equilibrium composition if argon gas is heated in an arc to 10 000 K, 1 kPa, assuming the plasma to consist of Ar, Ar^+, e^-. The ionization–equilibrium constant for the reaction

$$Ar \rightleftharpoons Ar^+ + e^-$$

at this temperature is 0.000 42.

> *Control volume:* Heating arc.
>
> *Exit state:* P, T known.
>
> *Model:* Ideal gas mixture at equilibrium.

Analysis and Solution

Consider an initial composition of 1 kmol neutral argon, and let z be the number of kilomoles ionized during the heating process. Therefore,

$$Ar \rightleftharpoons Ar^+ + e^-$$

Initial:	1	0	0
Change:	$-z$	$+z$	$+z$
Equilibrium:	$(1-z)$	z	z

and

$$n = (1-z) + z + z = 1 + z$$

Since the number of kilomoles of each component must be positive, the variable z is restricted to the range

$$0 \leq z \leq 1$$

The equilibrium mole fractions are

$$y_{Ar} = \frac{n_{Ar}}{n} = \frac{1-z}{1+z}$$

$$y_{Ar^+} = \frac{n_{Ar^+}}{n} = \frac{z}{1+z}$$

$$y_{e^-} = \frac{n_{e^-}}{n} = \frac{z}{1+z}$$

The equilibrium equation is

$$K = \frac{y_{Ar^+} y_{e^-}}{y_{Ar}} \left(\frac{P}{P^\circ}\right)^{1+1-1} = \frac{\left(\dfrac{z}{1+z}\right)\left(\dfrac{z}{1+z}\right)}{\left(\dfrac{1-z}{1+z}\right)} \left(\frac{P}{P^\circ}\right)$$

so that, at 10 000 K, 1 kPa,

$$0.000\,42 = \left(\frac{z^2}{1-z^2}\right)(0.01)$$

Solving,

$$z = 0.2008$$

and the composition is found to be

$$y_{Ar} = 0.6656$$
$$y_{Ar^+} = 0.1672$$
$$y_{e^-} = 0.1672$$

Simultaneous reactions, such as simultaneous molecular dissociation and ionization reactions or multiple ionization reactions, can be analyzed in the same manner as the ordinary simultaneous chemical reactions of Section 15.5. In doing so, we again make the assumption of thermal equilibrium in the plasma, which, as mentioned before, is, in many cases, a reasonable approximation. Figure 15.10 shows the equilibrium composition of air at high temperature and low density, and indicates the overlapping regions of the various dissociation and ionization processes.

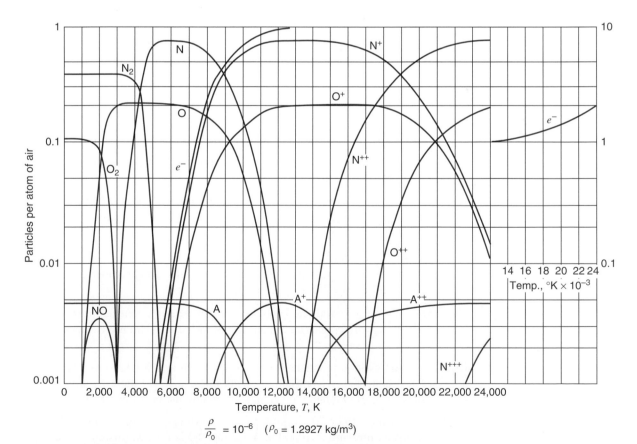

$$\frac{\rho}{\rho_0} = 10^{-6} \quad (\rho_0 = 1.2927 \text{ kg/m}^3)$$

FIGURE 15.10 Equilibrium composition of air [W. E. Moeckel and K. C. Weston, NACA TN 4265 (1958)].

PROBLEMS

15.1 Carbon dioxide at 15 MPa is injected into the top of a 5-km-deep well in connection with an enhanced oil-recovery process. The fluid column standing in the well is at a uniform temperature of 40°C. What is the pressure at the bottom of the well, assuming ideal gas behavior?

15.2 Consider a 2-km-deep gas well containing a gas mixture of methane and ethane at a uniform temperature of 30°C. The pressure at the top of the well is 14 MPa, and the composition on a mole basis is 90% methane, 10% ethane. Determine the pressure and composition at the bottom of the well, assuming an ideal gas mixture.

15.3 Using the same assumptions as those in developing Eq. d in Example 15.1, develop an expression for pressure at the bottom of a deep column of liquid in terms of the isothermal compressibility, β_T. For liquid water at 20°C, we know that $\beta_T = 0.0005$ [1/MPa]. Use the result of the first question to estimate the pressure in the Pacific Ocean at the depth of 3 km.

15.4 Calculate the equilibrium constant for the reaction $O_2 \rightleftharpoons 2O$ at temperatures of 298 K and 6000 K.

15.5 Calculate the equilibrium constant for the reaction $H_2 \rightleftharpoons 2H$ at a temperature of 2000 K, using properties from Table A.8. Compare the result with the value listed in Table A.10.

15.6 Plot to scale the values of $\ln K$ versus $1/T$ for the reaction $2CO_2 \rightleftharpoons 2CO + O_2$. Write an equation for $\ln K$ as a function of temperature.

15.7 Calculate the equilibrium constant for the reaction $2CO_2 \rightleftharpoons 2CO + O_2$ at 3000 K using values from Table A.8 and compare the result to Table A.10.

15.8 Pure oxygen is heated from 25°C to 3200 K in a SSSF process at a constant pressure of 200 kPa. Find the exit composition and the heat transfer.

15.9 Pure oxygen is heated from 25°C, 100 kPa to 3200 K in a constant volume container. Find the final pressure, composition, and the heat transfer.

15.10 Nitrogen gas, N_2, is heated to 4000 K, 10 kPa. What fraction of the N_2 is dissociated to N at this state?

15.11 Hydrogen gas is heated from room temperature to 4000 K, 500 kPa, at which state the diatomic species has partially dissociated to the monatomic form. Determine the equilibrium composition at this state.

15.12 Consider the chemical equilibrium involving H_2O, H_2, CO, and CO_2, and no other substances. Show that the equilibrium constant at any temperature can be found using values from Table A.10 only.

15.13 One kilomole Ar and one kilomole O_2 is heated at a constant pressure of 100 kPa to 3200 K, where it comes to equilibrium. Find the final mole fractions for Ar, O_2, and O.

15.14 A piston/cylinder contains 0.1 kmol hydrogen and 0.1 kmol Ar gas at 25°C, 200 kPa. It is heated in a constant pressure process so the mole fraction of atomic hydrogen is 10%. Find the final temperature and the heat transfer needed.

15.15 Air (assumed to be 79% nitrogen and 21% oxygen) is heated in a SSSF process at a constant pressure of 100 kPa, and some NO is formed (disregard dissociations of N_2 and O_2). At what temperature will the mole fraction of ṄO be 0.001?

15.16 Saturated liquid butane enters an insulated constant-pressure combustion chamber at 25°C, and x times theoretical oxygen gas enters at the same pressure and temperature. The combustion products exit at 3400 K. Assuming that the products are a chemical equilibrium gas mixture that includes CO, what is x?

15.17 The combustion products from burning pentane, C_5H_{12}, with pure oxygen in a stoichiometric ratio exit at 2400 K, 100 kPa. Consider the dissociation of only CO_2 and find the equilibrium mole fraction of CO.

15.18 Find the equilibrium constant for the reaction $2NO + O_2 \rightleftharpoons 2NO_2$ from the elementary reactions in Table A.10 to answer which of the nitrogen oxides, NO or NO_2, is the more stable at ambient conditions? What about at 2000 K?

15.19 Methane exists in equilibrium with carbon and hydrogen where $CH_4 \rightleftharpoons C + 2H_2$, $\ln K = -0.3362$ at 800 K. For a mixture at 100 kPa, 800 K, find the equilibrium mole fractions of all components (CH_4, C, and H_2, neglect hydrogen dissociation). Redo the mole fractions for a mixture state of 200 kPa, 800 K.

15.20 A mixture of 1 kmol carbon dioxide, 2 kmol carbon monoxide, and 2 kmol oxygen, at 25°C, 150 kPa, is heated in a constant pressure SSSF process to 3000 K. Assuming that only these same substances are present in the exiting chemical equilibrium mixture, determine the composition of that mixture.

15.21 Repeat the previous problem for an initial mixture that also includes 2 kmol of nitrogen, which does not dissociate during the process.

15.22 Complete combustion of hydrogen and pure oxygen in a stoichiometric ratio at P_0, T_0 to form water would result in a computed adiabatic flame temperature of 4990 K for a SSSF setup.

  a. How should the adiabatic flame temperature be found if the equilibrium reaction $2H_2 + O_2 \rightleftharpoons H_2O$ is considered? Disregard all other possible reactions (dissociations) and show the final equation(s) to be solved.

  b. Find the equilibrium composition at 3800 K, again disregarding all other reactions.

  c. Which other reactions should be considered and which components will be present in the final mixture?

15.23 Gasification of char (primarily carbon) with steam following coal pyrolysis yields a gas mixture of 1 kmol CO and 1 kmol H_2. We wish to upgrade the hydrogen content of this syngas fuel mixture, so it is fed to an appropriate catalytic reactor along with 1 kmol of H_2O. Exiting the reactor is a chemical equilibrium gas mixture of CO, H_2, H_2O, and CO_2 at 600 K, 500 kPa. Determine the equilibrium composition. Note: see Problem 15.12.

15.24 A gas mixture of 1 kmol carbon monoxide, 1 kmol nitrogen, and 1 kmol oxygen at 25°C, 150 kPa, is heated in a constant pressure SSSF process. The exit mixture can be assumed to be in chemical equilibrium with CO_2, CO, O_2, and N_2 present. The mole fraction of CO_2 at this point is 0.176. Calculate the heat transfer for the process.

15.25 A rigid container initially contains 2 kmol of carbon monoxide and 2 kmol of oxygen at 25°C, 100 kPa. The content is then heated to 3000 K at which point an equilibrium mixture of CO_2, CO, and O_2 exists. Disregard other possible species and determine the final pressure, the equilibrium composition, and the heat transfer for the process.

15.26 One approach to using hydrocarbon fuels in a fuel cell is to "reform" the hydrocarbon to obtain hydrogen, which is then fed to the fuel cell. As a part of the analysis of such a procedure, consider the reaction $CH_4 + H_2O \rightleftharpoons 3H_2 + CO$.

a. Determine the equilibrium constant for this reaction at a temperature of 800 K.

b. One kilomole each of methane and water are fed to a catalytic reformer. A mixture of CH_4, H_2O, H_2, and CO exits in chemical equilibrium at 800 K, 100 kPa; determine the equilibrium composition of this mixture.

15.27 In a test of a gas-turbine combustor, saturated-liquid methane at 115 K is burned with excess air to hold the adiabatic flame temperature to 1600 K. It is assumed that the products consist of a mixture of CO_2, H_2O, N_2, O_2, and NO in chemical equilibrium. Determine the percent excess air used in the combustion, and the percentage of NO in the products.

15.28 The van't Hoff equation

$$d \ln K = \frac{\Delta H^\circ}{\overline{R} T^2} dT \Big|_{p^\circ}$$

relates the chemical equilibrium constant K to the enthalpy of reaction ΔH°. From the value of K in Table A.10 for the dissociation of hydrogen at 2000 K and the value of ΔH° calculated from Table A.8 at 2000 K, use the van't Hoff equation to predict the equilibrium constant at 2400 K.

15.29 Catalytic gas generators are frequently used to decompose a liquid, providing a desired gas mixture (spacecraft control systems, fuel cell gas supply, and so forth). Consider feeding pure liquid hydrazine, N_2H_4, to a gas generator, from which exits a gas mixture of N_2, H_2, and NH_3 in chemical equilibrium at 100°C, 350 kPa. Calculate the mole fractions of the species in the equilibrium mixture.

15.30 Acetylene gas at 25°C is burned with 140% theoretical air, which enters the burner at 25°C, 100 kPa, 80% relative humidity. The combustion products form a mixture of CO_2, H_2O, N_2, O_2, and NO in chemical equilibrium at 2200 K, 100 kPa. This mixture is then cooled to 1000 K very rapidly, so that the composition does not change. Determine the mole fraction of NO in the products and the heat transfer for the overall process.

15.31 The equilibrium reaction with methane as $CH_4 \rightleftharpoons C + 2H_2$ has $\ln K = -0.3362$ at 800 K and $\ln K = -4.607$ at 600 K. By noting the relation of K to temperature show how you would interpolate $\ln K$ in $(1/T)$ to find K at 700 K and compare that to a linear interpolation.

15.32 Use the information in Problem 15.31 to estimate the enthalpy of reaction, ΔH°, at 700 K using the van't Hoff equation (see Problem 15.28) with finite differences for the derivatives.

15.33 A step in the production of a synthetic liquid fuel from organic waste matter is the following conversion process: 1 kmol of ethylene gas (converted from the waste) at 25°C, 5 MPa, and 2 kmol of steam at 300°C, 5 MPa, enter a catalytic reactor. An ideal gas mixture of ethanol, ethylene, and water in chemical equilibrium leaves the reactor at 700 K, 5 MPa. Determine the composition of the mixture and the heat transfer for the reactor.

15.34 Methane at 25°C, 100 kPa is burned with 200% theoretical oxygen at 400 K, 100 kPa in an adiabatic SSSF process, and the products of combustion exit at 100 kPa.

Assume that the only significant dissociation reaction in the products is that of carbon dioxide going to carbon monoxide and oxygen. Determine the equilibrium composition of the products and also their temperature at the combustor exit.

15.35 Calculate the irreversibility for the adiabatic combustion process described in the previous problem.

15.36 In rich (too much fuel) combustion the excess fuel may be broken down to give H_2, and CO may form. In the products at 1200 K, 200 kPa the reaction called the water gas reaction may take place: $CO_2 + H_2 \rightleftharpoons H_2O + CO$. Find the equilibrium constant for this reaction from the elementary reactions.

15.37 An important step in the manufacture of chemical fertilizer is the production of ammonia, according to the reaction: $N_2 + 3H_2 \rightleftharpoons 2NH_3$

a. Calculate the equilibrium constant for this reaction at 150°C.

b. For an initial composition of 25% nitrogen, 75% hydrogen, on a mole basis, calculate the equilibrium composition at 150°C, 5 MPa.

15.38 A space heating unit in Alaska uses propane combustion as the heat supply. Liquid propane comes from an outside tank at −44°C, and the air supply is also taken in from the outside at −44°D. The airflow reguator is misadjusted, such that only 90% of the theoretical air enters the combustion chamber, resulting in incomplete combustion. The products exit at 1000 K as a chemical equilibrium gas mixture including only CO_2, CO, H_2O, H_2, and N_2. Find the composition of the products. Hint: use the water gas reaction in Problem 15.36.

15.39 One kilomole of carbon dioxide, CO_2, and 1 kmol of hydrogen, H_2, at room temperature and 200 kPa is heated to 1200 K, 200 kPa. Use the water gas reaction (see Problem 15.36) to determine the mole fraction of CO. Neglect dissociations of H_2 and O_2.

15.40 Consider the production of a synthetic fuel (methanol) from coal. A gas mixture of 50% CO and 50% H_2 leaves a coal gasifier at 500 K, 1 MPa, and enters a catalytic converter. A gas mixture of methanol, CO and H_2 in chemical equilibrium with the reaction $CO + 2H_2 \rightleftharpoons CH_3OH$ leaves the converter at the same temperature and pressure, where it is known that $\ln K = -5.119$.

a. Calculate the equilibrium composition of the mixture leaving the converter.

b. Would it be more desirable to operate the converter at ambient pressure?

15.41 Consider the following coal gasifier proposed for supplying a syngas fuel to a gas turbine power plant. Fifty kilograms per second of dry coal (represented as 48 kg C plus 2 kg H) enter the gasifier, along with 4.76 kmol/s of air and 2 kmol/s of steam. The output stream from this unit is a gas mixture containing H_2, CO, N_2, CH_4, and CO_2 in chemical equilibrium at 900 K, 1 MPa.

a. Set up the reaction and equilibrium equation(s) for this system, and calculate the appropriate equilibrium constant(s).

b. Determine the composition of the gas mixture leaving the gasifier.

15.42 Ethane is burned with 150% theoretical air in a gas turbine combustor. The products exiting consist of a mixture of CO_2, H_2O, O_2, N_2, and NO in chemical equilibrium at 1800 K, 1 MPa. Determine the mole fraction of NO in the products. Is it reasonable to ignore CO in the products?

15.43 One kilomole of liquid oxygen, O_2, at 93 K, and x kmol of gaseous hydrogen, H_2, at 25°C, are fed to an SSSF combustion chamber. x is greater than 2, such that there is excess hydrogen for the combustion process. There is a heat loss from the chamber of 1000 kJ per kmol of reactants. Products exit the chamber at chemical equilibrium at 3800 K, 400 kPa, and are assumed to include only H_2O, H_2, and O.

　a. Determine the equilibrium composition of the products and also x, the amount of H_2 entering the combustion chamber.

　b. Should another substance(s) have been included in part (a) as being present in the products? Justify your answer.

15.44 Butane is burned with 200% theoretical air, and the products of combustion, an equilibrium mixture containing only CO_2, H_2O, O_2, N_2, NO, and NO_2, exits from the combustion chamber at 1400 K, 2 MPa. Determine the equilibrium composition at this state.

15.45 A mixture of 1 kmol water and 1 kmol oxygen at 400 K is heated to 3000 K, 200 kPa, in a SSSF process. Determine the equilibrium composition at the outlet of the heat exchanger, assuming that the mixture consists of H_2O, H_2, O_2, and OH.

15.46 One kilomole of air (assumed to be 78% nitrogen, 21% oxygen, and 1% argon) at room temperature is heated to 4000 K, 200 kPa. Find the equilibrium composition at this state, assuming that only N_2, O_2, NO, O, and Ar are present.

15.47 Acetylene gas and x times theoretical air ($x > 1$) at room temperature and 500 kPa are burned at constant pressure in an adiabatic SSSF process. The flame temperature is 2600 K, and the combustion products are assumed to consist of N_2, O_2, CO_2, H_2O, CO, and NO. Determine the value of x.

15.48 One kilomole of water vapor at 100 kPa, 400 K is heated to 3000 K in a constant pressure SSSF process. Determine the final composition, assuming that H_2O, H_2, H, O_2, and OH are present at equilibrium.

15.49 Operation of an MHD converter requires an electrically conducting gas. A helium gas "seeded" with 1.0 mole percent cesium, as shown in Fig. P15.49, is used where the cesium is partly ionized ($Cs \rightleftharpoons Cs^+ + e^-$) by heating the mixture to 1800 K, 1 MPa in a nuclear reactor to provide free electrons. No helium is ionized in this process, so that the mixture entering the converter consists of He, Cs, Cs^+, and e^-. Determine the mole fraction of electrons in the mixture at 1800 K, where $\ln K = 1.402$ for the cesium ionization reaction described.

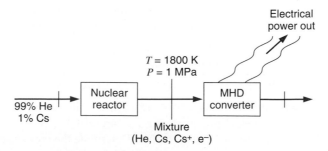

FIGURE P15.49

15.50 One kilomole of argon gas at room temperature is heated to 20 000 K, 100 kPa. Assume that the plasma in this condition consists of an equilibrium mixture of Ar, Ar^+, Ar^{++}, and e^- according to the simultaneous reactions

(1) $Ar \rightleftharpoons Ar^+ + e^-$ (2) $Ar^+ \rightleftharpoons Ar^{++} + e^-$

The ionization equilibrium constants for these reactions at 20 000 K have been calculated from spectroscopic data as $\ln K_1 = 3.11$ and $\ln K_2 = -4.92$. Determine the equilibrium composition of the plasma.

15.51 Plot to scale the equilibrium composition of nitrogen at 10 kPa over the temperature range 5000 K to 15 000 K, assuming that N_2, N, N^+, and e^- are present. For the ionization reaction $N \rightleftharpoons N^+ + e^-$, the ionization equilibrium constant K has been calculated from spectroscopic data as

T [K]	10000	12 000	14 000	16 000
$100K$	$6.26\ 10^{-2}$	1.51	15.1	92

15.52 Hydrides are rare earth metals, M, that have the ability to react with hydrogen to form a different substance MH_x with a release of energy. The hydrogen can then be released, the reaction reversed, by heat addition to the MH_x. In this reaction only the hydrogen is a gas so the formula developed for the chemical equilibrium is inappropriate. Show that the proper expression to be used instead of Eq. 15.34 is

$$\ln (P_{H2} / P_o) = \Delta G^o / RT$$

when the reaction is scaled to 1 kmol of H_2.

ADVANCED
PROBLEMS

15.53 Repeat Problem 15.1 using the generalized charts, instead of ideal gas behavior.

15.54 Derive the van't Hoff equation given in Problem 15.28, using Eqs. 15.12 and 15.15. Note: the $d(\bar{g}/T)$ at constant P^o for each component can be expressed using the relations in Eqs. 13.18 and 13.19.

15.55 A coal gasifier produces a mixture of 1 CO and $2H_2$ that is then fed to a catalytic converter to produce methane. A chemical-equilibrium gas mixture containing CH_4, CO, H_2, and H_2O exits the reactor at 600 K, 600 kPa. Determine the mole fraction of methane in the mixture.

15.56 Dry air is heated from 25°C to 4000 K in a 100-kPa constant-pressure process. List the possible reactions that may take place and determine the equilibrium composition. Find the required heat transfer.

15.57 Methane is burned with theoretical oxygen in a SSSF process, and the products exit the combustion chamber at 3200 K, 700 kPa. Calculate the equilibrium composition at this state, assuming that only CO_2, CO, H_2O, H_2, O_2, and OH are present.

ENGLISH UNIT
PROBLEMS

15.58E Carbon dioxide at 2200 lbf/in.^{2} is injected into the top of a 3-mi-deep well in connection with an enhanced oil recovery process. The fluid column standing in the well is at a uniform temperature of 100 F. What is the pressure at the bottom of the well assuming ideal gas behavior?

15.59E Calculate the equilibrium constant for the reaction $O_2 \rightleftharpoons 2O$ at temperatures of 537R and 10 800 R.

15.60E Pure oxygen is heated from 77 F to 5300 F in a SSSF process at a constant pressure of 30 lbf/in.^{2}. Find the exit composition and the heat transfer.

15.61E Pure oxygen is heated from 77 F, 14.7 lbf/in.^{2} to 5300 F in a constant volume container. Find the final pressure, composition, and the heat transfer.

15.62E Air (assumed to be 79% nitrogen and 21% oxygen) is heated in a SSSF process at a constant pressure of 14.7 lbf/in.^{2}, and some NO is formed. At what temperature will the mole fraction of NO be 0.001?

15.63E The combustion products from burning pentane, C_5H_{12}, with pure oxygen in a stoichiometric ratio exit at 4400 R. Consider the dissociation of only CO_2 and find the equilibrium mole fraction of CO.

15.64E A gas mixture of 1 pound mol carbon monoxide, 1 pound mol nitrogen, and 1 pound mol oxygen at 77 F, 20 lbf/in.^{2}, is heated in a constant pressure SSSF process. The exit mixture can be assumed to be in chemical equilibrium with CO_2, CO, O_2, and N_2 present. The mole fraction of CO_2 at this point is 0.176. Calculate the heat transfer for the process.

15.65E In a test of a gas-turbine combustor, saturated-liquid methane at 210 R is to be burned with excess air to hold the adiabatic flame temperature to 2880 R. It is assumed that the products consist of a mixture of CO_2, H_2O, N_2, O_2, and NO in chemical equilibrium. Determine the percent excess air used in the combustion, and the percentage of NO in the products.

15.66E Acetylene gas at 77 F is burned with 140% theoretical air, which enters the burner at 77 F, 14.7 lbf/in.^{2}, 80% relative humidity. The combustion products form a mixture of CO_2, H_2O, N_2, O_2, and NO in chemical equilibrium at 3500 F, 14.7 lbf/in.^{2}. This mixture is then cooled to 1340 F very rapidly, so that the composition does not change. Determine the mole fraction of NO in the products and the heat transfer for the overall process.

15.67E The equilibrium reaction with methane as $CH_4 \rightleftharpoons C + 2H_2$ has $\ln K = -0.3362$ at 1440 R and $\ln K = -4.607$ at 1080 R. By noting the relation of K to temperature, show how you would interpolate $\ln K$ in $(1/T)$ to find K at 1260 R and compare that to a linear interpolation.

15.68E Use the information in Problem 15.67 to estimate the enthalpy of reaction, $\Delta H°$, at 1260 R using the van't Hoff equation (see Problem 15.28) with finite differences for the derivatives.

15.69E An important step in the manufacture of chemical fertilizer is the production of ammonia, according to the reaction $N_2 + 3H_2 \rightleftharpoons 2NH_3$

a. Calculate the equilibrium constant for this reaction at 300 F.

b. For an initial composition of 25% nitrogen, 75% hydrogen, on a mole basis, calculate the equilibrium composition at 300 F, 750 lbf/in.^{2}.

15.70E Ethane is burned with 150% theoretical air in a gas turbine combustor. The products exiting consist of a mixture of CO_2, H_2O, O_2, N_2, and NO in chemical equilibrium at 2800 F, 150 lbf/in.^{2}. Determine the mole fraction of NO in the products. Is it reasonable to ignore CO in the products?

15.71E One pound mole of air (assumed to be 78% nitrogen, 21% oxygen, and 1% argon) at room temperature is heated to 7200 R, 30 lbf/in.^{2}. Find the equilibrium composition at this state, assuming that only N_2, O_2, NO, O, and Ar are present.

15.72E Dry air is heated from 77 F to 7200 R in a 14.7 lbf/in.^{2} constant-pressure process. List the possible reactions that may take place and determine the equilibrium composition. Find the required heat transfer.

15.73E Acetylene gas and x times theoretical air ($x > 1$) at room temperature and 75 lbf/in.^{2} are burned at constant pressure in an adiabatic SSSF process. The flame temperature is 4600 R, and the combustion products are assumed to consist of N_2, O_2, CO_2, H_2O, CO, and NO. Determine the value of x.

15.74E One pound mole of water vapor at 14.7 lbf/in.^{2}, 720 R is heated to 5400 R in a constant pressure SSSF process. Determine the final composition, assuming that H_2O, H_2, H, O_2, and OH are present at equilibrium.

15.75E Methane is burned with theoretical oxygen in a SSSF process, and the products exit the combustion chamber at 5300 F, 100 lbf/in.^{2}. Calculate the equilibrium composition at this state, assuming that only CO_2, CO, H_2O, H_2, O_2, and OH are present.

COMPUTER, DESIGN, AND OPEN-ENDED PROBLEMS

15.76 Use the menu-driven equilibrium program to solve for the adiabatic flame temperature including chemical equilibrium in Problem 15.22.

15.77 Write a program to solve the general case of Problem 15.33, in which the relative amount of steam input and the reactor temperature and pressure are program input variables and use constant specific heats.

15.78 Write a program to solve the following problem. One kmol of carbon at 25°C is burned with b kmol of oxygen in a constant pressure adiabatic process. The products consist of an equilibrium mixture of CO_2, CO, and O_2. We wish to determine the flame temperature for various combinations of b and the pressure P, assuming constant specific heat for the components from Table A.5.

15.79 Use the supplied menu-driven equilibrium program to do Problem 15.45 for a range of exit temperatures.

15.80 Use the supplied menu-driven equilibrium program to calculate the adiabatic flame temperature in Problem 15.57, assuming all the reactions included in the program are activated.

15.81 Study the chemical reactions that take place when CFC-type refrigerants are released into the the atmosphere. The chlorine may create compounds as HCl and $ClONO_2$ that react with the ozone O_3.

15.82 Examine the chemical equilibrium that takes place in an engine where CO and various nitrogen oxygen compounds summarized as NO_x may be formed. Study the processes for a range of air–fuel ratios and temperatures for typical fuels. Are there important reactions not listed in the book?

15.83 A number of products may be produced from the conversion of organic waste that can be used as fuel (see Problem 15.33). Study the subject and make a list of the major products that are formed and the conditions at which they are formed in desirable concentrations.

15.84 Using the supplied menu-driven software for chemical equilibrium calculations, study the formation of CO as a function of temperature as done in Problem 15.26 for a single temperature.

15.85 The hydrides as explained in Problem 15.52 can store large amounts of hydrogen. The penalty for the storage is that energy must be supplied when the hydrogen is released. Investigate the literature for quantitative information about the quantities and energy involved in such a hydrogen storage.

15.86 The hydrides explained in Problem 15.52 can be used in a chemical heat pump. The energy involved in the chemical reaction can be added and removed at different temperatures. For some hydrides these temperatures are low enough to make them feasible for heat pumps for heat upgrade, refrigerators, and air conditioners. Investigate the literature for such applications and give some typical values for these systems.

15.87 Power plants and engines have high peak temperatures in the combustion products where NO is produced. The equilibrium NO level at the high temperature is frozen at that level during the rapid drop in temperature with the expansion. The final exhaust therefore contains NO at a level much higher than the equilibrium value at the exhaust temperature. Study the NO level at equilibrium when natural gas, CH_4, is burned adiabatically with air (at T_0) in various ratios.

15.88 Excess air or steam addition is often used to lower the peak temperature in combustion to limit formation of pollutants like NO. Study the steam addition to the combustion of natural gas as in the Cheng cycle (see Problem 12.109), assuming the steam is added before the combustion. How does this affect the peak temperature and the NO concentration?

COMPRESSIBLE
FLOW 16

This chapter deals with the thermodynamic aspects of one-dimensional flow through nozzles and passages. In addition, the momentum equation for the control volume is developed and applied to these problems. The sonic velocity is defined in terms of thermodynamic properties, and the importance of the Mach number as a variable in compressible flow is noted.

16.1 STAGNATION PROPERTIES

In dealing with problems involving flow, many discussions and equations can be simplified by introducing the concept of the isentropic stagnation state and the properties associated with it. The isentropic stagnation state is the state a flowing fluid would attain if it underwent a reversible adiabatic deceleration to zero velocity. This state is designated in this chapter with the subscript 0. From the first law for a steady-state steady-flow process we conclude that

$$h + \frac{\mathbf{V}^2}{2} = h_0 \qquad (16.1)$$

The actual and the isentropic stagnation states for a typical gas or vapor are shown on the h–s diagram of Fig. 16.1. Sometimes it is advantageous to make a distinction between the actual and the isentropic stagnation states. The actual stagnation state is the state achieved after an actual deceleration to zero velocity (as at the nose of a body placed in a fluid stream), and there may be irreversibilities associated with the deceleration process. Therefore, the term stagnation property is sometimes reserved for the properties associated with the actual state, and the term total property is used for the isentropic stagnation state.

It is evident from Fig. 16.1 that the enthalpy is the same for both the actual and isentropic stagnation states (assuming that the actual process is adiabatic). Therefore, for an ideal gas, the actual stagnation temperature is the same as the isentropic stagnation temperature. However, the actual stagnation pressure may be less than the isentropic stagnation pressure and for this reason the term total pressure (meaning isentropic stagnation pressure) has particular meaning compared to the actual stagnation pressure.

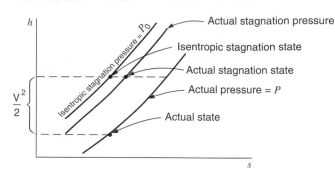

FIGURE 16.1
Enthalpy–entropy diagram illustrating the definition of stagnation state.

EXAMPLE 16.1   Air flows in a duct at a pressure of 150 kPa with a velocity of 200 m/s. The temperature of the air is 300 K. Determine the isentropic stagnation pressure and temperature.

Analysis and Solution

If we assume that the air is an ideal gas with contant specific heat as given in Table A.5, the calculation is as follows. From Eq. 16.1

$$\frac{\mathbf{V}^2}{2} = h_0 - h = C_{p0}(T_0 - T)$$

$$\frac{(200)^2}{2 \times 1000} = 1.0035(T_0 - 300)$$

$$T_0 = 319.9 \text{ K}$$

The stagnation pressure can be found from the relation

$$\frac{T_0}{T} = \left(\frac{P_0}{P}\right)^{(k-1)/k}$$

$$\frac{319.9}{300} = \left(\frac{P_0}{150}\right)^{0.286}$$

$$P_0 = 187.8 \text{ kPa}$$

The Air Tables, Table A.7, which are calculated from Table A.8, could also have been used, and then the variation of specific heat with temperature would have been taken into account. Since the actual and stagnation states have the same entropy, we proceed as follows: Using Table A.7,

$$T = 300 \text{ K} \qquad h = 300.47 \qquad P_r = 1.1146$$

$$h_0 = h + \frac{\mathbf{V}^2}{2} = 300.47 + \frac{(200)^2}{2 \times 1000} = 320.47$$

$$T_0 = 320 \text{ K} \qquad P_{r0} = 1.3956$$

$$P_0 = 150 \times \frac{1.3956}{1.1146} = 187.8 \text{ kPa}$$

16.2 THE MOMENTUM EQUATION FOR A CONTROL VOLUME

Before proceeding it will be advantageous to develop the momentum equation for the control volume. Newton's second law states that the sum of the external forces acting on a body in a given direction is proportional to the rate of change of momentum in the given direction. Writing this in equation form for the x-direction we have

$$\frac{d(m\mathbf{V}_x)}{dt} \propto \sum F_x$$

For the system of units used in this book, this proportionality can be written directly as an equality.

$$\frac{d(m\mathbf{V}_x)}{dt} = \sum F_x \tag{16.2}$$

Equation 16.2 has been written for a body of fixed mass, or in thermodynamic parlance, for a control mass. We now proceed to write the momentum equation for a control volume, and follow a procedure similar to that used in writing the continuity equation and the first and second laws of thermodynamics for a control volume.

Consider the control volume shown in Fig. 16.2 to be fixed relative to its coordinate frame. Each flow that enters or leaves the control volume possesses an amount of momentum per unit mass, so that it adds or subtracts a rate of momentum to or from the control volume.

Writing the momentum equation in a rate form similar to the balance equations for mass, energy, and entropy, Eqs. 6.1, 6.7, and 9.2, respectively, results in an expression of the form

$$\text{Rate of change} = \sum F_x + \text{in} - \text{out} \tag{16.3}$$

Only forces acting on the mass inside the control volume (for example, gravity) or on the control volume surface (for example, friction or piston forces) and the flow of mass carrying momentum can contribute to a change of momentum. Momentum is conserved, so that it cannot be created or destroyed, as was previously stated for the other control volume developments.

The momentum equation in the x-direction from the form of Eq. 16.3 becomes

$$\frac{d(m\mathbf{V}_x)}{dt} = \sum F_x + \sum \dot{m}_i \mathbf{V}_{ix} - \sum \dot{m}_e \mathbf{V}_{ex} \tag{16.4}$$

Similarly, for the y- and z-directions,

$$\frac{d(m\mathbf{V}_y)}{dt} = \sum F_y + \sum \dot{m}_i \mathbf{V}_{iy} - \sum \dot{m}_e \mathbf{V}_{ey} \tag{16.5}$$

and

$$\frac{d(m\mathbf{V}_z)}{dt} = \sum F_z + \sum \dot{m}_i \mathbf{V}_{iz} - \sum \dot{m}_e \mathbf{V}_{ez} \tag{16.6}$$

In the case of a control volume with no mass flow rates in or out (i.e., a control mass), these equations reduce to the form of Eq. 16.2 for each direction.

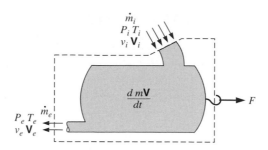

FIGURE 16.2 Development of the momentum equation for a control volume.

In this chapter we will be concerned primarily with steady-state, steady-flow processes in which there is a single flow with uniform properties into the control volume, and a single flow with uniform properties out of the control volume. The SSSF assumption means that the rate of momentum change for the control volume terms in Eqs. 16.4, 16.5, and 16.6 are equal to zero. That is,

$$\frac{d(m\mathbf{V}_x)_{\text{c.v.}}}{dt} = 0 \qquad \frac{d(m\mathbf{V}_y)_{\text{c.v.}}}{dt} = 0 \qquad \frac{d(m\mathbf{V}_z)_{\text{c.v.}}}{dt} = 0 \qquad (16.7)$$

Therefore for the SSSF process the momentum equation for the control volume, assuming uniform properties at each state, reduces to the form

$$\sum F_x = \sum \dot{m}_e(\mathbf{V}_e)_x - \sum \dot{m}_i(\mathbf{V}_i)_x \qquad (16.8)$$

$$\sum F_y = \sum \dot{m}_e(\mathbf{V}_e)_y - \sum \dot{m}_i(\mathbf{V}_i)_y \qquad (16.9)$$

$$\sum F_z = \sum \dot{m}_e(\mathbf{V}_e)_z - \sum \dot{m}_i(\mathbf{V}_i)_z \qquad (16.10)$$

Further, for the special case in which there is a single flow into and out of the control volume, these equations reduce to

$$\sum F_x = \dot{m}[(\mathbf{V}_e)_x - (\mathbf{V}_i)_x] \qquad (16.11)$$

$$\sum F_y = \dot{m}[(\mathbf{V}_e)_y - (\mathbf{V}_i)_y] \qquad (16.12)$$

$$\sum F_z = \dot{m}[(\mathbf{V}_e)_z - (\mathbf{V}_i)_z] \qquad (16.13)$$

EXAMPLE 16.2   On a level floor a man is pushing a wheelbarrow (Fig. 16.3) into which sand is falling at the rate of 1 kg/s. The man is walking at the rate of 1 m/s and the sand has a velocity of 10 m/s as it falls into the wheelbarrow. Determine the force the man must exert on the wheelbarrow and the force the floor exerts on the wheelbarrow due to the falling sand.

Analysis and Solution

Consider a control surface around the wheelbarrow. Consider first the x-direction. From Eq. 16.4

$$\sum F_x = \frac{d(m\mathbf{V}_x)_{\text{c.v.}}}{dt} + \sum \dot{m}_e(\mathbf{V}_e)_x - \sum \dot{m}_i(\mathbf{V}_i)_x$$

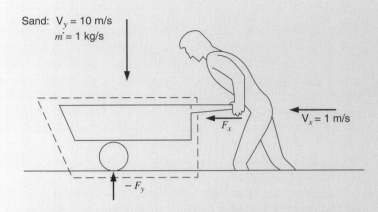

Sand: $V_y = 10$ m/s
$\dot{m} = 1$ kg/s

F_x

$V_x = 1$ m/s

$-F_y$

FIGURE 16.3 Sketch for Example 16.2.

Let us analyze this problem from the point of view of an observer riding on the wheelbarrow. For this observer, $\mathbf{V}_x$ of the material in the wheelbarrow is zero and therefore,

$$\frac{d(m\mathbf{V}_x)_{\text{c.v.}}}{dt} = 0$$

However, for this observer the sand crossing the control surface has an x-component velocity of -1 m/s, and $\dot{m}$, the mass flow out of the control volume, is -1 kg/s. Therefore,

$$F_x = (1\,\text{kg}/\text{s}) \times (1\,\text{m}/\text{s}) = 1\,\text{N}$$

If one considers this from the point of view of an observer who is stationary on the earth's surface we conclude that $\mathbf{V}_x$ of the falling sand is zero and therefore

$$\sum \dot{m}_e (\mathbf{V}_e)_x - \sum \dot{m}_i (\mathbf{V}_i)_x = 0$$

However, for this observer there is a change of momentum within the control volume, namely,

$$\sum F_x = \frac{d(m\mathbf{V}_x)_{\text{c.v.}}}{dt} = (1\,\text{m}/\text{s}) \times (1\,\text{kg}/\text{s}) = 1\,\text{N}$$

Next consider the vertical (y) direction.

$$\sum F_y = \frac{d(m\mathbf{V}_y)_{\text{c.v.}}}{dt} + \sum \dot{m}_e (\mathbf{V}_e)_y - \sum \dot{m}_i (\mathbf{V}_i)_y$$

For both the stationary and moving observer the first term drops out because $\mathbf{V}_y$ of the mass within the control volume is zero. However, for the mass crossing the control surface, $\mathbf{V}_y = 10$ m/s and

$$\dot{m} = -1\,\text{kg}/\text{s}$$

Therefore

$$F_y = (10\,\text{m}/\text{s}) \times (-1\,\text{kg}/\text{s}) = -10\,\text{N}$$

The minus sign indicates that the force is in the opposite direction to $\mathbf{V}_y$

16.3 FORCES ACTING ON A CONTROL SURFACE

In the last section we considered the momentum equation for the control volume. We now wish to evaluate the net force on a control surface that causes this change in momentum. Let us do this by considering the control mass shown in Fig. 16.4, which involves a pipe bend. The control surface is designated by the dotted lines, and is so chosen that at the point where the fluid crosses the boundary of the control mass are assumed to be negligible. Figure 16.4a shows the velocities and Fig. 16.4b shows the forces involved. The force R is the result of all external forces on the control mass, except for the pressure of all surroundings. The pressure of the surroundings, P_0, acts on the entire boundary except at $\mathscr{A}_i$ and $\mathscr{A}_e$, where the fluid crosses the control surface, P_i and P_e represent the absolute pressures at these points.

The net forces acting on the system in the x- and y-directions, F_x and F_y, are the sum of the pressure forces and the external force R in their respective directions. The influence of the pressure of the surroundings, P_0, is most easily taken into account by noting that it acts over the entire control mass boundary except at $\mathscr{A}_i$ and $\mathscr{A}_e$. Therefore, we can write

$$\sum F_x = (P_i \mathscr{A}_i)_x - (P_0 \mathscr{A}_i)_x + (P_e \mathscr{A}_e)_x - (P_0 \mathscr{A}_e)_x + R_x$$

$$\sum F_y = (P_i \mathscr{A}_i)_y - (P_0 \mathscr{A}_i)_y + (P_e \mathscr{A}_e)_y - (P_0 \mathscr{A}_e)_y + R_y$$

This equation may be simplified by combining the pressure terms.

$$\sum F_x = [(P_i - P_0)\mathscr{A}_i]_x + [(P_e - P_0)\mathscr{A}_e]_x + R_x$$

$$\sum F_y = [(P_i - P_0)\mathscr{A}_i]_y + [(P_e - P_0)\mathscr{A}_e]_y + R_y \qquad (16.14)$$

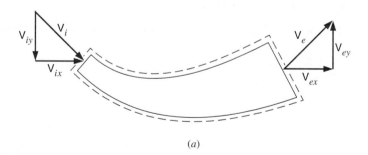

(a)

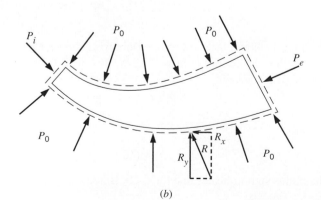

(b)

FIGURE 16.4 Forces acting on a control surface.

The proper sign for each pressure and force must of course be used in all calculations. Equations 16.8, 16.9, and 16.14 may be combined to give

$$\sum F_x = \sum \dot{m}_e (\mathbf{V}_e)_x - \sum \dot{m}_i (\mathbf{V}_i)_x$$
$$= \sum [(P_i - P_0)\mathcal{A}_i]_x + \sum [(P_e - P_0)\mathcal{A}_e]_x + R_x$$
$$\sum F_y = \sum \dot{m}_e (\mathbf{V}_e)_y - \sum \dot{m}_i (\mathbf{V}_i)_y$$
$$= \sum [(P_i - P_0)\mathcal{A}_i]_y + \sum [(P_e - P_0)\mathcal{A}_e]_y + R_y \qquad (16.15)$$

If there is a single flow across the control surface, Eqs. 16.11, 16.12, and 16.14 can be combined to give

$$\sum F_x = \dot{m}(\mathbf{V}_e - \mathbf{V}_i)_x = [(P_i - P_0)\mathcal{A}_i]_x + [(P_e - P_0)\mathcal{A}_e]_x + R_x$$
$$\sum F_y = \dot{m}(\mathbf{V}_e - \mathbf{V}_i)_y = [(P_i - P_0)\mathcal{A}_i]_y + [(P_e - P_0)\mathcal{A}_e]_y + R_y \qquad (16.16)$$

A similar equation could be written for the z-direction. These equations are very useful in analyzing the forces involved in a control volume analysis.

EXAMPLE 16.3

A jet engine is being tested on a test stand (Fig. 16.5). The inlet area to the compressor is 0.2 m^2 and air enters the compressor at 95 kPa, 100 m/s. The pressure of the atmosphere is 100 kPa. The exit area of the engine is 0.1 m^2, and the products of combustion leave the exit plane at a pressure of 125 kPa and a velocity of 450 m/s. The air–fuel ratio is 50 kg air/kg fuel, and the fuel enters with a low velocity. The rate of air flow entering the engine is 20 kg/s. Determine the thrust on the engine.

Analysis and Solution

In the solution that follows it is assumed that forces and velocities to the right are positive.

Using Eq. 16.16

$$R_x + [(P_i - P_0)\mathcal{A}_i]_x + [(P_e - P_0)\mathcal{A}_e]_x = (\dot{m}_e \mathbf{V}_e - \dot{m}_i \mathbf{V}_i)_x$$
$$R_x + [(95 - 100) \times 0.2] - [(125 - 100) \times 0.1] = \frac{20.4 \times 450 - 20 \times 100}{1000}$$
$$R_x = 10.68 \text{ kN}$$

(Note that the momentum of the fuel entering has been neglected.)

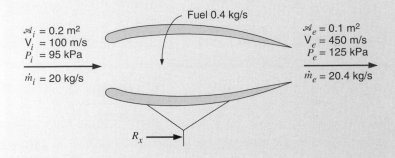

FIGURE 16.5 Sketch for Example 16.3.

16.4 ADIABATIC, ONE-DIMENSIONAL, STEADY-STATE STEADY FLOW OF AN INCOMPRESSIBLE FLUID THROUGH A NOZZLE

A nozzle is a device in which the kinetic energy of a fluid is increased in an adiabatic process. This increase involves a decrease in pressure and is accomplished by the proper change in flow area. A diffuser is a device that has the opposite function, namely, to increase the pressure by decelerating the fluid. In this section we discuss both nozzles and diffusers, but to minimize words we shall use only the term nozzle.

Consider the nozzle shown in Fig. 16.6, and assume an adiabatic, one-dimensional, steady-state steady-flow process of an incompressible fluid. From the continuity equation we conclude that

$$\dot{m}_e = \dot{m}_i = \rho \mathscr{A}_i \mathbf{V}_i = \rho \mathscr{A}_e \mathbf{V}_e$$

or

$$\frac{\mathscr{A}_i}{\mathscr{A}_e} = \frac{\mathbf{V}_e}{\mathbf{V}_i} \tag{16.17}$$

The first law for this process is

$$h_e - h_i + \frac{\mathbf{V}_e^2 - \mathbf{V}_i^2}{2} + (Z_e - Z_i)g = 0 \tag{16.18}$$

From the second law we conclude that $s_e \geq s_i$, where the equality holds for a reversible process. Therefore, from the relation

$$T\,ds = dh - v\,dP$$

we conclude that for the reversible process

$$h_e - h_i = \int_i^e v\,dP \tag{16.19}$$

If we assume that the fluid is incompressible, Eq. 16.19 can be integrated to give

$$h_e - h_i = v(P_e - P_i) \tag{16.20}$$

Substituting this in Eq. 16.18 we have

$$v(P_e - P_i) + \frac{\mathbf{V}_e^2 - \mathbf{V}_i^2}{2} + (Z_e - Z_i)g = 0 \tag{16.21}$$

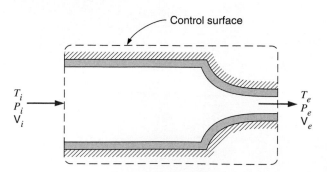

FIGURE 16.6 Schematic sketch of a nozzle.

This is of course the Bernoulli equation, which was derived in Section 9.3, Eq. 9.18, and for the reversible, adiabatic, one-dimensional, steady-state steady flow of an incompressible fluid through a nozzle the Bernoulli equation represents a combined statement of the first and second laws of thermodynamics.

EXAMPLE 16.4 Water enters the diffuser in a pump casing with a velocity of 30 m/s, a pressure of 350 kPa, and a temperature of 25°C. It leaves the diffuser with a velocity of 7 m/s and a pressure of 600 kPa. Determine the exit pressure for a reversible diffuser with these inlet conditions and exit velocity. Determine the increase in enthalpy, internal energy, and entropy for the actual diffuser.

Analysis and Solution

Consider first a control surface around a reversible diffuser with the given inlet conditions and exit velocity. Equation 16.21, the Bernoulli equation, is a statement of the first and second laws of thermodynamics for this process. Since there is no change in elevation this equation reduces to

$$v[(P_e)_s - P_i] + \frac{\mathbf{V}_e^2 - \mathbf{V}_i^2}{2} = 0$$

where $(P_e)_s$ represents the exit pressure for the reversible diffuser. From the steam tables, $v = 0.001\,003$ m³/kg.

$$P_{es} - P_i = \frac{(30)^2 - (7)^2}{0.001\,003 \times 2 \times 1000} = 424 \text{ kPa}$$

$$P_{es} = 774 \text{ kPa}$$

Next consider a control surface around the actual diffuser. The change in enthalpy can be found from the first law for this process, Eq. 16.18.

$$h_e - h_i = \frac{\mathbf{V}_i^2 - \mathbf{V}_e^2}{2} = \frac{(30)^2 - (7)^2}{2 \times 1000} = 0.4255 \text{ kJ/kg}$$

The change in internal energy can be found from the definition of enthalpy, $h_e - h_i = (u_e - u_i) + (P_e v_e - P_i v_i)$.

Thus, for an incompressible fluid

$$u_e - u_i = h_e - h_i - v(P_e - P_i)$$

$$= 0.4255 - 0.001\,003(600 - 350)$$

$$= 0.174\,75 \text{ kJ/kg}$$

The change of entropy can be approximated from the familiar relation

$$T\,ds = du + P\,dv$$

by assuming that the temperature is constant (which is approximately true in this case) and noting that for an incompressible fluid $dv = 0$. With these assumptions

$$s_e - s_i = \frac{u_e - u_i}{T} = \frac{0.174\,75}{298.2} = 0.000\,586 \text{ kJ/kg K}$$

Since this is an irreversible adiabatic process, the entropy will increase, as the above calculation indicates.

16.5 VELOCITY OF SOUND IN AN IDEAL GAS

When a pressure disturbance occurs in a compressible fluid, the disturbance travels with a velocity that depends on the state of the fluid. A sound wave is a very small pressure disturbance; the velocity of sound, also called the sonic velocity, is an important parameter in compressible-fluid flow. We proceed now to determine an expression for the sonic velocity of an ideal gas in terms of the properties of the gas.

Let a disturbance be set up by the movement of the piston at the end of the tube, Fig. 16.7a. A wave travels down the tube with a velocity c, which is the sonic velocity. Assume that after the wave has passed the properties of the gas have changed an infinitesimal amount and that the gas is moving with the velocity $d\mathbf{V}$ toward the wave front.

In Fig. 16.7b this process is shown from the point of view of an observer who travels with the wave front. Consider the control surface shown in Fig. 16.7b. From the first law for this steady-state steady-flow process we can write

$$h + \frac{c^2}{2} = (h + dh) + \frac{(c - d\mathbf{V})^2}{2}$$
$$dh - c\,d\mathbf{V} = 0 \qquad (16.22)$$

From the continuity equation we can write

$$\rho \mathscr{A}c = (\rho + d\rho)\mathscr{A}(c - d\mathbf{V})$$
$$c\,d\rho - \rho\,d\mathbf{V} = 0 \qquad (16.23)$$

Consider also the relation between properties

$$T\,ds = dh - \frac{dP}{\rho}$$

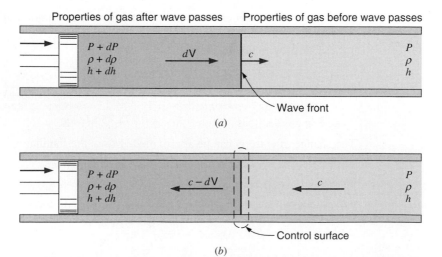

Properties of gas after wave passes   Properties of gas before wave passes

(a)

(b)

FIGURE 16.7 Diagram illustrating sonic velocity. (a) Stationary observer. (b) Observer traveling with wave front.

If the process is isentropic, $ds = 0$, and this equation can be combined with Eq. 16.22 to give the relation

$$\frac{dP}{\rho} - c \, d\mathbf{V} = 0 \qquad (16.24)$$

This can be combined with Eq. 16.23 to give the relation

$$\frac{dP}{d\rho} = c^2$$

Since we have assumed the process to be isentropic this is better written as a partial derivative.

$$\left(\frac{\partial P}{\partial \rho}\right)_s = c^2 \qquad (16.25)$$

An alternate derivation is to introduce the momentum equation. For the control volume of Fig. 16.7b the momentum equation is

$$P\mathcal{A} - (P + dP)\mathcal{A} = \dot{m}(c - d\mathbf{V} - c) = \rho\mathcal{A}c(c - d\mathbf{V} - c)$$
$$dP = \rho c \, d\mathbf{V} \qquad (16.26)$$

On combining this with Eq. 16.23 we obtain Eq. 16.25.

$$\left(\frac{\partial P}{\partial \rho}\right)_s = c^2$$

It will be of particular advantage to solve Eq. 16.25 for the velocity of sound in an ideal gas.

When an ideal gas undergoes an isentropic change of state, we found in Chapter 8 that, for this process, assuming constant specific heat

$$\frac{dP}{P} - k\frac{d\rho}{\rho} = 0$$

or

$$\left(\frac{\partial P}{\partial \rho}\right)_s = \frac{kP}{\rho}$$

Substituting this equation in Eq. 16.25 we have an equation for the velocity of sound in an ideal gas,

$$c^2 = \frac{kP}{\rho} \qquad (16.27)$$

Since for an ideal gas

$$\frac{P}{\rho} = RT$$

this equation may also be written

$$c^2 = kRT \qquad (16.28)$$

EXAMPLE 16.5   Determine the velocity of sound in air at 300 K and at 1000 K.

Analysis and Solution

Using Eq. 16.28

$$c = \sqrt{kRT}$$

$$= \sqrt{1.4 \times 0.287 \times 300 \times 1000} = 347.2 \text{ m/s}$$

Similarly, at 1000 K, using $k = 1.4$,

$$c = \sqrt{1.4 \times 0.287 \times 1000 \times 1000} = 633.9 \text{ m/s}$$

Note the significant increase in sonic velocity as the temperature increases.

The Mach number, M, is defined as the ratio of the actual velocity **V** to the sonic velocity c.

$$M = \frac{\mathbf{V}}{c} \tag{16.29}$$

When $M > 1$ the flow is supersonic; when $M < 1$ the flow is subsonic; and when $M = 1$ the flow is sonic. The importance of the Mach number as a parameter in fluid-flow problems will be evident in the paragraphs that follow.

16.6 REVERSIBLE, ADIABATIC, ONE-DIMENSIONAL FLOW OF AN IDEAL GAS THROUGH A NOZZLE

A nozzle or diffuser with both a converging and diverging section is shown in Fig. 16.8. The minimum cross-sectional area is called the throat.

Our first consideration concerns the conditions that determine whether a nozzle or diffuser should be converging or diverging, and the conditions that prevail at the throat. For the control volume shown the following relations can be written.

First law:

$$dh + \mathbf{V}\,d\mathbf{V} = 0 \tag{16.30}$$

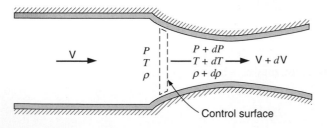

FIGURE 16.8 One-dimensional, reversible, adiabatic steady flow through a nozzle.

Property relation:

$$T \, ds = dh - \frac{dP}{\rho} = 0 \tag{16.31}$$

Continuity equation:

$$\rho \mathscr{A} \mathbf{V} = \dot{m} = \text{constant}$$

$$\frac{d\rho}{\rho} + \frac{d\mathscr{A}}{\mathscr{A}} + \frac{d\mathbf{V}}{\mathbf{V}} = 0 \tag{16.32}$$

Combining Eqs. 16.30 and 16.31 we have

$$dh = \frac{dP}{\rho} = -\mathbf{V} \, d\mathbf{V}$$

$$d\mathbf{V} = -\frac{1}{\rho \mathbf{V}} dP$$

Substituting this in Eq. 16.32

$$\frac{d\mathscr{A}}{\mathscr{A}} = \left(-\frac{d\rho}{\rho} - \frac{d\mathbf{V}}{\mathbf{V}} \right) = -\frac{d\rho}{\rho} \left(\frac{dP}{dP} \right) + \frac{1}{\rho \mathbf{V}^2} dP$$

$$= \frac{-dP}{\rho} \left(\frac{d\rho}{dP} - \frac{1}{\mathbf{V}^2} \right) = \frac{dP}{\rho} \left(-\frac{1}{(dP/d\rho)} + \frac{1}{\mathbf{V}^2} \right)$$

Since the flow is isentropic

$$\frac{dP}{d\rho} = c^2 = \frac{\mathbf{V}^2}{M^2}$$

and therefore

$$\frac{d\mathscr{A}}{\mathscr{A}} = \frac{dP}{\rho \mathbf{V}^2} (1 - M^2) \tag{16.33}$$

This is a very significant equation, for from it we can draw the following conclusions about the proper shape for nozzles and diffusers:

For a nozzle, $dP < 0$. Therefore,

> for a subsonic nozzle, $M < 1$, $d\mathscr{A} < 0$, and the nozzle is converging;

> for a supersonic nozzle, $M > 1$, $d\mathscr{A} > 0$, and the nozzle is diverging.

For a diffuser, $dP > 0$. Therefore,

> for a subsonic diffuser, $M < 1$, $\delta\mathscr{A} > 0$, and the diffuser is diverging;

> for a supersonic diffuser, $M > 1$, $d\mathscr{A} < 0$, and the diffuser is converging.

When $M = 1$, $d\mathscr{A} = 0$, which means that sonic velocity can be achieved only at the throat of a nozzle or diffuser. These conclusions are summarized in Fig. 16.9.

We will now develop a number of relations between the actual properties, stagnation properties, and Mach number. These relations are very useful in dealing with isentropic flow of an ideal gas in a nozzle.

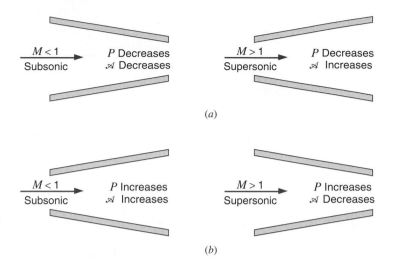

FIGURE 16.9 Required area changes for (*a*) nozzles and (*b*) diffusers.

Equation 16.1 gives the relation between enthalpy, stagnation enthalpy, and kinetic energy.

$$h + \frac{\mathbf{V}^2}{2} = h_0$$

For an ideal gas with constant specific heat Eq. 16.1 can be written

$$\mathbf{V}^2 = 2C_{po}(T_0 - T) = 2\frac{kRT}{k-1}\left(\frac{T_0}{T} - 1\right)$$

Since

$$c^2 = kRT$$

$$\mathbf{V}^2 = \frac{2c^2}{k-1}\left(\frac{T_0}{T} - 1\right)$$

$$\frac{\mathbf{V}^2}{c^2} = M^2 = \frac{2}{k-1}\left(\frac{T_0}{T} - 1\right)$$

$$\frac{T_0}{T} = 1 + \frac{(k-1)}{2}M^2 \tag{16.34}$$

For an isentropic process,

$$\left(\frac{T_0}{T}\right)^{k/(k-1)} = \frac{P_0}{P} \qquad \left(\frac{T_0}{T}\right)^{1/(k-1)} = \frac{\rho_0}{\rho}$$

Therefore,

$$\frac{P_0}{P} = \left[1 + \frac{(k-1)}{2}M^2\right]^{k/(k-1)} \tag{16.35}$$

$$\frac{\rho_0}{\rho} = \left[1 + \frac{(k-1)}{2}M^2\right]^{1/(k-1)} \tag{16.36}$$

TABLE 16.1
Critical Pressure, Density, and Temperature Ratios
for Isentropic Flow of an Ideal Gas

	k = 1.1	k = 1.2	k = 1.3	k = 1.4	k = 1.67
P^*/P_0	0.5847	0.5644	0.5457	0.5283	0.4867
ρ^*/ρ_0	0.6139	0.6209	0.6276	0.6340	0.6497
T^*/T_0	0.9524	0.9091	0.8696	0.8333	0.7491

Values of P/P_0, ρ/ρ_0, and T/T_0 are given as a function of M in Table A.11 for the value $k = 1.40$. In the software version of Table A.11, both M and k are input variables.

The conditions at the throat of the nozzle can be found by noting that $M = 1$ at the throat. The properties at the throat are denoted by an asterisk (*). Therefore,

$$\frac{T^*}{T_0} = \frac{2}{k+1} \tag{16.37}$$

$$\frac{P^*}{P_0} = \left(\frac{2}{k+1}\right)^{k/(k-1)} \tag{16.38}$$

$$\frac{\rho^*}{\rho_0} = \left(\frac{2}{k+1}\right)^{1/(k-1)} \tag{16.39}$$

These properties at the throat of a nozzle when $M = 1$ are frequently referred to as critical pressure, critical temperature, and critical density and the ratios given by Eqs. 16.37, 16.38, and 16.39 are referred to as the critical-temperature ratio, critical-pressure ratio, and critical-density ratio. Table 16.1 gives these ratios for various values of k.

16.7 MASS RATE OF FLOW OF AN IDEAL GAS THROUGH AN ISENTROPIC NOZZLE

We now turn our attention to a consideration of the mass rate of flow per unit area, $\dot{m}/\mathscr{A}$, in a nozzle. From the continuity equation we proceed as follows:

$$\frac{\dot{m}}{\mathscr{A}} = \rho \mathbf{V} = \frac{P\mathbf{V}}{RT}\sqrt{\frac{kT_0}{kT_0}}$$

$$= \frac{P\mathbf{V}}{\sqrt{kRT}}\sqrt{\frac{k}{R}}\sqrt{\frac{T_0}{T}}\sqrt{\frac{1}{T_0}}$$

$$= \frac{PM}{\sqrt{T_0}}\sqrt{\frac{k}{R}}\sqrt{1+\frac{k-1}{2}M^2} \tag{16.40}$$

By substituting Eq. 16.35 into Eq. 16.39 the flow per unit area can be expressed in terms of stagnation pressure, stagnation temperature, Mach number, and gas properties.

$$\frac{\dot{m}}{\mathscr{A}} = \frac{P_0}{\sqrt{T_0}}\sqrt{\frac{k}{R}} \times \frac{M}{\left(1+\frac{k-1}{2}M^2\right)^{(k+1)/2(k-1)}} \tag{16.41}$$

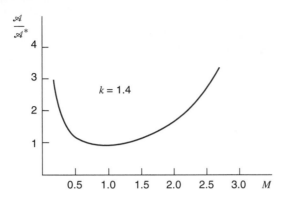

FIGURE 16.10 Area ratio as a function of Mach number for a reversible, adiabatic nozzle.

At the throat, $M = 1$, and therefore the flow per unit area at the throat, $\dot{m}/\mathscr{A}^*$, can be found by setting $M = 1$ in Eq. 16.41.

$$\frac{\dot{m}}{\mathscr{A}^*} = \frac{P_0}{\sqrt{T_0}}\sqrt{\frac{k}{R}} \times \frac{1}{\left(\dfrac{k+1}{2}\right)^{(k+1)/2(k-1)}} \tag{16.42}$$

The area ratio $\mathscr{A}/\mathscr{A}^*$ can be obtained by dividing Eq. 16.42 by Eq. 16.41.

$$\frac{\mathscr{A}}{\mathscr{A}^*} = \frac{1}{M}\left[\left(\frac{2}{k+1}\right)\left(1+\frac{k-1}{2}M^2\right)\right]^{(k+1)/2(k-1)} \tag{16.43}$$

The area ratio $\mathscr{A}/\mathscr{A}^*$ is the ratio of the area at the point where the Mach number is M to the throat area, and values of $\mathscr{A}/\mathscr{A}^*$ as a function of Mach number are given in Table A.16 in the Appendix. Figure 16.10 shows a plot of $\mathscr{A}/\mathscr{A}^*$ vs. M, which is in accordance with our previous conclusion that a subsonic nozzle is converging and a supersonic nozzle is diverging.

The final point to be made regarding the isentropic flow of an ideal gas through a nozzle involves the effect of varying the back pressure (the pressure outside the nozzle exit) on the mass rate of flow.

Consider first a convergent nozzle as shown in Fig. 16.11, which also shows the pressure ratio P/P_0 along the length of the nozzle. The conditions upstream are the stagnation conditions, which are assumed to be constant. The pressure at the exit plane of the nozzle is designated P_E, and the back pressure P_B. Let us consider how the mass rate of

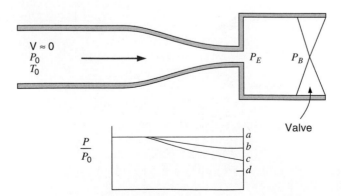

FIGURE 16.11 Pressure ratio as a function of back pressure for a convergent nozzle.

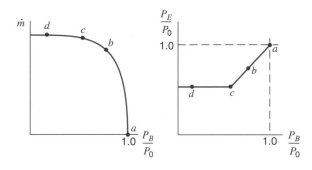

FIGURE 16.12 Mass rate of flow and exit pressure as a function of back pressure for a convergent nozzle.

flow $\dot{m}$ and the exit plane pressure P_E/P_0 vary as the back pressure P_B is decreased. These quantities are plotted in Fig. 16.12.

When $P_B/P_0 = 1$ there is of course no flow, and $P_E/P_0 = 1$ as designated by point a. Next let the back pressure P_B be lowered to that designated by point b, so that P_B/P_0 is greater than the critical-pressure ratio. The mass rate of flow has a certain value and $P_E = P_B$. The exit Mach number is less than 1. Next let the back pressure be lowered to the critical pressure, designated by point c. The Mach number at the exit is now unity, and P_E is equal to P_B. When P_B is decreased below the critical pressure, designated by point d, there is no further increase in the mass rate of flow, and P_E remains constant at a value equal to the critical pressure, and the exit Mach number is unity. The drop in pressure from P_E to P_B takes place outside the nozzle exit. Under these conditions the nozzle is said to be choked, which means that for given stagnation conditions the nozzle is passing the maximum possible mass flow.

Consider next a convergent–divergent nozzle in a similar arrangement, Fig. 16.13. Point a designates the condition when $P_B = P_0$ and there is no flow. When P_B is decreased to the pressure indicated by point b, so that P_B/P_0 is less than 1 but considerably greater than the critical-pressure ratio, the velocity increases in the convergent section, but $M < 1$ at the throat. Therefore, the diverging section acts as a subsonic diffuser in which the pressure increases and velocity decreases. Point c designates the back pressure at which $M = 1$ at the throat, but the diverging section acts as a subsonic diffuser (with $M = 1$ at the inlet) in which the pressure increases and velocity decreases. Point d designates one other back pressure that permits isentropic flow, and in this case the diverging section acts as a supersonic nozzle, with a decrease in pressure and an increase in velocity. Between the back pressures designated by points c and d, an isentropic solution is not possible, and shock waves will be present. This matter is discussed in the section that fol-

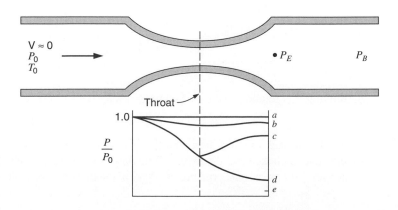

FIGURE 16.13 Nozzle pressure ratio as a function of back pressure for a reversible, convergent–divergent nozzle.

lows. When the back pressure is decreased below that designated by point d, the exit-plane pressure P_E remains constant, the drop in pressure from P_E to P_B takes place outside the nozzle. This is designated by point e.

EXAMPLE 16.6   A convergent nozzle has an exit area of 500 mm^2. Air enters the nozzle with a stagnation pressure of 1000 kPa and a stagnation temperature of 360 K. Determine the mass rate of flow for back pressures of 800 kPa, 528 kPa, and 300 kPa, assuming isentropic flow.

Analysis and Solution

For air $k = 1.4$ and Table A.11 may be used. The critical-pressure ratio, P^*/P_0, is 0.528. Therefore, for a back pressure of 528 kPa, $M = 1$ at the nozzle exit and the nozzle is choked. Decreasing the back pressure below 528 kPa will not increase the flow.

For a back pressure of 528 kPa,

$$\frac{T^*}{T_o} = 0.8333 \qquad T^* = 300 \text{ K}$$

At the exit

$$\mathbf{V} = c = \sqrt{kRT}$$

$$= \sqrt{1.4 \times 0.287 \times 300 \times 1000} = 347.2 \text{ m/s}$$

$$\rho^* = \frac{P^*}{RT^*} = \frac{528}{0.287 \times 300} = 6.1324 \text{ kg/m}^3$$

$$\dot{m} = \rho \mathscr{A} \mathbf{V}$$

Applying this relation to the throat section

$$\dot{m} = 6.1324 \times 500 \times 10^{-6} \times 347.2 = 1.0646 \text{ kg/s}$$

For a back pressure of 800 kPa, $P_E/P_0 = 0.8$ (subscript E designates the properties in the exit plane). From Table A.11

$$M_E = 0.573 \qquad T_E/T_0 = 0.9381$$

$$T_E = 337.7 \text{ K}$$

$$c_E = \sqrt{kRT_E} = \sqrt{1.4 \times 0.287 \times 337.7 \times 1000} = 368.4 \text{ m/s}$$

$$\mathbf{V}_E = M_E c_E = 211.1 \text{ m/s}$$

$$\rho_E = \frac{P_E}{RT_E} = \frac{800}{0.287 \times 337.7} = 8.2542 \text{ kg/m}^3$$

$$\dot{m} = \rho \mathscr{A} \mathbf{V}$$

Applying this relation to the exit section,

$$\dot{m} = 8.2542 \times 500 \times 10^{-6} \times 211.1 = 0.8712 \text{ kg/s}$$

For a back pressure less than the critical pressure, which in this case is 528 kPa, the nozzle is choked and the mass rate of flow is the same as that for the critical pressure. Therefore, for an exhaust pressure of 300 kPa, the mass rate of flow is 1.0646 kg/s.

EXAMPLE 16.7   A converging–diverging nozzle has an exit area to throat area ratio of 2. Air enters this nozzle with a stagnation pressure of 1000 kPa and a stagnation temperature of 360 K. The throat area is 500 mm². Determine the mass rate of flow, exit pressure, exit temperature, exit Mach number, and exit velocity for the following conditions:

(a) Sonic velocity at the throat, diverging section acting as a nozzle. (Corresponds to point d in Fig. 16.13.)

(b) Sonic velocity at the throat, diverging section acting as a diffuser. (Corresponds to point c in Fig. 16.13.)

Analysis and Solution

(a) In Table A.11 of the Appendix we find that there are two Mach numbers listed for $\mathscr{A}/\mathscr{A}^* = 2$. One of these is greater than unity and one is less than unity. When the diverging section acts as a supersonic nozzle we use the value for $M > 1$. The following are from Table A.11.

$$\frac{\mathscr{A}_E}{\mathscr{A}^*} = 2.0 \qquad M_E = 2.197 \qquad \frac{P_E}{P_0} = 0.0939 \qquad \frac{T_E}{T_0} = 0.5089$$

Therefore,

$$P_E = 0.0939(1000) = 93.9 \text{ kPa}$$
$$T_E = 0.5089(360) = 183.2 \text{ K}$$
$$c_E = \sqrt{kRT_E} = \sqrt{1.4 \times 0.287 \times 183.2 \times 1000} = 271.3 \text{ m/s}$$
$$\mathbf{V}_E = M_E c_E = 2.197(271.3) = 596.1 \text{ m/s}$$

The mass rate of flow can be determined by considering either the throat section or the exit section. However, in general it is preferable to determine the mass rate of flow from conditions at the throat. Since in this case $M = 1$ at the throat, the calculation is identical to the calculation for the flow in the convergent nozzle of Example 16.6 when it is choked.

(b) The following are from Table A.11.

$$\frac{\mathscr{A}_E}{\mathscr{A}^*} = 2.0 \qquad M = 0.308 \qquad \frac{P_E}{P_0} = 0.936 \qquad \frac{T_E}{T_0} = 0.9812$$

$$P_E = 0.936(1000) = 936 \text{ kPa}$$
$$T_E = 0.9812(360) = 353.3 \text{ K}$$
$$c_E = \sqrt{kRT_E} = \sqrt{1.4 \times 0.287 \times 353.3 \times 1000} = 376.8 \text{ m/s}$$
$$\mathbf{V}_E = M_E c_E = 0.308(376.8) = 116 \text{ m/s}$$

Since $M = 1$ at the throat, the mass rate of flow is the same as in (a), which is also equal to the flow in the convergent nozzle of Example 16.6 when it is choked.

In the example above a solution assuming isentropic flow is not possible if the back pressure is between 936 kPa and 93.9 kPa. If the back pressure is in this range there will be either a normal shock in the nozzle or oblique shock waves outside the nozzle. The matter of normal shock waves is considered in the following section.

16.8 NORMAL SHOCK IN AN IDEAL GAS FLOWING THROUGH A NOZZLE

A shock wave involves an extremely rapid and abrupt change of state. In a normal shock this change of state takes place across a plane normal to the direction of the flow. Figure 16.14 shows a control surface that includes such a normal shock. We can now determine the relations that govern the flow. Assuming steady-state, steady-flow we can write the following relations, where subscripts x and y denote the conditions upstream and downstream of the shock, respectively. Note that no heat and work across the control surface.

First law:

$$h_x + \frac{\mathbf{V}_x^2}{2} = h_y + \frac{\mathbf{V}_y^2}{2} = h_{0x} = h_{0y} \tag{16.44}$$

Continuity equation:

$$\frac{\dot{m}}{\mathscr{A}} = \rho_x \mathbf{V}_x = \rho_y \mathbf{V}_y \tag{16.45}$$

Momentum equation:

$$\mathscr{A}(P_x - P_y) = \dot{m}(\mathbf{V}_y - \mathbf{V}_x) \tag{16.46}$$

Second law: Since the process is adiabatic

$$s_y - s_x \geq 0 \tag{16.47}$$

The energy and continuity equations can be combined to give an equation that when plotted on the h–s diagram is called the Fanno line. Similarly, the momentum and continuity equations can be combined to give an equation the plot of which on the h–s diagram is known as the Rayleigh line. Both of these lines are shown on the h–s diagram of Fig. 16.15. It can be shown that the point of maximum entropy on each line, points a and b, corresponds to $M = 1$. The lower part of each line corresponds to supersonic velocities, and the upper part to subsonic velocities.

The two points where all three equations are satisfied are points x and y, x being in the supersonic region and y in the subsonic region. Since the second law requires that $s_y - s_x \geq 0$ in an adiabatic process, we conclude that the normal shock can proceed only from x to y. This means that the velocity changes from supersonic ($M > 1$) before the shock to subsonic ($M < 1$) after the shock.

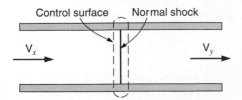

FIGURE 16.14 One-dimensional normal shock.

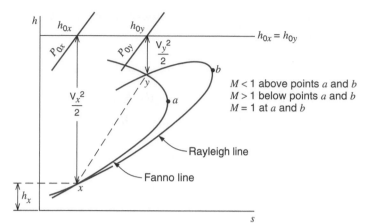

FIGURE 16.15 End states for a one-dimensional normal shock on an enthalpy–entropy diagram.

The equations governing normal shock waves will now be developed. If we assume constant specific heats we conclude from Eq. 16.44, the energy equation, that

$$T_{0x} = T_{0y} \qquad (16.48)$$

That is, there is no change in stagnation temperature across a normal shock. Introducing Eq. 16.34

$$\frac{T_{0x}}{T_x} = 1 + \frac{k-1}{2} M_x^2 \qquad \frac{T_{0y}}{T_y} = 1 + \frac{k-1}{2} M_y^2$$

and substituting into Eq. 16.48 we have

$$\frac{T_y}{T_x} = \frac{1 + \dfrac{k-1}{2} M_x^2}{1 + \dfrac{k-1}{2} M_y^2} \qquad (16.49)$$

The equation of state, the definition of Mach number, and the relation $c = \sqrt{kRT}$ can be introduced into the continuity equation as follows:

$$\rho_x \mathbf{V}_x = \rho_y \mathbf{V}_y$$

But

$$\rho_x = \frac{P_x}{RT_x} \qquad \rho_y = \frac{P_y}{RT_y}$$

$$\frac{T_y}{T_x} = \frac{P_y \mathbf{V}_y}{P_x \mathbf{V}_x} = \frac{P_y M_y c_y}{P_x M_x c_x} = \frac{P_y M_y \sqrt{T_y}}{P_x M_x \sqrt{T_x}}$$

$$= \left(\frac{P_y}{P_x}\right)^2 \left(\frac{M_y}{M_x}\right)^2 \qquad (16.50)$$

Combining Eqs. 16.49 and 16.50, which involves combining the energy equation and the continuity equation, gives the equation of the Fanno line.

$$\frac{P_y}{P_x} = \frac{M_x \sqrt{1 + \dfrac{k-1}{2} M_x^2}}{M_y \sqrt{1 + \dfrac{k-1}{2} M_y^2}} \qquad (16.51)$$

The momentum and continuity equations can be combined as follows to give the equation of the Rayleigh line.

$$P_x - P_y = \frac{\dot{m}}{\mathcal{A}}(\mathbf{V}_y - \mathbf{V}_x) = \rho_y \mathbf{V}_y^2 - \rho_x \mathbf{V}_x^2$$

$$P_x + \rho_x \mathbf{V}_x^2 = P_y + \rho_y \mathbf{V}_y^2$$

$$P_x + \rho_x M_x^2 c_x^2 = P_y + \rho_y M_y^2 c_y^2$$

$$P_x + \frac{P_x M_x^2}{RT_x}(kRT_x) = P_y + \frac{P_y M_y^2}{RT_y}(kRT_y)$$

$$P_x(1 + kM_x^2) = P_y(1 + kM_y^2)$$

$$\frac{P_y}{P_x} = \frac{1 + kM_x^2}{1 + kM_y^2} \tag{16.52}$$

Equations 16.51 and 16.52 can be combined to give the following equation relating M_x and M_y.

$$M_y^2 = \frac{M_x^2 + \dfrac{2}{k-1}}{\dfrac{2k}{k-1}M_x^2 - 1} \tag{16.53}$$

Table A.12 gives the normal shock functions, which include M_y as a function of M_x. This table applies to an ideal gas with a value $k = 1.40$. In the software version of Table A.12, both M and k are input variables. Note that M_x is always supersonic and M_y is always subsonic, which agrees with the previous statement that in a normal shock the velocity changes from supersonic to subsonic. These tables also give the pressure, density, temperature, and stagnation pressure ratios across a normal shock as a function of M_x. These are found from Eqs. 16.49 and 16.50 and the equation of state. Note that there is always a drop in stagnation pressure across a normal shock and an increase in the static pressure.

EXAMPLE 16.8

Consider the convergent–divergent nozzle of Example 16.7 in which the diverging section acts as a supersonic nozzle (Fig. 16.16). Assume that a normal shock stands in the exit plane of the nozzle. Determine the static pressure and temperature and the stagnation pressure just downstream of the normal shock.

Sketch: Fig. 16.16.

Analysis and Solution

From Table A.12

$$M_x = 2.197 \qquad M_y = 0.547 \qquad \frac{P_y}{P_x} = 5.46 \qquad \frac{T_y}{T_x} = 1.854 \qquad \frac{P_{0y}}{P_{0x}} = 0.630$$

$$P_y = 5.46 \times P_x = 5.46(93.9) = 512.7 \text{ kPa}$$

$$T_y = 1.854 \times T_x = 1.854(183.2) = 339.7 \text{ K}$$

$$P_{0y} = 0.630 \times P_{0x} = 0.630(1000) = 630 \text{ kPa}$$

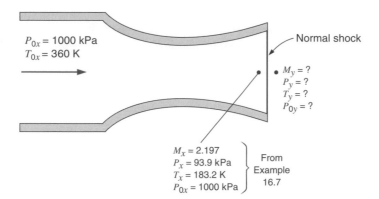

$P_{0x} = 1000$ kPa
$T_{0x} = 360$ K

Normal shock

$M_y = ?$
$P_y = ?$
$T_y = ?$
$P_{0y} = ?$

$M_x = 2.197$
$P_x = 93.9$ kPa
$T_x = 183.2$ K
$P_{0x} = 1000$ kPa

From
Example
16.7

FIGURE 16.16 Sketch for Example 16.8.

In the light of this example we can conclude the discussion concerning the flow through a convergent–divergent nozzle. Figure 16.13 is repeated here as Fig. 16.17 for convenience, except that points f, g, and h have been added. Consider point d. We have already noted that with this back pressure the exit plane pressure P_E is just equal to the back pressure P_B, and isentropic flow is maintained in the nozzle. Let the back pressure be raised to that designated by point f. The exit-plane pressure P_E is not influenced by this increase in back pressure, and the increase in pressure from P_E to P_B takes place outside the nozzle. Let the back pressure be raised to that designated by point g, which is just sufficient to cause a normal shock to stand in the exit plane of the nozzle. The exit-plane pressure P_E (downstream of the shock) is equal to the back pressure P_B, and $M < 1$ leaving the nozzle. This is the case in Example 16.8. Now let the back pressure be raised to that corresponding to point h. As the back pressure is raised from g to h the normal shock moves into the nozzle as indicated. Since $M < 1$ downstream of the normal shock, the diverging part of the nozzle that is downstream of the shock acts as a subsonic diffuser. As the back pressure is increased from h to c the shock moves further upstream and disappears at the nozzle throat where the back pressure corresponds to c. This is reasonable since there are no supersonic velocities involved when the back pressure corresponds to c, and hence no shock waves are possible.

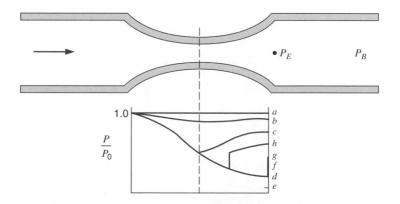

$\bullet P_E$ P_B

$\dfrac{P}{P_0}$ 1.0

a
b
c
h
g
f
d
e

FIGURE 16.17 Nozzle pressure ratio as a function of back pressure for a convergent–divergent nozzle.

EXAMPLE 16.9   Consider the convergent–divergent nozzle of Examples 16.7 and 16.8. Assume that there is a normal shock wave standing at the point where $M = 1.5$. Determine the exit-plane pressure, temperature, and Mach number. Assume isentropic flow except for the normal shock (Fig. 16.18).

Sketch: Fig. 16.18.

Analysis and Solution

The properties at point x can be determined from Table A.11, because the flow is isentropic to point x.

$$M_x = 1.5 \qquad \frac{P_x}{P_{0x}} = 0.2724 \qquad \frac{T_x}{T_{0x}} = 0.6897 \qquad \frac{\mathscr{A}_x}{\mathscr{A}_x^*} = 1.1762$$

Therefore,

$$P_x = 0.2724(1000) = 272.4 \text{ kPa}$$

$$T_x = 0.6897(360) = 248.3 \text{ K}$$

The properties at point y can be determined from the normal shock functions, Table A.12.

$$M_y = 0.7011 \qquad \frac{P_y}{P_x} = 2.4583 \qquad \frac{T_y}{T_x} = 1.320 \qquad \frac{P_{0y}}{P_{0x}} = 0.9298$$

$$P_y = 2.4583 P_x = 2.4583(272.4) = 669.6 \text{ kPa}$$

$$T_y = 1.320 T_x = 1.320(248.3) = 327.8 \text{ K}$$

$$P_{0y} = 0.9298 P_{0x} = 0.9298(1000) = 929.8 \text{ kPa}$$

Since there is no change in stagnation temperature across a normal shock,

$$T_{0x} = T_{0y} = 360 \text{ K}$$

From y to E the diverging section acts as a subsonic diffuser. In solving this problem it is convenient to think of the flow at y as having come from an isentropic nozzle having a throat area $\mathscr{A}_y^*$. Such a hypothetical nozzle is shown by the dotted line. From the table of isentropic flow functions, Table A.11, we find the following for $M_y = 0.7011$.

$$M_y = 0.7011 \qquad \frac{\mathscr{A}_y}{\mathscr{A}_y^*} = 1.0938 \qquad \frac{P_y}{P_{0y}} = 0.7202 \qquad \frac{T_y}{T_{0y}} = 0.9105$$

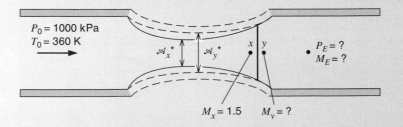

FIGURE 16.18 Sketch for Example 16.9.

From the statement of the problem

$$\frac{\mathscr{A}_E}{\mathscr{A}_x^*} = 2.0$$

Also, since the flow from y to E is isentropic,

$$\frac{\mathscr{A}_E}{\mathscr{A}_E^*} = \frac{\mathscr{A}_E}{\mathscr{A}_y^*} = \frac{\mathscr{A}_E}{\mathscr{A}_x^*} \times \frac{\mathscr{A}_x^*}{\mathscr{A}_x} \times \frac{\mathscr{A}_x}{\mathscr{A}_y} \times \frac{\mathscr{A}_y}{\mathscr{A}_y^*}$$

$$= \frac{\mathscr{A}_E}{\mathscr{A}_y^*} = 2.0 \times \frac{1}{1.1762} \times 1 \times 1.0938 = 1.860$$

From the table of isentropic flow functions for $\mathscr{A}/\mathscr{A}^* = 1.860$ and $M < 1$

$$M_E = 0.339 \qquad \frac{P_E}{P_{0E}} = 0.9222 \qquad \frac{T_E}{T_{0E}} = 0.9771$$

$$\frac{P_E}{P_{0E}} = \frac{P_E}{P_{0y}} = 0.9222$$

$$P_E = 0.9222(P_{0y}) = 0.9222(929.8) = 857.5 \text{ kPa}$$

$$T_E = 0.9771(T_{0E}) = 0.9771(360) = 351.7 \text{ K}$$

In conclusion it should be pointed out that in considering the normal shock we have ignored the effect of viscosity and thermal conductivity, which are certain to be present. The actual shock wave will occur over some finite thickness. However, the development as given here gives a very good qualitative picture of normal shocks, and also provides a basis for fairly accurate quantitative results.

16.9 NOZZLE AND DIFFUSER COEFFICIENTS

Up to this point we have considered only isentropic flow and normal shocks. As was pointed out in Chapter 9, isentropic flow through a nozzle provides a standard to which the performance of an actual nozzle can be compared. For nozzles, the three important parameters by which actual flow can be compared to the ideal flow are nozzle efficiency, velocity coefficient, and discharge coefficient. These are defined as follows:

The nozzle efficiency η_N is defined as

$$\eta_N = \frac{\text{Actual kinetic energy at nozzle exit}}{\text{Kinetic energy at nozzle exit with isentropic flow to same exit pressure}} \qquad (16.54)$$

The efficiency can be defined in terms of properties. On the h–s diagram of Fig. 16.19 state $0i$ represents the stagnation state of the fluid entering the nozzle; state e represents the actual state at the nozzle exit; and state s represents the state that would have been achieved at the nozzle exit if the flow had been reversible and adiabatic to the same exit pressure. Therefore, in terms of these states the nozzle efficiency is

$$\eta_N = \frac{h_{0i} - h_e}{h_{0i} - h_s}$$

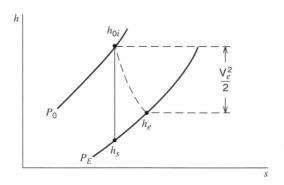

FIGURE 16.19 Enthalpy–entropy diagram showing the effects of irreversibility in a nozzle.

Nozzle efficiencies vary in general from 90 to 99%. Large nozzles usually have higher efficiencies than small nozzles, and nozzles with straight axes have higher efficiencies than nozzles with curved axes. The irreversibilities, which cause the departure from isentropic flow, are primarily due to frictional effects, and are confined largely to the boundary layer. The rate of change of cross-sectional area along the nozzle axis (that is, the nozzle contour) is an important parameter in the design of an efficient nozzle, particularly in the divergent section. Detailed consideration of this matter is beyond the scope of this text, and the reader is referred to standard references on the subject.

The velocity coefficient C_V is defined as

$$C_V = \frac{\text{Actual velocity at nozzle exit}}{\text{Velocity at nozzle exit with isentropic flow and same exit pressure}} \quad (16.55)$$

It follows that the velocity coefficient is equal to the square root of the nozzle efficiency

$$C_V = \sqrt{\eta_N} \quad (16.56)$$

The coefficient of discharge C_D is defined by the relation

$$C_D = \frac{\text{Actual mass rate of flow}}{\text{Mass rate of flow with isentropic flow}}$$

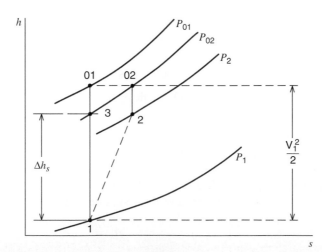

FIGURE 16.20 Enthalpy–entropy diagram showing the definition of diffuser efficiency.

In determining the mass rate of flow with isentropic conditions, the actual back pressure is used if the nozzle is not choked. If the nozzle is choked, the isentropic mass rate of flow is based on isentropic flow and sonic velocity at the minimum section (that is, sonic velocity at the exit of a convergent nozzle and at the throat of a convergent–divergent nozzle).

The performance of a diffuser is usually given in terms of diffuser efficiency, which is best defined with the aid of an h–s diagram. On the h–s diagram of Fig. 16.20 states 1 and 01 are the actual and stagnation states of the fluid entering the diffuser. States 2 and 02 are the actual and stagnation states of the fluid leaving the diffuser. State 3 is not attained in the diffuser, but it is the state that has the same entropy as the initial state and the pressure of the isentropic stagnation state leaving the diffuser. The efficiency of the diffuser η_D is defined as

$$\eta_D = \frac{\Delta h_s}{\mathbf{V}_1^2/2} = \frac{h_3 - h_1}{h_{01} - h_1} = \frac{h_3 - h_1}{h_{02} - h_1} \qquad (16.57)$$

If we assume an ideal gas with constant specific heat this reduces to

$$\eta_D = \frac{T_3 - T_1}{T_{02} - T_1} = \frac{\dfrac{(T_3 - T_1)}{T_1} T_1}{\dfrac{\mathbf{V}_1^2}{2\, C_{po}}}$$

$$C_{po} = \frac{kR}{k-1} \qquad T_1 = \frac{c_1^2}{kR} \qquad \mathbf{V}_1^2 = M_1^2 c_1^2 \qquad \frac{T_3}{T_1} = \left(\frac{P_{02}}{P_1}\right)^{(k-1)/k}$$

Therefore,

$$\eta_D = \frac{\left(\dfrac{P_{02}}{P_1}\right)^{(k-1)/k} - 1}{\dfrac{k-1}{2} M_1^2}$$

$$\left(\frac{P_{02}}{P_1}\right)^{(k-1)/k} = \left(\frac{P_{01}}{P_1}\right)^{(k-1)/k} \times \left(\frac{P_{02}}{P_{01}}\right)^{(k-1)/k}$$

$$\left(\frac{P_{02}}{P_1}\right)^{(k-1)/k} = \left(1 + \frac{k-1}{2} M_1^2\right)\left(\frac{P_{02}}{P_{01}}\right)^{(k-1)/k}$$

$$\eta_D = \frac{\left(1 + \dfrac{k-1}{2} M_1^2\right)\left(\dfrac{P_{02}}{P_{01}}\right)^{(k-1)/k} - 1}{\dfrac{k-1}{2} M_1^2} \qquad (16.58)$$

16.10 Nozzles and Orifices as Flow-Measuring Devices

The mass rate of flow of a fluid flowing in a pipe is frequently determined by measuring the pressure drop across a nozzle or orifice in the line, as shown in Fig. 16.21. The ideal process for such a nozzle or orifice is assumed to be isentropic flow through a nozzle that

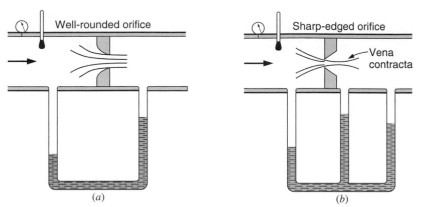

FIGURE 16.21 Nozzles and orifices as flow-measuring devices.

has the measured pressure drop from inlet to exit and a minimum cross-sectional area equal to the minimum area of the nozzle or orifice. The actual flow is related to the ideal flow by the coefficient of discharge, which is defined by Eq. 16.57.

The pressure difference measured across an orifice depends upon the location of the pressure taps as indicated in Fig. 16.21. Since the ideal flow is based on the measured pressure difference, it follows that the coefficient of discharge depends on the locations of the pressure taps. Also, the coefficient of discharge for a sharp-edged orifice is considerably less than that for a well-rounded nozzle, primarily due to a contraction of the stream, known as the vena contracta, as it flows through a sharpedged orifice.

There are two approaches to determining the discharge coefficient of a nozzle or orifice. One is to follow a standard design procedure, such as the ones established by the American Society of Mechanical Engineers,[1] and use the coefficient of discharge given for a particular design. A more accurate method is to calibrate a given nozzle or orifice, and determine the discharge coefficient for a given installation by accurately measuring the actual mass rate of flow. The procedure to be followed will depend on the accuracy desired and other factors involved (such as time, expense, availability of calibration facilities) in a given situation.

For incompressible fluids flowing through an orifice the ideal flow for a given pressure drop can be found by the procedure outlined in Section 16.4. Actually, it is advantageous to combine Eqs. 16.21 and 16.17 to give the following relation, which is valid for reversible flow.

$$v(P_2 - P_1) + \frac{\mathbf{V}_2^2 - \mathbf{V}_1^2}{2} = v(P_2 - P_1) + \frac{\mathbf{V}_2^2 - (\mathcal{A}_2 / \mathcal{A}_1)^2 \mathbf{V}_2^2}{2} = 0 \qquad (16.59)$$

or

$$v(P_2 - P_1) + \frac{\mathbf{V}_2^2}{2}\left[1 - \left(\frac{\mathcal{A}_2}{\mathcal{A}_1}\right)^2\right] = 0$$

$$\mathbf{V}_2 = \sqrt{\frac{2v(P_1 - P_2)}{[1 - (\mathcal{A}_2 / \mathcal{A}_1)^2]}} \qquad (16.60)$$

---

[1] *Fluid Meters, Their Theory and Application,* ASME, 1959, *Flow Measurement,* ASME, 1959.

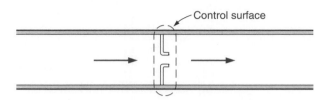

FIGURE 16.22 Analysis of a nozzle as a flow-measuring device.

For an ideal gas it is frequently advantageous to use the following simplified procedure when the pressure drop across an orifice or nozzle is small. Consider the nozzle shown in Fig. 16.22. From the first law we conclude that

$$h_i + \frac{\mathbf{V}_i^2}{2} = h_e + \frac{\mathbf{V}_e^2}{2}$$

Assuming constant specific heat, this reduces to

$$\frac{\mathbf{V}_e^2 - \mathbf{V}_i^2}{2} = h_i - h_e = C_{po}(T_i - T_e)$$

Let ΔP and ΔT be the decrease in pressure and temperature across the nozzle. Since we are considering reversible adiabatic flow we note that

$$\frac{T_e}{T_i} = \left(\frac{P_e}{P_i}\right)^{(k-1)/k}$$

or

$$\frac{T_i - \Delta T}{T_i} = \left(\frac{P_i - \Delta P}{P_i}\right)^{(k-1)/k}$$

$$1 - \frac{\Delta T}{T_i} = \left(1 - \frac{\Delta P}{P_i}\right)^{(k-1)/k}$$

Using the binomial expansion on the right side of the equation we have

$$1 - \frac{\Delta T}{T_i} = 1 - \frac{k-1}{k}\frac{\Delta P}{P_i} - \frac{k-1}{2k^2}\frac{\Delta P^2}{P_i^2}\cdots$$

If $\Delta P/P_i$ is small this reduces to

$$\frac{\Delta T}{T_i} = \frac{k-1}{k}\frac{\Delta P}{P_i}$$

Substituting this into the first-law equation we have

$$\frac{\mathbf{V}_e^2 - \mathbf{V}_i^2}{2} = C_{po}\frac{k-1}{k}\Delta P\frac{T_i}{P_i}$$

But for an ideal gas

$$C_{po} = \frac{kR}{k-1} \qquad \text{and} \qquad v_i = R\frac{T_i}{P_i}$$

Therefore,

$$\frac{\mathbf{V}_e^2 - \mathbf{V}_i^2}{2} = v_i\Delta P$$

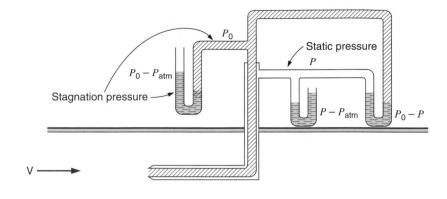

FIGURE 16.23 Schematic arrangement of a Pitot tube.

which is the same as Eq. 16.59, which was developed for incompressible flow. Therefore, when the pressure drop across a nozzle or orifice is small, the flow can be calculated with high accuracy by assuming incompressible flow.

The Pitot tube, Fig. 16.23, is an important instrument for measuring the velocity of a fluid. In calculating the flow with a Pitot tube it is assumed that the fluid is decelerated isentropically in front of the Pitot tube, and therefore the stagnation pressure of the free stream can be measured.

Applying the first law to this process we have

$$h + \frac{\mathbf{V}^2}{2} = h_0$$

If we assume incompressible flow for this isentropic process, the first law reduces to (because $T\,ds = dh - v\,dp$)

$$\frac{\mathbf{V}^2}{2} = h_0 - h = v(P_0 - P)$$

or

$$\mathbf{V} = \sqrt{2v(P_0 - P)} \tag{16.61}$$

If we consider the compressible flow of an ideal gas with constant specific heat, the velocity can be found from the relation

$$\frac{\mathbf{V}^2}{2} = h_0 - h = C_{po}(T_0 - T) = C_{po}T\left(\frac{T_0}{T} - 1\right)$$

$$= C_{po}T\left[\left(\frac{P_0}{P}\right)^{(k-1)/k} - 1\right] \tag{16.62}$$

It is of interest to know the error introduced by assuming incompressible flow when using the Pitot tube to measure the velocity of an ideal gas. To do so we introduce Eq. 16.35 and rearrange it as follows:

$$\frac{P_0}{P} = \left(1 + \frac{k-1}{2}M^2\right)^{k/(k-1)} = \left[1 + \left(\frac{k-1}{2}\right)\left(\frac{\mathbf{V}^2}{c^2}\right)\right]^{k/(k-1)} \tag{16.63}$$

TABLE 16.2

V/c_0	Approximate Room-Temperature Velocity, m/s	Error in Pressure for a Given Velocity, %	Error in Velocity for a Given Pressure, %
0.0	0	0	0
0.1	35	0.25	−0.13
0.2	70	1.0	−0.5
0.3	105	2.25	−1.2
0.4	140	4.0	−2.1
0.5	175	6.25	−3.3

But

$$\frac{\mathbf{V}^2}{2} + C_{po}T = C_{po}T_0$$

$$\frac{\mathbf{V}^2}{2} + \frac{kRc^2}{(k-1)kR} = \frac{kRc_0^2}{(k-1)kR}$$

$$1 + \frac{2c^2}{(k-1)\mathbf{V}^2} = \frac{2c_0^2}{(k-1)\mathbf{V}^2} \quad \text{where} \quad c_0 = \sqrt{kRT_0}$$

$$\frac{c^2}{\mathbf{V}^2} = \frac{k-1}{2}\left[\left(\frac{2}{k-1}\right)\left(\frac{c_0^2}{\mathbf{V}^2}\right) - 1\right] = \frac{c_0^2}{\mathbf{V}^2} - \frac{k-1}{2}$$

or

$$\frac{c^2}{\mathbf{V}^2} = \frac{c_0^2}{\mathbf{V}^2} - \frac{k-1}{2} \tag{16.64}$$

Substituting this into Eq. 16.63 and rearranging

$$\frac{P}{P_0} = \left[1 - \frac{k-1}{2}\left(\frac{\mathbf{V}}{c_0}\right)^2\right]^{k/(k-1)} \tag{16.65}$$

Expanding this equation by the binomial theorem, and including terms through $(\mathbf{V}/c_0)^4$, we have

$$\frac{P}{P_0} = 1 - \frac{k}{2}\left(\frac{\mathbf{V}}{c_0}\right)^2 + \frac{k}{8}\left(\frac{\mathbf{V}}{c_0}\right)^4$$

On rearranging this we have

$$\frac{P_0 - P}{\rho_0 \mathbf{V}^2/2} = 1 - \frac{1}{4}\left(\frac{\mathbf{V}}{c_0}\right)^2 \tag{16.66}$$

For incompressible flow the corresponding equation is

$$\frac{P_0 - P}{\rho_0 \mathbf{V}^2/2} = 1$$

Therefore, the second term on the right side of Eq. 16.66 represents the error involved if incompressible flow is assumed. The error in pressure for a given velocity and the error in velocity for a given pressure that would result from assuming incompressible flow are given in Table 16.2.

PROBLEMS

16.1 Steam leaves a nozzle with a pressure of 500 kPa, a temperature of 350°C, and a velocity of 250 m/s. What is the isentropic stagnation pressure and temperature?

16.2 An object from space enters the earth's upper atmosphere at 5 kPa, 100 K with a relative velocity of 2000 m/s or more. Estimate the object's surface temperature.

16.3 The products of combustion of a jet engine leave the engine with a velocity relative to the plane of 400 m/s, a temperature of 480°C, and a pressure of 75 kPa. Assuming that $k = 1.32$, $C_p = 1.15$ kJ/kg K for the products, determine the stagnation pressure and temperature of the products relative to the airplane.

16.4 A meteorite melts and burns up at temperatures of 3000 K. If it hits air at 5 kPa, 50 K, how high a velocity should it have to experience such a temperature?

16.5 I drive down the highway at 110 km/h on a day with 25°C, 101.3 kPa. I put my hand, cross-sectional area 0.01 m², flat out the window. What is the force on my hand and what temperature do I feel?

16.6 Air leaves a compressor in a pipe with a stagnation temperature and pressure of 150°C, 300 kPa, and a velocity of 125 m/s. The pipe has a cross-sectional area of 0.02 m². Determine the static temperature and pressure and the mass flow rate.

16.7 A stagnation pressure of 108 kPa is measured for an air flow where the pressure is 100 kPa and 20°C in the approach flow. What is the incoming velocity?

16.8 A jet engine receives a flow of 150 m/s air at 75 kPa, 5°C across an area of 0.6 m² with an exit flow at 450 m/s, 75 kPa, 600 K. Find the mass flow rate and thrust.

16.9 A water cannon sprays 1 kg/s liquid water at a velocity of 100 m/s horizontally out from a nozzle. It is driven by a pump that receives the water from a tank at 15°C, 100 kPa. Neglect elevation differences and the kinetic energy of the water flow in the pump and hose to the nozzle. Find the nozzle exit area, the required pressure out of the pump, and the horizontal force needed to hold the cannon.

16.10 An irrigation pump takes water from a lake and discharges it through a nozzle, as shown in Fig. P16.10. At the pump exit the pressure is 700 kPa, and the tempera-

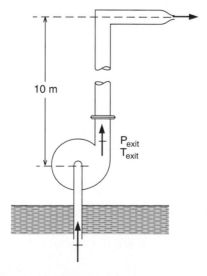

FIGURE P16.10

ture is 20°C. The nozzle is located 10 m above the pump, and the atmospheric pressure is 100 kPa. Assuming reversible flow through the system, determine the velocity of the water leaving the nozzle.

16.11 A water turbine using nozzles is located at the bottom of Hoover Dam 175 m below the surface of Lake Mead. The water enters the nozzles at a stagnation pressure corresponding to the column of water above it minus 20% due to friction. The temperature is 15°C, and the water leaves at standard atmospheric pressure. If the flow through the nozzle is reversible and adiabatic, determine the velocity and kinetic energy per kilogram of water leaving the nozzle.

16.12 A water tower on a farm holds 1 m³ liquid water at 20°C, 100 kPa in a tank on top of a 5-m-tall tower. A pipe leads to the ground level with a tap that can open a 1.5 cm-diameter hole. Neglect friction and pipe losses and estimate the time it will take to empty the tank of water.

16.13 Find the speed of sound for air at 100 kPa at the two temperatures 0°C and 30°C. Repeat the answer for carbon dioxide and argon gases.

16.14 If the sound of thunder is heard 5 seconds after seeing the lightning and the weather is 20°C, how far away is the lightning taking place?

16.15 Estimate the speed of sound for steam directly from Eq. 16.28 and the steam tables for a state of 6 MPa, 400°C. Use table values at 5 and 7 MPa at the same entropy as the wanted state. Equation 16.28 is then done by finite difference. Find also the answer for the speed of sound assuming steam is an ideal gas.

16.16 A convergent-divergent nozzle has a throat diameter of 0.05 m and an exit diameter of 0.1 m. The inlet stagnation state is 500 kPa, 500 K. Find the back pressure that will lead to the maximum possible flow rate and the mass flow rate for three different gases: air, hydrogen, or carbon dioxide.

16.17 Air is expanded in a nozzle from 2 MPa, 600 K, to 200 kPa. The mass flow rate through the nozzle is 5 kg/s. Assume the flow is reversible and adiabatic and determine the throat and exit areas for the nozzle.

16.18 Consider the nozzle of Problem 16.17 and determine what back pressure will cause a normal shock to stand in the exit plane of the nozzle. This is case g in Fig. 16.17. What is the mass flow rate under these conditions?

16.19 At what Mach number will the normal shock occur in the nozzle of Problem 16.17 if the back pressure is 1.4 MPa? (Trial and error on M_x.)

16.20 Consider the nozzle of Problem 16.17. What back pressure will be required to cause subsonic flow throughout the entire nozzle with $M = 1$ at the throat?

16.21 Determine the mass flow rate through the nozzle of Problem 16.17 for a back pressure of 1.9 MPa.

16.22 At what Mach number will the normal shock occur in the nozzle of Problem 16.16 flowing with air if the back pressure is halfway between the pressures at c and d in Fig. 16.17?

16.23 A convergent nozzle has minimum area of 0.1 m² and receives air at 175 kPa, 1000 K flowing with 100 m/s. What is the back pressure that will produce the maximum flow rate? Find that flow rate.

16.24 A convergent-divergent nozzle has a throat area of 100 mm^2 and an exit area of 175 mm^2. The inlet flow is helium at a total pressure of 1 MPa, stagnation temperature of 375 K. What is the back pressure that will give sonic condition at the throat, but subsonic everywhere else?

16.25 A nozzle is designed assuming reversible adiabatic flow with an exit Mach number of 2.6 while flowing air with a stagnation pressure and temperature of 2 MPa and 150°C, respectively. The mass flow rate is 5 kg/s, and k may be assumed to be 1.40 and constant.

    a. Determine the exit pressure, temperature, and area, and the throat area.

    b. Suppose that the back pressure at the nozzle exit is raised to 1.4 MPa, and that the flow remains isentropic except for a normal shock wave. Determine the exit Mach number and temperature, and the mass flow rate through the nozzle.

16.26 A jet plane travels through the air with a speed of 1000 km/h at an altitude of 6 km, where the pressure is 40 kPa and the temperature is −12°C. Consider the inlet diffuser of the engine where air leaves with a velocity of 100 m/s. Determine the pressure and temperature leaving the diffuser, and the ratio of inlet to exit area of the diffuser, assuming the flow to be reversible and adiabatic.

16.27 A 1-m^3 insulated tank contains air at 1 MPa, 560 K. The tank is now discharged through a small convergent nozzle to the atmosphere at 100 kPa. The nozzle has an exit area of 2×10^{-5} m^2.

    a. Find the initial mass flow rate out of the tank.

    b. Find the mass flow rate when half the mass has been discharged.

    c. Find the mass of air in the tank and the mass flow rate out of the tank when the nozzle flow changes to become subsonic.

16.28 A 1-m^3 uninsulated tank contains air at 1 MPa, 560 K. The tank is now discharged through a small convergent nozzle to the atmosphere at 100 kPa while heat transfer from some source keeps the air temperature in the tank at 560 K. The nozzle has an exit area of 2×10^{-5} m^2.

    a. Find the initial mass flow rate out of the tank.

    b. Find the mass flow rate when half the mass has been discharged.

    c. Find the mass of air in the tank and the mass flow rate out of the tank when the nozzle flow changes to become subsonic.

16.29 The products of combustion enter a convergent nozzle of a jet engine at a total pressure of 125 kPa, and a total temperature of 650°C. The atmospheric pressure is 45 kPa, and the flow is adiabatic with a rate of 25 kg/s. Determine the exit area of the nozzle.

16.30 Air is expanded in a nozzle from 700 kPa, 200°C, to 150 kPa in a nozzle having an efficiency of 90%. The mass flow rate is 4 kg/s. Determine the exit area of the nozzle, the exit velocity, and the increase of entropy per kilogram of air. Compare these results with those of a reversible adiabatic nozzle.

16.31 Repeat Problem 16.26, assuming a diffuser efficiency of 80%.

16.32 Consider the diffuser of a supersonic aircraft flying at $M = 1.4$ at such an altitude that the temperature is −20°C, and the atmospheric pressure is 50 kPa. Consider two possible ways in which the diffuser might operate, and for each case calculate the throat area required for a flow of 50 kg/s.

a. The diffuser operates as reversible adiabatic with subsonic exit velocity.

b. A normal shock stands at the entrance to the diffuser. Except for the normal shock the flow is reversible and adiabatic, and the exit velocity is subsonic. This is shown in Fig. P16.32. Assume a convergent-divergent diffuser with $M = 1$ at the throat.

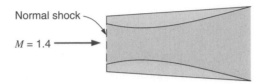

FIGURE P16.32

16.33 Air enters a diffuser with a velocity of 200 m/s, a static pressure of 70 kPa, and a temperature of −6°C. The velocity leaving the diffuser is 60 m/s, and the static pressure at the diffuser exit is 80 kPa. Determine the static temperature at the diffuser exit and the diffuser efficiency. Compare the stagnation pressures at the inlet and the exit.

16.34 Steam at a pressure of 1 MPa and temperature of 400°C expands in a nozzle to a pressure of 200 kPa. The nozzle efficiency is 90%, and the mass flow rate is 10 kg/s. Determine the nozzle exit area and the exit velocity.

16.35 Steam at 800 kPa, 350°C flows through a convergent-divergent nozzle that has a throat area of 350 mm². The pressure at the exit plane is 150 kPa, and the exit velocity is 800 m/s. The flow from the nozzle entrance to the throat is reversible and adiabatic. Determine the exit area of the nozzle, the overall nozzle efficiency, and the entropy generation in the process.

16.36 Air at 150 kPa, 290 K expands to the atmosphere at 100 kPa through a convergent nozzle with exit area of 0.01 m². Assume an ideal nozzle. What is the percent error in mass flow rate if the flow is assumed incompressible?

16.37 A sharp-edged orifice is used to measure the flow of air in a pipe. The pipe diameter is 100 mm, and the diameter of the orifice is 25 mm. Upstream of the orifice, the absolute pressure is 150 kPa, and the temperature is 35°C. The pressure drop across the orifice is 15 kPa, and the coefficient of discharge is 0.62. Determine the mass flow rate in the pipeline.

16.38 A critical nozzle is used for the accurate measurement of the flow rate of air. Exhaust from a car engine is diluted with air so its temperature is 50°C at a total pressure of 100 kPa. It flows through the nozzle with throat area of 700 mm² by suction from a blower. Find the needed suction pressure that will lead to critical flow in the nozzle, the mass flow rate and the blower work, assuming the blower exit is at atmospheric pressure 100 kPa.

16.39 A convergent nozzle is used to measure the flow of air to an engine. The atmosphere is at 100 kPa, 25°C. The nozzle used has a minimum area of 2000 mm², and the coefficient of discharge is 0.95. A pressure difference across the nozzle is measured to be 2.5 kPa. Find the mass flow rate assuming incompressible flow. Also find the mass flow rate assuming compressible adiabatic flow.

16.40 A convergent nozzle with exit diameter of 2 cm has an air inlet flow of 20°C, 101 kPa (stagnation conditions). The nozzle has an isentropic efficiency of 95%, and the pressure drop is measured to be 50 cm water column. Find the mass flow rate, assuming compressible adiabatic flow. Repeat this calculation for incompressible flow.

16.41 Steam at 600 kPa, 300°C is fed to a set of convergent nozzles in a steam turbine. The total nozzle exit area is 0.005 m², and they have a discharge coefficient of 0.94. The mass flow rate should be estimated from the measurement of the pressure drop across the nozzles, which is measured to be 200 kPa. Determine the mass flow rate.

16.42 The coefficient of discharge of a sharp-edged orifice is determined at one set of conditions by use of an accurately calibrated gasometer. The orifice has a diameter of 20 mm, and the pipe diameter is 50 mm. The absolute upstream pressure is 200 kPa, and the pressure drop across the orifice is 82 mm of mercury. The temperature of the air entering the orifice is 25°C, and the mass flow rate measured with the gasometer is 2.4 kg/min. What is the coefficient of discharge of the orifice at these conditions?

16.43 (Adv.) Atmospheric air is at 20°C, 100 kPa with zero velocity. An adiabatic reversible compressor takes atmospheric air in through a pipe with cross-sectional area of 0.1 m² at a rate of 1 kg/s. It is compressed up to a measured stagnation pressure of 500 kPa and leaves through a pipe with cross-sectional area of 0.01 m². What is the required compressor work and the air velocity, static pressure and temperature in the exit pipeline?

ENGLISH UNIT PROBLEMS

16.44E Steam leaves a nozzle with a velocity of 800 ft/s. The stagnation pressure is 100 lbf/in.², and the stagnation temperature is 500 F. What is the static pressure and temperature?

16.45E Air leaves the compressor of a jet engine at a temperature of 300 F, a pressure of 45 lbf/in.², and a velocity of 400 ft/s. Determine the isentropic stagnation temperature and pressure.

16.46E A meteorite melts and burns up at temperatures of 5500 R. If it hits air at 0.75 lbf/in.², 90 R, how high a velocity should it have to reach such a temperature?

16.47E A jet engine receives a flow of 500 ft/s air at 10 lbf/in.², 40 F, inlet area of 7 ft² with an exit at 1500 ft/s, 10 lbf/in.², 1100 R. Find the mass flow rate and thrust.

16.48E A water turbine using nozzles is located at the bottom of Hoover Dam 575 ft below the surface of Lake Mead. The water enters the nozzles at a stagnation pressure corresponding to the column of water above it minus 20% due to friction. The temperature is 60 F, and the water leaves at standard atmospheric pressure. If the flow through the nozzle is reversible and adiabatic, determine the velocity and kinetic energy per kilogram of water leaving the nozzle.

16.49E Find the speed of sound for air at 15 lbf/in.², at the two temperatures of 32 F and 90 F. Repeat the answer for carbon dioxide and argon gases.

16.50E Air is expanded in a nozzle from 300 lbf/in.^{2}, 1100 R to 30 lbf/in.^{2}. The mass flow rate through the nozzle is 10 lbm/s. Assume the flow is reversible and adiabatic and determine the throat and exit areas for the nozzle.

16.51E A convergent nozzle has a minimum area of 1 ft^2 and receives air at 25 lbf/in.^{2}, 1800 R flowing with 330 ft/s. What is the back pressure that will produce the maximum flow rate? Find that flow rate.

16.52E A jet plane travels through the air with a speed of 600 mi/h at an altitude of 20 000 ft, where the pressure is 5.75 lbf/in.^{2} and the temperature is 25 F. Consider the diffuser of the engine where air leaves with a velocity of 300 ft/s. Determine the pressure and temperature leaving the diffuser, and the ratio of inlet to exit area of the diffuser, assuming the flow to be reversible and adiabatic.

16.53E The products of combustion enter a nozzle of a jet engine at a total pressure of 18 lbf/in.^{2}, and a total temperature of 1200 F. The atmospheric pressure is 6.75 lbf/in.^{2}. The nozzle is convergent, and the mass flow rate is 50 lbm/s. Assume the flow is adiabatic. Determine the exit area of the nozzle.

16.54E Repeat Problem 16.52, assuming a diffuser efficiency of 80%.

16.55E A 50-ft^3 uninsulated tank contains air at 150 lbf/in.^{2}, 1000 R. The tank is now discharged through a small convergent nozzle to the atmosphere at 14.7 lbf/in.^{2} while heat transfer from some source keeps the air temperature in the tank at 1000 R. The nozzle has an exit area of 2×10^{-4} ft^2.

a. Find the initial mass flow rate out of the tank.

b. Find the mass flow rate when half the mass has been discharged.

c. Find the mass of air in the tank and the mass flow rate out of the tank when the nozzle flow changes to become subsonic.

16.56E Air enters a diffuser with a velocity of 600 ft/s, a static pressure of 10 lbf/in.^{2}, and a temperature of 20 F. The velocity leaving the diffuser is 200 ft/s, and the static pressure at the diffuser exit is 11.7 lbf/in.^{2}. Determine the static temperature at the diffuser exit and the diffuser efficiency. Compare the stagnation pressures at the inlet and the exit.

16.57E A convergent nozzle with exit diameter of 1 in. has an air inlet flow of 68 F, 14.7 lbf/in.^{2} (stagnation conditions). The nozzle has an isentropic efficiency of 95%, and the pressure drop is measured to be 20 in. water column. Find the mass flow rate assuming compressible adiabatic flow. Repeat this calculation for incompressible flow.

COMPUTER, DESIGN, AND OPEN-ENDED PROBLEMS

16.58 Make a program that calculates the stagnation pressure and temperature from a static pressure, temperature, and velocity. Assume the fluid is air with constant specific heats. If the inverse relation is sought, one of the three properties in the flow must be given. Include that case also.

16.59 Use the menu-driven software to solve Problem 16.35. Find from the menu-driven steam tables the ratio of specific heats at the inlet and the speed of sound from its definition in Eq. 16.28.

16.60 (Adv.) Make a program that will track the process in time as described in Problems 16.27 and 16.28. Investigate the time it takes to bring the tank pressure to 125 kPa as a function of the size of the nozzle exit area. Plot several of the key variables as functions of time.

16.61 A pump can deliver liquid water at an exit pressure of 400 kPa using 0.5 kW of power. Assume the inlet is water at 100 kPa, 15°C and that the pipe size is the same for the inlet and exit. Design a nozzle to be mounted on the exit line so the water exit velocity is at least 20 m/s. Show the exit velocity and mass flow rate as a function of the nozzle exit area with the same power to the pump.

16.62 In all the problems in the text the efficiency of a pump or compressor has been given as a constant. In reality it is a function of the mass flow rate and the fluid state through the device. Examine the literature for the characteristics of a real air compressor (blower).

16.63 The throttle plate in a carburetor gives a severe restriction to the air flow where at idle it is critical flow. For normal atmospheric conditions estimate then the inlet temperature and pressure to the cylinder of the engine.

16.64 For an experiment in the laboratory an air flow rate should be measured. The range should be from 0.05 to 0.10 kg/s, and the flow should be delivered to the experiment at 110 kPa. Size one (or two in parallel) convergent nozzle(s) that sits in a plate. The air is drawn through the nozzle(s) by suction of a blower that delivers the air at 110 kPa. What should be measured and what accuracy can be expected?

16.65 An afterburner in a jet engine adds fuel that is burned after the turbine but before the exit nozzle that accelerates the gases. Examine the effect on a nozzle exit velocity of having a higher inlet temperature, but same pressure as without the afterburner. Are these nozzles operating with subsonic or supersonic flow?

CONTENTS OF APPENDIX

TABLE A.1 *Conversion Factors*

Area

$1 \text{ mm}^2 = 1.0 \times 10^{-6} \text{ m}^2$ $1 \text{ ft}^2 = 144 \text{ in.}^2$

$1 \text{ cm}^2 = 1.0 \times 10^{-4} \text{ m}^2 = 0.1550 \text{ in.}^2$ $1 \text{ in.}^2 = 6.4516 \text{ cm}^2 = 6.4516 \times 10^{-4} \text{ m}^2$

$1 \text{ m}^2 = 10.7639 \text{ ft}^2$ $1 \text{ ft}^2 = 0.092\,903 \text{ m}^2$

Conductivity

$1 \text{ W/m-K} = 1 \text{ J/s-m-K}$

        $= 0.577\,789 \text{ Btu/h-ft-R}$ $1 \text{ Btu/h-ft-R} = 1.730\,735 \text{ W/m-K}$

Density

$1 \text{ kg/m}^3 = 0.06242797 \text{ lbm/ft}^3$ $1 \text{ lbm/ft}^3 = 16.018\,46 \text{ kg/m}^3$

$1 \text{ g/cm}^3 = 1000 \text{ kg/m}^3$

$1 \text{ g/cm}^3 = 1 \text{ kg/L}$

Energy

$1 \text{ J} \qquad = 1 \text{ N-m} = 1 \text{ kg-m}^2/\text{s}^2$

$1 \text{ J} \qquad = 0.737\,562 \text{ lbf-ft}$ $1 \text{ lbf-ft} = 1.355\,818 \text{ J}$

$1 \text{ cal (Int.)} = 4.1868 \text{ J}$ $= 1.28507 \times 10^{-3} \text{ Btu}$

                                   $1 \text{ Btu (Int.)} = 1.055\,056 \text{ kJ}$

$1 \text{ erg} \qquad = 1.0 \times 10^{-7} \text{ J}$ $= 778.1693 \text{ lbf-ft}$

$1 \text{ eV} \qquad = 1.602\,177\,33 \times 10^{-19} \text{ J}$

Force

$1 \text{ N} = 0.224809 \text{ lbf}$ $1 \text{ lbf} = 4.448\,222 \text{ N}$

$1 \text{ kp} = 9.80665 \text{ N} \; (1 \text{ kgf})$

Gravitation

$g = 9.80665 \text{ m/s}^2$ $g = 32.17405 \text{ ft/s}^2$

Heat capacity, specific entropy

$1 \text{ kJ/kg-K} = 0.238\,846 \text{ Btu/lbm-R}$ $1 \text{ Btu/lbm-R} = 4.1868 \text{ kJ/kg-K}$

Heat flux (per unit area)

$1 \text{ W/m}^2 = 0.316\,998 \text{ Btu/h-ft}^2$ $1 \text{ Btu/h-ft}^2 = 3.15459 \text{ W/m}^2$

Heat transfer coefficient

$1 \text{ W/m}^2\text{-K} = 0.176\,11 \text{ Btu/h-ft}^2\text{-R}$ $1 \text{ Btu/h-ft}^2\text{-R} = 5.67826 \text{ W/m}^2\text{-K}$

Length

$1 \text{ mm} = 0.001 \text{ m} = 0.1 \text{ cm}$ $1 \text{ ft} = 12 \text{ in.}$

$1 \text{ cm} = 0.01 \text{ m} = 10 \text{ mm} = 0.3970 \text{ in.}$ $1 \text{ in.} = 2.54 \text{ cm} = 0.0254 \text{ m}$

$1 \text{ m} = 3.28084 \text{ ft} = 39.370 \text{ in.}$ $1 \text{ ft} = 0.3048 \text{ m}$

$1 \text{ km} = 0.621\,371 \text{ mi}$ $1 \text{ mi} = 1.609344 \text{ km}$

$1 \text{ mi} = 1609.3 \text{ m (US statute)}$ $1 \text{ yd} = 0.9144 \text{ m}$

TABLE A.1 (Continued) *Conversion Factors*

Mass

$\quad$ 1 kg $\quad$ = 2.204 623 lbm $\qquad\qquad$ 1 lbm = 0.453 592 kg

$\quad$ 1 tonne = 1000 kg $\qquad\qquad\qquad$ 1 slug = 14.5939 kg

$\quad$ 1 grain = 6.47989×10^{-5} kg $\qquad$ 1 ton $\;$ = 2000 lbm

Moment (torque)

$\quad$ 1 N-m = 0.737 562 lbf-ft $\qquad\qquad$ 1 lbf-ft = 1.355 818 N-m

Momentum (mV)

$\quad$ 1 kg-m/s = 7.232 94 lbm-ft/s $\qquad\;$ 1 lbm-ft/s = 0.138 256 kg-m/s

$\qquad\qquad$ = 0.224809 lbf-s

Power

1 W	= 1 J/s = 1 N-m/s	1 lbf-ft/s	= 1.355 818 W
	= 0.737 562 lbf-ft/s		= 4.626 24 Btu/h
1 kW	= 3412.14 Btu/h	1 Btu/s	= 1.055 056 kW
1 hp (metric)	= 0.735 499 kW	1 hp (UK)	= 0.7457 kW
			= 550 lbf-ft/s
			= 2544.43 Btu/h

$\quad$ 1 ton of $\qquad\qquad\qquad\qquad\qquad$ 1 ton of

$\quad$ refrigeration $\;$ = 3.516 85 kW $\qquad$ refrigeration $\;$ = 12 000 Btu/h

Pressure

1 Pa	= 1 N/m^2 = 1 kg/m-s^2	1 lbf/in.^{2}	= 6.894 757 kPa
1 bar	= 1.0×10^5 Pa = 100 kPa		
1 atm	= 101.325 kPa	1 atm	= 14.695 94 lbf/in.^{2}
	= 1.01325 bar		= 29.921 in. Hg [32 F]
	= 760 mm Hg [0°C]		= 33.899 5 ft H_2O [4°C]
	= 10.332 56 m H_2O [4°C]		
1 torr	= 1 mm Hg [0°C]		
1 mm Hg [0°C] = 0.133 322 kPa		1 in. Hg [0°C] $\;$ = 0.49115 lbf/in.^{2}	
1 m H_2O [4°C] $\;$ = 9.806 38 kPa		1 in. H_2O [4°C] = 0.036126 lbf/in.^{2}	

Specific energy

$\quad$ 1 kJ/kg $\;$ = 0.42992 Btu/lbm $\qquad\;$ 1 Btu/lbm $\;$ = 2.326 kJ/kg

$\qquad\qquad$ = 334.55 lbf-ft/lbm $\qquad\;$ 1 lbf-ft/lbm = 2.98907×10^{-3} kJ/kg

$\qquad\qquad\qquad\qquad\qquad\qquad\qquad\qquad$ = 1.28507×10^{-3} Btu/lbm

---

TABLE A.1 (Continued) *Conversion Factors*

Specific kinetic energy (V^2)

$1 \ m^2/s^2 = 0.001 \ kJ/kg$ $1 \ ft^2/s^2 \quad = 3.9941 \times 10^{-5} \ Btu/lbm$

$1 \ kJ/kg = 1000 \ m^2/s^2$ $1 \ Btu/lbm = 25037 \ ft^2/s^2$

Specific potential energy (Zg)

$1 \ m\text{-}g_{std} = 9.80665 \times 10^{-3} \ kJ/kg$ $1 \ ft\text{-}g_{std} = 1.0 \ lbf\text{-}ft/lbm$

$\qquad\quad = 4.21607 \times 10^{-3} \ Btu/lbm$ $\qquad\quad = 0.001285 \ Btu/lbm$

$\qquad\qquad\qquad\qquad\qquad\qquad\qquad\qquad\quad = 0.002989 \ kJ/kg$

Specific volume

$1 \ cm^3/g = 0.001 \ m^3/kg$

$1 \ cm^3/g = 1 \ L/kg$

$1 \ m^3/kg = 16.018 \ 46 \ ft^3/lbm$ $1 \ ft^3/lbm = 0.062 \ 428 \ m^3/kg$

Temperature

$1 \ K = 1 \ °C = 1.8 \ R = 1.8 \ F$ $1 \ R = (5/9) \ K$

$TC = TK - 273.15$ $TF = TR - 459.67$

$\quad = (TF - 32)/1.8$ $\quad = 1.8 \ TC + 32$

$TK = TR/1.8$ $TR = 1.8 \ TK$

Universal Gas Constant

$R = N_0 \ k = 8.31451 \ kJ/kmol\text{-}K$ $R = 1.98589 \ Btu/lbmol\text{-}R$

$\qquad\quad = 1.98589 \ kcal/kmol\text{-}K$ $\quad = 1545.36 \ lbf\text{-}ft/lbmol\text{-}R$

$\qquad\quad = 82.0578 \ atm\text{-}L/kmol\text{-}K$ $\quad = 0.73024 \ atm\text{-}ft^3/lbmol\text{-}R$

$\qquad\qquad\qquad\qquad\qquad\qquad\qquad\quad = 10.7317 \ (lbf/in.^2)\text{-}ft^3/lbmol\text{-}R$

Velocity

$1 \ m/s \quad = 3.6 \ km/h$ $1 \ ft/s \ = 0.681818 \ mi/h$

$\qquad\quad = 3.28084 \ ft/s$ $\qquad = 0.3048 \ m/s$

$\qquad\quad = 2.23694 \ mi/h$ $\qquad = 1.09728 \ km/h$

$1 \ km/h \quad = 0.27778 \ m/s$ $1 \ mi/h = 1.46667 \ ft/s$

$\qquad\quad = 0.91134 \ ft/s$ $\qquad = 0.44704 \ m/s$

$\qquad\quad = 0.62137 \ mi/h$ $\qquad = 1.609344 \ km/h$

Volume

$1 \ m^3 \qquad = 35.3147 \ ft^3$ $1 \ ft^3 \qquad = 2.831 \ 685 \times 10^{-2} \ m^3$

$1 \ L \qquad = 1 \ dm^3 = 0.001 \ m^3$ $1 \ in.^3 \qquad = 1.6387 \times 10^{-5} \ m^3$

$1 \ Gal \ (US) = 3.785 \ 412 \ L$ $1 \ Gal \ (UK) = 4.546 \ 090 \ L$

$\qquad\qquad = 3.785 \ 412 \times 10^{-3} \ m^3$ $1 \ Gal \ (US) = 231.00 \ in.^3$

TABLE A.2 *Critical Constants (SI Units)*

Substance	Formula	Molec. Mass	Temp. K	Press. MPa	Vol. m³/kmol
Ammonia	NH_3	17.031	405.5	11.35	0.0725
Argon	Ar	39.948	150.8	4.87	0.0749
Bromine	Br_2	159.808	588	10.30	0.1272
Carbon dioxide	CO_2	44.01	304.1	7.38	0.0939
Carbon monoxide	CO	28.01	132.9	3.50	0.0932
Chlorine	Cl_2	70.906	416.9	7.98	0.1238
Fluorine	F_2	37.997	144.3	5.22	0.0663
Helium	He	4.003	5.19	0.227	0.0574
Hydrogen (normal)	H_2	2.016	33.2	1.30	0.0651
Krypton	Kr	83.80	209.4	5.50	0.0912
Neon	Ne	20.183	44.4	2.76	0.0416
Nitric oxide	NO	30.006	180	6.48	0.0577
Nitrogen	N_2	28.013	126.2	3.39	0.0898
Nitrogen dioxide	NO_2	46.006	431	10.1	0.1678
Nitrous oxide	N_2O	44.013	309.6	7.24	0.0974
Oxygen	O_2	31.999	154.6	5.04	0.0734
Sulfur dioxide	SO_2	64.063	430.8	7.88	0.1222
Water	H_2O	18.015	647.3	22.12	0.0571
Xenon	Xe	131.30	289.7	5.84	0.1184
Acetylene	C_2H_2	26.038	308.3	6.14	0.1127
Benzene	C_6H_6	78.114	562.2	4.89	0.2590
n-Butane	C_4H_{10}	58.124	425.2	3.80	0.2550
Chlorodifluoroethane (142b)	CH_3CCLF_2	100.495	410.3	4.25	0.2310
Chlorodifluoromethane (22)	$CHCLF_2$	86.469	369.3	4.97	0.1656
Dichlorodifluoromethane (12)	CCL_2F_2	120.914	385.0	4.14	0.2167
Dichlorofluoroethane (141)	CH_3CCL_2F	116.95	481.5	4.54	0.2520
Dichlorofluoromethane (21)	$CHCL_2F$	102.923	451.6	5.18	0.1964
Dichlorotrifluoroethane (123)	$CHCL_2CF_3$	152.93	456.9	3.66	0.2781
Difluoroethane (152a)	CHF_2CH_3	66.05	386.4	4.52	0.1795
Ethane	C_2H_6	30.070	305.4	4.88	0.1483
Ethyl alcohol	C_2H_5OH	46.069	513.9	6.14	0.1671
Ethylene	C_2H_4	28.054	282.4	5.04	0.1304
n-Heptane	C_7H_{16}	100.205	540.3	2.74	0.4320
n-Hexane	C_6H_{14}	86.178	507.5	3.01	0.3700
Methane	CH_4	16.043	190.4	4.60	0.0992
Methyl alcohol	CH_3OH	32.042	512.6	8.09	0.1180
n-Octane	C_8H_{18}	114.232	568.8	2.49	0.4920
n-Pentane	C_5H_{12}	72.151	469.7	3.37	0.3040
Propane	C_3H_8	44.094	369.8	4.25	0.2030
Propene	C_3H_6	42.081	364.9	4.60	0.1810
Tetrafluoroethane (134a)	CF_3CH_2F	102.03	374.2	4.06	0.2008

TABLE A.3 *Properties of Selected Solids at 25° C*

Substance	ρ kg/m³	C_p kJ/kg-K
Asphalt	2120	0.92
Brick, common	1800	0.84
Carbon, diamond	3250	0.51
Carbon, graphite	2000-2500	0.61
Coal	1200-1500	1.26
Concrete	2200	0.88
Glass, plate	2500	0.80
Glass, wool	200	0.66
Granite	2750	0.89
Ice (0 C)	917	2.04
Paper	700	1.2
Plexiglas	1180	1.44
Polystyrene	920	2.3
Polyvinyl chloride	1380	0.96
Rubber, soft	1100	1.67
Salt, rock	2100-2500	0.92
Sand, dry	1500	0.8
Silicon	2330	0.70
Snow, firm	560	2.1
Wood, hard (oak)	720	1.26
Wood, soft (pine)	510	1.38
Wool	100	1.72
Metals		
Aluminum	2700	0.90
Copper, commercial	8300	0.42
Brass, 60-40	8400	0.38
Gold	19300	0.13
Iron, cast	7272	0.42
Iron. 304 St Steel	7820	0.46
Lead	11340	0.13
Magnesium, 2% Mn	1778	1.00
Nickel, 10% Cr	8666	0.44
Silver, 99.9% Ag	10524	0.24
Sodium	971	1.21
Tin	7304	0.22
Tungsten	19300	0.13
Zinc	7144	0.39

TABLE A.4 *Properties of Some Liquids at 25° C*

Substance	ρ kg/m³	C_p kJ/kg-K
Ammonia	604	4.84
Benzene	879	1.72
Butane	556	2.47
CCL_4	1584	0.83
CO_2	680	2.9
Ethanol	783	2.46
Gasoline	750	2.08
Glycerine	1260	2.42
Kerosene	815	2.0
Methanol	787	2.55
n-octane	692	2.23
Oil engine	885	1.9
Oil light	910	1.8
Propane	510	2.54
R-12	1310	0.97
R-22	1190	1.26
R-134a	1206	1.43
Water	997	4.18
Liquid metals		
Bismuth, Bi	10040	0.14
Lead, Pb	10660	0.16
Mercury, Hg	13580	0.14
Potassium, K	828	0.81
Sodium, Na	929	1.38
Tin, Sn	6950	0.24
Zinc, Zn	6570	0.50
NaK (56/44)	887	1.13

TABLE A.5 *Properties of Various Ideal Gases at 25° C, 100 kPa* (SI Units)*

Gas	Chemical Formula	Molecular Mass	R kJ/kg-K	ρ kg/m^3	C_{po} kJ/kg-K	C_{vo} kJ/kg-K	k
Steam	H_2O	18.015	0.4615	0.0231	1.872	1.410	1.327
Acetylene	C_2H_2	26.038	0.3193	1.05	1.699	1.380	1.231
Air	--	28.97	0.287	1.169	1.004	0.717	1.400
Ammonia	NH_3	17.031	0.4882	0.694	2.130	1.642	1.297
Argon	Ar	39.948	0.2081	1.613	0.520	0.312	1.667
Butane	C_4H_{10}	58.124	0.1430	2.407	1.716	1.573	1.091
Carbon monoxide	CO	28.01	0.2968	1.13	1.041	0.744	1.399
Carbon dioxide	CO_2	44.01	0.1889	1.775	0.842	0.653	1.289
Ethane	C_2H_6	30.07	0.2765	1.222	1.766	1.490	1.186
Ethanol	C_2H_5OH	46.069	0.1805	1.883	1.427	1.246	1.145
Ethylene	C_2H_4	28.054	0.2964	1.138	1.548	1.252	1.237
Helium	He	4.003	2.0771	0.1615	5.193	3.116	1.667
Hydrogen	H_2	2.016	4.1243	0.0813	14.209	1.008	1.409
Methane	CH_4	16.043	0.5183	0.648	2.254	1.736	1.299
Methanol	CH_3OH	32.042	0.2595	1.31	1.405	1.146	1.227
Neon	Ne	20.183	0.4120	0.814	1.03	0.618	1.667
Nitric oxide	NO	30.006	0.2771	1.21	0.993	0.716	1.387
Nitrogen	N_2	28.013	0.2968	1.13	1.042	0.745	1.400
Nitrous oxide	N_2O	44.013	0.1889	1.775	0.879	0.690	1.274
n-octane	C_8H_{18}	114.23	0.07279	0.092	1.711	1.638	1.044
Oxygen	O_2	31.999	0.2598	1.292	0.922	0.662	1.393
Propane	C_3H_8	44.094	0.1886	1.808	1.679	1.490	1.126
R-12	CCL_2F_2	120.914	0.06876	4.98	0.616	0.547	1.126
R-22	$CHCLF_2$	86.469	0.09616	3.54	0.658	0.562	1.171
R-134a	CF_3CH_2F	102.03	0.08149	4.20	0.852	0.771	1.106
Sulfur dioxide	SO_2	64.059	0.1298	2.618	0.624	0.494	1.263
Sulfur trioxide	SO_3	80.053	0.10386	3.272	0.635	0.531	1.196

*Or saturation pressure if it is less than 100 kPa.

TABLE A.6 *Constant-Pressure Specific Heats of Various Ideal Gases (SI Units)*

$$C_{p0} = kJ/kmol\ K \qquad \theta = T(Kelvin)/100$$

Gas		Range K	Max Error %
N_2	$\overline{C}_{p0} = 39.060 - 512.79\,\theta^{-1.5} + 1072.7\,\theta^{-2} - 820.40\,\theta^{-3}$	300–3500	0.43
O_2	$\overline{C}_{p0} = 37.432 + 0.020\,102\,\theta^{1.5} - 178.57\,\theta^{-1.5} + 236.88\,\theta^{-2}$	300–3500	0.30
H_2	$\overline{C}_{p0} = 56.505 - 702.74\,\theta^{-0.75} + 1165.0\,\theta^{-1} - 560.70\,\theta^{-1.5}$	300–3500	0.60
CO	$\overline{C}_{p0} = 69.145 - 0.704\,63\,\theta^{0.75} - 200.77\,\theta^{-0.5} + 176.76\,\theta^{-0.75}$	300–3500	0.42
OH	$\overline{C}_{p0} = 81.546 - 59.350\,\theta^{0.25} + 17.329\,\theta^{0.75} - 4.2660\,\theta$	300–3500	0.43
NO	$\overline{C}_{p0} = 59.283 - 1.7096\,\theta^{0.5} - 70.613\,\theta^{-0.5} + 74.889\,\theta^{-1.5}$	300–3500	0.34
H_2O	$\overline{C}_{p0} = 143.05 - 183.54\,\theta^{0.25} + 82.751\,\theta^{0.5} - 3.6989\,\theta$	300–3500	0.43
CO_2	$\overline{C}_{p0} = -3.7357 + 30.529\,\theta^{0.5} - 4.1034\,\theta + 0.024\,198\,\theta^2$	300–3500	0.19
NO_2	$\overline{C}_{p0} = 46.045 + 216.10\,\theta^{-0.5} - 363.66\,\theta^{-0.75} + 232.550\,\theta^{-2}$	300–3500	0.26
CH_4	$\overline{C}_{p0} = -672.87 + 439.74\,\theta^{0.25} - 24.875\,\theta^{0.75} + 323.88\,\theta^{-0.5}$	300–2000	0.15
C_2H_4	$\overline{C}_{p0} = -95.395 + 123.15\,\theta^{0.5} - 35.641\,\theta^{0.75} + 182.77\,\theta^{-3}$	300–2000	0.07
C_2H_6	$\overline{C}_{p0} = 6.895 + 17.26\,\theta - 0.6402\,\theta^2 + 0.007\,28\,\theta^3$	300–1500	0.83
C_3H_8	$\overline{C}_{p0} = -4.042 + 30.46\,\theta - 1.571\,\theta^2 + 0.031\,71\,\theta^3$	300–1500	0.40

Source: From T.C. Scott and R.E. Sonntag. University of Michigan, unpublished 1971, except C_2H_6, C_3H_8, and C_4H_{10} from K.A. Kobe, *Petroleum Refiner,* 28, No. 2, 113 (1949).

TABLE A.7 *Ideal-Gas Properties of Air, Standard Entropy at 0.1 MPa (1 bar) Pressure*

T K	u kJ/kg	h kJ/kg	s^o kJ/kg K	P_r	v_r
200	142.768	200.174	6.46260	0.27027	493.466
220	157.071	220.218	6.55812	0.37700	389.150
240	171.379	240.267	6.64535	0.51088	313.274
260	185.695	260.323	6.72562	0.67573	256.584
280	200.022	280.390	6.79998	0.87556	213.257
290	207.191	290.430	6.83521	0.98990	195.361
298.15	213.036	298.615	6.86305	1.09071	182.288
300	214.364	300.473	6.86926	1.11458	179.491
320	228.726	320.576	6.93413	1.39722	152.728
340	243.113	340.704	6.99515	1.72814	131.200
360	257.532	360.863	7.05276	2.11226	113.654
380	271.988	381.060	7.10735	2.55479	99.1882
400	286.487	401.299	7.15926	3.06119	87.1367
420	301.035	421.589	7.20875	3.63727	77.0025
440	315.640	441.934	7.25607	4.28916	68.4088
460	330.306	462.340	7.30142	5.02333	61.0658
480	345.039	482.814	7.34499	5.84663	54.7479
500	359.844	503.360	7.38692	6.76629	49.2777
520	374.726	523.982	7.42736	7.78997	44.5143
540	389.689	544.686	7.46642	8.92569	40.3444
560	404.736	565.474	7.50422	10.18197	36.6765
580	419.871	586.350	7.54084	11.56771	33.4358
600	435.097	607.316	7.57638	13.09232	30.5609
620	450.415	628.375	7.61090	14.76564	28.0008
640	465.828	649.528	7.64448	16.59801	25.7132
660	481.335	670.776	7.67717	18.60025	23.6623
680	496.939	692.120	7.70903	20.78367	21.8182
700	512.639	713.561	7.74010	23.16010	20.1553
720	528.435	735.098	7.77044	25.74188	18.6519
740	544.328	756.731	7.80008	28.54188	17.2894
760	560.316	778.460	7.82905	31.57347	16.0518
780	576.400	800.284	7.85740	34.85061	14.9250
800	592.577	822.202	7.88514	38.38777	13.8972
850	633.422	877.397	7.95207	48.46828	11.6948
900	674.824	933.152	8.01581	60.51977	9.91692
950	716.756	989.436	8.07667	74.81519	8.46770
1000	759.189	1046.221	8.13493	91.65077	7.27604
1050	802.095	1103.478	8.19081	111.3467	6.28845
1100	845.445	1161.180	8.24449	134.2478	5.46408

TABLE A.7 (Continued) *Ideal-Gas Properties of Air, Standard Entropy at 0.1 MPa (1 bar) Pressure*

T K	u kJ/kg	h kJ/kg	s^o kJ/kg K	P_r	v_r
1100	845.445	1161.180	8.24449	134.2478	5.46408
1150	889.211	1219.298	8.29616	160.7245	4.77141
1200	933.367	1277.805	8.34596	191.1736	4.18586
1250	977.888	1336.677	8.39402	226.0192	3.68804
1300	1022.751	1395.892	8.44046	265.7145	3.26257
1350	1067.936	1455.429	8.48539	310.7426	2.89711
1400	1113.426	1515.270	8.52891	361.6192	2.58171
1450	1159.202	1575.398	8.57111	418.8942	2.30831
1500	1205.253	1635.800	8.61208	483.1554	2.07031
1550	1251.547	1696.446	8.65185	554.9577	1.86253
1600	1298.079	1757.329	8.69051	634.9670	1.68035
1650	1344.834	1818.436	8.72811	723.8560	1.52007
1700	1391.801	1879.755	8.76472	822.3320	1.37858
1750	1438.970	1941.275	8.80039	931.1376	1.25330
1800	1486.331	2002.987	8.83516	1051.051	1.14204
1850	1533.873	2064.882	8.86908	1182.888	1.04294
1900	1581.591	2126.951	8.90219	1327.498	0.95445
1950	1629.474	2189.186	8.93452	1485.772	0.87521
2000	1677.518	2251.581	8.96611	1658.635	0.80410
2050	1725.714	2314.128	8.99699	1847.077	0.74012
2100	1774.057	2376.823	9.02721	2052.109	0.68242
2150	1822.541	2439.659	9.05678	2274.789	0.63027
2200	1871.161	2502.630	9.08573	2516.217	0.58305
2250	1919.912	2565.733	9.11409	2777.537	0.54020
2300	1968.790	2628.962	9.14189	3059.939	0.50124
2350	2017.789	2692.313	9.16913	3364.658	0.46576
2400	2066.907	2755.782	9.19586	3692.974	0.43338
2450	2116.138	2819.366	9.22208	4046.215	0.40378
2500	2165.480	2883.059	9.24781	4425.759	0.37669
2550	2214.929	2946.859	9.27308	4833.031	0.35185
2600	2264.481	3010.763	9.29790	5269.505	0.32903
2650	2314.133	3074.767	9.32228	5736.707	0.30805
2700	2363.883	3138.868	9.34625	6236.215	0.28872
2750	2413.727	3203.064	9.36980	6769.657	0.27089
2800	2463.663	3267.351	9.39297	7338.715	0.25443
2850	2513.687	3331.726	9.41576	7945.124	0.23921
2900	2563.797	3396.188	9.43818	8590.676	0.22511
2950	2613.990	3460.733	9.46025	9277.216	0.21205
3000	2664.265	3525.359	9.48198	10006.645	0.19992

TABLE A.8 *Ideal-Gas Properties of Various Substances (SI Units), Entropies at 0.1-MPa (1-bar) Pressure*

	Nitrogen, Diatomic (N_2) $\bar{h}^o_{f,\,298} = 0$ kJ/kmol $M = 28.013$		Nitrogen, Monatomic (N) $\bar{h}^o_{f,\,298} = 472\,680$ kJ/kmol $M = 14.007$	
T K	$(\bar{h}-\bar{h}^o_{298})$ kJ/kmol	$\bar{s}^o$ kJ/kmol K	$(\bar{h}-\bar{h}^o_{298})$ kJ/kmol	$\bar{s}^o$ kJ/kmol
0	−8670	0	−6197	0
100	−5768	159.812	−4119	130.593
200	−2857	179.985	−2040	145.001
298	0	191.609	0	153.300
300	54	191.789	38	153.429
400	2971	200.181	2117	159.409
500	5911	206.740	4196	164.047
600	8894	212.177	6274	167.837
700	11937	216.865	8353	171.041
800	15046	221.016	10431	173.816
900	18223	224.757	12510	176.265
1000	21463	228.171	14589	178.455
1100	24760	231.314	16667	180.436
1200	28109	234.227	18746	182.244
1300	31503	236.943	20825	183.908
1400	34936	239.487	22903	185.448
1500	38405	241.881	24982	186.883
1600	41904	244.139	27060	188.224
1700	45430	246.276	29139	189.484
1800	48979	248.304	31218	190.672
1900	52549	250.234	33296	191.796
2000	56137	252.075	35375	192.863
2200	63362	255.518	39534	194.845
2400	70640	258.684	43695	196.655
2600	77963	261.615	47860	198.322
2800	85323	264.342	52033	199.868
3000	92715	266.892	56218	201.311
3200	100134	269.286	60420	202.667
3400	107577	271.542	64646	203.948
3600	115042	273.675	68902	205.164
3800	122526	275.698	73194	206.325
4000	130027	277.622	77532	207.437
4400	145078	281.209	86367	209.542
4800	160188	284.495	95457	211.519
5200	175352	287.530	104843	213.397
5600	190572	290.349	114550	215.195
6000	205848	292.984	124590	216.926

TABLE A.8 (Continued) *Ideal-Gas Properties of Various Substances (SI Units), Entropies at 0.1-MPa (1-bar) Pressure*

	Oxygen, Diatomic (O_2) $\bar{h}_{f,298}^o = 0$ kJ/kmol $M = 31.999$		Oxygen, Monatomic (O) $\bar{h}_{f,298}^o = 249\ 170$ kJ/kmol $M = 16.00$	
T K	$(\bar{h}-\bar{h}_{298}^o)$ kJ/kmol	$\bar{s}^o$ kJ/kmol K	$(\bar{h}-\bar{h}_{298}^o)$ kJ/kmol	$\bar{s}^o$ kJ/kmol K
0	−8683	0	−6725	0
100	−5777	173.308	−4518	135.947
200	−2868	193.483	−2186	152.153
298	0	205.148	0	161.059
300	54	205.329	41	161.194
400	3027	213.873	2207	167.431
500	6086	220.693	4343	172.198
600	9245	226.450	6462	176.060
700	12499	231.465	8570	179.310
800	15836	235.920	10671	182.116
900	19241	239.931	12767	184.585
1000	22703	243.579	14860	186.790
1100	26212	246.923	16950	188.783
1200	29761	250.011	19039	190.600
1300	33345	252.878	21126	192.270
1400	36958	255.556	23212	193.816
1500	40600	258.068	25296	195.254
1600	44267	260.434	27381	196.599
1700	47959	262.673	29464	197.862
1800	51674	264.797	31547	199.053
1900	55414	266.819	33630	200.179
2000	59176	268.748	35713	201.247
2200	66770	272.366	39878	203.232
2400	74453	275.708	44045	205.045
2600	82225	278.818	48216	206.714
2800	90080	281.729	52391	208.262
3000	98013	284.466	56574	209.705
3200	106022	287.050	60767	211.058
3400	114101	289.499	64971	212.332
3600	122245	291.826	69190	213.538
3800	130447	294.043	73424	214.682
4000	138705	296.161	77675	215.773
4400	155374	300.133	86234	217.812
4800	172240	303.801	94873	219.691
5200	189312	307.217	103592	221.435
5600	206618	310.423	112391	223.066
6000	224210	313.457	121264	224.597

TABLE A.8 (Continued) *Ideal-Gas Properties of Various Substances (SI Units), Entropies at 0.1-MPa (1-bar) Pressure*

T	Carbon Dioxide (CO_2) $\bar{h}^o_{f,298}$ = -393 522 kJ/kmol M = 44.01		Carbon Monoxide (CO) $\bar{h}^o_{f,298}$ = -110 527 kJ/kmol M = 28.01	
K	$(\bar{h}-\bar{h}^o_{298})$ **kJ/kmol**	$\bar{s}^o$ **kJ/kmol K**	$(\bar{h}-\bar{h}^o_{298})$ **kJ/kmol**	$\bar{s}^o$ **kJ/kmol K**
0	−9364	0	−8671	0
100	−6457	179.010	−5772	165.852
200	−3413	199.976	−2860	186.024
298	0	213.794	0	197.651
300	69	214.024	54	197.831
400	4003	225.314	2977	206.240
500	8305	234.902	5932	212.833
600	12906	243.284	8942	218.321
700	17754	250.752	12021	223.067
800	22806	257.496	15174	227.277
900	28030	263.646	18397	231.074
1000	33397	269.299	21686	234.538
1100	38885	274.528	25031	237.726
1200	44473	279.390	28427	240.679
1300	50148	283.931	31867	243.431
1400	55895	288.190	35343	246.006
1500	61705	292.199	38852	248.426
1600	67569	295.984	42388	250.707
1700	73480	299.567	45948	252.866
1800	79432	302.969	49529	254.913
1900	85420	306.207	53128	256.860
2000	91439	309.294	56743	258.716
2200	103562	315.070	64012	262.182
2400	115779	320.384	71326	265.361
2600	128074	325.307	78679	268.302
2800	140435	329.887	86070	271.044
3000	152853	334.170	93504	273.607
3200	165321	338.194	100962	276.012
3400	177836	341.988	108440	278.279
3600	190394	345.576	115938	280.422
3800	202990	348.981	123454	282.454
4000	215624	352.221	130989	284.387
4400	240992	358.266	146108	287.989
4800	266488	363.812	161285	291.290
5200	292112	368.939	176510	294.337
5600	317870	373.711	191782	297.167
6000	343782	378.180	207105	299.809

TABLE A.8 (Continued) *Ideal-Gas Properties of Various Substances (SI Units), Entropies at 0.1-MPa (1-bar) Pressure*

T K	Water (H_2O) $\bar{h}_{f,298}^\circ = -241\ 826$ kJ/kmol $M = 18.015$		Hydroxyl (OH) $\bar{h}_{f,298}^\circ = 38\ 987$ kJ/kmol $M = 17.007$	
	$(\bar{h}-\bar{h}_{298}^\circ)$ kJ/kmol	$\bar{s}^\circ$ kJ/kmol K	$(\bar{h}-\bar{h}_{298}^\circ)$ kJ/kmol	$\bar{s}^\circ$ kJ/kmol
0	−9904	0	−9172	0
100	−6617	152.386	−6140	149.591
200	−3282	175.488	−2975	171.592
298	0	188.835	0	183.709
300	62	189.043	55	183.894
400	3450	198.787	3034	192.466
500	6922	206.532	5991	199.066
600	10499	213.051	8943	204.448
700	14190	218.739	11902	209.008
800	18002	223.826	14881	212.984
900	21937	228.460	17889	216.526
1000	26000	232.739	20935	219.735
1100	30190	236.732	24024	222.680
1200	34506	240.485	27159	225.408
1300	38941	244.035	30340	227.955
1400	43491	247.406	33567	230.347
1500	48149	250.620	36838	232.604
1600	52907	253.690	40151	234.741
1700	57757	256.631	43502	236.772
1800	62693	259.452	46890	238.707
1900	67706	262.162	50311	240.556
2000	72788	264.769	53763	242.328
2200	83153	269.706	60751	245.659
2400	93741	274.312	67840	248.743
2600	104520	278.625	75018	251.614
2800	115463	282.680	82268	254.301
3000	126548	286.504	89585	256.825
3200	137756	290.120	96960	259.205
3400	149073	293.550	104388	261.456
3600	160484	296.812	111864	263.592
3800	171981	299.919	119382	265.625
4000	183552	302.887	126940	267.563
4400	206892	308.448	142165	271.191
4800	230456	313.573	157522	274.531
5200	254216	318.328	173002	277.629
5600	278161	322.764	188598	280.518
6000	302295	326.926	204309	283.227

TABLE A.8 (Continued) *Ideal-Gas Properties of Various Substances (SI Units),*
Entropies at 0.1-MPa (1-bar) Pressure

	Hydrogen (H_2) $\bar{h}^o_{f, 298} = 0$ kJ/kmol $M = 2.016$		Hydrogen, Monatomic (H) $\bar{h}^o_{f, 298} = 217\ 999$ kJ/kmol $M = 1.008$	
T K	$(\bar{h}-\bar{h}^o_{298})$ kJ/kmol	$\bar{s}^o$ kJ/kmol K	$(\bar{h}-\bar{h}^o_{298})$ kJ/kmol	$\bar{s}^o$ kJ/kmol K
0	−8467	0	−6197	0
100	−5467	100.727	−4119	92.009
200	−2774	119.410	−2040	106.417
298	0	130.678	0	114.716
300	53	130.856	38	114.845
400	2961	139.219	2117	120.825
500	5883	145.738	4196	125.463
600	8799	151.078	6274	129.253
700	11730	155.609	8353	132.457
800	14681	159.554	10431	135.233
900	17657	163.060	12510	137.681
1000	20663	166.225	14589	139.871
1100	23704	169.121	16667	141.852
1200	26785	171.798	18746	143.661
1300	29907	174.294	20825	145.324
1400	33073	176.637	22903	146.865
1500	36281	178.849	24982	148.299
1600	39533	180.946	27060	149.640
1700	42826	182.941	29139	150.900
1800	46160	184.846	31218	152.089
1900	49532	186.670	33296	153.212
2000	52942	188.419	35375	154.279
2200	59865	191.719	39532	156.260
2400	66915	194.789	43689	158.069
2600	74082	197.659	47847	159.732
2800	81355	200.355	52004	161.273
3000	88725	202.898	56161	162.707
3200	96187	205.306	60318	164.048
3400	103736	207.593	64475	165.308
3600	111367	209.773	68633	166.497
3800	119077	211.856	72790	167.620
4000	126864	213.851	76947	168.687
4400	142658	217.612	85261	170.668
4800	158730	221.109	93576	172.476
5200	175057	224.379	101890	174.140
5600	191607	227.447	110205	175.681
6000	208332	230.322	118519	177.114

TABLE A.8 (Continued) *Ideal-Gas Properties of Various Substances (SI Units), Entropies at 0.1-MPa (1-bar) Pressure*

| T | Nitric Oxide (NO) $\bar{h}_{f,298}^{o}$ = 90 291 kJ/kmol M = 30.006 | | Nitrogen Dioxide (NO_2) $\bar{h}_{f,298}^{o}$ = 33 100 kJ/kmol M = 46.005 | |
| | $(\bar{h}-\bar{h}_{298}^{o})$ | $\bar{s}^{o}$ | $(\bar{h}-\bar{h}_{298}^{o})$ | $\bar{s}^{o}$ |
K	kJ/kmol	kJ/kmol K	kJ/kmol	kJ/kmol K
0	−9192	0	−10186	0
100	−6073	177.031	−6861	202.563
200	−2951	198.747	−3495	225.852
298	0	210.759	0	240.034
300	55	210.943	68	240.263
400	3040	219.529	3927	251.342
500	6059	226.263	8099	260.638
600	9144	231.886	12555	268.755
700	12308	236.762	17250	275.988
800	15548	241.088	22138	282.513
900	18858	244.985	27180	288.450
1000	22229	248.536	32344	293.889
1100	25653	251.799	37606	298.904
1200	29120	254.816	42946	303.551
1300	32626	257.621	48351	307.876
1400	36164	260.243	53808	311.920
1500	39729	262.703	59309	315.715
1600	43319	265.019	64846	319.289
1700	46929	267.208	70414	322.664
1800	50557	269.282	76008	325.861
1900	54201	271.252	81624	328.898
2000	57859	273.128	87259	331.788
2200	65212	276.632	98578	337.182
2400	72606	279.849	109948	342.128
2600	80034	282.822	121358	346.695
2800	87491	285.585	132800	350.934
3000	94973	288.165	144267	354.890
3200	102477	290.587	155756	358.597
3400	110000	292.867	167262	362.085
3600	117541	295.022	178783	365.378
3800	125099	297.065	190316	368.495
4000	132671	299.007	201860	371.456
4400	147857	302.626	224973	376.963
4800	163094	305.940	248114	381.997
5200	178377	308.998	271276	386.632
5600	193703	311.838	294455	390.926
6000	209070	314.488	317648	394.926

TABLE A.9 *Enthalpy of Formation, and Absolute Entropy of Various Substances at 25°C, 100 kPa Pressure.*

Substance	Formula	M	State	$\bar{h}_f^0$ kJ/kmol	$\bar{s}_f^0$ kJ/kmol K
Water	H_2O	18.015	gas	−241 826	188.834
Water	H_2O	18.015	liq	−285 830	69.950
Hydrogen peroxide	H_2O_2	34.015	gas	−136 106	232.991
Ozone	O_3	47.998	gas	+142 674	238.932
Carbon (graphite)	C	12.011	solid	0	5.740
Carbon monoxide	CO	28.011	gas	−110 527	197.653
Carbon dioxide	CO_2	44.010	gas	−393 522	213.795
Methane	CH_4	16.043	gas	−74 873	186.251
Acetylene	C_2H_2	26.038	gas	+226 731	200.958
Ethene	C_2H_4	28.054	gas	+52 467	219.330
Ethane	C_2H_6	30.070	gas	−84 740	229.597
Propene	C_3H_6	42.081	gas	+20 430	267.066
Propane	C_3H_8	44.094	gas	−103 900	269.917
Butane	C_4H_{10}	58.124	gas	−126 200	306.647
Pentane	C_5H_{12}	72.151	gas	−146 500	348.945
Benzene	C_6H_6	78.114	gas	+82 980	269.562
Hexane	C_6H_{14}	86.178	gas	−167 300	387.979
Heptane	C_7H_{16}	100.205	gas	−187 900	427.805
n-Octane	C_8H_{18}	114.232	gas	−208 600	466.514
n-Octane	C_8H_{18}	114.232	liq	−250 105	360.575
Methanol	CH_3OH	32.042	gas	−201 300	239.709
Ethanol	C_2H_5OH	46.069	gas	−235 000	282.444
Ammonia	NH_3	17.031	gas	−45 720	192.572
T-T-Diesel	$C_{14.4}H_{24.9}$	198.06	liq	−174 000	525.90
Sulfur	S	32.06	solid	0	32.056
Sulfur dioxide	SO_2	64.059	gas	−296 842	248.212
Sulfur trioxide	SO_3	80.058	gas	−395 765	256.769
Nitrogen oxide	N_2O	44.013	gas	+82 050	219.957
Nitromethane	CH_3NO_2	61.04	liq	−113 100	171.80

TABLE A.10 *Logarithms to the Base e of the Equilibrium Constant K*

For the reaction $\nu_A A + \nu_B B \rightleftharpoons \nu_C C + \nu_D D$, the equilibrium constant K is defined as

$$K = \frac{y_C^{\nu_C} y_D^{\nu_D}}{y_A^{\nu_A} y_B^{\nu_B}} \left(\frac{P}{P°}\right)^{\nu_C + \nu_D - \nu_A - \nu_B}, \quad P° = 0.1 \text{ MPa}$$

Temp K	$H_2 \rightleftharpoons 2H$	$O_2 \rightleftharpoons 2O$	$N_2 \rightleftharpoons 2N$	$2H_2O \rightleftharpoons 2H_2 + O_2$	$2H_2O \rightleftharpoons H_2 + 2OH$	$2CO_2 \rightleftharpoons 2CO + O_2$	$N_2 + O_2 \rightleftharpoons 2NO$	$N_2 + 2O_2 \rightleftharpoons 2NO_2$
298	−164.003	−186.963	−367.528	−184.420	−212.075	−207.529	−69.868	−41.355
500	−92.830	−105.623	−213.405	−105.385	−120.331	−115.234	−40.449	−30.725
1000	−39.810	−45.146	−99.146	−46.321	−51.951	−47.052	−18.709	−23.039
1200	−30.878	−35.003	−80.025	−36.363	−40.467	−35.736	−15.082	−21.752
1400	−24.467	−27.741	−66.345	−29.222	−32.244	−27.679	−12.491	−20.826
1600	−19.638	−22.282	−56.069	−23.849	−26.067	−21.656	−10.547	−20.126
1800	−15.868	−18.028	−48.066	−19.658	−21.258	−16.987	−9.035	−19.577
2000	−12.841	−14.619	−41.655	−16.299	−17.406	−13.266	−7.825	−19.136
2200	−10.356	−11.826	−36.404	−13.546	−14.253	−10.232	−6.836	−18.773
2400	−8.280	−9.495	−32.023	−11.249	−11.625	−7.715	−6.012	−18.470
2600	−6.519	−7.520	−28.313	−9.303	−9.402	−5.594	−5.316	−18.214
2800	−5.005	−5.826	−25.129	−7.633	−7.496	−3.781	−4.720	−17.994
3000	−3.690	−4.356	−22.367	−6.184	−5.845	−2.217	−4.205	−17.805
3200	−2.538	−3.069	−19.947	−4.916	−4.401	−0.853	−3.755	−17.640
3400	−1.519	−1.932	−17.810	−3.795	−3.128	0.346	−3.359	−17.496
3600	−0.611	−0.922	−15.909	−2.799	−1.996	1.408	−3.008	−17.369
3800	0.201	−0.017	−14.205	−1.906	−0.984	2.355	−2.694	−17.257
4000	0.934	0.798	−12.671	−1.101	−0.074	3.204	−2.413	−17.157
4500	2.483	2.520	−9.423	0.602	1.847	4.985	−1.824	−16.953
5000	3.724	3.898	−6.816	1.972	3.383	6.397	−1.358	−16.797
5500	4.739	5.027	−4.672	3.098	4.639	7.542	−0.980	−16.678
6000	5.587	5.969	−2.876	4.040	5.684	8.488	−0.671	−16.588

Source: Consistent with thermodynamic data in *JANAF Thermochemical Tables*, third edition, Thermal Group, Dow Chemical U.S.A., Midland, MI 1985.

TABLE A.11 *One-Dimensional Isentropic Compressible-Flow Functions for an Ideal Gas with Constant Specific Heat and Molecular Weight and k = 1.4*

M	M*	A/A*	P/P_o	ρ/ρ_o	T/T_o
0.0	0.00000	∞	1.00000	1.00000	1.00000
0.1	0.10944	5.82183	0.99303	0.99502	0.99800
0.2	0.21822	2.96352	0.97250	0.98028	0.99206
0.3	0.32572	2.03506	0.93947	0.95638	0.98232
0.4	0.43133	1.59014	0.89561	0.92427	0.96899
0.5	0.53452	1.33984	0.84302	0.88517	0.95238
0.6	0.63481	1.18820	0.78400	0.84045	0.93284
0.7	0.73179	1.09437	0.72093	0.79158	0.91075
0.8	0.82514	1.03823	0.65602	0.73999	0.88652
0.9	0.91460	1.00886	0.59126	0.68704	0.86059
1.0	1.0000	1.00000	0.52828	0.63394	0.83333
1.1	1.0812	1.00793	0.46835	0.58170	0.80515
1.2	1.1583	1.03044	0.41238	0.53114	0.77640
1.3	1.2311	1.06630	0.36091	0.48290	0.74738
1.4	1.2999	1.11493	0.31424	0.43742	0.71839
1.5	1.3646	1.17617	0.27240	0.39498	0.68966
1.6	1.4254	1.25023	0.23527	0.35573	0.66138
1.7	1.4825	1.33761	0.20259	0.31969	0.63371
1.8	1.5360	1.43898	0.17404	0.28682	0.60680
1.9	1.5861	1.55526	0.14924	0.25699	0.58072
2.0	1.6330	1.68750	0.12780	0.23005	0.55556
2.1	1.6769	1.83694	0.10935	0.20580	0.53135
2.2	1.7179	2.00497	0.93522E-01	0.18405	0.50813
2.3	1.7563	2.19313	0.79973E-01	0.16458	0.48591
2.4	1.7922	2.40310	0.68399E-01	0.14720	0.46468
2.5	1.8257	2.63672	0.58528E-01	0.13169	0.44444
2.6	1.8571	2.89598	0.50115E-01	0.11787	0.42517
2.7	1.8865	3.18301	0.42950E-01	0.10557	0.40683
2.8	1.9140	3.50012	0.36848E-01	0.94626E-01	0.38941
2.9	1.9398	3.84977	0.31651E-01	0.84889E-01	0.37286
3.0	1.9640	4.23457	0.27224E-01	0.76226E-01	0.35714
3.5	2.0642	6.78962	0.13111E-01	0.45233E-01	0.28986
4.0	2.1381	10.7188	0.65861E-02	0.27662E-01	0.23810
4.5	2.1936	16.5622	0.34553E-02	0.17449E-01	0.19802
5.0	2.2361	25.0000	0.18900E-02	0.11340E-01	0.16667
6.0	2.2953	53.1798	0.63336E-03	0.51936E-02	0.12195
7.0	2.3333	104.143	0.24156E-03	0.26088E-02	0.09259
8.0	2.3591	190.109	0.10243E-03	0.14135E-02	0.07246
9.0	2.3772	327.189	0.47386E-04	0.81504E-03	0.05814
10.0	2.3905	535.938	0.23563E-04	0.49482E-03	0.04762
∞	2.4495	∞	0.0	0.0	0.0

TABLE A.12 *One-Dimensional Normal Shock Functions for an Ideal Gas with Constant Specific Heat and Molecular Weight and k = 1.4*

M_x	M_y	P_y/P_x	ρ_y/ρ_x	T_y/T_x	P_{0y}/P_{0x}	P_{0y}/P_x
1.00	1.00000	1.0000	1.0000	1.0000	1.00000	1.8929
1.05	0.95313	1.1196	1.0840	1.0328	0.99985	2.0083
1.10	0.91177	1.2450	1.1691	1.0649	0.99893	2.1328
1.15	0.87502	1.3763	1.2550	1.0966	0.99669	2.2661
1.20	0.84217	1.5133	1.3416	1.1280	0.99280	2.4075
1.25	0.81264	1.6563	1.4286	1.1594	0.98706	2.5568
1.30	0.78596	1.8050	1.5157	1.1909	0.97937	2.7136
1.35	0.76175	1.9596	1.6028	1.2226	0.96974	2.8778
1.40	0.73971	2.1200	1.6897	1.2547	0.95819	3.0492
1.45	0.71956	2.2863	1.7761	1.2872	0.94484	3.2278
1.50	0.70109	2.4583	1.8621	1.3202	0.92979	3.4133
1.55	0.68410	2.6362	1.9473	1.3538	0.91319	3.6057
1.60	0.66844	2.8200	2.0317	1.3880	0.89520	3.8050
1.65	0.65396	3.0096	2.1152	1.4228	0.87599	4.0110
1.70	0.64054	3.2050	2.1977	1.4583	0.85572	4.2238
1.75	0.62809	3.4063	2.2791	1.4946	0.83457	4.4433
1.80	0.61650	3.6133	2.3592	1.5316	0.81268	4.6695
1.85	0.60570	3.8263	2.4381	1.5693	0.79023	4.9023
1.90	0.59562	4.0450	2.5157	1.6079	0.76736	5.1418
1.95	0.58618	4.2696	2.5919	1.6473	0.74420	5.3878
2.00	0.57735	4.5000	2.6667	1.6875	0.72087	5.6404
2.05	0.56906	4.7362	2.7400	1.7285	0.69751	5.8996
2.10	0.56128	4.9783	2.8119	1.7705	0.67420	6.1654
2.15	0.55395	5.2263	2.8823	1.8132	0.65105	6.4377
2.20	0.54706	5.4800	2.9512	1.8569	0.62814	6.7165
2.25	0.54055	5.7396	3.0186	1.9014	0.60553	7.0018
2.30	0.53441	6.0050	3.0845	1.9468	0.58329	7.2937
2.35	0.52861	6.2762	3.1490	1.9931	0.56148	7.5920
2.40	0.52312	6.5533	3.2119	2.0403	0.54014	7.8969
2.45	0.51792	6.8363	3.2733	2.0885	0.51931	8.2083
2.50	0.51299	7.1250	3.3333	2.1375	0.49901	8.5261
2.55	0.50831	7.4196	3.3919	2.1875	0.47928	8.8505
2.60	0.50387	7.7200	3.4490	2.2383	0.46012	9.1813
2.70	0.49563	8.3383	3.5590	2.3429	0.42359	9.8624
2.80	0.48817	8.9800	3.6636	2.4512	0.38946	10.569
2.90	0.48138	9.6450	3.7629	2.5632	0.35773	11.302
3.00	0.47519	10.333	3.8571	2.6790	0.32834	12.061
4.00	0.43496	18.500	4.5714	4.0469	0.13876	21.068
5.00	0.41523	29.000	5.0000	5.8000	0.06172	32.653
10.00	0.38758	116.50	5.7143	20.387	0.00304	129.22

TABLE B.1 SI *Thermodynamic Properties of Water*
TABLE B.1.1 SI *Saturated Water*

Temp.	Press.	SpecificVolume, m³/kg			Internal Energy, kJ/kg		
C T	kPa P	Sat. Liquid v_f	Evap. v_{fg}	Sat. Vapor v_g	Sat. Liquid u_f	Evap. u_{fg}	Sat. Vapor u_g
0.01	0.6113	0.001000	206.131	206.132	0	2375.33	2375.33
5	0.8721	0.001000	147.117	147.118	20.97	2361.27	2382.24
10	1.2276	0.001000	106.376	106.377	41.99	2347.16	2389.15
15	1.705	0.001001	77.924	77.925	62.98	2333.06	2396.04
20	2.339	0.001002	57.7887	57.7897	83.94	2318.98	2402.91
25	3.169	0.001003	43.3583	43.3593	104.86	2304.90	2409.76
30	4.246	0.001004	32.8922	32.8932	125.77	2290.81	2416.58
35	5.628	0.001006	25.2148	25.2158	146.65	2276.71	2423.36
40	7.384	0.001008	19.5219	19.5229	167.53	2262.57	2430.11
45	9.593	0.001010	15.2571	15.2581	188.41	2248.40	2436.81
50	12.350	0.001012	12.0308	12.0318	209.30	2234.17	2443.47
55	15.758	0.001015	9.56734	9.56835	230.19	2219.89	2450.08
60	19.941	0.001017	7.66969	7.67071	251.09	2205.54	2456.63
65	25.03	0.001020	6.19554	6.19656	272.00	2191.12	2463.12
70	31.19	0.001023	5.04114	5.04217	292.93	2176.62	2469.55
75	38.58	0.001026	4.13021	4.13123	313.87	2162.03	2475.91
80	47.39	0.001029	3.40612	3.40715	334.84	2147.36	2482.19
85	57.83	0.001032	2.82654	2.82757	355.82	2132.58	2488.40
90	70.14	0.001036	2.35953	2.36056	376.82	2117.70	2494.52
95	84.55	0.001040	1.98082	1.98186	397.86	2102.70	2500.56
100	101.3	0.001044	1.67185	1.67290	418.91	2087.58	2506.50
105	120.8	0.001047	1.41831	1.41936	440.00	2072.34	2512.34
110	143.3	0.001052	1.20909	1.21014	461.12	2056.96	2518.09
115	169.1	0.001056	1.03552	1.03658	482.28	2041.44	2523.72
120	198.5	0.001060	0.89080	0.89186	503.48	2025.76	2529.24
125	232.1	0.001065	0.76953	0.77059	524.72	2009.91	2534.63
130	270.1	0.001070	0.66744	0.66850	546.00	1993.90	2539.90
135	313.0	0.001075	0.58110	0.58217	567.34	1977.69	2545.03
140	361.3	0.001080	0.50777	0.50885	588.72	1961.30	2550.02
145	415.4	0.001085	0.44524	0.44632	610.16	1944.69	2554.86
150	475.9	0.001090	0.39169	0.39278	631.66	1927.87	2559.54
155	543.1	0.001096	0.34566	0.34676	653.23	1910.82	2564.04
160	617.8	0.001102	0.30596	0.30706	674.85	1893.52	2568.37
165	700.5	0.001108	0.27158	0.27269	696.55	1875.97	2572.51
170	791.7	0.001114	0.24171	0.24283	718.31	1858.14	2576.46
175	892.0	0.001121	0.21568	0.21680	740.16	1840.03	2580.19
180	1002.2	0.001127	0.19292	0.19405	762.08	1821.62	2583.70
185	1122.7	0.001134	0.17295	0.17409	784.08	1802.90	2586.98
190	1254.4	0.001141	0.15539	0.15654	806.17	1783.84	2590.01

(Continued)

TABLE B.1.1 SI (Continued) *Saturated Water*

Temp.	Press.	Enthalpy, kJ/kg			Entropy, kJ/kg K		
C T	kPa P	Sat. Liquid h_f	Evap. h_{fg}	Sat. Vapor h_g	Sat. Liquid s_f	Evap. s_{fg}	Sat. Vapor s_g
0.01	0.6113	0.00	2501.35	2501.35	0	9.1562	9.1562
5	0.8721	20.98	2489.57	2510.54	0.0761	8.9496	9.0257
10	1.2276	41.99	2477.75	2519.74	0.1510	8.7498	8.9007
15	1.705	62.98	2465.93	2528.91	0.2245	8.5569	8.7813
20	2.339	83.94	2454.12	2538.06	0.2966	8.3706	8.6671
25	3.169	104.87	2442.30	2547.17	0.3673	8.1905	8.5579
30	4.246	125.77	2430.48	2556.25	0.4369	8.0164	8.4533
35	5.628	146.66	2418.62	2565.28	0.5052	7.8478	8.3530
40	7.384	167.54	2406.72	2574.26	0.5724	7.6845	8.2569
45	9.593	188.42	2394.77	2583.19	0.6386	7.5261	8.1647
50	12.350	209.31	2382.75	2592.06	0.7037	7.3725	8.0762
55	15.758	230.20	2370.66	2600.86	0.7679	7.2234	7.9912
60	19.941	251.11	2358.48	2609.59	0.8311	7.0784	7.9095
65	25.03	272.03	2346.21	2618.24	0.8934	6.9375	7.8309
70	31.19	292.96	2333.85	2626.80	0.9548	6.8004	7.7552
75	38.58	313.91	2321.37	2635.28	1.0154	6.6670	7.6824
80	47.39	334.88	2308.77	2643.66	1.0752	6.5369	7.6121
85	57.83	355.88	2296.05	2651.93	1.1342	6.4102	7.5444
90	70.14	376.90	2283.19	2660.09	1.1924	6.2866	7.4790
95	84.55	397.94	2270.19	2668.13	1.2500	6.1659	7.4158
100	101.3	419.02	2257.03	2676.05	1.3068	6.0480	7.3548
105	120.8	440.13	2243.70	2683.83	1.3629	5.9328	7.2958
110	143.3	461.27	2230.20	2691.47	1.4184	5.8202	7.2386
115	169.1	482.46	2216.50	2698.96	1.4733	5.7100	7.1832
120	198.5	503.69	2202.61	2706.30	1.5275	5.6020	7.1295
125	232.1	524.96	2188.50	2713.46	1.5812	5.4962	7.0774
130	270.1	546.29	2174.16	2720.46	1.6343	5.3925	7.0269
135	313.0	567.67	2159.59	2727.26	1.6869	5.2907	6.9777
140	361.3	589.11	2144.75	2733.87	1.7390	5.1908	6.9298
145	415.4	610.61	2129.65	2740.26	1.7906	5.0926	6.8832
150	475.9	632.18	2114.26	2746.44	1.8417	4.9960	6.8378
155	543.1	653.82	2098.56	2752.39	1.8924	4.9010	6.7934
160	617.8	675.53	2082.55	2758.09	1.9426	4.8075	6.7501
165	700.5	697.32	2066.20	2763.53	1.9924	4.7153	6.7078
170	791.7	719.20	2049.50	2768.70	2.0418	4.6244	6.6663
175	892.0	741.16	2032.42	2773.58	2.0909	4.5347	6.6256
180	1002.2	763.21	2014.96	2778.16	2.1395	4.4461	6.5857
185	1122.7	785.36	1997.07	2782.43	2.1878	4.3586	6.5464
190	1254.4	807.61	1978.76	2786.37	2.2358	4.2720	6.5078

(Continued)

TABLE B.1.1 SI (Continued) *Saturated Water*

Temp.	Press.	Specific Volume, m³/kg			Internal Energy, kJ/kg		
C T	kPa P	Sat. Liquid v_f	Evap. v_{fg}	Sat. Vapor v_g	Sat. Liquid u_f	Evap. u_{fg}	Sat. Vapor u_g
195	1397.8	0.001149	0.13990	0.14105	828.36	1764.43	2592.79
200	1553.8	0.001156	0.12620	0.12736	850.64	1744.66	2595.29
205	1723.0	0.001164	0.11405	0.11521	873.02	1724.49	2597.52
210	1906.3	0.001173	0.10324	0.10441	895.51	1703.93	2599.44
215	2104.2	0.001181	0.09361	0.09479	918.12	1682.94	2601.06
220	2317.8	0.001190	0.08500	0.08619	940.85	1661.49	2602.35
225	2547.7	0.001199	0.07729	0.07849	963.72	1639.58	2603.30
230	2794.9	0.001209	0.07037	0.07158	986.72	1617.17	2603.89
235	3060.1	0.001219	0.06415	0.06536	1009.88	1594.24	2604.11
240	3344.2	0.001229	0.05853	0.05976	1033.19	1570.75	2603.95
245	3648.2	0.001240	0.05346	0.05470	1056.69	1546.68	2603.37
250	3973.0	0.001251	0.04887	0.05013	1080.37	1522.00	2602.37
255	4319.5	0.001263	0.04471	0.04598	1104.26	1496.66	2600.93
260	4688.6	0.001276	0.04093	0.04220	1128.37	1470.64	2599.01
265	5081.3	0.001289	0.03748	0.03877	1152.72	1443.87	2596.60
270	5498.7	0.001302	0.03434	0.03564	1177.33	1416.33	2593.66
275	5941.8	0.001317	0.03147	0.03279	1202.23	1387.94	2590.17
280	6411.7	0.001332	0.02884	0.03017	1227.43	1358.66	2586.09
285	6909.4	0.001348	0.02642	0.02777	1252.98	1328.41	2581.38
290	7436.0	0.001366	0.02420	0.02557	1278.89	1297.11	2575.99
295	7992.8	0.001384	0.02216	0.02354	1305.21	1264.67	2569.87
300	8581.0	0.001404	0.02027	0.02167	1331.97	1230.99	2562.96
305	9201.8	0.001425	0.01852	0.01995	1359.22	1195.94	2555.16
310	9856.6	0.001447	0.01690	0.01835	1387.03	1159.37	2546.40
315	10547	0.001472	0.01539	0.01687	1415.44	1121.11	2536.55
320	11274	0.001499	0.01399	0.01549	1444.55	1080.93	2525.48
325	12040	0.001528	0.01267	0.01420	1474.44	1038.57	2513.01
330	12845	0.001561	0.01144	0.01300	1505.24	993.66	2498.91
335	13694	0.001597	0.01027	0.01186	1537.11	945.77	2482.88
340	14586	0.001638	0.00916	0.01080	1570.26	894.26	2464.53
345	15525	0.001685	0.00810	0.00978	1605.01	838.29	2443.30
350	16514	0.001740	0.00707	0.00881	1641.81	776.58	2418.39
355	17554	0.001807	0.00607	0.00787	1681.41	707.11	2388.52
360	18651	0.001892	0.00505	0.00694	1725.19	626.29	2351.47
365	19807	0.002011	0.00398	0.00599	1776.13	526.54	2302.67
370	21028	0.002213	0.00271	0.00493	1843.84	384.69	2228.53
374.1	22089	0.003155	0	0.00315	2029.58	0	2029.58

(Continued)

TABLE B.1.1 SI (Continued) *Saturated Water*

Temp.	Press.	Enthalpy, kJ/kg			Entropy, kJ/kg K		
C T	kPa P	Sat. Liquid h_f	Evap. h_{fg}	Sat. Vapor h_g	Sat. Liquid s_f	Evap. s_{fg}	Sat. Vapor s_g
195	1397.8	829.96	1959.99	2789.96	2.2835	4.1863	6.4697
200	1553.8	852.43	1940.75	2793.18	2.3308	4.1014	6.4322
205	1723.0	875.03	1921.00	2796.03	2.3779	4.0172	6.3951
210	1906.3	897.75	1900.73	2798.48	2.4247	3.9337	6.3584
215	2104.2	920.61	1879.91	2800.51	2.4713	3.8507	6.3221
220	2317.8	943.61	1858.51	2802.12	2.5177	3.7683	6.2860
225	2547.7	966.77	1836.50	2803.27	2.5639	3.6863	6.2502
230	2794.9	990.10	1813.85	2803.95	2.6099	3.6047	6.2146
235	3060.1	1013.61	1790.53	2804.13	2.6557	3.5233	6.1791
240	3344.2	1037.31	1766.50	2803.81	2.7015	3.4422	6.1436
245	3648.2	1061.21	1741.73	2802.95	2.7471	3.3612	6.1083
250	3973.0	1085.34	1716.18	2801.52	2.7927	3.2802	6.0729
255	4319.5	1109.72	1689.80	2799.51	2.8382	3.1992	6.0374
260	4688.6	1134.35	1662.54	2796.89	2.8837	3.1181	6.0018
265	5081.3	1159.27	1634.34	2793.61	2.9293	3.0368	5.9661
270	5498.7	1184.49	1605.16	2789.65	2.9750	2.9551	5.9301
275	5941.8	1210.05	1574.92	2784.97	3.0208	2.8730	5.8937
280	6411.7	1235.97	1543.55	2779.53	3.0667	2.7903	5.8570
285	6909.4	1262.29	1510.97	2773.27	3.1129	2.7069	5.8198
290	7436.0	1289.04	1477.08	2766.13	3.1593	2.6227	5.7821
295	7992.8	1316.27	1441.78	2758.05	3.2061	2.5375	5.7436
300	8581.0	1344.01	1404.93	2748.94	3.2533	2.4511	5.7044
305	9201.8	1372.33	1366.38	2738.72	3.3009	2.3633	5.6642
310	9856.6	1401.29	1325.97	2727.27	3.3492	2.2737	5.6229
315	10547	1430.97	1283.48	2714.44	3.3981	2.1821	5.5803
320	11274	1461.45	1238.64	2700.08	3.4479	2.0882	5.5361
325	12040	1492.84	1191.13	2683.97	3.4987	1.9913	5.4900
330	12845	1525.29	1140.56	2665.85	3.5506	1.8909	5.4416
335	13694	1558.98	1086.37	2645.35	3.6040	1.7863	5.3903
340	14586	1594.15	1027.86	2622.01	3.6593	1.6763	5.3356
345	15525	1631.17	964.02	2595.19	3.7169	1.5594	5.2763
350	16514	1670.54	893.38	2563.92	3.7776	1.4336	5.2111
355	17554	1713.13	813.59	2526.72	3.8427	1.2951	5.1378
360	18651	1760.48	720.52	2481.00	3.9146	1.1379	5.0525
365	19807	1815.96	605.44	2421.40	3.9983	0.9487	4.9470
370	21028	1890.37	441.75	2332.12	4.1104	0.6868	4.7972
374.1	22089	2099.26	0	2099.26	4.4297	0	4.4297

TABLE B.1.2 SI *Saturated Water Pressure Entry*

Press..	Temp.	SpecificVolume, m³/kg			Internal Energy, kJ/kg		
kPa P	C T	Sat. Liquid v_f	Evap. v_{fg}	Sat. Vapor v_g	Sat. Liquid u_f	Evap. u_{fg}	Sat. Vapor u_g
0.6113	0.01	0.001000	206.131	206.132	0	2375.3	2375.3
1	6.98	0.001000	129.20702	129.20802	29.29	2355.69	2384.98
1.5	13.03	0.001001	87.97913	87.98013	54.70	2338.63	2393.32
2	17.50	0.001001	67.00285	67.00385	73.47	2326.02	2399.48
2.5	21.08	0.001002	54.25285	54.25385	88.47	2315.93	2404.40
3	24.08	0.001003	45.66402	45.66502	101.03	2307.48	2408.51
4	28.96	0.001004	34.79915	34.80015	121.44	2293.73	2415.17
5	32.88	0.001005	28.19150	28.19251	137.79	2282.70	2420.49
7.5	40.29	0.001008	19.23674	19.23775	168.76	2261.74	2430.50
10	45.81	0.001010	14.67254	14.67355	191.79	2246.10	2437.89
15	53.97	0.001014	10.02117	10.02218	225.90	2222.83	2448.73
20	60.06	0.001017	7.64835	7.64937	251.35	2205.36	2456.71
25	64.97	0.001020	6.20322	6.20424	271.88	2191.21	2463.08
30	69.10	0.001022	5.22816	5.22918	289.18	2179.22	2468.40
40	75.87	0.001026	3.99243	3.99345	317.51	2159.49	2477.00
50	81.33	0.001030	3.23931	3.24034	340.42	2143.43	2483.85
75	91.77	0.001037	2.21607	2.21711	384.29	2112.39	2496.67
100	99.62	0.001043	1.69296	1.69400	417.33	2088.72	2506.06
125	105.99	0.001048	1.37385	1.37490	444.16	2069.32	2513.48
150	111.37	0.001053	1.15828	1.15933	466.92	2052.72	2519.64
175	116.06	0.001057	1.00257	1.00363	486.78	2038.12	2524.90
200	120.23	0.001061	0.88467	0.88573	504.47	2025.02	2529.49
225	124.00	0.001064	0.79219	0.79325	520.45	2013.10	2533.56
250	127.43	0.001067	0.71765	0.71871	535.08	2002.14	2537.21
275	130.60	0.001070	0.65624	0.65731	548.57	1991.95	2540.53
300	133.55	0.001073	0.60475	0.60582	561.13	1982.43	2543.55
325	136.30	0.001076	0.56093	0.56201	572.88	1973.46	2546.34
350	138.88	0.001079	0.52317	0.52425	583.93	1964.98	2548.92
375	141.32	0.001081	0.49029	0.49137	594.38	1956.93	2551.31
400	143.63	0.001084	0.46138	0.46246	604.29	1949.26	2553.55
450	147.93	0.001088	0.41289	0.41398	622.75	1934.87	2557.62
500	151.86	0.001093	0.37380	0.37489	639.66	1921.57	2561.23
550	155.48	0.001097	0.34159	0.34268	655.30	1909.17	2564.47
600	158.85	0.001101	0.31457	0.31567	669.88	1897.52	2567.40
650	162.01	0.001104	0.29158	0.29268	683.55	1886.51	2570.06
700	164.97	0.001108	0.27176	0.27286	696.43	1876.07	2572.49
750	167.77	0.001111	0.25449	0.25560	708.62	1866.11	2574.73
800	170.43	0.001115	0.23931	0.24043	720.20	1856.58	2576.79

(Continued)

TABLE B.1.2 SI (Continued) *Saturated Water Pressure Entry*

Press.	Temp.	Enthalpy, kJ/kg			Entropy, kJ/kg K		
kPa P	C T	Sat. Liquid h_f	Evap. h_{fg}	Sat. Vapor h_g	Sat. Liquid s_f	Evap. s_{fg}	Sat. Vapor s_g
0.6113	0.01	0.00	2501.3	2501.3	0	9.1562	9.1562
1.0	6.98	29.29	2484.89	2514.18	0.1059	8.8697	8.9756
1.5	13.03	54.70	2470.59	2525.30	0.1956	8.6322	8.8278
2.0	17.50	73.47	2460.02	2533.49	0.2607	8.4629	8.7236
2.5	21.08	88.47	2451.56	2540.03	0.3120	8.3311	8.6431
3.0	24.08	101.03	2444.47	2545.50	0.3545	8.2231	8.5775
4.0	28.96	121.44	2432.93	2554.37	0.4226	8.0520	8.4746
5.0	32.88	137.79	2423.66	2561.45	0.4763	7.9187	8.3950
7.5	40.29	168.77	2406.02	2574.79	0.5763	7.6751	8.2514
10	45.81	191.81	2392.82	2584.63	0.6492	7.5010	8.1501
15	53.97	225.91	2373.14	2599.06	0.7548	7.2536	8.0084
20	60.06	251.38	2358.33	2609.70	0.8319	7.0766	7.9085
25	64.97	271.90	2346.29	2618.19	0.8930	6.9383	7.8313
30	69.10	289.21	2336.07	2625.28	0.9439	6.8247	7.7686
40	75.87	317.55	2319.19	2636.74	1.0258	6.6441	7.6700
50	81.33	340.47	2305.40	2645.87	1.0910	6.5029	7.5939
75	91.77	384.36	2278.59	2662.96	1.2129	6.2434	7.4563
100	99.62	417.44	2258.02	2675.46	1.3025	6.0568	7.3593
125	105.99	444.30	2241.05	2685.35	1.3739	5.9104	7.2843
150	111.37	467.08	2226.46	2693.54	1.4335	5.7897	7.2232
175	116.06	486.97	2213.57	2700.53	1.4848	5.6868	7.1717
200	120.23	504.68	2201.96	2706.63	1.5300	5.5970	7.1271
225	124.00	520.69	2191.35	2712.04	1.5705	5.5173	7.0878
250	127.43	535.34	2181.55	2716.89	1.6072	5.4455	7.0526
275	130.60	548.87	2172.42	2721.29	1.6407	5.3801	7.0208
300	133.55	561.45	2163.85	2725.30	1.6717	5.3201	6.9918
325	136.30	573.23	2155.76	2728.99	1.7005	5.2646	6.9651
350	138.88	584.31	2148.10	2732.40	1.7274	5.2130	6.9404
375	141.32	594.79	2140.79	2735.58	1.7527	5.1647	6.9174
400	143.63	604.73	2133.81	2738.53	1.7766	5.1193	6.8958
450	147.93	623.24	2120.67	2743.91	1.8206	5.0359	6.8565
500	151.86	640.21	2108.47	2748.67	1.8606	4.9606	6.8212
550	155.48	655.91	2097.04	2752.94	1.8972	4.8920	6.7892
600	158.85	670.54	2086.26	2756.80	1.9311	4.8289	6.7600
650	162.01	684.26	2076.04	2760.30	1.9627	4.7704	6.7330
700	164.97	697.20	2066.30	2763.50	1.9922	4.7158	6.7080
750	167.77	709.45	2056.98	2766.43	2.0199	4.6647	6.6846
800	170.43	721.10	2048.04	2769.13	2.0461	4.6166	6.6627

(Continued)

TABLE B.1.2 SI (Continued) *Saturated Water Pressure Entry*

Press..	Temp.	SpecificVolume, m³/kg			Internal Energy, kJ/kg		
kPa	C	Sat. Liquid	Evap.	Sat. Vapor	Sat. Liquid	Evap.	Sat. Vapor
P	T	v_f	v_{fg}	v_g	u_f	u_{fg}	u_g
850	172.96	0.001118	0.22586	0.22698	731.25	1847.45	2578.69
900	175.38	0.001121	0.21385	0.21497	741.81	1838.65	2580.46
950	177.69	0.001124	0.20306	0.20419	751.94	1830.17	2582.11
1000	179.91	0.001127	0.19332	0.19444	761.67	1821.97	2583.64
1100	184.09	0.001133	0.17639	0.17753	780.08	1806.32	2586.40
1200	187.99	0.001139	0.16220	0.16333	797.27	1791.55	2588.82
1300	191.64	0.001144	0.15011	0.15125	813.42	1777.53	2590.95
1400	195.07	0.001149	0.13969	0.14084	828.68	1764.15	2592.83
1500	198.32	0.001154	0.13062	0.13177	843.14	1751.3	2594.5
1750	205.76	0.001166	0.11232	0.11349	876.44	1721.39	2597.83
2000	212.42	0.001177	0.09845	0.09963	906.42	1693.84	2600.26
2250	218.45	0.001187	0.08756	0.08875	933.81	1668.18	2601.98
2500	223.99	0.001197	0.07878	0.07998	959.09	1644.04	2603.13
2750	229.12	0.001207	0.07154	0.07275	982.65	1621.16	2603.81
3000	233.90	0.001216	0.06546	0.06668	1004.76	1599.34	2604.10
3250	238.38	0.001226	0.06029	0.06152	1025.62	1578.43	2604.04
3500	242.60	0.001235	0.05583	0.05707	1045.41	1558.29	2603.70
4000	250.40	0.001252	0.04853	0.04978	1082.28	1519.99	2602.27
5000	263.99	0.001286	0.03815	0.03944	1147.78	1449.34	2597.12
6000	275.64	0.001319	0.03112	0.03244	1205.41	1384.27	2589.69
7000	285.88	0.001351	0.02602	0.02737	1257.51	1322.97	2580.48
8000	295.06	0.001384	0.02213	0.02352	1305.54	1264.25	2569.79
9000	303.40	0.001418	0.01907	0.02048	1350.47	1207.28	2557.75
10000	311.06	0.001452	0.01657	0.01803	1393.00	1151.40	2544.41
11000	318.15	0.001489	0.01450	0.01599	1433.68	1096.06	2529.74
12000	324.75	0.001527	0.01274	0.01426	1472.92	1040.76	2513.67
13000	330.93	0.001567	0.01121	0.01278	1511.09	984.99	2496.08
14000	336.75	0.001611	0.00987	0.01149	1548.53	928.23	2476.76
15000	342.24	0.001658	0.00868	0.01034	1585.58	869.85	2455.43
16000	347.43	0.001711	0.00760	0.00931	1622.63	809.07	2431.70
17000	352.37	0.001770	0.00659	0.00836	1660.16	744.80	2404.96
18000	357.06	0.001840	0.00565	0.00749	1698.86	675.42	2374.28
19000	361.54	0.001924	0.00473	0.00666	1739.87	598.18	2338.05
20000	365.81	0.002035	0.00380	0.00583	1785.47	507.58	2293.05
21000	369.89	0.002206	0.00275	0.00495	1841.97	388.74	2230.71
22000	373.80	0.002808	0.00072	0.00353	1973.16	108.24	2081.39
22089	374.14	0.003155	0	0.00315	2029.58	0	2029.58

(Continued)

TABLE B.1.2 SI (Continued) *Saturated Water Pressure Entry*

Press.	Temp.	Enthalpy, kJ/kg			Entropy, kJ/kg K		
kPa P	C T	Sat. Liquid h_f	Evap. h_{fg}	Sat. Vapor h_g	Sat. Liquid s_f	Evap. s_{fg}	Sat. Vapor s_g
850	172.96	732.20	2039.43	2771.63	2.0709	4.5711	6.6421
900	175.38	742.82	2031.12	2773.94	2.0946	4.5280	6.6225
950	177.69	753.00	2023.08	2776.08	2.1171	4.4869	6.6040
1000	179.91	762.79	2015.29	2778.08	2.1386	4.4478	6.5864
1100	184.09	781.32	2000.36	2781.68	2.1791	4.3744	6.5535
1200	187.99	798.64	1986.19	2784.82	2.2165	4.3067	6.5233
1300	191.64	814.91	1972.67	2787.58	2.2514	4.2438	6.4953
1400	195.07	830.29	1959.72	2790.00	2.2842	4.1850	6.4692
1500	198.32	844.87	1947.28	2792.15	2.3150	4.1298	6.4448
1750	205.76	878.48	1917.95	2796.43	2.3851	4.0044	6.3895
2000	212.42	908.77	1890.74	2799.51	2.4473	3.8935	6.3408
2250	218.45	936.48	1865.19	2801.67	2.5034	3.7938	6.2971
2500	223.99	962.09	1840.98	2803.07	2.5546	3.7028	6.2574
2750	229.12	985.97	1817.89	2803.86	2.6018	3.6190	6.2208
3000	233.90	1008.41	1795.73	2804.14	2.6456	3.5412	6.1869
3250	238.38	1029.60	1774.37	2803.97	2.6866	3.4685	6.1551
3500	242.60	1049.73	1753.70	2803.43	2.7252	3.4000	6.1252
4000	250.40	1087.29	1714.09	2801.38	2.7963	3.2737	6.0700
5000	263.99	1154.21	1640.12	2794.33	2.9201	3.0532	5.9733
6000	275.64	1213.32	1571.00	2784.33	3.0266	2.8625	5.8891
7000	285.88	1266.97	1505.10	2772.07	3.1210	2.6922	5.8132
8000	295.06	1316.61	1441.33	2757.94	3.2067	2.5365	5.7431
9000	303.40	1363.23	1378.88	2742.11	3.2857	2.3915	5.6771
10000	311.06	1407.53	1317.14	2724.67	3.3595	2.2545	5.6140
11000	318.15	1450.05	1255.55	2705.60	3.4294	2.1233	5.5527
12000	324.75	1491.24	1193.59	2684.83	3.4961	1.9962	5.4923
13000	330.93	1531.46	1130.76	2662.22	3.5604	1.8718	5.4323
14000	336.75	1571.08	1066.47	2637.55	3.6231	1.7485	5.3716
15000	342.24	1610.45	1000.04	2610.49	3.6847	1.6250	5.3097
16000	347.43	1650.00	930.59	2580.59	3.7460	1.4995	5.2454
17000	352.37	1690.25	856.90	2547.15	3.8078	1.3698	5.1776
18000	357.06	1731.97	777.13	2509.09	3.8713	1.2330	5.1044
19000	361.54	1776.43	688.11	2464.54	3.9387	1.0841	5.0227
20000	365.81	1826.18	583.56	2409.74	4.0137	0.9132	4.9269
21000	369.89	1888.30	446.42	2334.72	4.1073	0.6942	4.8015
22000	373.80	2034.92	124.04	2158.97	4.3307	0.1917	4.5224
22089	374.14	2099.26	0	2099.26	4.4297	0	4.4297

(Continued)

TABLE B.1.3 SI *Superheated Vapor Water*

Temp. C	v m³/kg	u kJ/kg	h kJ/kg	s kJ/kg K	v m³/kg	u kJ/kg	h kJ/kg	s kJ/kg K
	P = 10 kPa (45.81)				P = 50 kPa (81.33)			
Sat.	14.67355	2437.89	2584.63	8.1501	3.24034	2483.85	2645.87	7.5939
50	14.86920	2443.87	2592.56	8.1749	--	--	--	--
100	17.19561	2515.50	2687.46	8.4479	3.41833	2511.61	2682.52	7.6947
150	19.51251	2587.86	2782.99	8.6881	3.88937	2585.61	2780.08	7.9400
200	21.82507	2661.27	2879.52	8.9037	4.35595	2659.85	2877.64	8.1579
250	24.13559	2735.95	2977.31	9.1002	4.82045	2734.97	2975.99	8.3555
300	26.44508	2812.06	3076.51	9.2812	5.28391	2811.33	3075.52	8.5372
400	31.06252	2968.89	3279.51	9.6076	6.20929	2968.43	3278.89	8.8641
500	35.67896	3132.26	3489.05	9.8977	7.13364	3131.94	3488.62	9.1545
600	40.29488	3302.45	3705.40	10.1608	8.05748	3302.22	3705.10	9.4177
700	44.91052	3479.63	3928.73	10.4028	8.98104	3479.45	3928.51	9.6599
800	49.52599	3663.84	4159.10	10.6281	9.90444	3663.70	4158.92	9.8852
900	54.14137	3855.03	4396.44	10.8395	10.82773	3854.91	4396.30	10.0967
1000	58.75669	4053.01	4640.58	11.0392	11.75097	4052.91	4640.46	10.2964
1100	63.37198	4257.47	4891.19	11.2287	12.67418	4257.37	4891.08	10.4858
1200	67.98724	4467.91	5147.78	11.4090	13.59737	4467.82	5147.69	10.6662
1300	72.60250	4683.68	5409.70	11.5810	14.52054	4683.58	5409.61	10.8382
	100 kPa (99.62)				200 kPa (120.23)			
Sat.	1.69400	2506.06	2675.46	7.3593	0.88573	2529.49	2706.63	7.1271
150	1.93636	2582.75	2776.38	7.6133	0.95964	2576.87	2768.80	7.2795
200	2.17226	2658.05	2875.27	7.8342	1.08034	2654.39	2870.46	7.5066
250	2.40604	2733.73	2974.33	8.0332	1.19880	2731.22	2970.98	7.7085
300	2.63876	2810.41	3074.28	8.2157	1.31616	2808.55	3071.79	7.8926
400	3.10263	2967.85	3278.11	8.5434	1.54930	2966.69	3276.55	8.2217
500	3.56547	3131.54	3488.09	8.8341	1.78139	3130.75	3487.03	8.5132
600	4.02781	3301.94	3704.72	9.0975	2.01297	3301.36	3703.96	8.7769
700	4.48986	3479.24	3928.23	9.3398	2.24426	3478.81	3927.66	9.0194
800	4.95174	3663.53	4158.71	9.5652	2.47539	3663.19	4158.27	9.2450
900	5.41353	3854.77	4396.12	9.7767	2.70643	3854.49	4395.77	9.4565
1000	5.87526	4052.78	4640.31	9.9764	2.93740	4052.53	4640.01	9.6563
1100	6.33696	4257.25	4890.95	10.1658	3.16834	4257.01	4890.68	9.8458
1200	6.79863	4467.70	5147.56	10.3462	3.39927	4467.46	5147.32	10.0262
1300	7.26030	4683.47	5409.49	10.5182	3.63018	4683.23	5409.26	10.1982
	300 kPa (133.55)				400 kPa (143.63)			
Sat.	0.60582	2543.55	2725.30	6.9918	0.46246	2553.55	2738.53	6.8958
150	0.63388	2570.79	2760.95	7.0778	0.47084	2564.48	2752.82	6.9299
200	0.71629	2650.65	2865.54	7.3115	0.53422	2646.83	2860.51	7.1706
250	0.79636	2728.69	2967.59	7.5165	0.59512	2726.11	2964.16	7.3788
300	0.87529	2806.69	3069.28	7.7022	0.65484	2804.81	3066.75	7.5661
400	1.03151	2965.53	3274.98	8.0329	0.77262	2964.36	3273.41	7.8984

(Continued)

TABLE B.1.3 SI (Continued) *Superheated Vapor Water*

Temp. C	v m³/kg	u kJ/kg	h kJ/kg	s kJ/kg K	v m³/kg	u kJ/kg	h kJ/kg	s kJ/kg K
	300 kPa (133.55)				400 kPa (143.63)			
500	1.18669	3129.95	3485.96	8.3250	0.88934	3129.15	3484.89	8.1912
600	1.34136	3300.79	3703.20	8.5892	1.00555	3300.22	3702.44	8.4557
700	1.49573	3478.38	3927.10	8.8319	1.12147	3477.95	3926.53	8.6987
800	1.64994	3662.85	4157.83	9.0575	1.23722	3662.51	4157.40	8.9244
900	1.80406	3854.20	4395.42	9.2691	1.35288	3853.91	4395.06	9.1361
1000	1.95812	4052.27	4639.71	9.4689	1.46847	4052.02	4639.41	9.3360
1100	2.11214	4256.77	4890.41	9.6585	1.58404	4256.53	4890.15	9.5255
1200	2.26614	4467.23	5147.07	9.8389	1.69958	4466.99	5146.83	9.7059
1300	2.42013	4682.99	5409.03	10.0109	1.81511	4682.75	5408.80	9.8780
	500 kPa (151.86)				600 kPa (158.85)			
Sat.	0.37489	2561.23	2748.67	6.8212	0.31567	2567.40	2756.80	6.7600
200	0.42492	2642.91	2855.37	7.0592	0.35202	2638.91	2850.12	6.9665
250	0.47436	2723.50	2960.68	7.2708	0.39383	2720.86	2957.16	7.1816
300	0.52256	2802.91	3064.20	7.4598	0.43437	2801.00	3061.63	7.3723
350	0.57012	2882.59	3167.65	7.6328	0.47424	2881.12	3165.66	7.5463
400	0.61728	2963.19	3271.83	7.7937	0.51372	2962.02	3270.25	7.7078
500	0.71093	3128.35	3483.82	8.0872	0.59199	3127.55	3482.75	8.0020
600	0.80406	3299.64	3701.67	8.3521	0.66974	3299.07	3700.91	8.2673
700	0.89691	3477.52	3925.97	8.5952	0.74720	3477.08	3925.41	8.5107
800	0.98959	3662.17	4156.96	8.8211	0.82450	3661.83	4156.52	8.7367
900	1.08217	3853.63	4394.71	9.0329	0.90169	3853.34	4394.36	8.9485
1000	1.17469	4051.76	4639.11	9.2328	0.97883	4051.51	4638.81	9.1484
1100	1.26718	4256.29	4889.88	9.4224	1.05594	4256.05	4889.61	9.3381
1200	1.35964	4466.76	5146.58	9.6028	1.13302	4466.52	5146.34	9.5185
1300	1.45210	4682.52	5408.57	9.7749	1.21009	4682.28	5408.34	9.6906
	800 kPa (170.43)				1000 kPa (179.91)			
Sat.	0.24043	2576.79	2769.13	6.6627	0.19444	2583.64	2778.08	6.5864
200	0.26080	2630.61	2839.25	6.8158	0.20596	2621.90	2827.86	6.6939
250	0.29314	2715.46	2949.97	7.0384	0.23268	2709.91	2942.59	6.9246
300	0.32411	2797.14	3056.43	7.2327	0.25794	2793.21	3051.15	7.1228
350	0.35439	2878.16	3161.68	7.4088	0.28247	2875.18	3157.65	7.3010
400	0.38426	2959.66	3267.07	7.5715	0.30659	2957.29	3263.88	7.4650
500	0.44331	3125.95	3480.60	7.8672	0.35411	3124.34	3478.44	7.7621
600	0.50184	3297.91	3699.38	8.1332	0.40109	3296.76	3697.85	8.0289
700	0.56007	3476.22	3924.27	8.3770	0.44779	3475.35	3923.14	8.2731
800	0.61813	3661.14	4155.65	8.6033	0.49432	3660.46	4154.78	8.4996
900	0.67610	3852.77	4393.65	8.8153	0.54075	3852.19	4392.94	8.7118
1000	0.73401	4051.00	4638.20	9.0153	0.58712	4050.49	4637.60	8.9119
1100	0.79188	4255.57	4889.08	9.2049	0.63345	4255.09	4888.55	9.1016
1200	0.84974	4466.05	5145.85	9.3854	0.67977	4465.58	5145.36	9.2821
1300	0.90758	4681.81	5407.87	9.5575	0.72608	4681.33	5407.41	9.4542

(Continued)

TABLE B.1.3 SI (Continued) *Superheated Vapor Water*

Temp. C	v m³/kg	u kJ/kg	h kJ/kg	s kJ/kg K	v m³/kg	u kJ/kg	h kJ/kg	s kJ/kg K
	1200 kPa (187.99)				1400 kPa (195.07)			
Sat.	0.16333	2588.82	2784.82	6.5233	0.14084	2592.83	2790.00	6.4692
200	0.16930	2612.74	2815.90	6.5898	0.14302	2603.09	2803.32	6.4975
250	0.19235	2704.20	2935.01	6.8293	0.16350	2698.32	2927.22	6.7467
300	0.21382	2789.22	3045.80	7.0316	0.18228	2785.16	3040.35	6.9533
350	0.23452	2872.16	3153.59	7.2120	0.20026	2869.12	3149.49	7.1359
400	0.25480	2954.90	3260.66	7.3773	0.21780	2952.50	3257.42	7.3025
500	0.29463	3122.72	3476.28	7.6758	0.25215	3121.10	3474.11	7.6026
600	0.33393	3295.60	3696.32	7.9434	0.28596	3294.44	3694.78	7.8710
700	0.37294	3474.48	3922.01	8.1881	0.31947	3473.61	3920.87	8.1160
800	0.41177	3659.77	4153.90	8.4149	0.35281	3659.09	4153.03	8.3431
900	0.45051	3851.62	4392.23	8.6272	0.38606	3851.05	4391.53	8.5555
1000	0.48919	4049.98	4637.00	8.8274	0.41924	4049.47	4636.41	8.7558
1100	0.52783	4254.61	4888.02	9.0171	0.45239	4254.14	4887.49	8.9456
1200	0.56646	4465.12	5144.87	9.1977	0.48552	4464.65	5144.38	9.1262
1300	0.60507	4680.86	5406.95	9.3698	0.51864	4680.39	5406.49	9.2983
	1600 kPa (201.40)				1800 kPa (207.15)			
Sat.	0.12380	2595.95	2794.02	6.4217	0.11042	2598.38	2797.13	6.3793
250	0.14184	2692.26	2919.20	6.6732	0.12497	2686.02	2910.96	6.6066
300	0.15862	2781.03	3034.83	6.8844	0.14021	2776.83	3029.21	6.8226
350	0.17456	2866.05	3145.35	7.0693	0.15457	2862.95	3141.18	7.0099
400	0.19005	2950.09	3254.17	7.2373	0.16847	2947.66	3250.90	7.1793
500	0.22029	3119.47	3471.93	7.5389	0.19550	3117.84	3469.75	7.4824
600	0.24998	3293.27	3693.23	7.8080	0.22199	3292.10	3691.69	7.7523
700	0.27937	3472.74	3919.73	8.0535	0.24818	3471.87	3918.59	7.9983
800	0.30859	3658.40	4152.15	8.2808	0.27420	3657.71	4151.27	8.2258
900	0.33772	3850.47	4390.82	8.4934	0.30012	3849.90	4390.11	8.4386
1000	0.36678	4048.96	4635.81	8.6938	0.32598	4048.45	4635.21	8.6390
1100	0.39581	4253.66	4886.95	8.8837	0.35180	4253.18	4886.42	8.8290
1200	0.42482	4464.18	5143.89	9.0642	0.37761	4463.71	5143.40	9.0096
1300	0.45382	4679.92	5406.02	9.2364	0.40340	4679.44	5405.56	9.1817

(Continued)

TABLE B.1.3 SI (Continued) *Superheated Vapor Water*

Temp. C	v m³/kg	u kJ/kg	h kJ/kg	s kJ/kg K	v m³/kg	u kJ/kg	h kJ/kg	s kJ/kg K
	2000 kPa (212.42)				2500 kPa (223.99)			
Sat.	0.09963	2600.26	2799.51	6.3408	0.07998	2603.13	2803.07	6.2574
250	0.11144	2679.58	2902.46	6.5452	0.08700	2662.55	2880.06	6.4084
300	0.12547	2772.56	3023.50	6.7663	0.09890	2761.56	3008.81	6.6437
350	0.13857	2859.81	3136.96	6.9562	0.10976	2851.84	3126.24	6.8402
400	0.15120	2945.21	3247.60	7.1270	0.12010	2939.03	3239.28	7.0147
450	0.16353	3030.41	3357.48	7.2844	0.13014	3025.43	3350.77	7.1745
500	0.17568	3116.20	3467.55	7.4316	0.13998	3112.08	3462.04	7.3233
600	0.19960	3290.93	3690.14	7.7023	0.15930	3287.99	3686.25	7.5960
700	0.22323	3470.99	3917.45	7.9487	0.17832	3468.80	3914.59	7.8435
800	0.24668	3657.03	4150.40	8.1766	0.19716	3655.30	4148.20	8.0720
900	0.27004	3849.33	4389.40	8.3895	0.21590	3847.89	4387.64	8.2853
1000	0.29333	4047.94	4634.61	8.5900	0.23458	4046.67	4633.12	8.4860
1100	0.31659	4252.71	4885.89	8.7800	0.25322	4251.52	4884.57	8.6761
1200	0.33984	4463.25	5142.92	8.9606	0.27185	4462.08	5141.70	8.8569
1300	0.36306	4678.97	5405.10	9.1328	0.29046	4677.80	5403.95	9.0291
	3000 kPa (233.90)				3500 kPa (242.60)			
Sat.	0.06668	2604.10	2804.14	6.1869	0.05707	2603.70	2803.43	6.1252
250	0.07058	2644.00	2855.75	6.2871	0.05873	2623.65	2829.19	6.1748
300	0.08114	2750.05	2993.48	6.5389	0.06842	2737.99	2977.46	6.4460
350	0.09053	2843.66	3115.25	6.7427	0.07678	2835.27	3103.99	6.6578
400	0.09936	2932.75	3230.82	6.9211	0.08453	2926.37	3222.24	6.8404
450	0.10787	3020.38	3344.00	7.0833	0.09196	3015.28	3337.15	7.0051
500	0.11619	3107.92	3456.48	7.2337	0.09918	3103.73	3450.87	7.1571
600	0.13243	3285.03	3682.34	7.5084	0.11324	3282.06	3678.40	7.4338
700	0.14838	3466.59	3911.72	7.7571	0.12699	3464.37	3908.84	7.6837
800	0.16414	3653.58	4146.00	7.9862	0.14056	3651.84	4143.80	7.9135
900	0.17980	3846.46	4385.87	8.1999	0.15402	3845.02	4384.11	8.1275
1000	0.19541	4045.40	4631.63	8.4009	0.16743	4044.14	4630.14	8.3288
1100	0.21098	4250.33	4883.26	8.5911	0.18080	4249.14	4881.94	8.5191
1200	0.22652	4460.92	5140.49	8.7719	0.19415	4459.76	5139.28	8.7000
1300	0.24206	4676.63	5402.81	8.9442	0.20749	4675.45	5401.66	8.8723

(Continued)

TABLE B.1.3 SI (Continued) *Superheated Vapor Water*

Temp. C	v m³/kg	u kJ/kg	h kJ/kg	s kJ/kg K	v m³/kg	u kJ/kg	h kJ/kg	s kJ/kg K
		4000 kPa (250.40)				4500 kPa (257.48)		
Sat.	0.04978	2602.27	2801.38	6.0700	0.04406	2600.03	2798.29	6.0198
300	0.05884	2725.33	2960.68	6.3614	0.05135	2712.00	2943.07	6.2827
350	0.06645	2826.65	3092.43	6.5820	0.05840	2817.78	3080.57	6.5130
400	0.07341	2919.88	3213.51	6.7689	0.06475	2913.29	3204.65	6.7046
450	0.08003	3010.13	3330.23	6.9362	0.07074	3004.91	3323.23	6.8745
500	0.08643	3099.49	3445.21	7.0900	0.07651	3095.23	3439.51	7.0300
600	0.09885	3279.06	3674.44	7.3688	0.08765	3276.04	3670.47	7.3109
700	0.11095	3462.15	3905.94	7.6198	0.09847	3459.91	3903.04	7.5631
800	0.12287	3650.11	4141.59	7.8502	0.10911	3648.37	4139.38	7.7942
900	0.13469	3843.59	4382.34	8.0647	0.11965	3842.15	4380.58	8.0091
1000	0.14645	4042.87	4628.65	8.2661	0.13013	4041.61	4627.17	8.2108
1100	0.15817	4247.96	4880.63	8.4566	0.14056	4246.78	4879.32	8.4014
1200	0.16987	4458.60	5138.07	8.6376	0.15098	4457.45	5136.87	8.5824
1300	0.18156	4674.29	5400.52	8.8099	0.16139	4673.12	5399.38	8.7548
		5000 kPa (263.99)				6000 kPa (275.64)		
Sat.	0.03944	2597.12	2794.33	5.9733	0.03244	2589.69	2784.33	5.8891
300	0.04532	2697.94	2924.53	6.2083	0.03616	2667.22	2884.19	6.0673
350	0.05194	2808.67	3068.39	6.4492	0.04223	2789.61	3042.97	6.3334
400	0.05781	2906.58	3195.64	6.6458	0.04739	2892.81	3177.17	6.5407
450	0.06330	2999.64	3316.15	6.8185	0.05214	2988.90	3301.76	6.7192
500	0.06857	3090.92	3433.76	6.9758	0.05665	3082.20	3422.12	6.8802
550	0.07368	3181.82	3550.23	7.1217	0.06101	3174.57	3540.62	7.0287
600	0.07869	3273.01	3666.47	7.2588	0.06525	3266.89	3658.40	7.1676
700	0.08849	3457.67	3900.13	7.5122	0.07352	3453.15	3894.28	7.4234
800	0.09811	3646.62	4137.17	7.7440	0.08160	3643.12	4132.74	7.6566
900	0.10762	3840.71	4378.82	7.9593	0.08958	3837.84	4375.29	7.8727
1000	0.11707	4040.35	4625.69	8.1612	0.09749	4037.83	4622.74	8.0751
1100	0.12648	4245.61	4878.02	8.3519	0.10536	4243.26	4875.42	8.2661
1200	0.13587	4456.30	5135.67	8.5330	0.11321	4454.00	5133.28	8.4473
1300	0.14526	4671.96	5398.24	8.7055	0.12106	4669.64	5395.97	8.6199

(Continued)

TABLE B.1.3 SI (Continued) *Superheated Vapor Water*

Temp. C	v m³/kg	u kJ/kg	h kJ/kg	s kJ/kg K	v m³/kg	u kJ/kg	h kJ/kg	s kJ/kg K
			7000 kPa (285.88)				8000 kPa (295.06)	
Sat.	0.02737	2580.48	2772.07	5.8132	0.02352	2569.79	2757.94	5.7431
300	0.02947	2632.13	2838.40	5.9304	0.02426	2590.93	2784.98	5.7905
350	0.03524	2769.34	3016.02	6.2282	0.02995	2747.67	2987.30	6.1300
400	0.03993	2878.55	3158.07	6.4477	0.03432	2863.75	3138.28	6.3633
450	0.04416	2977.91	3287.04	6.6326	0.03817	2966.66	3271.99	6.5550
500	0.04814	3073.33	3410.29	6.7974	0.04175	3064.30	3398.27	6.7239
550	0.05195	3167.21	3530.87	6.9486	0.04516	3159.76	3521.01	6.8778
600	0.05565	3260.69	3650.26	7.0894	0.04845	3254.43	3642.03	7.0205
700	0.06283	3448.60	3888.39	7.3476	0.05481	3444.00	3882.47	7.2812
800	0.06981	3639.61	4128.30	7.5822	0.06097	3636.08	4123.84	7.5173
900	0.07669	3834.96	4371.77	7.7991	0.06702	3832.08	4368.26	7.7350
1000	0.08350	4035.31	4619.80	8.0020	0.07301	4032.81	4616.87	7.9384
1100	0.09027	4240.92	4872.83	8.1933	0.07896	4238.60	4870.25	8.1299
1200	0.09703	4451.72	5130.90	8.3747	0.08489	4449.45	5128.54	8.3115
1300	0.10377	4667.33	5393.71	8.5472	0.09080	4665.02	5391.46	8.4842
			9000 kPa (303.40)				10000 kPa (311.06)	
Sat.	0.02048	2557.75	2742.11	5.6771	0.01803	2544.41	2724.67	5.6140
350	0.02580	2724.38	2956.55	6.0361	0.02242	2699.16	2923.39	5.9442
400	0.02993	2848.38	3117.76	6.2853	0.02641	2832.38	3096.46	6.2119
450	0.03350	2955.13	3256.59	6.4843	0.02975	2943.32	3240.83	6.4189
500	0.03677	3055.12	3386.05	6.6575	0.03279	3045.77	3373.63	6.5965
550	0.03987	3152.20	3511.02	6.8141	0.03564	3144.54	3500.92	6.7561
600	0.04285	3248.09	3633.73	6.9588	0.03837	3241.68	3625.34	6.9028
650	0.04574	3343.65	3755.32	7.0943	0.04101	3338.22	3748.27	7.0397
700	0.04857	3439.38	3876.51	7.2221	0.04358	3434.72	3870.52	7.1687
800	0.05409	3632.53	4119.38	7.4597	0.04859	3628.97	4114.91	7.4077
900	0.05950	3829.20	4364.74	7.6782	0.05349	3826.32	4361.24	7.6272
1000	0.06485	4030.30	4613.95	7.8821	0.05832	4027.81	4611.04	7.8315
1100	0.07016	4236.28	4867.69	8.0739	0.06312	4233.97	4865.14	8.0236
1200	0.07544	4447.18	5126.18	8.2556	0.06789	4444.93	5123.84	8.2054
1300	0.08072	4662.73	5389.22	8.4283	0.07265	4660.44	5386.99	8.3783

(Continued)

TABLE B.1.3 SI (Continued) *Superheated Vapor Water*

Temp. C	v m³/kg	u kJ/kg	h kJ/kg	s kJ/kg K	v m³/kg	u kJ/kg	h kJ/kg	s kJ/kg K
		12500 kPa (327.89)				15000 kPa (342.24)		
Sat.	0.01350	2505.08	2673.77	5.4623	0.01034	2455.43	2610.49	5.3097
350	0.01613	2624.57	2826.15	5.7117	0.01147	2520.36	2692.41	5.4420
400	0.02000	2789.25	3039.30	6.0416	0.01565	2740.70	2975.44	5.8810
450	0.02299	2912.44	3199.78	6.2718	0.01845	2879.47	3156.15	6.1403
500	0.02560	3021.68	3341.72	6.4617	0.02080	2996.52	3308.53	6.3442
550	0.02801	3124.94	3475.13	6.6289	0.02293	3104.71	3448.61	6.5198
600	0.03029	3225.37	3604.05	6.7810	0.02491	3208.64	3582.30	6.6775
650	0.03248	3324.43	3730.44	6.9218	0.02680	3310.37	3712.32	6.8223
700	0.03460	3422.93	3855.41	7.0536	0.02861	3410.94	3840.12	6.9572
800	0.03869	3620.02	4103.69	7.2965	0.03210	3610.99	4092.43	7.2040
900	0.04267	3819.11	4352.48	7.5181	0.03546	3811.89	4343.75	7.4279
1000	0.04658	4021.59	4603.81	7.7237	0.03875	4015.41	4596.63	7.6347
1100	0.05045	4228.23	4858.82	7.9165	0.04200	4222.55	4852.56	7.8282
1200	0.05430	4439.33	5118.02	8.0987	0.04523	4433.78	5112.27	8.0108
1300	0.05813	4654.76	5381.44	8.2717	0.04845	4649.12	5375.94	8.1839
		17500 kPa (354.75)				20000 kPa (365.81)		
Sat.	0.00792	2390.19	2528.79	5.1418	0.00583	2293.05	2409.74	4.9269
400	0.01245	2684.98	2902.82	5.7212	0.00994	2619.22	2818.07	5.5539
450	0.01517	2844.15	3109.69	6.0182	0.01270	2806.16	3060.06	5.9016
500	0.01736	2970.25	3274.02	6.2382	0.01477	2942.82	3238.18	6.1400
550	0.01929	3083.84	3421.37	6.4229	0.01656	3062.34	3393.45	6.3347
600	0.02106	3191.51	3560.13	6.5866	0.01818	3174.00	3537.57	6.5048
650	0.02274	3296.04	3693.94	6.7356	0.01969	3281.46	3675.32	6.6582
700	0.02434	3398.78	3824.67	6.8736	0.02113	3386.46	3809.09	6.7993
750	0.02588	3500.56	3953.48	7.0026	0.02251	3490.01	3940.27	6.9308
800	0.02738	3601.89	4081.13	7.1245	0.02385	3592.73	4069.80	7.0544
900	0.03031	3804.67	4335.05	7.3507	0.02645	3797.44	4326.37	7.2830
1000	0.03316	4009.25	4589.52	7.5588	0.02897	4003.12	4582.45	7.4925
1100	0.03597	4216.90	4846.37	7.7530	0.03145	4211.30	4840.24	7.6874
1200	0.03876	4428.28	5106.59	7.9359	0.03391	4422.81	5100.96	7.8706
1300	0.04154	4643.52	5370.50	8.1093	0.03636	4637.95	5365.10	8.0441
		25000 kPa				30000 kPa		
375	0.001973	1798.60	1847.93	4.0319	0.001789	1737.75	1791.43	3.9303
400	0.006004	2430.05	2580.16	5.1418	0.002790	2067.34	2151.04	4.4728
425	0.007882	2609.21	2806.25	5.4722	0.005304	2455.06	2614.17	5.1503
450	0.009162	2720.65	2949.70	5.6743	0.006735	2619.30	2821.35	5.4423
500	0.011124	2884.29	3162.39	5.9592	0.008679	2820.67	3081.03	5.7904
550	0.012724	3017.51	3335.62	6.1764	0.010168	2970.31	3275.36	6.0342
600	0.014138	3137.92	3491.36	6.3602	0.011446	3100.53	3443.91	6.2330
650	0.015433	3251.64	3637.46	6.5229	0.012596	3221.04	3598.93	6.4057
700	0.016647	3361.39	3777.56	6.6707	0.013661	3335.84	3745.67	6.5606

(Continued)

TABLE B.1.3 SI (Continued) *Superheated Vapor Water*

Temp. C	v m³/kg	u kJ/kg	h kJ/kg	s kJ/kg K	v m³/kg	u kJ/kg	h kJ/kg	s kJ/kg K
			25000 kPa				30000 kPa	
800	0.018913	3574.26	4047.08	6.9345	0.015623	3555.60	4024.31	6.8332
900	0.021045	3782.97	4309.09	7.1679	0.017448	3768.48	4291.93	7.0717
1000	0.023102	3990.92	4568.47	7.3801	0.019196	3978.79	4554.68	7.2867
1100	0.025119	4200.18	4828.15	7.5765	0.020903	4189.18	4816.28	7.4845
1200	0.027115	4412.00	5089.86	7.7604	0.022589	4401.29	5078.97	7.6691
1300	0.029101	4626.91	5354.44	7.9342	0.024266	4615.96	5343.95	7.8432
			35000 kPa				40000 kPa	
375	0.001700	1702.86	1762.37	3.8721	0.001641	1677.09	1742.71	3.8289
400	0.002100	1914.02	1987.52	4.2124	0.001908	1854.52	1930.83	4.1134
425	0.003428	2253.42	2373.41	4.7747	0.002532	2096.83	2198.11	4.5028
450	0.004962	2498.71	2672.36	5.1962	0.003693	2365.07	2512.79	4.9459
500	0.006927	2751.88	2994.34	5.6281	0.005623	2678.36	2903.26	5.4699
550	0.008345	2920.94	3213.01	5.9025	0.006984	2869.69	3149.05	5.7784
600	0.009527	3062.03	3395.49	6.1178	0.008094	3022.61	3346.38	6.0113
650	0.010575	3189.79	3559.91	6.3010	0.009064	3158.04	3520.58	6.2054
700	0.011533	3309.89	3713.54	6.4631	0.009942	3283.63	3681.29	6.3750
800	0.013278	3536.81	4001.54	6.7450	0.011523	3517.89	3978.80	6.6662
900	0.014883	3753.96	4274.87	6.9886	0.012963	3739.42	4257.93	6.9150
1000	0.016410	3966.70	4541.05	7.2063	0.014324	3954.64	4527.59	7.1356
1100	0.017895	4178.25	4804.59	7.4056	0.015643	4167.38	4793.08	7.3364
1200	0.019360	4390.67	5068.26	7.5910	0.016940	4380.11	5057.72	7.5224
1300	0.020815	4605.09	5333.62	7.7652	0.018229	4594.28	5323.45	7.6969
			50000 kPa				60000 kPa	
375	0.001559	1638.55	1716.52	3.7638	0.001503	1609.34	1699.51	3.7140
400	0.001731	1788.04	1874.58	4.0030	0.001633	1745.34	1843.35	3.9317
425	0.002007	1959.63	2059.98	4.2733	0.001817	1892.66	2001.65	4.1625
450	0.002486	2159.60	2283.91	4.5883	0.002085	2053.86	2178.96	4.4119
500	0.003892	2525.45	2720.07	5.1725	0.002956	2390.53	2567.88	4.9320
550	0.005118	2763.61	3019.51	5.5485	0.003957	2658.76	2896.16	5.3440
600	0.006112	2941.98	3247.59	5.8177	0.004835	2861.14	3151.21	5.6451
650	0.006966	3093.56	3441.84	6.0342	0.005595	3028.83	3364.55	5.8829
700	0.007727	3230.54	3616.91	6.2189	0.006272	3177.25	3553.56	6.0824
800	0.009076	3479.82	3933.62	6.5290	0.007459	3441.60	3889.12	6.4110
900	0.010283	3710.26	4224.41	6.7882	0.008508	3680.97	4191.47	6.6805
1000	0.011411	3930.53	4501.09	7.0146	0.009480	3906.36	4475.16	6.9126
1100	0.012497	4145.72	4770.55	7.2183	0.010409	4124.07	4748.61	7.1194
1200	0.013561	4359.12	5037.15	7.4058	0.011317	4338.18	5017.19	7.3082
1300	0.014616	4572.77	5303.56	7.5807	0.012215	4551.35	5284.28	7.4837

(Continued)

TABLE B.1.4 SI *Compressed Liquid Water*

Temp. C	v m³/kg	u kJ/kg	h kJ/kg	s kJ/kg K	v m³/kg	u kJ/kg	h kJ/kg	s kJ/kg K
		5000 kPa (263.99)				10000 kPa (311.06)		
Sat.	0.001286	1147.78	1154.21	2.9201	0.001452	1393.00	1407.53	3.3595
0	0.000998	0.03	5.02	0.0001	0.000995	0.10	10.05	0.0003
20	0.00100	83.64	88.64	0.2955	0.000997	83.35	93.32	0.2945
40	0.001006	166.93	171.95	0.5705	0.001003	166.33	176.36	0.5685
60	0.001015	250.21	255.28	0.8284	0.001013	249.34	259.47	0.8258
80	0.001027	333.69	338.83	1.0719	0.001025	332.56	342.81	1.0687
100	0.001041	417.50	422.71	1.3030	0.001039	416.09	426.48	1.2992
120	0.001058	501.79	507.07	1.5232	0.001055	500.07	510.61	1.5188
140	0.001077	586.74	592.13	1.7342	0.001074	584.67	595.40	1.7291
160	0.001099	672.61	678.10	1.9374	0.001195	670.11	681.07	1.9316
180	0.001124	759.62	765.24	2.1341	0.001120	756.63	767.83	2.1274
200	0.001153	848.08	853.85	2.3254	0.001148	844.49	855.97	2.3178
220	0.001187	938.43	944.36	2.5128	0.001181	934.07	945.88	2.5038
240	0.001226	1031.34	1037.47	2.6978	0.001219	1025.94	1038.13	2.6872
260	0.001275	1127.92	1134.30	2.8829	0.001265	1121.03	1133.68	2.8698
280					0.001322	1220.90	1234.11	3.0547
300					0.001397	1328.34	1342.31	3.2468
		15000 kPa (342.24)				20000 kPa (365.81)		
Sat.	0.001658	1585.58	1610.45	3.6847	0.002035	1785.47	1826.18	4.0137
0	0.000993	0.15	15.04	0.0004	0.000990	0.20	20.00	0.0004
20	0.000995	83.05	97.97	0.2934	0.000993	82.75	102.61	0.2922
40	0.001001	165.73	180.75	0.5665	0.000999	165.15	185.14	0.5646
60	0.001011	248.49	263.65	0.8231	0.001008	247.66	267.82	0.8205
80	0.001022	331.46	346.79	1.0655	0.001020	330.38	350.78	1.0623
100	0.001036	414.72	430.26	1.2954	0.001034	413.37	434.04	1.2917
120	0.001052	498.39	514.17	1.5144	0.001050	496.75	517.74	1.5101
140	0.001071	582.64	598.70	1.7241	0.001068	580.67	602.03	1.7192
160	0.001092	667.69	684.07	1.9259	0.001089	665.34	687.11	1.9203
180	0.001116	753.74	770.48	2.1209	0.001112	750.94	773.18	2.1146
200	0.001143	841.04	858.18	2.3103	0.001139	837.70	860.47	2.3031
220	0.001175	929.89	947.52	2.4952	0.001169	925.89	949.27	2.4869
240	0.001211	1020.82	1038.99	2.6770	0.001205	1015.94	1040.04	2.6673
260	0.001255	1114.59	1133.41	2.8575	0.001246	1108.53	1133.45	2.8459
280	0.001308	1212.47	1232.09	3.0392	0.001297	1204.69	1230.62	3.0248
300	0.001377	1316.58	1337.23	3.2259	0.001360	1306.10	1333.29	3.2071
320	0.001472	1431.05	1453.13	3.4246	0.001444	1415.66	1444.53	3.3978
340	0.001631	1567.42	1591.88	3.6545	0.001568	1539.64	1571.01	3.6074
360					0.001823	1702.78	1739.23	3.8770

(Continued)

TABLE B.1.4 SI (Continued) *Compressed Liquid Water*

Temp. C	v m³/kg	u kJ/kg	h kJ/kg	s kJ/kg K	v m³/kg	u kJ/kg	h kJ/kg	s kJ/kg K
	30000 kPa				50000 kPa			
0	0.000986	0.25	29.82	0.0001	0.000977	0.20	49.03	-0.0014
20	0.000989	82.16	111.82	0.2898	0.000980	80.98	130.00	0.2847
40	0.000995	164.01	193.87	0.5606	0.000987	161.84	211.20	0.5526
60	0.001004	246.03	276.16	0.8153	0.000996	242.96	292.77	0.8051
80	0.001016	328.28	358.75	1.0561	0.001007	324.32	374.68	1.0439
100	0.001029	410.76	441.63	1.2844	0.001020	405.86	456.87	1.2703
120	0.001044	493.58	524.91	1.5017	0.001035	487.63	539.37	1.4857
140	0.001062	576.86	608.73	1.7097	0.001052	569.76	622.33	1.6915
160	0.001082	660.81	693.27	1.9095	0.001070	652.39	705.91	1.8890
180	0.001105	745.57	778.71	2.1024	0.001091	735.68	790.24	2.0793
200	0.001130	831.34	865.24	2.2892	0.001115	819.73	875.46	2.2634
220	0.001159	918.32	953.09	2.4710	0.001141	904.67	961.71	2.4419
240	0.001192	1006.84	1042.60	2.6489	0.001170	990.69	1049.20	2.6158
260	0.001230	1097.38	1134.29	2.8242	0.001203	1078.06	1138.23	2.7860
280	0.001275	1190.69	1228.96	2.9985	0.001242	1167.19	1229.26	2.9536
300	0.001330	1287.89	1327.80	3.1740	0.001286	1258.66	1322.95	3.1200
320	0.001400	1390.64	1432.63	3.3538	0.001339	1353.23	1420.17	3.2867
340	0.001492	1501.71	1546.47	3.5425	0.001403	1451.91	1522.07	3.4556
360	0.001627	1626.57	1675.36	3.7492	0.001484	1555.97	1630.16	3.6290
380	0.001869	1781.35	1837.43	4.0010	0.001588	1667.13	1746.54	3.8100

(Continued)

TABLE B.1.5 SI *Saturated Solid-Saturated Vapor Water*

Temp.	Press.	SpecificVolume, m³/kg			Internal Energy, kJ/kg		
		Sat. Solid	Evap.	Sat. Vapor	Sat. Solid	Evap.	Sat. Vapor
C T	kPa P	v_i	v_{ig}	v_g	u_i	u_{ig}	u_g
0.01	0.6113	0.0010908	206.152	206.153	−333.40	2708.7	2375.3
0	0.6108	0.0010908	206.314	206.315	−333.42	2708.7	2375.3
−2	0.5177	0.0010905	241.662	241.663	−337.61	2710.2	2372.5
−4	0.4376	0.0010901	283.798	283.799	−341.78	2711.5	2369.8
−6	0.3689	0.0010898	334.138	334.139	−345.91	2712.9	2367.0
−8	0.3102	0.0010894	394.413	394.414	−350.02	2714.2	2364.2
−10	0.2601	0.0010891	466.756	466.757	−354.09	2715.5	2361.4
−12	0.2176	0.0010888	553.802	553.803	−358.14	2716.8	2358.7
−14	0.1815	0.0010884	658.824	658.824	−362.16	2718.0	2355.9
−16	0.1510	0.0010881	785.906	785.907	−366.14	2719.2	2353.1
−18	0.1252	0.0010878	940.182	940.183	−370.10	2720.4	2350.3
−20	0.10355	0.0010874	1128.112	1128.113	−374.03	2721.6	2347.5
−22	0.08535	0.0010871	1357.863	1357.864	−377.93	2722.7	2344.7
−24	0.07012	0.0010868	1639.752	1639.753	−381.80	2723.7	2342.0
−26	0.05741	0.0010864	1986.775	1986.776	−385.64	2724.8	2339.2
−28	0.04684	0.0010861	2415.200	2415.201	−389.45	2725.8	2336.4
−30	0.03810	0.0010858	2945.227	2945.228	−393.23	2726.8	2333.6
−32	0.03090	0.0010854	3601.822	3601.823	−396.98	2727.8	2330.8
−34	0.02499	0.0010851	4416.252	4416.253	−400.71	2728.7	2328.0
−36	0.02016	0.0010848	5430.115	5430.116	−404.40	2729.6	2325.2
−38	0.01618	0.0010844	6707.021	6707.022	−408.06	2730.5	2322.4
−40	0.01286	0.0010841	8366.395	8366.396	−411.70	2731.3	2319.6

(Continued)

TABLE B.1.5 SI (Continued) *Saturated Solid-Saturated Vapor Water*

Temp.	Press.	Enthalpy, kJ/kg			Entropy, kJ/kg K		
C T	kPa P	Sat. Solid h_i	Evap. h_{ig}	Sat. Vapor h_g	Sat. Solid s_i	Evap. s_{ig}	Sat. Vapor s_g
0.01	0.6113	−333.40	2834.7	2501.3	−1.2210	10.3772	9.1562
0	0.6108	−333.42	2834.8	2501.3	−1.2211	10.3776	9.1565
−2	0.5177	−337.61	2835.3	2497.6	−1.2369	10.4562	9.2193
−4	0.4376	−341.78	2835.7	2494.0	−1.2526	10.5358	9.2832
−6	0.3689	−345.91	2836.2	2490.3	−1.2683	10.6165	9.3482
−8	0.3102	−350.02	2836.6	2486.6	−1.2839	10.6982	9.4143
−10	0.2601	−354.09	2837.0	2482.9	−1.2995	10.7809	9.4815
−12	0.2176	−358.14	2837.3	2479.2	−1.3150	10.8648	9.5498
−14	0.1815	−362.16	2837.6	2475.5	−1.3306	10.9498	9.6192
−16	0.1510	−366.14	2837.9	2471.8	−1.3461	11.0359	9.6898
−18	0.1252	−370.10	2838.2	2468.1	−1.3617	11.1233	9.7616
−20	0.10355	−374.03	2838.4	2464.3	−1.3772	11.2120	9.8348
−22	0.08535	−377.93	2838.6	2460.6	−1.3928	11.3020	9.9093
−24	0.07012	−381.80	2838.7	2456.9	−1.4083	11.3935	9.9852
−26	0.05741	−385.64	2838.9	2453.2	−1.4239	11.4864	10.0625
−28	0.04684	−389.45	2839.0	2449.5	−1.4394	11.5808	10.1413
−30	0.03810	−393.23	2839.0	2445.8	−1.4550	11.6765	10.2215
−32	0.03090	−396.98	2839.1	2442.1	−1.4705	11.7733	10.3028
−34	0.02499	−400.71	2839.1	2438.4	−1.4860	11.8713	10.3853
−36	0.02016	−404.40	2839.1	2434.7	−1.5014	11.9704	10.4690
−38	0.01618	−408.06	2839.0	2431.0	−1.5168	12.0714	10.5546
−40	0.01286	−411.70	2838.9	2427.2	−1.5321	12.1768	10.6447

TABLE B.2 SI *Thermodynamic Properties of Ammonia*
TABLE B.2.1 SI *Saturated Ammonia*

Temp.	Press.	SpecificVolume, m³/kg			Internal Energy, kJ/kg		
C	kPa	Sat. Liquid	Evap.	Sat. Vapor	Sat. Liquid	Evap.	Sat. Vapor
T	P	v_f	v_{fg}	v_g	u_f	u_{fg}	u_g
-50	40.9	0.001424	2.62557	2.62700	-43.82	1309.1	1265.2
-45	54.5	0.001437	2.00489	2.00632	-22.01	1293.5	1271.4
-40	71.7	0.001450	1.55111	1.55256	-0.10	1277.6	1277.4
-35	93.2	0.001463	1.21466	1.21613	21.93	1261.3	1283.3
-30	119.5	0.001476	0.96192	0.96339	44.08	1244.8	1288.9
-25	151.6	0.001490	0.76970	0.77119	66.36	1227.9	1294.3
-20	190.2	0.001504	0.62184	0.62334	88.76	1210.7	1299.5
-15	236.3	0.001519	0.50686	0.50838	111.30	1193.2	1304.5
-10	290.9	0.001534	0.41655	0.41808	133.96	1175.2	1309.2
-5	354.9	0.001550	0.34493	0.34648	156.76	1157.0	1313.7
0	429.6	0.001566	0.28763	0.28920	179.69	1138.3	1318.0
5	515.9	0.001583	0.24140	0.24299	202.77	1119.2	1322.0
10	615.2	0.001600	0.20381	0.20541	225.99	1099.7	1325.7
15	728.6	0.001619	0.17300	0.17462	249.36	1079.7	1329.1
20	857.5	0.001638	0.14758	0.14922	272.89	1059.3	1332.2
25	1003.2	0.001658	0.12647	0.12813	296.59	1038.4	1335.0
30	1167.0	0.001680	0.10881	0.11049	320.46	1016.9	1337.4
35	1350.4	0.001702	0.09397	0.09567	344.50	994.9	1339.4
40	1554.9	0.001725	0.08141	0.08313	368.74	972.2	1341.0
45	1782.0	0.001750	0.07073	0.07248	393.19	948.9	1342.1
50	2033.1	0.001777	0.06159	0.06337	417.87	924.8	1342.7
55	2310.1	0.001804	0.05375	0.05555	442.79	899.9	1342.7
60	2614.4	0.001834	0.04697	0.04880	467.99	874.2	1342.1
65	2947.8	0.001866	0.04109	0.04296	493.51	847.4	1340.9
70	3312.0	0.001900	0.03597	0.03787	519.39	819.5	1338.9
75	3709.0	0.001937	0.03148	0.03341	545.70	790.4	1336.1
80	4140.5	0.001978	0.02753	0.02951	572.50	759.9	1332.4
85	4608.6	0.002022	0.02404	0.02606	599.90	727.8	1327.7
90	5115.3	0.002071	0.02093	0.02300	627.99	693.7	1321.7
95	5662.9	0.002126	0.01815	0.02028	656.95	657.4	1314.4
100	6253.7	0.002188	0.01565	0.01784	686.96	618.4	1305.3
105	6890.4	0.002261	0.01337	0.01564	718.30	575.9	1294.2
110	7575.7	0.002347	0.01128	0.01363	751.37	529.1	1280.5
115	8313.3	0.002452	0.00933	0.01178	786.82	476.2	1263.1
120	9107.2	0.002589	0.00744	0.01003	825.77	414.5	1240.3
125	9963.5	0.002783	0.00554	0.00833	870.69	337.7	1208.4
130	10891.6	0.003122	0.00337	0.00649	929.29	226.9	1156.2
132.3	11333.2	0.004255	0	0.00426	1037.62	0	1037.6

(Continued)

TABLE B.2.1 SI *Saturated Ammonia*

Temp.	Press.	Enthalpy, kJ/kg			Entropy, kJ/kg K		
C	kPa	Sat. Liquid	Evap.	Sat. Vapor	Sat. Liquid	Evap.	Sat. Vapor
T	P	h_f	h_{fg}	h_g	s_f	s_{fg}	s_g
-50	40.9	-43.76	1416.3	1372.6	-0.1916	6.3470	6.1554
-45	54.5	-21.94	1402.8	1380.8	-0.0950	6.1484	6.0534
-40	71.7	0	1388.8	1388.8	0	5.9567	5.9567
-35	93.2	22.06	1374.5	1396.5	0.0935	5.7715	5.8650
-30	119.5	44.26	1359.8	1404.0	0.1856	5.5922	5.7778
-25	151.6	66.58	1344.6	1411.2	0.2763	5.4185	5.6947
-20	190.2	89.05	1329.0	1418.0	0.3657	5.2498	5.6155
-15	236.3	111.66	1312.9	1424.6	0.4538	5.0859	5.5397
-10	290.9	134.41	1296.4	1430.8	0.5408	4.9265	5.4673
-5	354.9	157.31	1279.4	1436.7	0.6266	4.7711	5.3977
0	429.6	180.36	1261.8	1442.2	0.7114	4.6195	5.3309
5	515.9	203.58	1243.7	1447.3	0.7951	4.4715	5.2666
10	615.2	226.97	1225.1	1452.0	0.8779	4.3266	5.2045
15	728.6	250.54	1205.8	1456.3	0.9598	4.1846	5.1444
20	857.5	274.30	1185.9	1460.2	1.0408	4.0452	5.0860
25	1003.2	298.25	1165.2	1463.5	1.1210	3.9083	5.0293
30	1167.0	322.42	1143.9	1466.3	1.2005	3.7734	4.9738
35	1350.4	346.80	1121.8	1468.6	1.2792	3.6403	4.9196
40	1554.9	371.43	1098.8	1470.2	1.3574	3.5088	4.8662
45	1782.0	396.31	1074.9	1471.2	1.4350	3.3786	4.8136
50	2033.1	421.48	1050.0	1471.5	1.5121	3.2493	4.7614
55	2310.1	446.96	1024.1	1471.0	1.5888	3.1208	4.7095
60	2614.4	472.79	997.0	1469.7	1.6652	2.9925	4.6577
65	2947.8	499.01	968.5	1467.5	1.7415	2.8642	4.6057
70	3312.0	525.69	938.7	1464.4	1.8178	2.7354	4.5533
75	3709.0	552.88	907.2	1460.1	1.8943	2.6058	4.5001
80	4140.5	580.69	873.9	1454.6	1.9712	2.4746	4.4458
85	4608.6	609.21	838.6	1447.8	2.0488	2.3413	4.3901
90	5115.3	638.59	800.8	1439.4	2.1273	2.2051	4.3325
95	5662.9	668.99	760.2	1429.2	2.2073	2.0650	4.2723
100	6253.7	700.64	716.2	1416.9	2.2893	1.9195	4.2088
105	6890.4	733.87	668.1	1402.0	2.3740	1.7667	4.1407
110	7575.7	769.15	614.6	1383.7	2.4625	1.6040	4.0665
115	8313.3	807.21	553.8	1361.0	2.5566	1.4267	3.9833
120	9107.2	849.36	482.3	1331.7	2.6593	1.2268	3.8861
125	9963.5	898.42	393.0	1291.4	2.7775	0.9870	3.7645
130	10892	963.29	263.7	1227.0	2.9326	0.6540	3.5866
132.3	11333	1085.85	0	1085.9	3.2316	0	3.2316

(Continued)

TABLE B.2.2 SI *Superheated Ammonia*

Temp. C	v m³/kg	h kJ/kg	s kJ/kg K	v m³/kg	h kJ/kg	s kJ/kg K	v m³/kg	h kJ/kg	s kJ/kg K
	50 kPa (-46.53)			75 kPa (-39.16)			100 kPa (-33.60)		
Sat.	2.17521	1378.3	6.0839	1.48922	1390.1	5.9411	1.13806	1398.7	5.8401
-30	2.34484	1413.4	6.2333	1.55321	1410.1	6.0247	1.15727	1406.7	5.8734
-20	2.44631	1434.6	6.3187	1.62221	1431.7	6.1120	1.21007	1428.8	5.9626
-10	2.54711	1455.7	6.4006	1.69050	1453.3	6.1954	1.26213	1450.8	6.0477
0	2.64736	1476.9	6.4795	1.75823	1474.8	6.2756	1.31362	1472.6	6.1291
10	2.74716	1498.1	6.5556	1.82551	1496.2	6.3527	1.36465	1494.4	6.2073
20	2.84661	1519.3	6.6293	1.89243	1517.7	6.4272	1.41532	1516.1	6.2826
30	2.94578	1540.6	6.7008	1.95906	1539.2	6.4993	1.46569	1537.7	6.3553
40	3.04472	1562.0	6.7703	2.02547	1560.7	6.5693	1.51582	1559.5	6.4258
50	3.14348	1583.5	6.8379	2.09168	1582.4	6.6373	1.56577	1581.2	6.4943
60	3.24209	1605.1	6.9038	2.15775	1604.1	6.7036	1.61557	1603.1	6.5609
70	3.34058	1626.9	6.9682	2.22369	1626.0	6.7683	1.66525	1625.1	6.6258
80	3.43897	1648.8	7.0312	2.28954	1648.0	6.8315	1.71482	1647.1	6.6892
100	3.63551	1693.2	7.1533	2.42099	1692.4	6.9539	1.81373	1691.7	6.8120
120	3.83183	1738.2	7.2708	2.55221	1737.5	7.0716	1.91240	1736.9	6.9300
140	4.02797	1783.9	7.3842	2.68326	1783.4	7.1853	2.01091	1782.8	7.0439
160	4.22398	1830.4	7.4941	2.81418	1829.9	7.2953	2.10927	1829.4	7.1540
180	4.41988	1877.7	7.6008	2.94499	1877.2	7.4021	2.20754	1876.8	7.2609
200	4.61570	1925.7	7.7045	3.07571	1925.3	7.5059	2.30571	1924.9	7.3648
	125 kPa (-29.07)			150 kPa (-25.22)			200 kPa (-18.86)		
Sat.	0.92365	1405.4	5.7620	0.77870	1410.9	5.6983	0.59460	1419.6	5.5979
-20	0.96271	1425.9	5.8446	0.79774	1422.9	5.7465	—	—	—
-10	1.00506	1448.3	5.9314	0.83364	1445.7	5.8349	0.61926	1440.6	5.6791
0	1.04682	1470.5	6.0141	0.86892	1468.3	5.9189	0.64648	1463.8	5.7659
10	1.08811	1492.5	6.0933	0.90373	1490.6	5.9992	0.67319	1486.8	5.8484
20	1.12903	1514.4	6.1694	0.93815	1512.8	6.0761	0.69951	1509.4	5.9270
30	1.16964	1536.3	6.2428	0.97227	1534.8	6.1502	0.72553	1531.9	6.0025
40	1.21003	1558.2	6.3138	1.00615	1556.9	6.2217	0.75129	1554.3	6.0751
50	1.25022	1580.1	6.3827	1.03984	1578.9	6.2910	0.77685	1576.6	6.1453
60	1.29026	1602.1	6.4496	1.07338	1601.0	6.3583	0.80226	1598.9	6.2133
70	1.33017	1624.1	6.5149	1.10678	1623.2	6.4238	0.82754	1621.3	6.2794
80	1.36998	1646.3	6.5785	1.14009	1645.4	6.4877	0.85271	1643.7	6.3437
100	1.44937	1691.0	6.7017	1.20646	1690.2	6.6112	0.90282	1688.8	6.4679
120	1.52852	1736.3	6.8199	1.27259	1735.6	6.7297	0.95268	1734.4	6.5869
140	1.60749	1782.2	6.9339	1.33855	1781.7	6.8439	1.00237	1780.6	6.7015
160	1.68633	1828.9	7.0443	1.40437	1828.4	6.9544	1.05192	1827.4	6.8123
180	1.76507	1876.3	7.1513	1.47009	1875.9	7.0615	1.10136	1875.0	6.9196
200	1.84371	1924.5	7.2553	1.53572	1924.1	7.1656	1.15072	1923.3	7.0239
220	1.92229	1973.4	7.3566	1.60127	1973.1	7.2670	1.20000	1972.4	7.1255

(Continued)

TABLE B.2.2 SI (Continued) *Superheated Ammonia*

Temp. C	v m³/kg	h kJ/kg	s kJ/kg K	v m³/kg	h kJ/kg	s kJ/kg K	v m³/kg	h kJ/kg	s kJ/kg K
	250 kPa (-13.66)			300 kPa (-9.24)			350 kPa (-5.36)		
Sat.	0.48213	1426.3	5.5201	0.40607	1431.7	5.4565	0.35108	1436.3	5.4026
0	0.51293	1459.3	5.6441	0.42382	1454.7	5.5420	0.36011	1449.9	5.4532
10	0.53481	1482.9	5.7288	0.44251	1478.9	5.6290	0.37654	1474.9	5.5427
20	0.55629	1506.0	5.8093	0.46077	1502.6	5.7113	0.39251	1499.1	5.6270
30	0.57745	1529.0	5.8861	0.47870	1525.9	5.7896	0.40814	1522.9	5.7068
40	0.59835	1551.7	5.9599	0.49636	1549.0	5.8645	0.42350	1546.3	5.7828
50	0.61904	1574.3	6.0309	0.51382	1571.9	5.9365	0.43865	1569.5	5.8557
60	0.63958	1596.8	6.0997	0.53111	1594.7	6.0060	0.45362	1592.6	5.9259
70	0.65998	1619.4	6.1663	0.54827	1617.5	6.0732	0.46846	1615.5	5.9938
80	0.68028	1641.9	6.2312	0.56532	1640.2	6.1385	0.48319	1638.4	6.0596
100	0.72063	1687.3	6.3561	0.59916	1685.8	6.2642	0.51240	1684.3	6.1860
120	0.76073	1733.1	6.4756	0.63276	1731.8	6.3842	0.54135	1730.5	6.3066
140	0.80065	1779.4	6.5906	0.66618	1778.3	6.4996	0.57012	1777.2	6.4223
160	0.84044	1826.4	6.7016	0.69946	1825.4	6.6109	0.59876	1824.4	6.5340
180	0.88012	1874.1	6.8093	0.73263	1873.2	6.7188	0.62728	1872.3	6.6421
200	0.91972	1922.5	6.9138	0.76572	1921.7	6.8235	0.65571	1920.9	6.7470
220	0.95923	1971.6	7.0155	0.79872	1970.9	6.9254	0.68407	1970.2	6.8491
240	0.99868	2021.5	7.1147	0.83167	2020.9	7.0247	0.71237	2020.3	6.9486
260	1.03808	2072.2	7.2115	0.86455	2071.6	7.1217	0.74060	2071.0	7.0456
	400 kPa (-1.89)			500 kPa (4.13)			600 kPa (9.28)		
Sat.	0.30942	1440.2	5.3559	0.25035	1446.5	5.2776	0.21038	1451.4	5.2133
10	0.32701	1470.7	5.4663	0.25757	1462.3	5.3340	0.21115	1453.4	5.2205
20	0.34129	1495.6	5.5525	0.26949	1488.3	5.4244	0.22154	1480.8	5.3156
30	0.35520	1519.8	5.6338	0.28103	1513.5	5.5090	0.23152	1507.1	5.4037
40	0.36884	1543.6	5.7111	0.29227	1538.1	5.5889	0.24118	1532.5	5.4862
50	0.38226	1567.1	5.7850	0.30328	1562.3	5.6647	0.25059	1557.3	5.5641
60	0.39550	1590.4	5.8560	0.31410	1586.1	5.7373	0.25981	1581.6	5.6383
70	0.40860	1613.6	5.9244	0.32478	1609.6	5.8070	0.26888	1605.7	5.7094
80	0.42160	1636.7	5.9907	0.33535	1633.1	5.8744	0.27783	1629.5	5.7778
100	0.44732	1682.8	6.1179	0.35621	1679.8	6.0031	0.29545	1676.8	5.9081
120	0.47279	1729.2	6.2390	0.37681	1726.6	6.1253	0.31281	1724.0	6.0314
140	0.49808	1776.0	6.3552	0.39722	1773.8	6.2422	0.32997	1771.5	6.1491
160	0.52323	1823.4	6.4671	0.41748	1821.4	6.3548	0.34699	1819.4	6.2623
180	0.54827	1871.4	6.5755	0.43764	1869.6	6.4636	0.36389	1867.8	6.3717
200	0.57321	1920.1	6.6806	0.45771	1918.5	6.5691	0.38071	1916.9	6.4776
220	0.59809	1969.5	6.7828	0.47770	1968.1	6.6717	0.39745	1966.6	6.5806
240	0.62289	2019.6	6.8825	0.49763	2018.3	6.7717	0.41412	2017.1	6.6808
260	0.64764	2070.5	6.9797	0.51749	2069.3	6.8692	0.43073	2068.2	6.7786
280	0.67234	2122.1	7.0747	0.53731	2121.1	6.9644	0.44729	2120.1	6.8741

(Continued)

TABLE B.2.2 SI (Continued) *Superheated Ammonia*

Temp. C	v m³/kg	h kJ/kg	s kJ/kg K	v m³/kg	h kJ/kg	s kJ/kg K	v m³/kg	h kJ/kg	s kJ/kg K
		700 kPa (13.80)			800 kPa (17.85)			900 kPa (21.52)	
Sat.	0.18148	1455.3	5.1586	0.15958	1458.6	5.1110	0.14239	1461.2	5.0686
20	0.18721	1473.0	5.2196	0.16138	1464.9	5.1328	—	—	—
30	0.19610	1500.4	5.3115	0.16947	1493.5	5.2287	0.14872	1486.5	5.1530
40	0.20464	1526.7	5.3968	0.17720	1520.8	5.3171	0.15582	1514.7	5.2447
50	0.21293	1552.2	5.4770	0.18465	1547.0	5.3996	0.16263	1541.7	5.3296
60	0.22101	1577.1	5.5529	0.19189	1572.5	5.4774	0.16922	1567.9	5.4093
70	0.22894	1601.6	5.6254	0.19896	1597.5	5.5513	0.17563	1593.3	5.4847
80	0.23674	1625.8	5.6949	0.20590	1622.1	5.6219	0.18191	1618.4	5.5565
100	0.25205	1673.7	5.8268	0.21949	1670.6	5.7555	0.19416	1667.5	5.6919
120	0.26709	1721.4	5.9512	0.23280	1718.7	5.8811	0.20612	1716.1	5.8187
140	0.28193	1769.2	6.0698	0.24590	1766.9	6.0006	0.21787	1764.5	5.9389
160	0.29663	1817.3	6.1837	0.25886	1815.3	6.1150	0.22948	1813.2	6.0541
180	0.31121	1866.0	6.2935	0.27170	1864.2	6.2254	0.24097	1862.4	6.1649
200	0.32570	1915.3	6.3999	0.28445	1913.6	6.3322	0.25236	1912.0	6.2721
220	0.34012	1965.2	6.5032	0.29712	1963.7	6.4358	0.26368	1962.3	6.3762
240	0.35447	2015.8	6.6037	0.30973	2014.5	6.5367	0.27493	2013.2	6.4774
260	0.36876	2067.1	6.7018	0.32228	2065.9	6.6350	0.28612	2064.8	6.5760
280	0.38299	2119.1	6.7975	0.33477	2118.0	6.7310	0.29726	2117.0	6.6722
300	0.39718	2171.8	6.8911	0.34722	2170.9	6.8248	0.30835	2170.0	6.7662
		1000 kPa (24.90)			1200 kPa (30.94)			1400 kPa (36.26)	
Sat.	0.12852	1463.4	5.0304	0.10751	1466.8	4.9635	0.09231	1469.0	4.9060
30	0.13206	1479.1	5.0826	—	—	—	—	—	—
40	0.13868	1508.5	5.1778	0.11287	1495.4	5.0564	0.09432	1481.6	4.9463
50	0.14499	1536.3	5.2654	0.11846	1525.1	5.1497	0.09942	1513.4	5.0462
60	0.15106	1563.1	5.3471	0.12378	1553.3	5.2357	0.10423	1543.1	5.1370
70	0.15695	1589.1	5.4240	0.12890	1580.5	5.3159	0.10882	1571.5	5.2209
80	0.16270	1614.6	5.4971	0.13387	1606.8	5.3916	0.11324	1598.8	5.2994
100	0.17389	1664.3	5.6342	0.14347	1658.0	5.5325	0.12172	1651.4	5.4443
120	0.18477	1713.4	5.7622	0.15275	1708.0	5.6631	0.12986	1702.5	5.5775
140	0.19545	1762.2	5.8834	0.16181	1757.5	5.7860	0.13777	1752.8	5.7023
160	0.20597	1811.2	5.9992	0.17071	1807.1	5.9031	0.14552	1802.9	5.8208
180	0.21638	1860.5	6.1105	0.17950	1856.9	6.0156	0.15315	1853.2	5.9343
200	0.22669	1910.4	6.2182	0.18819	1907.1	6.1241	0.16068	1903.8	6.0437
220	0.23693	1960.8	6.3226	0.19680	1957.9	6.2292	0.16813	1955.0	6.1495
240	0.24710	2011.9	6.4241	0.20534	2009.3	6.3313	0.17551	2006.7	6.2523
260	0.25720	2063.6	6.5229	0.21382	2061.3	6.4308	0.18283	2059.0	6.3523
280	0.26726	2116.0	6.6194	0.22225	2114.0	6.5278	0.19010	2111.9	6.4498
300	0.27726	2169.1	6.7137	0.23063	2167.3	6.6225	0.19732	2165.5	6.5450
320	0.28723	2222.9	6.8059	0.23897	2221.3	6.7151	0.20450	2219.8	6.6380

(Continued)

TABLE B.2.2 SI (Continued) *Superheated Ammonia*

Temp. C	v m³/kg	h kJ/kg	s kJ/kg K	v m³/kg	h kJ/kg	s kJ/kg K	v m³/kg	h kJ/kg	s kJ/kg K
	1600 kPa (41.03)			1800 kPa (45.38)			2000 kPa (49.37)		
Sat.	0.08079	1470.5	4.8553	0.07174	1471.3	4.8096	0.06444	1471.5	4.7680
50	0.08506	1501.0	4.9510	0.07381	1487.9	4.8614	0.06471	1473.9	4.7754
60	0.08951	1532.5	5.0472	0.07801	1521.4	4.9637	0.06875	1509.8	4.8848
70	0.09372	1562.3	5.1351	0.08193	1552.7	5.0561	0.07246	1542.7	4.9821
80	0.09774	1590.7	5.2167	0.08565	1582.2	5.1410	0.07595	1573.5	5.0707
100	0.10539	1644.8	5.3659	0.09267	1638.0	5.2948	0.08248	1631.1	5.2294
120	0.11268	1696.9	5.5018	0.09931	1691.2	5.4337	0.08861	1685.5	5.3714
140	0.11974	1748.0	5.6286	0.10570	1743.1	5.5624	0.09447	1738.2	5.5022
160	0.12662	1798.7	5.7485	0.11192	1794.5	5.6838	0.10016	1790.2	5.6251
180	0.13339	1849.5	5.8631	0.11801	1845.7	5.7995	0.10571	1842.0	5.7420
200	0.14005	1900.5	5.9734	0.12400	1897.2	5.9107	0.11116	1893.9	5.8540
220	0.14663	1952.0	6.0800	0.12990	1949.1	6.0180	0.11652	1946.1	5.9621
240	0.15314	2004.1	6.1834	0.13574	2001.4	6.1221	0.12182	1998.8	6.0668
260	0.15959	2056.7	6.2839	0.14152	2054.3	6.2232	0.12705	2052.0	6.1685
280	0.16599	2109.9	6.3819	0.14724	2107.8	6.3217	0.13224	2105.8	6.2675
300	0.17234	2163.7	6.4775	0.15291	2161.9	6.4178	0.13737	2160.1	6.3641
320	0.17865	2218.2	6.5710	0.15854	2216.7	6.5116	0.14246	2215.1	6.4583
340	0.18492	2273.3	6.6624	0.16414	2272.0	6.6034	0.14751	2270.7	6.5505
360	0.19115	2329.1	6.7519	0.16969	2328.0	6.6932	0.15253	2326.8	6.6406
	5000 kPa (88.90)			10000 kPa (125.20)			20000 kPa		
Sat.	0.02365	1441.4	4.3454	0.00826	1289.4	3.7587	—	—	—
100	0.02636	1501.5	4.5091	—	—	—	—	—	—
120	0.03024	1586.3	4.7306	—	—	—	—	—	—
140	0.03350	1657.3	4.9068	0.01195	1461.3	4.1839	0.00251	918.9	2.7630
160	0.03643	1721.7	5.0591	0.01461	1578.3	4.4610	0.00323	1097.2	3.1838
180	0.03916	1782.7	5.1968	0.01666	1667.2	4.6617	0.00490	1329.7	3.7087
200	0.04174	1841.8	5.3245	0.01842	1744.5	4.8287	0.00653	1497.7	4.0721
220	0.04422	1900.0	5.4450	0.02001	1816.0	4.9767	0.00782	1618.7	4.3228
240	0.04662	1957.9	5.5600	0.02150	1884.2	5.1123	0.00891	1718.6	4.5214
260	0.04895	2015.6	5.6704	0.02290	1950.6	5.2392	0.00988	1807.6	4.6916
280	0.05123	2073.6	5.7771	0.02424	2015.9	5.3596	0.01077	1890.5	4.8442
300	0.05346	2131.8	5.8805	0.02552	2080.7	5.4746	0.01159	1969.6	4.9847
320	0.05565	2190.3	5.9809	0.02676	2145.2	5.5852	0.01237	2046.3	5.1164
340	0.05779	2249.2	6.0786	0.02796	2209.6	5.6921	0.01312	2121.6	5.2412
360	0.05990	2308.6	6.1738	0.02913	2274.1	5.7955	0.01382	2195.8	5.3603
380	0.06198	2368.4	6.2668	0.03026	2338.7	5.8960	0.01450	2269.4	5.4748
400	0.06403	2428.6	6.3576	0.03137	2403.5	5.9937	0.01516	2342.6	5.5851
420	0.06606	2489.3	6.4464	0.03245	2468.5	6.0888	0.01579	2415.4	5.6917
440	0.06806	2550.4	6.5334	0.03351	2533.7	6.1815	0.01641	2488.1	5.7950

TABLE B.3 SI *Thermodynamic Properties of R-12*
TABLE B.3.1 SI *Saturated R-12*

Temp.	Press.	SpecificVolume, m³/kg			Internal Energy, kJ/kg		
		Sat. Liquid	Evap.	Sat. Vapor	Sat. Liquid	Evap.	Sat. Vapor
C	kPa	v_f	v_{fg}	v_g	u_f	u_{fg}	u_g
T	P						
-90	2.8	0.000608	4.41494	4.41555	-43.29	177.20	133.91
-80	6.2	0.000617	2.13773	2.13835	-34.73	172.54	137.82
-70	12.3	0.000627	1.12665	1.12728	-26.14	167.94	141.81
-60	22.6	0.000637	0.63727	0.63791	-17.50	163.36	145.86
-50	39.1	0.000648	0.38246	0.38310	-8.80	158.76	149.95
-45	50.4	0.000654	0.30203	0.30268	-4.43	156.44	152.01
-40	64.2	0.000659	0.24125	0.24191	-0.04	154.11	154.07
-35	80.7	0.000666	0.19473	0.19540	4.37	151.77	156.13
-30	100.4	0.000672	0.15870	0.15937	8.79	149.40	158.19
-29.8	101.3	0.000672	0.15736	0.15803	8.98	149.30	158.28
-25	123.7	0.000679	0.13049	0.13117	13.24	147.01	160.25
-20	150.9	0.000685	0.10816	0.10885	17.71	144.59	162.31
-15	182.6	0.000693	0.09033	0.09102	22.20	142.15	164.35
-10	219.1	0.000700	0.07595	0.07665	26.72	139.67	166.39
-5	261.0	0.000708	0.06426	0.06496	31.26	137.16	168.42
0	308.6	0.000716	0.05467	0.05539	35.83	134.61	170.44
5	362.6	0.000724	0.04676	0.04749	40.43	132.01	172.44
10	423.3	0.000733	0.04018	0.04091	45.06	129.36	174.42
15	491.4	0.000743	0.03467	0.03541	49.73	126.65	176.38
20	567.3	0.000752	0.03003	0.03078	54.45	123.87	178.32
25	651.6	0.000763	0.02609	0.02685	59.21	121.03	180.23
30	744.9	0.000774	0.02273	0.02351	64.02	118.09	182.11
35	847.7	0.000786	0.01986	0.02064	68.88	115.06	183.95
40	960.7	0.000798	0.01737	0.01817	73.82	111.92	185.74
45	1084.3	0.000811	0.01522	0.01603	78.83	108.66	187.49
50	1219.3	0.000826	0.01334	0.01417	83.93	105.24	189.17
55	1366.3	0.000841	0.01170	0.01254	89.12	101.66	190.78
60	1525.9	0.000858	0.01025	0.01111	94.43	97.88	192.31
65	1698.8	0.000877	0.00897	0.00985	99.87	93.86	193.73
70	1885.8	0.000897	0.00783	0.00873	105.46	89.56	195.03
75	2087.5	0.000920	0.00680	0.00772	111.23	84.94	196.17
80	2304.6	0.000946	0.00588	0.00682	117.21	79.90	197.11
85	2538.0	0.000976	0.00503	0.00600	123.45	74.34	197.80
90	2788.5	0.001012	0.00425	0.00526	130.02	68.12	198.14
95	3056.9	0.001056	0.00351	0.00456	137.01	60.98	197.99
100	3344.1	0.001113	0.00279	0.00390	144.59	52.48	197.07
105	3650.9	0.001197	0.00205	0.00324	153.15	41.58	194.73
110	3978.5	0.001364	0.00110	0.00246	164.12	24.08	188.20
112.0	4116.8	0.001792	0	0.00179	176.06	0	176.06

(Continued)

TABLE B.3.1 SI (Continued) *Saturated R-12*

Temp.	Press.	Enthalpy, kJ/kg			Entropy, kJ/kg K		
C T	kPa P	Sat. Liquid h_f	Evap. h_{fg}	Sat. Vapor h_g	Sat. Liquid s_f	Evap. s_{fg}	Sat. Vapor s_g
-90	2.8	-43.28	189.75	146.46	-0.2086	1.0359	0.8273
-80	6.2	-34.72	185.74	151.02	-0.1631	0.9616	0.7984
-70	12.3	-26.13	181.76	155.64	-0.1198	0.8947	0.7749
-60	22.6	-17.49	177.77	160.29	-0.0783	0.8340	0.7557
-50	39.1	-8.78	173.73	164.95	-0.0384	0.7785	0.7401
-45	50.4	-4.40	171.68	167.28	-0.0190	0.7524	0.7334
-40	64.2	0	169.59	169.59	0	0.7274	0.7274
-35	80.7	4.42	167.48	171.90	0.0187	0.7032	0.7219
-30	100.4	8.86	165.34	174.20	0.0371	0.6799	0.7170
-29.8	101.3	9.05	165.24	174.29	0.0379	0.6790	0.7168
-25	123.7	13.33	163.15	176.48	0.0552	0.6574	0.7126
-20	150.9	17.82	160.92	178.74	0.0731	0.6356	0.7087
-15	182.6	22.33	158.64	180.97	0.0906	0.6145	0.7051
-10	219.1	26.87	156.31	183.19	0.1080	0.5940	0.7019
-5	261.0	31.45	153.93	185.37	0.1251	0.5740	0.6991
0	308.6	36.05	151.48	187.53	0.1420	0.5545	0.6965
5	362.6	40.69	148.96	189.65	0.1587	0.5355	0.6942
10	423.3	45.37	146.37	191.74	0.1752	0.5169	0.6921
15	491.4	50.10	143.68	193.78	0.1915	0.4986	0.6902
20	567.3	54.87	140.91	195.78	0.2078	0.4806	0.6884
25	651.6	59.70	138.03	197.73	0.2239	0.4629	0.6868
30	744.9	64.59	135.03	199.62	0.2399	0.4454	0.6853
35	847.7	69.55	131.90	201.45	0.2559	0.4280	0.6839
40	960.7	74.59	128.61	203.20	0.2718	0.4107	0.6825
45	1084.3	79.71	125.16	204.87	0.2877	0.3934	0.6811
50	1219.3	84.94	121.51	206.45	0.3037	0.3760	0.6797
55	1366.3	90.27	117.65	207.92	0.3197	0.3585	0.6782
60	1525.9	95.74	113.52	209.26	0.3358	0.3407	0.6765
65	1698.8	101.36	109.10	210.46	0.3521	0.3226	0.6747
70	1885.8	107.15	104.33	211.48	0.3686	0.3040	0.6726
75	2087.5	113.15	99.14	212.29	0.3854	0.2847	0.6702
80	2304.6	119.39	93.44	212.83	0.4027	0.2646	0.6672
85	2538.0	125.93	87.11	213.04	0.4204	0.2432	0.6636
90	2788.5	132.84	79.96	212.80	0.4389	0.2202	0.6590
95	3056.9	140.23	71.71	211.94	0.4583	0.1948	0.6531
100	3344.1	148.31	61.81	210.12	0.4793	0.1656	0.6449
105	3650.9	157.52	49.05	206.57	0.5028	0.1297	0.6325
110	3978.5	169.55	28.44	197.99	0.5333	0.0742	0.6076
112.0	4116.8	183.43	0	183.43	0.5689	0	0.5689

(Continued)

TABLE B.3.2 SI *Superheated R-12*

Temp. C	v m³/kg	h kJ/kg	s kJ/kg K	v m³/kg	h kJ/kg	s kJ/kg K	v m³/kg	h kJ/kg	s kJ/kg K
	25 kPa (-58.26)			50 kPa (-45.18)			100 kPa (-30.10)		
Sat.	0.58130	161.10	0.7527	0.30515	167.19	0.7336	0.15999	174.15	0.7171
-30	0.66179	176.19	0.8187	0.32738	175.55	0.7691	0.16006	174.21	0.7174
-20	0.69001	181.74	0.8410	0.34186	181.17	0.7917	0.16770	179.99	0.7406
-10	0.71811	187.40	0.8630	0.35623	186.89	0.8139	0.17522	185.84	0.7633
0	0.74613	193.17	0.8844	0.37051	192.70	0.8356	0.18265	191.77	0.7854
10	0.77409	199.03	0.9055	0.38472	198.61	0.8568	0.18999	197.77	0.8070
20	0.80198	204.99	0.9262	0.39886	204.62	0.8776	0.19728	203.85	0.8281
30	0.82982	211.05	0.9465	0.41296	210.71	0.8981	0.20451	210.02	0.8488
40	0.85762	217.20	0.9665	0.42701	216.89	0.9181	0.21169	216.26	0.8691
50	0.88538	223.45	0.9861	0.44103	223.16	0.9378	0.21884	222.58	0.8889
60	0.91312	229.77	1.0054	0.45502	229.51	0.9572	0.22596	228.98	0.9084
70	0.94083	236.19	1.0244	0.46898	235.95	0.9762	0.23305	235.46	0.9276
80	0.96852	242.68	1.0430	0.48292	242.46	0.9949	0.24011	242.01	0.9464
90	0.99618	249.26	1.0614	0.49684	249.05	1.0133	0.24716	248.63	0.9649
100	1.02384	255.91	1.0795	0.51074	255.71	1.0314	0.25419	255.32	0.9831
110	1.05148	262.63	1.0972	0.52463	262.45	1.0493	0.26121	262.08	1.0009
120	1.07910	269.43	1.1148	0.53851	269.26	1.0668	0.26821	268.91	1.0185
	200 kPa (-12.53)			300 kPa (-0.86)			400 kPa (8.15)		
Sat.	0.08354	182.07	0.7035	0.05690	187.16	0.6969	0.04321	190.97	0.6928
0	0.08861	189.80	0.7325	0.05715	187.72	0.6989	—	—	—
10	0.09255	196.02	0.7548	0.05998	194.17	0.7222	0.04363	192.21	0.6972
20	0.09642	202.28	0.7766	0.06273	200.64	0.7446	0.04584	198.91	0.7204
30	0.10023	208.60	0.7978	0.06542	207.12	0.7663	0.04797	205.58	0.7428
40	0.10399	214.97	0.8184	0.06805	213.64	0.7875	0.05005	212.25	0.7645
50	0.10771	221.41	0.8387	0.07064	220.19	0.8081	0.05207	218.94	0.7855
60	0.11140	227.90	0.8585	0.07319	226.79	0.8282	0.05406	225.65	0.8060
70	0.11506	234.46	0.8779	0.07571	233.44	0.8479	0.05601	232.40	0.8259
80	0.11869	241.09	0.8969	0.07820	240.15	0.8671	0.05794	239.19	0.8454
90	0.12230	247.77	0.9156	0.08067	246.90	0.8860	0.05985	246.02	0.8645
100	0.12590	254.53	0.9339	0.08313	253.72	0.9045	0.06173	252.89	0.8831
110	0.12948	261.34	0.9519	0.08557	260.58	0.9226	0.06360	259.81	0.9015
120	0.13305	268.21	0.9696	0.08799	267.50	0.9405	0.06546	266.79	0.9194
130	0.13661	275.15	0.9870	0.09041	274.48	0.9580	0.06730	273.81	0.9370
140	0.14016	282.14	1.0042	0.09281	281.51	0.9752	0.06913	280.88	0.9544
150	0.14370	289.19	1.0210	0.09520	288.59	0.9922	0.07095	287.99	0.9714

(Continued)

TABLE B.3.2 SI (Continued) *Superheated R-12*

Temp. C	v m³/kg	h kJ/kg	s kJ/kg K	v m³/kg	h kJ/kg	s kJ/kg K	v m³/kg	h kJ/kg	s kJ/kg K
	500 kPa (15.60)			750 kPa (30.26)			1000 kPa (41.64)		
Sat.	0.03482	194.03	0.6899	0.02335	199.72	0.6852	0.01744	203.76	0.6820
30	0.03746	203.96	0.7235	—	—	—	—	—	—
40	0.03921	210.81	0.7457	0.02467	206.91	0.7086	—	—	—
50	0.04091	217.64	0.7672	0.02595	214.18	0.7314	0.01837	210.32	0.7026
60	0.04257	224.48	0.7881	0.02718	221.37	0.7533	0.01941	217.97	0.7259
70	0.04418	231.33	0.8083	0.02837	228.52	0.7745	0.02040	225.49	0.7481
80	0.04577	238.21	0.8281	0.02952	235.65	0.7949	0.02134	232.91	0.7695
90	0.04734	245.11	0.8473	0.03064	242.76	0.8148	0.02225	240.28	0.7900
100	0.04889	252.05	0.8662	0.03174	249.89	0.8342	0.02313	247.61	0.8100
110	0.05041	259.03	0.8847	0.03282	257.03	0.8530	0.02399	254.93	0.8293
120	0.05193	266.06	0.9028	0.03388	264.19	0.8715	0.02483	262.25	0.8482
130	0.05343	273.12	0.9205	0.03493	271.38	0.8895	0.02566	269.57	0.8665
140	0.05492	280.23	0.9379	0.03596	278.59	0.9072	0.02647	276.90	0.8845
150	0.05640	287.39	0.9550	0.03699	285.84	0.9246	0.02728	284.26	0.9021
160	0.05788	294.59	0.9718	0.03801	293.13	0.9416	0.02807	291.63	0.9193
170	0.05934	301.83	0.9884	0.03902	300.45	0.9583	0.02885	299.04	0.9362
180	0.06080	309.12	1.0046	0.04002	307.81	0.9747	0.02963	306.47	0.9528
	1500 kPa (59.22)			2000 kPa (72.88)			4000 kPa (110.32)		
Sat.	0.01132	209.06	0.6768	0.00813	211.97	0.6713	0.00239	196.90	0.6046
80	0.01305	226.73	0.7284	0.00870	219.02	0.6914	—	—	—
90	0.01377	234.77	0.7508	0.00941	228.23	0.7171	—	—	—
100	0.01446	242.65	0.7722	0.01003	236.94	0.7408	—	—	—
110	0.01512	250.41	0.7928	0.01061	245.34	0.7630	—	—	—
120	0.01575	258.10	0.8126	0.01116	253.53	0.7841	0.00374	225.18	0.6777
130	0.01636	265.74	0.8318	0.01168	261.58	0.8043	0.00433	238.69	0.7116
140	0.01696	273.35	0.8504	0.01217	269.53	0.8238	0.00478	249.93	0.7392
150	0.01754	280.94	0.8686	0.01265	277.41	0.8426	0.00517	260.12	0.7636
160	0.01811	288.52	0.8863	0.01312	285.24	0.8609	0.00552	269.71	0.7860
170	0.01867	296.11	0.9036	0.01357	293.04	0.8787	0.00585	278.90	0.8069
180	0.01922	303.70	0.9205	0.01401	300.82	0.8961	0.00615	287.82	0.8269
190	0.01977	311.31	0.9371	0.01445	308.59	0.9131	0.00643	296.55	0.8459
200	0.02031	318.93	0.9534	0.01488	316.36	0.9297	0.00671	305.14	0.8642
210	0.02084	326.58	0.9694	0.01530	324.14	0.9459	0.00697	313.61	0.8820
220	0.02137	334.24	0.9851	0.01572	331.92	0.9619	0.00723	322.01	0.8992

TABLE B.4 SI *Thermodynamic Properties of R-22*
TABLE B.4.1 SI *Saturated R-22*

Temp.	Press.	SpecificVolume, m³/kg			Internal Energy, kJ/kg		
		Sat. Liquid	Evap.	Sat. Vapor	Sat. Liquid	Evap.	Sat. Vapor
C T	kPa P	v_f	v_{fg}	v_g	u_f	u_{fg}	u_g
-70	20.5	0.000670	0.94027	0.94094	-30.62	230.13	199.51
-65	28.0	0.000676	0.70480	0.70547	-25.68	227.21	201.54
-60	37.5	0.000682	0.53647	0.53715	-20.68	224.25	203.57
-55	49.5	0.000689	0.41414	0.41483	-15.62	221.21	205.59
-50	64.4	0.000695	0.32386	0.32456	-10.50	218.11	207.61
-45	82.7	0.000702	0.25629	0.25699	-5.32	214.94	209.62
-40.8	101.3	0.000708	0.21191	0.21261	-0.87	212.18	211.31
-40	104.9	0.000709	0.20504	0.20575	-0.07	211.68	211.60
-35	131.7	0.000717	0.16568	0.16640	5.23	208.34	213.57
-30	163.5	0.000725	0.13512	0.13584	10.61	204.91	215.52
-25	201.0	0.000733	0.11113	0.11186	16.04	201.39	217.44
-20	244.8	0.000741	0.09210	0.09284	21.55	197.78	219.32
-15	295.7	0.000750	0.07688	0.07763	27.11	194.07	221.18
-10	354.3	0.000759	0.06458	0.06534	32.74	190.25	222.99
-5	421.3	0.000768	0.05457	0.05534	38.44	186.33	224.77
0	497.6	0.000778	0.04636	0.04714	44.20	182.30	226.50
5	583.8	0.000789	0.03957	0.04036	50.03	178.15	228.17
10	680.7	0.000800	0.03391	0.03471	55.92	173.87	229.79
15	789.1	0.000812	0.02918	0.02999	61.88	169.47	231.35
20	909.9	0.000824	0.02518	0.02600	67.92	164.92	232.85
25	1043.9	0.000838	0.02179	0.02262	74.04	160.22	234.26
30	1191.9	0.000852	0.01889	0.01974	80.23	155.35	235.59
35	1354.8	0.000867	0.01640	0.01727	86.53	150.30	236.82
40	1533.5	0.000884	0.01425	0.01514	92.92	145.02	237.94
45	1729.0	0.000902	0.01238	0.01328	99.42	139.50	238.93
50	1942.3	0.000922	0.01075	0.01167	106.06	133.70	239.76
55	2174.4	0.000944	0.00931	0.01025	112.85	127.56	240.41
60	2426.6	0.000969	0.00803	0.00900	119.83	121.01	240.84
65	2699.9	0.000997	0.00689	0.00789	127.04	113.94	240.98
70	2995.9	0.001030	0.00586	0.00689	134.54	106.22	240.76
75	3316.1	0.001069	0.00491	0.00598	142.44	97.61	240.05
80	3662.3	0.001118	0.00403	0.00515	150.92	87.71	238.63
85	4036.8	0.001183	0.00317	0.00436	160.32	75.78	236.10
90	4442.5	0.001282	0.00228	0.00356	171.51	59.90	231.41
95	4883.5	0.001521	0.00103	0.00255	188.93	29.89	218.83
96.0	4969.0	0.001906	0	0.00191	203.07	0	203.07

(Continued)

TABLE B.4.1 SI (Continued) *Saturated R-22*

Temp.	Press.	Enthalpy, kJ/kg			Entropy, kJ/kg K		
C	kPa	Sat. Liquid	Evap.	Sat. Vapor	Sat. Liquid	Evap.	Sat. Vapor
T	P	h_f	h_{fg}	h_g	s_f	s_{fg}	s_g
-70	20.5	-30.61	249.43	218.82	-0.1401	1.2277	1.0876
-65	28.0	-25.66	246.93	221.27	-0.1161	1.1862	1.0701
-60	37.5	-20.65	244.35	223.70	-0.0924	1.1463	1.0540
-55	49.5	-15.59	241.70	226.12	-0.0689	1.1079	1.0390
-50	64.4	-10.46	238.96	228.51	-0.0457	1.0708	1.0251
-45	82.7	-5.26	236.13	230.87	-0.0227	1.0349	1.0122
-40.8	101.3	-0.80	233.65	232.85	-0.0034	1.0053	1.0019
-40	104.9	0	233.20	233.20	0	1.0002	1.0002
-35	131.7	5.33	230.16	235.48	0.0225	0.9664	0.9889
-30	163.5	10.73	227.00	237.73	0.0449	0.9335	0.9784
-25	201.0	16.19	223.73	239.92	0.0670	0.9015	0.9685
-20	244.8	21.73	220.33	242.06	0.0890	0.8703	0.9593
-15	295.7	27.33	216.80	244.13	0.1107	0.8398	0.9505
-10	354.3	33.01	213.13	246.14	0.1324	0.8099	0.9422
-5	421.3	38.76	209.32	248.09	0.1538	0.7806	0.9344
0	497.6	44.59	205.36	249.95	0.1751	0.7518	0.9269
5	583.8	50.49	201.25	251.73	0.1963	0.7235	0.9197
10	680.7	56.46	196.96	253.42	0.2173	0.6956	0.9129
15	789.1	62.52	192.49	255.02	0.2382	0.6680	0.9062
20	909.9	68.67	187.84	256.51	0.2590	0.6407	0.8997
25	1043.9	74.91	182.97	257.88	0.2797	0.6137	0.8934
30	1191.9	81.25	177.87	259.12	0.3004	0.5867	0.8871
35	1354.8	87.70	172.52	260.22	0.3210	0.5598	0.8809
40	1533.5	94.27	166.88	261.15	0.3417	0.5329	0.8746
45	1729.0	100.98	160.91	261.90	0.3624	0.5058	0.8682
50	1942.3	107.85	154.58	262.43	0.3832	0.4783	0.8615
55	2174.4	114.91	147.80	262.71	0.4042	0.4504	0.8546
60	2426.6	122.18	140.50	262.68	0.4255	0.4217	0.8472
65	2699.9	129.73	132.55	262.28	0.4472	0.3920	0.8391
70	2995.9	137.63	123.77	261.40	0.4695	0.3607	0.8302
75	3316.1	145.99	113.90	259.89	0.4927	0.3272	0.8198
80	3662.3	155.01	102.47	257.49	0.5173	0.2902	0.8075
85	4036.8	165.09	88.60	253.69	0.5445	0.2474	0.7918
90	4442.5	177.20	70.04	247.24	0.5767	0.1929	0.7695
95	4883.5	196.36	34.93	231.28	0.6273	0.0949	0.7222
96.0	4969.0	212.54	0	212.54	0.6708	0	0.6708

(Continued)

TABLE B.4.2 SI *Superheated R-22*

Temp. C	v m³/kg	h kJ/kg	s kJ/kg K	v m³/kg	h kJ/kg	s kJ/kg K	v m³/kg	h kJ/kg	s kJ/kg K
	50 kPa (-54.80)			100 kPa (-41.03)			150 kPa (-32.02)		
Sat.	0.41077	226.21	1.0384	0.21525	232.72	1.0026	0.14727	236.83	0.9826
-40	0.44063	234.72	1.0762	0.21633	233.34	1.0052	—	—	—
-30	0.46064	240.60	1.1008	0.22675	239.36	1.0305	0.14872	238.08	0.9877
-20	0.48054	246.59	1.1250	0.23706	245.47	1.0551	0.15585	244.32	1.0129
-10	0.50036	252.68	1.1485	0.24728	251.67	1.0791	0.16288	250.63	1.0373
0	0.52010	258.87	1.1717	0.25742	257.96	1.1026	0.16982	257.02	1.0612
10	0.53977	265.18	1.1943	0.26749	264.35	1.1256	0.17670	263.50	1.0844
20	0.55939	271.59	1.2166	0.27750	270.83	1.1481	0.18352	270.06	1.1072
30	0.57897	278.12	1.2385	0.28747	277.42	1.1702	0.19028	276.71	1.1295
40	0.59851	284.74	1.2600	0.29739	284.10	1.1919	0.19701	283.45	1.1514
50	0.61801	291.48	1.2811	0.30729	290.89	1.2132	0.20370	290.29	1.1729
60	0.63749	298.32	1.3020	0.31715	297.77	1.2342	0.21036	297.22	1.1940
70	0.65694	305.26	1.3225	0.32699	304.76	1.2548	0.21700	304.25	1.2148
80	0.67636	312.31	1.3428	0.33680	311.84	1.2752	0.22361	311.37	1.2353
90	0.69577	319.47	1.3627	0.34660	319.03	1.2952	0.23020	318.58	1.2554
100	0.71516	326.72	1.3824	0.35637	326.31	1.3150	0.23678	325.90	1.2753
110	0.73454	334.07	1.4019	0.36614	333.69	1.3345	0.24333	333.30	1.2948
	200 kPa (-25.12)			250 kPa (-19.46)			300 kPa (-14.61)		
Sat.	0.11237	239.87	0.9688	0.09102	242.28	0.9583	0.07657	244.29	0.9499
-20	0.11520	243.14	0.9818	—	—	—	—	—	—
-10	0.12065	249.57	1.0068	0.09528	248.49	0.9823	0.07834	247.38	0.9617
0	0.12600	256.07	1.0310	0.09969	255.10	1.0069	0.08213	254.10	0.9868
10	0.13129	262.63	1.0546	0.10402	261.76	1.0309	0.08583	260.86	1.0111
20	0.13651	269.27	1.0776	0.10829	268.48	1.0542	0.08947	267.67	1.0347
30	0.14168	275.99	1.1002	0.11251	275.27	1.0770	0.09305	274.53	1.0577
40	0.14681	282.80	1.1222	0.11668	282.13	1.0993	0.09659	281.46	1.0802
50	0.15190	289.69	1.1439	0.12082	289.08	1.1211	0.10009	288.46	1.1022
60	0.15696	296.66	1.1652	0.12492	296.10	1.1425	0.10355	295.54	1.1238
70	0.16200	303.73	1.1861	0.12899	303.21	1.1635	0.10699	302.69	1.1449
80	0.16701	310.89	1.2066	0.13304	310.41	1.1842	0.11040	309.92	1.1657
90	0.17200	318.14	1.2269	0.13708	317.69	1.2045	0.11379	317.24	1.1861
100	0.17697	325.48	1.2468	0.14109	325.06	1.2246	0.11716	324.64	1.2062
110	0.18193	332.91	1.2665	0.14509	332.52	1.2443	0.12052	332.13	1.2260
120	0.18688	340.44	1.2858	0.14907	340.07	1.2637	0.12387	339.70	1.2455
130	0.19181	348.05	1.3050	0.15304	347.70	1.2829	0.12720	347.36	1.2648

(Continued)

TABLE B.4.2 SI (Continued) *Superheated R-22*

Temp. C	v m³/kg	h kJ/kg	s kJ/kg K	v m³/kg	h kJ/kg	s kJ/kg K	v m³/kg	h kJ/kg	s kJ/kg K
	400 kPa (-6.52)			500 kPa (0.15)			600 kPa (5.88)		
Sat.	0.05817	247.50	0.9367	0.04692	250.00	0.9267	0.03929	252.04	0.9185
0	0.06013	252.05	0.9536	—	—	—	—	—	—
10	0.06306	259.02	0.9787	0.04936	257.11	0.9522	0.04018	255.11	0.9295
20	0.06591	266.01	1.0029	0.05175	264.30	0.9772	0.04228	262.52	0.9552
30	0.06871	273.03	1.0265	0.05408	271.48	1.0013	0.04431	269.89	0.9799
40	0.07146	280.09	1.0494	0.05636	278.69	1.0247	0.04628	277.25	1.0038
50	0.07416	287.21	1.0717	0.05859	285.93	1.0474	0.04820	284.62	1.0270
60	0.07683	294.39	1.0936	0.06079	293.22	1.0696	0.05008	292.02	1.0495
70	0.07947	301.63	1.1150	0.06295	300.55	1.0913	0.05193	299.46	1.0715
80	0.08209	308.94	1.1361	0.06509	307.95	1.1126	0.05375	306.94	1.0930
90	0.08468	316.33	1.1567	0.06721	315.41	1.1334	0.05555	314.48	1.1140
100	0.08725	323.80	1.1770	0.06930	322.94	1.1539	0.05733	322.07	1.1347
110	0.08981	331.34	1.1969	0.07138	330.54	1.1740	0.05909	329.73	1.1549
120	0.09236	338.96	1.2165	0.07345	338.21	1.1937	0.06084	337.46	1.1748
130	0.09489	346.66	1.2359	0.07550	345.96	1.2132	0.06258	345.26	1.1944
140	0.09741	354.45	1.2550	0.07755	353.79	1.2324	0.06430	353.12	1.2137
150	0.09992	362.31	1.2738	0.07958	361.69	1.2513	0.06601	361.07	1.2327
	700 kPa (10.93)			800 kPa (15.47)			900 kPa (19.61)		
Sat.	0.03377	253.73	0.9116	0.02958	255.16	0.9056	0.02629	256.39	0.9002
20	0.03549	260.67	0.9357	0.03037	258.74	0.9179	0.02636	256.71	0.9013
30	0.03731	268.24	0.9611	0.03203	266.53	0.9440	0.02792	264.76	0.9283
40	0.03906	275.77	0.9855	0.03363	274.24	0.9690	0.02940	272.67	0.9540
50	0.04076	283.28	1.0091	0.03518	281.91	0.9931	0.03082	280.50	0.9786
60	0.04242	290.80	1.0320	0.03667	289.55	1.0164	0.03219	288.28	1.0023
70	0.04405	298.34	1.0543	0.03814	297.20	1.0391	0.03353	296.04	1.0253
80	0.04565	305.91	1.0761	0.03957	304.87	1.0611	0.03483	303.81	1.0476
90	0.04722	313.53	1.0973	0.04097	312.57	1.0826	0.03611	311.59	1.0693
100	0.04878	321.19	1.1181	0.04236	320.30	1.1036	0.03736	319.40	1.0905
110	0.05031	328.91	1.1386	0.04373	328.09	1.1242	0.03860	327.25	1.1113
120	0.05183	336.70	1.1586	0.04508	335.93	1.1444	0.03982	335.15	1.1316
130	0.05334	344.54	1.1783	0.04641	343.82	1.1642	0.04102	343.09	1.1516
140	0.05484	352.45	1.1977	0.04774	351.78	1.1837	0.04222	351.10	1.1712
150	0.05632	360.44	1.2168	0.04905	359.80	1.2029	0.04340	359.16	1.1905
160	0.05780	368.49	1.2356	0.05036	367.89	1.2218	0.04457	367.28	1.2094
170	0.05927	376.61	1.2541	0.05166	376.04	1.2404	0.04574	375.47	1.2281

(Continued)

TABLE B.4.2 SI (Continued) *Superheated R-22*

Temp. C	v m³/kg	h kJ/kg	s kJ/kg K	v m³/kg	h kJ/kg	s kJ/kg K	v m³/kg	h kJ/kg	s kJ/kg K
	1000 kPa (23.42)			1200 kPa (30.26)			1400 kPa (36.31)		
Sat.	0.02364	257.46	0.8954	0.01960	259.18	0.8868	0.01668	260.48	0.8792
30	0.02460	262.91	0.9136	—	—	—	—	—	—
40	0.02599	271.04	0.9400	0.02085	267.60	0.9141	0.01712	263.86	0.8901
50	0.02732	279.05	0.9651	0.02205	276.01	0.9405	0.01825	272.77	0.9181
60	0.02860	286.97	0.9893	0.02319	284.26	0.9657	0.01930	281.40	0.9444
70	0.02984	294.86	1.0126	0.02428	292.42	0.9898	0.02029	289.86	0.9694
80	0.03104	302.73	1.0352	0.02534	300.51	1.0131	0.02125	298.20	0.9934
90	0.03221	310.60	1.0572	0.02636	308.57	1.0356	0.02217	306.47	1.0165
100	0.03336	318.49	1.0786	0.02736	316.62	1.0574	0.02306	314.70	1.0388
110	0.03449	326.41	1.0996	0.02833	324.68	1.0788	0.02393	322.92	1.0606
120	0.03561	334.36	1.1200	0.02929	332.76	1.0996	0.02477	331.13	1.0817
130	0.03671	342.36	1.1401	0.03024	340.87	1.1199	0.02561	339.35	1.1024
140	0.03780	350.41	1.1599	0.03117	349.02	1.1399	0.02643	347.60	1.1226
150	0.03887	358.51	1.1792	0.03208	357.21	1.1595	0.02723	355.89	1.1424
160	0.03994	366.68	1.1983	0.03299	365.45	1.1787	0.02803	364.21	1.1618
170	0.04100	374.90	1.2171	0.03389	373.74	1.1977	0.02882	372.57	1.1809
180	0.04205	383.19	1.2356	0.03479	382.09	1.2163	0.02960	380.99	1.1997
	1600 kPa (41.75)			1800 kPa (46.71)			2000 kPa (51.28)		
Sat.	0.01446	261.43	0.8724	0.01271	262.10	0.8659	0.01129	262.53	0.8598
50	0.01535	269.26	0.8969	0.01305	265.42	0.8763	—	—	—
60	0.01635	278.36	0.9246	0.01403	275.10	0.9057	0.01213	271.56	0.8873
70	0.01728	287.17	0.9507	0.01492	284.33	0.9330	0.01301	281.31	0.9161
80	0.01817	295.80	0.9755	0.01576	293.28	0.9588	0.01381	290.64	0.9429
90	0.01901	304.30	0.9992	0.01655	302.05	0.9832	0.01456	299.70	0.9682
100	0.01983	312.73	1.0221	0.01730	310.68	1.0067	0.01528	308.57	0.9923
110	0.02061	321.10	1.0442	0.01803	319.24	1.0293	0.01596	317.32	1.0155
120	0.02138	329.46	1.0658	0.01874	327.75	1.0512	0.01662	325.99	1.0378
130	0.02213	337.81	1.0867	0.01943	336.22	1.0725	0.01726	334.61	1.0594
140	0.02287	346.16	1.1072	0.02010	344.70	1.0933	0.01788	343.20	1.0805
150	0.02359	354.54	1.1272	0.02076	353.17	1.1136	0.01849	351.78	1.1010
160	0.02430	362.95	1.1469	0.02141	361.67	1.1334	0.01909	360.37	1.1211
170	0.02501	371.39	1.1661	0.02204	370.19	1.1528	0.01967	368.97	1.1407
180	0.02570	379.87	1.1851	0.02267	378.74	1.1719	0.02025	377.60	1.1600
190	0.02639	388.40	1.2037	0.02330	387.33	1.1907	0.02082	386.25	1.1788
200	0.02707	396.97	1.2220	0.02391	395.96	1.2091	0.02138	394.94	1.1974

(Continued)

TABLE B.4.2 SI (Continued) *Superheated R-22*

Temp. C	v m³/kg	h kJ/kg	s kJ/kg K	v m³/kg	h kJ/kg	s kJ/kg K	v m³/kg	h kJ/kg	s kJ/kg K
	2500 kPa (61.38)			3000 kPa (70.07)			3500 kPa (77.70)		
Sat.	0.00868	262.61	0.8450	0.00688	261.38	0.8300	0.00552	258.72	0.8135
70	0.00946	272.68	0.8748	—	—	—	—	—	—
80	0.01024	283.33	0.9054	0.00775	274.53	0.8678	0.00576	262.74	0.8249
90	0.01095	293.34	0.9333	0.00847	286.04	0.9000	0.00660	277.27	0.8655
100	0.01160	302.94	0.9594	0.00910	296.66	0.9288	0.00726	289.50	0.8987
110	0.01221	312.26	0.9841	0.00967	306.74	0.9555	0.00783	300.64	0.9282
120	0.01279	321.40	1.0076	0.01021	316.47	0.9805	0.00835	311.13	0.9552
130	0.01334	330.41	1.0302	0.01072	325.96	1.0044	0.00883	321.20	0.9805
140	0.01388	339.34	1.0521	0.01120	335.27	1.0272	0.00928	330.98	1.0045
150	0.01440	348.21	1.0733	0.01166	344.47	1.0492	0.00970	340.55	1.0274
160	0.01491	357.04	1.0939	0.01211	353.58	1.0705	0.01011	349.99	1.0494
170	0.01540	365.86	1.1141	0.01255	362.65	1.0912	0.01051	359.32	1.0707
180	0.01589	374.68	1.1338	0.01298	371.68	1.1113	0.01089	368.59	1.0914
190	0.01636	383.51	1.1530	0.01339	380.70	1.1310	0.01127	377.81	1.1115
200	0.01683	392.35	1.1719	0.01380	389.71	1.1502	0.01163	387.00	1.1311
210	0.01730	401.23	1.1905	0.01420	398.73	1.1691	0.01199	396.19	1.1503
220	0.01776	410.13	1.2087	0.01460	407.77	1.1876	0.01234	405.37	1.1692
	4000 kPa (84.53)			5000 kPa			6000 kPa		
Sat.	0.00443	254.13	0.7935	—	—	—	—	—	—
110	0.00641	293.75	0.9009	0.00425	275.92	0.8406	0.00243	243.28	0.7467
120	0.00692	305.27	0.9306	0.00485	291.36	0.8804	0.00333	272.39	0.8218
130	0.00739	316.08	0.9578	0.00533	304.47	0.9134	0.00390	290.25	0.8668
140	0.00782	326.42	0.9831	0.00576	316.38	0.9426	0.00435	304.76	0.9023
150	0.00823	336.45	1.0071	0.00614	327.56	0.9693	0.00473	317.63	0.9331
160	0.00861	346.25	1.0300	0.00649	338.27	0.9943	0.00507	329.55	0.9609
170	0.00898	355.89	1.0520	0.00683	348.63	1.0180	0.00539	340.85	0.9867
180	0.00933	365.41	1.0732	0.00714	358.76	1.0406	0.00568	351.72	1.0110
190	0.00968	374.85	1.0939	0.00744	368.71	1.0623	0.00596	362.27	1.0340
200	0.01001	384.24	1.1139	0.00774	378.54	1.0833	0.00622	372.60	1.0561
210	0.01033	393.59	1.1335	0.00802	388.27	1.1036	0.00648	382.76	1.0773
220	0.01065	402.93	1.1526	0.00829	397.93	1.1234	0.00672	392.80	1.0979
230	0.01097	412.25	1.1713	0.00856	407.55	1.1427	0.00696	402.75	1.1179
240	0.01127	421.58	1.1897	0.00882	417.14	1.1616	0.00719	412.62	1.1373
250	0.01158	430.91	1.2077	0.00908	426.71	1.1801	0.00742	422.45	1.1563
260	0.01188	440.27	1.2254	0.00933	436.28	1.1982	0.00764	432.25	1.1748

TABLE B.5 SI *Thermodynamic Properties of R-134a*
TABLE B.5.1 SI *Saturated R-134a*

Temp.	Press.	SpecificVolume, m³/kg			Internal Energy, kJ/kg		
C T	kPa P	Sat. Liquid v_f	Evap. v_{fg}	Sat. Vapor v_g	Sat. Liquid u_f	Evap. u_{fg}	Sat. Vapor u_g
-70	8.3	0.000675	1.97207	1.97274	119.46	218.74	338.20
-65	11.7	0.000679	1.42915	1.42983	123.18	217.76	340.94
-60	16.3	0.000684	1.05199	1.05268	127.52	216.19	343.71
-55	22.2	0.000689	0.78609	0.78678	132.36	214.14	346.50
-50	29.9	0.000695	0.59587	0.59657	137.60	211.71	349.31
-45	39.6	0.000701	0.45783	0.45853	143.15	208.99	352.15
-40	51.8	0.000708	0.35625	0.35696	148.95	206.05	355.00
-35	66.8	0.000715	0.28051	0.28122	154.93	202.93	357.86
-30	85.1	0.000722	0.22330	0.22402	161.06	199.67	360.73
-26.3	101.3	0.000728	0.18947	0.19020	165.73	197.16	362.89
-25	107.2	0.000730	0.17957	0.18030	167.30	196.31	363.61
-20	133.7	0.000738	0.14576	0.14649	173.65	192.85	366.50
-15	165.0	0.000746	0.11932	0.12007	180.07	189.32	369.39
-10	201.7	0.000755	0.09845	0.09921	186.57	185.70	372.27
-5	244.5	0.000764	0.08181	0.08257	193.14	182.01	375.15
0	294.0	0.000773	0.06842	0.06919	199.77	178.24	378.01
5	350.9	0.000783	0.05755	0.05833	206.48	174.38	380.85
10	415.8	0.000794	0.04866	0.04945	213.25	170.42	383.67
15	489.5	0.000805	0.04133	0.04213	220.10	166.35	386.45
20	572.8	0.000817	0.03524	0.03606	227.03	162.16	389.19
25	666.3	0.000829	0.03015	0.03098	234.04	157.83	391.87
30	771.0	0.000843	0.02587	0.02671	241.14	153.34	394.48
35	887.6	0.000857	0.02224	0.02310	248.34	148.68	397.02
40	1017.0	0.000873	0.01915	0.02002	255.65	143.81	399.46
45	1160.2	0.000890	0.01650	0.01739	263.08	138.71	401.79
50	1318.1	0.000908	0.01422	0.01512	270.63	133.35	403.98
55	1491.6	0.000928	0.01224	0.01316	278.33	127.68	406.01
60	1681.8	0.000951	0.01051	0.01146	286.19	121.66	407.85
65	1889.9	0.000976	0.00899	0.00997	294.24	115.22	409.46
70	2117.0	0.001005	0.00765	0.00866	302.51	108.27	410.78
75	2364.4	0.001038	0.00645	0.00749	311.06	100.68	411.74
80	2633.6	0.001078	0.00537	0.00645	319.96	92.26	412.22
85	2926.2	0.001128	0.00437	0.00550	329.35	82.67	412.01
90	3244.5	0.001195	0.00341	0.00461	339.51	71.24	410.75
95	3591.5	0.001297	0.00243	0.00373	351.17	56.25	407.42
100	3973.2	0.001557	0.00108	0.00264	368.55	28.19	396.74
101.2	4064.0	0.001969	0	0.00197	382.97	0	382.97

(Continued)

TABLE B.5.1 SI (Continued) *Saturated R-134s*

Temp.	Press.	Enthalpy, kJ/kg			Entropy, kJ/kg K		
		Sat. Liquid	Evap.	Sat. Vapor	Sat. Liquid	Evap.	Sat. Vapor
C T	kPa P	h_f	h_{fg}	h_g	s_f	s_{fg}	s_g
-70	8.3	119.47	235.15	354.62	0.6645	1.1575	1.8220
-65	11.7	123.18	234.55	357.73	0.6825	1.1268	1.8094
-60	16.3	127.53	233.33	360.86	0.7031	1.0947	1.7978
-55	22.2	132.37	231.63	364.00	0.7256	1.0618	1.7874
-50	29.9	137.62	229.54	367.16	0.7493	1.0286	1.7780
-45	39.6	143.18	227.14	370.32	0.7740	0.9956	1.7695
-40	51.8	148.98	224.50	373.48	0.7991	0.9629	1.7620
-35	66.8	154.98	221.67	376.64	0.8245	0.9308	1.7553
-30	85.1	161.12	218.68	379.80	0.8499	0.8994	1.7493
-26.3	101.3	165.80	216.36	382.16	0.8690	0.8763	1.7453
-25	107.2	167.38	215.57	382.95	0.8754	0.8687	1.7441
-20	133.7	173.74	212.34	386.08	0.9007	0.8388	1.7395
-15	165.0	180.19	209.00	389.20	0.9258	0.8096	1.7354
-10	201.7	186.72	205.56	392.28	0.9507	0.7812	1.7319
-5	244.5	193.32	202.02	395.34	0.9755	0.7534	1.7288
0	294.0	200.00	198.36	398.36	1.0000	0.7262	1.7262
5	350.9	206.75	194.57	401.32	1.0243	0.6995	1.7239
10	415.8	213.58	190.65	404.23	1.0485	0.6733	1.7218
15	489.5	220.49	186.58	407.07	1.0725	0.6475	1.7200
20	572.8	227.49	182.35	409.84	1.0963	0.6220	1.7183
25	666.3	234.59	177.92	412.51	1.1201	0.5967	1.7168
30	771.0	241.79	173.29	415.08	1.1437	0.5716	1.7153
35	887.6	249.10	168.42	417.52	1.1673	0.5465	1.7139
40	1017.0	256.54	163.28	419.82	1.1909	0.5214	1.7123
45	1160.2	264.11	157.85	421.96	1.2145	0.4962	1.7106
50	1318.1	271.83	152.08	423.91	1.2381	0.4706	1.7088
55	1491.6	279.72	145.93	425.65	1.2619	0.4447	1.7066
60	1681.8	287.79	139.33	427.13	1.2857	0.4182	1.7040
65	1889.9	296.09	132.21	428.30	1.3099	0.3910	1.7008
70	2117.0	304.64	124.47	429.11	1.3343	0.3627	1.6970
75	2364.4	313.51	115.94	429.45	1.3592	0.3330	1.6923
80	2633.6	322.79	106.40	429.19	1.3849	0.3013	1.6862
85	2926.2	332.65	95.45	428.10	1.4117	0.2665	1.6782
90	3244.5	343.38	82.31	425.70	1.4404	0.2267	1.6671
95	3591.5	355.83	64.98	420.81	1.4733	0.1765	1.6498
100	3973.2	374.74	32.47	407.21	1.5228	0.0870	1.6098
101.2	4064.0	390.98	0	390.98	1.5658	0	1.5658

(Continued)

TABLE B.5.2 SI *Superheated R-134a*

Temp. C	v m³/kg	h kJ/kg	s kJ/kg K	v m³/kg	h kJ/kg	s kJ/kg K	v m³/kg	h kJ/kg	s kJ/kg K
		50 kPa (-40.67)			100 kPa (-26.54)			150 kPa (-17.29)	
Sat.	0.36889	373.06	1.7629	0.19257	381.98	1.7456	0.13139	387.77	1.7372
-20	0.40507	388.82	1.8279	0.19860	387.22	1.7665	—	—	—
-10	0.42222	396.64	1.8582	0.20765	395.27	1.7978	0.13602	393.84	1.7606
0	0.43921	404.59	1.8878	0.21652	403.41	1.8281	0.14222	402.19	1.7917
10	0.45608	412.70	1.9170	0.22527	411.67	1.8578	0.14828	410.60	1.8220
20	0.47287	420.96	1.9456	0.23392	420.05	1.8869	0.15424	419.11	1.8515
30	0.48958	429.38	1.9739	0.24250	428.56	1.9155	0.16011	427.73	1.8804
40	0.50623	437.96	2.0017	0.25101	437.22	1.9436	0.16592	436.47	1.9088
50	0.52284	446.70	2.0292	0.25948	446.03	1.9712	0.17168	445.35	1.9367
60	0.53941	455.60	2.0563	0.26791	454.99	1.9985	0.17740	454.37	1.9642
70	0.55595	464.66	2.0831	0.27631	464.10	2.0255	0.18308	463.53	1.9913
80	0.57247	473.88	2.1096	0.28468	473.36	2.0521	0.18874	472.83	2.0180
90	0.58896	483.26	2.1358	0.29302	482.78	2.0784	0.19437	482.28	2.0444
100	0.60544	492.81	2.1617	0.30135	492.35	2.1044	0.19999	491.89	2.0705
110	0.62190	502.50	2.1874	0.30967	502.07	2.1301	0.20559	501.64	2.0963
120	0.63835	512.36	2.2128	0.31797	511.95	2.1555	0.21117	511.54	2.1218
130	0.65479	522.37	2.2379	0.32626	521.98	2.1807	0.21675	521.60	2.1470
		200 kPa (-10.22)			300 kPa (0.56)			400 kPa (8.84)	
Sat.	0.10002	392.15	1.7320	0.06787	398.69	1.7259	0.05136	403.56	1.7223
0	0.10501	400.91	1.7647	—	—	—	—	—	—
10	0.10974	409.50	1.7956	0.07111	407.17	1.7564	0.05168	404.65	1.7261
20	0.11436	418.15	1.8256	0.07441	416.12	1.7874	0.05436	413.97	1.7584
30	0.11889	426.87	1.8549	0.07762	425.10	1.8175	0.05693	423.22	1.7895
40	0.12335	435.71	1.8836	0.08075	434.12	1.8468	0.05940	432.46	1.8195
50	0.12776	444.66	1.9117	0.08382	443.23	1.8755	0.06181	441.75	1.8487
60	0.13213	453.74	1.9394	0.08684	452.44	1.9035	0.06417	451.10	1.8772
70	0.13646	462.95	1.9666	0.08982	461.76	1.9311	0.06648	460.55	1.9051
80	0.14076	472.30	1.9935	0.09277	471.21	1.9582	0.06877	470.09	1.9325
90	0.14504	481.79	2.0200	0.09570	480.78	1.9850	0.07102	479.75	1.9595
100	0.14930	491.42	2.0461	0.09861	490.48	2.0113	0.07325	489.52	1.9860
110	0.15355	501.21	2.0720	0.10150	500.32	2.0373	0.07547	499.43	2.0122
120	0.15777	511.13	2.0976	0.10437	510.30	2.0631	0.07767	509.46	2.0381
130	0.16199	521.21	2.1229	0.10723	520.43	2.0885	0.07985	519.63	2.0636
140	0.16620	531.43	2.1479	0.11008	530.69	2.1136	0.08202	529.94	2.0889
150	0.17039	541.80	2.1727	0.11292	541.09	2.1385	0.08418	540.38	2.1139

(Continued)

TABLE B.5.2 SI (Continued) *Superheated R-134a*

Temp. C	v m³/kg	h kJ/kg	s kJ/kg K	v m³/kg	h kJ/kg	s kJ/kg K	v m³/kg	h kJ/kg	s kJ/kg K
	500 kPa (15.66)			600 kPa (21.52)			700 kPa (26.67)		
Sat.	0.04126	407.45	1.7198	0.03442	410.66	1.7179	0.02947	413.38	1.7163
20	0.04226	411.65	1.7342	—	—	—	—	—	—
30	0.04446	421.22	1.7663	0.03609	419.09	1.7461	0.03007	416.81	1.7277
40	0.04656	430.72	1.7971	0.03796	428.88	1.7779	0.03178	426.93	1.7606
50	0.04858	440.20	1.8270	0.03974	438.59	1.8084	0.03339	436.89	1.7919
60	0.05055	449.72	1.8560	0.04145	448.28	1.8379	0.03493	446.78	1.8220
70	0.05247	459.29	1.8843	0.04311	457.99	1.8666	0.03641	456.66	1.8512
80	0.05435	468.94	1.9120	0.04473	467.76	1.8947	0.03785	466.55	1.8796
90	0.05620	478.69	1.9392	0.04632	477.61	1.9222	0.03925	476.51	1.9074
100	0.05804	488.55	1.9660	0.04788	487.55	1.9492	0.04063	486.53	1.9347
110	0.05985	498.52	1.9924	0.04943	497.59	1.9758	0.04198	496.65	1.9614
120	0.06164	508.61	2.0184	0.05095	507.75	2.0019	0.04331	506.88	1.9878
130	0.06342	518.83	2.0440	0.05246	518.03	2.0277	0.04463	517.21	2.0137
140	0.06518	529.19	2.0694	0.05396	528.43	2.0532	0.04594	527.66	2.0393
150	0.06694	539.67	2.0945	0.05544	538.95	2.0784	0.04723	538.23	2.0646
160	0.06869	550.29	2.1193	0.05692	549.61	2.1033	0.04851	548.92	2.0896
170	0.07043	561.04	2.1438	0.05839	560.40	2.1279	0.04979	559.75	2.1143
	800 kPa (31.30)			900 kPa (35.50)			1000 kPa (39.37)		
Sat.	0.02571	415.72	1.7150	0.02276	417.76	1.7137	0.02038	419.54	1.7125
40	0.02711	424.86	1.7446	0.02345	422.64	1.7294	0.02047	420.25	1.7148
50	0.02861	435.11	1.7768	0.02487	433.23	1.7627	0.02185	431.24	1.7494
60	0.03002	445.22	1.8076	0.02619	443.60	1.7943	0.02311	441.89	1.7818
70	0.03137	455.27	1.8373	0.02745	453.83	1.8246	0.02429	452.34	1.8127
80	0.03268	465.31	1.8662	0.02865	464.03	1.8539	0.02542	462.70	1.8425
90	0.03394	475.38	1.8943	0.02981	474.22	1.8823	0.02650	473.03	1.8713
100	0.03518	485.50	1.9218	0.03094	484.44	1.9101	0.02754	483.36	1.8994
110	0.03639	495.70	1.9487	0.03204	494.73	1.9373	0.02856	493.74	1.9268
120	0.03758	505.99	1.9753	0.03313	505.09	1.9640	0.02956	504.17	1.9537
130	0.03876	516.38	2.0014	0.03419	515.54	1.9902	0.03053	514.69	1.9801
140	0.03992	526.88	2.0271	0.03524	526.10	2.0161	0.03150	525.30	2.0061
150	0.04107	537.50	2.0525	0.03628	536.76	2.0416	0.03244	536.02	2.0318
160	0.04221	548.23	2.0775	0.03731	547.54	2.0668	0.03338	546.84	2.0570
170	0.04334	559.09	2.1023	0.03832	558.44	2.0917	0.03431	557.77	2.0820
180	0.04446	570.08	2.1268	0.03933	569.45	2.1162	0.03523	568.83	2.1067

(Continued)

TABLE B.5.2 SI (Continued) *Superheated R-134a*

Temp. C	v m³/kg	h kJ/kg	s kJ/kg K	v m³/kg	h kJ/kg	s kJ/kg K	v m³/kg	h kJ/kg	s kJ/kg K
	1200 kPa (46.31)			1400 kPa (52.42)			1600 kPa (57.90)		
Sat.	0.01676	422.49	1.7102	0.01414	424.78	1.7077	0.01215	426.54	1.7051
50	0.01724	426.84	1.7237	—	—	—	—	—	—
60	0.01844	438.21	1.7584	0.01503	434.08	1.7360	0.01239	429.32	1.7135
70	0.01953	449.18	1.7908	0.01608	445.72	1.7704	0.01345	441.89	1.7507
80	0.02055	459.92	1.8217	0.01704	456.94	1.8026	0.01438	453.72	1.7847
90	0.02151	470.55	1.8514	0.01793	467.93	1.8333	0.01522	465.15	1.8166
100	0.02244	481.13	1.8801	0.01878	478.79	1.8628	0.01601	476.33	1.8469
110	0.02333	491.70	1.9081	0.01958	489.59	1.8914	0.01676	487.39	1.8762
120	0.02420	502.31	1.9354	0.02036	500.38	1.9192	0.01748	498.39	1.9045
130	0.02504	512.97	1.9621	0.02112	511.19	1.9463	0.01817	509.37	1.9321
140	0.02587	523.70	1.9884	0.02186	522.05	1.9730	0.01884	520.38	1.9591
150	0.02669	534.51	2.0143	0.02258	532.98	1.9991	0.01949	531.43	1.9855
160	0.02750	545.43	2.0398	0.02329	543.99	2.0248	0.02013	542.54	2.0115
170	0.02829	556.44	2.0649	0.02399	555.10	2.0502	0.02076	553.73	2.0370
180	0.02907	567.57	2.0898	0.02468	566.30	2.0752	0.02138	565.02	2.0622
	1800 kPa (62.89)			2000 kPa (67.48)			2500 kPa (77.57)		
Sat.	0.01057	427.85	1.7022	0.00930	428.75	1.6991	0.00694	429.41	1.6893
70	0.01134	437.56	1.7309	0.00958	432.53	1.7101	—	—	—
80	0.01227	450.20	1.7672	0.01055	446.30	1.7497	0.00722	433.80	1.7018
90	0.01310	462.16	1.8006	0.01137	458.95	1.7850	0.00816	449.50	1.7457
100	0.01385	473.74	1.8320	0.01211	471.00	1.8177	0.00891	463.28	1.7831
110	0.01456	485.09	1.8620	0.01279	482.69	1.8487	0.00956	476.13	1.8171
120	0.01523	496.32	1.8910	0.01342	494.19	1.8783	0.01015	488.46	1.8489
130	0.01587	507.50	1.9190	0.01403	505.57	1.9069	0.01069	500.47	1.8790
140	0.01649	518.66	1.9464	0.01461	516.90	1.9346	0.01121	512.31	1.9080
150	0.01709	529.84	1.9731	0.01517	528.22	1.9617	0.01170	524.04	1.9361
160	0.01768	541.07	1.9994	0.01571	539.57	1.9882	0.01217	535.72	1.9634
170	0.01825	552.36	2.0251	0.01624	550.96	2.0142	0.01262	547.40	1.9900
180	0.01881	563.72	2.0505	0.01676	562.42	2.0398	0.01307	559.10	2.0161

(Continued)

TABLE B.5.2 SI (Continued) *Superheated R-134a*

Temp. C	v m³/kg	h kJ/kg	s kJ/kg K	v m³/kg	h kJ/kg	s kJ/kg K	v m³/kg	h kJ/kg	s kJ/kg K
	3000 kPa (86.20)			3500 kPa (93.72)			4000 kPa (100.33)		
Sat.	0.00528	427.67	1.6759	0.00396	422.43	1.6552	0.00252	404.94	1.6036
90	0.00575	436.19	1.6995	—	—	—	—	—	—
100	0.00665	453.73	1.7472	0.00484	440.43	1.7039	—	—	—
110	0.00734	468.50	1.7862	0.00567	459.21	1.7535	0.00428	446.84	1.7148
120	0.00792	482.04	1.8211	0.00629	474.70	1.7935	0.00500	465.99	1.7642
130	0.00845	494.91	1.8535	0.00681	488.77	1.8288	0.00556	481.87	1.8040
140	0.00893	507.39	1.8840	0.00728	502.08	1.8614	0.00603	496.29	1.8394
150	0.00937	519.62	1.9133	0.00771	514.93	1.8922	0.00644	509.92	1.8720
160	0.00980	531.70	1.9415	0.00810	527.50	1.9215	0.00683	523.07	1.9027
170	0.01021	543.71	1.9689	0.00848	539.89	1.9498	0.00718	535.92	1.9320
180	0.01060	555.69	1.9956	0.00884	552.18	1.9772	0.00752	548.57	1.9603
	5000 kPa			6000 kPa			7000 kPa		
90	0.001089	336.61	1.4163	0.001059	334.70	1.4081	0.001037	333.29	1.4013
100	0.001216	357.68	1.4735	0.001150	353.61	1.4595	0.001110	351.10	1.4497
110	0.001659	392.10	1.5644	0.001307	375.90	1.5184	0.001215	370.68	1.5015
120	0.002969	440.47	1.6892	0.001698	406.78	1.5979	0.001393	393.45	1.5601
130	0.003705	464.63	1.7499	0.002396	441.18	1.6843	0.001720	420.73	1.6286
140	0.004226	482.86	1.7946	0.002985	466.25	1.7458	0.002169	448.28	1.6961
150	0.004652	498.77	1.8327	0.003439	485.82	1.7926	0.002599	471.55	1.7518
160	0.005023	513.48	1.8670	0.003814	502.77	1.8322	0.002968	491.16	1.7976
170	0.005357	527.47	1.8990	0.004141	518.30	1.8676	0.003287	508.52	1.8373
180	0.005665	541.00	1.9292	0.004435	532.96	1.9004	0.003569	524.51	1.8729
	8000 kPa			10000 kPa			20000 kPa		
90	0.001019	332.20	1.3955	0.000991	330.62	1.3856	0.000912	327.89	1.3520
100	0.001081	349.30	1.4420	0.001040	346.85	1.4297	0.000939	342.49	1.3917
110	0.001163	367.57	1.4903	0.001100	363.73	1.4744	0.000969	357.33	1.4309
120	0.001282	387.56	1.5417	0.001175	381.44	1.5200	0.001002	372.38	1.4697
130	0.001465	409.98	1.5981	0.001272	400.16	1.5670	0.001037	387.65	1.5081
140	0.001736	434.40	1.6579	0.001400	419.98	1.6155	0.001076	403.13	1.5460
150	0.002061	458.21	1.7148	0.001564	440.63	1.6649	0.001118	418.83	1.5836
160	0.002384	479.59	1.7648	0.001758	461.34	1.7133	0.001164	434.72	1.6207
170	0.002680	498.61	1.8082	0.001965	481.30	1.7589	0.001214	450.80	1.6574
180	0.002946	515.93	1.8469	0.002172	500.12	1.8009	0.001268	467.03	1.6936

TABLE B.6 SI *Thermodynamic Properties of Nitrogen*
TABLE B.6.1 SI *Saturated Nitrogen*

Temp.	Press.	Specific Volume, m³/kg			Internal Energy, kJ/kg		
K T	kPa P	Sat. Liquid v_f	Evap. v_{fg}	Sat. Vapor v_g	Sat. Liquid u_f	Evap. u_{fg}	Sat. Vapor u_g
63.1	12.5	0.001150	1.48074	1.48189	-150.92	196.86	45.94
65	17.4	0.001160	1.09231	1.09347	-147.19	194.37	47.17
70	38.6	0.001191	0.52513	0.52632	-137.13	187.54	50.40
75	76.1	0.001223	0.28052	0.28174	-127.04	180.47	53.43
77.3	101.3	0.001240	0.21515	0.21639	-122.27	177.04	54.76
80	137.0	0.001259	0.16249	0.16375	-116.86	173.06	56.20
85	229.1	0.001299	0.10018	0.10148	-106.55	165.20	58.65
90	360.8	0.001343	0.06477	0.06611	-96.06	156.76	60.70
95	541.1	0.001393	0.04337	0.04476	-85.35	147.60	62.25
100	779.2	0.001452	0.02975	0.03120	-74.33	137.50	63.17
105	1084.6	0.001522	0.02066	0.02218	-62.89	126.18	63.29
110	1467.6	0.001610	0.01434	0.01595	-50.81	113.11	62.31
115	1939.3	0.001729	0.00971	0.01144	-37.66	97.36	59.70
120	2513.0	0.001915	0.00608	0.00799	-22.42	76.63	54.21
125	3208.0	0.002355	0.00254	0.00490	-0.83	40.73	39.90
126.2	3397.8	0.003194	0	0.00319	18.94	0	18.94

TABLE B.6.2 SI *Superheated Nitrogen*

Temp. K	v m³/kg	h kJ/kg	s kJ/kg K	v m³/kg	h kJ/kg	s kJ/kg K	v m³/kg	h kJ/kg	s kJ/kg K
	100 kPa (77.24)			200 kPa (83.62)			500 kPa (93.98)		
Sat.	0.21903	76.61	5.4059	0.11520	81.05	5.2673	0.04834	86.15	5.0802
100	0.29103	101.94	5.6944	0.14252	100.24	5.4775	0.05306	94.46	5.1660
120	0.35208	123.15	5.8878	0.17397	121.93	5.6753	0.06701	118.12	5.3821
140	0.41253	144.20	6.0501	0.20476	143.28	5.8399	0.08007	140.44	5.5541
160	0.47263	165.17	6.1901	0.23519	164.44	5.9812	0.09272	162.22	5.6996
180	0.53254	186.09	6.3132	0.26542	185.49	6.1052	0.10515	183.70	5.8261
200	0.59231	206.97	6.4232	0.29551	206.48	6.2157	0.11744	205.00	5.9383
220	0.65199	227.83	6.5227	0.32552	227.41	6.3155	0.12964	226.18	6.0392
240	0.71161	248.67	6.6133	0.35546	248.32	6.4064	0.14177	247.27	6.1310
260	0.77118	269.51	6.6967	0.38535	269.21	6.4900	0.15385	268.31	6.2152
280	0.83072	290.33	6.7739	0.41520	290.08	6.5674	0.16590	289.31	6.2930

(Continued)

TABLE B.6.1 SI (Continued) *Saturated Nitrogen*

Temp.	Press.	Enthalpy, kJ/kg			Entropy, kJ/kg K		
K T	kPa P	Sat. Liquid h_f	Evap. h_{fg}	Sat. Vapor h_g	Sat. Liquid s_f	Evap. s_{fg}	Sat. Vapor s_g
63.1	12.5	-150.91	215.39	64.48	2.4234	3.4109	5.8343
65	17.4	-147.17	213.38	66.21	2.4816	3.2828	5.7645
70	38.6	-137.09	207.79	70.70	2.6307	2.9684	5.5991
75	76.1	-126.95	201.82	74.87	2.7700	2.6909	5.4609
77.3	101.3	-122.15	198.84	76.69	2.8326	2.5707	5.4033
80	137.0	-116.69	195.32	78.63	2.9014	2.4415	5.3429
85	229.1	-106.25	188.15	81.90	3.0266	2.2135	5.2401
90	360.8	-95.58	180.13	84.55	3.1466	2.0015	5.1480
95	541.1	-84.59	171.07	86.47	3.2627	1.8007	5.0634
100	779.2	-73.20	160.68	87.48	3.3761	1.6068	4.9829
105	1084.6	-61.24	148.59	87.35	3.4883	1.4151	4.9034
110	1467.6	-48.45	134.15	85.71	3.6017	1.2196	4.8213
115	1939.3	-34.31	116.19	81.88	3.7204	1.0104	4.7307
120	2513.0	-17.61	91.91	74.30	3.8536	0.7659	4.6195
125	3208.0	6.73	48.88	55.60	4.0399	0.3910	4.4309
126.2	3397.8	29.79	0	29.79	4.2193	0	4.2193

TABLE B.6.2 SI (Continued) *Superheated Nitrogen*

Temp. K	v m³/kg	h kJ/kg	s kJ/kg K	v m³/kg	h kJ/kg	s kJ/kg K	v m³/kg	h kJ/kg	s kJ/kg K
	100 kPa (77.24)			200 kPa (83.62)			500 kPa (93.98)		
300	0.89023	311.16	6.8457	0.44503	310.94	6.6393	0.17792	310.28	6.3653
350	1.03891	363.24	7.0063	0.51952	363.09	6.8001	0.20788	362.63	6.5267
400	1.18752	415.41	7.1456	0.59392	415.31	6.9396	0.23777	414.99	6.6666
450	1.33607	467.77	7.2690	0.66827	467.70	7.0630	0.26759	467.49	6.7902
500	1.48458	520.41	7.3799	0.74258	520.37	7.1740	0.29739	520.24	6.9014
600	1.78154	626.94	7.5741	0.89114	626.94	7.3682	0.35691	626.93	7.0959
700	2.07845	735.58	7.7415	1.03965	735.61	7.5357	0.41637	735.68	7.2635
800	2.37532	846.60	7.8897	1.18812	846.64	7.6839	0.47581	846.78	7.4118
900	2.67217	960.01	8.0232	1.33657	960.07	7.8175	0.53522	960.24	7.5454
1000	2.96900	1075.68	8.1451	1.48501	1075.75	7.9393	0.59462	1075.96	7.6673

(Continued)

TABLE B.6.2 SI (Continued) *Superheated Nitrogen*

Temp. K	v m³/kg	h kJ/kg	s kJ/kg K	v m³/kg	h kJ/kg	s kJ/kg K	v m³/kg	h kJ/kg	s kJ/kg K
	600 kPa (96.37)			800 kPa (100.38)			1000 kPa (103.73)		
Sat.	0.04046	86.85	5.0411	0.03038	87.52	4.9768	0.02416	87.51	4.9237
120	0.05510	116.79	5.3204	0.04017	114.02	5.2191	0.03117	111.08	5.1357
140	0.06620	139.47	5.4953	0.04886	137.50	5.4002	0.03845	135.47	5.3239
160	0.07689	161.47	5.6422	0.05710	159.95	5.5501	0.04522	158.42	5.4772
180	0.08734	183.10	5.7696	0.06509	181.89	5.6793	0.05173	180.67	5.6082
200	0.09766	204.50	5.8823	0.07293	203.51	5.7933	0.05809	202.52	5.7234
220	0.10788	225.76	5.9837	0.08067	224.94	5.8954	0.06436	224.11	5.8263
240	0.11803	246.92	6.0757	0.08835	246.23	5.9880	0.07055	245.53	5.9194
260	0.12813	268.01	6.1601	0.09599	267.42	6.0728	0.07670	266.83	6.0047
280	0.13820	289.05	6.2381	0.10358	288.54	6.1511	0.08281	288.04	6.0833
300	0.14824	310.06	6.3105	0.11115	309.62	6.2238	0.08889	309.18	6.1562
350	0.17326	362.48	6.4722	0.12998	362.17	6.3858	0.10401	361.87	6.3187
400	0.19819	414.89	6.6121	0.14873	414.68	6.5260	0.11905	414.47	6.4591
450	0.22308	467.42	6.7359	0.16743	467.28	6.6500	0.13404	467.15	6.5832
500	0.24792	520.20	6.8471	0.18609	520.12	6.7613	0.14899	520.04	6.6947
600	0.29755	626.93	7.0416	0.22335	626.93	6.9560	0.17883	626.92	6.8895
700	0.34712	735.70	7.2093	0.26056	735.76	7.1237	0.20862	735.81	7.0573
800	0.39666	846.82	7.3576	0.29773	846.91	7.2721	0.23837	847.00	7.2057
900	0.44618	960.30	7.4912	0.33488	960.42	7.4058	0.26810	960.54	7.3394
1000	0.49568	1076.02	7.6131	0.37202	1076.16	7.5277	0.29782	1076.30	7.4614
	1500 kPa (110.38)			2000 kPa (115.58)			3000 kPa (123.61)		
Sat.	0.01555	85.51	4.8148	0.01100	81.25	4.7193	0.00582	63.47	4.5032
120	0.01899	102.75	4.9650	0.01260	92.10	4.8116	—	—	—
140	0.02452	130.15	5.1767	0.01752	124.40	5.0618	0.01038	111.13	4.8706
160	0.02937	154.50	5.3394	0.02144	150.43	5.2358	0.01350	141.85	5.0763
180	0.03393	177.60	5.4755	0.02503	174.48	5.3775	0.01614	168.09	5.2310
200	0.03832	200.03	5.5937	0.02844	197.53	5.4989	0.01857	192.49	5.3596
220	0.04260	222.05	5.6987	0.03174	219.99	5.6060	0.02088	215.88	5.4711
240	0.04682	243.80	5.7933	0.03496	242.08	5.7021	0.02312	238.66	5.5702
260	0.05099	265.36	5.8796	0.03814	263.90	5.7894	0.02531	261.02	5.6597
280	0.05512	286.78	5.9590	0.04128	285.53	5.8696	0.02746	283.09	5.7414
300	0.05922	308.10	6.0325	0.04440	307.03	5.9438	0.02958	304.94	5.8168
350	0.06940	361.13	6.1960	0.05209	360.39	6.1083	0.03480	358.96	5.9834
400	0.07949	413.96	6.3371	0.05971	413.47	6.2500	0.03993	412.50	6.1264
450	0.08953	466.82	6.4616	0.06727	466.49	6.3750	0.04502	465.87	6.2521
500	0.09953	519.84	6.5733	0.07480	519.65	6.4870	0.05008	519.29	6.3647
600	0.11948	626.92	6.7685	0.08980	626.93	6.6825	0.06013	626.95	6.5609
700	0.13937	735.94	6.9365	0.10474	736.07	6.8507	0.07012	736.35	6.7295
800	0.15923	847.22	7.0851	0.11965	847.45	6.9994	0.08008	847.92	6.8785
900	0.17906	960.83	7.2189	0.13454	961.13	7.1333	0.09003	961.73	7.0125
1000	0.19889	1076.65	7.3409	0.14942	1077.01	7.2553	0.09996	1077.72	7.1347

(Continued)

TABLE B.6.2 SI *Superheated Nitrogen*

Temp. K	v m³/kg	h kJ/kg	s kJ/kg K	v m³/kg	h kJ/kg	s kJ/kg K	v m³/kg	h kJ/kg	s kJ/kg K
		6000 kPa			8000 kPa			10000 kPa	
140	0.002941	47.44	4.2926	0.002224	27.78	4.1167	0.002003	20.87	4.0373
160	0.005556	112.16	4.7292	0.003748	91.80	4.5453	0.002908	76.52	4.4088
180	0.007309	148.02	4.9411	0.005193	134.69	4.7988	0.004021	122.65	4.6813
200	0.008771	177.29	5.0955	0.006387	167.47	4.9717	0.005014	158.35	4.8697
220	0.010095	203.77	5.2217	0.007449	196.07	5.1082	0.005902	188.88	5.0153
240	0.011337	228.73	5.3303	0.008433	222.48	5.2231	0.006721	216.64	5.1362
260	0.012526	252.73	5.4264	0.009367	247.55	5.3235	0.007495	242.72	5.2406
280	0.013678	276.09	5.5130	0.010264	271.74	5.4131	0.008235	267.69	5.3331
300	0.014803	298.99	5.5920	0.011135	295.32	5.4945	0.008952	291.90	5.4167
350	0.017532	354.95	5.7646	0.013236	352.51	5.6709	0.010670	350.26	5.5967
400	0.020187	409.83	5.9111	0.015264	408.24	5.8197	0.012320	406.79	5.7477
450	0.022794	464.19	6.0392	0.017248	463.22	5.9492	0.013927	462.36	5.8786
500	0.025370	518.37	6.1534	0.019202	517.88	6.0644	0.015507	517.48	5.9948
600	0.030463	627.12	6.3516	0.023053	627.32	6.2639	0.018611	627.58	6.1955
700	0.035506	737.27	6.5214	0.026856	737.94	6.4344	0.021669	738.65	6.3667
800	0.040519	849.37	6.6710	0.030631	850.38	6.5845	0.024700	851.43	6.5172
900	0.045514	963.59	6.8055	0.034388	964.86	6.7194	0.027714	966.15	6.6523
1000	0.050495	1079.88	6.9281	0.038132	1081.35	6.8421	0.030715	1082.84	6.7753
		15000 kPa			20000 kPa			50000 kPa	
140	0.001770	14.81	3.9273	0.001655	13.75	3.8587	0.001391	28.05	3.6405
160	0.002183	59.14	4.2232	0.001929	53.63	4.1250	0.001497	61.62	3.8647
180	0.002749	102.34	4.4778	0.002281	93.02	4.3570	0.001612	94.31	4.0573
200	0.003365	140.60	4.6796	0.002687	130.17	4.5529	0.001736	126.15	4.2250
220	0.003964	174.10	4.8394	0.003108	164.26	4.7154	0.001867	157.12	4.3726
240	0.004531	204.33	4.9710	0.003525	195.59	4.8518	0.002003	187.24	4.5037
260	0.005071	232.41	5.0834	0.003930	224.82	4.9689	0.002143	216.53	4.6209
280	0.005589	259.01	5.1820	0.004323	252.50	5.0714	0.002285	245.02	4.7266
300	0.006088	284.56	5.2702	0.004704	279.01	5.1629	0.002428	272.78	4.8223
350	0.007280	345.47	5.4581	0.005617	341.86	5.3568	0.002786	339.44	5.0280
400	0.008416	403.79	5.6139	0.006487	401.65	5.5166	0.003138	403.08	5.1980
450	0.009517	460.71	5.7480	0.007329	459.70	5.6534	0.003484	464.64	5.3431
500	0.010593	516.88	5.8664	0.008149	516.78	5.7737	0.003823	524.82	5.4699
600	0.012697	628.50	6.0699	0.009748	629.76	5.9797	0.004484	642.94	5.6853
700	0.014759	740.63	6.2427	0.011310	742.85	6.1540	0.005129	760.04	5.8658
800	0.016797	854.18	6.3943	0.012849	857.11	6.3065	0.005762	877.47	6.0226
900	0.018818	969.50	6.5301	0.014374	972.98	6.4430	0.006385	995.87	6.1621
1000	0.020828	1086.64	6.6535	0.015887	1090.55	6.5668	0.007001	1115.51	6.2881

TABLE B.7 SI *Thermodynamic Properties of Methane*
TABLE B.7.1 SI *Saturated Methane*

Temp.	Press.	Specific Volume, m³/kg			Internal Energy, kJ/kg		
		Sat. Liquid	Evap.	Sat. Vapor	Sat. Liquid	Evap.	Sat. Vapor
K T	kPa P	v_f	v_{fg}	v_g	u_f	u_{fg}	u_g
90.7	11.7	0.002215	3.97941	3.98163	-358.10	496.59	138.49
95	19.8	0.002243	2.44845	2.45069	-343.79	488.62	144.83
100	34.4	0.002278	1.47657	1.47885	-326.90	478.96	152.06
105	56.4	0.002315	0.93780	0.94012	-309.79	468.89	159.11
110	88.2	0.002353	0.62208	0.62443	-292.50	458.41	165.91
111.7	101.3	0.002367	0.54760	0.54997	-286.74	454.85	168.10
115	132.3	0.002395	0.42800	0.43040	-275.05	447.48	172.42
120	191.6	0.002439	0.30367	0.30610	-257.45	436.02	178.57
125	269.0	0.002486	0.22108	0.22357	-239.66	423.97	184.32
130	367.6	0.002537	0.16448	0.16701	-221.65	411.25	189.60
135	490.7	0.002592	0.12458	0.12717	-203.40	397.77	194.37
140	641.6	0.002653	0.09575	0.09841	-184.86	383.42	198.56
145	823.7	0.002719	0.07445	0.07717	-165.97	368.06	202.09
150	1040.5	0.002794	0.05839	0.06118	-146.65	351.53	204.88
155	1295.6	0.002877	0.04605	0.04892	-126.82	333.61	206.79
160	1592.8	0.002974	0.03638	0.03936	-106.35	314.01	207.66
165	1935.9	0.003086	0.02868	0.03177	-85.06	292.30	207.24
170	2329.3	0.003222	0.02241	0.02563	-62.67	267.81	205.14
175	2777.6	0.003393	0.01718	0.02058	-38.75	239.47	200.72
180	3286.4	0.003623	0.01266	0.01629	-12.43	205.16	192.73
185	3863.2	0.003977	0.00846	0.01243	18.47	159.49	177.96
190	4520.5	0.004968	0.00300	0.00797	69.10	67.01	136.11
190.6	4599.2	0.006148	0	0.00615	101.46	0	101.46

TABLE B.6.2 SI *Superheated Methane*

Temp. K	v m³/kg	h kJ/kg	s kJ/kg K	v m³/kg	h kJ/kg	s kJ/kg K	v m³/kg	h kJ/kg	s kJ/kg K
	50 kPa (103.73)			100 kPa (111.50)			200 kPa (120.61)		
Sat.	1.05026	209.85	9.7295	0.55665	223.56	9.5084	0.29422	238.14	9.2918
125	1.27928	255.82	10.1331	0.63126	253.33	9.7606	0.30695	248.19	9.3736
150	1.54333	308.45	10.5170	0.76586	306.77	10.1504	0.37700	303.31	9.7759
175	1.80540	360.81	10.8399	0.89840	359.56	10.4759	0.44486	357.02	10.1071
200	2.06648	413.17	11.1196	1.02994	412.19	10.7570	0.51165	410.21	10.3912
225	2.32702	465.79	11.3674	1.16092	464.99	11.0058	0.57786	463.38	10.6417
250	2.58720	518.94	11.5914	1.29154	518.27	11.2303	0.64370	516.93	10.8674

(Continued)

TABLE B.7.1 SI (Continued) *Saturated Methane*

Temp.	Press.	Enthalpy, kJ/kg			Entropy, kJ/kg K		
K	kPa	Sat. Liquid	Evap.	Sat. Vapor	Sat. Liquid	Evap.	Sat. Vapor
T	P	h_f	h_{fg}	h_g	s_f	s_{fg}	s_g
90.7	11.7	-358.07	543.12	185.05	4.2264	5.9891	10.2155
95	19.8	-343.75	537.18	193.43	4.3805	5.6545	10.0350
100	34.4	-326.83	529.77	202.94	4.5538	5.2977	9.8514
105	56.4	-309.66	521.82	212.16	4.7208	4.9697	9.6905
110	88.2	-292.29	513.29	221.00	4.8817	4.6663	9.5480
111.7	101.3	-286.50	510.33	223.83	4.9336	4.5706	9.5042
115	132.3	-274.74	504.12	229.38	5.0368	4.3836	9.4205
120	191.6	-256.98	494.20	237.23	5.1867	4.1184	9.3051
125	269.0	-238.99	483.44	244.45	5.3321	3.8675	9.1996
130	367.6	-220.72	471.72	251.00	5.4734	3.6286	9.1020
135	490.7	-202.13	458.90	256.77	5.6113	3.3993	9.0106
140	641.6	-183.16	444.85	261.69	5.7464	3.1775	8.9239
145	823.7	-163.73	429.38	265.66	5.8794	2.9613	8.8406
150	1040.5	-143.74	412.29	268.54	6.0108	2.7486	8.7594
155	1295.6	-123.09	393.27	270.18	6.1415	2.5372	8.6787
160	1592.8	-101.61	371.96	270.35	6.2724	2.3248	8.5971
165	1935.9	-79.08	347.82	268.74	6.4046	2.1080	8.5126
170	2329.3	-55.17	320.02	264.85	6.5399	1.8824	8.4224
175	2777.6	-29.33	287.20	257.87	6.6811	1.6411	8.3223
180	3286.4	-0.53	246.77	246.25	6.8333	1.3710	8.2043
185	3863.2	33.83	192.16	226.00	7.0095	1.0387	8.0483
190	4520.5	91.56	80.58	172.14	7.3015	0.4241	7.7256
190.6	4599.2	129.74	0	129.74	7.4999	0	7.4999

TABLE B.7.2 SI (Continued) *Superheated Methane*

Temp.	v	h	s	v	h	s	v	h	s
K	m³/kg	kJ/kg	kJ/kg K	m³/kg	kJ/kg	kJ/kg K	m³/kg	kJ/kg	kJ/kg K
	50 kPa (103.73)			100 kPa (111.50)			200 kPa (120.61)		
275	2.84715	572.92	11.7972	1.42193	572.36	11.4365	0.70931	571.22	11.0743
300	3.10694	628.06	11.9891	1.55215	627.58	11.6286	0.77475	626.60	11.2670
325	3.36661	684.65	12.1702	1.68225	684.23	11.8100	0.84008	683.38	11.4488
350	3.62619	742.94	12.3429	1.81226	742.57	11.9829	0.90530	741.83	11.6220
375	3.88569	803.13	12.5090	1.94220	802.80	12.1491	0.97046	802.16	11.7885
400	4.14515	865.38	12.6697	2.07209	865.10	12.3099	1.03557	864.53	11.9495
425	4.40455	929.80	12.8259	2.20193	929.55	12.4661	1.10062	929.05	12.1059

(Continued)

TABLE B.7.2 SI (Continued) *Superheated Methane*

Temp. K	v m³/kg	h kJ/kg	s kJ/kg K	v m³/kg	h kJ/kg	s kJ/kg K	v m³/kg	h kJ/kg	s kJ/kg K
	400 kPa (131.42)			600 kPa (138.72)			800 kPa (144.40)		
Sat.	0.15427	252.72	9.0754	0.10496	260.51	8.9458	0.07941	265.23	8.8505
150	0.18233	296.09	9.3843	0.11717	288.38	9.1390	0.08434	280.00	8.9509
175	0.21799	351.81	9.7280	0.14227	346.39	9.4970	0.10433	340.76	9.3260
200	0.25246	406.18	10.0185	0.16603	402.06	9.7944	0.12278	397.85	9.6310
225	0.28631	460.13	10.2726	0.18911	456.84	10.0525	0.14050	453.50	9.8932
250	0.31978	514.23	10.5007	0.21180	511.52	10.2830	0.15781	508.78	10.1262
275	0.35301	568.94	10.7092	0.23424	566.66	10.4931	0.17485	564.35	10.3381
300	0.38606	624.65	10.9031	0.25650	622.69	10.6882	0.19172	620.73	10.5343
325	0.41899	681.69	11.0857	0.27863	680.00	10.8716	0.20845	678.31	10.7186
350	0.45183	740.36	11.2595	0.30067	738.88	11.0461	0.22510	737.41	10.8938
375	0.48460	800.87	11.4265	0.32264	799.57	11.2136	0.24167	798.28	11.0617
400	0.51731	863.39	11.5879	0.34456	862.25	11.3754	0.25818	861.12	11.2239
425	0.54997	928.04	11.7446	0.36643	927.04	11.5324	0.27465	926.03	11.3813
450	0.58260	994.89	11.8974	0.38826	994.00	11.6855	0.29109	993.11	11.5346
475	0.61520	1063.97	12.0468	0.41006	1063.18	11.8351	0.30749	1062.40	11.6845
500	0.64778	1135.29	12.1931	0.43184	1134.59	11.9816	0.32387	1133.89	11.8311
525	0.68033	1208.81	12.3366	0.45360	1208.18	12.1252	0.34023	1207.56	11.9749
	1000 kPa (149.13)			1500 kPa (158.52)			2000 kPa (165.86)		
Sat.	0.06367	268.12	8.7735	0.04196	270.47	8.6215	0.03062	268.25	8.4975
175	0.08149	334.87	9.1871	0.05078	318.81	8.9121	0.03504	299.97	8.6839
200	0.09681	393.53	9.5006	0.06209	382.26	9.2514	0.04463	370.17	9.0596
225	0.11132	450.11	9.7672	0.07239	441.44	9.5303	0.05289	432.43	9.3532
250	0.12541	506.01	10.0028	0.08220	499.00	9.7730	0.06059	491.84	9.6036
275	0.13922	562.04	10.2164	0.09171	556.21	9.9911	0.06796	550.31	9.8266
300	0.15285	618.76	10.4138	0.10103	613.82	10.1916	0.07513	608.85	10.0303
325	0.16635	676.61	10.5990	0.11022	672.37	10.3790	0.08216	668.12	10.2200
350	0.17976	735.94	10.7748	0.11931	732.26	10.5565	0.08909	728.58	10.3992
375	0.19309	797.00	10.9433	0.12832	793.78	10.7263	0.09594	790.57	10.5703
400	0.20636	859.98	11.1059	0.13728	857.16	10.8899	0.10274	854.34	10.7349
425	0.21959	925.03	11.2636	0.14619	922.54	11.0484	0.10949	920.06	10.8942
450	0.23279	992.23	11.4172	0.15506	990.02	11.2027	0.11620	987.84	11.0491
475	0.24595	1061.61	11.5672	0.16391	1059.66	11.3532	0.12289	1057.72	11.2003
500	0.25909	1133.19	11.7141	0.17273	1131.46	11.5005	0.12955	1129.74	11.3480
525	0.27221	1206.95	11.8580	0.18152	1205.41	11.6448	0.13619	1203.88	11.4927
550	0.28531	1282.84	11.9992	0.19031	1281.48	11.7864	0.14281	1280.13	11.6346

(Continued)

TABLE B.7.2 SI (Continued) *Superheated Methane*

Temp. K	v m³/kg	h kJ/kg	s kJ/kg K	v m³/kg	h kJ/kg	s kJ/kg K	v m³/kg	h kJ/kg	s kJ/kg K
		3000 kPa (177.26)			4000 kPa (186.10)			5000 kPa	
Sat.	0.01856	253.33	8.2718	0.01160	219.34	8.0035	—	—	—
200	0.02690	342.70	8.7492	0.01763	308.23	8.4675	0.01142	258.30	8.1459
225	0.03333	413.29	9.0823	0.02347	392.39	8.8653	0.01749	369.34	8.6728
250	0.03896	477.06	9.3512	0.02814	461.63	9.1574	0.02165	445.55	8.9945
275	0.04421	538.32	9.5848	0.03235	526.07	9.4031	0.02525	513.60	9.2540
300	0.04924	598.83	9.7954	0.03631	588.73	9.6212	0.02857	578.57	9.4802
325	0.05412	659.59	9.9899	0.04011	651.07	9.8208	0.03173	642.56	9.6851
350	0.05889	721.23	10.1726	0.04381	713.93	10.0071	0.03477	706.67	9.8751
375	0.06358	784.19	10.3464	0.04742	777.86	10.1835	0.03774	771.60	10.0543
400	0.06822	848.76	10.5130	0.05097	843.24	10.3523	0.04064	837.78	10.2251
425	0.07281	915.15	10.6740	0.05448	910.31	10.5149	0.04350	905.53	10.3894
450	0.07736	983.50	10.8303	0.05795	979.23	10.6725	0.04632	975.04	10.5483
475	0.08188	1053.89	10.9825	0.06139	1050.12	10.8258	0.04911	1046.42	10.7026
500	0.08638	1126.34	11.1311	0.06481	1123.01	10.9753	0.05187	1119.74	10.8531
525	0.09086	1200.88	11.2765	0.06820	1197.93	11.1215	0.05462	1195.04	11.0000
550	0.09532	1277.47	11.4190	0.07158	1274.86	11.2646	0.05735	1272.30	11.1437
575	0.09976	1356.07	11.5588	0.07495	1353.77	11.4049	0.06006	1351.51	11.2846
		6000 kPa			8000 kPa			10000 kPa	
200	0.006109	160.30	7.6125	0.004120	88.54	7.2069	0.003756	72.22	7.0862
225	0.013466	343.74	8.4907	0.008460	284.98	8.1344	0.005945	229.32	7.8245
250	0.017325	428.84	8.8502	0.011983	393.92	8.5954	0.008915	358.60	8.3716
275	0.020529	500.95	9.1253	0.014688	475.39	8.9064	0.011272	450.09	8.7210
300	0.023430	568.39	9.3601	0.017055	548.15	9.1598	0.013297	528.37	8.9936
325	0.026156	634.09	9.5705	0.019235	617.40	9.3815	0.015137	601.22	9.2270
350	0.028767	699.48	9.7643	0.021297	685.39	9.5831	0.016861	671.80	9.4362
375	0.031295	765.41	9.9463	0.023277	753.34	9.7706	0.018505	741.72	9.6292
400	0.033763	832.42	10.1192	0.025197	821.95	9.9477	0.020090	811.92	9.8104
425	0.036185	900.84	10.2851	0.027072	891.71	10.1169	0.021632	882.98	9.9827
450	0.038571	970.91	10.4453	0.028911	962.92	10.2796	0.023140	955.27	10.1480
475	0.040927	1042.79	10.6007	0.030721	1035.75	10.4372	0.024620	1029.04	10.3075
500	0.043259	1116.54	10.7520	0.032508	1110.34	10.5902	0.026077	1104.44	10.4622
525	0.045571	1192.21	10.8997	0.034276	1186.74	10.7393	0.027516	1181.55	10.6126
550	0.047866	1269.81	11.0441	0.036027	1264.99	10.8849	0.028939	1260.42	10.7594
575	0.050147	1349.31	11.1854	0.037765	1345.07	11.0272	0.030350	1341.06	10.9028

TABLE C.1 *Critical Constants (English Units)*

Substance	Formula	Molec. Weight	Temp. R	Pressure lbf/in.^{2}	Volume ft^3/lb-mole
Ammonia	NH_3	17.031	729.9	1646	1.1613
Argon	Ar	39.948	271.4	706	1.1998
Bromine	Br_2	159.808	1058.4	1494	2.0375
Carbon dioxide	CO_2	44.010	547.4	1070	1.5041
Carbon monoxide	CO	28.010	239.2	508	1.4929
Chlorine	Cl_2	70.906	750.4	1157	1.9831
Fluorine	F_2	37.997	259.7	757	1.0620
Helium	He	4.003	9.34	32.9	0.9195
Hydrogen (normal)	H_2	2.016	59.76	188.6	1.0428
Krypton	Kr	83.800	376.9	798	1.4609
Neon	Ne	20.183	79.92	400	0.6664
Nitric oxide	NO	30.006	324.0	940	0.9243
Nitrogen	N_2	28.013	227.2	492	1.4385
Nitrogen dioxide	NO_2	46.006	775.8	1465	2.6879
Nitrous oxide	N_2O	44.013	557.3	1050	1.5602
Oxygen	O_2	31.999	278.3	731	1.1758
Sulfur dioxide	SO_2	64.063	775.4	1143	1.9575
Water	H_2O	18.015	1165.1	3208	0.9147
Xenon	Xe	131.300	521.5	847	1.8966
Acetylene	C_2H_2	26.038	554.9	891	1.8053
Benzene	C_6H_6	78.114	1012.0	709	4.1488
n-Butane	C_4H_{10}	58.124	765.4	551	4.0847
Chlorodifluoroethane (142b)	CH_3CClF_2	100.495	738.5	616	3.7003
Chlorodifluoromethane (22)	$CHClF_2$	86.469	664.7	721	2.6527
Dichlorodifluoromethane (12)	CCl_2F_2	120.914	693.0	600	3.4712
Dichlorofluoroethane (141)	CH_3CCl_2F	116.950	866.7	658	4.0367
Dichlorofluoromethane (21)	$CHCl_2F$	102.923	812.9	751	3.1460
Dichlorotrifluoroethane(123)	$CHCl_2CF_3$	152.930	822.4	532	4.4547
Difluoroethane (152a)	CHF_2CH_3	66.050	695.5	656	2.8753
Ethane	C_2H_6	30.070	549.7	708	2.3755
Ethyl alcohol	C_2H_5OH	46.069	925.0	891	2.6767
Ethylene	C_2H_4	28.054	508.3	731	2.0888
n-Heptane	C_7H_{16}	100.205	972.5	397	6.9200
n-Hexane	C_6H_{14}	86.178	913.5	437	5.9268
Methane	CH_4	16.043	342.7	667	1.5890
Methyl alcohol	CH_3OH	32.042	922.7	1173	1.8902
n-Octane	C_8H_{18}	114.232	1023.8	361	7.8811
n-Pentane	C_5H_{12}	72.151	845.5	489	4.8696
Propane	C_3H_8	44.094	665.6	616	4.2517
Propene	C_3H_6	42.081	656.8	667	2.8993
Tetrafluoroethane (134a)	CF_3CH_2F	102.030	673.6	589	3.1717

TABLE C.2 *Properties of Selected Solids at 77 F*

Substance	ρ lbm/ft^3	C_p Btu/lbm-R
Asphalt	132.3	0.225
Brick, common	112.4	0.20
Carbon, diamond	202.9	0.122
Carbon, graphite	125-156	0.146
Coal	75-95	0.305
Concrete	137	0.21
Glass, plate	156	0.191
Glass, wool	12.5	0.158
Granite	172	0.212
Ice (32 F)	57.2	0.487
Paper	43.7	0.287
Plexiglas	73.7	0.344
Polystyrene	57.4	0.549
Polyvinyl chloride	86.1	0.229
Rubber, soft	68.7	0.399
Salt, rock	130-156	0.2196
Sand, dry	93.6	0.191
Silicon	145.5	0.167
Snow, firm	35	0.501
Wood, hard (oak)	44.9	0.301
Wood, soft (pine)	31.8	0.33
Wool	6.24	0.411
Metals		
Aluminum, duralumin	170	0.215
Copper, commercial	518	0.100
Brass, 60-40	524	0.0898
Gold	1205	0.03082
Iron, cast	454	0.100
Iron. 304 St Steel	488	0.110
Lead	708	0.031
Magnesium, 2% Mn	111	0.239
Nickel, 10% Cr	541	0.1066
Silver, 99.9% Ag	657	0.0564
Sodium	60.6	0.288
Tin	456	0.0525
Tungsten	1205	0.032
Zinc	446	0.0927

TABLE C.3 *Properties of Some Liquids at 77 F*

Substance	ρ $\dfrac{\text{lbm}}{\text{ft}^3}$	C_p $\dfrac{\text{Btu}}{\text{lbm-R}}$
Ammonia	37.7	1.151
Benzene	54.9	0.41
Butane	34.7	0.60
CCL$_4$	98.9	0.20
CO$_2$	42.5	0.69
Ethanol	48.9	0.59
Gasoline	46.8	0.50
Glycerine	78.7	0.58
Kerosene	50.9	0.48
Methanol	49.1	0.61
n-octane	43.2	0.53
Oil light	57	0.43
Oil engine	55.2	0.46
Propane	31.8	0.61
R-12	81.8	0.232
R-22	74.3	0.30
R-134a	75.3	0.34
Water	62.2	1.00
Liquid Metals		
Bismuth, Bi	627	0.033
Lead, Pb	665	0.038
Mercury, Hg	848	0.033
Potassium, K	51.7	0.193
Sodium, Na	58	0.33
Tin, Sn	434	0.057
Zinc, Zn	410	0.12
NaK (56/44)	55.4	0.27

TABLE C.4 *Properties of Various Ideal Gases at 77 F, 1 atm* (English Units)*

Gas	Chemical Formula	Molecular Mass	R ft-lbf/lbm-R	$\rho \times 10^3$ lbm/ft^3	C_{po} Btu/lbm-R	C_{vo} Btu/lbm-R	k C_{po}/C_{vo}
Steam	H_2O	18.015	85.76	1.442	0.447	0.337	1.327
Acetylene	C_2H_2	26.038	59.34	65.55	0.406	0.330	1.231
Air	--	28.97	53.34	72.98	0.240	0.171	1.400
Ammonia	NH_3	17.031	90.72	43.325	0.509	0.392	1.297
Argon	Ar	39.948	38.68	100.7	0.124	0.0745	1.667
Butane	C_4H_{10}	58.124	26.58	150.3	0.410	0.376	1.091
Carbon monoxide	CO	28.01	55.16	70.5	0.249	0.178	1.399
Carbon dioxide	CO_2	44.01	35.10	110.8	0.201	0.156	1.289
Ethane	C_2H_6	30.07	51.38	76.29	0.422	0.356	1.186
Ethanol	C_2H_5OH	46.069	33.54	117.6	0.341	0.298	1.145
Ethylene	C_2H_4	28.054	55.07	71.04	0.370	0.299	1.237
Helium	He	4.003	386.0	10.08	1.240	0.744	1.667
Hydrogen	H_2	2.016	766.5	5.075	3.394	0.241	1.409
Methane	CH_4	16.043	96.35	40.52	0.538	0.415	1.299
Methanol	CH_3OH	32.042	48.22	81.78	0.336	0.274	1.227
Neon	Ne	20.183	76.55	50.81	0.246	0.148	1.667
Nitric oxide	NO	30.006	51.50	75.54	0.237	0.171	1.387
Nitrogen	N_2	28.013	55.15	70.61	0.249	0.178	1.400
Nitrous oxide	N_2O	44.013	35.10	110.8	0.210	0.165	1.274
n-octane	C_8H_{18}	114.23	13.53	5.74	0.409	0.391	1.044
Oxygen	O_2	31.999	48.28	80.66	0.220	0.158	1.393
Propane	C_3H_8	44.094	35.04	112.9	0.401	0.356	1.126
R-12	CCL_2F_2	120.914	12.78	310.9	0.147	0.131	1.126
R-22	$CHCLF_2$	86.469	17.87	221.0	0.157	0.134	1.171
R-134a	CF_3CH_2F	102.03	15.15	262.2	0.203	0.184	1.106
Sulfur dioxide	SO_2	64.059	24.12	163.4	0.149	0.118	1.263
Sulfur trioxide	SO_3	80.053	19.30	204.3	0.152	0.127	1.196

*Or saturation pressure if it is less than 1 atm.

TABLE C.5 *Constant-Pressure Specific Heats of Various Ideal Gases (English Units)*

$$C_{p0} = \frac{\text{Btu}}{\text{lb mole R}} \qquad \theta = \frac{T(\text{Rankine})}{180}$$

Gas		Range R	Max Error %
N_2	$\overline{C}_{po} = 9.3355 - 122.56\,\theta^{-1.5} + 256.38\,\theta^{-2} - 196.08\,\theta^{-3}$	540–6300	0.43
O_2	$\overline{C}_{po} = 8.9465 + 4.8044 \times 10^{-3}\,\theta^{1.5} - 42.679\,\theta^{-1.5} + 56.615\,\theta^{-2}$	540–6300	0.30
H_2	$\overline{C}_{po} = 13.505 - 167.96\,\theta^{-0.75} + 278.44\,\theta^{-1} - 134.01\,\theta^{-1.5}$	540–6300	0.60
CO	$\overline{C}_{po} = 16.526 - 0.16841\,\theta^{0.75} - 47.985\,\theta^{-0.5} + 42.246\,\theta^{-0.75}$	540–6300	0.42
OH	$\overline{C}_{po} = 19.490 - 14.185\,\theta^{0.25} + 4.1418\,\theta^{0.75} - 1.0196\,\theta$	540–6300	0.43
NO	$\overline{C}_{po} = 14.169 - 0.40861\,\theta^{0.5} - 16877\,\theta^{-0.5} + 17.889\,\theta^{-1.5}$	540–6300	0.34
H_2O	$\overline{C}_{po} = 34.190 - 43.868\,\theta^{0.25} + 19.778\,\theta^{0.5} - 0.88407\,\theta$	540–6300	0.43
CO_2	$\overline{C}_{po} = -0.89286 + 7.2967\,\theta^{0.5} - 0.98074\,\theta + 5.7835 \times 10^{-3}\,\theta^2$	540–6300	0.19
NO_2	$\overline{C}_{po} = 11.005 + 51.650\,\theta^{-0.5} - 86.916\,\theta^{-0.75} + 55.580\,\theta^{-2}$	540–6300	0.26
CH_4	$\overline{C}_{po} = -160.82 + 105.10\,\theta^{0.25} - 5.9452\,\theta^{0.75} + 77.408\,\theta^{-0.5}$	540–3600	0.15
C_2H_4	$\overline{C}_{po} = -22.800 + 29.433\,\theta^{0.5} - 8.5185\,\theta^{0.75} + 43.683\,\theta^{-3}$	540–3600	0.07
C_2H_6	$\overline{C}_{po} = 1.648 + 4.124\,\theta - 0.153\,\theta^2 + 1.74 \times 10^{-3}\,\theta^3$	540–2700	0.83
C_3H_8	$\overline{C}_{po} = -0.966 + 7.2790 - 0.3755\,\theta^2 + 7.58 \times 10^{-3}\,\theta^3$	540–2700	0.40
C_4H_{10}	$\overline{C}_{po} = 0.945 + 8.873\,\theta - 0.438\,\theta^2 + 8.36 \times 10^{-3}\,\theta^3$	540–2700	0.54

Source: From T.C. Scott and R.E. Sonntag. University of Michigan, unpublished 1971, except C_2H_6, C_3H_8, and C_4H_{10} from K.A. Kobe, Petroleum Refiner, 28, No. 2, 113 (1949).

TABLE C.6 *Ideal-Gas Properties of Air, English Units Standard Entropy at*
1 atm = 101.325 kPa = 14.696 lbf/in.^{2}

T R	u Btu/lbm	h Btu/lbm	s^0 Btu/lbm R	P_r	v_r
400	68.212	95.634	1.56788	0.39046	379.523
440	75.047	105.212	1.59071	0.54470	299.264
480	81.887	114.794	1.61155	0.73825	240.877
520	88.733	124.383	1.63074	0.97670	197.244
536.67	91.589	128.381	1.63831	1.09071	182.288
540	92.160	129.180	1.63979	1.11458	179.491
560	95.589	133.980	1.64852	1.26592	163.885
600	102.457	143.590	1.66510	1.61217	137.880
640	109.340	153.216	1.68063	2.02204	117.260
680	116.242	162.860	1.69524	2.50257	100.666
720	123.167	172.528	1.70906	3.06119	87.1367
760	130.118	182.221	1.72216	3.70585	75.9775
800	137.099	191.944	1.73463	4.44496	66.6778
840	144.114	201.701	1.74653	5.28751	58.8555
880	151.165	211.494	1.75791	6.24303	52.2211
920	158.255	221.327	1.76884	7.32166	46.5519
960	165.388	231.202	1.77935	8.53415	41.6744
1000	172.564	241.121	1.78947	9.89193	37.4523
1040	179.787	251.086	1.79924	11.40706	33.7768
1080	187.058	261.099	1.80868	13.09232	30.5609
1120	194.378	271.161	1.81783	14.96119	27.7339
1160	201.748	281.273	1.82670	17.02788	25.2381
1200	209.168	291.436	1.83532	19.30735	23.0259
1240	216.640	301.650	1.84369	21.81531	21.0581
1280	224.163	311.915	1.85184	24.56826	19.3017
1320	231.737	322.231	1.85977	27.58348	17.7290
1360	239.362	332.598	1.86751	30.87907	16.3168
1400	247.037	343.016	1.87506	34.47392	15.0451
1440	254.762	353.483	1.88243	38.38777	13.8972
1480	262.537	364.000	1.88964	42.64121	12.8585
1520	270.359	374.565	1.89668	47.25567	11.9165
1560	278.230	385.177	1.90357	52.25344	11.0603
1600	286.146	395.837	1.91032	57.65771	10.2807
1650	296.106	409.224	1.91856	65.02144	9.40126
1700	306.136	422.681	1.92659	73.10700	8.61487
1750	316.232	436.205	1.93444	81.96560	7.90980
1800	326.393	449.794	1.94209	91.65077	7.27604
1850	336.616	463.445	1.94957	102.2183	6.70505

(Continued)

TABLE C.6 (Continued) *Ideal-Gas Properties of Air, English Units Standard Entropy at 1 atm = 101.325 kPa = 14.696 lbf/in.^{2}*

T R	u Btu/lbm	h Btu/lbm	s^0 Btu/lbm R	P_r	v_r
1900	346.901	477.158	1.95689	113.7264	6.18944
1950	357.243	490.928	1.96404	126.2356	5.72284
2000	367.642	504.755	1.97104	139.8090	5.29973
2050	378.096	518.636	1.97790	154.5119	4.91531
2100	388.602	532.570	1.98461	170.4125	4.56538
2150	399.158	546.554	1.99119	187.5812	4.24627
2200	409.764	560.588	1.99765	206.0915	3.95477
2300	431.114	588.793	2.01018	247.4432	3.44359
2400	452.640	617.175	2.02226	295.1096	3.01292
2500	474.330	645.721	2.03391	349.7802	2.64791
2600	496.175	674.421	2.04517	412.1964	2.33683
2700	518.165	703.267	2.05606	483.1554	2.07031
2800	540.286	732.244	2.06659	563.4304	1.84110
2900	562.532	761.345	2.07681	653.9284	1.64296
3000	584.895	790.564	2.08671	755.5802	1.47096
3100	607.369	819.894	2.09633	869.3694	1.32104
3200	629.948	849.328	2.10567	996.3336	1.18988
3300	652.625	878.861	2.11476	1137.566	1.07472
3400	675.396	908.488	2.12361	1294.214	0.97327
3500	698.257	938.204	2.13222	1467.483	0.88360
3600	721.203	968.005	2.14062	1658.635	0.80410
3700	744.230	997.888	2.14880	1869.020	0.73341
3800	767.334	1027.848	2.15679	2100.030	0.67037
3900	790.513	1057.882	2.16459	2353.126	0.61401
4000	813.763	1087.988	2.17221	2629.834	0.56350
4100	837.081	1118.162	2.17967	2931.747	0.51810
4200	860.466	1148.402	2.18695	3260.527	0.47722
4300	883.913	1178.705	2.19408	3617.908	0.44032
4400	907.422	1209.069	2.20106	4005.693	0.40694
4500	930.989	1239.492	2.20790	4425.759	0.37669
4600	954.613	1269.972	2.21460	4880.058	0.34921
4700	978.292	1300.506	2.22117	5370.617	0.32421
4800	1002.023	1331.093	2.22761	5899.541	0.30143
4900	1025.806	1361.732	2.23392	6469.012	0.28062
5000	1049.638	1392.419	2.24012	7081.293	0.26159
5100	1073.518	1423.155	2.24621	7738.728	0.24415
5200	1097.444	1453.936	2.25219	8443.744	0.22815
5300	1121.414	1484.762	2.25806	9198.851	0.21345
5400	1145.428	1515.632	2.26383	10006.645	0.19992

TABLE C.7 *Ideal-Gas Properties of Various Substances (English Units), Entropies at 1 atm Pressure*

	Nitrogen, Diatomic (N_2) $\bar{h}^o_{f,537} = 0$ Btu/lb mol $M = 28.013$		Nitrogen, Monatomic (N) $\bar{h}^o_{f,537} = 203\ 216$ Btu/lb mol $M = 14.007$	
T R	$\bar{h}^o - \bar{h}^o_{537}$ Btu/lb mol	$\bar{s}^o$ Btu/lbmol/R	$\bar{h}^o - \bar{h}^o_{537}$ Btu/lb mol	$\bar{s}^o$ Btu/lbmol/R
0	−3727	0	−2664	0
200	−2341	38.877	−1671	31.689
400	−950	43.695	−679	35.130
537	0	45.739	0	36.589
600	441	46.515	314	37.143
800	1837	48.524	1307	38.571
1000	3251	50.100	2300	39.679
1200	4693	51.414	3293	40.584
1400	6169	52.552	4286	41.349
1600	7681	53.561	5279	42.012
1800	9227	54.472	6272	42.597
2000	10804	55.302	7265	43.120
2200	12407	56.066	8258	43.593
2400	14034	56.774	9251	44.025
2600	15681	57.433	10244	44.423
2800	17345	58.049	11237	44.791
3000	19025	58.629	12230	45.133
3200	20717	59.175	13223	45.454
3400	22421	59.691	14216	45.755
3600	24135	60.181	15209	46.038
3800	25857	60.647	16202	46.307
4000	27587	61.090	17195	46.562
4200	29324	61.514	18189	46.804
4400	31068	61.920	19183	47.035
4600	32817	62.308	20178	47.256
4800	34571	62.682	21174	47.468
5000	36330	63.041	22171	47.672
5500	40745	63.882	24670	48.148
6000	45182	64.654	27186	48.586
6500	49638	65.368	29724	48.992
7000	54109	66.030	32294	49.373
7500	58595	66.649	34903	49.733
8000	63093	67.230	37559	50.076
8500	67603	67.777	40270	50.405
9000	72125	68.294	43040	50.721
9500	76658	68.784	45875	51.028
10000	81203	69.250	48777	51.325

(Continued)

TABLE C.7 (Continued) *Ideal-Gas Properties of Various Substances (English Units), Entropies at 1 atm Pressure*

T R	Oxygen, Diatomic (O_2) $\bar{h}^o_{f,\,537} = 0$ Btu/lb mol $M = 31.999$		Oxygen, Monatomic (O) $\bar{h}^o_{f,\,537} = 107\ 124$ Btu/lb mol $M = 16.00$	
	$\bar{h}^o - \bar{h}^o_{537}$ Btu/lb mol	$\bar{s}^o$ Btu/lbmol/R	$\bar{h}^o - \bar{h}^o_{537}$ Btu/lb mol	$\bar{s}^o$ Btu/lbmol/R
0	−3733	0	−2891	0
200	−2345	42.100	−1829	33.041
400	−955	46.920	−724	36.884
537	0	48.973	0	38.442
600	446	49.758	330	39.023
800	1881	51.819	1358	40.503
1000	3366	53.475	2374	41.636
1200	4903	54.876	3383	42.556
1400	6487	56.096	4387	43.330
1600	8108	57.179	5389	43.999
1800	9761	58.152	6389	44.588
2000	11438	59.035	7387	45.114
2200	13136	59.844	8385	45.589
2400	14852	60.591	9381	46.023
2600	16584	61.284	10378	46.422
2800	18329	61.930	11373	46.791
3000	20088	62.537	12369	47.134
3200	21860	63.109	13364	47.455
3400	23644	63.650	14359	47.757
3600	25441	64.163	15354	48.041
3800	27250	64.652	16349	48.310
4000	29071	65.119	17344	48.565
4200	30904	65.566	18339	48.808
4400	32748	65.995	19334	49.039
4600	34605	66.408	20330	49.261
4800	36472	66.805	21327	49.473
5000	38350	67.189	22325	49.677
5500	43091	68.092	24823	50.153
6000	47894	68.928	27329	50.589
6500	52751	69.705	29847	50.992
7000	57657	70.433	32378	51.367
7500	62608	71.116	34924	51.718
8000	67600	71.760	37485	52.049
8500	72633	72.370	40063	52.362
9000	77708	72.950	42658	52.658
9500	82828	73.504	45270	52.941
10000	87997	74.034	47897	53.210

(Continued)

TABLE C.7 (Continued) *Ideal-Gas Properties of Various Substances (English Units), Entropies at 1 atm Pressure*

T R	Carbon Dioxide (CO_2) $\bar{h}^o_{f,\,537} = -169\,184$ Btu/lb mol $M = 44.01$		Carbon Monoxide (CO) $\bar{h}^o_{f,\,537} = -47\,518$ Btu/lb mol $M = 28.01$	
	$\bar{h}^o - \bar{h}^o_{537}$ Btu/lb mol	$\bar{s}^o$ Btu/lbmol/R	$\bar{h}^o - \bar{h}^o_{537}$ Btu/lb mol	$\bar{s}^o$ Btu/lb
0	−4026	0	−3728	0
200	−2636	43.466	−2343	40.319
400	−1153	48.565	−951	45.137
537	0	51.038	0	47.182
600	573	52.047	441	47.959
800	2525	54.848	1842	49.974
1000	4655	57.222	3266	51.562
1200	6927	59.291	4723	52.891
1400	9315	61.131	6220	54.044
1600	11798	62.788	7754	55.068
1800	14358	64.295	9323	55.992
2000	16982	65.677	10923	56.835
2200	19659	66.952	12549	57.609
2400	22380	68.136	14197	58.326
2600	25138	69.239	15864	58.993
2800	27926	70.273	17547	59.616
3000	30741	71.244	19243	60.201
3200	33579	72.160	20951	60.752
3400	36437	73.026	22669	61.273
3600	39312	73.847	24395	61.767
3800	42202	74.629	26128	62.236
4000	45105	75.373	27869	62.683
4200	48021	76.084	29614	63.108
4400	50948	76.765	31366	63.515
4600	53885	77.418	33122	63.905
4800	56830	78.045	34883	64.280
5000	59784	78.648	36650	64.641
5500	55739	68.649	39393	61.477
6000	74660	81.360	45548	66.263
6500	82155	82.560	50023	66.979
7000	89682	83.675	54514	67.645
7500	97239	84.718	59020	68.267
8000	104823	85.697	63539	68.850
8500	112434	86.620	68069	69.399
9000	120071	87.493	72610	69.918
9500	127734	88.321	77161	70.410
10000	135426	89.110	81721	70.878

(Continued)

TABLE C.7 (Continued) *Ideal-Gas Properties of Various Substances (English Units), Entropies at 1 atm Pressure*

	Water (H_2O) $\bar{h}^o_{f,537} = -103\,966$ Btu/lb mol $M = 18.015$		Hydroxyl (OH) $\bar{h}^o_{f,537} = 16\,761$ Btu/lb mol $M = 17.007$	
T R	$\bar{h}^o - \bar{h}^o_{537}$ Btu/lb mol	$\bar{s}^o$ Btu/lbmol/R	$\bar{h}^o - \bar{h}^o_{537}$ Btu/lb mol	$\bar{s}^o$ Btu/lbmol/R
0	−4258	0	−3943	0
200	−2686	37.209	−2484	36.521
400	−1092	42.728	−986	41.729
537	0	45.076	0	43.852
600	509	45.973	452	44.649
800	2142	48.320	1870	46.689
1000	3824	50.197	3280	48.263
1200	5566	51.784	4692	49.549
1400	7371	53.174	6112	50.643
1600	9241	54.422	7547	51.601
1800	11178	55.563	9001	52.457
2000	13183	56.619	10477	53.235
2200	15254	57.605	11978	53.950
2400	17388	58.533	13504	54.614
2600	19582	59.411	15054	55.235
2800	21832	60.245	16627	55.817
3000	24132	61.038	18220	56.367
3200	26479	61.796	19834	56.887
3400	28867	62.520	21466	57.382
3600	31293	63.213	23114	57.853
3800	33756	63.878	24777	58.303
4000	36251	64.518	26455	58.733
4200	38774	65.134	28145	59.145
4400	41325	65.727	29849	59.542
4600	43899	66.299	31563	59.922
4800	46496	66.852	33287	60.289
5000	49114	67.386	35021	60.643
5500	55739	68.649	39393	61.477
6000	62463	69.819	43812	62.246
6500	69270	70.908	48272	62.959
7000	76146	71.927	52767	63.626
7500	83081	72.884	57294	64.250
8000	90069	73.786	61851	64.838
8500	97101	74.639	66434	65.394
9000	104176	75.448	71043	65.921
9500	111289	76.217	75677	66.422
10000	118440	76.950	80335	66.900

(Continued)

TABLE C.7 (Continued) *Ideal-Gas Properties of Various Substances (English Units), Entropies at 1 atm Pressure*

	Hydrogen (H_2) $\bar{h}^o_{f,\,537} = 0$ Btu/lb mol $M = 2.016$		Hydrogen, Monatomic (H) $\bar{h}^o_{f,\,537} = 93\ 723$ Btu/lb mol $M = 1.008$	
T R	$\bar{h}^o\text{-}\bar{h}^o_{537}$ Btu/lb mol	$\bar{s}^o$ Btu/lbmol/R	$\bar{h}^o\text{-}\bar{h}^o_{537}$ Btu/lb mol	$\bar{s}^o$ Btu/lbmol/R
0	−3640	0	−2664	0
200	−2224	24.703	−1672	22.473
400	−927	29.193	−679	25.914
537	0	31.186	0	27.373
600	438	31.957	314	27.927
800	1831	33.960	1307	29.355
1000	3225	35.519	2300	30.463
1200	4622	36.797	3293	31.368
1400	6029	37.883	4286	32.134
1600	7448	38.831	5279	32.797
1800	8884	39.676	6272	33.381
2000	10337	40.441	7265	33.905
2200	11812	41.143	8258	34.378
2400	13309	41.794	9251	34.810
2600	14829	42.401	10244	35.207
2800	16372	42.973	11237	35.575
3000	17938	43.512	12230	35.917
3200	19525	44.024	13223	36.238
3400	21133	44.512	14215	36.539
3600	22761	44.977	15208	36.823
3800	24407	45.422	16201	37.091
4000	26071	45.849	17194	37.346
4200	27752	46.260	18187	37.588
4400	29449	46.655	19180	37.819
4600	31161	47.035	20173	38.040
4800	32887	47.403	21166	38.251
5000	34627	47.758	22159	38.454
5500	39032	48.598	24641	38.927
6000	43513	49.378	27124	39.359
6500	48062	50.105	29606	39.756
7000	52678	50.789	32088	40.124
7500	57356	51.434	34571	40.467
8000	62094	52.045	37053	40.787
8500	66889	52.627	39535	41.088
9000	71738	53.182	42018	41.372
9500	76638	53.712	44500	41.640
10000	81581	54.220	46982	41.895

(Continued)

TABLE C.7 (Continued) *Ideal-Gas Properties of Various Substances (English Units), Entropies at 1 atm Pressure*

	Nitric Oxide (NO) $\bar{h}^o_{f, 537} = 38\ 818$ Btu/lb mol $M = 30.006$		Nitrogen Dioxide (NO$_2$) $\bar{h}^o_{f, 537} = 14\ 230$ Btu/lb mol $M = 46.005$	
T R	$\bar{h}^o - \bar{h}^o_{537}$ Btu/lb mol	$\bar{s}^o$ Btu/lbmol/R	$\bar{h}^o - \bar{h}^o_{537}$ Btu/lb mol	$\bar{s}^o$ Btu/lbmol/R
0	−3952	0	−4379	0
200	−2224	24.703	−1672	22.473
400	−927	29.193	−679	25.914
537	0	31.186	0	27.373
600	438	31.957	314	27.927
800	1831	33.960	1307	29.355
1000	3225	35.519	2300	30.463
1200	4622	36.797	3293	31.368
1400	6029	37.883	4286	32.134
1600	7448	38.831	5279	32.797
1800	8884	39.676	6272	33.381
2000	10337	40.441	7265	33.905
2200	11812	41.143	8258	34.378
2400	13309	41.794	9251	34.810
2600	14829	42.401	10244	35.207
2800	16372	42.973	11237	35.575
3000	17938	43.512	12230	35.917
3200	19525	44.024	13223	36.238
3400	21133	44.512	14215	36.539
3600	22761	44.977	15208	36.823
3800	24407	45.422	16201	37.091
4000	26071	45.849	17194	37.346
4200	27752	46.260	18187	37.588
4400	29449	46.655	19180	37.819
4600	31161	47.035	20173	38.040
4800	32887	47.403	21166	38.251
5000	34627	47.758	22159	38.454
5500	41726	68.965	63395	84.990
6000	43513	49.378	27124	39.359
6500	48062	50.105	29606	39.756
7000	52678	50.789	32088	40.124
7500	57356	51.434	34571	40.467
8000	62094	52.045	37053	40.787
8500	66889	52.627	39535	41.088
9000	71738	53.182	42018	41.372
9500	76638	53.712	44500	41.640
10000	81581	54.220	46982	41.895

TABLE C.8 ENG *Thermodynamic Properties of Water*
TABLE C.8.1 ENG *Saturated Water*

Temp.	Press.	SpecificVolume, ft³/lbm			Internal Energy, Btu/lbm		
F	psia	Sat. Liquid	Evap.	Sat. Vapor	Sat. Liquid	Evap.	Sat. Vapor
T	P	v_f	v_{fg}	v_g	u_f	u_{fg}	u_g
32	0.0887	0.01602	3301.6545	3301.6705	0	1021.21	1021.21
35	0.100	0.01602	2947.5021	2947.5181	2.99	1019.20	1022.19
40	0.122	0.01602	2445.0713	2445.0873	8.01	1015.84	1023.85
45	0.147	0.01602	2036.9527	2036.9687	13.03	1012.47	1025.50
50	0.178	0.01602	1703.9867	1704.0027	18.05	1009.10	1027.15
60	0.256	0.01603	1206.7283	1206.7443	28.08	1002.36	1030.44
70	0.363	0.01605	867.5791	867.5952	38.09	995.64	1033.72
80	0.507	0.01607	632.6739	632.6900	48.08	988.91	1036.99
90	0.699	0.01610	467.5865	467.6026	58.06	982.18	1040.24
100	0.950	0.01613	349.9602	349.9764	68.04	975.43	1043.47
110	1.276	0.01617	265.0548	265.0709	78.01	968.67	1046.68
120	1.695	0.01620	203.0105	203.0267	87.99	961.88	1049.87
130	2.225	0.01625	157.1419	157.1582	97.96	955.07	1053.03
140	2.892	0.01629	122.8567	122.8730	107.95	948.21	1056.16
150	3.722	0.01634	96.9611	96.9774	117.94	941.32	1059.26
160	4.745	0.01639	77.2079	77.2243	127.94	934.39	1062.32
170	5.997	0.01645	61.9983	62.0148	137.94	927.41	1065.35
180	7.515	0.01651	50.1826	50.1991	147.96	920.38	1068.34
190	9.344	0.01657	40.9255	40.9421	157.99	913.29	1071.29
200	11.530	0.01663	33.6146	33.6312	168.03	906.15	1074.18
210	14.126	0.01670	27.7964	27.8131	178.09	898.95	1077.04
212.0	14.696	0.01672	26.7864	26.8032	180.09	897.51	1077.60
220	17.189	0.01677	23.1325	23.1492	188.16	891.68	1079.84
230	20.781	0.01685	19.3677	19.3846	198.25	884.33	1082.58
240	24.968	0.01692	16.3088	16.3257	208.36	876.91	1085.27
250	29.823	0.01700	13.8077	13.8247	218.48	869.41	1087.90
260	35.422	0.01708	11.7503	11.7674	228.64	861.82	1090.46
270	41.848	0.01717	10.0483	10.0655	238.81	854.14	1092.95
280	49.189	0.01726	8.6325	8.6498	249.02	846.35	1095.37
290	57.535	0.01735	7.4486	7.4660	259.25	838.46	1097.71
300	66.985	0.01745	6.4537	6.4712	269.51	830.45	1099.96
310	77.641	0.01755	5.6136	5.6312	279.80	822.32	1102.13
320	89.609	0.01765	4.9010	4.9186	290.13	814.07	1104.20
330	103.00	0.01776	4.2938	4.3115	300.50	805.68	1106.17
340	117.94	0.01787	3.7742	3.7921	310.90	797.14	1108.04
350	134.54	0.01799	3.3279	3.3459	321.35	788.45	1109.80

TABLE C.8.1 ENG (Continued) *Saturated Water*

Temp.	Press.	Specific Volume, ft³/lbm			Internal Energy, Btu/lbm		
F	psia	Sat. Liquid	Evap.	Sat. Vapor	Sat. Liquid	Evap.	Sat. Vapor
T	P	v_f	v_{fg}	v_g	u_f	u_{fg}	u_g
32	0.0887	0	1075.38	1075.39	0	2.1869	2.1869
35	0.100	2.99	1073.71	1076.70	0.0061	2.1703	2.1764
40	0.122	8.01	1070.89	1078.90	0.0162	2.1430	2.1591
45	0.147	13.03	1068.06	1081.10	0.0262	2.1161	2.1423
50	0.178	18.05	1065.24	1083.29	0.0361	2.0898	2.1259
60	0.256	28.08	1059.59	1087.67	0.0555	2.0388	2.0943
70	0.363	38.09	1053.95	1092.04	0.0746	1.9896	2.0642
80	0.507	48.08	1048.31	1096.39	0.0933	1.9423	2.0356
90	0.699	58.06	1042.65	1100.72	0.1116	1.8966	2.0083
100	0.950	68.04	1036.98	1105.02	0.1296	1.8526	1.9822
110	1.276	78.01	1031.28	1109.29	0.1473	1.8101	1.9574
120	1.695	87.99	1025.55	1113.54	0.1646	1.7690	1.9336
130	2.225	97.97	1019.78	1117.75	0.1817	1.7292	1.9109
140	2.892	107.96	1013.96	1121.92	0.1985	1.6907	1.8892
150	3.722	117.95	1008.10	1126.05	0.2150	1.6533	1.8683
160	4.745	127.95	1002.18	1130.14	0.2313	1.6171	1.8484
170	5.997	137.96	996.21	1134.17	0.2473	1.5819	1.8292
180	7.515	147.98	990.17	1138.15	0.2631	1.5478	1.8109
190	9.344	158.02	984.06	1142.08	0.2786	1.5146	1.7932
200	11.530	168.07	977.87	1145.94	0.2940	1.4822	1.7762
210	14.126	178.13	971.61	1149.74	0.3091	1.4507	1.7599
212.0	14.696	180.13	970.35	1150.49	0.3121	1.4446	1.7567
220	17.189	188.21	965.26	1153.47	0.3240	1.4201	1.7441
230	20.781	198.31	958.81	1157.12	0.3388	1.3901	1.7289
240	24.968	208.43	952.27	1160.70	0.3533	1.3609	1.7142
250	29.823	218.58	945.61	1164.19	0.3677	1.3324	1.7001
260	35.422	228.75	938.84	1167.59	0.3819	1.3044	1.6864
270	41.848	238.95	931.95	1170.90	0.3960	1.2771	1.6731
280	49.189	249.17	924.93	1174.10	0.4098	1.2504	1.6602
290	57.535	259.43	917.76	1177.19	0.4236	1.2241	1.6477
300	66.985	269.73	910.45	1180.18	0.4372	1.1984	1.6356
310	77.641	280.06	902.98	1183.03	0.4507	1.1731	1.6238
320	89.609	290.43	895.34	1185.76	0.4640	1.1483	1.6122
330	103.00	300.84	887.52	1188.36	0.4772	1.1238	1.6010
340	117.94	311.29	879.51	1190.80	0.4903	1.0997	1.5900
350	134.54	321.80	871.30	1193.10	0.5033	1.0760	1.5793

TABLE C.8.1 ENG (Continued) *Saturated Water*

Temp.	Press.	SpecificVolume, ft³/lbm			Internal Energy, Btu/lbm		
F	psia	Sat. Liquid	Evap.	Sat. Vapor	Sat. Liquid	Evap.	Sat. Vapor
T	P	v_f	v_{fg}	v_g	u_f	u_{fg}	u_g
360	152.93	0.01811	2.9430	2.9611	331.83	779.60	1111.43
370	173.24	0.01823	2.6098	2.6280	342.37	770.57	1112.94
380	195.61	0.01836	2.3203	2.3387	352.95	761.37	1114.31
390	220.17	0.01850	2.0680	2.0865	363.58	751.97	1115.55
400	247.08	0.01864	1.8474	1.8660	374.26	742.37	1116.63
410	276.48	0.01878	1.6537	1.6725	385.00	732.56	1117.56
420	308.52	0.01894	1.4833	1.5023	395.80	722.52	1118.32
430	343.37	0.01909	1.3329	1.3520	406.67	712.24	1118.91
440	381.18	0.01926	1.1998	1.2191	417.61	701.71	1119.32
450	422.13	0.01943	1.0816	1.1011	428.63	690.90	1119.53
460	466.38	0.01961	0.9764	0.9961	439.73	679.82	1119.55
470	514.11	0.01980	0.8826	0.9024	450.92	668.43	1119.35
480	565.50	0.02000	0.7986	0.8186	462.21	656.72	1118.93
490	620.74	0.02021	0.7233	0.7435	473.60	644.67	1118.28
500	680.02	0.02043	0.6556	0.6761	485.11	632.26	1117.37
510	743.53	0.02066	0.5946	0.6153	496.75	619.46	1116.21
520	811.48	0.02091	0.5395	0.5604	508.53	606.23	1114.76
530	884.07	0.02117	0.4896	0.5108	520.46	592.56	1113.02
540	961.51	0.02145	0.4443	0.4658	532.56	578.39	1110.95
550	1044.02	0.02175	0.4031	0.4249	544.85	563.69	1108.54
560	1131.85	0.02207	0.3656	0.3876	557.35	548.42	1105.76
570	1225.21	0.02241	0.3312	0.3536	570.07	532.50	1102.56
580	1324.37	0.02278	0.2997	0.3225	583.05	515.87	1098.91
590	1429.58	0.02318	0.2707	0.2939	596.31	498.44	1094.76
600	1541.13	0.02362	0.2440	0.2676	609.91	480.11	1090.02
610	1659.32	0.02411	0.2193	0.2434	623.87	460.76	1084.63
620	1784.48	0.02465	0.1963	0.2209	638.26	440.20	1078.46
630	1916.96	0.02525	0.1747	0.2000	653.17	418.22	1071.38
640	2057.17	0.02593	0.1545	0.1804	668.68	394.52	1063.20
650	2205.54	0.02673	0.1353	0.1620	684.96	368.66	1053.63
660	2362.59	0.02766	0.1169	0.1446	702.24	340.02	1042.26
670	2528.88	0.02882	0.0990	0.1278	720.91	307.52	1028.43
680	2705.09	0.03031	0.0809	0.1112	741.70	269.26	1010.95
690	2891.99	0.03248	0.0618	0.0943	766.34	220.82	987.16
700	3090.47	0.03665	0.0377	0.0743	801.66	145.92	947.57
705.4	3203.79	0.05053	0	0.0505	872.56	0	872.56

TABLE C.8.1 ENG (Continued) *Saturated Water*

Temp.	Press.	Specific Volume, ft³/lbm			Internal Energy, Btu/lbm		
F	psia	Sat. Liquid	Evap.	Sat. Vapor	Sat. Liquid	Evap.	Sat. Vapor
T	P	v_f	v_{fg}	v_g	u_f	u_{fg}	u_g
360	152.93	332.35	862.88	1195.23	0.5162	1.0526	1.5688
370	173.24	342.95	854.24	1197.19	0.5289	1.0295	1.5584
380	195.61	353.61	845.36	1198.97	0.5416	1.0067	1.5483
390	220.17	364.33	836.23	1200.56	0.5542	0.9841	1.5383
400	247.08	375.11	826.84	1201.95	0.5667	0.9617	1.5284
410	276.48	385.96	817.17	1203.13	0.5791	0.9395	1.5187
420	308.52	396.89	807.20	1204.09	0.5915	0.9175	1.5090
430	343.37	407.89	796.93	1204.82	0.6038	0.8957	1.4995
440	381.18	418.97	786.34	1205.31	0.6160	0.8740	1.4900
450	422.13	430.15	775.40	1205.54	0.6282	0.8523	1.4805
460	466.38	441.42	764.09	1205.51	0.6404	0.8308	1.4711
470	514.11	452.80	752.40	1205.20	0.6525	0.8093	1.4618
480	565.50	464.30	740.30	1204.60	0.6646	0.7878	1.4524
490	620.74	475.92	727.76	1203.68	0.6767	0.7663	1.4430
500	680.02	487.68	714.76	1202.44	0.6888	0.7447	1.4335
510	743.53	499.59	701.27	1200.86	0.7009	0.7232	1.4240
520	811.48	511.67	687.25	1198.92	0.7130	0.7015	1.4144
530	884.07	523.93	672.66	1196.58	0.7251	0.6796	1.4048
540	961.51	536.38	657.45	1193.83	0.7374	0.6576	1.3950
550	1044.02	549.05	641.58	1190.63	0.7496	0.6354	1.3850
560	1131.85	561.97	624.98	1186.95	0.7620	0.6129	1.3749
570	1225.21	575.15	607.59	1182.74	0.7745	0.5901	1.3646
580	1324.37	588.63	589.32	1177.95	0.7871	0.5668	1.3539
590	1429.58	602.45	570.06	1172.51	0.7999	0.5431	1.3430
600	1541.13	616.64	549.71	1166.35	0.8129	0.5187	1.3317
610	1659.32	631.27	528.08	1159.36	0.8262	0.4937	1.3199
620	1784.48	646.40	505.00	1151.41	0.8397	0.4677	1.3075
630	1916.96	662.12	480.21	1142.33	0.8537	0.4407	1.2943
640	2057.17	678.55	453.33	1131.89	0.8681	0.4122	1.2803
650	2205.54	695.87	423.89	1119.76	0.8831	0.3820	1.2651
660	2362.59	714.34	391.13	1105.47	0.8990	0.3493	1.2483
670	2528.88	734.39	353.83	1088.23	0.9160	0.3132	1.2292
680	2705.09	756.87	309.77	1066.64	0.9350	0.2718	1.2068
690	2891.99	783.72	253.88	1037.60	0.9575	0.2208	1.1783
700	3090.47	822.61	167.47	990.09	0.9901	0.1444	1.1345
705.4	3203.79	902.52	0	902.52	1.0580	0	1.0580

TABLE C.8.2 ENG *Superheated Vapor Water*

Temp. F	v ft³/lbm	u Btu/lbm	h Btu/lbm	s Btu/lbm R	v ft³/lbm	u Btu/lbm	h Btu/lbm	s Btu/lbm R
		1 psia (101.70)				5 psia (162.20)		
Sat.	333.58	1044.02	1105.75	1.9779	73.531	1062.99	1131.03	1.8441
200	392.51	1077.49	1150.12	2.0507	78.147	1076.25	1148.55	1.8715
240	416.42	1091.22	1168.28	2.0775	83.001	1090.25	1167.05	1.8987
280	440.32	1105.02	1186.50	2.1028	87.831	1104.27	1185.53	1.9244
320	464.19	1118.92	1204.82	2.1269	92.645	1118.32	1204.04	1.9487
360	488.05	1132.92	1223.23	2.1499	97.447	1132.42	1222.59	1.9719
400	511.91	1147.02	1241.75	2.1720	102.24	1146.61	1241.21	1.9941
440	535.76	1161.23	1260.37	2.1932	107.03	1160.89	1259.92	2.0154
500	571.53	1182.77	1288.53	2.2235	114.21	1182.50	1288.17	2.0458
600	631.13	1219.30	1336.09	2.2706	126.15	1219.10	1335.82	2.0930
700	690.72	1256.65	1384.47	2.3142	138.08	1256.50	1384.26	2.1367
800	750.30	1294.86	1433.70	2.3549	150.01	1294.73	1433.53	2.1774
900	809.88	1333.94	1483.81	2.3932	161.94	1333.84	1483.68	2.2157
1000	869.45	1373.93	1534.82	2.4294	173.86	1373.85	1534.71	2.2520
1100	929.03	1414.83	1586.75	2.4638	185.78	1414.77	1586.66	2.2864
1200	988.60	1456.67	1639.61	2.4967	197.70	1456.61	1639.53	2.3192
1300	1048.17	1499.43	1693.40	2.5281	209.62	1499.38	1693.33	2.3507
1400	1107.74	1543.13	1748.12	2.5584	221.53	1543.09	1748.06	2.3809
		10 psia (193.19)				14.696 psia (211.99)		
Sat.	38.424	1072.21	1143.32	1.7877	26.803	1077.60	1150.49	1.7567
200	38.848	1074.67	1146.56	1.7927	—	—	—	—
240	41.320	1089.03	1165.50	1.8205	27.999	1087.87	1164.02	1.7764
280	43.768	1103.31	1184.31	1.8467	29.687	1102.40	1183.14	1.8030
320	46.200	1117.56	1203.05	1.8713	31.359	1116.83	1202.11	1.8280
360	48.620	1131.81	1221.78	1.8948	33.018	1131.22	1221.01	1.8516
400	51.032	1146.10	1240.53	1.9171	34.668	1145.62	1239.90	1.8741
440	53.438	1160.46	1259.34	1.9385	36.313	1160.05	1258.80	1.8956
500	57.039	1182.16	1287.71	1.9690	38.772	1181.83	1287.27	1.9262
600	63.027	1218.85	1335.48	2.0164	42.857	1218.61	1335.16	1.9737
700	69.006	1256.30	1384.00	2.0601	46.932	1256.12	1383.75	2.0175
800	74.978	1294.58	1433.32	2.1009	51.001	1294.43	1433.13	2.0584
900	80.946	1333.72	1483.51	2.1392	55.066	1333.60	1483.35	2.0967
1000	86.912	1373.74	1534.57	2.1755	59.128	1373.65	1534.44	2.1330
1100	92.875	1414.68	1586.54	2.2099	63.188	1414.60	1586.44	2.1674
1200	98.837	1456.53	1639.43	2.2428	67.247	1456.47	1639.34	2.2003
1300	104.798	1499.32	1693.25	2.2743	71.304	1499.26	1693.17	2.2318
1400	110.759	1543.03	1747.99	2.3045	75.361	1542.98	1747.92	2.2620
1500	116.718	1587.67	1803.66	2.3337	79.417	1587.63	1803.60	2.2912
1600	122.678	1633.24	1860.25	2.3618	83.473	1633.20	1860.20	2.3194

(Continued)

TABLE C.8.2 ENG (Continued) *Superheated Vapor Water*

Temp. F	v ft³/lbm	u Btu/lbm	h Btu/lbm	s Btu/lbm R	v ft³/lbm	u Btu/lbm	h Btu/lbm	s Btu/lbm R
		20 psia (227.96)				40 psia (267.26)		
Sat.	20.091	1082.02	1156.38	1.7320	10.501	1092.27	1170.00	1.6767
240	20.475	1086.54	1162.32	1.7405	—	—	—	—
280	21.734	1101.36	1181.80	1.7676	10.711	1097.31	1176.59	1.6857
320	22.976	1116.01	1201.04	1.7929	11.360	1112.81	1196.90	1.7124
360	24.206	1130.55	1220.14	1.8168	11.996	1127.98	1216.77	1.7373
400	25.427	1145.06	1239.17	1.8395	12.623	1142.95	1236.38	1.7606
440	26.642	1159.59	1258.19	1.8611	13.243	1157.82	1255.84	1.7827
500	28.456	1181.46	1286.78	1.8919	14.164	1180.06	1284.91	1.8140
600	31.466	1218.35	1334.80	1.9395	15.685	1217.33	1333.43	1.8621
700	34.466	1255.91	1383.47	1.9834	17.196	1255.14	1382.42	1.9063
800	37.460	1294.27	1432.91	2.0243	18.701	1293.65	1432.08	1.9474
900	40.450	1333.47	1483.17	2.0626	20.202	1332.96	1482.50	1.9859
1000	43.437	1373.54	1534.30	2.0989	21.700	1373.12	1533.74	2.0222
1100	46.422	1414.51	1586.32	2.1334	23.196	1414.16	1585.86	2.0568
1200	49.406	1456.39	1639.24	2.1663	24.690	1456.09	1638.85	2.0897
1300	52.389	1499.19	1693.08	2.1978	26.184	1498.94	1692.75	2.1212
1400	55.371	1542.92	1747.85	2.2280	27.677	1542.70	1747.56	2.1515
1500	58.352	1587.58	1803.54	2.2572	29.169	1587.38	1803.29	2.1807
1600	61.333	1633.15	1860.14	2.2854	30.660	1632.97	1859.92	2.2089
		60 psia (292.73)				80 psia (312.06)		
Sat.	7.177	1098.33	1178.02	1.6444	5.474	1102.56	1183.61	1.6214
320	7.485	1109.46	1192.56	1.6633	5.544	1105.95	1188.02	1.6270
360	7.924	1125.31	1213.29	1.6893	5.886	1122.53	1209.67	1.6541
400	8.353	1140.77	1233.52	1.7134	6.217	1138.53	1230.56	1.6790
440	8.775	1156.01	1253.44	1.7360	6.541	1154.15	1250.98	1.7022
500	9.399	1178.64	1283.00	1.7678	7.017	1177.19	1281.07	1.7346
600	10.425	1216.31	1332.06	1.8165	7.794	1215.28	1330.66	1.7838
700	11.440	1254.35	1381.37	1.8609	8.561	1253.57	1380.31	1.8285
800	12.448	1293.03	1431.24	1.9022	9.322	1292.41	1430.40	1.8700
900	13.452	1332.46	1481.82	1.9408	10.078	1331.95	1481.14	1.9087
1000	14.454	1372.71	1533.19	1.9773	10.831	1372.29	1532.63	1.9453
1100	15.454	1413.81	1585.39	2.0119	11.583	1413.46	1584.93	1.9799
1200	16.452	1455.80	1638.46	2.0448	12.333	1455.51	1638.08	2.0129
1300	17.449	1498.69	1692.42	2.0764	13.082	1498.43	1692.09	2.0445
1400	18.445	1542.48	1747.28	2.1067	13.830	1542.26	1746.99	2.0749
1500	19.441	1587.18	1803.04	2.1359	14.577	1586.99	1802.79	2.1041
1600	20.436	1632.79	1859.70	2.1641	15.324	1632.62	1859.48	2.1323
1800	22.426	1726.69	1975.69	2.2178	16.818	1726.54	1975.50	2.1861
2000	24.415	1824.02	2095.10	2.2685	18.310	1823.88	2094.94	2.2367

(Continued)

TABLE C.8.2 ENG (Continued) *Superheated Vapor Water*

Temp. F	v ft³/lbm	u Btu/lbm	h Btu/lbm	s Btu/lbm R	v ft³/lbm	u Btu/lbm	h Btu/lbm	s Btu/lbm R
		100 psia (327.85)				120 psia (341.30)		
Sat.	4.4340	1105.76	1187.81	1.6034	3.7302	1108.28	1191.11	1.5886
350	4.5917	1115.39	1200.36	1.6191	3.7835	1112.20	1196.22	1.5950
400	4.9344	1136.21	1227.53	1.6517	4.0785	1133.83	1224.39	1.6287
450	5.2646	1156.20	1253.62	1.6812	4.3600	1154.34	1251.16	1.6590
500	5.5866	1175.72	1279.10	1.7085	4.6330	1174.22	1277.10	1.6868
550	5.9032	1195.02	1304.25	1.7340	4.9002	1193.78	1302.59	1.7127
600	6.2160	1214.23	1329.26	1.7582	5.1636	1213.18	1327.84	1.7371
700	6.8340	1252.78	1379.24	1.8033	5.6825	1251.98	1378.17	1.7825
800	7.4455	1291.78	1429.56	1.8449	6.1948	1291.15	1428.72	1.8243
900	8.0528	1331.45	1480.47	1.8838	6.7029	1330.94	1479.78	1.8633
1000	8.6574	1371.87	1532.08	1.9204	7.2082	1371.46	1531.52	1.9000
1100	9.2599	1413.12	1584.47	1.9551	7.7114	1412.77	1584.01	1.9348
1200	9.8610	1455.21	1637.69	1.9882	8.2132	1454.92	1637.30	1.9679
1300	10.4610	1498.18	1691.76	2.0198	8.7140	1497.93	1691.43	1.9996
1400	11.0602	1542.04	1746.71	2.0502	9.2139	1541.82	1746.42	2.0300
1500	11.6588	1586.79	1802.54	2.0794	9.7133	1586.60	1802.29	2.0592
1600	12.2570	1632.44	1859.25	2.1076	10.2122	1632.26	1859.03	2.0875
1800	13.4525	1726.38	1975.32	2.1614	11.2091	1726.23	1975.14	2.1412
2000	14.6472	1823.74	2094.78	2.2120	12.2052	1823.59	2094.62	2.1919
		140 psia (353.08)				160 psia (363.59)		
Sat.	3.2214	1110.31	1193.77	1.5760	2.8359	1111.99	1195.95	1.5650
400	3.4664	1131.36	1221.16	1.6088	3.0066	1128.81	1217.83	1.5910
450	3.7135	1152.44	1248.64	1.6399	3.2282	1150.49	1246.07	1.6230
500	3.9515	1172.70	1275.07	1.6682	3.4402	1171.15	1273.01	1.6518
550	4.1837	1192.52	1300.90	1.6944	3.6461	1191.25	1299.20	1.6784
600	4.4118	1212.12	1326.41	1.7191	3.8478	1211.05	1324.97	1.7033
700	4.8599	1251.18	1377.09	1.7648	4.2430	1250.38	1376.00	1.7494
800	5.3014	1290.53	1427.87	1.8068	4.6314	1289.89	1427.02	1.7916
900	5.7387	1330.43	1479.10	1.8459	5.0155	1329.92	1478.42	1.8308
1000	6.1730	1371.04	1530.96	1.8827	5.3967	1370.62	1530.40	1.8677
1100	6.6054	1412.42	1583.54	1.9176	5.7759	1412.07	1583.08	1.9026
1200	7.0363	1454.62	1636.91	1.9507	6.1536	1454.32	1636.52	1.9358
1300	7.4661	1497.67	1691.10	1.9824	6.5302	1497.42	1690.77	1.9676
1400	7.8952	1541.60	1746.14	2.0128	6.9061	1541.37	1745.85	1.9980
1500	8.3236	1586.40	1802.04	2.0421	7.2813	1586.20	1801.79	2.0273
1600	8.7516	1632.08	1858.81	2.0704	7.6561	1631.91	1858.59	2.0556
1800	9.6067	1726.08	1974.96	2.1242	8.4049	1725.92	1974.77	2.1094
2000	10.4610	1823.45	2094.46	2.1748	9.1528	1823.31	2094.31	2.1601

(Continued)

TABLE C.8.2 ENG (Continued) *Superheated Vapor Water*

Temp. F	v ft³/lbm	u Btu/lbm	h Btu/lbm	s Btu/lbm R	v ft³/lbm	u Btu/lbm	h Btu/lbm	s Btu/lbm R
	180 psia (373.12)				200 psia (381.86)			
Sat.	2.5333	1113.38	1197.76	1.5553	2.2892	1114.55	1199.28	1.5464
400	2.6482	1126.18	1214.39	1.5749	2.3609	1123.45	1210.83	1.5600
450	2.8504	1148.49	1243.43	1.6078	2.5477	1146.44	1240.73	1.5938
500	3.0424	1169.57	1270.91	1.6372	2.7238	1167.96	1268.77	1.6238
550	3.2279	1189.96	1297.47	1.6641	2.8932	1188.65	1295.72	1.6512
600	3.4091	1209.96	1323.52	1.6893	3.0580	1208.87	1322.05	1.6767
700	3.7631	1249.57	1374.92	1.7357	3.3792	1248.76	1373.82	1.7234
800	4.1102	1289.26	1426.17	1.7781	3.6932	1288.62	1425.31	1.7659
900	4.4529	1329.41	1477.73	1.8175	4.0029	1328.90	1477.04	1.8055
1000	4.7928	1370.20	1529.84	1.8544	4.3097	1369.77	1529.28	1.8425
1100	5.1307	1411.72	1582.61	1.8894	4.6145	1411.36	1582.15	1.8776
1200	5.4670	1454.03	1636.13	1.9227	4.9178	1453.73	1635.74	1.9109
1300	5.8023	1497.16	1690.43	1.9544	5.2200	1496.91	1690.10	1.9427
1400	6.1368	1541.15	1745.56	1.9849	5.5214	1540.93	1745.28	1.9732
1500	6.4707	1586.01	1801.54	2.0142	5.8222	1585.81	1801.29	2.0025
1600	6.8041	1631.73	1858.37	2.0425	6.1225	1631.55	1858.15	2.0308
1800	7.4701	1725.77	1974.59	2.0963	6.7223	1725.62	1974.41	2.0847
2000	8.1353	1823.17	2094.15	2.1470	7.3214	1823.02	2093.99	2.1354
	250 psia (401.03)				300 psia (417.42)			
Sat.	1.8448	1116.73	1202.08	1.5274	1.5441	1118.14	1203.86	1.5115
450	2.0018	1141.09	1233.70	1.5632	1.6361	1135.37	1226.20	1.5365
500	2.1498	1163.81	1263.27	1.5948	1.7662	1159.47	1257.52	1.5701
550	2.2903	1185.30	1291.26	1.6233	1.8878	1181.85	1286.65	1.5997
600	2.4258	1206.09	1318.32	1.6494	2.0041	1203.24	1314.50	1.6266
650	2.5581	1226.49	1344.84	1.6739	2.1168	1224.08	1341.60	1.6516
700	2.6879	1246.71	1371.06	1.6970	2.2269	1244.63	1368.26	1.6751
800	2.9426	1287.02	1423.16	1.7401	2.4421	1285.41	1420.99	1.7187
900	3.1929	1327.61	1475.32	1.7799	2.6528	1326.31	1473.58	1.7589
1000	3.4402	1368.72	1527.87	1.8172	2.8604	1367.65	1526.45	1.7964
1100	3.6854	1410.48	1580.98	1.8524	3.0660	1409.60	1579.80	1.8317
1200	3.9291	1452.98	1634.76	1.8858	3.2700	1452.24	1633.77	1.8653
1300	4.1718	1496.27	1689.27	1.9177	3.4730	1495.63	1688.43	1.8972
1400	4.4136	1540.37	1744.56	1.9483	3.6751	1539.82	1743.84	1.9279
1500	4.6549	1585.32	1800.66	1.9777	3.8767	1584.82	1800.03	1.9573
1600	4.8957	1631.11	1857.59	2.0060	4.0777	1630.66	1857.04	1.9857
1800	5.3763	1725.23	1973.95	2.0599	4.4790	1724.85	1973.50	2.0396
2000	5.8562	1822.67	2093.59	2.1106	4.8794	1822.32	2093.20	2.0904

(Continued)

TABLE C.8.2 ENG (Continued) *Superheated Vapor Water*

Temp. F	v ft³/lbm	u Btu/lbm	h Btu/lbm	s Btu/lbm R	v ft³/lbm	u Btu/lbm	h Btu/lbm	s Btu/lbm R
		350 psia (431.81)				400 psia (444.69)		
Sat.	1.3267	1119.00	1204.93	1.4978	1.1619	1119.44	1205.45	1.4856
450	1.3733	1129.24	1218.18	1.5125	1.1745	1122.63	1209.57	1.4901
500	1.4913	1154.91	1251.49	1.5481	1.2843	1150.11	1245.17	1.5282
550	1.5998	1178.27	1281.88	1.5790	1.3834	1174.56	1276.95	1.5605
600	1.7025	1200.32	1310.59	1.6068	1.4760	1197.33	1306.58	1.5892
700	1.8975	1242.52	1365.42	1.6562	1.6503	1240.38	1362.54	1.6396
800	2.0846	1283.78	1418.80	1.7004	1.8163	1282.14	1416.59	1.6844
900	2.2670	1325.01	1471.83	1.7409	1.9776	1323.69	1470.07	1.7252
1000	2.4463	1366.58	1525.02	1.7787	2.1357	1365.51	1523.59	1.7632
1100	2.6235	1408.71	1578.63	1.8142	2.2917	1407.81	1577.44	1.7989
1200	2.7993	1451.48	1632.78	1.8478	2.4462	1450.73	1631.79	1.8327
1300	2.9739	1494.99	1687.59	1.8799	2.5995	1494.34	1686.76	1.8648
1400	3.1476	1539.26	1743.12	1.9106	2.7520	1538.70	1742.40	1.8956
1500	3.3208	1584.33	1799.41	1.9401	2.9039	1583.83	1798.78	1.9251
1600	3.4935	1630.22	1856.48	1.9685	3.0553	1629.77	1855.93	1.9535
1700	3.6659	1676.93	1914.36	1.9959	3.2064	1676.52	1913.86	1.9810
1800	3.8380	1724.47	1973.04	2.0225	3.3573	1724.08	1972.59	2.0076
1900	4.0099	1772.82	2032.53	2.0482	3.5080	1772.45	2032.11	2.0333
2000	4.1817	1821.96	2092.80	2.0732	3.6585	1821.61	2092.41	2.0584
		500 psia (467.12)				600 psia (486.33)		
Sat.	0.9283	1119.43	1205.32	1.4645	0.7702	1118.54	1204.06	1.4464
500	0.9924	1139.69	1231.51	1.4922	0.7947	1127.97	1216.21	1.4592
550	1.0792	1166.71	1266.56	1.5279	0.8749	1158.23	1255.36	1.4990
600	1.1583	1191.09	1298.26	1.5585	0.9456	1184.50	1289.49	1.5320
650	1.2327	1213.98	1328.04	1.5860	1.0109	1208.63	1320.87	1.5609
700	1.3040	1236.01	1356.66	1.6112	1.0728	1231.51	1350.62	1.5871
800	1.4407	1278.81	1412.11	1.6571	1.1900	1275.42	1407.55	1.6343
900	1.5723	1321.04	1466.52	1.6986	1.3021	1318.36	1462.92	1.6766
1000	1.7008	1363.34	1520.71	1.7371	1.4108	1361.15	1517.79	1.7155
1100	1.8271	1406.01	1575.06	1.7731	1.5173	1404.20	1572.66	1.7519
1200	1.9518	1449.21	1629.80	1.8071	1.6222	1447.68	1627.80	1.7861
1300	2.0754	1493.05	1685.07	1.8395	1.7260	1491.74	1683.38	1.8186
1400	2.1981	1537.57	1740.96	1.8704	1.8289	1536.44	1739.51	1.8497
1500	2.3203	1582.84	1797.52	1.9000	1.9312	1581.84	1796.26	1.8794
1600	2.4419	1628.88	1854.82	1.9285	2.0330	1627.98	1853.71	1.9080
1700	2.5632	1675.70	1912.87	1.9560	2.1345	1674.88	1911.87	1.9355
1800	2.6843	1723.32	1971.68	1.9826	2.2357	1722.55	1970.78	1.9622
1900	2.8052	1771.72	2031.27	2.0084	2.3367	1771.00	2030.44	1.9880

(Continued)

TABLE C.8.2 ENG (Continued) *Superheated Vapor Water*

Temp. F	v ft³/lbm	u Btu/lbm	h Btu/lbm	s Btu/lbm R	v ft³/lbm	u Btu/lbm	h Btu/lbm	s Btu/lbm R
	800 psia (518.36)				1000 psia (544.74)			
Sat.	0.5691	1115.02	1199.26	1.4160	0.4459	1109.86	1192.37	1.3903
550	0.6154	1138.83	1229.93	1.4469	0.4534	1114.77	1198.67	1.3965
600	0.6776	1170.10	1270.41	1.4861	0.5140	1153.66	1248.76	1.4450
650	0.7324	1197.22	1305.64	1.5186	0.5637	1184.74	1289.06	1.4822
700	0.7829	1222.08	1337.98	1.5471	0.6080	1212.03	1324.54	1.5135
750	0.8306	1245.65	1368.62	1.5729	0.6490	1237.23	1357.33	1.5412
800	0.8764	1268.45	1398.19	1.5969	0.6878	1261.21	1388.49	1.5664
900	0.9640	1312.88	1455.60	1.6408	0.7610	1307.26	1448.08	1.6120
1000	1.0482	1356.71	1511.88	1.6807	0.8305	1352.17	1505.86	1.6530
1100	1.1300	1400.52	1567.81	1.7178	0.8976	1396.77	1562.88	1.6908
1200	1.2102	1444.60	1623.76	1.7525	0.9630	1441.46	1619.67	1.7260
1300	1.2892	1489.11	1679.97	1.7854	1.0272	1486.45	1676.53	1.7593
1400	1.3674	1534.17	1736.59	1.8167	1.0905	1531.88	1733.67	1.7909
1500	1.4448	1579.85	1793.74	1.8467	1.1531	1577.84	1791.21	1.8210
1600	1.5218	1626.19	1851.49	1.8754	1.2152	1624.40	1849.27	1.8499
1700	1.5985	1673.25	1909.89	1.9031	1.2769	1671.61	1907.91	1.8777
1800	1.6749	1721.03	1968.98	1.9298	1.3384	1719.51	1967.18	1.9046
1900	1.7510	1769.55	2028.77	1.9557	1.3997	1768.11	2027.12	1.9305
2000	1.8271	1818.80	2089.28	1.9808	1.4608	1817.41	2087.74	1.9557
	1250 psia (572.56)				1500 psia (596.38)			
Sat.	0.3454	1101.68	1181.57	1.3619	0.2769	1091.81	1168.67	1.3358
600	0.3786	1129.00	1216.58	1.3954	0.2816	1096.61	1174.78	1.3416
650	0.4267	1167.24	1265.95	1.4409	0.3329	1146.95	1239.34	1.4012
700	0.4670	1198.42	1306.44	1.4766	0.3716	1183.44	1286.60	1.4429
750	0.5030	1226.08	1342.42	1.5070	0.4049	1214.13	1326.52	1.4766
800	0.5364	1251.75	1375.82	1.5341	0.4350	1241.79	1362.53	1.5058
850	0.5680	1276.25	1407.64	1.5589	0.4631	1267.69	1396.23	1.5321
900	0.5984	1300.02	1438.42	1.5819	0.4897	1292.53	1428.46	1.5562
1000	0.6563	1346.37	1498.18	1.6244	0.5400	1340.43	1490.32	1.6001
1100	0.7116	1392.01	1556.62	1.6631	0.5876	1387.16	1550.26	1.6398
1200	0.7652	1437.49	1614.49	1.6990	0.6334	1433.45	1609.25	1.6765
1300	0.8176	1483.08	1672.19	1.7328	0.6778	1479.68	1667.82	1.7108
1400	0.8690	1528.98	1729.98	1.7647	0.7213	1526.06	1726.28	1.7431
1500	0.9197	1575.31	1788.04	1.7952	0.7641	1572.77	1784.86	1.7738
1600	0.9699	1622.15	1846.49	1.8242	0.8064	1619.90	1843.72	1.8031
1700	1.0197	1669.57	1905.44	1.8522	0.8482	1667.53	1902.98	1.8312
1800	1.0693	1717.62	1964.95	1.8791	0.8899	1715.73	1962.73	1.8582
1900	1.1186	1766.32	2025.07	1.9052	0.9313	1764.53	2023.03	1.8843

(Continued)

TABLE C.8.2 ENG (Continued) *Superheated Vapor Water*

Temp. F	v ft³/lbm	u Btu/lbm	h Btu/lbm	s Btu/lbm R	v ft³/lbm	u Btu/lbm	h Btu/lbm	s Btu/lbm R
	1750 psia (617.30)				2000 psia (635.99)			
Sat.	0.2268	1080.21	1153.65	1.3109	0.1881	1066.63	1136.25	1.2861
650	0.2627	1122.53	1207.61	1.3603	0.2057	1091.06	1167.18	1.3141
700	0.3022	1166.72	1264.60	1.4106	0.2487	1147.74	1239.79	1.3782
750	0.3341	1201.27	1309.47	1.4485	0.2803	1187.32	1291.07	1.4216
800	0.3622	1231.27	1348.55	1.4801	0.3071	1220.13	1333.80	1.4562
850	0.3878	1258.77	1384.37	1.5080	0.3312	1249.46	1372.03	1.4860
900	0.4119	1284.79	1418.18	1.5334	0.3534	1276.78	1407.58	1.5126
1000	0.4569	1334.34	1482.29	1.5789	0.3945	1328.10	1474.09	1.5598
1100	0.4990	1382.21	1543.79	1.6197	0.4325	1377.17	1537.23	1.6017
1200	0.5392	1429.35	1603.95	1.6570	0.4685	1425.19	1598.58	1.6398
1300	0.5780	1476.23	1663.40	1.6918	0.5031	1472.74	1658.95	1.6751
1400	0.6158	1523.12	1722.55	1.7245	0.5368	1520.15	1718.81	1.7082
1500	0.6530	1570.21	1781.67	1.7555	0.5697	1567.64	1778.48	1.7395
1600	0.6896	1617.64	1840.95	1.7850	0.6020	1615.37	1838.18	1.7692
1700	0.7258	1665.49	1900.53	1.8132	0.6340	1663.45	1898.08	1.7976
1800	0.7617	1713.85	1960.52	1.8404	0.6656	1711.97	1958.32	1.8248
1900	0.7974	1762.76	2021.00	1.8666	0.6971	1760.99	2018.99	1.8511
2000	0.8330	1812.26	2082.02	1.8919	0.7284	1810.56	2080.15	1.8765
	2500 psia (668.30)				3000 psia (695.52)			
Sat.	0.1306	1030.99	1091.41	1.2326	0.0840	968.77	1015.42	1.1575
700	0.1684	1098.70	1176.61	1.3073	0.0977	1003.88	1058.13	1.1944
750	0.2030	1155.21	1249.13	1.3686	0.1483	1114.74	1197.08	1.3122
800	0.2291	1195.69	1301.66	1.4112	0.1757	1167.64	1265.19	1.3675
850	0.2513	1229.54	1345.78	1.4456	0.1973	1207.64	1317.18	1.4080
900	0.2712	1259.90	1385.35	1.4752	0.2160	1241.80	1361.69	1.4413
950	0.2896	1288.20	1422.17	1.5018	0.2328	1272.72	1401.99	1.4705
1000	0.3069	1315.19	1457.19	1.5262	0.2485	1301.69	1439.63	1.4967
1100	0.3393	1366.83	1523.81	1.5704	0.2772	1356.16	1510.05	1.5434
1200	0.3696	1416.71	1587.69	1.6101	0.3036	1408.02	1576.59	1.5847
1300	0.3984	1465.66	1649.95	1.6465	0.3285	1458.45	1640.85	1.6223
1400	0.4261	1514.15	1711.28	1.6804	0.3524	1508.08	1703.69	1.6571
1500	0.4531	1562.47	1772.07	1.7123	0.3754	1557.26	1765.65	1.6895
1600	0.4795	1610.83	1832.64	1.7424	0.3978	1606.27	1827.12	1.7201
1700	0.5055	1659.38	1893.22	1.7711	0.4198	1655.30	1888.38	1.7492
1800	0.5312	1708.23	1953.96	1.7986	0.4416	1704.51	1949.64	1.7769
1900	0.5567	1757.48	2015.01	1.8251	0.4631	1754.00	2011.07	1.8035
2000	0.5820	1807.20	2076.44	1.8506	0.4844	1803.86	2072.78	1.8291

(Continued)

TABLE C.8.2 ENG (Continued) *Superheated Vapor Water*

Temp. F	v ft³/lbm	u Btu/lbm	h Btu/lbm	s Btu/lbm R	v ft³/lbm	u Btu/lbm	h Btu/lbm	s Btu/lbm R
	3500 psia				4000 psia			
650	0.02491	663.52	679.65	0.8629	0.02447	657.71	675.82	0.8574
700	0.03058	759.52	779.32	0.9506	0.02867	742.13	763.35	0.9345
750	0.10460	1058.38	1126.13	1.2440	0.06332	960.69	1007.56	1.1395
800	0.13626	1134.74	1222.99	1.3226	0.10523	1095.04	1172.93	1.2740
850	0.15819	1183.43	1285.88	1.3716	0.12833	1156.47	1251.46	1.3352
900	0.17625	1222.36	1336.51	1.4095	0.14623	1201.47	1309.71	1.3789
950	0.19214	1256.40	1380.85	1.4416	0.16152	1239.20	1358.75	1.4143
1000	0.20663	1287.60	1421.43	1.4699	0.17520	1272.94	1402.62	1.4449
1100	0.23282	1345.17	1495.97	1.5193	0.19954	1333.90	1481.60	1.4973
1200	0.25657	1399.15	1565.32	1.5624	0.22129	1390.11	1553.91	1.5423
1300	0.27871	1451.13	1631.65	1.6012	0.24137	1443.72	1622.38	1.5823
1400	0.29972	1501.93	1696.05	1.6368	0.26029	1495.73	1688.39	1.6188
1500	0.31992	1552.01	1759.22	1.6699	0.27837	1546.73	1752.78	1.6525
1600	0.33953	1601.70	1821.61	1.7010	0.29586	1597.12	1816.11	1.6841
1700	0.35872	1651.24	1883.57	1.7303	0.31291	1647.17	1878.79	1.7138
1800	0.37759	1700.80	1945.36	1.7583	0.32964	1697.11	1941.11	1.7420
1900	0.39624	1750.54	2007.17	1.7851	0.34616	1747.10	2003.32	1.7689
2000	0.41474	1800.56	2069.17	1.8108	0.36251	1797.27	2065.60	1.7948
	6000 psia				8000 psia			
650	0.02322	639.99	665.77	0.8404	0.02239	627.01	660.16	0.8278
700	0.02563	708.08	736.53	0.9028	0.02418	688.59	724.39	0.8844
750	0.02978	788.60	821.66	0.9746	0.02671	755.67	795.21	0.9441
800	0.03942	896.87	940.64	1.0708	0.03061	830.67	875.99	1.0095
850	0.05818	1018.83	1083.42	1.1820	0.03706	915.81	970.67	1.0832
900	0.07588	1102.93	1187.18	1.2599	0.04657	1003.68	1072.63	1.1596
950	0.09009	1162.00	1262.02	1.3140	0.05721	1079.59	1164.28	1.2259
1000	0.10207	1209.11	1322.44	1.3561	0.06722	1141.04	1240.55	1.2791
1100	0.12219	1286.42	1422.08	1.4222	0.08445	1236.84	1361.85	1.3595
1200	0.13928	1352.69	1507.33	1.4752	0.09892	1314.18	1460.62	1.4210
1300	0.15453	1413.30	1584.87	1.5206	0.11161	1382.27	1547.50	1.4718
1400	0.16854	1470.48	1657.61	1.5608	0.12309	1444.85	1627.08	1.5158
1500	0.18169	1525.37	1727.10	1.5972	0.13372	1503.78	1701.74	1.5549
1600	0.19421	1578.68	1794.31	1.6307	0.14373	1560.12	1772.89	1.5904
1700	0.20627	1630.90	1859.93	1.6618	0.15328	1614.58	1841.49	1.6229
1800	0.21801	1682.40	1924.45	1.6910	0.16251	1667.69	1908.27	1.6531
1900	0.22952	1733.45	1988.28	1.7186	0.17151	1719.85	1973.75	1.6815
2000	0.24087	1784.28	2051.72	1.7450	0.18034	1771.38	2038.36	1.7083

TABLE C.8.3 ENG *Compressed Liquid Water*

Temp. F	v ft³/lbm	u Btu/lbm	h Btu/lbm	s Btu/lbm R	v ft³/lbm	u Btu/lbm	h Btu/lbm	s Btu/lbm R
	250 psia (401.03)				500 psia (467.12)			
Sat.	0.01865	375.37	376.23	0.5680	0.01975	447.69	449.51	0.6490
32	0.0160	0.00	0.74	0.0000	0.01599	0.00	1.48	0.0000
50	0.0160	18.04	18.78	0.0360	0.01599	18.02	19.50	0.0360
100	0.0161	67.95	68.70	0.1295	0.0161	67.87	69.36	0.1293
125	0.0162	92.86	93.61	0.1730	0.0162	92.75	94.24	0.1728
150	0.0163	117.80	118.55	0.2148	0.0163	117.66	119.17	0.2146
175	0.0165	142.79	143.55	0.2550	0.0165	142.62	144.14	0.2547
200	0.0166	167.84	168.61	0.2937	0.0166	167.64	169.18	0.2934
225	0.0168	192.99	193.76	0.3311	0.0168	192.76	194.31	0.3308
250	0.0170	218.25	219.04	0.3674	0.0170	217.99	219.56	0.3670
275	0.0172	243.66	244.46	0.4026	0.0172	243.36	244.95	0.4022
300	0.0174	269.26	270.06	0.4369	0.0174	268.91	270.52	0.4364
325	0.0177	295.07	295.89	0.4703	0.0177	294.68	296.32	0.4698
350	0.0180	321.14	321.97	0.5030	0.0180	320.70	322.36	0.5025
375	0.0183	347.52	348.36	0.5351	0.0183	347.01	348.70	0.5345
400	0.0186	374.25	375.12	0.5667	0.0186	373.68	375.40	0.5660
450	—	—	—	—	0.0194	428.39	430.19	0.6280
	1000 psia (544.74)				1500 psia (596.38)			
Sat.	0.02159	538.37	542.36	0.74318	0.02346	604.95	611.46	0.80821
32	0.01597	0.02	2.98	0.0000	0.01594	0.04	4.47	0.0001
50	0.0160	17.98	20.94	0.0359	0.0159	17.95	22.37	0.0358
100	0.0161	67.70	70.67	0.1290	0.0161	67.53	71.99	0.1287
125	0.0162	92.52	95.51	0.1724	0.0162	92.30	96.78	0.1720
150	0.0163	117.37	120.39	0.2141	0.0163	117.10	121.61	0.2136
175	0.0164	142.28	145.32	0.2542	0.0164	141.95	146.50	0.2536
200	0.0166	167.25	170.32	0.2928	0.0166	166.86	171.46	0.2922
225	0.0168	192.30	195.40	0.3301	0.0167	191.86	196.50	0.3295
250	0.0169	217.46	220.60	0.3663	0.0169	216.95	221.65	0.3655
275	0.0171	242.77	245.94	0.4014	0.0171	242.18	246.93	0.4005
300	0.0174	268.24	271.45	0.4355	0.0173	267.57	272.39	0.4346
325	0.0176	293.91	297.17	0.4688	0.0176	293.16	298.04	0.4678
350	0.0179	319.83	323.14	0.5014	0.0179	318.97	323.93	0.5003
375	0.0182	346.02	349.39	0.5333	0.0182	345.05	350.10	0.5321
400	0.0185	372.55	375.98	0.5647	0.0185	371.45	376.58	0.5634
425	0.0189	399.47	402.97	0.5957	0.0189	398.21	403.45	0.5942
450	0.0193	426.89	430.47	0.6263	0.0193	425.43	430.78	0.6247
500	0.0204	483.77	487.54	0.6874	0.0202	481.76	487.38	0.6852
550	—	—	—	—	0.0216	542.08	548.07	0.7469

(Continued)

TABLE C.8.3 ENG (Continued) *Compressed Liquid Water*

Temp. F	v ft³/lbm	u Btu/lbm	h Btu/lbm	s Btu/lbm R	v ft³/lbm	u Btu/lbm	h Btu/lbm	s Btu/lbm R
		2000 psia (635.99)				4000 psia		
Sat.	0.02565	662.38	671.87	0.8622	—	—	—	—
32	0.0159	0.06	5.95	0.0001	0.0158	0.10	11.79	0.0000
50	0.0159	17.91	23.80	0.0357	0.0158	17.75	29.46	0.0353
100	0.0160	67.36	73.30	0.1284	0.0159	66.71	78.51	0.1271
125	0.0161	92.07	98.04	0.1716	0.0160	91.22	103.09	0.1701
150	0.0162	116.82	122.84	0.2132	0.0161	115.77	127.72	0.2113
175	0.0164	141.62	147.68	0.2531	0.0163	140.36	152.41	0.2510
200	0.0165	166.48	172.60	0.2916	0.0164	165.01	177.17	0.2893
225	0.0167	191.42	197.59	0.3288	0.0166	189.72	202.00	0.3262
250	0.0169	216.45	222.69	0.3648	0.0168	214.51	226.92	0.3620
275	0.0171	241.61	247.93	0.3998	0.0170	239.40	251.96	0.3967
300	0.0173	266.92	273.33	0.4337	0.0172	264.43	277.14	0.4304
325	0.0176	292.42	298.92	0.4669	0.0174	289.61	302.49	0.4632
350	0.0178	318.14	324.74	0.4993	0.0177	314.97	328.04	0.4952
375	0.0181	344.11	350.82	0.5310	0.0179	340.54	353.81	0.5266
400	0.0184	370.38	377.20	0.5621	0.0182	366.34	379.84	0.5573
450	0.0192	424.03	431.13	0.6231	0.0189	418.83	432.83	0.6172
500	0.0201	479.84	487.29	0.6832	0.0198	472.90	487.53	0.6758
550	0.0214	539.24	547.16	0.7440	0.0208	529.44	544.87	0.7340
600	0.0233	605.37	613.99	0.8086	0.0223	590.01	606.51	0.7935
		6000 psia				8000 psia		
50	0.01573	17.57	35.03	0.0348	0.01563	17.38	40.52	0.0342
100	0.01585	66.09	83.69	0.1259	0.01577	65.49	88.83	0.1246
125	0.01595	90.40	108.11	0.1685	0.01586	89.62	113.10	0.1670
150	0.01606	114.76	132.59	0.2096	0.01597	113.81	137.45	0.2078
175	0.01619	139.17	157.14	0.2490	0.01610	138.04	161.87	0.2471
200	0.01633	163.62	181.75	0.2871	0.01623	162.31	186.34	0.2849
225	0.01648	188.12	206.43	0.3238	0.01639	186.61	210.87	0.3214
250	0.01666	212.69	231.18	0.3593	0.01655	210.97	235.47	0.3567
275	0.01684	237.34	256.04	0.3937	0.01673	235.39	260.16	0.3909
300	0.01705	262.10	281.02	0.4271	0.01693	259.91	284.97	0.4241
325	0.01727	286.99	306.16	0.4597	0.01714	284.53	309.91	0.4564
350	0.01751	312.03	331.47	0.4915	0.01737	309.29	335.01	0.4878
375	0.01777	337.25	356.98	0.5225	0.01762	334.19	360.28	0.5186
400	0.01805	362.66	382.70	0.5528	0.01788	359.26	385.73	0.5486
450	0.01869	414.17	434.92	0.6119	0.01848	409.94	437.30	0.6069
500	0.01945	466.89	488.48	0.6692	0.01918	461.56	489.95	0.6633
550	0.02039	521.38	544.02	0.7256	0.02002	514.49	544.13	0.7183
600	0.02159	578.59	602.57	0.7822	0.02106	569.36	600.53	0.7728

TABLE C.8.4 *Saturated Solid-Saturated Vapor Water (English Units)*

Temp. F T	Press. lbf/in.^{2} P	Specific Volume ft^3/lbm		Internal Energy Btu/lbm		
		Sat. Solid v_i	Sat. Vapor $v_g \times 10^{-3}$	Sat. Solid u_i	Evap. u_{ig}	Sat. Vapor u_g
32.02	0.08866	0.017473	3.302	−143.34	1164.5	1021.2
32	0.08859	0.01747	3.305	−143.35	1164.5	1021.2
30	0.08083	0.01747	3.607	−144.35	1164.9	1020.5
25	0.06406	0.01746	4.505	−146.84	1165.7	1018.9
20	0.05051	0.01745	5.655	−149.31	1166.5	1017.2
15	0.03963	0.01745	7.133	−151.75	1167.3	1015.6
10	0.03093	0.01744	9.043	−154.16	1168.1	1013.9
5	0.02402	0.01743	11.522	−156.56	1168.8	1012.2
0	0.01855	0.01742	14.761	−158.93	1169.5	1010.6
−5	0.01424	0.01742	19.019	−161.27	1170.2	1008.9
−10	0.01086	0.01741	24.657	−163.59	1170.8	1007.3
−15	0.00823	0.01740	32.169	−165.89	1171.5	1005.6
−20	0.00620	0.01740	42.238	−168.16	1172.1	1003.9
−25	0.00464	0.01739	55.782	−170.40	1172.7	1002.3
−30	0.00346	0.01738	74.046	−172.63	1173.2	1000.6
−35	0.00256	0.01737	98.890	−174.82	1173.8	998.9
−40	0.00187	0.01737	134.017	−177.00	1174.3	997.3

(Continued)

TABLE C.8.4 (Continued) *Saturated Solid-Saturated Vapor Water (English Units)*

Temp. F T	Press. lbf/in.² P	Enthalpy Btu/lbm			Entropy Btu/lbm R		
		Sat. Solid h_i	Evap. h_{ig}	Sat. Vapor h_g	Sat. Solid s_i	Evap. s_{ig}	Sat. Vapor s_g
32.02	0.08866	−143.34	1218.7	1075.4	−0.2916	2.4786	2.1869
32	0.08859	−143.35	1218.7	1075.4	−0.2917	2.4787	2.1870
30	0.08083	−144.35	1218.8	1074.5	−0.2938	2.4891	2.1953
25	0.06406	−146.84	1219.1	1072.3	−0.2990	2.5154	2.2164
20	0.05051	−149.31	1219.4	1070.1	−0.3042	2.5422	2.2380
15	0.03963	−151.75	1219.6	1067.9	−0.3093	2.5695	2.2601
10	0.03093	−154.16	1219.8	1065.7	−0.3145	2.5973	2.2827
5	0.02402	−156.56	1220.0	1063.5	−0.3197	2.6256	2.3059
0	0.01855	−158.93	1220.2	1061.2	−0.3248	2.6544	2.3296
−5	0.01424	−161.27	1220.3	1059.0	−0.3300	2.6839	2.3539
−10	0.01086	−163.59	1220.4	1056.8	−0.3351	2.7140	2.3788
−15	0.00823	−165.89	1220.5	1054.6	−0.3403	2.7447	2.4044
−20	0.00620	−168.16	1220.5	1052.4	−0.3455	2.7761	2.4307
−25	0.00464	−170.40	1220.6	1050.2	−0.3506	2.8081	2.4575
−30	0.00346	−172.63	1220.6	1048.0	−0.3557	2.8406	2.4849
−35	0.00256	−174.82	1220.6	1045.7	−0.3608	2.8737	2.5129
−40	0.00187	−177.00	1220.5	1043.5	−0.3659	2.9084	2.5425

TABLE C.9 ENG *Thermodynamic Properties of Ammonia*
TABLE C.9.1 ENG *Saturated Ammonia*

Temp.	Press.	SpecificVolume, ft³/lbm			Internal Energy, Btu/lbm		
F	psia	Sat. Liquid	Evap.	Sat. Vapor	Sat. Liquid	Evap.	Sat. Vapor
T	P	v_f	v_{fg}	v_g	u_f	u_{fg}	u_g
-60	5.547	0.02277	44.7397	44.7625	-20.92	564.27	543.36
-50	7.663	0.02299	33.0702	33.0932	-10.51	556.84	546.33
-40	10.404	0.02322	24.8464	24.8696	-0.04	549.25	549.20
-30	13.898	0.02345	18.9490	18.9724	10.48	541.50	551.98
-28.0	14.696	0.02350	17.9833	18.0068	12.59	539.93	552.52
-20	18.289	0.02369	14.6510	14.6747	21.07	533.57	554.64
-10	23.737	0.02394	11.4714	11.4953	31.73	525.47	557.20
0	30.415	0.02420	9.0861	9.1103	42.46	517.18	559.64
10	38.508	0.02446	7.2734	7.2979	53.26	508.71	561.96
20	48.218	0.02474	5.8792	5.9039	64.12	500.04	564.16
30	59.756	0.02502	4.7945	4.8195	75.06	491.17	566.23
40	73.346	0.02532	3.9418	3.9671	86.07	482.09	568.15
50	89.226	0.02564	3.2647	3.2903	97.16	472.78	569.94
60	107.641	0.02597	2.7221	2.7481	108.33	463.24	571.56
70	128.849	0.02631	2.2835	2.3098	119.58	453.44	573.02
80	153.116	0.02668	1.9260	1.9526	130.92	443.37	574.30
90	180.721	0.02706	1.6323	1.6594	142.36	433.01	575.37
100	211.949	0.02747	1.3894	1.4168	153.89	422.34	576.23
110	247.098	0.02790	1.1870	1.2149	165.53	411.32	576.85
120	286.473	0.02836	1.0172	1.0456	177.28	399.92	577.20
130	330.392	0.02885	0.8740	0.9028	189.17	388.10	577.27
140	379.181	0.02938	0.7524	0.7818	201.20	375.82	577.02
150	433.181	0.02995	0.6485	0.6785	213.40	363.01	576.41
160	492.742	0.03057	0.5593	0.5899	225.80	349.61	575.41
170	558.231	0.03124	0.4822	0.5135	238.42	335.53	573.95
180	630.029	0.03199	0.4153	0.4472	251.33	320.66	571.99
190	708.538	0.03281	0.3567	0.3895	264.58	304.87	569.45
200	794.183	0.03375	0.3051	0.3388	278.24	287.96	566.20
210	887.424	0.03482	0.2592	0.2941	292.43	269.70	562.13
220	988.761	0.03608	0.2181	0.2542	307.28	249.72	557.00
230	1098.766	0.03759	0.1807	0.2183	323.03	227.47	550.50
240	1218.113	0.03950	0.1460	0.1855	340.05	202.02	542.06
250	1347.668	0.04206	0.1126	0.1547	359.03	171.57	530.60
260	1488.694	0.04599	0.0781	0.1241	381.74	131.74	513.48
270.1	1643.742	0.06816	0	0.0682	446.09	0	446.09

(Continued)

TABLE C.9.1 ENG (Continued) *Saturated Ammonia*

Temp.	Press.	Enthalpy, Btu/lbm			Entropy, Btu/lbm R		
F	psia	Sat. Liquid	Evap.	Sat. Vapor	Sat. Liquid	Evap.	Sat. Vapor
T	P	h_f	h_{fg}	h_g	s_f	s_{fg}	s_g
-60	5.547	-20.89	610.19	589.30	-0.0510	1.5267	1.4758
-50	7.663	-10.48	603.73	593.26	-0.0252	1.4737	1.4485
-40	10.404	0	597.08	597.08	0	1.4227	1.4227
-30	13.898	10.54	590.23	600.77	0.0248	1.3737	1.3985
-28.0	14.696	12.65	588.84	601.49	0.0297	1.3641	1.3938
-20	18.289	21.15	583.15	604.31	0.0492	1.3263	1.3755
-10	23.737	31.84	575.85	607.69	0.0731	1.2806	1.3538
0	30.415	42.60	568.32	610.92	0.0967	1.2364	1.3331
10	38.508	53.43	560.54	613.97	0.1200	1.1935	1.3134
20	48.218	64.34	552.50	616.84	0.1429	1.1518	1.2947
30	59.756	75.33	544.18	619.52	0.1654	1.1113	1.2768
40	73.346	86.41	535.59	622.00	0.1877	1.0719	1.2596
50	89.226	97.58	526.68	624.26	0.2097	1.0334	1.2431
60	107.641	108.84	517.46	626.30	0.2314	0.9957	1.2271
70	128.849	120.21	507.89	628.09	0.2529	0.9589	1.2117
80	153.116	131.68	497.94	629.62	0.2741	0.9227	1.1968
90	180.721	143.26	487.60	630.86	0.2951	0.8871	1.1822
100	211.949	154.97	476.83	631.80	0.3159	0.8520	1.1679
110	247.098	166.80	465.59	632.40	0.3366	0.8173	1.1539
120	286.473	178.79	453.84	632.63	0.3571	0.7829	1.1400
130	330.392	190.93	441.54	632.47	0.3774	0.7488	1.1262
140	379.181	203.26	428.61	631.87	0.3977	0.7147	1.1125
150	433.181	215.80	415.00	630.80	0.4180	0.6807	1.0987
160	492.742	228.58	400.61	629.19	0.4382	0.6465	1.0847
170	558.231	241.65	385.35	627.00	0.4586	0.6120	1.0705
180	630.029	255.06	369.08	624.14	0.4790	0.5770	1.0560
190	708.538	268.88	351.63	620.51	0.4997	0.5412	1.0410
200	794.183	283.20	332.80	616.00	0.5208	0.5045	1.0253
210	887.424	298.14	312.27	610.42	0.5424	0.4663	1.0087
220	988.761	313.88	289.63	603.51	0.5647	0.4261	0.9909
230	1098.766	330.67	264.21	594.89	0.5882	0.3831	0.9713
240	1218.113	348.95	234.93	583.87	0.6132	0.3358	0.9490
250	1347.668	369.52	199.65	569.17	0.6410	0.2813	0.9224
260	1488.694	394.41	153.25	547.66	0.6743	0.2129	0.8872
270.1	1643.742	466.83	0	466.83	0.7718	0	0.7718

TABLE C.9.2 ENG *Superheated Ammonia*

Temp. F	v ft³/lbm	h Btu/lbm	s Btu/lbm R	v ft³/lbm	h Btu/lbm	s Btu/lbm R	v ft³/lbm	h Btu/lbm	s Btu/lbm R
	\multicolumn 5 psia (-63.09)			10 psia (-41.33)			15 psia (-27.27)		
Sat.	49.32002	588.05	1.4846	25.80648	596.58	1.4261	17.66533	601.75	1.3921
-40	52.3487	599.56	1.5128	25.8962	597.27	1.4277	—	—	—
-20	54.9506	609.53	1.5360	27.2401	607.60	1.4518	17.9999	605.63	1.4010
0	57.5366	619.51	1.5582	28.5674	617.88	1.4746	18.9086	616.22	1.4245
20	60.1099	629.50	1.5795	29.8814	628.12	1.4964	19.8036	626.72	1.4469
40	62.6732	639.52	1.5999	31.1852	638.34	1.5173	20.6880	637.15	1.4682
60	65.2288	649.57	1.6197	32.4809	648.56	1.5374	21.5641	647.54	1.4886
80	67.7782	659.67	1.6387	33.7703	658.80	1.5567	22.4338	657.91	1.5082
100	70.3228	669.84	1.6572	35.0549	669.07	1.5754	23.2985	668.29	1.5271
120	72.8637	680.06	1.6752	36.3356	679.38	1.5935	24.1593	678.70	1.5453
140	75.4015	690.36	1.6926	37.6133	689.75	1.6111	25.0170	689.14	1.5630
160	77.9370	700.74	1.7097	38.8886	700.19	1.6282	25.8723	699.64	1.5803
180	80.4706	711.20	1.7263	40.1620	710.70	1.6449	26.7256	710.21	1.5970
200	83.0026	721.75	1.7425	41.4338	721.30	1.6612	27.5774	720.84	1.6134
220	85.5334	732.39	1.7584	42.7043	731.98	1.6771	28.4278	731.56	1.6294
240	88.0631	743.13	1.7740	43.9737	742.74	1.6928	29.2772	742.36	1.6451
260	90.5918	753.96	1.7892	45.2422	753.61	1.7081	30.1256	753.24	1.6604
280	93.1199	764.90	1.8042	46.5100	764.56	1.7231	30.9733	764.23	1.6755
	20 psia (-16.63)			25 psia (-7.95)			30 psia (-0.57)		
Sat.	13.49628	605.47	1.3680	10.95013	608.37	1.3494	9.22850	610.74	1.3342
0	14.0774	614.54	1.3881	11.1771	612.82	1.3592	9.2423	611.06	1.3349
20	14.7635	625.30	1.4111	11.7383	623.86	1.3827	9.7206	622.39	1.3591
40	15.4385	635.94	1.4328	12.2881	634.72	1.4049	10.1872	633.49	1.3817
60	16.1051	646.51	1.4535	12.8291	645.46	1.4260	10.6447	644.41	1.4032
80	16.7651	657.02	1.4734	13.3634	656.12	1.4461	11.0954	655.21	1.4236
100	17.4200	667.51	1.4925	13.8926	666.73	1.4654	11.5407	665.93	1.4431
120	18.0709	678.01	1.5109	14.4176	677.32	1.4840	11.9820	676.62	1.4618
140	18.7187	688.53	1.5287	14.9395	687.91	1.5020	12.4200	687.29	1.4799
160	19.3640	699.09	1.5461	15.4589	698.54	1.5194	12.8554	697.98	1.4975
180	20.0073	709.71	1.5629	15.9763	709.20	1.5363	13.2888	708.70	1.5145
200	20.6491	720.39	1.5794	16.4920	719.93	1.5528	13.7206	719.47	1.5311
220	21.2895	731.14	1.5954	17.0065	730.72	1.5689	14.1511	730.29	1.5472
240	21.9288	741.97	1.6111	17.5198	741.58	1.5847	14.5804	741.19	1.5630
260	22.5673	752.88	1.6265	18.0322	752.52	1.6001	15.0088	752.16	1.5785
280	23.2049	763.89	1.6416	18.5439	763.55	1.6152	15.4365	763.21	1.5936
300	23.8419	774.99	1.6564	19.0548	774.67	1.6301	15.8634	774.36	1.6085
320	24.4783	786.18	1.6709	19.5652	785.89	1.6446	16.2898	785.59	1.6231

(Continued)

TABLE C.9.2 ENG (Continued) *Superheated Ammonia*

Temp. F	v ft³/lbm	h Btu/lbm	s Btu/lbm R	v ft³/lbm	h Btu/lbm	s Btu/lbm R	v ft³/lbm	h Btu/lbm	s Btu/lbm R
	35 psia (5.89)			40 psia (11.66)			50 psia (21.66)		
Sat.	7.98414	612.73	1.3214	7.04135	614.45	1.3103	5.70491	617.30	1.2917
20	8.2786	620.90	1.3387	7.1964	619.39	1.3206	—	—	—
40	8.6860	632.23	1.3618	7.5596	630.96	1.3443	5.9814	628.37	1.3142
60	9.0841	643.34	1.3836	7.9132	642.26	1.3665	6.2731	640.07	1.3372
80	9.4751	654.29	1.4043	8.2596	653.37	1.3874	6.5573	651.49	1.3588
100	9.8606	665.14	1.4240	8.6004	664.33	1.4074	6.8356	662.70	1.3792
120	10.2420	675.92	1.4430	8.9370	675.21	1.4265	7.1096	673.79	1.3986
140	10.6202	686.67	1.4612	9.2702	686.04	1.4449	7.3800	684.78	1.4173
160	10.9957	697.42	1.4788	9.6008	696.86	1.4626	7.6478	695.73	1.4352
180	11.3692	708.19	1.4959	9.9294	707.69	1.4798	7.9135	706.67	1.4526
200	11.7410	719.01	1.5126	10.2562	718.54	1.4965	8.1775	717.61	1.4695
220	12.1115	729.87	1.5288	10.5817	729.44	1.5128	8.4400	728.59	1.4859
240	12.4808	740.80	1.5447	10.9061	740.40	1.5287	8.7014	739.62	1.5018
260	12.8493	751.80	1.5602	11.2296	751.43	1.5442	8.9619	750.70	1.5175
280	13.2169	762.88	1.5753	11.5522	762.54	1.5594	9.2216	761.86	1.5327
300	13.5838	774.04	1.5902	11.8741	773.72	1.5744	9.4805	773.09	1.5477
320	13.9502	785.29	1.6049	12.1955	785.00	1.5890	9.7389	784.40	1.5624
340	14.3160	796.64	1.6192	12.5163	796.36	1.6034	9.9967	795.80	1.5769
	60 psia (30.19)			70 psia (37.68)			80 psia (44.38)		
Sat.	4.80091	619.57	1.2764	4.14732	621.44	1.2635	3.65200	623.02	1.2523
40	4.9277	625.69	1.2888	4.1738	622.94	1.2665	—	—	—
60	5.1787	637.82	1.3126	4.3961	635.52	1.2912	3.8083	633.16	1.2721
80	5.4217	649.57	1.3348	4.6099	647.62	1.3140	4.0005	645.63	1.2956
100	5.6586	661.05	1.3557	4.8174	659.37	1.3354	4.1861	657.66	1.3175
120	5.8909	672.34	1.3755	5.0201	670.88	1.3556	4.3667	669.39	1.3381
140	6.1197	683.50	1.3944	5.2191	682.21	1.3749	4.5435	680.90	1.3577
160	6.3456	694.59	1.4126	5.4153	693.44	1.3933	4.7174	692.27	1.3763
180	6.5694	705.64	1.4302	5.6093	704.60	1.4110	4.8890	703.55	1.3942
200	6.7915	716.68	1.4472	5.8014	715.73	1.4281	5.0588	714.79	1.4115
220	7.0121	727.73	1.4637	5.9921	726.87	1.4448	5.2270	726.00	1.4283
240	7.2316	738.83	1.4798	6.1816	738.03	1.4610	5.3941	737.23	1.4446
260	7.4501	749.97	1.4955	6.3702	749.23	1.4767	5.5602	748.50	1.4604
280	7.6678	761.17	1.5108	6.5579	760.49	1.4922	5.7254	759.80	1.4759
300	7.8848	772.45	1.5259	6.7449	771.81	1.5073	5.8900	771.17	1.4911
320	8.1011	783.80	1.5406	6.9313	783.21	1.5221	6.0538	782.61	1.5059
340	8.3169	795.24	1.5551	7.1171	794.68	1.5366	6.2172	794.12	1.5205
360	8.5323	806.77	1.5693	7.3025	806.24	1.5509	6.3801	805.71	1.5348

(Continued)

TABLE C.9.2 ENG (Continued) *Superheated Ammonia*

Temp. F	v ft³/lbm	h Btu/lbm	s Btu/lbm R	v ft³/lbm	h Btu/lbm	s Btu/lbm R	v ft³/lbm	h Btu/lbm	s Btu/lbm R
	90 psia (50.45)			100 psia (56.02)			125 psia (68.28)		
Sat.	3.26324	624.36	1.2423	2.94969	625.52	1.2334	2.37866	627.80	1.2143
60	3.3503	630.74	1.2547	2.9831	628.25	1.2387	—	—	—
80	3.5260	643.59	1.2790	3.1459	641.51	1.2637	2.4597	636.11	1.2299
100	3.6947	655.92	1.3014	3.3013	654.16	1.2867	2.5917	649.59	1.2544
120	3.8583	667.88	1.3224	3.4513	666.36	1.3082	2.7177	662.44	1.2770
140	4.0179	679.58	1.3423	3.5972	678.24	1.3283	2.8392	674.83	1.2980
160	4.1745	691.10	1.3612	3.7400	689.91	1.3475	2.9574	686.90	1.3178
180	4.3287	702.50	1.3793	3.8804	701.44	1.3658	3.0730	698.74	1.3366
200	4.4811	713.83	1.3967	4.0188	712.87	1.3834	3.1865	710.44	1.3546
220	4.6319	725.13	1.4136	4.1558	724.25	1.4004	3.2985	722.04	1.3720
240	4.7816	736.43	1.4300	4.2915	735.63	1.4169	3.4091	733.59	1.3887
260	4.9302	747.75	1.4459	4.4261	747.01	1.4329	3.5187	745.13	1.4050
280	5.0779	759.11	1.4615	4.5599	758.42	1.4485	3.6274	756.68	1.4208
300	5.2250	770.53	1.4767	4.6930	769.88	1.4638	3.7353	768.27	1.4362
320	5.3714	782.01	1.4916	4.8254	781.40	1.4788	3.8426	779.89	1.4514
340	5.5173	793.56	1.5063	4.9573	792.99	1.4935	3.9493	791.58	1.4662
360	5.6626	805.18	1.5206	5.0887	804.66	1.5079	4.0555	803.33	1.4807
380	5.8076	816.90	1.5348	5.2196	816.40	1.5220	4.1613	815.15	1.4949
	150 psia (78.79)			175 psia (88.03)			200 psia (96.31)		
Sat.	1.99226	629.45	1.1986	1.71282	630.64	1.1850	1.50102	631.49	1.1731
80	1.9997	630.36	1.2003	—	—	—	—	—	—
100	2.1170	644.81	1.2265	1.7762	639.77	1.2015	1.5190	634.45	1.1785
120	2.2275	658.37	1.2504	1.8762	654.13	1.2267	1.6117	649.71	1.2052
140	2.3331	671.31	1.2723	1.9708	667.67	1.2497	1.6984	663.90	1.2293
160	2.4351	683.80	1.2928	2.0614	680.62	1.2710	1.7807	677.36	1.2514
180	2.5343	695.99	1.3122	2.1491	693.17	1.2909	1.8598	690.30	1.2719
200	2.6313	707.96	1.3306	2.2345	705.44	1.3098	1.9365	702.87	1.2913
220	2.7267	719.79	1.3483	2.3181	717.51	1.3278	2.0114	715.20	1.3097
240	2.8207	731.54	1.3653	2.4002	729.46	1.3451	2.0847	727.35	1.3273
260	2.9136	743.24	1.3818	2.4813	741.33	1.3619	2.1569	739.39	1.3443
280	3.0056	754.93	1.3978	2.5613	753.16	1.3781	2.2280	751.38	1.3607
300	3.0968	766.63	1.4134	2.6406	764.99	1.3939	2.2984	763.33	1.3767
320	3.1873	778.37	1.4287	2.7192	776.84	1.4092	2.3680	775.30	1.3922
340	3.2772	790.15	1.4436	2.7972	788.72	1.4243	2.4370	787.28	1.4074
360	3.3667	801.99	1.4582	2.8746	800.65	1.4390	2.5056	799.30	1.4223
380	3.4557	813.90	1.4726	2.9516	812.64	1.4535	2.5736	811.38	1.4368
400	3.5442	825.88	1.4867	3.0282	824.70	1.4677	2.6412	823.51	1.4511

(Continued)

TABLE C.9.2 ENG (Continued) *Superheated Ammonia*

Temp. F	v ft³/lbm	h Btu/lbm	s Btu/lbm R	v ft³/lbm	h Btu/lbm	s Btu/lbm R	v ft³/lbm	h Btu/lbm	s Btu/lbm R
	250 psia (110.78)			300 psia (123.20)			350 psia (134.14)		
Sat.	1.20063	632.43	1.1528	0.99733	632.63	1.1356	0.85027	632.28	1.1205
120	1.2384	640.21	1.1663	—	—	—	—	—	—
140	1.3150	655.95	1.1930	1.0568	647.32	1.1605	0.8696	637.87	1.1299
160	1.3863	670.53	1.2170	1.1217	663.27	1.1866	0.9309	655.48	1.1588
180	1.4539	684.34	1.2389	1.1821	678.07	1.2101	0.9868	671.46	1.1842
200	1.5188	697.59	1.2593	1.2394	692.08	1.2317	1.0391	686.34	1.2071
220	1.5815	710.45	1.2785	1.2943	705.55	1.2518	1.0886	700.47	1.2282
240	1.6426	723.05	1.2968	1.3474	718.63	1.2708	1.1362	714.08	1.2479
260	1.7024	735.46	1.3142	1.3991	731.44	1.2888	1.1822	727.32	1.2666
280	1.7612	747.76	1.3311	1.4497	744.07	1.3062	1.2270	740.31	1.2844
300	1.8191	759.98	1.3474	1.4994	756.58	1.3228	1.2708	753.12	1.3015
320	1.8762	772.18	1.3633	1.5482	769.02	1.3390	1.3138	765.82	1.3180
340	1.9328	784.37	1.3787	1.5965	781.43	1.3547	1.3561	778.46	1.3340
360	1.9887	796.59	1.3938	1.6441	793.84	1.3701	1.3979	791.07	1.3496
380	2.0442	808.83	1.4085	1.6913	806.27	1.3850	1.4391	803.67	1.3648
400	2.0993	821.13	1.4230	1.7380	818.72	1.3997	1.4798	816.30	1.3796
420	2.1540	833.48	1.4372	1.7843	831.23	1.4141	1.5202	828.95	1.3942
440	2.2083	845.90	1.4512	1.8302	843.78	1.4282	1.5602	841.65	1.4085
	400 psia (143.97)			600 psia (175.93)			800 psia (200.65)		
Sat.	0.73876	631.50	1.1070	0.47311	625.39	1.0620	0.33575	615.67	1.0242
160	0.7860	647.06	1.1324	—	—	—	—	—	—
180	0.8392	664.44	1.1601	0.4834	630.48	1.0700	—	—	—
200	0.8880	680.32	1.1845	0.5287	652.67	1.1041	—	—	—
220	0.9338	695.21	1.2067	0.5680	671.78	1.1327	0.3769	642.62	1.0645
240	0.9773	709.40	1.2273	0.6035	689.03	1.1577	0.4115	665.08	1.0971
260	1.0192	723.10	1.2466	0.6366	705.06	1.1803	0.4419	684.62	1.1246
280	1.0597	736.47	1.2650	0.6678	720.26	1.2011	0.4694	702.36	1.1489
300	1.0992	749.60	1.2825	0.6976	734.88	1.2206	0.4951	718.93	1.1710
320	1.1379	762.58	1.2993	0.7264	749.09	1.2391	0.5193	734.69	1.1915
340	1.1758	775.45	1.3156	0.7542	763.02	1.2567	0.5425	749.89	1.2108
360	1.2131	788.27	1.3315	0.7814	776.75	1.2737	0.5648	764.68	1.2290
380	1.2499	801.06	1.3469	0.8079	790.34	1.2901	0.5864	779.19	1.2465
400	1.2862	813.85	1.3619	0.8340	803.86	1.3060	0.6074	793.50	1.2634
420	1.3221	826.66	1.3767	0.8595	817.32	1.3215	0.6279	807.68	1.2797
440	1.3576	839.51	1.3911	0.8847	830.76	1.3366	0.6480	821.76	1.2955
460	1.3928	852.39	1.4053	0.9095	844.21	1.3514	0.6677	835.80	1.3109
480	1.4277	865.34	1.4192	0.9340	857.67	1.3658	0.6871	849.80	1.3260

TABLE C 10 ENG *Thermodynamic Properties of R-22*
TABLE C.10.1 ENG *Saturated R-22*

Temp.	Press.	SpecificVolume, ft³/lbm			Internal Energy, Btu/lbm		
F T	psia P	Sat. Liquid v_f	Evap. v_{fg}	Sat. Vapor v_g	Sat. Liquid u_f	Evap. u_{fg}	Sat. Vapor u_g
-100	2.398	0.01066	18.4219	18.4326	-14.57	99.76	85.19
-90	3.423	0.01077	13.2243	13.2351	-12.22	98.38	86.16
-80	4.782	0.01088	9.6840	9.6949	-9.85	96.98	87.13
-70	6.552	0.01099	7.2208	7.2318	-7.44	95.54	88.10
-60	8.818	0.01111	5.4733	5.4844	-5.01	94.07	89.06
-50	11.674	0.01124	4.2111	4.2224	-2.54	92.56	90.02
-41.4	14.696	0.01135	3.3944	3.4058	-0.37	91.22	90.85
-40	15.222	0.01136	3.2844	3.2957	-0.03	91.01	90.97
-30	19.573	0.01150	2.5934	2.6049	2.51	89.41	91.91
-20	24.845	0.01163	2.0709	2.0826	5.08	87.76	92.84
-10	31.162	0.01178	1.6707	1.6825	7.68	86.07	93.75
0	38.657	0.01193	1.3603	1.3723	10.32	84.33	94.65
10	47.464	0.01209	1.1170	1.1290	13.00	82.53	95.53
20	57.727	0.01226	0.9241	0.9363	15.71	80.67	96.38
30	69.591	0.01243	0.7697	0.7821	18.45	78.76	97.21
40	83.206	0.01262	0.6449	0.6575	21.23	76.79	98.02
50	98.727	0.01282	0.5432	0.5561	24.04	74.75	98.79
60	116.312	0.01303	0.4597	0.4727	26.89	72.65	99.54
70	136.123	0.01325	0.3905	0.4037	29.78	70.46	100.24
80	158.326	0.01349	0.3327	0.3462	32.71	68.19	100.91
90	183.094	0.01375	0.2841	0.2979	35.69	65.83	101.53
100	210.604	0.01404	0.2430	0.2570	38.72	63.37	102.09
110	241.042	0.01435	0.2079	0.2222	41.81	60.78	102.59
120	274.604	0.01469	0.1777	0.1924	44.96	58.05	103.01
130	311.496	0.01508	0.1515	0.1666	48.19	55.14	103.33
140	351.944	0.01552	0.1287	0.1442	51.52	52.02	103.54
150	396.194	0.01603	0.1085	0.1245	54.97	48.63	103.60
160	444.525	0.01663	0.0904	0.1070	58.58	44.88	103.46
170	497.259	0.01737	0.0739	0.0913	62.42	40.62	103.04
180	554.783	0.01833	0.0585	0.0768	66.62	35.57	102.18
190	617.590	0.01973	0.0431	0.0628	71.46	29.10	100.55
200	686.356	0.02244	0.0250	0.0474	78.01	18.81	96.83
204.8	720.698	0.03053	0	0.0305	87.30	0	87.30

(Continued)

TABLE C.10.1 ENG (Continued) *Saturated R-22*

Temp.	Press.	Enthalpy, Btu/lbm			Entropy, Btu/lbm R		
F T	psia P	Sat. Liquid h_f	Evap. h_{fg}	Sat. Vapor h_g	Sat. Liquid s_f	Evap. s_{fg}	Sat. Vapor s_g
-100	2.398	-14.56	107.94	93.37	-0.0373	0.3001	0.2627
-90	3.423	-12.22	106.76	94.54	-0.0309	0.2888	0.2579
-80	4.782	-9.84	105.55	95.71	-0.0246	0.2780	0.2534
-70	6.552	-7.43	104.30	96.87	-0.0183	0.2676	0.2493
-60	8.818	-4.99	103.00	98.01	-0.0121	0.2577	0.2456
-50	11.674	-2.51	101.66	99.14	-0.0060	0.2481	0.2421
-41.4	14.696	-0.34	100.45	100.11	-0.0008	0.2401	0.2393
-40	15.222	0	100.26	100.26	0	0.2389	0.2389
-30	19.573	2.55	98.80	101.35	0.0060	0.2299	0.2359
-20	24.845	5.13	97.28	102.42	0.0119	0.2213	0.2332
-10	31.162	7.75	95.70	103.46	0.0178	0.2128	0.2306
0	38.657	10.41	94.06	104.47	0.0236	0.2046	0.2282
10	47.464	13.10	92.34	105.44	0.0293	0.1966	0.2259
20	57.727	15.84	90.55	106.38	0.0350	0.1888	0.2238
30	69.591	18.61	88.67	107.28	0.0407	0.1811	0.2218
40	83.206	21.42	86.72	108.14	0.0463	0.1735	0.2199
50	98.727	24.27	84.68	108.95	0.0519	0.1661	0.2180
60	116.312	27.17	82.54	109.71	0.0574	0.1588	0.2163
70	136.123	30.12	80.30	110.41	0.0630	0.1516	0.2146
80	158.326	33.11	77.94	111.05	0.0685	0.1444	0.2129
90	183.094	36.16	75.46	111.62	0.0739	0.1373	0.2112
100	210.604	39.27	72.84	112.11	0.0794	0.1301	0.2096
110	241.042	42.45	70.05	112.50	0.0849	0.1230	0.2079
120	274.604	45.71	67.08	112.78	0.0904	0.1157	0.2061
130	311.496	49.06	63.88	112.94	0.0960	0.1083	0.2043
140	351.944	52.53	60.40	112.93	0.1016	0.1007	0.2023
150	396.194	56.14	56.58	112.73	0.1074	0.0928	0.2002
160	444.525	59.95	52.32	112.26	0.1133	0.0844	0.1978
170	497.259	64.02	47.42	111.44	0.1196	0.0753	0.1949
180	554.783	68.50	41.57	110.07	0.1263	0.0650	0.1913
190	617.590	73.71	34.02	107.73	0.1341	0.0524	0.1865
200	686.356	80.86	21.99	102.85	0.1446	0.0333	0.1779
204.8	720.698	91.38	0	91.38	0.1602	0	0.1602

TABLE C.10.2 ENG *Superheated R-22*

Temp. F	v ft³/lbm	h Btu/lbm	s Btu/lbm R	v ft³/lbm	h Btu/lbm	s Btu/lbm R	v ft³/lbm	h Btu/lbm	s Btu/lbm R
	5 psia (-78.62)			10 psia (-55.59)			15 psia (-40.57)		
Sat.	9.30117	95.87	0.2528	4.87779	98.52	0.2440	3.34121	100.19	0.2391
-40	10.2935	101.09	0.2659	5.0838	100.69	0.2493	3.3463	100.28	0.2393
-20	10.8034	103.89	0.2724	5.3460	103.53	0.2559	3.5261	103.16	0.2460
0	11.3114	106.73	0.2787	5.6060	106.41	0.2623	3.7037	106.09	0.2525
20	11.8177	109.64	0.2849	5.8643	109.36	0.2686	3.8794	109.06	0.2588
40	12.3227	112.61	0.2910	6.1212	112.35	0.2747	4.0537	112.09	0.2650
60	12.8265	115.64	0.2969	6.3769	115.40	0.2807	4.2268	115.17	0.2710
80	13.3293	118.72	0.3027	6.6316	118.51	0.2865	4.3989	118.30	0.2769
100	13.8313	121.87	0.3085	6.8855	121.67	0.2923	4.5701	121.48	0.2827
120	14.3327	125.07	0.3141	7.1387	124.89	0.2979	4.7406	124.72	0.2884
140	14.8335	128.33	0.3196	7.3913	128.17	0.3035	4.9105	128.00	0.2940
160	15.3337	131.64	0.3250	7.6434	131.49	0.3089	5.0799	131.34	0.2995
180	15.8336	135.01	0.3304	7.8951	134.87	0.3143	5.2489	134.74	0.3049
200	16.3331	138.44	0.3357	8.1464	138.31	0.3196	5.4174	138.18	0.3102
220	16.8323	141.92	0.3409	8.3974	141.80	0.3248	5.5857	141.68	0.3154
240	17.3312	145.45	0.3460	8.6481	145.34	0.3300	5.7537	145.23	0.3205
260	17.8298	149.03	0.3510	8.8986	148.93	0.3350	5.9215	148.83	0.3256
280	18.3283	152.67	0.3560	9.1489	152.57	0.3400	6.0890	152.48	0.3306
	20 psia (-29.12)			25 psia (-19.73)			30 psia (-11.71)		
Sat.	2.55270	101.44	0.2357	2.07040	102.44	0.2331	1.74388	103.28	0.2310
0	2.7521	105.76	0.2454	2.1808	105.42	0.2397	1.7997	105.08	0.2350
20	2.8867	108.77	0.2518	2.2908	108.47	0.2462	1.8933	108.16	0.2415
40	3.0198	111.83	0.2580	2.3992	111.56	0.2525	1.9853	111.29	0.2479
60	3.1516	114.93	0.2641	2.5063	114.69	0.2586	2.0760	114.45	0.2541
80	3.2823	118.08	0.2701	2.6123	117.86	0.2646	2.1655	117.64	0.2602
100	3.4122	121.28	0.2759	2.7175	121.09	0.2705	2.2542	120.89	0.2661
120	3.5414	124.54	0.2816	2.8219	124.36	0.2762	2.3421	124.18	0.2718
140	3.6700	127.84	0.2872	2.9257	127.68	0.2819	2.4294	127.51	0.2775
160	3.7981	131.19	0.2927	3.0289	131.04	0.2874	2.5162	130.89	0.2830
180	3.9257	134.60	0.2981	3.1318	134.46	0.2928	2.6025	134.32	0.2885
200	4.0529	138.05	0.3034	3.2342	137.93	0.2982	2.6884	137.80	0.2938
220	4.1799	141.56	0.3087	3.3363	141.44	0.3034	2.7739	141.32	0.2991
240	4.3065	145.12	0.3138	3.4382	145.01	0.3086	2.8592	144.90	0.3043
260	4.4329	148.73	0.3189	3.5397	148.62	0.3137	2.9443	148.52	0.3094
280	4.5591	152.38	0.3239	3.6411	152.28	0.3187	3.0291	152.19	0.3144
300	4.6851	156.09	0.3288	3.7423	155.99	0.3236	3.1138	155.90	0.3194
320	4.8109	159.84	0.3337	3.8434	159.75	0.3285	3.1983	159.67	0.3243

(Continued)

TABLE C.10.2 ENG (Continued) *Superheated R-22*

Temp. F	v ft³/lbm	h Btu/lbm	s Btu/lbm R	v ft³/lbm	h Btu/lbm	s Btu/lbm R	v ft³/lbm	h Btu/lbm	s Btu/lbm R
	40 psia (1.63)			50 psia (12.61)			60 psia (22.03)		
Sat.	1.32853	104.63	0.2278	1.07436	105.69	0.2253	0.90223	106.57	0.2234
20	1.3959	107.54	0.2340	1.0968	106.90	0.2279	—	—	—
40	1.4676	110.73	0.2405	1.1564	110.16	0.2346	0.9486	109.58	0.2295
60	1.5378	113.95	0.2468	1.2145	113.44	0.2410	0.9987	112.92	0.2361
80	1.6068	117.20	0.2530	1.2714	116.74	0.2472	1.0475	116.28	0.2424
100	1.6749	120.48	0.2589	1.3272	120.08	0.2533	1.0952	119.66	0.2486
120	1.7423	123.81	0.2648	1.3822	123.44	0.2592	1.1420	123.06	0.2545
140	1.8090	127.18	0.2705	1.4366	126.84	0.2649	1.1882	126.50	0.2604
160	1.8751	130.59	0.2761	1.4903	130.28	0.2706	1.2338	129.96	0.2660
180	1.9407	134.04	0.2816	1.5436	133.75	0.2761	1.2788	133.47	0.2716
200	2.0060	137.54	0.2869	1.5965	137.28	0.2815	1.3235	137.01	0.2771
220	2.0709	141.08	0.2922	1.6491	140.84	0.2869	1.3678	140.60	0.2824
240	2.1356	144.67	0.2974	1.7013	144.45	0.2921	1.4118	144.22	0.2877
260	2.2000	148.31	0.3026	1.7533	148.10	0.2972	1.4556	147.89	0.2928
280	2.2641	151.99	0.3076	1.8051	151.80	0.3023	1.4991	151.60	0.2979
300	2.3281	155.72	0.3126	1.8567	155.54	0.3073	1.5424	155.35	0.3029
320	2.3919	159.49	0.3175	1.9081	159.32	0.3122	1.5856	159.15	0.3079
340	2.4556	163.31	0.3223	1.9594	163.15	0.3171	1.6286	162.99	0.3127
	70 psia (30.32)			80 psia (37.76)			100 psia (50.77)		
Sat.	0.77766	107.31	0.2217	0.68319	107.95	0.2203	0.54908	109.01	0.2179
40	0.7998	108.97	0.2251	0.6878	108.35	0.2211	—	—	—
60	0.8443	112.39	0.2318	0.7282	111.84	0.2279	0.5650	110.70	0.2212
80	0.8874	115.81	0.2382	0.7671	115.32	0.2345	0.5982	114.32	0.2280
100	0.9293	119.23	0.2445	0.8048	118.80	0.2408	0.6300	117.91	0.2345
120	0.9704	122.68	0.2505	0.8415	122.29	0.2470	0.6608	121.49	0.2408
140	1.0107	126.15	0.2564	0.8775	125.80	0.2529	0.6908	125.08	0.2469
160	1.0504	129.65	0.2621	0.9129	129.33	0.2587	0.7201	128.67	0.2528
180	1.0896	133.18	0.2677	0.9477	132.89	0.2643	0.7488	132.29	0.2586
200	1.1284	136.75	0.2732	0.9821	136.48	0.2699	0.7771	135.93	0.2642
220	1.1669	140.35	0.2786	1.0161	140.10	0.2753	0.8050	139.60	0.2696
240	1.2050	143.99	0.2839	1.0498	143.76	0.2806	0.8326	143.30	0.2750
260	1.2428	147.68	0.2891	1.0833	147.46	0.2858	0.8599	147.03	0.2803
280	1.2805	151.40	0.2942	1.1165	151.20	0.2909	0.8869	150.80	0.2854
300	1.3179	155.17	0.2992	1.1495	154.98	0.2960	0.9137	154.61	0.2905
320	1.3552	158.97	0.3042	1.1823	158.80	0.3009	0.9404	158.45	0.2955
340	1.3923	162.82	0.3090	1.2150	162.66	0.3058	0.9669	162.33	0.3004
360	1.4292	166.72	0.3138	1.2476	166.56	0.3106	0.9932	166.25	0.3052

(Continued)

TABLE C.10.2 ENG (Continued) *Superheated R-22*

Temp. F	v ft³/lbm	h Btu/lbm	s Btu/lbm R	v ft³/lbm	h Btu/lbm	s Btu/lbm R	v ft³/lbm	h Btu/lbm	s Btu/lbm R
	125 psia (64.53)			150 psia (76.38)			175 psia (86.85)		
Sat.	0.43988	110.04	0.2155	0.36587	110.83	0.2135	0.31224	111.45	0.2117
80	0.4622	112.99	0.2210	0.3705	111.56	0.2148	—	—	—
100	0.4896	116.74	0.2279	0.3953	115.50	0.2220	0.3273	114.18	0.2167
120	0.5158	120.46	0.2344	0.4187	119.37	0.2288	0.3488	118.22	0.2238
140	0.5411	124.15	0.2406	0.4410	123.18	0.2353	0.3691	122.17	0.2305
160	0.5656	127.84	0.2467	0.4624	126.97	0.2415	0.3884	126.07	0.2369
180	0.5896	131.53	0.2526	0.4832	130.74	0.2475	0.4070	129.94	0.2430
200	0.6130	135.23	0.2583	0.5034	134.52	0.2533	0.4250	133.79	0.2489
220	0.6360	138.96	0.2638	0.5232	138.31	0.2589	0.4426	137.64	0.2547
240	0.6587	142.71	0.2693	0.5426	142.11	0.2645	0.4597	141.49	0.2603
260	0.6810	146.48	0.2746	0.5618	145.93	0.2698	0.4765	145.36	0.2657
280	0.7032	150.29	0.2798	0.5806	149.78	0.2751	0.4930	149.25	0.2711
300	0.7251	154.13	0.2849	0.5993	153.65	0.2803	0.5094	153.16	0.2763
320	0.7468	158.00	0.2900	0.6177	157.55	0.2854	0.5255	157.10	0.2814
340	0.7683	161.91	0.2949	0.6360	161.49	0.2903	0.5414	161.06	0.2864
360	0.7898	165.85	0.2998	0.6541	165.46	0.2952	0.5572	165.06	0.2913
380	0.8111	169.83	0.3046	0.6721	169.46	0.3001	0.5728	169.08	0.2962
400	0.8322	173.85	0.3093	0.6900	173.50	0.3048	0.5884	173.14	0.3010
	200 psia (96.27)			250 psia (112.76)			300 psia (126.98)		
Sat.	0.27150	111.93	0.2102	0.21352	112.59	0.2074	0.17400	112.90	0.2049
100	0.2755	112.75	0.2116	—	—	—	—	—	—
120	0.2959	117.00	0.2191	0.2204	114.30	0.2104	—	—	—
140	0.3149	121.11	0.2261	0.2379	118.82	0.2180	0.1852	116.20	0.2104
160	0.3327	125.13	0.2327	0.2540	123.14	0.2251	0.2006	120.94	0.2182
180	0.3497	129.10	0.2390	0.2690	127.34	0.2318	0.2146	125.44	0.2253
200	0.3661	133.03	0.2450	0.2833	131.46	0.2381	0.2276	129.78	0.2320
220	0.3820	136.95	0.2509	0.2969	135.53	0.2442	0.2398	134.04	0.2384
240	0.3974	140.87	0.2566	0.3100	139.58	0.2501	0.2515	138.22	0.2445
260	0.4125	144.79	0.2621	0.3228	143.60	0.2558	0.2628	142.38	0.2503
280	0.4273	148.72	0.2675	0.3352	147.63	0.2613	0.2737	146.50	0.2560
300	0.4419	152.67	0.2727	0.3474	151.66	0.2667	0.2843	150.62	0.2615
320	0.4563	156.64	0.2779	0.3593	155.70	0.2719	0.2947	154.74	0.2668
340	0.4705	160.63	0.2830	0.3711	159.76	0.2770	0.3048	158.86	0.2720
360	0.4845	164.65	0.2879	0.3827	163.83	0.2821	0.3148	163.00	0.2771
380	0.4984	168.70	0.2928	0.3942	167.93	0.2870	0.3247	167.15	0.2821
400	0.5122	172.78	0.2976	0.4055	172.05	0.2919	0.3344	171.31	0.2870
420	0.5259	176.89	0.3023	0.4167	176.20	0.2966	0.3440	175.51	0.2919

(Continued)

TABLE C.10.2 ENG (Continued) *Superheated R-22*

Temp. F	v ft³/lbm	h Btu/lbm	s Btu/lbm R	v ft³/lbm	h Btu/lbm	s Btu/lbm R	v ft³/lbm	h Btu/lbm	s Btu/lbm R
	400 psia (150.82)			500 psia (170.50)			600 psia (187.29)		
Sat.	0.12297	112.70	0.2000	0.09053	111.38	0.1947	0.06663	108.51	0.1880
160	0.1305	115.52	0.2046	—	—	—	—	—	—
180	0.1446	121.01	0.2133	0.0987	115.06	0.2005	—	—	—
200	0.1567	126.02	0.2210	0.1122	121.43	0.2103	0.0791	115.12	0.1981
220	0.1677	130.76	0.2281	0.1232	126.95	0.2186	0.0919	122.32	0.2089
240	0.1778	135.31	0.2347	0.1328	132.05	0.2260	0.1020	128.29	0.2175
260	0.1874	139.76	0.2410	0.1416	136.89	0.2328	0.1106	133.70	0.2252
280	0.1965	144.13	0.2470	0.1498	141.57	0.2392	0.1184	138.79	0.2321
300	0.2052	148.45	0.2527	0.1576	146.14	0.2453	0.1256	143.67	0.2386
320	0.2137	152.74	0.2583	0.1650	150.63	0.2512	0.1323	148.41	0.2448
340	0.2219	157.01	0.2637	0.1721	155.08	0.2568	0.1387	153.05	0.2507
360	0.2299	161.27	0.2690	0.1789	159.49	0.2622	0.1449	157.63	0.2563
380	0.2378	165.54	0.2741	0.1856	163.88	0.2675	0.1508	162.16	0.2618
400	0.2455	169.81	0.2791	0.1921	168.26	0.2727	0.1566	166.66	0.2671
420	0.2530	174.09	0.2841	0.1985	172.63	0.2777	0.1622	171.15	0.2722
440	0.2605	178.38	0.2889	0.2048	177.01	0.2826	0.1677	175.62	0.2773
460	0.2679	182.69	0.2936	0.2109	181.40	0.2875	0.1730	180.09	0.2822
480	0.2752	187.02	0.2983	0.2170	185.80	0.2922	0.1783	184.57	0.2870
	700 psia (201.88)			800 psia			900 psia		
Sat.	0.04365	101.02	0.1750	—	—	—	—	—	—
220	0.0671	116.05	0.1975	0.0422	104.39	0.1788	0.0305	95.43	0.1648
240	0.0788	123.79	0.2087	0.0600	118.02	0.1986	0.0434	109.91	0.1857
260	0.0879	130.09	0.2176	0.0702	125.88	0.2097	0.0557	120.85	0.2011
280	0.0956	135.73	0.2253	0.0782	132.34	0.2186	0.0643	128.53	0.2117
300	0.1025	141.01	0.2324	0.0851	138.13	0.2263	0.0713	135.01	0.2203
320	0.1089	146.05	0.2389	0.0913	143.54	0.2333	0.0774	140.87	0.2279
340	0.1149	150.93	0.2451	0.0970	148.70	0.2399	0.0830	146.36	0.2349
360	0.1206	155.69	0.2510	0.1023	153.69	0.2460	0.0881	151.60	0.2413
380	0.1260	160.39	0.2566	0.1074	158.56	0.2519	0.0929	156.67	0.2475
400	0.1312	165.02	0.2621	0.1122	163.34	0.2575	0.0975	161.62	0.2533
420	0.1363	169.62	0.2674	0.1169	168.07	0.2630	0.1018	166.49	0.2589
440	0.1412	174.20	0.2725	0.1214	172.76	0.2682	0.1060	171.29	0.2643
460	0.1460	178.76	0.2775	0.1258	177.41	0.2734	0.1101	176.05	0.2695
480	0.1507	183.31	0.2824	0.1301	182.05	0.2783	0.1141	180.77	0.2746
500	0.1553	187.87	0.2872	0.1343	186.67	0.2832	0.1179	185.47	0.2795
520	0.1599	192.42	0.2919	0.1384	191.30	0.2880	0.1217	190.17	0.2844

TABLE C.11 ENG *Thermodynamic Properties of R-134a*
TABLE C.11.1 ENG *Saturated R-134a*

Temp.	Press.	SpecificVolume, ft³/lbm			Internal Energy, Btu/lbm		
F	psia	Sat. Liquid	Evap.	Sat. Vapor	Sat. Liquid	Evap.	Sat. Vapor
T	P	v_f	v_{fg}	v_g	u_f	u_{fg}	u_g
-100	0.951	0.01077	39.5032	39.5139	50.47	94.15	144.62
-90	1.410	0.01083	27.3236	27.3345	52.03	93.89	145.92
-80	2.047	0.01091	19.2731	19.2840	53.96	93.27	147.24
-70	2.913	0.01101	13.8538	13.8648	56.19	92.38	148.57
-60	4.067	0.01111	10.1389	10.1501	58.64	91.26	149.91
-50	5.575	0.01122	7.5468	7.5580	61.27	89.99	151.26
-40	7.511	0.01134	5.7066	5.7179	64.04	88.58	152.62
-30	9.959	0.01146	4.3785	4.3900	66.90	87.09	153.99
-20	13.009	0.01159	3.4049	3.4165	69.83	85.53	155.36
-15.3	14.696	0.01166	3.0350	3.0466	71.25	84.76	156.02
-10	16.760	0.01173	2.6805	2.6922	72.83	83.91	156.74
0	21.315	0.01187	2.1340	2.1458	75.88	82.24	158.12
10	26.787	0.01202	1.7162	1.7282	78.96	80.53	159.50
20	33.294	0.01218	1.3928	1.4050	82.09	78.78	160.87
30	40.962	0.01235	1.1398	1.1521	85.25	76.99	162.24
40	49.922	0.01253	0.9395	0.9520	88.45	75.16	163.60
50	60.311	0.01271	0.7794	0.7921	91.68	73.27	164.95
60	72.271	0.01291	0.6503	0.6632	94.95	71.32	166.28
70	85.954	0.01313	0.5451	0.5582	98.27	69.31	167.58
80	101.515	0.01335	0.4588	0.4721	101.63	67.22	168.85
90	119.115	0.01360	0.3873	0.4009	105.04	65.04	170.09
100	138.926	0.01387	0.3278	0.3416	108.51	62.77	171.28
110	161.122	0.01416	0.2777	0.2919	112.03	60.38	172.41
120	185.890	0.01448	0.2354	0.2499	115.62	57.85	173.48
130	213.425	0.01483	0.1993	0.2142	119.29	55.17	174.46
140	243.932	0.01523	0.1684	0.1836	123.04	52.30	175.34
150	277.630	0.01568	0.1415	0.1572	126.89	49.21	176.11
160	314.758	0.01620	0.1181	0.1343	130.86	45.85	176.71
170	355.578	0.01683	0.0974	0.1142	134.99	42.12	177.11
180	400.392	0.01760	0.0787	0.0963	139.32	37.91	177.23
190	449.572	0.01862	0.0614	0.0801	143.97	32.94	176.90
200	503.624	0.02013	0.0444	0.0645	149.19	26.59	175.79
210	563.438	0.02334	0.0238	0.0471	156.18	16.17	172.34
214.1	589.953	0.03153	0	0.0315	164.65	0	164.65

(Continued)

TABLE C.11.1 ENG (Continued) *Saturated R-134a*

Temp.	Press.	Enthalpy, Btu/lbm			Entropy, Btu/lbm R		
F T	psia P	Sat. Liquid h_f	Evap. h_{fg}	Sat. Vapor h_g	Sat. Liquid s_f	Evap. s_{fg}	Sat. Vapor s_g
-100	0.951	50.47	101.10	151.57	0.1563	0.2811	0.4373
-90	1.410	52.04	101.02	153.05	0.1605	0.2733	0.4338
-80	2.047	53.97	100.58	154.54	0.1657	0.2649	0.4306
-70	2.913	56.19	99.85	156.04	0.1715	0.2562	0.4277
-60	4.067	58.65	98.90	157.55	0.1777	0.2474	0.4251
-50	5.575	61.29	97.77	159.06	0.1842	0.2387	0.4229
-40	7.511	64.05	96.52	160.57	0.1909	0.2300	0.4208
-30	9.959	66.92	95.16	162.08	0.1976	0.2215	0.4191
-20	13.009	69.86	93.72	163.59	0.2044	0.2132	0.4175
-15.3	14.696	71.28	93.02	164.30	0.2076	0.2093	0.4169
-10	16.760	72.87	92.22	165.09	0.2111	0.2051	0.4162
0	21.315	75.92	90.66	166.58	0.2178	0.1972	0.4150
10	26.787	79.02	89.04	168.06	0.2244	0.1896	0.4140
20	33.294	82.16	87.36	169.53	0.2310	0.1821	0.4132
30	40.962	85.34	85.63	170.98	0.2375	0.1749	0.4124
40	49.922	88.56	83.83	172.40	0.2440	0.1678	0.4118
50	60.311	91.82	81.97	173.79	0.2504	0.1608	0.4112
60	72.271	95.13	80.02	175.14	0.2568	0.1540	0.4108
70	85.954	98.48	77.98	176.46	0.2631	0.1472	0.4103
80	101.515	101.88	75.84	177.72	0.2694	0.1405	0.4099
90	119.115	105.34	73.58	178.92	0.2757	0.1339	0.4095
100	138.926	108.86	71.19	180.06	0.2819	0.1272	0.4091
110	161.122	112.46	68.66	181.11	0.2882	0.1205	0.4087
120	185.890	116.12	65.95	182.07	0.2945	0.1138	0.4082
130	213.425	119.88	63.04	182.92	0.3008	0.1069	0.4077
140	243.932	123.73	59.90	183.63	0.3071	0.0999	0.4070
150	277.630	127.70	56.49	184.18	0.3135	0.0926	0.4061
160	314.758	131.81	52.73	184.53	0.3200	0.0851	0.4051
170	355.578	136.09	48.53	184.63	0.3267	0.0771	0.4037
180	400.392	140.62	43.74	184.36	0.3336	0.0684	0.4020
190	449.572	145.52	38.05	183.56	0.3409	0.0586	0.3995
200	503.624	151.07	30.73	181.80	0.3491	0.0466	0.3957
210	563.438	158.61	18.65	177.26	0.3601	0.0278	0.3879
214.1	589.953	168.09	0	168.09	0.3740	0	0.3740

TABLE C.11.2 ENG *Superheated R-134a*

Temp. F	v ft³/lbm	h Btu/lbm	s Btu/lbm R	v ft³/lbm	h Btu/lbm	s Btu/lbm R	v ft³/lbm	h Btu/lbm	s Btu/lbm R
		5 psia (-53.51)			10 psia (-29.85)			15 psia (-14.44)	
Sat.	8.3676	158.53	0.4236	4.3732	162.10	0.4190	2.9885	164.42	0.4168
-20	9.1149	164.47	0.4377	4.4879	163.92	0.4232	—	—	—
0	9.5533	168.11	0.4458	4.7168	167.66	0.4315	3.1033	167.19	0.4229
20	9.9881	171.83	0.4537	4.9417	171.45	0.4396	3.2586	171.06	0.4311
40	10.4202	175.63	0.4615	5.1637	175.31	0.4475	3.4109	174.97	0.4391
60	10.8502	179.52	0.4691	5.3836	179.24	0.4552	3.5610	178.95	0.4469
80	11.2786	183.50	0.4766	5.6019	183.25	0.4628	3.7093	183.00	0.4545
100	11.7059	187.56	0.4840	5.8189	187.34	0.4702	3.8563	187.12	0.4620
120	12.1322	191.71	0.4913	6.0350	191.51	0.4775	4.0024	191.31	0.4694
140	12.5578	195.95	0.4985	6.2503	195.77	0.4848	4.1476	195.59	0.4767
160	12.9828	200.28	0.5056	6.4650	200.11	0.4919	4.2922	199.95	0.4838
180	13.4073	204.69	0.5126	6.6791	204.54	0.4989	4.4364	204.39	0.4909
200	13.8314	209.19	0.5195	6.8929	209.05	0.5059	4.5801	208.91	0.4978
220	14.2551	213.77	0.5263	7.1064	213.64	0.5127	4.7234	213.51	0.5047
240	14.6786	218.44	0.5331	7.3195	218.32	0.5195	4.8665	218.19	0.5115
260	15.1019	223.19	0.5398	7.5324	223.07	0.5262	5.0093	222.96	0.5182
280	15.5250	228.02	0.5464	7.7452	227.91	0.5328	5.1519	227.80	0.5248
300	15.9478	232.93	0.5530	7.9577	232.83	0.5394	5.2943	232.72	0.5314
320	16.3706	237.92	0.5595	8.1701	237.82	0.5459	5.4365	237.73	0.5379
		30 psia (15.15)			40 psia (28.83)			50 psia (40.08)	
Sat.	1.5517	168.82	0.4136	1.1787	170.81	0.4125	0.9506	172.41	0.4118
20	1.5725	169.82	0.4157	—	—	—	—	—	—
40	1.6559	173.93	0.4240	1.2157	173.18	0.4173	—	—	—
60	1.7367	178.05	0.4321	1.2796	177.42	0.4256	1.0045	176.76	0.4203
80	1.8155	182.21	0.4400	1.3413	181.67	0.4336	1.0563	181.10	0.4285
100	1.8929	186.43	0.4477	1.4015	185.95	0.4414	1.1062	185.45	0.4364
120	1.9691	190.70	0.4552	1.4604	190.27	0.4490	1.1549	189.83	0.4441
140	2.0445	195.03	0.4625	1.5184	194.65	0.4565	1.2026	194.26	0.4516
160	2.1192	199.44	0.4697	1.5757	199.09	0.4637	1.2495	198.74	0.4590
180	2.1933	203.92	0.4769	1.6324	203.60	0.4709	1.2957	203.28	0.4662
200	2.2670	208.48	0.4839	1.6886	208.18	0.4780	1.3415	207.89	0.4733
220	2.3403	213.11	0.4908	1.7444	212.84	0.4849	1.3869	212.56	0.4803
240	2.4133	217.82	0.4976	1.7999	217.57	0.4918	1.4319	217.31	0.4872
260	2.4860	222.61	0.5044	1.8552	222.37	0.4985	1.4766	222.13	0.4939
280	2.5585	227.47	0.5110	1.9102	227.25	0.5052	1.5211	227.02	0.5007
300	2.6309	232.41	0.5176	1.9650	232.20	0.5118	1.5655	231.99	0.5073
320	2.7030	237.43	0.5241	2.0196	237.23	0.5184	1.6096	237.03	0.5138
340	2.7750	242.53	0.5306	2.0741	242.34	0.5248	1.6536	242.15	0.5203
360	2.8469	247.70	0.5370	2.1285	247.52	0.5312	1.6974	247.34	0.5267

(Continued)

TABLE C.11.2 ENG (Continued) *Superheated R-134a*

Temp. F	v ft^3/lbm	h Btu/lbm	s Btu/lbm R	v ft^3/lbm	h Btu/lbm	s Btu/lbm R	v ft^3/lbm	h Btu/lbm	s Btu/lbm R
	60 psia (49.72)			70 psia (58.20)			80 psia (65.81)		
Sat.	0.7961	173.75	0.4113	0.6844	174.90	0.4108	0.5996	175.91	0.4105
60	0.8204	176.06	0.4157	0.6882	175.32	0.4116	—	—	—
80	0.8657	180.51	0.4241	0.7291	179.89	0.4203	0.6262	179.24	0.4168
100	0.9091	184.94	0.4322	0.7679	184.41	0.4285	0.6617	183.86	0.4252
120	0.9510	189.38	0.4400	0.8051	188.92	0.4364	0.6954	188.44	0.4332
140	0.9918	193.86	0.4476	0.8411	193.45	0.4441	0.7279	193.03	0.4410
160	1.0318	198.38	0.4550	0.8763	198.01	0.4516	0.7595	197.64	0.4485
180	1.0712	202.95	0.4623	0.9107	202.62	0.4589	0.7903	202.28	0.4559
200	1.1100	207.59	0.4694	0.9446	207.28	0.4661	0.8205	206.98	0.4632
220	1.1484	212.29	0.4764	0.9781	212.01	0.4731	0.8503	211.72	0.4702
240	1.1865	217.05	0.4833	1.0112	216.79	0.4801	0.8796	216.53	0.4772
260	1.2243	221.89	0.4902	1.0440	221.65	0.4869	0.9087	221.41	0.4841
280	1.2618	226.80	0.4969	1.0765	226.57	0.4937	0.9375	226.34	0.4909
300	1.2991	231.78	0.5035	1.1088	231.57	0.5003	0.9661	231.35	0.4975
320	1.3362	236.83	0.5101	1.1410	236.63	0.5069	0.9945	236.43	0.5041
340	1.3732	241.96	0.5166	1.1729	241.77	0.5134	1.0227	241.58	0.5107
360	1.4100	247.16	0.5230	1.2048	246.98	0.5199	1.0508	246.80	0.5171
380	1.4468	252.43	0.5294	1.2365	252.26	0.5262	1.0788	252.09	0.5235
400	1.4834	257.78	0.5357	1.2681	257.62	0.5325	1.1066	257.46	0.5298
	90 psia (72.72)			100 psia (79.08)			125 psia (93.09)		
Sat.	0.5331	176.81	0.4102	0.4794	177.61	0.4100	0.3814	179.28	0.4094
80	0.5457	178.55	0.4135	0.4809	177.83	0.4104	—	—	—
100	0.5788	183.28	0.4221	0.5122	182.68	0.4192	0.3910	181.06	0.4126
120	0.6100	187.95	0.4303	0.5414	187.44	0.4276	0.4171	186.08	0.4214
140	0.6398	192.60	0.4382	0.5691	192.15	0.4356	0.4413	190.98	0.4297
160	0.6685	197.25	0.4458	0.5957	196.86	0.4433	0.4642	195.84	0.4377
180	0.6965	201.94	0.4532	0.6215	201.58	0.4508	0.4861	200.68	0.4454
200	0.7239	206.66	0.4605	0.6466	206.34	0.4581	0.5073	205.52	0.4529
220	0.7508	211.44	0.4676	0.6712	211.15	0.4653	0.5278	210.40	0.4601
240	0.7773	216.27	0.4746	0.6954	216.00	0.4723	0.5480	215.32	0.4673
260	0.8035	221.16	0.4815	0.7193	220.91	0.4792	0.5677	220.28	0.4743
280	0.8294	226.12	0.4883	0.7429	225.88	0.4861	0.5872	225.30	0.4811
300	0.8551	231.14	0.4950	0.7663	230.92	0.4928	0.6064	230.38	0.4879
320	0.8806	236.23	0.5017	0.7895	236.03	0.4994	0.6254	235.51	0.4946
340	0.9059	241.39	0.5082	0.8125	241.20	0.5060	0.6442	240.71	0.5012
360	0.9311	246.62	0.5146	0.8353	246.44	0.5124	0.6629	245.98	0.5077
380	0.9561	251.92	0.5210	0.8580	251.75	0.5188	0.6814	251.32	0.5141
400	0.9811	257.29	0.5274	0.8806	257.13	0.5252	0.6998	256.72	0.5205

(Continued)

TABLE C.11.2 ENG (Continued) *Superheated R-134a*

Temp. F	v ft³/lbm	h Btu/lbm	s Btu/lbm R	v ft³/lbm	h Btu/lbm	s Btu/lbm R	v ft³/lbm	h Btu/lbm	s Btu/lbm R
	150 psia (105.13)			175 psia (115.73)			200 psia (125.25)		
Sat.	0.3150	180.61	0.4089	0.2669	181.68	0.4085	0.2304	182.53	0.4080
120	0.3332	184.57	0.4159	0.2719	182.88	0.4105	—	—	—
140	0.3554	189.72	0.4246	0.2933	188.34	0.4198	0.2459	186.82	0.4152
160	0.3761	194.75	0.4328	0.3126	193.58	0.4284	0.2645	192.33	0.4242
180	0.3955	199.72	0.4407	0.3305	198.71	0.4365	0.2814	197.64	0.4327
200	0.4141	204.67	0.4484	0.3474	203.78	0.4444	0.2971	202.85	0.4407
220	0.4321	209.63	0.4558	0.3636	208.84	0.4519	0.3120	208.01	0.4484
240	0.4496	214.62	0.4630	0.3791	213.90	0.4592	0.3262	213.15	0.4559
260	0.4666	219.64	0.4701	0.3943	218.98	0.4664	0.3400	218.31	0.4631
280	0.4833	224.70	0.4770	0.4091	224.10	0.4734	0.3534	223.48	0.4702
300	0.4998	229.82	0.4838	0.4236	229.26	0.4803	0.3664	228.69	0.4772
320	0.5160	235.00	0.4906	0.4379	234.47	0.4871	0.3792	233.94	0.4840
340	0.5320	240.23	0.4972	0.4519	239.74	0.4937	0.3918	239.24	0.4907
360	0.5479	245.52	0.5037	0.4658	245.06	0.5003	0.4042	244.60	0.4973
380	0.5636	250.88	0.5102	0.4795	250.45	0.5068	0.4165	250.01	0.5038
400	0.5792	256.31	0.5166	0.4931	255.90	0.5132	0.4286	255.48	0.5103
	250 psia (141.87)			300 psia (156.14)			350 psia (168.69)		
Sat.	0.1783	183.75	0.4068	0.1428	184.43	0.4055	0.1167	184.63	0.4039
160	0.1955	189.46	0.4162	0.1467	185.84	0.4078	—	—	—
180	0.2117	195.28	0.4255	0.1637	192.53	0.4184	0.1275	189.13	0.4110
200	0.2261	200.84	0.4340	0.1779	198.59	0.4278	0.1425	196.01	0.4216
220	0.2394	206.26	0.4421	0.1905	204.35	0.4364	0.1550	202.24	0.4309
240	0.2519	211.60	0.4498	0.2020	209.93	0.4445	0.1660	208.14	0.4395
260	0.2638	216.90	0.4573	0.2128	215.43	0.4522	0.1761	213.86	0.4476
280	0.2752	222.21	0.4646	0.2230	220.88	0.4597	0.1856	219.49	0.4553
300	0.2863	227.52	0.4717	0.2328	226.31	0.4669	0.1945	225.06	0.4627
320	0.2971	232.86	0.4786	0.2423	231.75	0.4740	0.2031	230.61	0.4699
340	0.3076	238.24	0.4854	0.2515	237.21	0.4809	0.2114	236.16	0.4769
360	0.3180	243.66	0.4921	0.2605	242.70	0.4877	0.2194	241.74	0.4838
380	0.3282	249.13	0.4987	0.2693	248.24	0.4944	0.2273	247.33	0.4906
400	0.3382	254.65	0.5052	0.2779	253.81	0.5009	0.2349	252.97	0.4972

(Continued)

TABLE C.11.2 ENG (Continued) *Superheated R-134a*

Temp. F	v ft³/lbm	h Btu/lbm	s Btu/lbm R	v ft³/lbm	h Btu/lbm	s Btu/lbm R	v ft³/lbm	h Btu/lbm	s Btu/lbm R
	400 psia (179.92)			500 psia (199.36)			600 psia		
Sat.	0.0965	184.37	0.4020	0.0655	181.96	0.3960	—	—	—
190	0.1066	189.00	0.4092	—	—	—	—	—	—
200	0.1146	192.92	0.4152	0.0666	182.54	0.3969	—	—	—
210	0.1215	196.50	0.4205	0.0786	189.06	0.4067	0.0211	155.64	0.3554
220	0.1277	199.86	0.4255	0.0867	193.80	0.4137	0.0507	182.23	0.3948
230	0.1333	203.08	0.4302	0.0933	197.89	0.4197	0.0627	190.23	0.4065
240	0.1386	206.19	0.4347	0.0990	201.62	0.4251	0.0703	195.56	0.4142
250	0.1436	209.22	0.4390	0.1042	205.13	0.4300	0.0763	200.02	0.4205
260	0.1484	212.20	0.4432	0.1089	208.47	0.4347	0.0815	204.02	0.4261
270	0.1530	215.13	0.4472	0.1133	211.71	0.4392	0.0861	207.74	0.4312
280	0.1573	218.03	0.4512	0.1174	214.86	0.4435	0.0903	211.27	0.4360
290	0.1616	220.90	0.4550	0.1214	217.95	0.4476	0.0942	214.67	0.4406
300	0.1657	223.76	0.4588	0.1252	220.99	0.4517	0.0978	217.96	0.4450
310	0.1697	226.60	0.4625	0.1288	224.00	0.4556	0.1013	221.18	0.4492
320	0.1737	229.44	0.4662	0.1323	226.98	0.4594	0.1046	224.34	0.4533
330	0.1775	232.27	0.4698	0.1357	229.93	0.4632	0.1077	227.46	0.4572
340	0.1813	235.09	0.4733	0.1390	232.87	0.4669	0.1108	230.54	0.4611
350	0.1850	237.92	0.4769	0.1423	235.80	0.4705	0.1138	233.59	0.4649
360	0.1886	240.75	0.4803	0.1454	238.73	0.4741	0.1166	236.62	0.4686
370	0.1922	243.58	0.4838	0.1485	241.64	0.4777	0.1194	239.64	0.4723
380	0.1957	246.42	0.4872	0.1516	244.56	0.4812	0.1222	242.64	0.4759
390	0.1992	249.27	0.4905	0.1546	247.48	0.4846	0.1249	245.64	0.4794
400	0.2027	252.12	0.4939	0.1575	250.39	0.4880	0.1275	248.63	0.4829

TABLE C.12 *Enthalpy of Formation, Gibbs Function of Formation, and Absolute Entropy of Various Substances at 77 F, 1 atm Pressure*

Substance	Formula	M	State	$\overline{h}_f^o$ Btu/lbmol	$\overline{s}_f^o$ Btu/lbmol R
Water	H_2O	18.015	gas	−103 966	45.076
Water	H_2O	18.015	liq	−122 885	16.707
Hydrogen peroxide	H_2O_2	34.015	gas	−58 515	55.623
Ozone	O_3	47.998	gas	+61 339	57.042
Carbon (graphite)	C	12.011	solid	0	1.371
Carbon monoxide	CO	28.011	gas	−47 518	47.182
Carbon dioxide	CO_2	44.010	gas	−169 184	51.038
Methane	CH_4	16.043	gas	−32 190	44.459
Acetylene	C_2H_2	26.038	gas	+97 477	47.972
Ethene	C_2H_4	28.054	gas	+22.557	52.360
Ethane	C_2H_6	30.070	gas	−36 432	54.812
Propene	C_3H_6	42.081	gas	+8 783	63.761
Propane	C_3H_8	44.094	gas	−44 669	64.442
Butane	C_4H_{10}	58.124	gas	−54 256	73.215
Pentane	C_5H_{12}	72.151	gas	−62 984	83.318
Benzene	C_6H_6	78.114	gas	+35 675	64.358
Hexane	C_6H_{14}	86.178	gas	−71 926	92.641
Heptane	C_7H_{16}	100.205	gas	−80 782	102.153
n-Octane	C_8H_{18}	114.232	gas	−89 682	111.399
n-Octane	C_8H_{18}	114.232	liq	−107 526	86.122
Methanol	CH_3OH	32.042	gas	−86 543	57.227
Ethanol	C_2H_5OH	46.069	gas	−101 032	67.434
Ammonia	NH_3	17.031	gas	−19 656	45.969
T-T-Diesel	$C_{14.4}H_{24.9}$	198.06	liq	−74 807	125.609
Sulfur	S	32.06	solid	0	7.656
Sulfur dioxide	SO_2	64.059	gas	−127 619	59.258
Sulfur trioxide	SO_3	80.058	gas	−170 148	61.302
Nitrogen oxide	N_2O	44.013	gas	+35 275	52.510
Nitromethane	CH_3NO_2	61.04	liq	−48 624	41.034

EQUATIONS OF STATE

Some of the most used pressure explicit equations of state can be shown in a form with two parameters. This form is known as a cubic equation of state and contains as a special case the ideal gas law.

$$P = \frac{RT}{v-b} - \frac{a}{v^2 + cbv + db^2}$$

where (a,b) are parameters and (c,d) defines the model as shown in the following table with the acentric factor (w) and

$$b = b_o RT_c/P_c \qquad \text{and} \qquad a = a_o R^2 T_c^2/P_c$$

TABLE D.1 Equations of State

Model	c	d	b_o	a_o
Ideal Gas	0	0	0	0
van der Waal	0	0	1/8	27/64
Redlich-Kwong	1	0	0.08664	$0.42748\, T_r^{-1/2}$
Soave	1	0	0.08664	$0.42748[1+f(1-T_r^{1/2})]^2$
Peng-Robinson	2	−1	0.0778	$0.45724[1+f(1-T_r^{1/2})]^2$

$$f = 0.48 + 1.574\,\omega - 0.176\,\omega^2 \qquad \text{for Soave}$$
$$f = 0.37464 + 1.54226\omega - 0.26992\,\omega^2 \qquad \text{for Peng-Robinson}$$

TABLE D.2 Empirical Constants for Benedict-Webb-Rubin Equation
Units: Atmospheres, Liters, Moles, K, Gas constant R = 0.08206.

Gas	Formula	A_0	B_0	$C_0 \times 10^{-6}$	a
Methane	CH_4	1.85500	0.042600	0.022570	0.49400
Ethylene	C_2H_4	3.33958	0.0556833	0.131140	0.25900
Ethane	C_2H_6	4.15556	0.0627724	0.179592	0.34516
Propylene	C_3H_6	6.11220	0.0850647	0.439182	0.774056
Propane	C_3H_8	6.872	0.097313	0.508256	0.94770
n-Butane	C_4H_{10}	10.0847	0.124361	0.992830	1.88231
n-Pentane	C_5H_{12}	12.1794	0.156751	2.12121	4.07480
n-Hexane	C_6H_{14}	14.4373	0.177813	3.31935	7.11671
n-Heptane	C_7H_{16}	17.5206	0.199005	4.74574	10.36475
Nitrogen	N_2	1.19250	0.04580	0.0058891	0.01490
Oxygen	O_2	1.49880	0.046524	0.0038617	−0.040507
Ammonia	NH_3	3.78928	0.0516461	0.178567	0.10354
Carbon dioxide	CO_2	2.67340	0.045628	0.11333	0.051689

TABLE D.2 (Continued)

Gas	Formula	b	$c \times 10^{-6}$	$\alpha \times 10^3$	$\gamma \times 10^2$
Methane	CH_4	0.00338004	0.002545	0.124359	0.600
Ethylene	C_2H_4	0.008600	0.021120	0.17800	0.923
Ethane	C_2H_6	0.011122	0.032767	0.243389	1.180
Propylene	C_3H_6	0.0187059	0.102611	0.455696	1.829
Propane	C_3H_8	0.022500	0.12900	0.607175	2.200
n-Butane	C_4H_{10}	0.0399983	0.316400	1.10132	3.400
n-Pentane	C_5H_{12}	0.066812	0.82417	1.81000	4.750
n-Hexane	C_6H_{14}	0.109131	1.51276	2.81086	6.66849
n-Heptane	C_7H_{16}	0.151954	2.47000	4.35611	9.000
Nitrogen	N_2	0.00198154	0.000548064	0.291545	0.750
Oxygen	O_2	−0.000027963	−0.00020376	0.008641	0.359
Ammonia	NH_3	0.000719561	0.000157536	0.00465189	1.980
Carbon dioxide	CO_2	0.0030819	0.0070672	0.11271	0.494

TABLE D.3 *The Lee-Kesler Equation of State, The Lee-Kesler Generalized Equation of State is*

$$Z = \frac{P_r v_r'}{T_r} = 1 + \frac{B}{v_r'} + \frac{C}{v_r'^2} + \frac{D}{v_r'^5} + \frac{c_4}{T_r^3 v_r'^2}\left(\beta + \frac{\gamma}{v_r'^2}\right)\exp\left(-\frac{\gamma}{v_r'^2}\right)$$

$$B = b_1 - \frac{b_2}{T_r} - \frac{b_3}{T_r^2} - \frac{b_4}{T_r^3}$$

$$C = c_1 - \frac{c_2}{T_r} + \frac{c_3}{T_r^3}$$

$$D - d_1 + \frac{d_2}{T_r}$$

in which

$$T_r = \frac{T}{T_c}, \qquad P_r = \frac{P}{P_c}, \qquad v_r' = \frac{v}{RT_c/P_c}$$

The set of constants is as follows:

Constant	Simple Fluids	Constant	Simple Fluids
b_1	0.118 119 3	c_3	0.0
b_2	0.265 728	c_4	0.042 724
b_3	0.154 790	$d_1 \times 10^4$	0.155 488
b_4	0.030 323	$d_2 \times 10^4$	0.623 689
c_1	0.023 674 4	β	0.653 92
c_2	0.018 698 4	γ	0.060 167

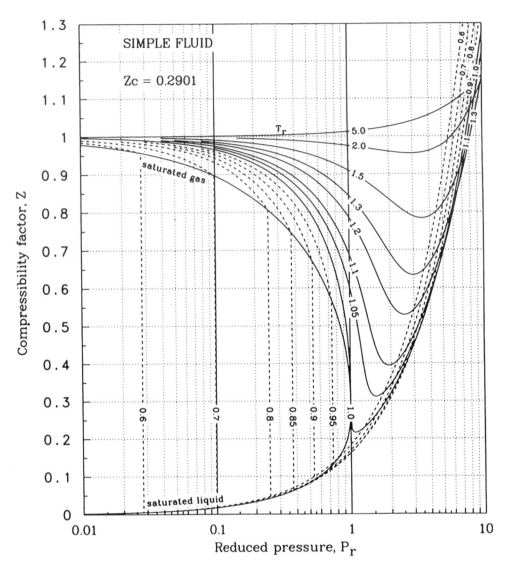

Figure D.1 Lee-Kesler Simple Fluid Compressibility Factor.

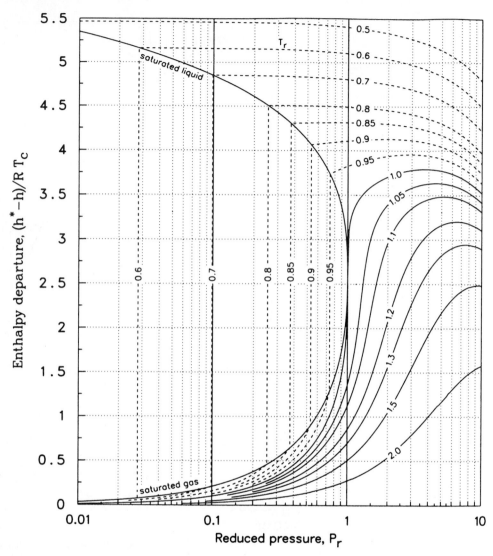

Figure D.2 Lee-Kesler Simple Fluid Enthalpy Departure.

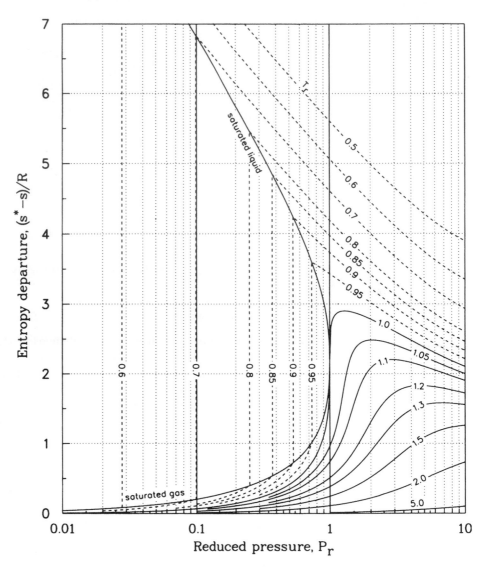

Figure D.3 Lee-Kesler Simple Fluid Entropy Departure.

APPENDIX E
IDEAL GAS SPECIFIC HEAT

Three types of energy storage or possession were identified in Section 2.6, of which two, translation and intramolecular energy, are associated with the individual molecules. These comprise the ideal gas model, with the third type, the system intermolecular potential energy, then accounting for the behavior of real (non-ideal gas) substances. This appendix deals with the ideal gas contributions. Since these contribute to the energy, and therefore also the enthalpy, they also contribute to the specific heat of each gas. The different possibilities can be grouped according to the intramolecular energy contributions as follows:

Monatomic Gases (inert gases Ar, He, Ne, Xe, Kr, also N, O, H, Cl, F, ...)

$$\bar{h} = \bar{h}_{\text{translation}} + \bar{h}_{\text{electronic}} = \bar{h}_t + \bar{h}_e$$

$$\frac{d\bar{h}}{dT} = \frac{d\bar{h}_t}{dT} + \frac{d\bar{h}_e}{dT}, \qquad \overline{C}_{P0} = \overline{C}_{P0t} + \overline{C}_{P0e} = \frac{5}{2}\overline{R} + f_e(T)$$

where the electronic contribution, $f_e(T)$, is usually small, except at very high T (common exceptions are O, Cl, F).

Diatomic and Linear Polyatomic Gases (N_2, O_2, CO, OH, ..., CO_2, N_2O, ...)

In addition to translational and electronic contributions to specific heat, these also have molecular rotation (about the center of mass of the molecule) and also $(3a - 5)$ independent modes of molecular vibration of the a atoms in the molecule relative to one another, such that

$$\overline{C}_{P0} = \overline{C}_{P0t} + \overline{C}_{P0r} + \overline{C}_{P0v} + \overline{C}_{P0e} = \frac{5}{2}\overline{R} + \overline{R} + f_v(T) + f_e(T)$$

where the vibrational contribution is

$$f_v(T) = \overline{R} \sum_{i=1}^{3a-5} \left[x_i^2 e_i^x / \left(e_i^x - 1 \right)^2 \right], \quad x_i = \frac{\theta_i}{T}$$

For each vibrational mode, θ_i is an experimentally observed constant, expressed in temperature units, that is indicative of the stiffness of the vibration. As with the monatomic gases, the electronic contribution, $f_e(T)$, is usually small, except at very high T (common exceptions are O_2, NO, OH).

Example N_2, $3a - 5 = 1$ vibrational mode, with $\theta_i = 3392$ K

At $T = 300$K, $\overline{C}_{P0} = 20.786 + 8.314 + 0.013 + \approx 0 = 29.113$ kJ/kmol K

At $T = 1000$K, $\overline{C}_{P0} = 20.786 + 8.314 + 3.446 + \approx 0 = 32.546$ kJ/kmol K
(an increase of 11.8% from 300 K)

Example CO_2, $3a - 5 = 4$ vibrational modes, with $\theta_i = 960$ K, 960 K, 1993 K, 3380 K

At $T = 300$K, $\overline{C}_{P0} = 20.786 + 8.314 + 8.036 + \approx 0 = 37.136$ kJ/kmol K

At $T = 1000$K, $\overline{C}_{P0} = 20.786 + 8.314 + 24.907 + \approx 0 = 54.007$ kJ/kmol K
(an increase of 45.4% from 300 K)

Non-Linear Polyatomic Molecules (H_2O, NH_3, CH_4, C_2H_6, ...)

Contributions to specific heat are similar to those for linear molecules, except that the rotational contribution is larger, and there are $3a - 6$ independent vibrational modes, such that

$$\overline{C}_{P0} = \overline{C}_{P0t} + \overline{C}_{P0r} + \overline{C}_{P0v} + \overline{C}_{P0e} = \frac{5}{2}\overline{R} + \frac{3}{2}\overline{R} + f_v(T) + f_e(T)$$

where the vibrational contribution is

$$f_v(T) = \overline{R}\sum_{i=1}^{3a-6}\left[x_i^2 e^{x_i} / \left(e^{x_i} - 1\right)^2\right], \quad x_i = \frac{\theta_i}{T}$$

and $f_e(T)$ is usually small, except at very high temperatures.

Example CH_4, $3a - 6 = 9$ vibrational modes, with θ_i = 4196 K, 2207 K (two modes), 1879 K (three), 4343 K (three)

At T = 300 K, $\overline{C}_{P0}$ = 20.786 + 12.472 + 2.450 + ≈ 0 = 35.708 kJ/kmol K

At T = 1000K, $\overline{C}_{P0}$ = 20.786 + 12.472 + 38.539 + ≈ 0 = 71.797 kJ/kmol K

(an increase of 101.1% from 300 K)

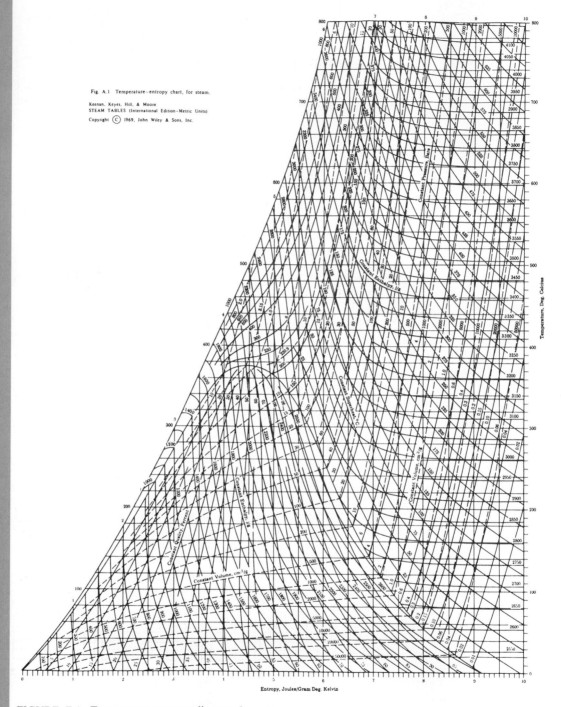

Fig. A.1 Temperature-entropy chart, for steam.

Keenan, Keyes, Hill, & Moore
STEAM TABLES (International Edition-Metric Units)
Copyright © 1969, John Wiley & Sons, Inc.

Entropy, Joules/Gram Deg. Kelvin

FIGURE F.1. Temperature entropy diagram for water.

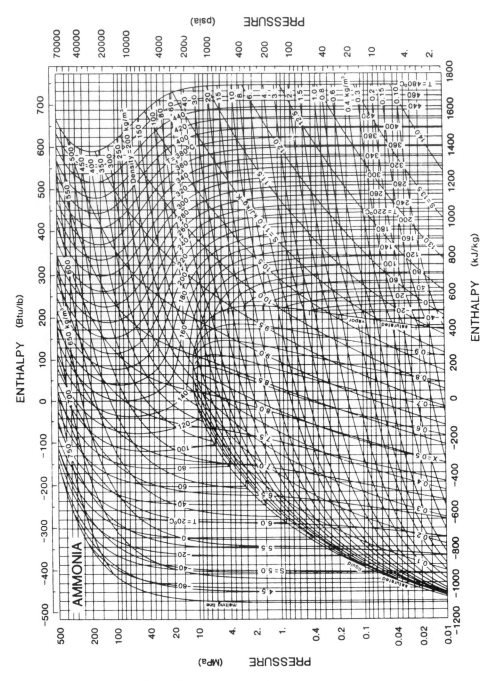

FIGURE F.2. Pressure enthalpy diagram for ammonia.

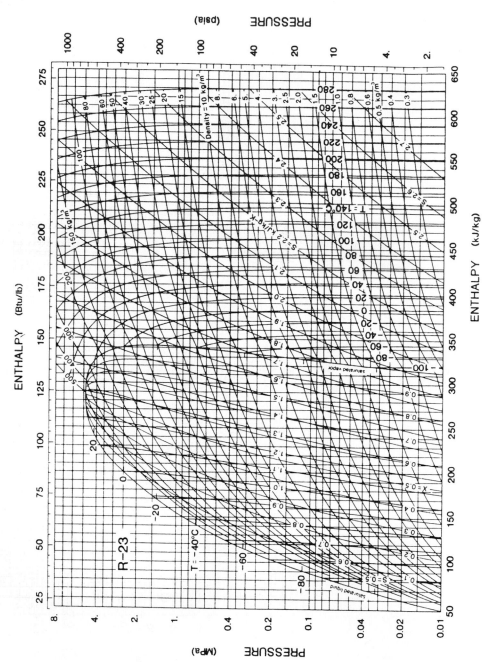

FIGURE F.3. Pressure enthalpy diagram for R-23.

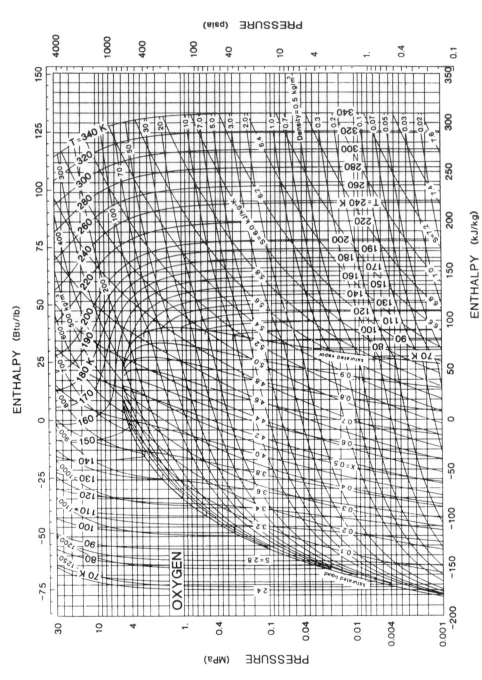

FIGURE F.4. Pressure enthalpy diagram for oxygen.

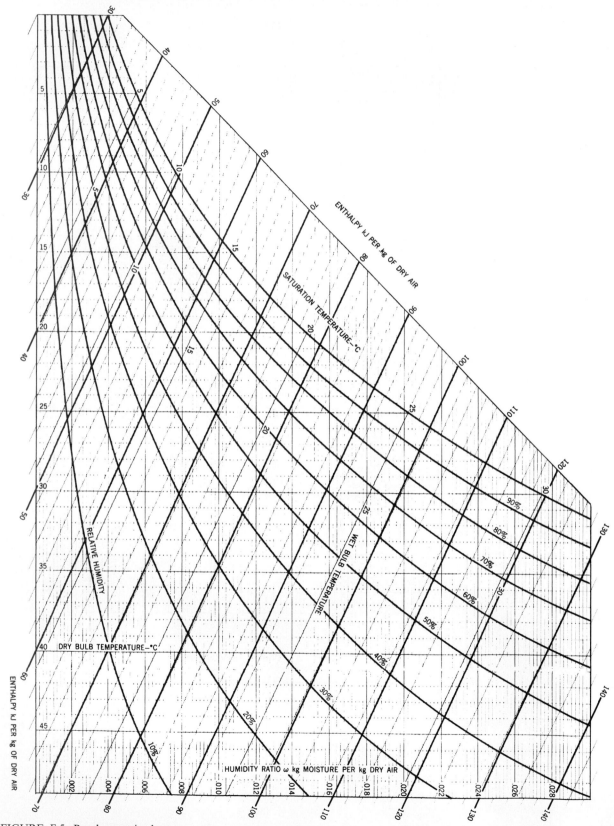

FIGURE F.5. Psychrometric chart.

SOME SELECTED REFERENCES

C. Borgnakke and R. E. Sonntag, *Thermodynamic and Transport Properties*, First Edition, John Wiley & Sons, New York, 1997.

H. B. Callen, *Thermodynamics and an Introduction to Thermostatics,* Second Edition, John Wiley & Sons, New York, 1985.

G. N. Hatsopoulos and J. H. Keenan, *Principles of General Thermodynamics,* John Wiley & Sons, New York, 1965, 1981.

J. R. Howell and R. O. Buckius, *Fundamentals of Engineering Thermodynamics,* Second Edition, McGraw-Hill Book Co., New York, 1992.

J. H. Keenan, *Thermodynamics,* M.I.T. Press, Cambridge, MA, 1970.

J. Kestin, *A Course in Thermodynamics,* Hemisphere Publishing Corp., Washington, D.C., 1966, 1979.

M. J. Moran and H. N. Shapiro, *Fundamentals of Engineering Thermodynamics,* Third Edition, John Wiley & Sons, New York, 1995.

H. Reiss, *Methods of Thermodynamics,* Blaisdell Publishing Co., Waltham, MA, 1965.

W. C. Reynolds and H. C. Perkins, *Engineering Thermodynamics,* Second Edition, McGraw-Hill Book Co., New York, 1977.

A. H. Shapiro, *The Dynamics and Thermodynamics of Compressible Fluid Flow,* The Ronald Press Co., New York, 1953.

R. E. Sonntag and G. J. Van Wylen, *Introduction to Thermodynamics, Classical and Statistical,* Third Edition, John Wiley & Sons, New York, 1991.

W. F. Stoecker, *Design of Thermal Systems,* Third Edition, McGraw-Hill Book Co., New York, 1989.

K. Wark, *Thermodynamics,* Fifth Edition, McGraw-Hill Book Co., New York, 1988.

M. W. Zemansky, M. M. Abbott, and H. C. Van Ness, *Basic Engineering Thermodynamics,* McGraw-Hill Book Co., New York, 1966, 1975.

ANSWERS TO SELECTED PROBLEMS

2.3   3583 N

2.6   2850 N; 30 m/s

2.9   2361 N

2.12   5.32 kg; 0.0055 m^3/kg

2.15   96.34 kPa

2.18   198 kPa

2.21   90 mm

2.24   1.77 kPa; 13.2 mm

2.27   6.9 mm

2.30   25.8 kPa

2.33   83.3 N

2.36   6.2 MPa

2.39   1165 lbf; 3.4 s

2.42   15 ft^3/lbm; 480 ft^3/lb mol

2.45   0.83 ft

2.48   24 lbf/in^2

3.3 $-1°$C

3.6   1.19 kg; 0.83 kg; 1.81 kg

3.9   8.7 kg; 38.7 kg; 3.3 MPa

3.12   79.7 kg; 8.4 m^3

3.15   3.98 m

3.18   5.93 kg; 142 kg; 114 kg

3.21   361 K

3.27   0.008; 0.045; 0.060; 0.0135 m^3/kg

3.30   a. 0.56; 198.5 kPa

   b. 121.6°C

   c. 0.43; 0.26 kPa

   d. 364 K

   e. 0.666; 857.5 kPa

3.33   4.5%; 1.4%

3.36   218 kPa

3.39   152 000 kg; 4.7×10^{-4}

3.42   18.9 kPa

3.45   641°C

3.54   1554 kPa; 0.118

3.57   1.3 MPa; 93 kg

3.60   6.8%

3.63   1.5 lbm

3.66   10%

3.69   a. 0.6832

   b. 0.01747

   c. 3.15

   d. 6.65

   e. 4.1 ft^3/lbm

3.72   5.3%; 0.03%

3.75   3200 lbf/in^2

3.78   0.983

3.81   8.6 ft^3/lbm

4.3   50 N; 2.5 J

4.6   131 kPa; -97 kJ/kg

4.9   1.55 MPa; 0.5 m^3; 80 kJ

4.12 -80.4 kJ

4.15   60 kJ

4.18   3.169 kPa; 0.0361; 0

4.21 -13.4 kJ

4.24   143.6°C; 143.6°C; 0.46 m^3; 2611 kJ

4.27   40 kJ

4.30 -49.4 kJ

4.33   1.77 kJ

4.36 -74.5 kJ/kg

4.39   3.93 J

4.45   14 661 kJ

4.48 15.8°C

4.51 1600 m^2

4.54 45°C

4.57 1000 K

4.63 7.7 Btu; 0.128 Btu/s

4.66 3.33 Btu

4.69 44.4 Btu

4.72 −45.5 Btu

4.75 18.2 Btu/lbm

4.78 0.22 ft

5.3 0.716 m^3

5.6 a. 3973 kPa; 1664 kJ/kg

    b. 264 kJ/kg

    c. −337.6 kJ/kg; 0.00109 m^3/kg

    d. 405.1 kJ/kg; 32.5°C

    e. 1436 kJ/kg; 0.2645 m^3/kg

5.9 a. 680.7 kPa; 0.0289 m^3/kg; 219.7 kJ/kg; 0.8286

    b. 1.375 MPa; 0.204 m^3/kg; 2869.5 kJ/kg

    c. 69.7°C; 0.3624 m^3/kg; 208.3 kJ/kg

    d. 500 kPa; 0.04656 m^3/kg; 430.7 kJ/kg

    e. 857 kPa; 978.4 kJ/kg; 1064.1 kJ/kg; 0.666

5.12 −274.6 kJ

5.15 −23.5 kJ

5.18 0.59 kg; 0.97 kg; −470 kJ; −264.8 kJ

5.21 995 kJ

5.24 2611 kJ

5.27 0.0326 m^3; 0.228 m^3; 605 kJ

5.30 146 kJ

5.33 37 kJ; 446 kJ

5.36 1.68 m^3; 400 kPa; 254 kJ; 8840 kJ

5.39 25 510 kJ

5.42 568.5 kPa

5.45 842 kPa; −1381 kJ

5.48 361 kPa; 2080 kJ; 60 kJ

5.51 288 m^3

5.54 27 kJ

5.57 1187 kPa; 1485 kJ/kg; 0; 737 kJ/kg

5.60 66°C

5.63 498 K; 188 kPa

5.66 845 kPa; 459 K; −0.015 kJ

5.69 172 kJ/kg; 670 kJ/kg; 0; −498 kJ/kg

5.72 233 kPa; −444 kJ; 31%

5.75 1345 K; 1.93 m^3; 321 kJ; 2020 kJ; 213 kJ

5.78 1239 kJ

5.81 14.2 kJ; 27.6 kJ

5.84 1491 kPa; 41.5 kJ; 1025 kJ

5.87 0.53°C/min

5.90 15 h

5.93 262 kPa; 6.1 kJ; 8.5 kJ

5.96 0.15 m^3; 30 kJ; 459 kJ

5.99 1732 K; 196 kPa; 2450 K; 290 kJ; 818 kJ

5.102 1.5 ft^3

5.105 a. 0.8155

     b. 200 lbf/in^2

     c. 100 F

     d. 55 lbf/in^2

     e. 129 lbf/in^2

5.108 751 Btu

5.111 121 Btu; 2451 Btu

5.114 4452 Btu

5.117 −766 Btu

5.120 87 F

5.123 122 lbf/in^2; 826 R; −0.014 Btu

5.126 2.55 lbm; 21.336 ft^3; 30 lbf/in^2; 687.3 R; 95.9 Btu

5.129 a. 10 in^3; 4.26 × 10^{-5} lbm

     b. 0.0362 Btu; 0.0142 Btu

     c. 0.022 Btu; 166 ft/s

5.132 1.0 Btu

5.135 2.73 hp

6.3 1.52 kg/s ± 2%

6.6 2.66 m^3/s; 4.33 m

6.9 1374 kJ/kg

6.12 8973 kW

6.15 12 850 kJ

6.18 382 m/s

6.21 360 K; 306 K; 330 K

6.24 131°C

6.27 0.964 kg/s

6.30 0.00125 kg/s

6.33 664 m^3/s; 663 000 kg/s

6.36 12 kg/s

6.39 24.8 MW; −56 MW

6.42 a. 0.9755

    b. 22.5 MW

c. 18.4 MW

d. 0.26

6.45 a. 119 MW

　　b. 128 kW

6.48 26.4 kJ

6.51 −380 MJ

6.54 39.9 kJ; −31.3 kJ

6.57 27.2 kg

6.60 520°C; 0.55 kg

6.63 3.08 kg; 225 kJ; −819 kJ

6.66 7.59 kg; 282 K; −5055 kJ

6.69 1081 kPa

6.72 0.516; 423 kPa; −89 kJ

6.75 1.45 Btu/s; 4.26 Btu/s

6.78 0.16

6.81 1116 ft/s

6.84 255 F

6.87 7.57 lbm/h

6.90 137 Btu/lbm; 127 Btu/lbm

6.93 222 540 lbm/h

6.96 115 lbf/in^2

6.99 69.6 lbf/in^2; 0.814; −62.4 Btu

6.102 47 479 Btu

7.6 0.595

7.9 0.7 MW

7.12 1.39 kW

7.15 2.56 kW

7.18 0.41 kW

7.21 0.051

7.24 62.3 kJ; 9.85 kJ

7.27 334.6 kJ; 48 kJ

7.30 10.4°C; 27°C

7.33 0.69

7.36 300 J; 3.3×10^{-8}

7.45 9.76 kW

7.51 3615 kPa; 0.269 m^3/kg

7.54 0.3

7.57 663 Btu/s

7.60 8035 Btu/h

7.69 47 ft^2

7.72 1591 Btu

8.3 637 kJ/kg; 0.454

8.6 0.93; 0.22; 7.83

8.9 364 kJ; 397 kJ

8.12 7.1 kJ; 59.6 kJ

8.15 2.05 MPa; 474 kJ

8.18 3214 kJ; 617 kJ

8.21 476°C

8.24 4910 kJ; 1290 kJ

8.33 26.3 kJ/K

8.36 0.395 kJ/kg K

8.39 10 kJ; 0.0048 kJ/K

8.42 91.94 MJ

8.45 661 kJ; 0.66 kJ/K

8.51 1.9 cm^3; 0.145 J

8.54 750 kPa; 1500 K; 333 kPa; 667 K; 1244 kJ

8.57 6.52 kg

8.60 400 K; 300 kPa; 0.52 kJ/K

8.63 a. −312 kJ

　　b. −316 kJ

　　c. −314.5 kJ

8.66 229 K; −14.7 kJ; 23 kJ

8.69 1.3034; 0.022 m^3; −21.3 kJ; −5.2 kJ; 0.0037 kJ/K

8.72 843 kJ

8.75 18 kJ; 0.106 kJ/K

8.81 3189 kPa; −291 kJ; −2376 kJ; 5.52 kJ/K

8.84 a. 1058 Btu/lbm; 152.9 F; 0.93

　　b. 1020 F; 1.608 Btu/lbm R

　　c. 0.2261 Btu/lbm R

　　d. 0.784 ft^3/lbm; 0.31 Btu/lbm R

　　e. 582.9 Btu/lbm

8.87 0.98; 0.253; 8.0

8.90 6.5 Btu; 54.5 Btu

8.93 −4.2 Btu; −5.05 Btu

8.99 14.2 Btu/R

8.102 1.3 Btu/R

8.105 23.9 in.; 0.46 Btu

8.108 45 lbf/in^2; 720 R; 0.32 Btu/R

8.111 422 R; −11.8 Btu; 19.4 Btu

9.3 358 kPa; 178 mm^2

9.6 1397 kJ/kg; −2328.5 kJ/kg

9.9 281°C; 0.72 kW/K

9.12 0.128 kJ/kg K

9.15 167.2 kW; 661 kW

9.18   0.42 kW/K

9.21   a. 706 K; 558 kJ/kg

      b. 662 K; 540 kJ/kg

      c. 706 K; 554 kJ/kg

9.24   793 K; 13 kW; 433 K; 349 kPa

9.27   6.96 MPa; 15.3 kJ/K

9.30   13 190 kJ; 12.4 kJ/K

9.33   12 295 kJ

9.36   −11.6 kJ; 0.063 kJ/K

9.39   2.36 kW

9.42   42.4 m/s

9.45   100.2 kPa; 290.3 K

9.48   a. 10.2°C; 10.2°C

      b. 98.5°C; 86.8°C

9.51   85%; 0.15 kJ/kg K

9.54   269 kPa; 143.5°C

9.57   a. 1343 kJ/kg; 92.6%

      b. 29.5 kJ/kg; 197 kJ/kg

      c. 36.7%

9.60   129 kPa; 313 K

9.63   1230 kJ/kg; 0.5 kJ/kg K; −2496 kJ/kg

9.66   18.7 MW; 3.6 kW/K

9.69   86%; 76%; 10.3 kW/K

9.72   45.4 kg/s; 20.9 kg/s; 0.52 kW/K

9.75   −180.3 kJ/kg; 56°C

9.78   461 kPa; 8 kW

9.81   813 K; 11 kW; 414 K; 245 kPa

9.84   0.095; −0.6 kJ/kg; 328 kJ/kg

9.87   613 m/s

9.90   0.99; 136.5°C

9.93   −60.6 Btu/lbm; 771 R; 1682 ft/s

9.96   21.6 lbf/in^2; 579 R

9.99   1.67 lbm/s; 8.33 lbm/s; 0.33 Btu/R s

9.102 1 003 187 Btu

9.105 a. −21.8 hp; 116 F

       b. −0.4 hp; 10.9 F

9.108 100 lbm/min; 4.4 Btu/R min

9.111 52 lbf/in^2; 11 hp

9.114 4.15 hp

9.117 −79.2 Btu/lbm; 136 F

10.3   −48.2 kJ/kg

10.6   312.7 kW

10.9   1483.9 kJ/kg; 1636.8 kJ/kg

10.12 26.4 kW; 7444 kW

10.15 790 K

10.18 60.2 kW

10.21 420.3 kJ

10.24 0.16 kJ/kg; 850 kJ/kg; 1118 kJ

10.27 64.6 kJ; 1285 kJ

10.30 2.46 kJ/kg

10.33 924 K

10.36 −1576 kJ; 1460 kJ

10.39 44.5 kJ/kg; 95%

10.42 146.8 kJ/kg; 84%

10.45 1 kg/s; 77%

10.48 85%

10.51 91%

10.54 −1.35 kJ; 4.53 kJ; −1.1 kJ; 4.78 kJ

10.57 299.5 K; 21 533 kJ; 300.3 K; 22 170 kJ

10.60 550 K; 31 kJ; 81.6%

10.63 961 Btu/s

10.66 1414 R

10.69 −1000 Btu; −1000 Btu; −537 Btu

10.72 261.7 Btu; 122.9 Btu; 152.3 Btu

10.75 171.25 Btu

10.78 61.3 Btu/lbm; 82.9%

10.81 33.5%; 70%

10.84 150 lbf/in^2; 0.057 Btu/lbm

10.87 337 Btu

11.3   0.356; 0.56

11.6   0.1034

11.9   a. 5280 kW; 156 kW

      b. 23 429 kg/s; 22 736 kg/s

      c. 0.026

11.12 3027 kJ/kg; 1055.5 kJ/kg; −1975 kJ/kg; −3.53 kJ/kg;
      0.3475

11.15 a. 21.568 kg/s

       b. −44 786 kW

       c. 0.307

       d. 0.526

11.18 0.366

11.21 0.2364; 5.04 kJ/kg; 4.51 kJ/kg

11.24 0.366; 915.7 kJ/kg

11.27 0.19; 4755 kW

11.30 571 kg/s; 1.37 m

11.33 3473 kJ/kg; 1155.1 kJ/kg; −2323 kJ/kg; −5.04 kJ/kg; 0.842; 3.05 kJ/kg K

11.36 2607.2 kJ/kg; 735.2 kJ/kg; −1876.4 kJ/kg; −4.4 kJ/kg; 0.28

11.39 0.178; 51.2°C; 183.9°C

11.42 400°C; 4409 kW; 4409 kW

11.45 7424 kW

11.48 0.565

11.51 166.33 MW; 0.399; 0.53

11.54 375 kPa; 442 kJ/kg; 0.339 kg/s; 958.8 K; 0.687

11.57 478.1 kJ/kg; 1235.5 kJ/kg; 0.613

11.60 141.5 kJ/kg; 141.5 kJ/kg 141.5 kJ/kg

11.63 138.2 kJ/kg; 600 kJ/kg; 0.77

11.66 412 kPa; 909 m/s

11.69 2118 K; 5366 kPa

11.72 5730 kPa; 2576 K; 0.475; 1105 kPa

11.75 9803 kPa; 3084 K; 0.602; 1255 kPa

11.78 11 796 kPa; 419.8 kJ/kg; 0.698

11.81 3040 K

11.84 −197.1 kJ/kg; −197.1 kJ/kg; 0; 698.6 kJ/kg; 841.4 kJ/kg; 841.4 kJ/kg; 0; −698.6 kJ/kg; 0.766

11.87 3.198; 3.172

11.90 106.9 kJ/kg; 134 kJ/kg; 3.945

11.93 5.201; 5.143; 5.428

11.96 15.55 kW

11.99 0.147

11.102 0.258

11.105 26.584 kg/s; 133.6°C

11.108 0.672; 1.0

11.111 0.438; 0.473; 0.488

11.114 1301.28 kJ/kg; 254.78 kJ/kg; 0.776

11.117 1108.5 Btu/lbm; 365.3 Btu/lbm; −744.7 Btu/lbm; − 1.8 Btu/lbm; 0.33

11.120 0.104

11.123 0.349; 0.08

11.126 0.357; 421.2 Btu/lbm

11.129 0.284; 0.0153 Btu/lbm R

11.132 165 600 hp; 0.396; 0.53

11.135 242.7 Btu/lbm; 449.5 Btu/lbm; 0.427

11.138 3817 R; 791 lbf/in^2

11.141 1033 lbf/in^2; 5789 R; 0.488; 169.5 lbf/in^2

11.144 5455 R

11.147 1250 R; 363 lbf/in^2

11.150 3.206

11.153 0.1485

12.3 0.525 N_2, 0.375 Ar, 0.10 O_2; 0.58 kg

12.6 0.021 H_2, 0.469 CO, 0.276 CO_2, 0.234 N_2; 0.00098 kg; 0.072 kJ

12.9 333.6 K; 303 kPa

12.12 679 K

12.15 308.7 K; 5.35 kJ/kmol K

12.18 1164 K

12.21 −1285 kJ/kg; 476 kJ/kg

12.24 242 kPa; 315.5 K

12.27 363 K; 152 kJ/kg

12.30 292 K; 1.09 m; 0.0276 kJ/K

12.33 0.2051 kmol; 0.4; 437 K; 700 kPa; −1038 kJ

12.36 0.51 kg; 0.0043; 1.4°C

12.39 0.56 kg

12.42 0.0189; 0.0108; 43.5 kJ/kg air

12.45 28.2°C; −2.77 kJ

12.48 1.7°C; 0.0288 kg/kg air; −1 kW

12.51 68%; 366 kW

12.54 a. 0.016; 22.3°C
       b. 0.0106; 15°C
       c. 57%; 14.4°C
       d. 86%; 0.017

12.57 21.5°C

12.60 a. 17%; 16 kJ/kg
       b. 100%; −15 kJ/kg

12.63 19.4 kg/h; −3.53 kW

12.66 3.77; 6.43 kJ/kg air

12.69 24.5%; 22 kJ/kg air; 0.0227 kg/s; 0.000 27 kg/s

12.72 0.06 kg/min; 0.016 kg/min; 32.5°C; 12%

12.75 532 kPa; 0.19 kJ/K

12.78 50%; 25.8 kJ/kg air

12.81 0.54 N_2, 0.21 H_2O, 0.25 O_2; 59.87 ft lbf/lbm R; 164 ft^3

12.84 989 Btu/s

12.87 1184 Btu/s

12.90 38 lbf/in^2; 565 R; 0.836 Btu/R

12.93 0.66 lbm; 0.00436; 35.5 F

12.96 −4.18 Btu; 0.0227 lbm; 20.8 lbf/in^2

12.99 95%

13.3 40 462 kPa

13.6   48 kPa; 0.00216 Pa

13.15 1415 m/s; 506 m/s

13.18 451.6 K; 116.6 kJ

13.21 244.6 K

13.30 −281.7 kJ/kmol

13.33 5.0°C; 4.7°C

13.36 −62.4 kJ/kg; −378.5 kJ/kg

13.39 0.474 kg; 36.3 kJ; 101.8 kJ

13.42 65.3 kJ/kg; −9.9 kJ/kg

13.45 0.1827; −10 939 kJ

13.48 934.2 kJ/kg; 368 K; 418.7 kJ/kg

13.51 −1651 kJ

13.54 5309 kW; 11.368 kW/K

13.57 −205.7 kJ; 0.0284 kJ/K

13.60 −8326 kW

13.66 1655 kPa

13.69 275 K; 0.727

13.72 200.9 kJ/kg; 0.153

13.75 0.2912 lbf/in^2; 132 ft^3/lbm

13.78 0.000 103 l/F; 0.000 0033 in^2/lbf; 0.000 143 l/F; 0.000 0108 in^2/lbf

13.81 48 Btu

13.84 443 R

13.87 43.7 F; 42.1 F

13.90 −895 Btu

13.93 −704 Btu

13.96 301.2; −52.6 Btu/lbm; −233.9 Btu/lbm

14.6   824.1 kg; 23.765 kmol; 32.778 kmol

14.9   0.461; 149.3%

14.12 2.95 kg air/kg

14.15 43.2°C; 0.0639 kg/kg fuel

14.18 −179 796 kJ/kmol; 1.352

14.21 2804 K

14.24 −1 234 583 kJ

14.27 38.66 kW; −83.27 kW

14.30 6.78; 9.567

14.33 2044 K

14.36 238%

14.39 2460.7 K; −393 522 kJ

14.42 16 666 kJ/m^3

14.45 20 986 kJ/kg coal

14.48 a. −24 746 kJ/kg

b. 4487 K

14.51 52.9°C; 1.303 kg/kg fuel

14.54 1839.14 kJ/K

14.60 3.536; 154 721 kJ/kmol C

14.63 0.328; 0.414

14.66 −100 949 kJ/kmol

14.69 238 kPa; −1 613 000 kJ; 4070 kJ/K

14.72 a. 140.697 kJ/kmol K

b. 0.7263 kmol liq; 3.2737 kmol gas

c. 433.4°C

14.75 1.232 lbm/lbm fuel; 1.492 lbm/lbm fuel

14.78 125.8 lbf/in^2; −194 945 Btu

14.81 985.6 R; 1 241 560 Btu/2 lb mol fuel

14.84 −369 746 Btu/lb mol; −337 570 Btu/lb mol

14.87 3628 R

14.90 3510 R

14.93 79.845 Btu/ft^3

14.96 −21 110 Btu/h; 0.137

15.3   29.682 MPa

15.9   1108 kPa; 6.3% O, 93.7% O_2; 97 681 kJ/kmol O_2

15.15 1444 K

15.18 2.4×10^{12}; 1.2×10^{-5}

15.21 16.4% CO, 29.5% CO_2, 23.5% O_2, 30.6% N_2

15.24 176 811 kJ

15.27 73.7%; 0.13% NO

15.30 0.006 55; −835 974 kJ

15.33 1.37% C_2H_5OH, 32.4% C_2H_4, 66.2% H_2O; 41 330 kJ

15.36 1.368

15.39 0.27

15.42 0.0024

15.45 40.6% H_2O, 1.9% H_2, 46.3% O_2, 11.2% OH

15.48 66.1% H_2O, 12.9% H_2, 5.4% O_2, 9.9% OH, 5.7% H

15.57 48.8% H_2O, 5.7% H_2, 7.6% O_2, 8.5% OH, 15.5% CO_2, 13.9% CO

15.60 14.1% O, 85.9% O_2; 1948 Btu/lbm

15.63 0.102

15.66 0.006 55; −359 465 Btu

15.69 6.826; 85.9% NH_3, 3.5% N_2, 10.6% H_2

15.72 0.0356 O_2, 0.3 O, 0.765 N_2, 0.00167 N, 0.049 NO; 92 135 Btu/lb mol air

15.75 48.7% H_2O, 5.7% H_2, 7.6% O_2, 8.6% OH, 15.5% CO_2, 13.9% CO

16.3   108 kPa; 823 K

16.6   415.4 K; 281 kPa; 5.9 kg/s

16.9   10 mm^2; 150 kPa; 100 N

16.12 9.53 min

16.15 603.8 m/s; 633.2 m/s

16.18 1055 kPa; 5 kg/s

16.21 3.56 kg/s

16.24 906 kPa

16.27 a. 0.0342 kg/s

      b. 0.0149 kg/s

      c. 1.895 kg; 0.0082 kg/s

16.30 4460 mm^2; 0.0539 kJ/kg K

16.33 285.3 K; 0.608; 88.3 kPa; 81.8 kPa

16.36 25%

16.39 0.1454 kg/s; 0.1433 kg/s

16.42 0.58

16.45 773 R; 47.8 lbf/in^2

16.48 192.4 ft/s; 0.739 Btu/lbm

16.51 13.26 lbf/in^2; 45.2 lbm/s

16.54 9.5 lbf/in^2; 1382 R; 0.414

16.57 0.116 lbm/s; 0.119 lbm/s

INDEX